Lecture Notes in Computer Science 4245

Commenced Publication in 1973
Founding and Former Series Editors:
Gerhard Goos, Juris Hartmanis, and Jan van Leeuwen

Preface

DGCI 2006, the 13th in a series of international conferences on Discrete Geometry for Computer Imagery, was held in Szeged, Hungary, October 25-27, 2006. DGCI 2006 attracted a large number of research contributions from academic and research institutions in this field. In fact, 99 papers were submitted from all around the world. After review, 55 contributions were accepted from which 28 were selected for oral and 27 for poster presentation. All accepted contributions were scheduled in single-track sessions. The program was enriched by three invited lectures, presented by internationally well-known speakers: Jean-Marc Chassery (Domaine Universitaire Grenoble, France), T. Yung Kong (City University of New York, USA), and László Lovász (Eötvös Loránd University, Budapest, Hungary).

We were pleased that DGCI got the sponsorship of the International Association of Pattern Recognition (IAPR). DGCI 2006 is also a conference associated with the IAPR Technical Committee on Discrete Geometry (TC18). Hereby, we would like to thank all contributors, the invited speakers, all reviewers and members of the Steering and Program Committees, and all supporting personnel who made the conference happen. We are also grateful to the Institute of Informatics, University of Szeged, for the financial and infrastructural help, which was essential to the organization of a successful conference. Finally, we thank all the participants and hope that they found interest in the scientific program and also that they had a pleasant stay in Szeged.

October 2006

Attila Kuba
László G. Nyúl
Kálmán Palágyi

DGCI meetings

Edition	Venue	Date	Proc.	Editors / Organizers
13th	Szeged, Hungary	Oct. 25–27, 2006	LNCS 4245	A. Kuba L.G. Nyúl K. Palágyi
12th	Poitiers, France	Apr. 13–15, 2005	LNCS 3429	E. Andres G. Damiand P. Lienhardt
11th	Naples, Italy	Nov. 19–21, 2003	LNCS 2886	I. Nyström G. Sanniti di Baja S. Svensson
10th	Bordeaux, France	Apr. 3–5, 2002	LNCS 2886	A. Braquelaire J.-O. Lauchaud A. Vialard
9th	Uppsala, Sweden	Dec. 13–15, 2000	LNCS 1953	G. Borgefors I. Nyström G. Sanniti di Baja
8th	Marne-la-Vallée, France	Mar. 17–19, 1999	LNCS 1568	G. Bertrand M. Couprie L. Perroton
7th	Montpellier, France	Dec. 3–5, 1997	LNCS 1347	E. Ahronovitz C. Fiorio
6th	Lyon, France	Nov. 13–15, 1996	LNCS 1176	S. Miguet A. Montanvert S. Ubeda
5th	Clermont-Ferrand, France	Sep. 25–27, 1995	–	D. Richard
4th	Grenoble, France	Sep. 19–20, 1994	–	J.-M. Chassery A. Montanvert
3rd	Strasbourg, France	Sep. 20–21, 1993	–	J. Françon J.-P. Reveillès
2nd	Grenoble, France	Sep. 17–18, 1992	–	J.-M. Chassery A. Montanvert
1st	Strasbourg, France	Sep. 26–27, 1991	–	J. Françon J.-P. Reveillès

Organization

DGCI 2006 was organized by the Department of Image Processing and Computer Graphics, University of Szeged, Hungary.

Organizing Committee

Attila Kuba	University of Szeged, Hungary
László G. Nyúl	University of Szeged, Hungary
Kálmán Palágyi	University of Szeged, Hungary

Steering Committee

Eric Andres	Université de Poitiers, France
Gunilla Borgefors	Swedish University of Agricultural Sciences, Uppsala, Sweden
Achille Braquelaire	Université Bordeaux 1, France
Jean-Marc Chassery	Domaine Universitaire Grenoble, France
Annick Montanvert	Domaine Universitaire Grenoble, France
Gabriella Sanniti di Baja	Istituto di Cibernetica "E. Caianiello" del CNR, Pozzuoli (Naples), Italy

Program Committee

Reneta Barneva	SUNY College at Fredonia, Fredonia, USA
Gilles Bertrand	Groupe ESIEE, Noisy-le-Grand, France
Valentin E. Brimkov	SUNY Buffalo State College, Buffalo, USA
David Coeurjolly	Université Claude Bernard Lyon 1, Villeurbanne, France
Michel Couprie	Groupe ESIEE, Noisy-le-Grand, France
Leila De Floriani	Dipartimento di Informatica e Scienze dell'Informazione, Genova, Italy
Isabelle Debled-Rennesson	LORIA Nancy, Vandœuvre-lès-Nancy Cedex, France
Ulrich Eckhardt	Universität Hamburg, Germany
Oscar Figueiredo	École Supérieure Chimie Physique Électronique de Lyon, France
Christophe Fiorio	Universitaire de Montpellier, France
Atsushi Imiya	Chiba University, Japan
Pieter Jonker	Delft University of Technology, The Netherlands

Ron Kimmel	Technion, Haifa, Israel
Christer O. Kiselman	Uppsala University, Sweden
Reinhard Klette	The University of Auckland, New Zealand
Walter G. Kropatsch	Vienna University of Technology, Austria
Jacques-Olivier Lachaud	Université Bordeaux 1, France
Gregoire Malandain	INRIA, Sophia-Antipolis, France
Remy Malgouyres	GREYC, ISMRA, Aubiere, France
Serge Miguet	Université Lumière Lyon 2, Bron, France
Ingela Nyström	Uppsala University, Sweden
Pierre Soille	Joint Research Centre of the European Commission, Ispra (VA), Italy
Edouard Thiel	Université de la Méditerranée Aix-Marseille 2, France

Referees

Eric Andres
Reneta Barneva
Gilles Bertrand
Gunilla Borgefors
Achille Braquelaire
Valentin E. Brimkov
Jean-Marc Chassery
Pierre Chatelier
David Coeurjolly
Michel Couprie
Jean Cousty
Emanuele Danovaro
Leila De Floriani
Isabelle Debled-Rennesson
Eric Domenjoud
Ulrich Eckhardt
Fabien Feschet
Oscar Figueiredo
Christophe Fiorio
Céline Fouard
Pierre-Marie Gandoin
Yan Gérard
Yll Haxhimusa
Atsushi Imiya
Adrian Ion
Pieter Jonker
Zoltán Kató
Yukiko Kenmochi
Bertrand Kerautret
Ron Kimmel
Christer O. Kiselman
Reinhard Klette
Walter G. Kropatsch
Attila Kuba
Jacques-Olivier Lachaud
Gaëlle Largeteau-Skapin
Gregoire Malandain
Remy Malgouyres
Serge Miguet
Annick Montanvert
Ingela Nyström
László G. Nyúl
Kálmán Palágyi
Laura Papaleo
Gabriella Sanniti di Baja
Isabelle Sivignon
Pierre Soille
Robin Strand
Stina Svensson
Hugues Talbot
Edouard Thiel
Laure Tougne
Jean-Luc Toutant

Table of Contents

Discrete Geometry

Invited Paper

Discrete Tomography

Discrete Topology

Invited Paper

Distance

Image Analysis

Shape Representation

Segmentation

Skeletonization

Surfaces and Volumes

Duality and Geometry
Straightness, Characterization and Envelope

Jean-Marc Chassery[1], David Coeurjolly[2], and Isabelle Sivignon[2]

[1] Laboratoire LIS
Domaine universitaire Grenoble - BP46
38402 St Martin d'Hères Cedex, France
jean-marc.chassery@lis.inpg.fr

[2] Laboratoire LIRIS - Université Claude Bernard Lyon 1
Bâtiment Nautibus - 8, boulevard Niels Bohr
69622 Villeurbanne cedex, France
{david.coeurjolly, isabelle.sivignon}@liris.cnrs.fr

Abstract. Duality applied to geometrical problems is widely used in many applications in computer vision or computational geometry. A classical example is the Hough Transform to detect linear structures in images. In this paper, we focus on two kinds of duality/polarity applied to geometrical problems: digital straightness detection and envelope computation.

Introduction

In domain of geometry, notion of duality is often used to represent the same structure in different domains like spatial domain or parametric one. The objective is to facilitate transformations like characterization, detection, recognition or classical ones such as intersection or union. A first example is illustrated with Voronoi partition in which polygonal regions are not homogeneous in terms of number of vertices. Nevertheless, the corresponding dual mesh, called Delaunay mesh, is composed of triangles. According to applications the choice of the alternative representations can be used on optimality criteria (computational cost, database structure, ...).

Following this first example, we focus in this paper on dual transformations illustrated by problems of digital straightness and envelope.

1 Example of the Hough Transform

The Hough transform (HT for short) is a very classical tool in image analysis to detect geometric features in images. These features may be line segments, circles, ellipses or any other parameterized curve. The HT, introduced in 1962 by Hough [1], is a dual transformation that enables to find a set of global structures,

A. Kuba, L.G. Nyúl, and K. Palágyi (Eds.): DGCI 2006, LNCS 4245, pp. 1–16, 2006.

without any a priori knowledge on the number of structures to be found. Note also that this method is robust to noise and disconnected features.

1.1 Definition of Hough Transform

The general idea of this transform is that every point of the image contributes to the definition of the solution set for a given parameterized structure. Consider for instance a point p_0 of coordinates (x_0, y_0) and the parameterization of lines $y = \alpha x + \beta$. Then the set of lines going through p_0 are the ones of parameters (α, β) fulfilling the equality $y_0 = \alpha x_0 + \beta$. This equality may be rewritten as $\beta = -\alpha x_0 + y_0$, and if a new geometrical space $(\alpha\beta)$, called *dual space*, or *parameter space*, is defined, this equation defines a line : in this dual space, each point of this line represents a line of the (xy) space going through the point p_0. An illustration of three points and the three corresponding lines in the dual space $(\alpha\beta)$ are represented in Figure 1 (a)-(b): note that the three lines in $(\alpha\beta)$ space are concurrent in one point, the coordinates of which defines a line going through the three points in (xy) space.

However, as noticed by Duda in [2], the linear parameterization of lines defined by $y = \alpha x + \beta$ is not the handiest one since the two parameters α and β are unbounded. Thus, another transform consists in using the polar parameterization of straight lines $\rho = x \cos\theta + y \sin\theta$. Any point in the (xy) space defines a sinusoidal curve in the $(\theta\rho)$ space, where only the parameter ρ has unbounded values (see Figure 1(c) for an illustration).

General properties fulfilled by these two representations, and suitable for straight line detection in images were expressed by Duda [2]:

Property 1

- A point in the (xy) space matches up with one curve in the dual space;
- A point in the dual space matches up with a straight line in the (xy) space;
- Points lying on a same line in the (xy) space match up with concurrent curves in the dual space;
- Points on a same curve in the dual space match up with concurrent straight lines in the (xy) space.

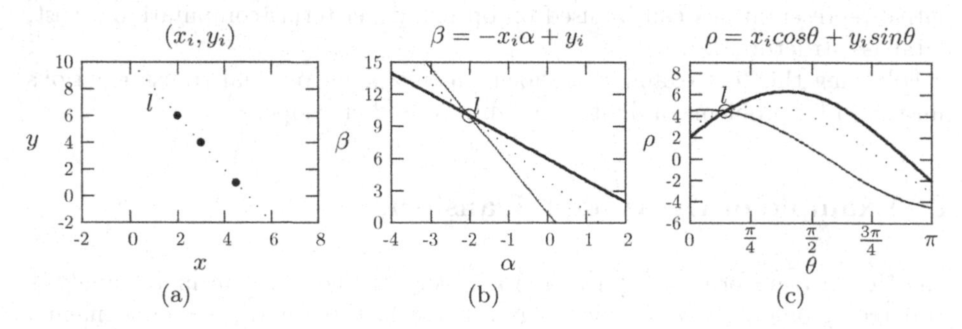

Fig. 1. (a) Three points in the (xy) space; (b) Dual representation in the $(\alpha\beta)$ space; (c) Dual representation in the $(\theta\rho)$ space

1.2 Recognition of Parameterized Structures

Line segment detection in images does not consist in finding the pixels lined up according to the Euclidean straight line definition, but a relaxation of this definition has to be used. To do so, the method generally used consists of, first, decomposing, or quantifying the dual space along the two axis, and second, defining a counter for each cell of the dual space. Algorithm 1 describes the general algorithm for finding parameterized curves in an image using HT. The quantization step is a trade-off between precision on one part, and memory/computation cost on the other hand. Moreover, a good quantization should provide constant densities for equally probable line parameters. An illustration of Algorithm 1 is proposed in Figure 2.

Algorithm 1. Hough transform for parameterized curve detection

Input: Set of pixels P
Quantify the dual space of the parameterized curve;
Set all the cell counters to zero;
for *every pixel p of P* **do**
  Compute $HT(p)$ and digitize it according to the quantization grid;
  Add one to the counters of $HT(p)$ digitization;
end
Look for local maxima among the cells counters: each maximum matches with the parameters of a curve found in P.

2 Duality in Discrete Geometry

During a HT, the discrete nature of the data processed is taken into account with a quantization of the dual space. On the contrary, we see in this section that the classical notion of dual space used in discrete geometry introduces the discrete nature of the data in the definition of the dual representation of a point.

2.1 Definition of the Dual Space

In digital geometry, pixels are said to be lined up if they belong to a digital straight line, which is the digitization of a straight line. In a general way, a digital straight line of parameters (a, b, μ) and bounds $\rho(a, b)$ and $\omega(a, b)$ is the set of pixels (x, y) such that $\rho(a, b) \leq ax - by + \mu \leq \omega(a, b)$. Without loss of generality, we suppose that $|b| > |a|$, and $b > 0$ in the following. With these conditions, the previous definition may be rewritten as $\rho'(\alpha, \beta) \leq \alpha x - y + \beta \leq \omega'(\alpha, \beta)$. Given a point p_0 of coordinates (x_0, y_0), the digital lines containing are the ones for which (x_0, y_0) fulfills the inequalities. Thus, we can once again define a dual space $(\alpha\beta)$ to represent the space of line parameters, but contrary to HT, a given point p_0 of coordinates (x_0, y_0) matches up with the intersection of two linear constraints defined by $E^+ : \beta \geq -\alpha x_0 + y_0 + \rho'(\alpha, \beta)$ and $E^- : \beta < -\alpha x_0 + y_0 + \omega'(\alpha, \beta)$.

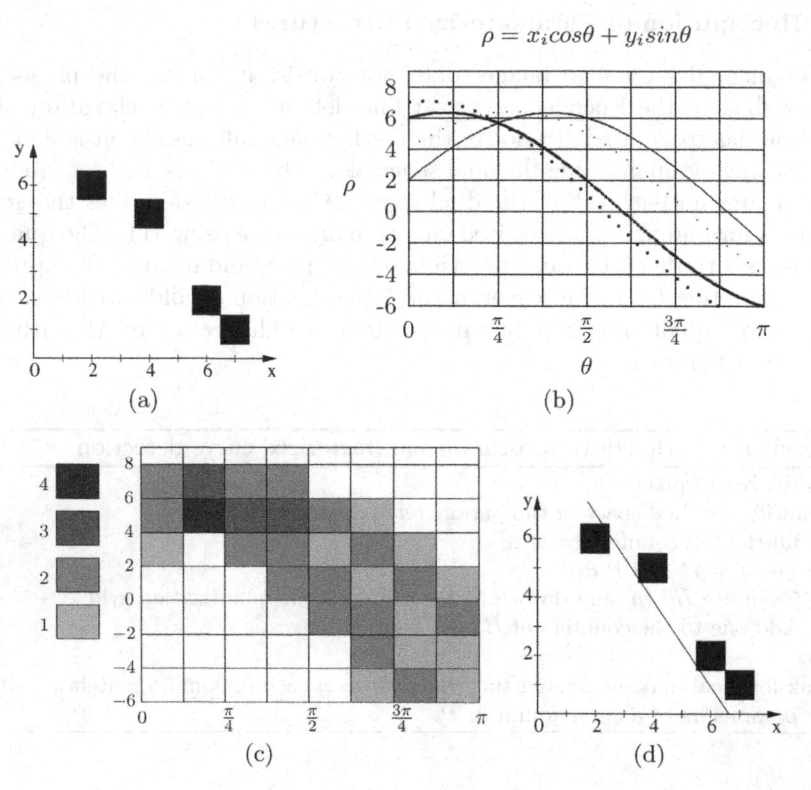

Fig. 2. Detection of a line segment with HT: (a) the four pixels of the set P; (b) dual representation in the quantified dual space; (c) result of the digitization of the sinusoidal curves; (d) straight line computed from the local maximum found

Definition 1. *Let P be a set of pixels. The preimage of P denoted by $\mathcal{P}(P)$ is defined as follows: $\mathcal{P}(P) = \{(\alpha, \beta),\ |\alpha| \leq 1 \mid \forall(x, y) \in P,\ \rho'(\alpha, \beta) \leq \alpha x - y + \beta < \omega'(\alpha, \beta)\}$. (See Figure 3).*

As we can see, in digital geometry, the linear parameterization of lines is used in order to define the dual space. Nevertheless, we pointed out that for the Hough transform, using a polar parameterization is more convenient in order to handle bounded parameters. Actually, the polar parameterization is not appropriate for preimage definition since intersection of sinusoidal curves would be involved. Thus, the handling of unbounded domains has to be tackled. First, the parameter β takes its values in an unbounded domain since it represents all the possible translation of a line. This problem is easy to solve, operating a translation of the set of pixels studied such that one particular pixel of the set is set to the origin. Next, the slope α of the lines also have unbounded values. The idea here is to use two dual spaces instead of one :

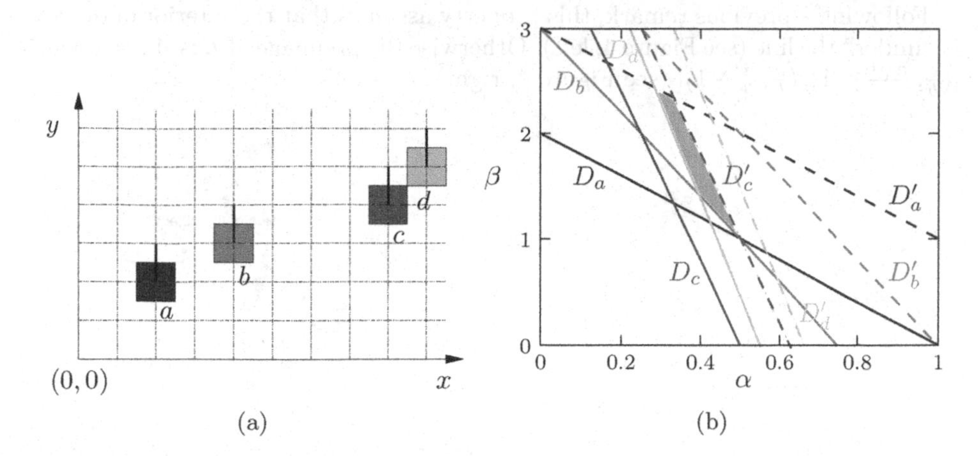

Fig. 3. Illustration of the preimage of a set of pixels (digitization process fixed): each point matches up with two linear constraints, and the preimage is the intersection of these constraints

Definition 2. *The dual space $\mathcal{P}_y$ is defined as the space where one point (α, β), $|\alpha| < 1$ stands for the line $\alpha x - y + \beta = 0$. In the same way, a point (α, β), $|\alpha| < 1$ of the dual space denoted $\mathcal{P}_x$ stands for the line $\alpha y - x + \beta = 0$.*

2.2 Preimages of Digital Lines and Line Segments

The definition of preimage depends on the values of $\rho'(\alpha, \beta)$ and $\omega'(\alpha, \beta)\}$, and in most applications, these values are defined according to the digitization process considered during the definition of the digital straight line. In this section, firstly we give some examples of preimages of digital straight lines in respect to the digitization process considered, and secondly, we emphasize on particular properties of the preimage of digital straight line segments (DSS for short) for one digitization process.

Digitization and Preimage. Let us first consider the *OBQ* (object boundary quantization) digitization scheme: given a straight line of equation $ax - by + \mu = 0$, its *OBQ* digitization is the set of pixels such that $0 \leq ax - by + \mu < b$ (see conditions over a and b previously defined). Since the *OBQ* digitization is based on the definition of the inside and this outside of an object, this definition assumes that the line $ax - by + \mu = 0$ is part of the boundary of an object the inside of which is given by the direction of the normal vector $(a, -b)$.

From this definition, we derive a characterization of the preimage of an infinite digital line according to the *OBQ* digitization process [3]:

Property 2. Let L be a digital straight line defined by $0 \leq ax - by + \mu < b$, with $0 \leq a < b$. Then the preimage of L according to the *OBQ* digitization process is the vertical segment $[(\frac{a}{b}, \frac{\mu}{b}), (\frac{a}{b}, \frac{\mu+1}{b})]$.

Following a previous remark, this property assumes that the interior of object is "under" the line (see Figure 4, left). Otherwise the preimage of L is the segment $[(\frac{a}{b}, \frac{\mu+1}{b} - 1), (\frac{a}{b}, \frac{\mu}{b} - 1)[$ (see Figure 4, right).

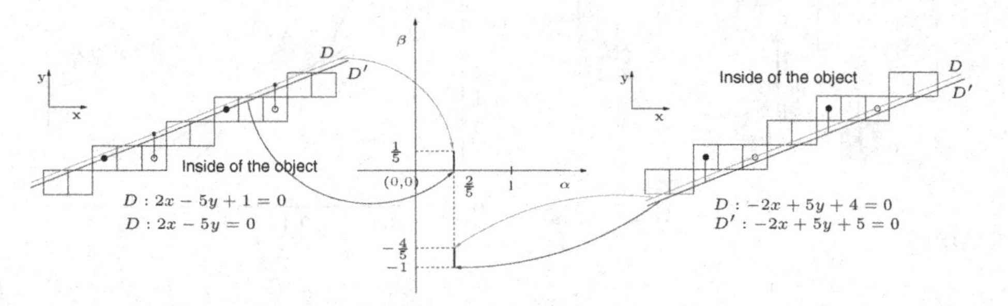

Fig. 4. Preimage of the digital straight line $0 \leq 2x - 5y < 5$: two solutions according to the direction of the solutions normal vector

Let us now consider the standard digitization process: given a straight line of equation $ax - by + \mu = 0$ such that $0 \leq a < b$, its standard digitization is the set of pixels such that $-\frac{|a|+|b|}{2} \leq ax - by + \mu < \frac{|a|+|b|}{2}$. Contrary to the OBQ digitization process, the standard digitization of a line does not depend on the direction of the normal vector of the line. However, we have the same kind of results on the characterization of the preimage (see Figure 5 for an illustration):

Property 3. Let L be a digital straight line defined by $-\frac{|a|+|b|}{2} \leq ax - by + \mu < \frac{|a|+|b|}{2}$, with $0 \leq a < b$. Then the preimage of L according to the standard digitization process is the vertical segment defined by:

- $[(\frac{a}{b}, \frac{\mu}{b}), (\frac{a}{b}, \frac{\mu+1}{b})[$ if $|a| + |b|$ is even;
- $[(\frac{a}{b}, \frac{\mu}{b} - \frac{1}{2b}), (\frac{a}{b}, \frac{\mu}{b} + \frac{1}{2b})[$ if $|a| + |b|$ is odd.

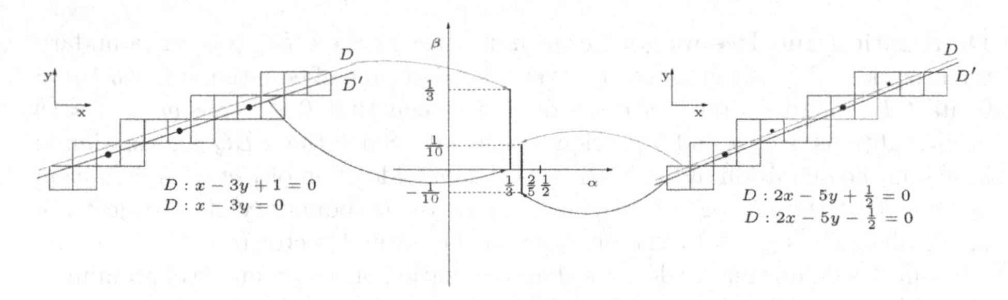

Fig. 5. Preimages of digital straight lines according to the standard digitization process: on the left, the sum $|a| + |b|$ is even, on the right, it is odd

DSS Preimage. Using a dual space is a common technique for digital straight line recognition, and thus many works have been carried out about the geometrical and arithmetical structure of the preimage of a digital straight segment. We simply recall here some of the main and classical properties on this structure [4, 5]:

Property 4. Let S be a set of $N+1$ 8-connected pixels, and x_0 be the minimum abscissa of this set of pixels. Then the preimage $\mathcal{P}(S)$ of S has the following properties:

1. $\mathcal{P}(S)$ is a convex polygon with at most four vertices;
2. two consecutive abscissa of the vertices are consecutive terms in the Farey series [6] of order $\max(x_0, N - x_0)$. Moreover, for a given abscissa equal to $\frac{p}{q}$, the corresponding ordinate is a multiple of $\frac{1}{q}$;
3. if this polygon has four vertices, then two out of the four vertices have the same abscissa.

This property shows that there is a strong connection between DSS preimages and Farey series. Actually, given a Farey series of order d, a Farey diagram may be defined ([5], see Figure 6, on the left for an example). An important property of this diagram is that their is a bijection between the cells of the Farey diagram of order q and the preimages of the DSS of length $q+1$ [7]. This property is illustrated in Figure 6 in the case of $q = 2$: there are only four DSS of length 3, and their preimages are the cells of the diagram. These strong arithmetical and geometrical features of DSS preimages are used to design efficient recognition algorithms [8, 9].

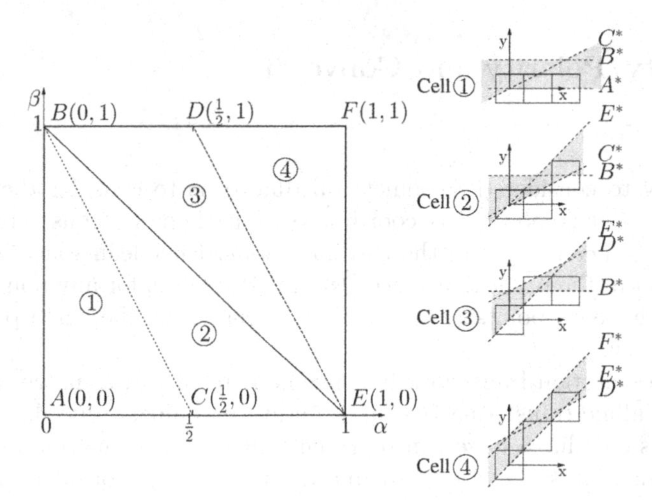

Fig. 6. Bijection between the cells of a Farey diagram of order 2 and the preimages of DSS of length 3

3 Generalizations and Applications

These definitions of dual space and preimage easily extend to higher dimensions for digital hyperplanes. Nevertheless, even if the characterization of the preimages of infinite hyperplanes is easy to handle, few works have been carried out concerning the structure of the preimage in 3D or more. In [10], the authors propose a first structural and arithmetical characterization of the preimage of a digital plane segment for particular cases. In [11], a theoretical and experimental study on the number of faces of the preimage in higher dimensions is proposed. Concerning the definition of dual space and preimage, recent work [12] proposes a generalization of the preimage which enables to define the dual of a polygon as the set of lines crossing this polygon. Together with the use of standard digital planes, this new dual structure enables to define nice algorithms for digital curve reconstruction.

Concerning the interest of using a dual space for applications, we already mentioned the digital line recognition problem, for which using a dual space offers a nice solution. This algorithm can also be extended for digital plane segments recognition [13]. This notion may also be used to study the properties of the intersection of two digital lines or two digital planes, as in [14]. In this case, the preimage of infinite digital lines and planes are involved in the characterization of the minimal parameters of the set of intersection grid points. Finally, let us also mention the work carried out by Veelaert in [15,16,17] concerning the detection of collinear, parallel or concurrent segments in an image. In these works, the dual representation of the studied properties enables to extract a graph in which particular structures (*e.g.* cliques) are sought. Then these structures represent sets of segments fulfilling the desired property.

4 Duality/Polarity and Convexity

4.1 Definitions

Another way to consider the geometrical duality is to consider the projective group on $(d+1)$ homogeneous coordinates. The homogeneous representation of a point $\boldsymbol{p} = (x_1, \dots, x_d)$ in the $d-$dimensional Euclidean space is the point $(x_1, \dots, x_d, 1)$ in the projective space [18]. Furthermore, for any non-zero scalar λ, the homogeneous points $(\lambda x_1, \dots, \lambda x_d, \lambda)$ represent the same point in the Euclidean space.

This representation framework is convenient to obtain a matrix representation of both affine transformations and duality mappings. Indeed, with $(d+1)$ homogeneous coordinates, we can represent Euclidean points and all linear varieties of dimension $k < d+1$. To define the polarity, we consider the following transformation:

$$\boldsymbol{\alpha} = \boldsymbol{\beta} B$$

where $\boldsymbol{\alpha}$ and $\boldsymbol{\beta}$ are vectors in $(d+1)$ homogeneous coordinates and B is a $(d+1) \times (d+1)$ matrix.

In the following, we focus on matrices B such that $|B| \neq 0$ (defining the projective group) and $B = B^T$. In dimension 2, this transformation maps points to lines and lines to points using the homogeneous coordinate system. Furthermore, this class of transformation preserves the incidence: $\boldsymbol{\alpha} \in \boldsymbol{\beta}$, then $\boldsymbol{\beta} B \in \boldsymbol{\alpha} B$. We thus obtain a geometric duality.

Note that the matrix B is such that $\boldsymbol{\alpha} B \boldsymbol{\alpha}^T = 0$ which corresponds to the equation of a conic in homogeneous coordinates [18]. In computational geometry, the mapping induced by such a conic is called polarity.

4.2 Polarity with Respect to Unit Circle

In the following, we focus on the polarity defined by the matrix:

$$B = \begin{bmatrix} 1 & 0 & 0 \\ 0 & 1 & 0 \\ 0 & 0 & -1 \end{bmatrix}$$

The conic defined by this mapping in homogeneous coordinates is a unit circle in dimension 2. This transformation maps a point $(x_1, x_2, 1)$ to the line $x_1 x + x_2 y - z = 0$ in homogeneous coordinates.

A final property of this special duality is that we have a kind of metric preservation. Given a point (or a line) $\boldsymbol{\alpha}$, we have:

$$distance(\boldsymbol{\alpha}, 0) \cdot distance(dual(\boldsymbol{\alpha}), 0) = 1$$

In the Cartesian space, we can thus consider the dual by polarity of a polygon. Note that the center of the unit circle (called the *pole*) needs to be specified. Figure 7 illustrates the dual transformation of each straight line defined by the polygon edges. Connecting the polar elements using the incidence property, we thus obtain a polygon (not necessarily simple) called the *dual polygon*.

Using the classical property of polarity that maps union to intersection and conversely we mention the property that in the general case, the convex hull

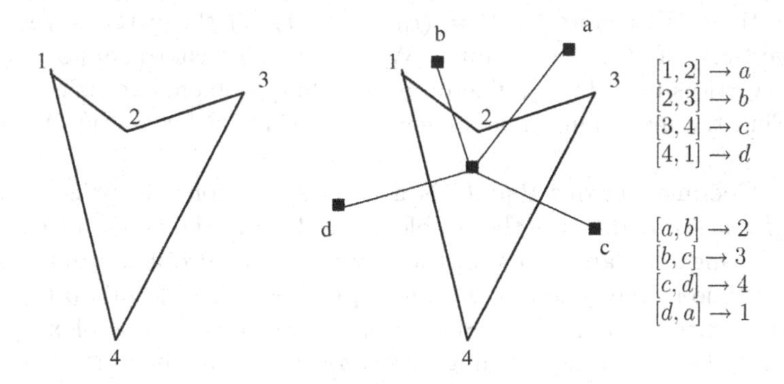

Fig. 7. A polygon and its dual by polarity

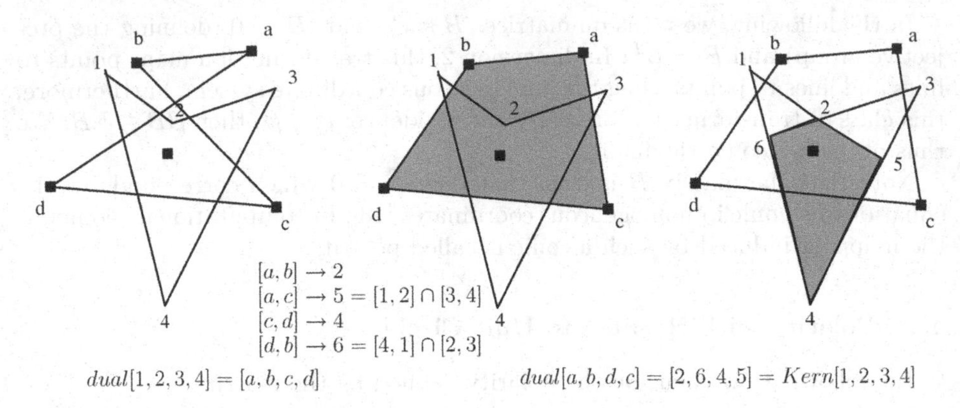

Fig. 8. A polygon, its kernel as the dual of the convex hull of the dual

of the dual polygon corresponds to the dual of a geometrical object called the *kernel*. This object corresponds to the set of all the points inside the polygon which are visible from any point of polygon contour (see below for a formal definition of the kernel). The class of polygons the kernel of which is not empty is called the *star-shaped* class of polygons. Figure 8 illustrates this construction (remind that the position of the pole is very important).

5 Application to Convex Optimization by Interior

This section addresses the problem of the access of the shape by its interior. Polygon inclusion problems are defined as follows: given a non-convex polygon, how to extract the maximum area subset included in that polygon ? The search of the maximum horizontal-vertical convex polygon included into a simple orthogonal polygon has been solved in the continuous case by Wood and Yap with complexity in $O(n^2)$ [19]. For more details about a lot of proposed solutions to inclusion problems, refer to [20].

For the rest of the presentation, we consider a polygon $P = (v_0, v_1, \ldots, v_{n-1})$ with n vertices. We denote by $R = (r_0, r_1, \ldots, r_{k-1})$ the k reflex vertices (or concave vertices) of P (maybe empty). We note by C_i a chord supported by two successive vertices v_i and v_{i+1}. The potato-peeling problem can be expressed as follows: Find the maximum area convex subset (MACS for short) Q contained in P.

In [21], Goodman proves that Q is a convex polygon. He presents explicit solutions for $n \leq 5$ and leaves the problem unsolved in the general case.

In [22], Chang and Yap prove that the potato-peeling problem can be solved in polynomial time in the general case. More precisely, they detail an $O(n^7)$ time algorithm to extract Q from P. Since this algorithm uses complex geometric concepts and dynamic programming in several key steps, it is not tractable in practical applications.

In the following we propose an approximation based algorithm to approach the MACS of a star-shaped polygon P. The proposed algorithm is an iterative process based on a kernel dilatation framework.

5.1 Fast Approximation Algorithm

In this section, we assume that P is a star-shaped polygon. P is a *star-shaped polygon* if there exist a point q in P such that $q\bar{v}_i$ lies inside P for all vertices v_i of P. The set of points q satisfying this property is called the *kernel* of P.

An extremal chord is a chord which contains two or more vertices of P. We note that an edge of P is always included in an extremal chord.

To end with definitions a chord is called single-pivot chord if it contains only one reflex vertex (chord C_1 in Figure 9) and double-pivot chord if it contains two distinct reflex points (chord C_2 in Figure 9).

The kernel of P can be seen as the intersection between P and the half-planes C_i^+ defined by all extremal chords C_i associated to all reflex vertices, as illustrated in Figure 10.

Figure 10 is an illustration of such proposition. We have the property:

Property 5. Let P be a star-shaped polygon, then its kernel is a subset of the maximum area convex subset of P.

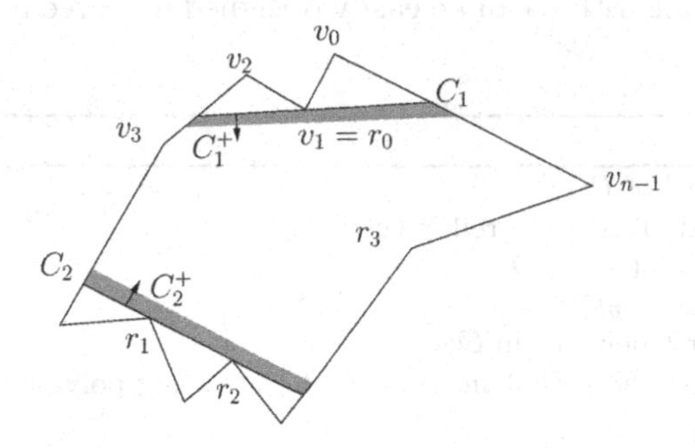

Fig. 9. Notations and illustrations of chords and half-planes generated by these chords

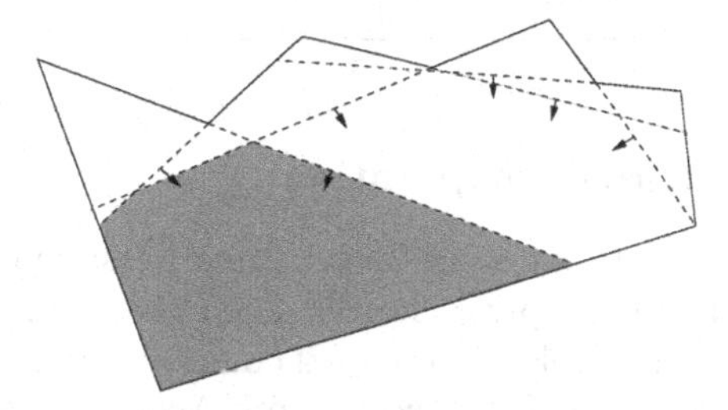

Fig. 10. Illustration of the kernel computation based on intersection of extremal chords

Proof. Details of the proof are given in [20].

In other words, there exists a continuous deformation that transforms the kernel to the MACS. In the following, the strategy we choose to approximate the MACS is to consider the deformation as an Euclidean dilatation of the kernel. Based on this heuristic, several observations can be made: the reflex vertices must be taken into account in the order in which they are reached by the dilatation wavefront. More formally, we consider the list $\mathcal{O}$ of reflex vertices such that the points are sorted according to their minimum distance to the kernel polygon. When a reflex vertex is analyzed, we fix the possible chords as follows:

- the chord may be an extremal one;
- the chord may be a single-pivot chord such that its slope is tangent to the wavefront (this point will be detailed in the next section);
- the chord may be a double-pivot chord. In that case, the second reflex vertex that belongs to the chord is necessary. It must correspond to the next reflex point in the order $\mathcal{O}$.

Furthermore, when a reflex vertex is analyzed, we choose the chord from this list that maximizes the area of the resulting polygon. If we denote by P' the polygon given by the intersection between P and the half-plane associated to the chosen chord, the chord must maximize the area of P'. In the algorithm, it is equivalent to minimize the area of the removed parts P/P'. Using these heuristics, the approximated MACS algorithm can be easily designed in a greedy process.

Algorithm 2

Compute the kernel of P;
Compute the ordered list $\mathcal{O}$ of reflex vertices;
Extract the first point r_1 in $\mathcal{O}$;
tant que $\mathcal{O}$ *is not empty* **faire**
    Extract the first point r_2 in $\mathcal{O}$;
    Choose the best chord that maximizes the resulting polygon area with the chords (r_1, r_2);
    Modify the polygon P accordingly;
    Update the list $\mathcal{O}$ removing reflex points excluded by the chord;
    $r_1 \leftarrow r_2$;
fin

5.2 Single-Pivot Chords Computation

Given a reflex point r_i of P, we have listed three possible classes of chord: extremal, single-pivot and double-pivot chords. The Figure 11 reminds the possibles chords. The extremal and double-pivot chord computation is direct. However, we have to detail the single-pivot chord extraction. According to our heuristic, the single-pivot chord associated to r_i must be tangent to the wavefront propagation of the kernel dilatation.

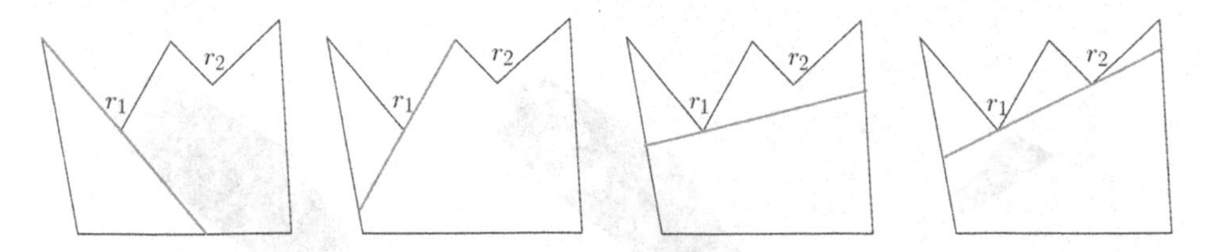

Fig. 11. All possible chords that can be associated to the reflex point r_1 (the two extremal chords, a single-pivot balanced chord and the double-pivot chord)

Using the exterior angular bisector structure issued from the computation of the generalized Voronoi diagram of $Kern(P)$, we can efficiently compute the slopes of such chords. In Figure 12, let e_1 and e_2 be two adjacent edges of $Kern(P)$ (e_1 and e_2 are incident to the vertex v). Let p (resp. q) be a point in the plane that belongs to the cell generated by e_1 (resp. e_2). We can distinguish two cases: p is closer to e_1 than to one of its extremities and q is closer to v than to e_2 (without the extremities). Hence the straight line going through p and tangent to the wave-front propagation is parallel to e_1. In the second case, the tangent to wavefront straight line going through q is tangent to the circle of center v and radius $\|\boldsymbol{vq}\|$ (see Figure 12).

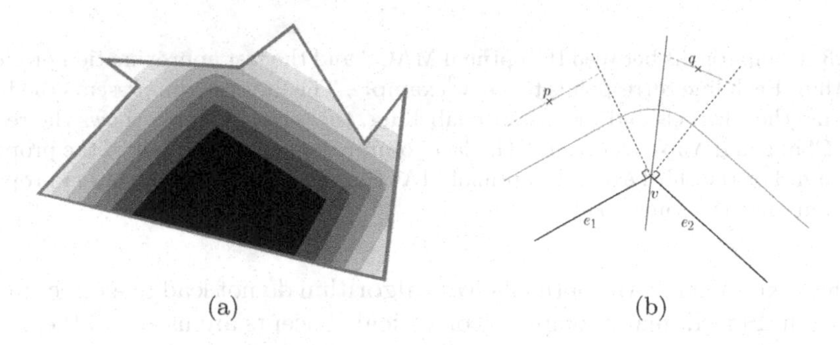

Fig. 12. Computing a chord parallel to the kernel dilatation wavefront: (a) illustration of the kernel dilatation, (b) single-pivot slope computation

Finally, if each reflex point r_i of P is labelled according to the closest edge e_i of $Kern(P)$ (extremities included), we can directly compute the single-pivot chord: if r_i is closer to e_j than one of its extremities, the chord is parallel to e_j, otherwise, the chord is tangent to a given circle. Computational cost analysis is developed in details in [20].

5.3 Experiments

In this section, we present some results of the proposed algorithm. First of all, Figure 13 compares the results between the optimal Chang and Yap's algorithm [22] and the approximated MACS extraction process on 3 examples. In

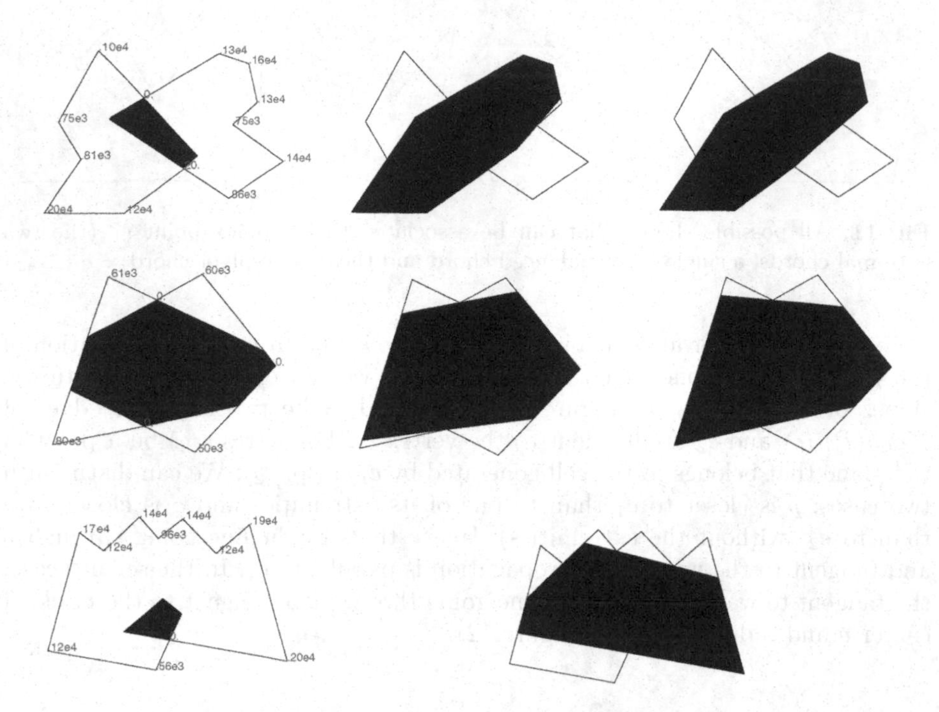

Fig. 13. Comparisons between the optimal MACS and the fast approximation proposed algorithm. Each line corresponds to a new example. The first column presents the input polygons, their kernels and the distance labelling, the second column shows the results of the Chang and Yap's algorithm. The last column presents the result of the proposed algorithm. For the third example, optimal MACS and the fast approximation proposed algorithm offer the same result.

practical experiments, the optimal $O(n^7)$ algorithm do not lead to a direct implementation. Indeed, many complex geometrical concepts are used and the overall algorithm is not really tractable. In Figure 13, the first column presents the polygon, its kernel and the distance labelling of all vertices, the second row contains the optimal MACS and the third one the fast approximation of the MACS presented in Section 5.1. Note that the results of the last row are identical. If we compute the area error between the optimal and the approximated MACS on these examples, the error is less than one percent.

6 Conclusion

Duality can be seen in two ways : a geometric or a parametric transform. Geometrical duality and graphs have been widely investigated with Voronoi diagrams and Delaunay triangulations. In this paper, we focused on transformation based duality. Two different processes have been investigated. The first one is based on characterization of digital straight lines using duality and preimage. The second one is based on polarity for kernel construction used as an initialization step for the problem of convex envelope of polygon by interior. The analysis of the geometry of the dual polygon seems to be very promising to have a direct computation of the MACS. Furthermore, a unification of both duality and polarity frameworks is a challenging future work. Many other examples or approaches could be studied and elaborated using this concept of duality. All these methods share the choice of alternative representations in order to offer optimality criteria.

References

1. Hough, P.: Method and means for recognizing complex patterns. United States Patent, n3, 069, 654 (1962)
2. Duda, R.O., Hart, P.E.: Use of the Hough transformation to detect lines and curves in pictures. Communications of the ACM **15**(1) (1972) 11–15
3. Sivignon, I.: De la caractérisation des primitives la reconstruction polyédrique de surfaces en géométrie discrète. Thèse de doctorat, Institut National Polytechnique de Grenoble, Grenoble, France (2004)
4. Dorst, L., Smeulders, A.N.M.: Discrete representation of straight lines. IEEE Trans. on Pattern Anal. and Mach. Intell. **6**(4) (1984) 450–463
5. McIlroy, M.D.: A note on discrete representation of lines. AT&T Technical Journal **64**(2) (1985) 481–490
6. Hardy, G.H., Wright, E.M.: An Introduction to the Theory of Numbers. Oxford Society (1989)
7. Vittone, J.: Caractérisation et reconnaissance de droites et de plans en géométrie discrète. Thèse de doctorat, Université Joseph Fourier, Grenoble, France (1999)
8. Lindenbaum, M., Bruckstein, A.: On recursive, $\mathcal{O}(n)$ partitioning of a digitized curve into digital straight segments. IEEE Trans. on Pattern Anal. and Mach. Intell. **15**(9) (1993) 949–953
9. Coeurjolly, D.: Algorithmique et géométrie discrte pour la caractrisation des courbes et surfaces. Thèse de doctorat, Université Lumière Lyon 2, Lyon, France (2002)
10. Coeurjolly, D., Sivignon, I., Dupont, F., Feschet, F., Chassery, J.M.: On digital plane preimage structure. Discrete Applied Math. **151**(1-3) (2005) 78–92
11. Coeurjolly, D., Brimkov, V.: Computational aspects of digital plane and hyperplane recognition. In: 11th International Workshop on Combinatorial Image Analysis. LNCS, Springer-Verlag (2006)
12. Dexet, M., Andres, E.: Linear discrete line recognition and reconstruction based on a generalized preimage. In Reulke, R., Eckhardt, U., B., F., U., K., Polthier, K., eds.: 11th International Workshop on Combinatorial Image Analysis. Volume 4040 of LNCS., Springer-Verlag (2006) 174–188
13. Vittone, J., Chassery, J.M.: Recognition of digital naive planes and polyhedrization. In Borgefors, G., Nystrm, I., Sanniti di Baja, G., eds.: Discrete Geometry for Computer Imagery. Volume 1953 of Lect. Notes on Comp. Sci., Springer-Verlag (2000) 296–307
14. Sivignon, I., Dupont, F., Chassery, J.M.: Digital intersections : minimal carrier, connectiviy and periodicity properties. Graphical Models **66**(4) (2004) 226–244
15. Veelaert, P.: Geometric constructions in the digital plane. Journal of Mathematical Imaging and Vision **11** (1999) 99–118

16. Veelaert, P.: Collinearity and weak collinearity in the digital plane. In Bertrand, G., Imiya, A., Klette, R., eds.: Digital and Image Geometry. Volume 2243 of Lect. Notes on Comp. Sci., Dagstuhl, Allemagne, Springer-Verlag (2001) 439–453
17. Veelaert, P.: Graph-theoretical properties of parallelism in the digital plane. Discrete Applied Mathematics **125** (2003) 135–160
18. Preparata, F.P., Shamos, M.I.: Computational Geometry : An Introduction. Springer-Verlag (1985)
19. Wood, D., Yap, C.K.: The orthogonal convex skull problem. Discrete and Computational Geometry **3** (1988) 349–365
20. Chassery, J.M., Coeurjolly, D.: Optimal shape and inclusion: open problems. In Ronse, C., Najman, L., Decencière, E., eds.: International Symposium on Mathematical Morphology. Volume 30 of Computational Imaging and Vision., Springer (2005) 229–248
21. Goodman, J.E.: On the largest convex polygon contained in a non-convex $n-$gon or how to peel a potato. Geometricae Dedicata **11** (1981) 99–106
22. Chang, J.S., Yap, C.K.: A polynomial solution for the potato-peeling problem. Discrete & Computational Geometry **1** (1986) 155–182

On Minimal Perimeter Polyminoes*

Yaniv Altshuler[1], Vladimir Yanovsky[1], Daniel Vainsencher[1], Israel A. Wagner[1,2], and Alfred M. Bruckstein[1]

[1] Computer Science Department, Technion, Haifa 32000 Israel
{yanival@cs, volodyan@cs, danielv@tx, wagner@cs, freddy@cs}.technion.ac.il
[2] IBM Haifa Labs, MATAM, Haifa 31905 Israel
wagner@il.ibm.com

Abstract. This paper explores proofs of the isoperimetric inequality for 4-connected shapes on the integer grid $\mathbb{Z}^2$, and its geometric meaning. Pictorially, we discuss ways to place a maximal number unit square tiles on a chess board so that the shape they form has a minimal number of unit square neighbors. Previous works have shown that "digital spheres" have a minimum of neighbors for their area. We here characterize all shapes that are optimal and show that they are all close to being digital spheres. In addition, we show a similar result when the 8-connectivity metric is assumed (i.e. connectivity through vertices or edges, instead of edge connectivity as in 4-connectivity).

1 Introduction

The isoperimetric inequality for $\mathbb{R}^2$ states that the area enclosed by a closed simple curve is at most that enclosed by a circle of the same length, with equality occurring only for curves that are circles. This implies two conclusions about circles that are equivalent in the continuous case, but distinct in discrete spaces. It is clear that among closed simple curves of a certain length, a circle encloses a maximal area, and on the other hand, that among curves enclosing a certain area, a circle has minimal length. For discrete spaces there are special shapes that have been proved to have minimal "perimeter", for various definitions of the perimeter, corresponding to the first conclusion. In the context of the $\mathbb{Z}^n$ grid, Wang and Wang [1] presented an ordering of grid points, such that every finite prefix of the sequence forms a set with minimal boundary size for that cardinality. Similar arguments have been applied to B^n (the hypercube of dimension n) and other classes of spaces, and are reviewed by Bezrukov [2]. More results appear in [3, 4, 5].

This paper is concerned with shapes that are optimal in both having minimal boundaries and having maximal areas given their boundary size. In this way, they are similar to disks. We limit our treatment to the 2 dimensional grid, and provide a characterization of shapes that are optimal in this "double" sense.

* This research was supported in part by the Israeli Ministry of Science Infrastructural Grant No. 3-942 and in part by the Devorah fund.

A. Kuba, L.G. Nyúl, and K. Palágyi (Eds.): DGCI 2006, LNCS 4245, pp. 17–28, 2006.

We will start with a useful result that illustrates the differences between the double optimality we require and the weak optimality that was imposed before. We seek a tight lower bound on the size of the neighborhood of a general subset of the $\mathbb{Z}^2$ grid for which we know the cardinality i.e. the area. This is clearly a form of isoperimetric inequality. Bounds of this sort were used to prove lower bounds on the efficiency of a multi-agent algorithm for sweeping dynamically growing shapes [6].

Let A be a finite subset of the $\mathbb{Z}^2$ grid. We define its neighborhood as $N(A) = \{p \in \mathbb{Z}^2 | d(p, A) = 1\}$, where d is the *Manhattan* metric $d((a, b), (x, y)) = |a - x| + |b - y|$. Then the lower bound we seek can be written in the form of an integer sequence $n(k) : \mathbb{N} \to \mathbb{N}$, defined via $n(k) = \min_{|A| \geq k} \{|N(A)|\}$.

Let us first look at the 2 dimensional case of the sequence described by Wang and Wang in [1]. Every prefix of this sequence is a set of tiles (a shape) that can be described as the union of a discrete sphere (all tiles whose coordinates sum to at most k) and part of the shell needed for the next largest sphere (some of the tiles whose coordinates sum to $k + 1$). The first elements of the Wang^2 sequence are (0,0), (0,1), (1,0), (-1,0), (0,-1), (1,1), (-1,1). The corresponding shapes can be seen in Figure 1. Because Wang^2 show that the shape formed by every such prefix has a minimal boundary size for its area, a formula to calculate the neighborhood size of every such shape would provide us with a way to calculate $n(k)$. Geometrically we can say that the boundary size changes whenever the expansion of the outer shell enters a new quadrant.

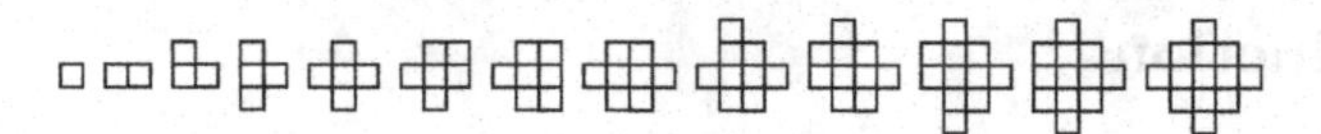

Fig. 1. The first few shapes in the 2D Wang sequence

In our approach, we first note that the function $n(k)$ is not affected by a shape that has a non minimal neighborhood size for its area (because it will not be chosen in the min), nor by a shape that has non maximal neighborhood area (since then the shape of maximal area can be used instead). Then at the beginning of the next section we provide an explicit expression for $n(k)$, which we later justify by characterizing the set of shapes that are simultaneously optimal in both having largest area for the given neighborhood size and having smallest neighborhood size given their area.

The rest of the paper is organized as follows — section 2 contains a detailed analysis of the above while section 3 presents an alternative method of producing similar results. This alternative approach is later used with slight modifications to derive similar results under the 8-connectivity metric.

2 The Isoperimetric Inequality Theorem

We shall next provide the promised explicit expression for $n(k)$, whose first few values are 0, 4, 6, 7, 8, 8, 9. This sequence already highlights the fact that the fourth shape in the Wang[2] sequence (Figure 1) obviously does not have maximal area for its boundary size, because $n(4) = n(5) = 8$.

Theorem 1. *If $k = 0$, then $n(k) = 0$. For $k > 0$ $n(k) = 4(m+1) + i$ where $(m, i) \in \mathbb{N} \times \mathbb{Z}_4$ is the first pair for which one of the following holds:*

1. $i = 0 \wedge k \leq 2m^2 + 2m + 1$
2. $i = 1 \wedge k \leq 2m^2 + 3m + 1$
3. $i = 2 \wedge k \leq 2m^2 + 4m + 2$
4. $i = 3 \wedge k \leq 2m^2 + 5m + 3$

where $\mathbb{N} \times \mathbb{Z}_4$ is ordered lexicographically (with priority to $\mathbb{N}$).

While the formula for $n(k)$ in the above theorem is explicit, one might find this expression somewhat difficult to grasp. However it is easy to understand as a way to fill a simple look up table, the first few columns of which are shown below and which is scanned column first for the first value above or equal to k. Then the column and row of that value provide i and m needed to calculate the perimeter.

$i\backslash m$	0	1	2	3	4	5	6	7	8	
0	1	5	13	25	41	61	85	113	145	
1	1	6	15	28	45	66	91	120	153	...
2	2	8	18	32	50	72	98	128	162	
3	3	10	21	36	55	78	105	136	171	

Note that the increasing sequence of values that appear in the successive rows of the table, i.e. 1 2 3 56 8 10 13 15 ... etc. are the areas of the double optimal shapes.

To gain some geometrical understanding of $n(k)$ in terms of m and i, we reconsider the Wang[2] sequence. This sequence includes among others, some optimal shapes, which are those whose area appears in the table above. As we mentioned, such a shape can be seen as a digital sphere enclosed by a shell of zero to three quadrants. It is easy to see by continuing the sequence that the radius of the digital sphere is $m+1$ and the number of quadrants is i. A complete correspondence of these areas to all optimal shapes, including small ones, will be proved in the next sections.

Our exploration of optimal shapes that yields theorem 1 consists of two phases. First we show that the optimal shapes belong to a class of simple shapes (section 2.1) and explore the structure common to all the shapes of this class (sections 2.2 and 2.3). Then we use this structure to find which shapes in that class are indeed optimal (section 2.4).

2.1 Simple Shapes

In this section we present an algorithm that allows us to cover every shape with a simple shape of the same neighborhood size and at least as much area. This will show that only simple shapes may be optimal. We will show how to calculate the neighborhood size of a simple shape, and later on its area.

Definition 1. *(x, y) is called a* 4 neighbor *of A if :*

$$(x, y) \notin A \text{ and } \{(x, y+1), (x+1, y), (x-1, y), (x, y-1)\} \cap A \neq \emptyset$$

The set of 4 neighbors of A is written $N(A)$.

Definition 2. *A shape A is called* optimal *if for every shape B :*

$$(|N(B)| \leq |N(A)| \Rightarrow |B| \leq |A|) \text{ and } (|B| \geq |A| \Rightarrow |N(B)| \geq |N(A)|)$$

Definition 3. *A shape is called* simple *if it can be written as :*

$$B = \{(x, y) \,|y - x \in [j_1, j_2] \text{ and } x + y \in [k_1, k_2]\}$$

We often refer to the sizes of a simple shape as $j = j_2 - j_1; k = k_2 - k_1$. Sometimes the specific directions do not matter, in which cases we denote w.l.o.g $a = \min\{j, k\}; b = \max\{j, k\}$. Note that there may be different shapes that have the same dimensions.

Theorem 2. *If A is optimal, then A is simple.*

Proof. Let $A = \{(x, y)\}$ be a set of tiles, $k_1 = \min\{k|\exists (x, y) \in B \wedge x - y = k\}$, $k_2 = \max\{k|\exists (x, y) \in B \wedge x - y = k\}$, $j_1 = \min\{j|\exists (x, y) \in B \wedge x + y = j\}$, and $j_2 = \max\{j|\exists (x, y) \in B \wedge x + y = j\}$.

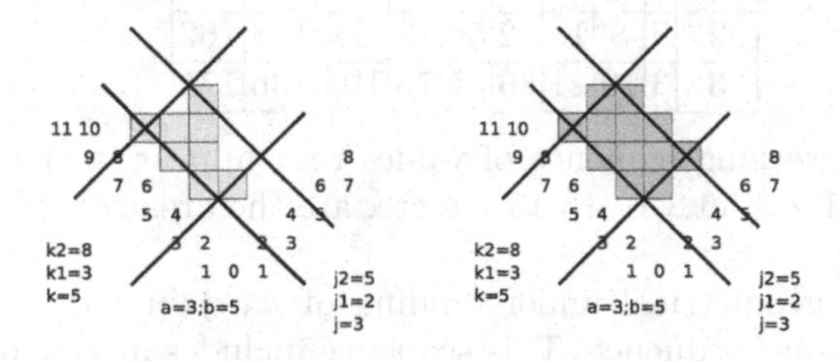

Fig. 2. A general shape A and the corresponding simple shape B

We look at the shape $B = \{(x, y) \mid x - y \in [j_1, j_2] \wedge x + y \in [k_1, k_2]\}$, then clearly $B \supset A$. We will show that if A is optimal, $B = A$. Since B is simple, this is sufficient. On each boundary line there is at least one point that is in A. Let $\{(x_1, y_1), (x_2, y_2), (x_3, y_3), (x_4, y_4)\}$ be on such a boundary, where $(x_i, y_i), (x_{i+1}, y_{i+1})$ are on non-opposite sides. Note that it is possible that $(x_i, y_i) = (x_{i+1}, y_{i+1})$, for example for $A' = \{(x_0, y_0)\}$ all the points are the same, and that we consider the indexes i modulo 4, so that $i = 4 \Rightarrow i + 1 = 1$.

W.l.o.g, we assume that $y_i - x_i = k_2 \wedge x_{i+1} + y_{i+1} = j_2$, then $x_i \leq x_{i+1}$. Since there are no vacant columns between x_i, x_{i+1}, A has at least $x_{i+1} - x_i + 1$ neighbors from above (in each column, the neighbor above the highest tile of A in that column — see Figure 3). Doing the same for the other 3 adjacent pairs of points, we find a lower bound on neighbors from the left, from below, and from the right. Note that this bound is tight for shape B, which has no other neighbors, and has all the possible tiles. Then if $A \neq B$, A is not optimal, because it has at least as many neighbors, and not as many tiles. □

Definition 4. *A simple shape A with $a = 0$ is denoted as* degenerate shape.

Note that degenerate shapes behave differently from other simple shapes (for example, in the degenerate case b cannot have odd values, because the Manhattan distance between two tiles on a diagonal is always even).

Lemma 1. *The only optimal degenerate shapes have an area of 0, 1 or 2.*

Proof. If A is degenerate, then $a = 0$. We assume by contradiction that $b \geq 4$ (see Figure 3) and A is optimal. But the shape B, created by placing all the tiles in the same column has exactly as many neighbors (two horizontal neighbors per tile, and two additional vertical neighbors), and the same area, but is not simple, therefore is not optimal. Then A cannot be optimal. The shapes with $b \leq 3$ have areas as described, as can be seen in Figure 7. □

Since we have seen that only a small and finite set of degenerate shapes is of interest to our discussion we shall assume that all simple shapes have $a \geq 1$.

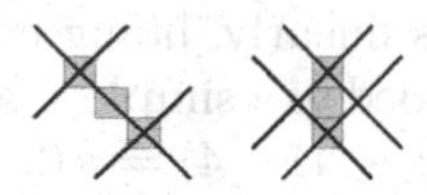

Fig. 3. A degenerate shape A, and a variation A' which clarifies that A is non optimal

Lemma 2. *Every simple shape has $j + k + 4$ 4-Neighbors*

Proof. By induction on j and k. This is true for the shape of two neighboring tiles (i.e $j = k = 1$), and 6 neighbors. In the induction step, we assume validity for j, k and prove it for $j+1$ (same reasoning applies to expansion in k). Adding 1 to j causes one diagonal side (having r tiles) to expand to some direction (an expansion up is illustrated in Figure 4). As a result, r neighbors in that direction become new tiles, and r vacant beyond those in the same direction become neighbors, not modifying the neighborhood size yet. However, the new tile that is last in the direction of advancement is exposed to a new neighbor from the side. Having been diagonal to an extreme tile in the shape, it was not a neighbor before (i.e. increasing j or k adds one neighbor), thus a shape of dimensions $(j + 1), k$ has $j+1+k+4$ neighbors, completing the induction step. □

2.2 Expansion

In this section we demonstrate how each simple shape can be described as a "*spine*" expanded by an iterative expansion process. This process and its effects on the area and neighborhood size of a simple shape is described.

Definition 5. *Let A be a simple shape of dimensions j, k. We call increasing each of j, k by two an* expansion step.

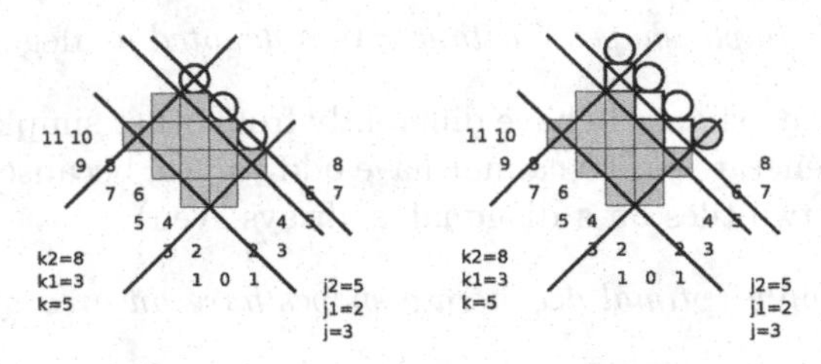

Fig. 4. Expanding a simple shape to one side

Note that each expansion step performed on a simple shape adds exactly all of its 4-neighbors. Thus, the number of tiles of the shape increases by $j + k + 4$, and the number of neighbors grows by 4. See Figure 5 for an example.

Lemma 3. *Let A be a simple shape with dimensions j, k. After s expansion steps, its neighborhood grows by $4s$ and its area grows by $E(j, k, s) = s(2 + j + k + 2s)$.*

Proof. The neighborhood grows linearly, being equal to $j + k + 4$. $E(j, k, s)$ is defined as the number of tiles added to a simple shape by s expansions, therefore : $E(j, k, s) = \sum_{i=0}^{s-1}((j + 2i) + (k + 2i) + 4) = s(2 + j + k + 2s)$. □

2.3 Spines

Definition 6. *A simple shape such that $a \in \{1, 2\}$ is called a* spine.

Theorem 3. *A simple shape A can be described as a spine, expanded some finite number (possibly zero) of times. This description is unique.*

We shall next show that there are only 4 kinds of spines. Thus, since we know the area added by each expansion step, we can calculate the areas of all simple shapes.

Proof. Of the theorem. If a of A is even, we say that A_s has dimensions $2, b-a+2$, otherwise $1, b - a + 1$. Either way, A_s is a spine and expanding it $s = \lceil \frac{a}{2} \rceil - 1$ times yields exactly A. Then the area of every simple shape is the sum of the area of its spine A_s and the area added in the expansions. We note that starting from any other spine will result in the wrong shape - a different initial width (or different number of expansions) results in wrong parity of the final width, and the same spine width but different different length results in a wrong difference between length and width. Therefore this description is unique. □

Lemma 4. *Let A_s be a spine of dimensions $a \leq b$, then its area is given by (See Figure 5):*

1. *If $a = 1$, the area is $b + 1$*
2. *If $a = 2$, then we have the following options:*

(a) *If b is odd, then* $|A_s| = \frac{3\cdot(b+1)}{2}$.
(b) *b is even, of type 1, then* $|A_s| = \frac{3\cdot b}{2} + 1$.
(c) *b is even, of type 2, then* $|A_s| = \frac{3\cdot b}{2} + 2$.

Proof. For $a = 1$, there are b tiles at distances 0 to $b-1$ from one line, and one more. For $a = 2$, there are $\lfloor \frac{b+1}{2} \rfloor$ triplets of tiles. Note that there are two ways of getting from an odd b to an even one, depending on which boundary is moved, resulting in different area increases. □

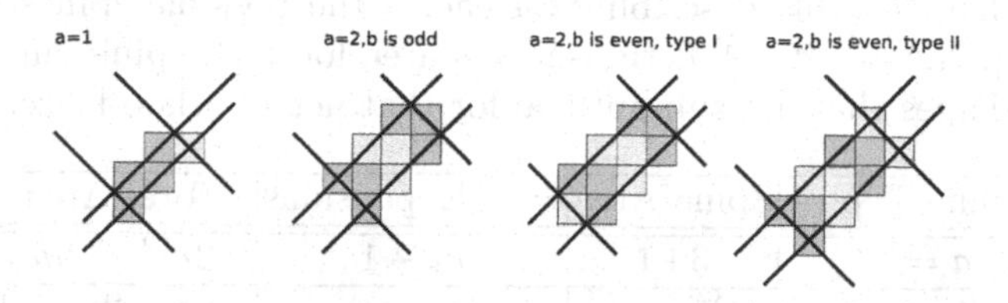

Fig. 5. Spine types and their areas

2.4 Spines of Optimal Shapes

Theorem 4. *Let A_s be a spine with dimension a, b of an optimal shape A, then $a + 4 > b$.*

Later we show that this result, while necessary in our construction, is not tight.

Proof. We assume by contradiction that A is an optimal shape with spine $a+4 \leq b$, extended s times. Then we take the same skeleton with b shortened by 4, and expanding it $s+1$ times we get B, such that $|N(A)| = |N(B)|$.

We will now show that $|B| > |A|$, contradicting the optimality of A. $|A|$ is the sum of spine size and $E(a, b, s)$.

First we note that $E(a, b-4, s+1) - (E(a, b, s)) = a + b$, then if $a = 1$, the area of the skeletons is $b - 4 + 1$ and $b + 1$, then $b - 4 + a \geq a + 4 - 4 + a > 0 \Rightarrow (b - 4 + 1) + (a + b) > b + 1$.

If $a = 2$, in all the variations, subtracting 4 from b reduces the skeleton area by precisely 2, but the expansions more than offset that because $a + b \geq 2a + 4 > 2$. □

Corollary 1. *The dimensions of spines of optimal shapes are a subset of:*

$$\{(1,1), (1,2), (1,3), (1,4), (2,2), (2,3)\,(2,4), (2,5)\}$$

Recalling Lemma 4, we note that spines of dimensions $\{(2,2), (2,4)\}$ mentioned above come in two types. As we saw then type 2 spines have strictly more area than those of type 1, with the same neighborhood. Therefore only type 2 spines can result in optimal shapes. In this context, each set of spine dimensions results in a certain spine area and neighborhood size.

Theorem 5. *Let A be a non degenerate optimal shape with dimensions j, k, so that $|N(A)| = 4(m+1) + i$, with $i \in \{0, 1, 2, 3\}$. Then:*

$$|A| = 2m^2 + (1+i)m + \max\{1, i\}$$

Proof. Let a, b be the dimensions of A's spine, then remembering each expansion increases the neighborhood size by 4, we see that $4(m+1) + i = j + k + 4 = a + b + 4(s+1)$. One conclusion is that $a + b \equiv i \bmod 4$, and another is that $s = \frac{4m+i-a-b}{4}$. Hence, denoting $|A_s|$ the area of the skeleton of dimensions a, b, the total area for such a shape is exactly $|A| = |A_s| + E\left(a, b, \frac{4m+i-a-b}{4}\right)$.

Below we have a table describing for each i the possible spines for optimal shapes with $|N(A)| = 4m + i$, the shape's area for each spine, and the spines resulting in shapes that are sub-optimal for that neighborhood size.

i	Spine	Spine Type	Spine Area	# of Expansions	Total Area	Sub-opt.
0	$(1,3)$	$a=1$	3+1	$m-1$	$2m^2+2m$	yes
0	$(2,2)$	$a=2$ b is even	$\frac{3\cdot 2}{2}+2$	$m-1$	$2m^2+2m+1$	
1	$(1,4)$	$a=1$	4+1	$m-1$	$2m^2+3m$	yes
1	$(2,3)$	$a=2$ b is odd	$\frac{3\cdot(3+1)}{2}$	$m-1$	$2m^2+3m+1$	
2	$(1,1)$	$a=1$	1+1	m	$2m^2+4m+2$	
2	$(2,4)$	$a=2$ b is even	$\frac{3\cdot 4}{2}+2$	$m-1$	$2m^2+4m+2$	
3	$(1,2)$	$a=1$	2+1	m	$2m^2+5m+3$	
3	$(2,5)$	$a=2$ b is odd	$\frac{3\cdot(5+1)}{2}$	$m-1$	$2m^2+5m+2$	yes

□

Theorem 5 provided a necessary condition for an optimal non degenerate simple shape A. However, although we have shown the optimal area for every specific neighborhood size, we are not done yet. We must show that no shape exists having larger area and smaller neighborhood. This should hold because $n(k)$ is defined so that it is a non-decreasing sequence.

First we note that for any specific m, $|A|$ is strictly monotonous in i. Furthermore, we see that $2(m+1)^2 + 2(m+1) + 1 = 2m^2 + 4m + 2 + 2m + 2 + 1 = 2m^2 + 6m + 5 > 2m^2 + 5m + 3$, then $|A|$ is strictly monotonous in $N(A)$. Thus, all size values in the above result are indeed areas of optimal shapes. Therefore:

Theorem 6. *The non-degenerate optimal shapes are those simple shapes that when decomposed into spine and expansion have a spine of one of the following forms: $(a, b) \in \{(1,1), (1,2), (2,2), (2,3)\ (2,4)\}$ (these spines appear in Figure 6).*

Corollary 2. *Let A_s be a spine with dimension a, b of an optimal shape A, then $a + 3 > b$.*

Corollary 2 is a tighter version of theorem 4, and can now be verified by inspection of the list of optimal spines.

Corollary 3. *The degenerate simple shapes with areas $0, 1, 2$ are all optimal.*

All these appear in Figure 7.

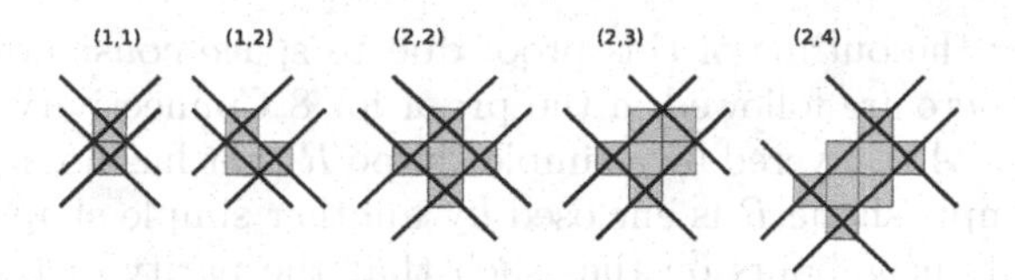

Fig. 6. The optimal spines

Proof. There are no other optimal shapes with neighborhoods sizes 0 or 4, and the other optimal shapes of neighborhood size 6 also have area equal to 2. □

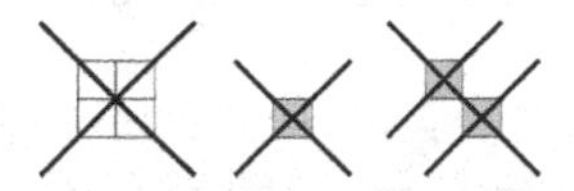

Fig. 7. The optimal degenerate shapes: two with dimensions (0,0) and one with (0,2)

We have now identified all the optimal shapes, degenerate and simple, with explicit expressions for their neighborhood sizes and areas. This allows us to state that every shape of area k has a neighborhood at least as large as that of (every) optimal shape with area $\leq k$. From this characterization, Theorem 1 immediately follows.

3 Alternative Analysis

This section describes an alternative approach to the grid isoperimetric inequality. Some results similar to those presented in section 2 are rederived, as well as a new result, concerning the 8-connectivity grid metric.

3.1 Four Connectivity in $\mathbb{Z}^2$

Let A be a finite subset of the $\mathbb{Z}^2$ grid, having the *neighborhood* $N(A) = \{p \in \mathbb{Z}^2 | d(p, A) = 1\}$, where d is the *Manhattan* metric $d((a,b),(c,d)) = |d-b| + |c-a|$. Let us denote $n(A) = |N(A)|$.

For some area $k \in \mathbb{N}$ let $A_{MIN}(k)$ be defined as the shape of area k whose neighborhood is the smallest, namely :

$$A_{MIN}(k) \subset \mathbb{Z}^2 \wedge |A_{MIN}(k)| = k \wedge$$
$$\forall A \subset \mathbb{Z}^2 \quad (|A| = k) \rightarrow (n(A) \geq n(A_{MIN}(k)))$$

Theorem 7. *For every positive k, the neighborhood of $A_{MIN}(k)$ is at least as large as this of the largest digital sphere (assuming 4 Connectivity) of size at most k, minus two, namely :*

$$\forall k \in \mathbb{N} \quad n(A_{MIN}(k)) \geq \max\{n(A_{SPHERE}) \mid |A_{SPHERE}| \leq k\} - 2$$

We provide only the outline of this proof due to space considerations, but the steps described here are followed in the proof for 8 Connectivity. As we proved before, any shape A is covered by a simple shape R that has no more neighbours than A. This simple shape R is enclosed by another simple shape denoted CR, that has all of its neighbours on tiles such that the parity of the sum of their coordinates is constant (if the plane is colored as a chess board, the neighbours are all of the same color) and such that CR has at most two more neighbours than C and A. We call such a shape a Canonical Rectangle. We derive formulae for the area and boundary size of CR based on the sizes of its sides a, b, then find a lower bound on the minimal boundary for a canonical rectangle of the same area by differentiating by b. This minimum is found when $a = b$, that is when the canonical rectangle is a Manhattan sphere.

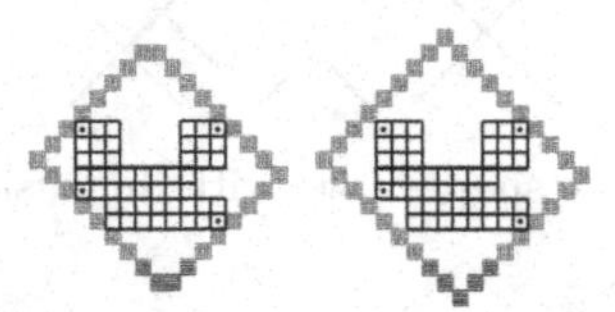

Fig. 8. For a shape A, the left chart demonstrates R while the right chart demonstrates the canonical rectangle CR

3.2 Eight Connectivity in $\mathbb{Z}^2$

Let B be a finite subset of the $\mathbb{Z}^2$ grid, having the *neighborhood* $N_8(B) = \{p \in \mathbb{Z}^2 | d_8(p, B) = 1\}$, where $d_8((a,b),(c,d)) = \max\{(d-b),(c-a)\}$. Let us denote $n_8(B) = |N_8(B)|$.

For some area $k \in \mathbb{N}$ let $B_{MIN}(k)$ be defined as the shape of area k whose neighborhood is the smallest, namely :

$$B_{MIN}(k) \subset \mathbb{Z}^2 \wedge |B_{MIN}(k)| = k \wedge$$
$$\forall B \subset \mathbb{Z}^2 \quad (|B| = k) \rightarrow (n_8(B) \geq n_8(B_{MIN}(k)))$$

Theorem 8. *For every positive k, the size of the neighborhood of $B_{MIN}(k)$ is at least as large as this of the largest digital sphere (assuming 8 Connectivity) of size at most k, namely :*

$$\forall k \in \mathbb{N} \quad n_8(B_{MIN}(k)) \geq \max\{n_8(B_{SPHERE}) \mid |B_{SPHERE}| \leq k\}$$

Note that a digital sphere of radius 3, for example, is a 5 by 5 square.

Proof. Let us denote the bounding rectangle of B by *bounding-rectangle*(B). For each of the four sides of *bounding-rectangle*(B) (i.e. *top, right, down, left*) let us denote the last tiles of B that are 4 neighbors of the four sides (assuming clockwise movement) by 1, 2, 3 and 4 respectively. See an example in Figure 9.

Let us project all the tiles of *bounding-rectangle*(B) between points 1 and 2 in 45° *down-left*, the points between 2 and 3 in 45° *up-left*, the points between

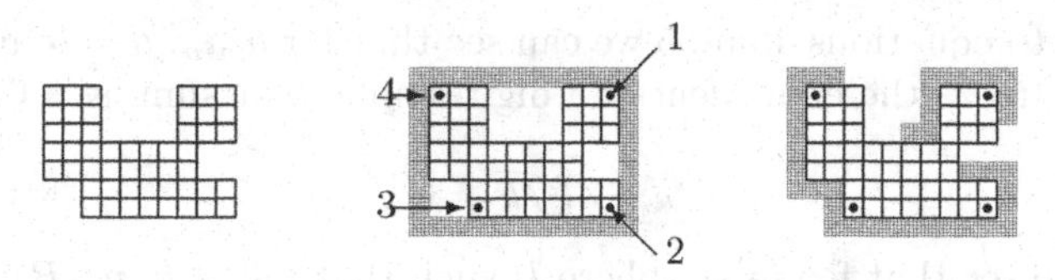

Fig. 9. An example of the *bounding-rectangle* and its projection

3 and 4 in 45° *up-right* and the points between 4 and 1 in 45° *down-right*. An example appears in Figure 9.

Clearly, after such projection each tile of *bounding-rectangle*(B) a 4 neighbor of at least a single tile of B. In addition, it is impossible that two tiles of *bounding-rectangle*(B) will merge in the same spot. Thus, $n_8(B)$ is at least the number of tiles in *bounding-rectangle*(B), namely :

$$\forall B \quad n_8(B) \geq |\textit{bounding-rectangle}(B)| \tag{1}$$

Let $R(k)$ be the smallest rectangle which contains at least k tiles. Let a and b denote the sides of $R(k)$ and let c denote the number of tiles $R(k)$ comprises. Then :

$$c = 2(a+b) - 4 \tag{2}$$

Let $f(a,b)$ denote the area of a rectangle of sides a and b :

$$f(a,b) = (a-2)(b-2) \tag{3}$$

We would like to find a solution for the following optimization problem :

$$\min \quad c \quad s.t \quad f(a,b) \geq k \qquad \wedge \qquad c = 2(a+b) - 4$$

After some arithmetics equation 3 can be written as :

$$a = \frac{f(a,b)}{b-2} + 2 \tag{4}$$

Combining this with 2 we get $c = \frac{2f(a,b)}{b-2} + 2b$. Since we require that $f(a,b) \geq k$ we can write the following :

$$c \geq \rho \triangleq \frac{2k}{b-2} + 2b \tag{5}$$

Note that while the minimizing b may not be an integer, it still gives a bound valid over the integer b. In order to minimize ρ we require that $\frac{\partial \rho}{\partial b} = 2 - \frac{2k}{(b-2)^2} = 0$ and after some arithmetics we get that :

$$b = \sqrt{k} + 2 \tag{6}$$

By examining the behavior of $\frac{\partial^2 \rho}{\partial b^2}$ we can see that for $b = \sqrt{k} + 2$ since $k \geq 1$ then $\frac{\partial^2 \rho}{\partial b^2} > 0$, meaning that ρ is indeed minimized at this point. By assigning the value of b_{min} to equations 4 and 5 we can see that for b_{min}, $a = b$ (meaning that $R(k)$ is a square — the equivalent of a digital sphere, assuming 8 Connectivity) and that :

$$c \geq 4(\sqrt{k} + 1) \tag{7}$$

It is easy to see that for some sphere B such that $|B| = k$, $n_8(B) = 4(\sqrt{k}+1)$ and therefore it is the shape that minimizes the neighborhood for shapes of given area k. The rest of the Theorem is implied. □

References

1. Wang, D.L., Wang, P.: Discrete Isoperimetric Problems. SIAM J. Appl. Math., **32**(4) (1977) 860–870
2. Bezrukov, S.L.: Isoperimetric Problems in Discrete Spaces. Extremal Problems for Finite Sets, Janos Bolyai Soc. Math. Stud. 3, Budapest, Hungary (1994) 59–91
3. Chung, F.: Discrete Isoperimetric Inequalities. Surveys in Differential Geometry IX, International Press (2004) 53–82
4. Bollob'as, B., Radcliffe, A.J.: Isoperimetric Inequalities for Faces of the Cube and the Grid. Europ. J. Combinatorics, **11** (1990) 323–333
5. Tiersma, H.J.: A Note on Hamming Spheres. Discr. Math., **54** (1985) 225–228
6. Altshuler, Y., Bruckstein, A.M., Wagner, I.A.: Swarm Robotics for a Dynamic Cleaning Problem. IEEE Swarm Intelligence Symposium (2005) 209–216

A Generic Approach for n-Dimensional Digital Lines

Fabien Feschet and Jean-Pierre Reveillès

LAIC Laboratory
IUT Clermont-Ferrand - Campus des Cézeaux
63172 Aubière Cedex - France
{feschet, reveil}@laic.u-clermont1.fr

Abstract. In this paper, we provide an unified view of two definitions of digital lines in 3D via the use of lattice theory and specific projections of the lattice $\mathbb{Z}^3$. We use this unified vision to explain the extension of the definition of Voss [1] to an arbitrary dimension and we show how to extend the definition of Figueiredo and Reveillès [2] to an arbitrary dimension.

1 Introduction

Digital lines are among the simplest primitives in Digital Geometry. Many definitions have been proposed by many authors [3], which are almost all equivalent in 2D. Several drawing algorithms are known as well as several recognition algorithms. All of this explains why digital lines are extremely central for a lot of digital algorithms. Thus, it is natural to look for an extension of the definition of 2D digital lines to 3D digital lines. Moreover, as the applications nowadays manipulate 3D, 4D and sometimes higher dimensional data, extensions of digital lines to n-D becomes also very important and critical.

Several extensions have been proposed to define 3D digital lines. First, Voss [1] recalled some previous works by Kim [4] and proposed a definition of n-D digital lines based on the integer part function $\lfloor . \rfloor$. Second, the work of Debled-Rennesson et al. [5,6] proposed to define 3D digital lines through their projections (two or three in the general case) on the planes defined by the axes of the standard basis of $\mathbb{Z}^3$, and used the arithmetical approach of Reveillès [7]. It must be noticed that [4] also used projections onto the coordinates planes. The work of Debled et al [6] also leads to a recognition algorithm. A third approach was also done by Figueiredo and Reveillès in [8, 2] using lattice theory and projections onto the orthogonal plane of a direction v in $\mathbb{Z}^3$. As it can be seen, only the definition given by Voss [1] extends to an arbitrary dimension. Beside this, we can note that there exist drawing algorithms of digital lines in n-D [9] based on displacement vectors. Moreover, the definition of n-D digital lines is related to the notion of digitization. Some models are presented by Klette [10] (with the important correction given in [11]).

The goal of this paper is to present a unified and generic view of the definition of Voss [1] and the definition of Figueiredo and Reveillès [2]. Moreover, due

A. Kuba, L.G. Nyúl, and K. Palágyi (Eds.): DGCI 2006, LNCS 4245, pp. 29–40, 2006.

to this unified viewpoint, we also extend the last definition to an arbitrary dimension. The main mathematical tools used in this paper are lattice theory and projections of lattices. We prove in the paper that both definition are obtained via projection of the lattice $\mathbb{Z}^3$ onto specific planes which are the xOy plane for the definition of Voss and the orthogonal plane - as it was already the case - in the definition of Figueiredo and Reveillès.

The structure of the paper is as follows. We first recall how to manage symmetries in 3D via the octaedral group in section 2. This is followed in section 3 by the construction of the definition of Voss in dimension 3, as well as a recall of the construction of Figueiredo and Reveillès. We end this section by the presentation of a drawing algorithm. Then in section 4, those approaches are extended to an arbitrary dimension. We also present some results concerning the basis of the lattice we manipulate. The paper ends in section 5 with some conclusions and perspectives.

2 Preliminaries

In 2D, it is usual to restrict the study of digital lines to the first octant where for each point (x, y), we have $0 \leq y \leq x$. In higher dimension, we can do the same following the approach of Reveillès [12]. Hence, we will use the group of the symmetries of the unit cube in $3D$. We denote this octaedral group by O_h. This group can be identified to the product of the group $(\frac{\mathbb{Z}}{2\mathbb{Z}})^3$ of order 8, and the group $\mathcal{S}_3$ of the permutations of the three letters a, b, c, whose order is 6. The order of O_h is thus 48. Its geometrical interpretation is easy using rotations and symmetries and is given on Fig. 1.

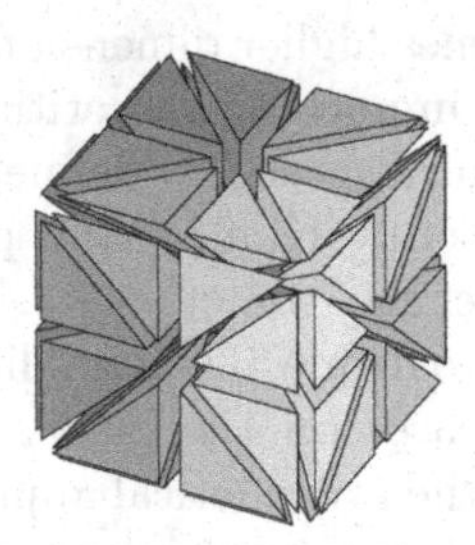

Fig. 1. The octaedral group O_h associated to the decomposition of a cube into 48 tetraedra, each being a transformation of the fundamental domain $0 \leq a \leq b \leq c$ by an element of the group O_h

Using the octaedral group, we could study only the fundamental domain which is the subset F of $\mathbb{Z}^3$ composed of the integer points (a, b, c) such that $0 \leq a \leq b \leq c$. To generate all possible cases, we simply study the action of O_h on a triple of signed symbols $(\pm a, \pm b, \pm c)$. Each of the eight elements of the subgroup $(\frac{\mathbb{Z}}{2\mathbb{Z}})^3$ of O_h modify the signs of the symbols and the other six, coming from S_3, permute them. Consequently, O_h can be identified to the group of 3×3 matrices where each row and column contains only one value being either +1 or −1. We can effectively find 48 of such matrices.

Let $n = (n_1, n_2, n_3) \in \mathbb{Z}^3$ be a vector and let us denote by Dom_n the domain of n. The element g_n of O_h which maps Dom_n onto F can be constructed as follows. We sort the matrix whose columns are the n_i such that the order of the first row is increasing. The resulting matrix is the element M_n. Let us now denote by $perm_n : \mathbb{Z}^3 \to \mathbb{Z}^3$ and by $sgn_n : \mathbb{Z}^3 \to \mathbb{Z}^3$ the applications

$$perm_n : \begin{pmatrix} x \\ y \\ z \end{pmatrix} \mapsto M_n \begin{pmatrix} x \\ y \\ z \end{pmatrix} \quad sgn_n : \begin{pmatrix} x \\ y \\ z \end{pmatrix} \mapsto \begin{pmatrix} sign(n_1)x \\ sign(n_2)y \\ sign(n_3)z \end{pmatrix}$$

where $sign(w)$ is the sign of w. We have $g_n = sgn_n \circ perm_n$.

3 3D Digital Lines

We present in this section our construction of 3D digital lines using lattices of $\mathbb{R}^n$ and arithmetics. Recall that if $v_1, v_2, \ldots, v_p$ is a collection of p vectors of $\mathbb{R}^n$, then the lattice generated by the collection is the set of all integral combinations

$$\sum a_i v_i, \quad \forall i = 1, 2 \ldots, p, \quad a_i \in \mathbb{Z}$$

Our approach is based on the study of the repartition of integer points of $\mathbb{Z}^3$ in the neighborhood of the integral direction given by the vector (a, b, c). We will define 3D digital lines based on the notion of 1D dotted lines as it is the case for the two dimensional lines [13]. Using the octaedral group O_h, we suppose that (a, b, c) belongs to the fundamental domain F. Moreover, a, b and c are supposed to be relatively prime.

Let us denote by $\mathcal{E}$ the set of all Euclidean lines whose direction vector is v and which contain integer points. We will call *1D dotted lines* with direction $v = (a, b, c)$, the intersection of the Euclidean lines - with direction v - with $\mathbb{Z}^3$. The plane (P) given by $ax + by + cz = 0$ is a subgroup of $\mathbb{R}^3$. The *orthogonal projection* of $\mathbb{Z}^3$ onto (P) is the intersection of $\mathcal{E}$ and (P) and is a lattice of (P) denoted by $\mathcal{E}_\mathcal{P}$. This lattice is clearly a rational lattice (see Fig. 4).

It is easy to verify that the intersection of the planes $-cx + az = 0$ and $-cy + bz = 0$ is the line directed by v and passing through the origin. The intersections of the planes $-cx + az = k$ and $-cy + bz = l$ where $k, l \in \mathbb{Z}$ also give a family of lines of direction v. We denote by $\mathcal{D}$ this family. It is clear that $\mathcal{E} \subset \mathcal{D}$, but the converse is false as the following system shows it,

$$\begin{cases} 13x - 3z &= 2 \quad (1) \\ 13y - 5z &= 3 \quad (2) \end{cases}$$

corresponds to the line whose direction is $(3, 5, 13)$ but this line does not intersect $\mathbb{Z}^3$. Indeed, the solution of (1) are given by $(2 + 3\mu, 0, 8 + 13\mu)$, $\mu \in \mathbb{Z}$, whereas the solutions of (2) are $(0, 6 + 5\nu, 15 + 13\nu)$, $\nu \in \mathbb{Z}$. To have an integer solution of the system, one must have: $\exists \mu, \nu \in \mathbb{Z},\ 8 + 13\mu = 15 + 13\nu$; this is clearly impossible.

Let us denote by $\mathcal{D}_{\mathcal{P}}$ the lattice given by the intersection of $\mathcal{D}$ and the plane (P). It is clear that $\mathcal{E}_P \subset \mathcal{D}_P$. The lattice $\mathcal{E}_P$ belonging to (P) is the projection of all *1D dotted lines*, but its use is not very easy. However, we can see it as a sublattice of $\mathcal{D}_P$. Since $\mathcal{D}_P$ is a Cartesian lattice, it is much easier to work with. These lattices were introduced in [2] to propose a new definition of $3D$ digital lines. Beside this definition, we can refer to definition 4.2.3 of Voss [1] of nD digital lines. In the sequel, we explain how to obtain Voss definition using two specific lattices whose construction is similar to the one of $\mathcal{E}_P$ and $\mathcal{D}_P$ of [2].

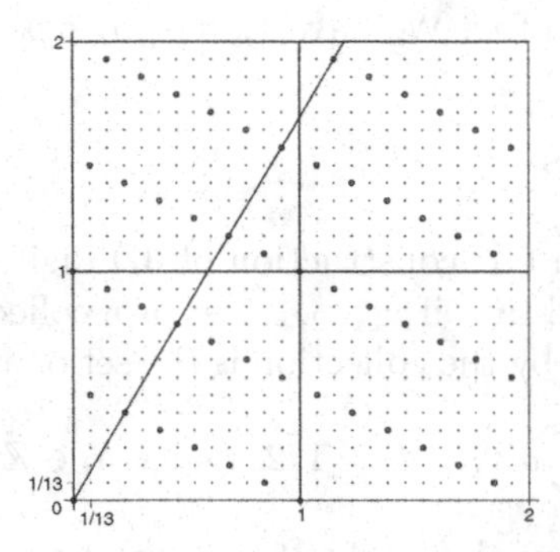

Fig. 2. The lattices $\mathcal{D}_{xy}$, $\mathcal{E}_{xy}$, the line l with direction $(a, b, c) = (3, 5, 13)$

3.1 The Lattices $\mathcal{E}_{xy}$ and $\mathcal{D}_{xy}$

Let us denote by $\mathcal{E}_{xy}$ and $\mathcal{D}_{xy}$ the lattices which are respectively the intersections of the collection of lines $\mathcal{D}$ and $\mathcal{E}$ with the xOy plane (see Fig. 2).

Proposition 1. *The lattice $\mathcal{D}_{xy}$ is the set of integer points of the xOy plane, given by $(\frac{k}{c}, \frac{l}{c})$ where k and l are arbitrary integers.*

Let $\mathcal{L}$ be the line with direction v containing the origin and let l be its projection onto the plane $z = 0$. We then have the following.

Proposition 2. *The lattice $\mathcal{E}_{xy}$ is the set of rationnal points $(x - \frac{az}{c}, y - \frac{bz}{c})$ of the plane $z = 0$ where x, y, z are arbitrary integers.*

To efficiently manipulate 3D digital lines, we must clearly understand the lattice $\mathcal{E}_{xy}$. To do this, we now give a modular generation of this last lattice.

When z varies in $\mathbb{Z}$, the points $(x + \frac{za}{c}, y + \frac{zb}{c})$ are located into unit squares given by $[k, k+1[\times[l, l+1[\subset \mathbb{R}^2$ where k and l are well chosen. We can consider the reduction of this series modulo $(1, 1)$, that is $(\frac{ka \mod c}{c}, \frac{kb \mod c}{c})$. We simplify the notation by denoting by $\{\frac{u}{v}\}$ the value of $u \mod v$, such that the previous couple is $\frac{1}{c}(\{ka/c\}, \{kb/c\})$.

The c points $\frac{1}{c}(\{ka/c\}, \{kb/c\}), k = 0, 1, 2, \ldots, c-1$ of the lattice $\mathcal{E}_{xy}$ all belong to the unit square $[0, 1[\times[0, 1[$. We denote by Π_{abc} this set. The whole lattice $\mathcal{E}_{xy}$ is obtained by periodic translations of Π_{abc}. The only consideration of the lattices $\mathcal{D}_{xy}$ and $\mathcal{E}_{xy}$ will lead to the notion of 3D digital lines.

First, the points $\frac{1}{c}(za, zb)$ of $\mathcal{E}_{xy}$, belonging to l, are the projections of the intersections of $\mathcal{L}$ with the planes $z = cste$. But l also intersects the unit squares of $\mathbb{R}^2$ given by $[x, x+1[\times[y, y+1[$. These squares are the projection over the plane $z = 0$ of the voxels $[x, x+1[\times[y, y+1[\times[z, z+1[$ of $\mathbb{R}^3$. Consequently, the study of the intersections of $\mathcal{L}$ with the unit cube of $\mathbb{R}^3$ is equivalent to the study of the intersections of l with the squares $[x, x+1[\times[y, y+1[$ and the study of the lattices $\mathcal{D}_{xy}$ and $\mathcal{E}_{xy}$.

If we consider the parallel lines to $\mathcal{L}$ containing a point of Π_{abc}, thus the *1D dotted lines* with direction v, we can verify that they intersect the same voxels or neighboor voxels than $\mathcal{L}$. Hence, the set of intersected voxels is a 26-connected structure we can call a 3D digital lines.

Among the points of $\mathcal{E}_{xy}$ belonging to l, which are generally rational points, the one given by $k(a, b) = \frac{kc}{c}(a, b)$ with $k \in \mathbb{Z}$ are *integer points*. These are the projections of the integer points $k(a, b, c),\ k \in \mathbb{Z}$ belonging to the line $\mathcal{L}$.

Theorem 1. *The projection of $k(a, b, c),\ k \in \mathbb{Z}$ of the line $\mathcal{L}$ is the series of integer points of the plane $z = 0$ given by $k(a, b),\ k \in \mathbb{Z}$. If $\frac{1}{c}(m, n)$ is a point different from $(0, 0)$ in Π_{abc}, either $i = -ma^{-1} \mod c$ or $i = -nb^{-1} \mod c$ where a^{-1} (resp. b^{-1}) is the inverse of a (resp. b) in the group $\frac{\mathbb{Z}}{c\mathbb{Z}}$, then $(\frac{1}{c}(m + ia), \frac{1}{c}(n + ib))$ is an integer point and is the projection of the point $(1 + \left[\frac{ia}{c}\right], 1 + \left[\frac{ib}{c}\right], i)$ of the dotted line parallel to $\mathcal{L}$ and containing the point (m, n).*

Proof. The first relation on $\mathcal{L}$ has already been given. For the second on an arbitray dotted lines, as the three integers a, b, c are relatively prime ($(a, b, c) = 1$ where (a, b, c) is the gcd of the three numbers), then a or b is invertible $\mod c$ (as $(m, n) \in \mathcal{E}_{xy}$, $ma^{-1} = nb^{-1} \mod c$ if both are invertible). Consequently, $m + ia$ and $n + ib$ are multiples of c. Using the Euclidean division between $m + ia$ and $n + ib$, we obtain the last relation of the theorem. □

We now describe the construction of the 3D digital lines Δ_{abc} directed by v and containing the origin. This line is an union of c 1D dotted lines and $\mathcal{L}$ is one of them. We then apply theorem 1 to add one by one 1D dotted lines to Δ_{abc}.

We consider the point $\frac{1}{c}(m, n) = \frac{1}{c}((c-1)a, (c-1)b)$ of Π_{abc} and the line δ directed by (a, b) and containing $\frac{1}{c}(m, n)$. The sum $\frac{1}{c}(m, n) + \frac{1}{c}(a, b))$ is equal to (a, b), hence the point following $\frac{1}{c}(m, n)$ on δ is an *integer* point which is the projection of the point $(1, 1, 1)$ (here $i = 1$). We obtain thus that the 3D line directed by v and containing the point $(1, 1, 1)$ is parallel to $\mathcal{L}$. It is also a 1D dotted line whose integer points are $(1, 1, 1) + k(a, b, c),\ k \in \mathbb{Z}$. We add it to Δ_{abc}.

By adding the vector $\frac{1}{c}(2a, 2b)$ to the point $\frac{1}{c}(m, n) = \frac{1}{c}((c-2)a, (c-2)b)$ of Π_{abc}, we also obtain an integer point (x, y) of $\mathcal{E}_{xy}$ (equal to $(1, 1)$ or $(1, 2)$ depending on the relative values of a, b, c). This point (x, y) is the projection of the point $(x, y, 2)$ of $\mathbb{Z}^3$ defining the 1D dotted line made by the points $(x, y, 2) + k(a, b, c),\ k \in \mathbb{Z}$. We also add it to Δ_{abc}.

We pursue this construction by adding all the dotted lines defined by the points $\frac{1}{c}((c-i)a,(c-i)b)$ of Π_{abc}. At the end, when $i=c$, we obtain an integer point of $\mathcal{L}$. Hence, Δ_{abc} is periodic with period (a,b,c). Its period is composed of the c first voxels described previously. As $(c-i)a = -ia \mod c$, it is easy to compute the x and y coordinates of the voxels.

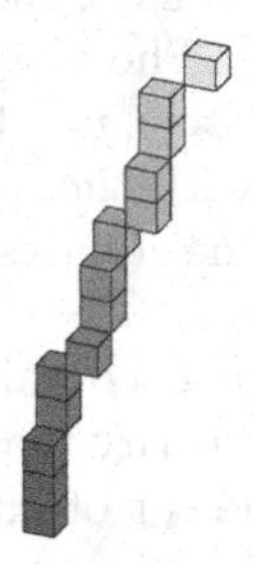

Fig. 3. The first 13 voxels of the first period followed by the first voxel of the second one. The direction vector is $(a,b,c)=(3,5,13)$.

We thus obtain Voss definition of a 3D digital lines (see Fig. 3).

Definition 1. *The 3D digital line with direction (a,b,c) and initial conditions m,n at the origin, where $0 \leq m < c$ and $0 \leq n < c$, is given by*

$$\begin{cases} x &= \left[\frac{az+m}{c}\right] \\ y &= \left[\frac{bz+n}{c}\right] \\ z &= z \end{cases} \quad \text{with} \quad z \in \mathbb{Z}.$$

Contrarily to the 2D case, choosing $m=n=\frac{c}{2}$ does not produce the approximation with rounding of the Euclidean line. In fact, we will recall that this Bresenham-like 3D digital line is generated via the lattices $\mathcal{E}_P$ and $\mathcal{D}_P$.

From the previous study, it becomes easy to find an algorithm to draw the 3D digital lines. Indeed, if we translate the point $\frac{1}{c}(m,n) \in \Pi_{xy}$ by the vector (a,b), 4 cases happen

$$\begin{cases} 0 \leq m < c-a \ \text{ and } 0 \leq n < c-b & \text{step } (0,0,1) \\ c-a \leq m \ \text{ and } 0 \leq n < c-b & \text{step } (1,0,1) \\ 0 \leq m < c-a \ \text{ and } c-b \leq n & \text{step } (0,1,1) \\ c-a \leq m \ \text{ and } c-b \leq n & \text{step } (1,1,1) \end{cases}$$

The $3D$ digital lines, previously defined, with direction vector $n=(a,b,c)$ such that $0 \leq a < b < c$ and a,b,c relatively prime, are given by the intersection of two particular digital planes.

Definition 2. *A $3D$ digital lines with direction vector (a,b,c) such that $0 \leq a < b < c$ and a,b,c relatively prime is the set of solutions of the linear systems of inequalities given by*

$$\begin{cases} \gamma \leq cx - az < \gamma + c \\ \gamma' \leq cy - bz < \gamma' + c \end{cases}$$

The *lower bounds* can be used to translate the digital line so that it can contain any given integer point of $\mathbb{Z}^3$. Their *arithmetical thickness* is the integer c. This number can be replaced by any couple ϵ, ϵ' of integers in orderto define $3D$ digital lines with prescribed thickness.

One on the fundamental properties of $3d$ digital lines is that they cover $\mathbb{Z}^3$. Moreover, the action of the octahedral group O_h can be used to define $3D$ digital lines with any direction vector.

3.2 The Lattices $\mathcal{D}_P$ and $\mathcal{E}_P$

Recall that for a direction vector $n = (a, b, c)$, the canonical lattices attached to the famillies $\mathcal{D}$ and $\mathcal{E}$ are the intersections of these famillies with the plane $(P): \ ax + by + cz = 0$. The lattices are respectively $\mathcal{D}_P$ and $\mathcal{E}_P$.

The lattice $\mathcal{E}_P$ is very interesting since it permits to measure Euclidean distances between the 1D dotted lines so to locate the integer points which are closest to an Euclidean lines with direction vector n. As it was the case for $\mathcal{D}$, $\mathcal{D}_P$ is a cartesian lattice which contains $\mathcal{E}_P$ and with which it is easier to work.

The coordinates of the points of $\mathcal{D}_P$ and $\mathcal{E}_P$ are more complex that for $\mathcal{D}_{xy}$ and $\mathcal{E}_{xy}$ but their dependances are algegraically similar. We refer to [2] for the computation. The coordinates of the points of $\mathcal{D}_P$ are given by:

$$\begin{cases} x &= \frac{(b^2+c^2)u-abv}{a^2+b^2+c^2} \\ y &= \frac{(a^2+c^2)v-abu}{a^2+b^2+c^2} \\ z &= \frac{-c(au+bv)}{a^2+b^2+c^2} \end{cases} \qquad u, v \in \mathbb{Z}$$

Hence, $\mathcal{D}_P$ is generated by the vectors $\alpha = \frac{1}{a^2+b^2+c^2}(b^2 + c^2, -ab, -ac)$ and $\beta = \frac{1}{a^2+b^2+c^2}(-ab, a^2 + c^2, -bc)$.

The lattice $\mathcal{E}_P$ is generated by the reductions modulo α and β of the vectors $\frac{k}{a^2+b^2+c^2}(ac, bc, -(a^2 + b^2)), \quad k \in \mathbb{Z}$. Fig. 4 shows a partial view of a lattice $\mathcal{E}_P$ as well as several 1D dotted lines of $\mathcal{E}$.

Both $\mathcal{D}_P$ and $\mathcal{E}_P$ are planar lattice with rank 2. Then, using a convenient isometry we can map them onto xOy. After some tedious calculus, the isometric lattice of $\mathcal{D}_P$ is generated by the vectors $U = \frac{1}{\sqrt{(a^2+b^2)}}(1, \frac{-ab}{c\sqrt{(a^2+b^2+c^2)}})$ and $V = \frac{a^2+c^2}{c\sqrt{(a^2+b^2)}\sqrt{(a^2+b^2+c^2)}}(0, 1)$. The image of the lattice $\mathcal{E}_P$ is the reduction modulo U and V of the vectors $k(aU, bV) \quad k \in \mathbb{Z}$. As it can be easily seen, this situation is the analoguous of the link between $\mathcal{D}_{xy}$ and $\mathcal{E}_{xy}$.

Given a point in $\mathcal{E}_P$, the closest points in $\mathcal{E}_P$ to the given points enables us to define the notion of closest $3D$ digital lines (see Fig. 5). This corresponds to a Bresenham-like $3D$ digital lines. To define it, one must *sort* the points in $\mathcal{E}_P$ around a given point in $\mathcal{E}_P$.

Let $\pi : \mathbb{Z}^3 \mapsto \mathcal{E}_P$ be the application which maps a 1D dotted line to its intersection with the plane (P), let ω be a point of $\mathcal{E}_P$ and let $\Delta_{\omega,\rho}$ be the set

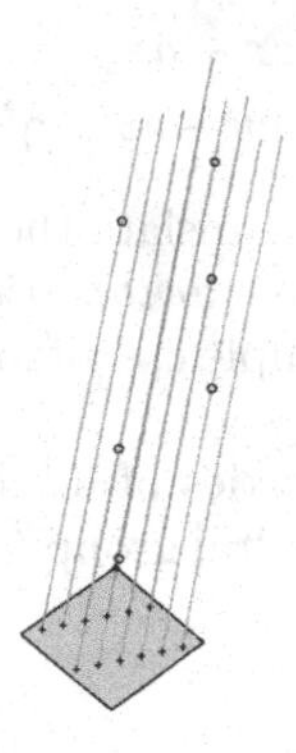

Fig. 4. A part of the lattice $\mathcal{E}_P$ where $(a, b, c) = (3, 5, 13)$ as well as some 1D dotted lines

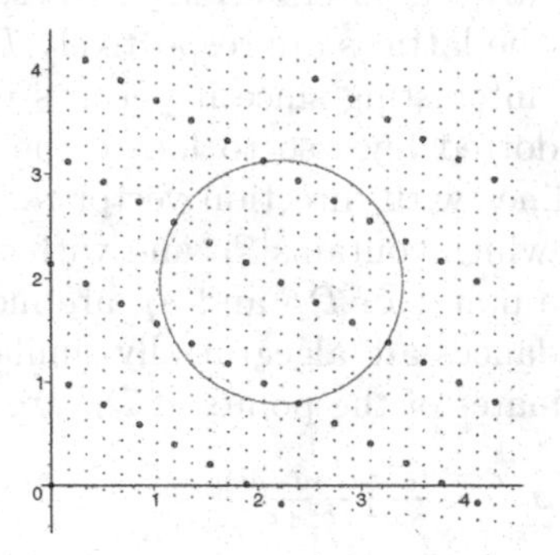

Fig. 5. The isometric images of $\mathcal{D}_P$ and $\mathcal{E}_P$, where $(a, b, c) = (3, 5, 13)$, in the plane xOy and a circle of radius 1.17 containing some closest points of one element of $\mathcal{E}_P$

of points in $\mathcal{E}_P$ belonging to a disk (in plane (P)) with center ω and radius ρ, then we have the following,

Definition 3. *The 3D digital line with best integer approximation of order* ρ *of the 1D dotted line containing the point* ω *of* $\mathcal{E}_P$ *is the reciprocical image* $\pi^{-1}(\Delta_{\omega,\rho})$.

Obviously, these $3D$ digital lines does not cover $\mathbb{Z}^3$, which could be a bad behaviour. Nevertheless, it guarantees that the digital lines is as closest as possible of the Euclidean corresponding line.

3.3 3D Segment Drawing

If any segment AB is given let us denote by OV the vector $B - A$ and by $n = (n_1, n_2, n_3)$ the components of OV divided by their greatest common divider, so that n_1, n_2, n_3 are relatively prime. Construction given in section 3 about the symmetry group O_h can be used to give an operator g_n mapping the domain of

n (or OC) to the fundamental domain F of O_h. Let $n' = g_n.n = (a, b, c)$ and $OC' = g_n.OC = (u'_1, u'_2, u'_3)$ in F.

Operator g_n^{-1} followed by the translation of OC to AB leads to a procedure $g(x, y, z)$ which maps the $3D$ discrete segment associated with OC' to the one associated to AB, (g_n being orthogonal g_n^{-1} is equal to the transposed of g_n). Drawing the $3D$ discrete segment associated to AB is thus reduced to the following algorithm giving the discrete approximation of OC' directed by $n' = (a, b, c)$.

```
Drawing of 3D segment OC'=(u'1,u'2,u'3) directed by n'=(a,b,c).
 //(a,b,c) satisfy 0<=a<=b<=c and gcd(a,b,c)=1
 x=y=0;
 // integer division so that line OC' is in the
 // middle of generated voxels
 rx=c/2;
 ry=rx
 for z = 0 to u'3
   draw g(x,y,z);
   if rx>=c-a then
      rx=rx+a-c;
      x=x+1;
   else
      rx=rx+a;
   end if;
   if ry>=c-b then
      ry=ry+b-c;
      y=y+1;
   else
      ry=ry+b
   end if;
 end for
```

4 nD Digital Lines

Let $v = (a_1, a_2, \ldots, a_n) \in F$ an integer point in the fondamental domain of the hyperoctaedral group $\mathcal{B}_n$. This group of order $2^n.n!$ can be identified with integer matrices of order n where each row and column contains one and only one non-zero term equals to ± 1.

Let P be the hyperplane whose equation is $a_1x_1 + a_2x_2 + \ldots a_nx_n = 0$, and E_P be the lattice obtained by projection of $\mathbb{Z}^n$ on P along direction v. We denote by $\{u_i\}$ $1 \leq i \leq n$ the canonical basis of $\mathbb{Z}^n$ and X_i the projection of u_i onto P along v. Vectors X_i belong to E_P and from equation of hyperplane P we have:

$$X_n = a_1(\frac{X_1}{a_n}) + a_2(\frac{X_2}{a_n}) + \cdots + a_{n-1}(\frac{X_{n-1}}{a_n})$$

We consider the lattice D_P generated in P by the $n-1$ vectors $\frac{X_1}{a_n}$, $\frac{X_2}{a_n}$, $\ldots$, $\frac{X_{n-1}}{a_n}$; of course E_P is a sublattice of D_P. Moreover, E_P is $n-1$ periodic, one period being given by $E_P \cap \delta$, δ denoting the fondamental domain of D_P. Any point of E_P is obtained by reduction modulo δ of integer multiples of X_n.

Let $\sigma^2 = a_1^2 + a_2^2 + \cdots + a_n^2$ and $\widehat{\sigma_i^2} = a_1^2 + a_2^2 + \cdots + \widehat{(a_i^2)} + \cdots + a_n^2$ where, in the sum $\widehat{u}$ means omission (so that $\widehat{\sigma_i^2} = \sigma^2 - a_i^2$), then components x_{ij} of $X_i = (x_{ij})$ can be computed and we have $x_{ij} = \frac{-a_i a_j}{\sigma^2}$ if $i \neq j$ and $x_{ii} = \frac{\widehat{\sigma_i^2}}{\sigma^2}$ for $i = j$.

From these expressions of vectors X_i the determinant of the Gram matrix of D_P can be evaluated giving $\det((X_i.X_j)) = a_n$ showing that domain δ contains a_n elements of E_P.

The hypothesis $v \in F$ leads to a very natural observation which will be helpful to define nD digital lines. Computation of the norm of $X_i - X_j$ gives $\| X_i - X_j \|^2 = \frac{\widehat{\sigma_i^2} + \widehat{\sigma_j^2} + (a_i + a_j)^2}{\sigma^2}$ from which inequalities $\| X_i \| \leq \| X_i - X_j \|$ and $\| X_j \| \leq \| X_i - X_j \|$ can be deduced showing that the set $\{X_1, X_2, \ldots X_{n-1}\}$ is *almost orthogonal* in D_P.

Let Π_i be the hyperplane generated by $u_1, \ldots, \widehat{u_i}, \ldots, u_{n-1}, v$ where, again, $\widehat{-}$ means omission. Intersection $\Pi_i \cap P$ is the subspace of P generated by $X_1, \ldots, \widehat{X_i}, \ldots, X_{n-1}$ so that these hyperplanes too are *almost orthogonal.* Definition of digital hyperplanes in $\mathbb{Z}^n$ being obvious we can define nD digital line through 0 and directed by v as the intersection of digital hyperplanes associated to $\Pi_i' s$.

Definition 4. *Digital line through* 0 *directed by* $v = (a_1, a_2, \ldots, a_n)$ *where* $0 \leq a_1 \leq a_2 \leq \cdots \leq a_n$ *is the set of integer points solution of the* $n-1$ *diophantine inequations*

$$\gamma_i \leq a_1 x_1 + \cdots + \widehat{(a_i x_i)} + \ldots a_{n-1} x_{n-1} + a_n x_n < \gamma_i + \epsilon_i \quad 1 \leq i \leq n-1$$

$\widehat{-}$ *meaning omission.*

Vector (γ_i) is the *lower bound* and vector (ϵ_i) the *arithmetical thickness.*

Algorithms can be given to draw digital nD lines defined in this way. They use a vector of errors $\rho = (r_1, r_2 \ldots r_{n-1}$ and the simplest one draws $2^n - 1$-connected lines when $\epsilon_i = a_n \quad \forall i = 1, 2, \ldots n-1$; again we suppose $v \in F$, the general case being solved with the help of operators of the Hyperoctaedral group $\mathcal{H}_\backslash$ in a similar way as what has been donne in $3D$.

Suppose $M = (m_1, m_2, \ldots, m_n)$ is a point in $\mathbb{Z}^n$ and $v = (a_1, a_2, \ldots, a_n) \in F$ and $\gcd(a_i) = 1$, then to obtain the first $nbPoints$ of the nD and 2^n-1-connected digital line through M and directed by v we have the following algorithm.

```
nD digital line drawing
------------------------
M=(m1,m2,...,mn); // starting point
x=(x1,x2,...,xn)=M; // initialization of x variable
v=(a1,a2,...,an); // line direction in F and gcd(ai)=1
rho=(an/2,an/2,...,an/2);//n-i components of rho are equal to an/2
```

```
for i=1 to NbPoints
  draw x;
  for j=1 to n-1
    if rj>=an-aj {
       rj=rj+aj-an;
       xj=xj+1;
     }
    else
       rj=rj+aj;
    end if;
  end for
end for
```

Initializing error vector ρ with half of thickness, that is setting $\frac{a_n}{2}$ for all ρ's component we are assured that integer points generated are *well* distributed around the euclidean line going through M and directed by v.

Following array shows an application of this algorithm for the drawing of the first 13 points of the $4D$ digital line going through origin and directed by vector $v = (3, 5, 7, 13)$. First 3 lines show evolution of $\rho = (r_1, r_2, r_3)$ error vector and the last ones are coordinates of approximating points. One period of this line is thus obtained; following ones are obtained by translating this one by integer multiples of vector v.

r_1	6	9	12	2	5	8	11	1	4	7	10	0	3	6
r_2	6	11	3	8	0	5	10	2	7	12	4	9	1	6
r_3	6	0	7	1	8	2	9	3	10	4	11	5	12	6

x_1	0	0	0	1	1	1	1	2	2	2	2	3	3
x_2	0	0	1	1	2	2	2	3	3	3	4	4	5
x_3	0	1	1	2	2	3	3	4	4	5	5	6	6
x_4	0	1	2	3	4	5	6	7	8	9	10	11	12

5 Conclusion

We have presented in this paper a unified view of the definitions of Voss [1] and Figueiredo and Reveillès [2]. This permits us, for instance, to give a short drawing algorithm in 3D. Moreover, the presentation is extended to an arbitrary dimension via the use of lattice theory and specific projections. We also give a 13-lines long drawing algorithm for nD digital lines. It should be very interesting to study the link between this approach and multi-dimensonal continued fraction given by Arnold [14] and this is a future work.

References

1. Voss, K.: Discrete Images, Objects and Functions in $\mathbb{Z}^n$. Springer-Verlag (1993)
2. Figueiredo, O., Reveillès, J.P.: New results about 3D digital lines. In Melter, R.A., Wu, A.Y., Latecki, L., eds.: Vision Geometry V. Volume 2826. (1996) 98–108

3. Klette, R., Rosenfeld, A.: Digital Geometry. Morgan-Kaufmann (2004)
4. Kim, C.: Three-dimensional digital line segments. IEEE Trans. Pattern Analysis and Machine Intelligence **5** (1983) 231–234
5. Debled-Rennesson, I.: Etude et reconnaissance des droites et plans discrets. PhD thesis, Université Louis Pasteur - Strasbourg (1995)
6. Cœurjolly, D., Debled-Rennesson, I., Teytaud, O.: Segmentation and length estimation of 3D discrete curves. In Bertrand, G., et al, eds.: Digital and Image Geometry. Volume 2243 of LNCS., Springer-Verlag (2001) 299–317
7. Reveillès, J.P.: Géométrie discrète, calcul en nombres entiers et algorithmique. Thèse d'etat, Université Louis Pasteur, Strasbourg (1991)
8. Figueiredo, O., Reveillès, J.P.: A contribution to 3D digital lines. In: Proc. Discrete Geometry for Computer Imagery, Université d'Auvergne - LLAIC (1995) 187–198
9. Ibanez, L., Hamitouche, C., Roux, C.: A Vectorial Algorithm for Tracing Discrete Straight Lines in N-Dimensional Generalized Grids. IEEE Trans. Vis. Comput. Graph. **7**(2) (2001) 97–108
10. Klette, R.: The m-dimensional grid point space. Computer Vision, Graphics, and Image Processing **30** (1985) 1–12
11. Stojmenovic, I., Tosic, R.: Digitization Schemes and the Recognition of Digital Straight Lines, Hyperplanes, and Flats in Arbitrary Dimensions. Contemporary Mathematics **119** (1991) 197–212
12. Reveillès, J.P.: The Geometry of the Intersection of Voxel Spaces. In Fourey, S., Herman, G., Kong, T.Y., eds.: IWCIA 2001. Volume 46 of Electronic Notes in Theoretical Computer Science., Elsevier (2001) 1–24
13. Debled, I., Reveillès, J.: A linear algorithm for segmentation of digital curves. In: 3rd IWPIA. (1994)
14. Arnold, V.: Higher dimensional Continued Fractions. Regular and Chaotics Dynamics **33** (1998) 10–17

Two Discrete-Euclidean Operations Based on the Scaling Transform

Gaëlle Largeteau-Skapin and Eric Andres

Laboratoire SIC,
Université de Poitiers,
BP 30179 86962 Futuroscope Chasseneuil cédex, France
{glargeteau, andres}@sic.univ-poitiers.fr

Abstract. In this paper we study the relationship between the Euclidean and the discrete world thru two operations based on the Euclidean scaling function: the discrete smooth scaling and the discrete based geometrical simplification.

Keywords: discrete geometry, operations, discrete scale, multi-representation modeller.

1 Introduction

The Euclidean and the discrete world are generally considered as antagonists. Both worlds have different properties and it is reflected in the operations. Operations might be trivial in one world and difficult to transpose in the other one. For instance, there isn't a satisfying discrete rotation that is at the same time one-to-one and commutative. Two primary properties of the Euclidean rotation. Boolean operations (intersection, union, difference) that are trivial in the discrete world become tedious to perform in the Euclidean world because of numerical errors. The goal of this paper is to show how the specificities of both worlds can be used to define operations with new interesting properties. To illustrate this we propose two operations: one in the discrete (discrete smooth scaling) and one in the Euclidean world (discrete based geometrical simplification). Each operation is partly performed in the other world with a digitization and/or an analytical reconstruction step. The digitization process allows us to move from the Euclidean world to the discrete world. The analytical reconstruction process allows us to move from the discrete to the Euclidean world.

The first operation that we are proposing is called "discrete smooth scaling". The idea behind this operation is to describe a discrete object in a smaller (finer) grid. We want to perform this operation without filtering or smoothing. The information in a discrete cell (pixel, voxel) can be a complex information that can't simply be smoothed. So far, discrete scaling didn't respect geometrical properties of the object (discrete edge slopes for instance) [1]. To solve this problem, we perform the dilation in the space best adapted: the Euclidean space. We perform an analytical reconstruction on the original image followed by a Euclidean scaling. The discretization provides us with the final "refined" image. This discrete

A. Kuba, L.G. Nyúl, and K. Palágyi (Eds.): DGCI 2006, LNCS 4245, pp. 41–52, 2006.

smooth scaling operation possesses a remarkable property: the almost stability by inverse scale. If we make a discrete smooth scale of factor $\alpha \geq 1$ followed by a discrete smooth scale of factor $\beta = \frac{1}{\alpha}$ we obtain the original discrete object with an error bounded by a factor proportional to $\frac{1}{\alpha}$.

The second operation is a discrete based geometrical simplification operation. The operation consists, this time, starting with a Euclidean object, to digitize with a given grid size and then to reconstruct it. When we reconstruct a discrete object, the "shape complexity" (resulting vertice and edge number) depends on the size of the object. The smaller the object, the less complex the reconstructed object. It is however difficult to assure a topological consistence between the initial object and the reconstructed object. An interesting property of this operation is that the Hausdorff distance between the original object and the simplified object is bounded by a factor proportional to the grid size.

The interest of these two operations is that they each make use of the properties of the other world. The discrete operation uses the properties of the Euclidean world and the Euclidean operation those of the discrete world. These operations show how the duality between the discrete and the Euclidean world can be used at our advantage.

In section two, we introduce the basic notions used in this paper such as discrete analytical models, the principle of the analytical reconstruction method and the notations used through out the paper. In the third section we introduce the discrete discrete smooth scaling operation. In section four we present the discrete geometrical simplification operation. We conclude and propose some extensions in section five.

2 Preliminaries

2.1 Basic Notations in Discrete Geometry

The following notations correspond to those given by Cohen and Kaufman in [2] and those given by Andres in [3]. We provide only a short recall of these notions.

A **discrete** (resp. **Euclidean**) **point** is an element of $\mathbb{Z}^n$ (resp. $\mathbb{R}^n$). A **discrete** (resp. **Euclidean**) **object** is a set of discrete (resp. Euclidean) points. We denote p_i the ith coordinate of a point p of $\mathbb{Z}^n$. The **voxel** $\mathbb{V}(p) \subset \mathbb{R}^n$ of a discrete nD point p is defined by $\mathbb{V}(p) = [p_1 - \frac{1}{2}, p_1 + \frac{1}{2}] \times ... \times [p_n - \frac{1}{2}, p_n + \frac{1}{2}]$. For a discrete object D, $\mathbb{V}(D) = \bigcup_{p \in D} \mathbb{V}(p)$

In this paper, we use the Hausdorff distance defined by:

Definition 1. *Let h be the direct Hausdorff distance: $A \subset \mathbb{R}^n$, $B \subset \mathbb{R}^n$, $h(A, B) = max_{a \in A} \left(min_{b \in B} \left(d_2(a, b) \right) \right)$, where d_2 is the Euclidean distance. The Hausdorff distance H between A and B is $H(A, B) = max \left(h(A, B), h(B, A) \right)$.*

This paper is based on the relations between the Euclidean and the discrete world and the way operations can benefit from this duality. We present two operations that are based on the Euclidean **scale** function noted Sc. We consider, without loss of generality, that the center of the scale function Sc is the origin.

2.2 Digitization and Reconstruction

The basic idea behind this paper is to profit from the possibility to travel between the discrete world $\mathbb{Z}^n$ and the Euclidean world $\mathbb{R}^n$. The transformation from the discrete to the Euclidean world is called *digitization*. The transformation from the Euclidean world to the discrete world is called *reconstruction*. The experiments presented in this paper have been conducted with the standard analytical model [3] (see also Fig. 1). The theoretical results are however not restricted to the standard analytical model and are also verified for a larger class of digitization schemes. Most of the digitization schemes commonly used seem actually to fit the definition that follows including the Bresenham algorithms, the supercover model, the naive digitization, the standard model, etc. Let us try to propose a characterisation of the digitization schemes that suit the purpose of this paper.

We consider digitization transforms defined by *narrow offset areas*. A narrow offset area $\mathbb{O}$ is defined for classes of Euclidean objects. It simply has to verify two fundamental conditions: A narrow offset area $\mathbb{O}(E) \subset \mathbb{R}^n$ of a Euclidean object E must be *narrow* meaning that if $x \in \mathbb{O}(E) \cap \mathbb{Z}^n \Rightarrow \mathbb{V}(x) \cap E \neq \varnothing$. It simply requires that the digitization of an Euclidean object E to be composed of pixels that are intersected by E. The second condition is a constructive condition. A narrow offset area must verify a stability property for the union: $\mathbb{O}(E \cup F) = \mathbb{O}(E) \cup \mathbb{O}(F)$.

Definition 2. *The digitization based on a narrow offset area is defined by:*

$$\mathcal{D} : \mathcal{P}(\mathbb{R}^n) \longrightarrow \mathcal{P}(\mathbb{Z}^n)$$
$$\mathcal{D}(E) = \{p \in \mathbb{Z}^n \mid p \in \mathbb{O}(E)\} = \mathbb{O}(E) \cap \mathbb{Z}^n.$$

A good way to define a wide class of digitization tranforms is to define the offset area with a distance d.

$$\mathbb{O}(E) = \left\{ x \in \mathbb{R}^n \,\middle|\, d(x, E) \leq \frac{1}{2} \right\}.$$

The best known discrete analytical model is called the supercover model [4,5,6] with an offset defined by the Chebyshev distance d_∞. The distance d_1 defines the closed naïve model and the distance d_2 defines the closed pythagorean model. All distances, of course, don't verify the narrowness property but many do. There exist also narrow offset areas that aren't defined with distances. This is the case for the Bresenham algorithms, the standard analytical model, the naive digitization, etc.

Digitization based on narrow offset areas verify, by construction, properties such as $\mathcal{D}(E \cup F) = \mathcal{D}(E) \cup \mathcal{D}(F)$; $\mathcal{D}(E \cap F) \subset \mathcal{D}(E) \cap \mathcal{D}(F)$ and $F \subset G \Longrightarrow \mathcal{D}(E) \subset \mathcal{D}(F)$. These properties ensure that we can build complex discrete objects out of a set of basic elements. We can, for instance, build all linear objects out of simplices.

Defining a reconstruction transform is much more difficult. If we want the reconstruction transform to make any sense we must define some properties that

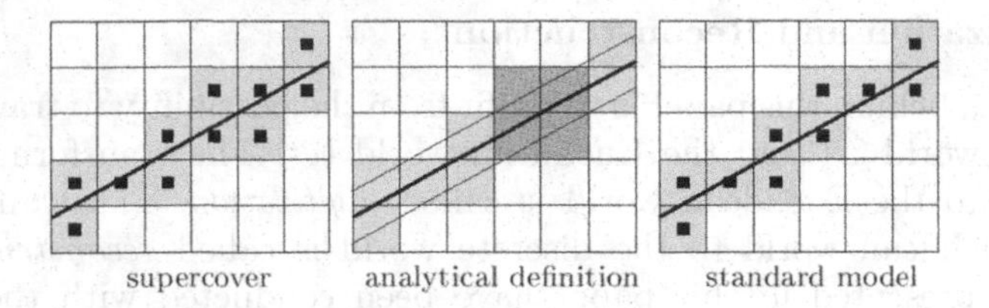

Fig. 1. Supercover and standard model examples

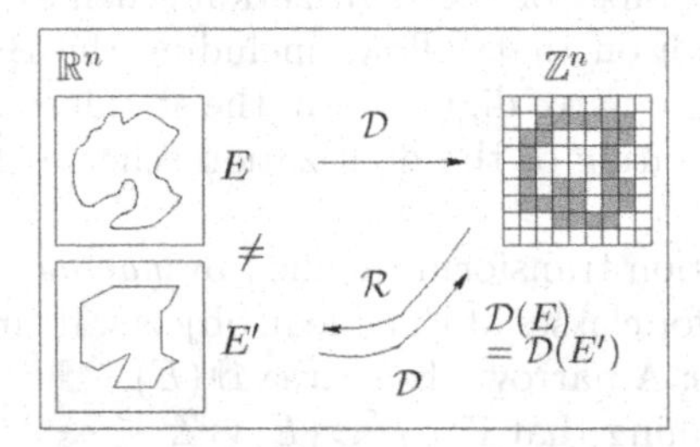

Fig. 2. Coherence between discrete and Euclidean world

have to be verified. For any given analytical digitization, we have an infinity of reconstruction operations [7, 8, 9, 10]. For instance, it's natural to associate a reconstruction transform to a digitization. Indeed, we can define a equivalence relation $\approx$ between two Euclidean objects E and F by $E \approx F$ iff $\mathcal{D}(E) = \mathcal{D}(F)$. There is a one-to-one mapping between the discrete objects and the equivalence classes defined by $\approx$. One of the properties of any reconstruction $\mathcal{R}$ is to stay in the equivalence class if we digitize and then reconstruct. Of course, in general, $\mathcal{R}(\mathcal{D}(E)) \neq E$ (see Fig. 2).

Definition 3. *Reconstruction*
A reconstruction operation $\mathcal{R} : \mathcal{P}(\mathbb{Z}^n) \longrightarrow \mathcal{P}(\mathbb{R}^n)$ *associated to an analytical digitization* $\mathcal{D}$ *is an operation verifying, for any Euclidean object* E*:*

$$\mathcal{R}(\mathcal{D}(E)) \approx E.$$

A property that we won't have systematically but that will be verified in many practical situations is: $\mathcal{D}(\mathcal{R}(A)) = A$ for a given discrete object A. This property will be verified if there isn't any *missing information* in A. For instance, if we reconstruct a Bresenham line segment, that isn't missing any pixels, the property will be verified.

3 Discrete-Euclidean Operations

In this part, we will study two operations linking the discrete and Euclidean world. The first is an operation from $\mathbb{Z}^n$ to $\mathbb{Z}^n$ that use the Euclidean scale properties to define a discrete smooth scale. The second, from $\mathbb{R}^n$ to $\mathbb{R}^n$, uses the digitization properties to erase details in Euclidean objects.

3.1 Discrete Smooth Scaling

The first operation that we are proposing is called *discrete smooth scaling.* The idea behind this operation is to describe a discrete object on a smaller grid. We want to perform this operation without filtering or smoothing (see Fig. 3). We therefore perform the dilation in the space best adapted: the Euclidean space.

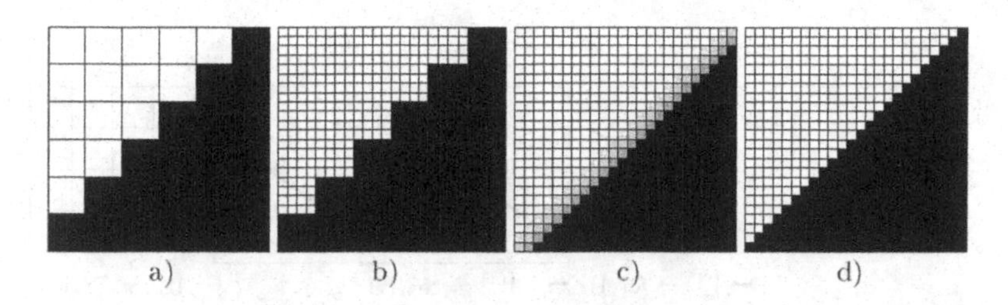

Fig. 3. a) original discrete object. b)reduced grid size. c) classical smoothing. d) discrete smooth scaling.

Definition 4. *We call discrete smooth scaling of a discrete object A of $\mathbb{Z}^n$ by a scale α, $\alpha \in \mathbb{R}^{+*}$, the following operation denoted $DSS_\alpha(A)$:*

$$DSS_\alpha(A) = \mathcal{D} \circ Sc_\alpha \circ R(A).$$

We can see in section 4 some examples of this operation on discrete objects. The operation is meant to work for $\alpha \geq 1$. We can consider scales smaller than 1 especially in order to define the inverse operation. However the intuitive $DSS_{\frac{1}{\alpha}}$ is actually not an exact inverse operation (see Fig. 4). We don't know for the moment how to define the exact inverse transform but we can estimate the error commited with $DSS_{\frac{1}{\alpha}}$. This error is due to the reconstruction part of the operation. We don't measure the error between two discrete objects A and $DSS_{\frac{1}{\alpha}}(DSS_\alpha(A))$ but between $\mathcal{R}(A)$ and $Sc_{\frac{1}{\alpha}}(\mathcal{R}(DSS_\alpha(A)))$. This error measure is translation independant.

Note that the error bound we are proposing makes sense for objects verifying $\mathcal{D}(\mathcal{R}(A)) = A$. In case of missing information and partial information reconstruction the result of the theorem that follows stands but it's not very meaningful. Measuring an error between an incomplete discrete object and its scaled and descaled reconstruction isn't, in our case, very interesting. So, let us suppose, for what follows, that $\mathcal{D}(\mathcal{R}(A)) = A$.

Let us introduce several notations: for a discrete object A, we note $A_{first} = \mathcal{R}(A)$ the reconstruction of the original discrete object and we note $A_{last} = Sc_{\frac{1}{\alpha}}(\mathcal{R}(DSS_\alpha(A)))$ the Euclidean object which discretization is $A_{last} \cap \mathbb{Z}^n = DSS_{\frac{1}{\alpha}}(DSS_\alpha(A))$. The error measure is a bound on the Hausdorff distance between both objects.

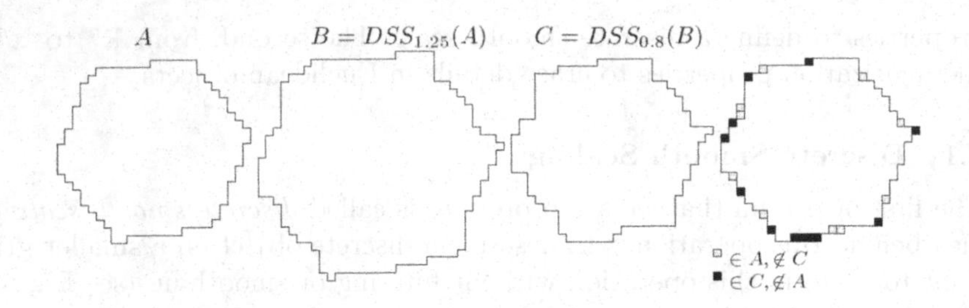

Fig. 4. Discrete smooth scaling: inversibility problem

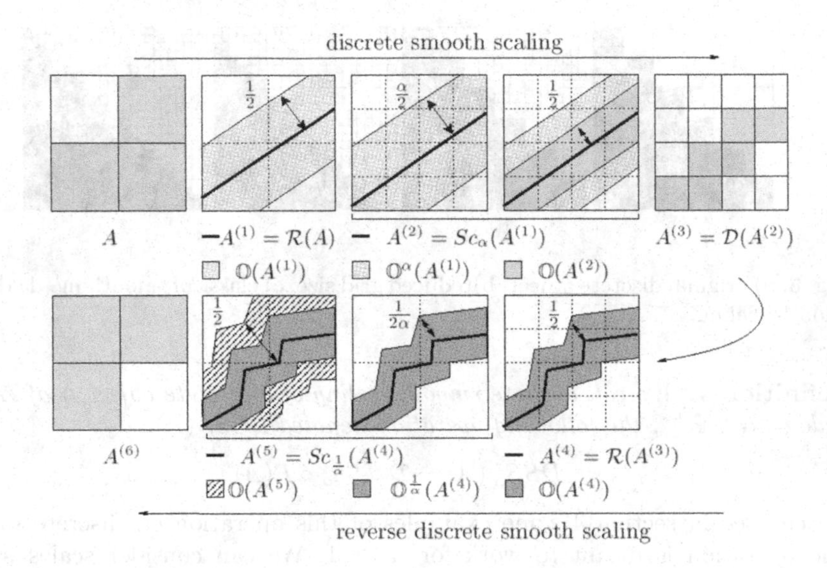

Fig. 5. Discrete smooth scaling inversibility

Theorem 1

$$H\left(A_{first}, A_{last}\right) = H\left(\mathcal{R}\left(A\right), Sc_{\frac{1}{\alpha}}\left(\mathcal{R}\left(\mathcal{D}\left(Sc_{\alpha}\left(\mathcal{R}(A)\right)\right)\right)\right)\right) \leq \frac{1}{\alpha}\sqrt{n}.$$

Proof (see Fig. 5): Let A be a discrete object and let: $A_{first} = A^{(1)} = \mathcal{R}(A) \subset \mathbb{R}^n$, $A^{(2)} = Sc_{\alpha}(A^{(1)}) \subset \mathbb{R}^n$, $A^{(3)} = \mathcal{D}(A^{(2)}) \subset \mathbb{Z}^n$, $A^{(4)} = \mathcal{R}(A^{(3)}) \subset \mathbb{R}^n$, $A_{last} = A^{(5)} = Sc_{\frac{1}{\alpha}}(A^{(4)}) \subset \mathbb{R}^n$ and $A^{(6)} = \mathcal{D}(A^{(5)}) \subset \mathbb{Z}^n$.According to our notations, we have $A_{first} = A^{(1)}$ and $A_{last} = A^{(5)}$.

Digitization and reconstruction definitions and properties provide the following result: $A^{(2)} \approx A^{(4)}$ and thus $A^{(3)} = (\mathbb{O}(A^{(2)})) \cap \mathbb{Z}^n = (\mathbb{O}(A^{(4)})) \cap \mathbb{Z}^n$. The narrowness property of the digitization tells us that $A^{(2)}$ and $A^{(4)}$ intersect each voxel of $A^{(3)}$ and therefore, each voxel of $A^{(3)}$ contains at least one point of $A^{(2)}$ and one of $A^{(4)}$. The Euclidean distance between these two points is bounded by the voxel diagonal length: $\sqrt{n}$. We can generalize: $\forall x \in A^{(2)}, \exists y \in D^{(4)} | d_2(x, y) \leq \sqrt{n}$. This implies the following result on the direct Hausdorff distance: $\forall x \in A^{(2)}$, $min_{y \in A^{(4)}}(d_2(x,y)) \leq \sqrt{n}$ and therefore $h(A^{(2)}, A^{(4)}) \leq \sqrt{n}$. The same reasoning stands for $h(A^{(4)}, A^{(2)})$ and leads to $h(A^{(4)}, A^{(2)}) \leq \sqrt{n}$. The result is : $H(A^{(2)}, A^{(4)}) \leq \sqrt{n}$. We then apply the scale operation $Sc_{\frac{1}{\alpha}}$.We have $H\left(Sc_{\frac{1}{\alpha}}\left(A^{(2)}\right), Sc_{\frac{1}{\alpha}}\left(A^{(4)}\right)\right) = H\left(A^{(1)}, A^{(5)}\right) \leq \frac{1}{\alpha}\sqrt{n}$ since $Sc_{\frac{1}{\alpha}}\left(A^{(2)}\right) = A^{(1)}$ by construction and $Sc_{\frac{1}{\alpha}}\left(A^{(4)}\right) = A^{(5)}$ by definition.

Finally: $H\left(A_{first}, A_{last}\right) \leq \frac{1}{\alpha}\sqrt{n}$. □

Here are some comments on these results. The first obvious comment is that the bigger the scaling factor, the smaller the possible difference between A and $DSS_{\frac{1}{\alpha}}\left(DSS_{\alpha}\left(A\right)\right)$ is (in case of $\mathcal{D}\left(\mathcal{R}\left(A\right)\right) = A$ as already stated). Since $\mathcal{R}(A) - Sc_{\frac{1}{\alpha}}\left(\mathcal{R}\left(DSS_{\alpha}\left(A\right)\right)\right)$ is smaller, there is a lesser chance that it contains a discrete point. The result of our theorem is a quite general bounding value. It doesn't take into account the fact that the reconstruction algorithms are deterministic and that it's often the case that if A and B are very similar then $\mathcal{R}(A)$ is similar to $\mathcal{R}(B)$. This occurs especially for small scale factors. We can thus suppose, and experimentation supports it, that in many cases the actual Hausdorff distance is much smaller than the theoretical bounding value we propose. For the case $\alpha = 1$ we have no difference between $A_{first} = A_{last}$ and thus $H\left(A_{first}, A_{last}\right) = 0$.

Corollary 1. $\lim_{\alpha \to \infty} H\left(A_{first}, A_{last}\right) = 0$.

The corollary tells us that the discrete smooth scaling is invertible when α tends to infinite. In fact, the discrete smooth scaling operation can be seen as a multi-scale digitization of the Euclidean scaling function with an approximation factor α. We can say that when α tends to infinite then DSS tends to Sc. Some more theoretical work needs to be done here. Non standard analysis is one way of looking at this problem [11].

3.2 Discrete Based Geometrical Simplification

The second operation we have studied and implemented is a discrete based geometrical simplification operation. This operation acts on a Euclidean object that is first digitized on a given grid size and then reconstructed. According to the grid size, details are gathered in the same voxel and thus do not appear in the reconstructed object. The bigger the voxel, the lesser details from the Euclidean object will remain after the reconstruction. The object is simplified and can be represented at different levels of details (see Fig. 6). In practice, it's not the voxel size that changes but the object size. The object is scaled with the Euclidean scaling function to fit the grid size. For a scaling factor x the voxel size is $\frac{1}{x}$.

Definition 5. *We call discrete based geometrical simplification of a Euclidean object E of $\mathbb{R}^n$ by a factor α, $\alpha \in \mathbb{R}^{+*}$, the following operation denoted $Sp_{\alpha}(E)$:*

$$Sp_{\alpha}(E) = Sc_{\frac{1}{\alpha}} \circ \mathcal{R} \circ \mathcal{D} \circ Sc_{\alpha}\left(E\right).$$

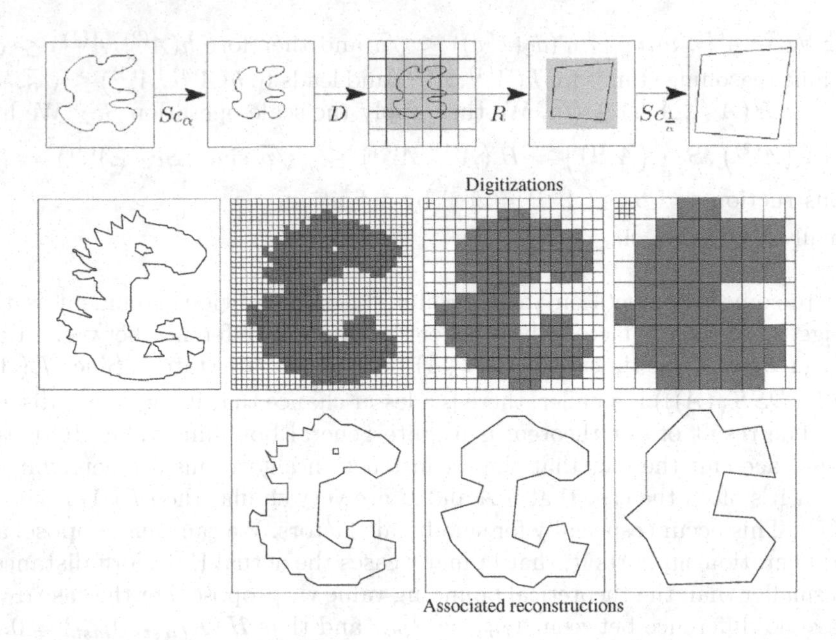

Fig. 6. Discrete based geometrical simplification principle : $\alpha = 1, \frac{1}{2}$ and $\frac{1}{4}$

We remark that the discrete based simplification has a similar property as the discrete smooth scaling operation: the Hausdorff distance between the original object and its simplification is bounded by a factor proportional to the grid size.

Theorem 2

$$\forall E \subset \mathbb{R}^n,\ H(E, Sp_\alpha(E)) \leq \frac{1}{\alpha}\sqrt{n}.$$

Proof
Theorem 2 is similar to theorem 1. We proved that $H\Big(\mathcal{R}(A), Sc_{\frac{1}{\alpha}}(\mathcal{R}(\mathcal{D}(Sc_\alpha(R(A)))))\Big) \leq \frac{1}{\alpha}\sqrt{n}$ for A a discrete object. Now, $\mathcal{R}(A)$ is a Euclidean object so, if we call $E = \mathcal{R}(A)$, we have $H\Big(E, Sc_{\frac{1}{\alpha}}(\mathcal{R}(\mathcal{D}(Sc_\alpha(E))))\Big) \leq \frac{1}{\alpha}\sqrt{n}$. By definition $Sc_{\frac{1}{\alpha}}(\mathcal{R}(\mathcal{D}(Sc_\alpha(E)))) = Sp_\alpha(E)$ which leads to $H(E, Sp_\alpha(E)) \leq \frac{1}{\alpha}\sqrt{n}$. □

The theorem tells us that the geometrical simplification process respects the general shape of an object. The error we commit by replacing the Euclidean object by its simplified version is bounded.

Corollary 2. $\lim_{\alpha\to\infty} Sp_\alpha(E) = E$.

4 Results: Implementation and Illustrations

Let us comment our implementation choices and present some images to illustrate the operations. The theoretical results we presented in this paper are valid in dimension n for a very large class of digitization and related reconstructions transforms. We implemented both operations in 2D. We present also an image of our first results in 3D.

4.1 Implementation

For several years our discrete geometry team develops a multi-representation modelling software intended to represent objects under four different embeddings (see Fig. 7): a Euclidean version, its analytical equivalent, the region representation and finally a discrete 2D pixel or 3D voxel representation. This allows us to choose the best adapted representation form depending on the type operation we want to realise.

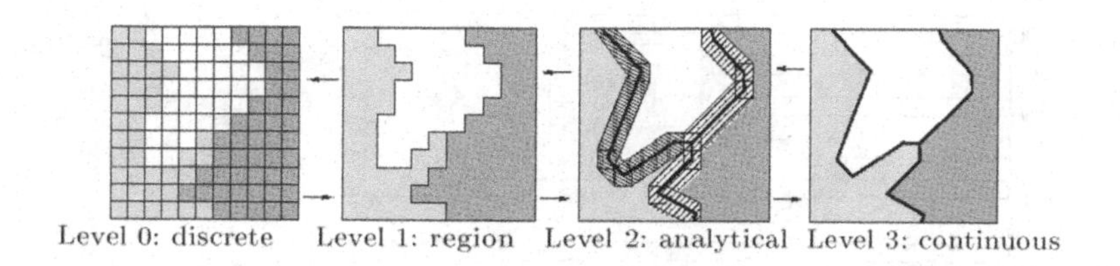

Fig. 7. Multi-representation modeller

In this modeller discrete object are defined using the standard analytic model [3] (see Fig. 1). The reconstruction implemented in the modeller was defined in [7,8] and is based on the *preimage* notion [12]. This algorithm computes the set of Euclidean hyperplane segments which digitization contains the original discrete object: $\mathcal{R}(A) \subset \mathbb{V}(A)$ (the standard model is a cover). This approach is based on discrete analytic geometry and is composed of two steps: the regognition of discrete analytical hyperplane segments (see [10] for an overview on recognition algorithms) and the analytical polygonalisation of the curve [9,8].

4.2 Illustrations

Here we present illustrations of the discrete smooth scaling transform with scaling factor $\alpha = 5$ and $\alpha = 10$.

The reconstruction operation we implemented [7,8] reconstruct objects with line segments, plane segments. The discrete smooth scaling is thus quite good on discrete objects with linear borders. The arrows in Fig. 8 show that on more *circular parts* the reconstruction creates less natural reconstruction shapes. This comes of course from the fact that a circle in low resolution will be reconstructed as a polygon.

The discrete based geometrical simplification operation decreases object detail level and therefore decreases its complexity. This operation can be used to simplify object when details are not perceptible by a human observer and when only the global pattern of the object is meaningful. Our simplification operation

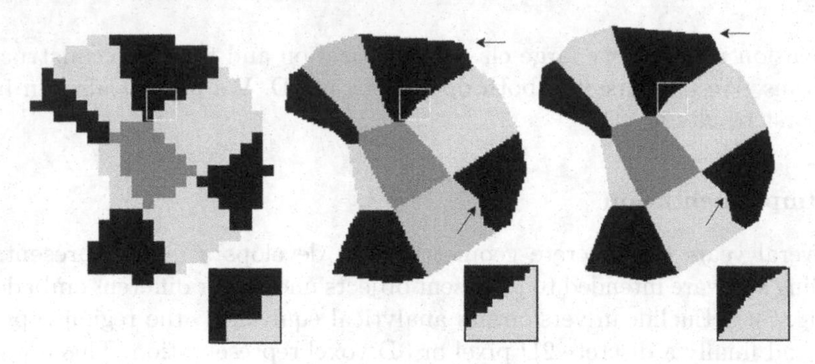

Fig. 8. Discrete smooth scale example ($\alpha = 5$ and $\alpha = 10$) with details

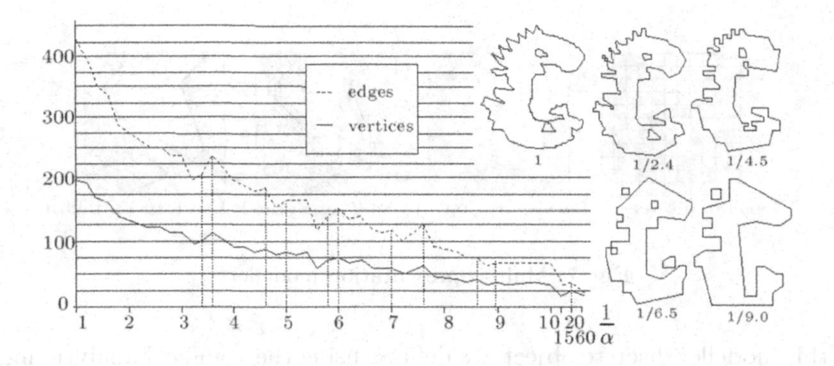

Fig. 9. Edges and vertices number evolution

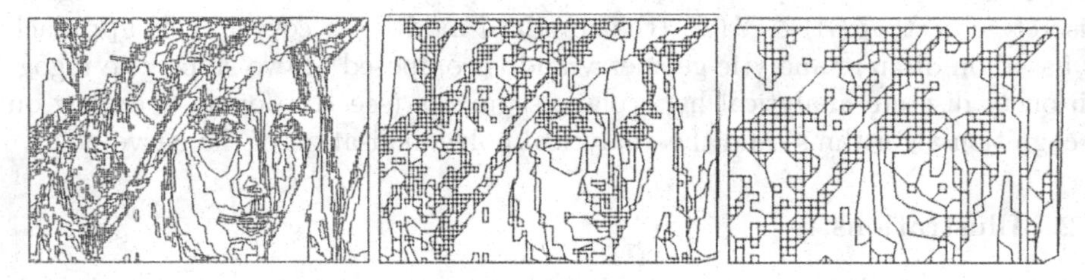

Fig. 10. Simplification example: $\alpha = 1$, $\frac{1}{2}$, and $\frac{1}{3.5}$

allows to decrease significantly the number of object element to be rendered (see Fig. 9).

However, as we can see on Fig. 9, for some coefficients (see dot lines), the number of object element increases. This is due to the instability of the digitization grid resulting from a simplification with coefficients in $\mathbb{R}$ and to the non determinism of the number of reconstructed edges. Figure 9 shows several resulting pictures. We can notice that the object topology is modified: a hole can appear and then disappear. In [13], authors provide a theorem linking topology modifications and grid size. This gives only a general bound because the recon-

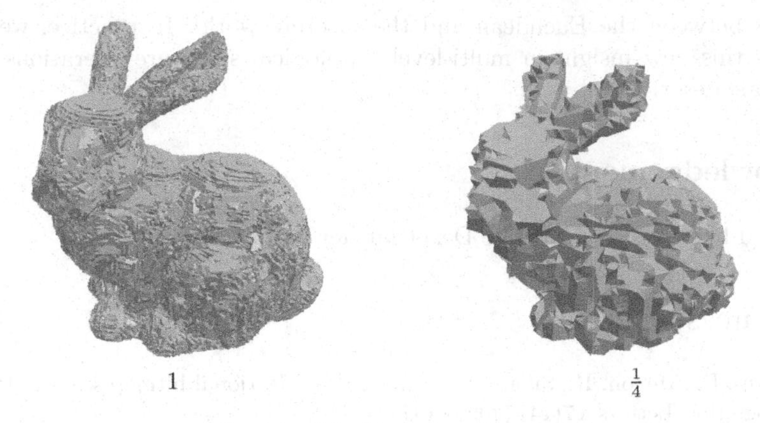

Fig. 11. $3D$ object simplification

struction process in not translation invariant. With a same grid size we can get different topologies. Object topology not only depends on the grid size but on its position. The center of the scaling function modifies the end result.

Figure 10 presents a discrete based simplification example and Fig. 11 shows its extention in $3D$.

5 Conclusion

In this paper, we have presented two operations that both use Euclidean and discrete world properties. Both operations are based on the Euclidean scaling transform. In the first case it scales the object, in the second one, it scales the grid. Both operations, while they seem quite different have a strong link and we obtained similar error bounds for them. The first operation is called the "discrete smooth scaling". We bounded the error done while trying to reverse this operation. The bigger the scale, the closer the discrete operation is to the Euclidean scaling transform. The discrete smooth scaling can be seen as a digitization of the Euclidean scale transform.

The second operation is a Euclidean operation that uses the discrete world properties. We define an operation that digitizes and recontructs Euclidean objects according to a given grid size. Depending on the grid size, a certain number of details are gathered in the same pixel and do therefore disappear during the reconstruction process. The result is a simplified Euclidean object that can be used in a multi-level representation form. The quality measure of a simplified object is a bound of the Hausdorff distance between the simplified and the original object proportional to the grid size.

In the future we are going to consider discrete-Euclidean transforms based on Euclidean operations such as rotations, translations and general affine transforms. We are also considering discrete-Euclidean transforms based on discrete operations such as boolean operations, mathematical morphology operations, etc. The long term theoretical goal of this study is to better understand the relations between the Euclidean and the discrete world. In practice, we hope to apply this new insight in multi-level topological structure operations or on multi-scale described objects.

Acknowledgements

We would like to thank Martine Dexet for her help.

References

1. Andres, E., Menon, R., Sibata, C., Acharya, R.: Rational bitmap scaling. Pattern Recognition Letters **17**(14) (1996) 1471–1475.
2. Cohen-Or, D., Kaufman, A.: Fundamentals of surface voxelization. Graphical models and image processing **57**(6) (1995) 453–461.
3. Andres, E., Menon, R., Acharya, R., Sibata, C.: Discrete linear objects in dimension n : The standard model. Graphical Models **65** (2003) 92–111.
4. Andres, E., Acharya, R., Sibata, C.: The supercover 3d polygon. In: Discrete Geometry for Computer Imagery. LNCS, Lyon, France (1996) 237–242.
5. Andres, E., Nehlig, P., Françon, J.: Supercover of straight lines, planes and triangles. In: Discrete Geometry for Computer Imagery. Volume LNCS n1347 of Lecture Notes on Computer Science., Springer Verlag (1997) 243–257.
6. Lincke, C., Wuthrich, W.: Towards a unified approach between digitization of linear objects and discrete analytical objects. In: WSCG 2000 Workshop, V. Skala (Ed.) (2000) 124–131.
7. Breton, R., Sivignon, I., Dupont, F., Andres, E.: Towards an invertible euclidean reconstruction of a discrete object. In: Discrete Geometry for Computer Imagery. Volume 2886 of Lecture Notes on Computer Science., Naples, Italie, Springer Verlag (2003) 246–256.
8. Dexet, M., Andres, E.: Linear discrete line recognition and reconstruction based on a generalized preimage. In: IWCIA Papers Proceedings. Volume LNCS 4040 of Lecture Notes on Computer Science., Berlin, Germany, Springer Verlag (2006) 174–188.
9. Françon, J., Schramm, J.M., Tajine, M.: Recognizing arithmetic straight lines and planes. In: DGCI'96, LNCS 1176, Springer-Verlag (1996) 141–150.
10. Rosenfeld, A., Klette, R.: Digital straightness. In: UMD. (2001)
11. Reveillès, J.P.: Géométrie discrète, Calcul en nombres entiers et algorithmique. PhD thesis, Université Louis Pasteur, Strasbourg, France (1991)
12. Dorst, L., Smeulders, A.: Discrete representation of straight lines. IEEE Transactions on Pattern Analysis and Machine Intelligence **6**(4) (1984) 450–463.
13. Tajine, M., Ronse, M.: Preservation of topolgy by hausdorff discretization and comparison to other discretization schemes. Theoretical Computer Science **283**(1) (2002) 243–268.

Geometry of Neighborhood Sequences in Hexagonal Grid*

Benedek Nagy

GRLMC, Rovira i Virgili University, Tarragona, Spain
Faculty of Informatics, University of Debrecen, Hungary
H-4010 Debrecen P.O. Box 12
nbenedek@inf.unideb.hu

Abstract. In this paper the nodes of the hexagonal grid are used as points. Three types of neighbors are used on this grid, therefore neighborhood sequences contain values 1, 2 and 3. The grid is coordinatized by three coordinates in a symmetric way. Digital circles are classified based on digital distances using neighborhood sequences. They can be triangle, hexagon, enneagon and dodecagon. The corners of the convex hulls of these polygons are computed.

1 Introduction

The classical digital geometry started by [1], where the authors defined the two basic neighborhood relations on the square grid. The topic is well developed due to people of image processing and computer graphics communities. We refer to [2] as a recent textbook on the topic. In [3] the authors used the so-called neighborhood sequences to vary the neighborhood criterion in a path. They used only periodic neighborhood sequences in their analysis. Some properties of distances based on neighborhood sequences are detailed in [4]. Nowadays, in many applications it is worth to consider other grids than the square one. The hexagonal grid has some nice properties and it is regular, therefore it is not too hard to handle it. The geometry of the hexagonal grid with a symmetric coordinate system is described in [5]. In [6] the neighborhood sequences were also defined for the hexagonal grid. In this paper we will analyse some properties of the distances based on neighborhood sequences on this grid.

The structure of the paper is as follows. In the second section we give our notation, and provide some properties of the concepts introduced. In the other sections we detail some former results of Das and Chatterji [4] on the hexagonal grid. We use only initial parts of the neighborhood sequences in our analysis, therefore we do not care about the periodic property of the whole neighborhood sequences. In the third section we describe the smallest digital circles of the hexagonal grid using only a step from the origin. In the fourth section changing and developments of wave-fronts and digital circles are analysed. We compute

* This research was supported by grants from the Hungarian Foundation for Scientific Research (OTKA F043090 and T049409).

A. Kuba, L.G. Nyúl, and K. Palágyi (Eds.): DGCI 2006, LNCS 4245, pp. 53–64, 2006.

the coordinate values of the corners of these polygons. In the fifth section we give a description of all the digital circles with neighborhood sequences in the hexagonal grid. We show a characterization of them. We present some properties, in which the hexagonal grid differs from the square grid. In the last section we summarize our results.

2 Basic Notation and Concepts

In this section we recall some definitions and notation from the literature mentioned earlier concerning neighborhood relations and sequences.

There are usually three types of neighbors defined, as Fig. 1 shows, among the nodes of the hexagonal grid.

In Figure 1 a node and its 12 neighbors are shown. Only the 1-neighbors are directly connected by an edge, the other 2- and 3-neighbors are at the positions of shorter and longer diagonals, respectively. These relations are reflexive (i.e., a node is a 1-, 2-, and a 3-neighbor of itself by definition) and symmetric. In

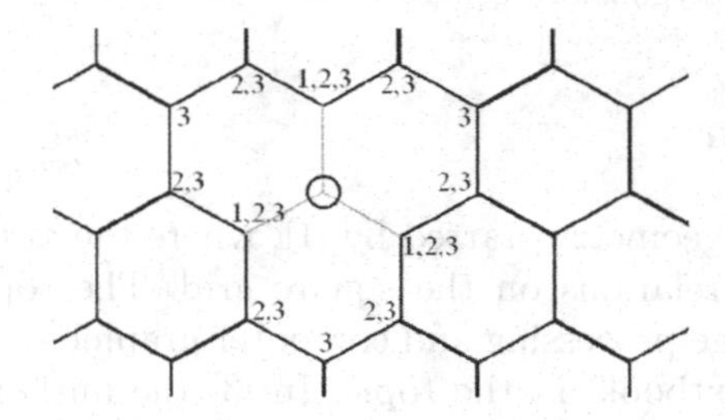

Fig. 1. Types of neighbors in the hexagonal grid

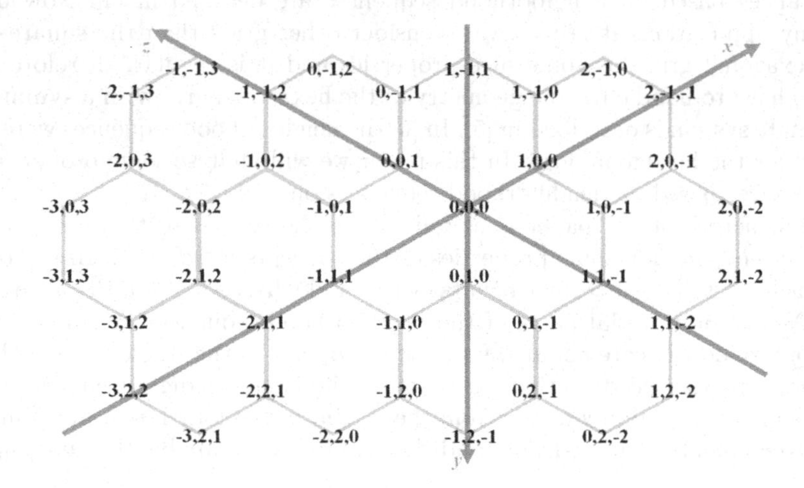

Fig. 2. Coordinate values of nodes

addition, all 1-neighbors of a point are its 2-neighbors and all 2-neighbors are 3-neighbors, as well (i.e., they have increasing and inclusion properties).

The coordinate values of the grid were introduced as it is shown in Fig. 2. The coordinate axes meet at a grid-point called Origin having triplet (0,0,0). They are the direction of grid-edges starting at the Origin. The coordinate values of each point can be computed as the sum of steps on the grid-edges taken into direction of the edges. A step by direction of axis x increases the first coordinate value by 1, while a step to inverse direction decreases the first coordinate. Similarly steps on the edges parallel to axis y and z modify the second and third coordinate values, respectively. Three coordinate values are used to address a point taking advantage of the symmetry of the grid.

With the help of the assigned coordinate values we are able to describe the grid in a mathematical way. By the presented coordinate system every node has a unique triplet which exactly shows the place of the node. The hexagonal grid contains exactly those triplets which have sum of coordinate values 0 or 1. We call the points with 0-sum value *even* (their connections have shape **Y** in the figure); the points with 1-sum are *odd* (opposite shape). We can write the neighborhood relations in the following formal form.

The points $P(p(1), p(2), p(3))$ and $Q(q(1), q(2), q(3))$ of the hexagonal grid are m-neighbors ($m = 1, 2, 3$), if the following two conditions hold:

1. $|p(i) - q(i)| \leq 1$, for $i = 1, 2, 3$,
2. $|p(1) - q(1)| + |p(2) - q(2)| + |p(3) - q(3)| \leq m$.

It is easy to check that the formal definition above with the presented coordinate values (Fig. 2) gives the neighborhood relations shown in Fig. 1.

Now, we are recalling some concepts about the theory of neighborhood sequences. In this paper, we are dealing only with neighborhood sequences in this grid. The sequence $B = (b(i))_{i=1}^{\infty}$, where $1 \leq b(i) \leq 3$ for all $i \in \mathbb{N}$, is called a *neighborhood sequence* (on the hexagonal grid). When we need only the initial part up to the l-th element, then we briefly write $B_l = (b(1), b(2), \ldots, b(l))$.

A movement is called a $b(i)$-step when we move from a point P to a point Q and they are $b(i)$-neighbors. Let P, Q be two points and B be a neighborhood sequence. The point-sequence $P = P_0, P_1, \ldots, P_k = Q$, in which we move from P_{i-1} to P_i by a $b(i)$-step ($1 \leq i \leq k$), is called a B-path from P to Q of length k. The B-distance $d(P, Q; B)$ from P to Q is defined as the length of the shortest B-path(s). In a B-path an initial sequence of B is used.

The sequence of 1-neighbor points, for which a coordinate value remains constant, forms a so-called *lane.* In Fig. 3 there are some examples; the black line shows the lane for which the third coordinate is 0 and the gray lane represents the lane for which the second value is 0.

Every lane is 'orthogonal' to one of the coordinate axes, especially that one of $\{x, y, z\}$, for which the coordinate value is fixed.

If a point is on an axis, then two of its coordinate-values coincide. For instance on axis x they are the second and third ones (meaning that the point is on the

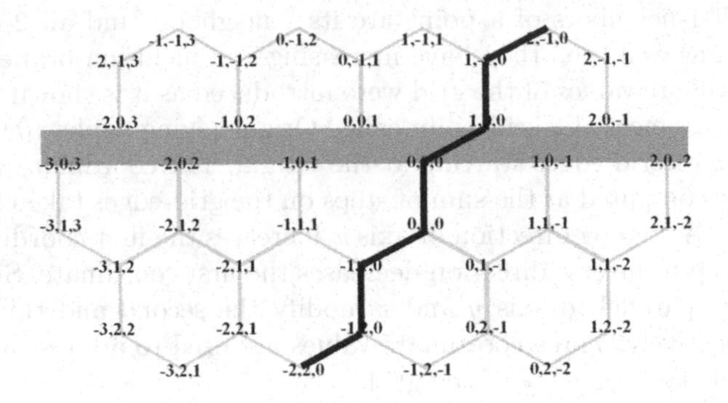

Fig. 3. Examples for lanes

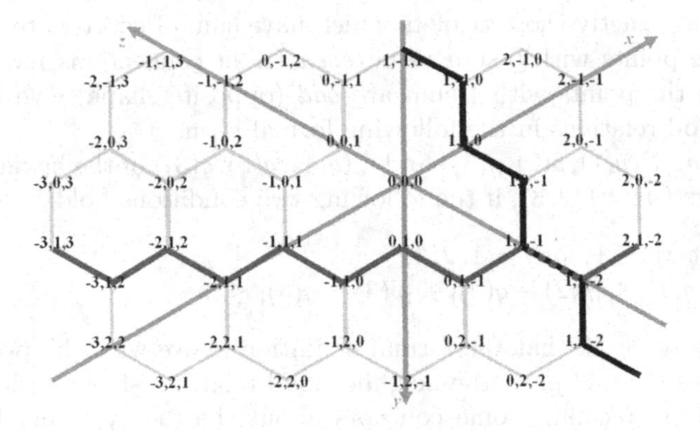

Fig. 4. Lanes and coordinate axes

lanes for which the same value is fixed on axes y and z). See Fig. 4, where the lanes $x = 1$ and $y = 1$ are shown with black and dark-gray, respectively. They are orthogonal to the axes x and y, respectively. The two points where they meet are on the axis z. It is a nice property of the assignment of coordinates to the grid, that a point and its symmetric pair mirroring it to an axis have the same coordinate values, but – if they are not on the symmetry axis, – in a different order. So a point and its mirror images are identical up to a permutation of their coordinates.

To use the line between the lanes $x = 0$ and $x = 1$ as a symmetry axis such that these two lanes are mirror images of each others we have the following formula. The mirror image of the point $P(x, y, z)$ is given as $P'(-(x-1), -z, -y)$.

In this paper we investigate the way of a neighborhood sequence spreads in the digital space starting from a point of the hexagonal grid. This spreading is translation-invariant among the points of the same parity and it is symmetric concerning points with different parities. So, for simplicity we may choose the Origin $\mathbf{O}(0, 0, 0)$ as the starting point.

Let

$$C_{B_k} = \{P : d(\mathbf{O}, P; B) \leq k\}.$$

It is the region (*digital circle*) occupied by B after the first k steps.

In [4] Das and Chatterji showed that for every initial part of a neighborhood sequence in the square grid (using Cartesian coordinates and two types of neighbors), the obtained digital circle is always an octagon. These octagons can be degenerated, so the digital circles are squares in the following cases: Using only 1-steps in the initial part of the neighborhood sequence we get only four edges (the corners will be $(0,k), (k,0), (0,-k), (-k,0)$), while using only 2-steps we get a square with corners $(k,k), (k,-k), (-k,-k), (-k,k)$. In the case when we use both 1-step(s) and 2-step(s) our result is a non-degenerated octagon.

The following observations hold in the square grid, and they are true in the hexagonal case as well:

Remark 1. *The convex hull of every C_{B_k} is digitally convex in the usual sense (see e.g. p. 171, Definition 4.3.4. in [7])*

On figures we will use the convex hull as the occupied polygon of the digital circle. These polygons has sides and corners in the usual sense. Since they are convex, the sets of the coordinate triplets of corners describe them. (We use the term corner only for angles less than π.)

Remark 2. *For any neighborhood sequence B, the sequence of regions $(C_{B_k})_{k=1}^{\infty}$ is a strictly monotone increasing sequence. That is, $k > l$ implies $C_{B_k} \supsetneq C_{B_l}$.*

In the following sections we will underline some properties which are different for the digital circles in square grid and in hexagonal grid. Now, we are moving to describe all digital circles of the hexagonal grid in details.

3 Description of Small Circles

Table 1 shows the three possible circles obtained by a step.

The three circles have four kinds of corners. Corner-type α signs corners with angle $\frac{\pi}{3}$. Corner-types β and γ refer for angles $\frac{2\pi}{3}$ (where the sides connected at a corner type γ are parallel to some edges and so to coordinate axes of the hexagonal grid, and the sides at a corner type β are orthogonal to some edges and so to coordinate axes of the grid.) At corners type δ the angles are $\frac{5\pi}{6}$. Note that the circle obtained by a 2-step from $\mathbf{O}$ is the same as the circle obtained by two 1-steps. This property (a circle can have more radii, depending on the used neighborhood sequences) is not present in the square grid.

Proposition 1. *Contrary to the square grid in the hexagonal grid it is possible for two neighborhood sequences B', B'' that $C_{B'_k} = C_{B''_l}$ with $k \neq l$.*

Let B' be given such a way that $B'_2 = (1,1)$ and let it hold for B'' that $B''_1 = (2)$ then $C_{(1,1)} = C_{(2)}$.

Table 1. Digital circles with radius 1 by various steps

Initial part of the neighbourhood sequence	The shape of the circle (figure)	Corner-types and coordinates
1 $C_{(1)}$		triangle with number 3 α-type: α(1,0,0) α(0,1,0) α(0,0,1)
11 2 $C_{(2)} = C_{(1,1)}$		hexagon with number 6 β-type: β(1,-1,0) β(1,0,-1) β(0,1,-1) β(0,-1,1) β(-1,1,0) β(-1,0,1)
3 $C_{(3)}$		enneagon with γ(1,-1,0) γ(1,0,-1) γ(0,1,-1) γ(0,-1,1) γ(-1,1,0) γ(-1,0,1) δ(1,1,-1) δ(1,-1,1) δ(-1,1,-1)

4 Development of Corners

In [3] the authors analysed the wavefront sets of neighborhood sequences in the square grid. In this part we are detailing how the wave-fronts are occupying the hexagonal grid. Because of symmetry it is sufficient to deal with a sixth of the plane. For our convenience we will use the region between the axes z and y, especially between the positive part of z and the negative part of y (we refer here to Fig. 2). One can check that the points $P(x, y, z)$ with $z \geq x \geq y$ are exactly those ones which are in this sixth of the plane. We will call this region as the *analysed section*.

It is sufficient to deal with this analysed section since we have the following statements. For any point P the points obtained by permutating the coordinate-values of P are exactly the same points as the mirror images of P obtained by mirroring it to some of the coordinate axes. Every point has the same parity as its symmetric pairs. Let P be given as $(p(1), p(2), p(3))$. It is easy to check that, for instance, the mirror image of P with axis x is $P'(p(1), p(3), p(2))$. Similar facts hold for the other mirror points. A point with its mirror images represent at most six points. In this case there are not equal coordinate values, i.e. the point is not on an axis. Every point has a mirror image in the analysed section of the grid (using only axial mirroring to the coordinate axes). It can be obtained by permutation of coordinate-values.

Let us check which corners of the small circles are in the analysed section. $C_{(1)}$ has only one corner in this section, namely $\alpha(0,0,1)$. $C_{(2)}$ has also only one corner in this area of the grid, namely $\beta(0,-1,1)$, while $C_{(3)}$ has two corners in the analysed section; they are $\gamma(1,-1,1)$ on axis y and $\delta(0,-1,1)$ with an orthogonal edge to axis z.

Now we analyze how the possible vertices change in the growing steps. Table 2 lists all kinds of corners occur in different digital circles. (We detail only the changing way of the coordinate values of those corners which are in the analysed sixth of the hexagonal plane.) The table shows all possibilities of the evolution of the corners by a step in the analysed section of the grid. It is due to the facts that all corners of Table 1 (which are in the analysed section) are also in Table 2, and all possible evolving corners are in Table 2, as well.

Based on Table 2 we summarize the evolution of corners via the growing procedure.

Proposition 2

- *(a) The new corner(s) are b-neighbors of the previous one when we obtain them by a b-step.*
- *(b) A corner type α occurs only at digital circle $C_{(1)}$.*
- *(c) A corner type γ occurs only when two coordinate values have the same value, i.e. the corner is on an axis.*
- *(d) In some steps a corner type β or a corner type γ is divided to two corners type δ.*
- *(e) If a corner type γ remains the same type after a step then it stays on the same axis as before (by changing the parity).*
- *(f) The corners type δ never change their type, moreover their position is fixed, i.e. when a side of the corner was orthogonal to an axis then this is also the case after any kind of step.*
- *(g) The corners type δ coming from a corner type γ have a side orthogonal to that axis on which the corner γ was.*
- *(h) In some cases the resulted digital object by a 3-step is the same as the one obtained by a 2-step.*

Proof. Most of the statements above are easy to check. We analyse only the statements (e) and (f). Let us start with (e).

If a corner $\gamma(x,y,z)$ on the axis y, then $x=z$. If it is even then with a 3-step we get $\gamma(x+1,y-1,z+1)$ which is also on the axis y. Similarly if γ is odd then with a 1-step the first and the third coordinate values do not change, therefore the new corner is on the axis y as well. The same analysis works when γ is on the axis z. It is evident that the parity of the corner is changing by these steps.

To prove (f) assume that $\delta(x,y,z)$ has an edge which is orthogonal to the axis y. Then the mirror images of δ are also corners. Let $\delta'(z,y,x)$ be its

Table 2. Development table of corners by taking a step

original corner: type and coordinates	corner after a 1-step	corner after a 2-step	corner after a 3-step
$\alpha(0,0,1)$	$\beta(0,-1,1)$	$\beta(0,-1,2)$	$\gamma(-1,-1,2)$ on axis z, $\delta(0,-1,2)$ with edge orthogonal to axis y
$\beta(x,y,z)$ even	$\beta(x,y,z+1)$	$\beta(x,y-1,z+1)$	$\delta(x,y-1,z+1)$ with edge orthogonal to axis z, $\delta(x+1,y-1,z+1)$ with edge orthogonal to axis z
$\beta(x,y,z)$ odd	$\beta(x,y-1,z)$	$\beta(x,y-1,z+1)$	$\delta(x,y-1,z+1)$ with edge orthogonal to axis y, $\delta(x-1,y-1,z+1)$ with edge orthogonal to axis z
$\gamma(x,y,z)$ even on axis y	$\delta(x,y,z+1)$, $\delta(x+1,y,z)$	$\delta(x,y-1,z+1)$, $\delta(x+1,y-1,z)$	$\gamma(x+1,y-1,z+1)$
$\gamma(x,y,z)$ odd on axis y	$\gamma(x,y-1,z)$	$\delta(x,y-1,z+1)$, $\delta(x+1,y-1,z)$	$\delta(x,y-1,z+1)$, $\delta(x+1,y-1,z)$
$\gamma(x,y,z)$ even on axis z	$\gamma(x,y,z+1)$	$\delta(x-1,y,z+1)$, $\delta(x,y-1,z+1)$	$\delta(x-1,y,z+1)$, $\delta(x,y-1,z+1)$
$\gamma(x,y,z)$ odd on axis z	$\delta(x-1,y,z)$, $\delta(x,y-1,z)$	$\delta(x-1,y,z+1)$, $\delta(x,y-1,z+1)$	$\gamma(x-1,y-1,z+1)$
$\delta(x,y,z)$ even with edge orthogonal to axis y	$\delta(x,y,z+1)$	$\delta(x,y-1,z+1)$	$\delta(x+1,y-1,z+1)$
$\delta(x,y,z)$ odd with edge orthogonal to axis y	$\delta(x,y-1,z)$	$\delta(x,y-1,z+1)$	$\delta(x,y-1,z+1)$
$\delta(x,y,z)$ even with edge orthogonal to axis z	$\delta(x,y,z+1)$	$\delta(x,y-1,z+1)$	$\delta(x,y-1,z+1)$
$\delta(x,y,z)$ odd with edge orthogonal to axis z	$\delta(x,y-1,z)$	$\delta(x,y-1,z+1)$	$\delta(x-1,y-1,z+1)$

symmetric pair to axis y. Then the side of the polygon connecting δ and δ' is orthogonal to the axis y. After any kind of step the new corners obtained from δ and δ' have the same property. The proof is similar if one of the edges connected at corner δ is orthogonal to axis z. □

As we can see the corner-types are in a closed set, i.e. we cannot step out from the above used set by the growing steps.

Now, we present a method which calculates the corners of any digital circle. We are using Table 1 and the transition table given in Table 2.

Let us calculate the corner of the digital circle with origin (x_0, y_0, z_0) using the initial part B_k of a neighborhood sequence B. First we compute the corners of C_{B_k} using $\mathbf{O}$ as origin, and after this we will translate the circle.

Algorithm

1. Initially start with corners of the circle $C_{(b(1))}$ from Table 1 which are in the analysed section of the grid, where $b(1)$ is the first element of B. Let $i = 1$.
2. While $i < k$ let $i := i + 1$ and the analysed corners of the new polygon are from Table 2 by a $b(i)$-step. (Keep only those ones which are in the analysed part of the grid.)
3. Get every point which coordinate values form a permutation of a computed point. Those are the corners of C_{B_k}.
4. If (x_0, y_0, z_0) is even then add the vector (x_0, y_0, z_0) to each corner of C_{B_k} in order to get the result.
5. If (x_0, y_0, z_0) is odd then then, first, let the new values of coordinates of the corners be given by the formula $(x, y, z) \rightarrow (-(x-1), -z, -y)$ (they represent the digital circle with origin $(1, 0, 0)$ generated by B with radius k); secondly, add the vector $(x_0 - 1, y_0, z_0)$ to each corner to obtain the final result.

Now we present an example. Let us determine the corners of the digital circle starting from point $(-5, 3, 3)$ with $B_4 = (1, 3, 1, 2)$.

1. $b(1) = 1$, therefore we start with $C_{(1)}$, so we have $\alpha(0, 0, 1)$ and $i = 1$.
2. $i < 4$ therefore $i := 2$, $b(i) = b(2) = 3$, the result: $\gamma(-1, -1, 2)$ and $\delta(0, -1, 2)$.
   $i < 4$ therefore $i := 3$, $b(i) = b(3) = 1$, $\gamma(-1, -1, 2)$ is even and it is on axis z (the first two coördinates have the same value): $\gamma(-1, -1, 3)$ and
   $\delta(0, -1, 2)$ is odd and it has edge orthogonal to the axis y so the new corner: $\delta(0, -2, 2)$.
   $i < 4$ therefore $i := 4$, $b(i) = b(4) = 2$, $\gamma(-1, -1, 3)$ is odd and it is on axis z, so we get $\delta(-2, -1, 4)$ and $\delta(-1, -2, 4)$, but the first one is outside of the analysed section of the plane
   $\delta(0, -2, 2)$ is even and it has edge orthogonal to the axis y so the new corner: $\delta(0, -3, 3)$.
   $i = 4$, the loop of Step 2 is finished.
3. The corners with permutations: all of them are type-δ: $(-1, -2, 4)$, $(-2, -1, 4)$, $(-1, 4, -2)$, $(-2, 4, -1)$, $(4, -1, -2)$, $(4, -2, -1)$ and $(0, -3, 3)$, $(0, 3, -3)$, $(-3, 0, 3)$, $(-3, 3, 0)$, $(3, 0, -3)$, $(3, -3, 0)$.
4. The given origin is odd, so
5. first we get: $(2, -4, 2)$, $(3, -4, 1)$, $(2, 2, -4)$, $(3, 1, -4)$, $(-3, 2, 1)$, $(-3, 1, 2)$, $(1, -3, 3)$, $(1, 3, -3)$, $(4, -3, 0)$, $(4, 0, -3)$, $(-2, 3, 0)$, $(-2, 0, 3)$.
   Secondly, adding $(-6, 3, 3)$ the final result: $(-4, -1, 5)$, $(-3, -1, 4)$, $(-4, 5, -1)$, $(-3, 4, -1)$, $(-9, 5, 4)$, $(-9, 4, 5)$, $(-5, 0, 6)$, $(-5, 6, 0)$, $(-2, 0, 3)$, $(-2, 3, 0)$, $(-8, 6, 3)$, $(-8, 3, 6)$ and all of them are type-δ.

Using the three digital circles with radius 1 and our growing table we get all possible digital circles of the hexagonal grid. In the next section we will list their types.

5 The Shapes of the Digital Circles

In this section - continuing our previous results - we characterize the digital circles by neighborhood sequences in the hexagonal grid.

Theorem 1. *The shape of the digital circle generated by the neighborhood sequence B in k steps is a triangle if and only if it is $C_{(1)}$. The shape is a hexagon if and only if there is no element 3 in the initial part of B up to the k-th element. (Except the previous case, in which $C_{(1)}$ is especially a triangle.) The shape is an enneagon if and only if there is not any element 2 and nor any consecutive 1,1 or 3,3 occur in the initial part B_k. (Except $C_{(1)}$.) In every other case the digital circle is a dodecagon.*

Proof. In the first case the triangle has three α corners. According to point (b) of Proposition 2 only other types of corners can occur at the other digital circles. With other types of corners it is impossible to get a triangle.

Now let us consider the other digital circles. It is easy to check in Table 1 that starting with an element 2 we get a hexagon with six β corners. Moreover by Table 2 we know that using 1-step and/or 2-step from corners type α and from corners type β the new corners will be type β as well. With only corners type β there must be six of them to make a polygon. Therefore without a 3-step the result is a hexagon with corners type β.

It is shown in Table 1 that $C_{(3)}$ is an enneagon with three γ and six δ corners. The corner $\gamma(1,-1,1)$ is odd and it is on the axis y. Therefore with a 1-step it grows to a γ_2 which is even and is on the same axis. (With a 2-step or a 3-step the corner would be divided to two δ vertices.) From γ_2 with a 3-step γ_3 is resulted; it is odd on axis y. (From γ_2 with a 1-step or a 2-step we would obtain two δ corners.) Since γ_3 is in the same 'class' (i.e. in the same row of Table 2) as $\gamma(1,-1,1)$ the computing cycle is starting again, and it is going on while the steps are 1-step and 3-step turn by turn. Observe in Table 2 that type γ corner can be obtained from α, but cannot from β. Therefore there is no way to get γ vertices from $C_{(2)}$ and so from any hexagons. From $C_{(1)}$ one can obtain an enneagon in one way, namely to get $C_{(1,3)}$. The obtained corner $\gamma(-1,-1,2)$ is even and it is on the axis z. One can check that there is a similar computing cycle for this γ to keep it with only 1-steps and 3-steps by turns. (Leaving this computing cycle two corners type δ are obtained instead of the type γ.)

Finally, in all other cases the polygons only have δ vertices. When twelve corners type δ are in a digital circle, then it never happens that they change to another type (see Table 2) and twelve of them are needed for a polygon. Therefore the last statement is proved. □

Analysing the digital circles on the square and on the hexagonal grid we have found the following important difference.

In the square grid the region occupied by k steps of a neighborhood sequence A is independent of the ordering of the first k element of A. (see [4])

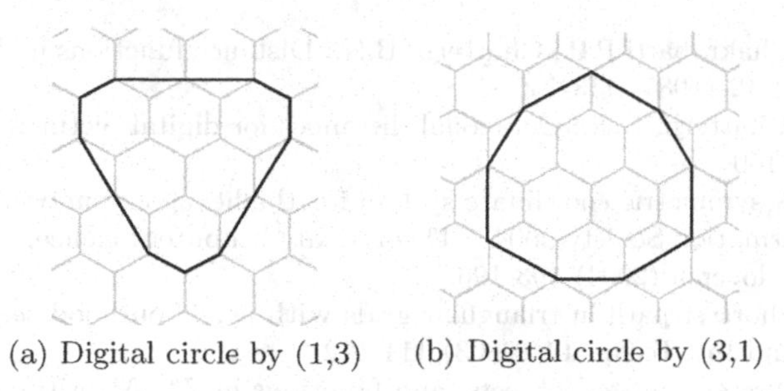

(a) Digital circle by (1,3) (b) Digital circle by (3,1)

Fig. 5. The elements of the neighborhood sequence on the hexagonal grid are not permutable, in general

Proposition 3. *Contrary to the case of the square grid, it is possible for a neighborhood sequence B and for a $k \in \mathbb{N}$, that the region C_{B_k} depends on the order of the first k elements of B.*

Assume that $B_2' = (1,3)$ and $B_2'' = (3,1)$ then our regions $C_{B_2'}$ and $C_{B_2''}$ differ as Fig. 5 shows.

6 Conclusions

In this paper we presented some results about neighborhood sequences on the hexagonal grid. We made a classification of the digital circles. We gave the possible types of corners of these digital polygons and studied their development in the growing procedure. Moreover an algorithm is presented to compute the coordinate-values of the corners of any digital circle (arbitrary origin, arbitrary neighborhood sequence and arbitrary radius) on the hexagonal grid. We listed the types of the digital circles occupied by neighborhood sequences, as well. Since the convex hulls of the digital circles are convex polygons, the lists of their corners determine them.

Our results can be used in digital image processing and in the field of networks as well. It is useful in region growing procedures. In grid-structured networks the non-common properties are useful. Some digital circles have several radii or the non permutability of the elements of the neighborhood sequence are exotic properties. In practice, it would also be interesting to analyse the development of the wave-front sets in the case of "barrels", or starting not from a point, but from other digital object, for instance from a lane. Another possible direction of future research is the further analysis of meeting waves, etc. It would be interesting if one mixed our method of region growing with the methods used in practice ([8,9,2]). Extensions to non-regular grids can be topics of further research, as well.

References

1. Rosenfeld, A., Pfaltz, J.L.: Distance functions on digital pictures. Pattern Recognition **1** (1968) 33-61
2. Klette, R., Rosenfeld, A.: Digital geometry. Geometric methods for digital picture analysis. Morgan Kaufmann Publishers, San Francisco, CA, Elsevier Science B.V., Amsterdam (2004)

3. Das, P.P., Chakrabarti P.P., Chatterji, B.N.: Distance functions in digital geometry. Inform. Sci. **42** (1987) 113-136
4. Das, P.P., Chatterji, B.N.: Octagonal distances for digital pictures. Inform. Sci. **50** (1990) 123-150
5. Nagy, B.: A symmetric coordinate system for the hexagonal networks. In: Proceedings of Information Society 2004 – Theoretical Computer Science, ACM conference, Ljubljana, Slovenia (2004) 193-196
6. Nagy, B.: Shortest path in triangular grids with neighbourhood sequences. Journal of Comp. and Inf. Techn. **11** (2003) 111-122
7. Voss, K.: Discrete images, objects, and functions in $\mathbb{Z}^n$. Algorithms and Combinatorics Vol. 11. Springer-Verlag, Berlin Heidelberg New York (1993)
8. Chen, C.H., Pau, L.F., Wang, P.S.P. (eds.): Handbook of pattern recognition & computer vision. Handbooks in Science and Technology. Academic Press, Orlando, FL (1986)
9. Gonzalez, R.C., Woods, R.E.: Digital image processing. Addison-Wesley, Reading, MA (1992)

Recognition of Blurred Pieces of Discrete Planes

Laurent Provot[1], Lilian Buzer[2], and Isabelle Debled-Rennesson[1]

[1] LORIA Nancy
Campus Scientifique - BP 239
54506 Vandœuvre-lès-Nancy Cedex, France
{provot, debled}@loria.fr

[2] Laboratory CNRS-UMLV-ESIEE, UMR 8049
ESIEE, 2, boulevard Blaise Pascal
Cité Descartes, BP 99
93162 Noisy le Grand Cedex, France
buzerl@esiee.fr

Abstract. We introduce a new discrete primitive, the blurred piece of a discrete plane, which relies on the arithmetic definition of discrete planes. It generalizes such planes, admitting that some points are missing and then permits to adapt to noisy discrete data. Two recognition algorithms of such primitives are proposed: the first one is a geometrical algorithm and minimizes the Euclidean distance and the second one relies on linear programming and minimizes the vertical distance.

1 Introduction

The recognition of discrete primitives as digital straight lines and digital planes is a deeply studied problem in digital geometry (see a review in the book [1]). This problem consists in determining if a set of discrete points corresponds to a known discrete primitive and, in such case, in identifying its characteristics. Three main classes of algorithms can be defined:

- Structural algorithms: based on geometric (convex hull, chords) or combinatorial (size of the steps) properties of discrete primitives. Indeed, the structural regularity of these primitives can lead to efficient algorithms.
- Arithmetic algorithms: based on the definition of discrete primitives as Diophantine inequalities, these algorithms make profit of the well defined arithmetical structure of discrete primitives.
- Dual space algorithms: the recognition problem is translated in a dual space where each grid point is represented by a double linear constraint. The recognition problem is then defined as a linear programming problem, optimized using particular knowledge on the constraints geometry.

Recently, a new discrete primitive, the blurred segment [2,3], was introduced to deal with the noise or artefacts due to the acquisition tools or methods. Relying on an arithmetic definition of discrete lines [4], it generalizes such lines, admitting that some points are missing. Efficient blurred segments recognition

A. Kuba, L.G. Nyúl, and K. Palágyi (Eds.): DGCI 2006, LNCS 4245, pp. 65–76, 2006.

algorithms were proposed [2, 3, 5] and they were used in applications in image analysis [6]. In the same framework, we introduce in the paper the new notion of blurred pieces of discrete planes, relying on the definition of arithmetic discrete plane [7] by considering a variable thickness.

Two recognition algorithms of blurred pieces of discrete planes are proposed. The first one is based on a structural approach: the computation of the convex hull of the given voxels is done while we search for the two parallel planes that mark out this convex hull and that minimize the Euclidean distance between themselves. An incremental algorithm is given. The second one is based on a dual space approach in the context of linear programming: the recognition problem is modelled by a system of linear constraints defined by the initial set of points. The simplex algorithm is then used to solve the problem by minimizing the vertical distance between two parallel planes containing all the points of the initial set. A geometrical interpretation of this method is also given. The codes of these algorithms and examples are available on `http://www.loria.fr/~debled/BlurredPlane`.

In section 2, after recalling definitions and basic properties of arithmetic discrete planes, we define the related notion of blurred pieces of discrete planes and optimal bounding planes. Then, in section 3, a geometrical method is proposed to solve the recognition problem by minimizing the Euclidean distance. The second method, based on linear programming, is presented in section 4 as well as a geometrical interpretation of the dual problem. The paper ends up with some conclusions and perspectives in section 5.

2 Blurred Pieces of Discrete Planes

An ***arithmetic discrete plane*** [7], named $\mathcal{P}(a, b, c, \mu, \omega)$, is a set of integer points (x, y, z) verifying $\mu \leq ax + by + cz < \mu + \omega$ where $(a, b, c) \in \mathbb{Z}^3$ is the normal vector. $\mu \in \mathbb{Z}$ is named the translation constant and $\omega \in \mathbb{Z}$ the arithmetical thickness.

The two real planes, defined by the following equations: $ax + by + cz = \mu$ and $ax + by + cz = \mu + \omega - 1$, are called the ***leaning planes*** of $\mathcal{P}(a, b, c, \mu, \omega)$. All the points of $\mathcal{P}$ are located between the leaning planes of $\mathcal{P}$.

We hereafter propose a generalization of the notion of discrete plane relying on the arithmetical definition and admitting that some points are missing. Consider a norm N on $\mathbb{R}^3$. We define the notion of bounding plane, relative to N, as follows:

Definition 1. *Let $\mathcal{E}$ be a set of points in $\mathbb{Z}^3$. We say that the discrete plane $\mathcal{P}(a, b, c, \mu, \omega)$ is a* ***bounding plane of*** *$\mathcal{E}$ if all the points of $\mathcal{E}$ belong to $\mathcal{P}$. We call* ***width of*** *$\mathcal{P}(a, b, c, \mu, \omega)$, the value $\frac{\omega - 1}{N(a,b,c)}$.*

Interpretation of the Width:

1. if $N = \|\cdot\|_2$, the width $\frac{\omega-1}{N(a,b,c)}$ represents the Euclidean distance between the two leaning planes of the bounding plane $\mathcal{P}(a, b, c, \mu, \omega)$. Indeed, let P_1 :

$ax+by+cz=\mu$ and $P_2 : ax+by+cz=\mu+\omega-1$ be the two leaning planes of $\mathcal{P}$. As P_1 and P_2 are parallel, the distance between P_1 and P_2 is equal to $\frac{|\mu+\omega-1-\mu|}{\sqrt{a^2+b^2+c^2}}$, i.e. $\frac{\omega-1}{\|(a,b,c)\|_2}$ since $\omega > 0$.

2. if $N = \|\cdot\|_\infty$, the width $\frac{\omega-1}{N(a,b,c)}$ represents the distance according to the main direction of the vector (a,b,c). Indeed and without loss of generality we can assume that $\max(|a|,|b|,|c|) = |c|$, which means the main direction is the Oz axis. Let $M_1(x_1,y_1,z_1) \in P_1$ and $M_2(x_2,y_2,z_2) \in P_2$ such that $x_1 = x_2$ and $y_1 = y_2$. The distance between P_1 and P_2 is equal to $|z_1 - z_2| = \frac{|c(z_1-z_2)|}{|c|} = \frac{|a(x_1-y_2)+b(y_1-y_2)+c(z_1-z_2)|}{|c|} = \frac{|\mu-(\mu+\omega-1)|}{|c|}$ because $M_1 \in P_1$ and $M_2 \in P_2$, i.e. $\frac{\omega-1}{\|(a,b,c)\|_\infty}$ since $\omega > 0$.

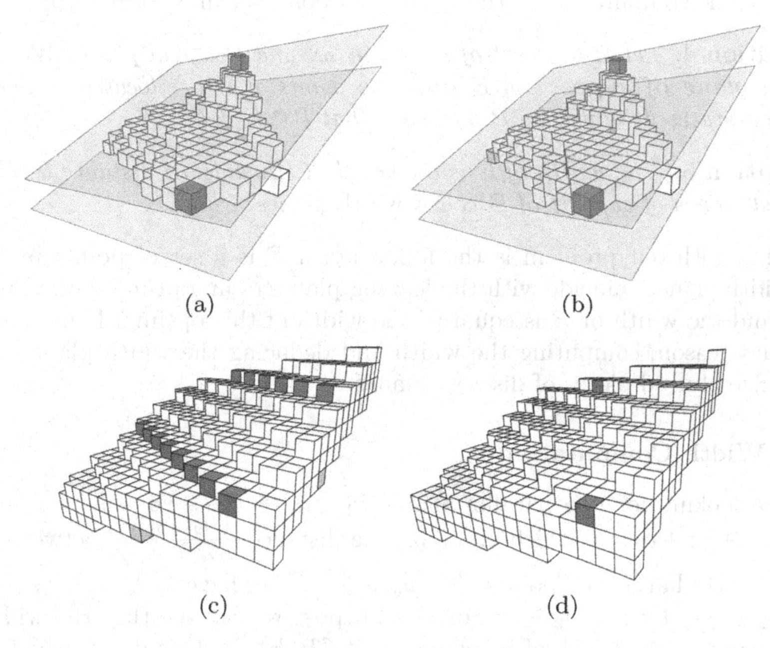

Fig. 1. A width-3 blurred piece of discrete plane (a and b), its optimal bounding planes (c) for Euclidean norm: $\mathcal{P}_2(4,8,19,-80,49)$ and the width of $\mathcal{P}_2 = 2.28$ (d) for infinity norm: $\mathcal{P}_\infty(31,65,157,-680,397)$ and the width of $\mathcal{P}_\infty = 2.52$. The leaning planes and corresponding leaning points of $\mathcal{P}_2$ and $\mathcal{P}_\infty$ are respectively drawn on (a,c) and (b,d).

Definition 2. *Let $\mathcal{E}$ be a point set in $\mathbb{Z}^3$. A bounding plane of $\mathcal{E}$ is said* ***optimal*** *if its width is minimal.*

This leads us to the definition of a blurred piece of discrete plane (Fig. 1).

Definition 3. *A point set $\mathcal{E}$ in $\mathbb{Z}^3$ is a* ***width-ν blurred piece of discrete plane*** *if and only if the width of its optimal bounding plane is less or equal to ν.*

In the following sections we propose two algorithms which solve the recognition problem of blurred pieces of discrete planes. For a given set of points $\mathcal{E}$ in $\mathbb{Z}^3$ and a width ν these algorithms decide whether $\mathcal{E}$ is a width-ν blurred piece of discrete plane. In addition, they give the characteristics of an optimal bounding plane of $\mathcal{E}$ for which the width is minimal. We also show how these algorithms can be made incremental.

3 Geometrical Method for the Recognition of Blurred Pieces of Discrete Planes

The first approach allows to solve the problem in terms of the norm $\|\cdot\|_2$. It relies on the computation of the width of a point set in 3-space [8,9].

Definition 4. *Let E be a set of points in $\mathbb{R}^3$ and P a real plane. We say that P is a* ***plane of support*** *of E if all the points of E are located in one of the two half-spaces delimited by P and such that $P \cap E \neq \emptyset$.*

Definition 5. *The* ***width*** *of E is the smallest (Euclidean) distance between two parallel planes of support of E called* width planes.

The link with our problem is the following: if E is a set of points in $\mathbb{Z}^3$ then the width planes coincide with the leaning planes of an optimal bounding plane of E and the width of E is equal to the width of this optimal bounding plane. For that reason, computing the width and deducing the width planes allow to recognize blurred pieces of discrete planes.

3.1 Width Computation

We are looking for two parallel planes $P_1 : \alpha x + \beta y + \gamma z + \delta_1 = 0$ and $P_2 : \alpha x + \beta y + \gamma z + \delta_2 = 0$ which minimize the distance $\frac{|\delta_2 - \delta_1|}{\sqrt{\alpha^2+\beta^2+\gamma^2}}$ between P_1 and P_2 and such that, for all points $p(p_x, p_y, p_z) \in E$, we have $p_x + \beta p_y + \gamma p_z + \delta_1 \leq 0$ and $p_x + \beta p_y + \gamma p_z + \delta_2 \geq 0$. For this purpose we can see that the width of E is the same as the width of its convex hull $\mathcal{CH}(E)$ [8]. It is due to the fact that $\mathcal{CH}(E)$ is the intersection of all the half-spaces containing all the points of E. We can then simplify the problem by introducing antipodal pairs. Consider the convex hull of a set of points E in 3-space. Two of its edges form an ***antipodal edge-edge (E-E) pair*** when two parallel planes of support of E contain these edges. Similarly, we define ***vertex-vertex (V-V)***, ***face-face (F-F)***, ***vertex-face (V-F)***, ***vertex-edge (V-E)*** and ***edge-face (E-F) pairs***.

In [8], M.E. Houle and G.T. Toussaint show that, to compute the width of E, it is sufficient to focus only on parallel planes which contain an E-E pair or a V-F pair. Therefore, we will enumerate all the E-E and V-F pairs of $\mathcal{CH}(E)$ and keep the ones whose distance is minimal.

In [9], B. Gärtner and T. Herrmann propose a direct approach relying on the geometry and combinatorial properties of the convex hull. The method is inspired from the *rotating calipers* [10] but generalized to the three-dimensional space.

They start with an arbitrary face f of $\mathcal{CH}(E)$ and determine its antipodal vertices $V = \{v_1, \ldots, v_k\}$ by exploring all the vertices of $\mathcal{CH}(E)$. Thus, they obtain an initial V-F pair and the two parallel planes P_1 and P_2 supporting V and f respectively. Next, they rotate the two planes about an incident edge e of f until P_2 supports the other facet f' incident to e. During this rotation the parallelism and the supporting property of the two planes are preserved and all E-E pairs belonging to e as well as the antipodal vertices of f' are reported.

The important part is as follows: given a V-E pair (w, e) and two parallel planes P_1 and P_2 supporting w and e respectively, two events of interest might happen during the rotation of P_2 about e:

1. P_2 supports a new face f' incident to e, a new V-F pair (w, f') is found.
2. P_1 supports an additional vertex v, a new E-E pair $((wv), a)$ is found.

Thus, a rotation about an edge e of $\mathcal{CH}(E)$ allows to get all E-E pairs belonging to e and all V-F belonging to the two incident faces of e. Hence, by rotating about all the edges of $\mathcal{CH}(E)$ we get all the possible E-E and V-F pairs of $\mathcal{CH}(E)$. At least one of them belongs to the width planes and the distance between these planes is the width $\mathcal{W}$ of E.

As $\mathcal{W}$ represents the width of an optimal bounding plane of E, if $\mathcal{W} \leq \nu$ then E is a width-ν blurred piece of discrete plane.

Furthermore, we can obtain the characteristics of this optimal bounding plane. As the width planes coincide with the leaning planes of the bounding plane $\mathcal{P}(a, b, c, \mu, \omega)$ of E, we have $a = \alpha$, $b = \beta$ and $c = \gamma$. Relying on the width interpretation in Section 2, we get $\omega = |\delta_2 - \delta_1| + 1$. Lastly, owing to the leaning planes equations, $\mu = \min(-\delta_1, -\delta_2)$.

3.2 Incremental Algorithm

Here we propose an incremental version, in order to get an algorithm which gives the characteristics of an optimal bounding plane of E each time we add a new point. A naive method consists in recomputing the width of E each time we add a point. Nevertheless some observations allow to improve this process.

On the one hand, only one point differs from one step to another. Thus, we can advantageously replace the computation of the convex hull of all the points of E by an incremental computation ([11] pp 235–246). Let us briefly recall

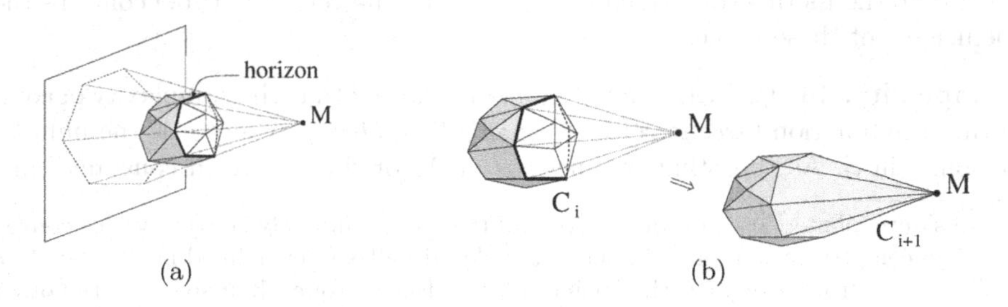

Fig. 2. (a) The horizon from M; (b) Adding a point to the convex hull

the procedure. At a general step i of the algorithm, a convex hull C_i is given and we add a new point M. If it lies inside C_i or on its boundary, then there is nothing to be done. Otherwise we look for all the visible[1] faces of C_i, standing from M. This set of faces is enclosed by a curve called ***horizon*** (Fig. 2(a)). All the visible faces are removed from C_i and replaced by new ones created by joining each vertex of the horizon to the point M (Fig. 2(b)). Some of them could be coplanar with non-visible faces so they have to be merged together. The resulting polytope is the new convex hull C_{i+1}.

On the other hand, we can observe that, at each step of the algorithm, we know the characteristics of an optimal bounding plane $\mathcal{P}(a, b, c, \mu, \omega)$ of $\mathcal{E}$. So, if we add a point $M(x_M, y_M, z_M)$, we can compute the remainder value of M relative to $\mathcal{P}$: $r_\mathcal{P}(M) = ax_M + by_M + cz_M - \mu$. According to a property of discrete planes, if $r_M \in [0, \omega - 1]$ then $M \in \mathcal{P}$, so it is useless to recompute the width of $\mathcal{E}$ since it does not change.

Algorithm 1. Incremental Recognition

Data: $\mathcal{E} \in \mathbb{Z}^3$, the convex hull C of $\mathcal{E}$, the characteristics a, b, c, μ and ω of the optimal bounding plane of $\mathcal{E}$
Input: A point $M \in \mathbb{Z}^3$
Result: The updated data after the addition of M

```
1  begin
2  |  E ⟵ E ∪ M
3  |  Update C using the incremental process
4  |  r_M ⟵ a x_M + b y_M + c z_M − μ
5  |  if r_M ∉ [0, ω − 1] then
6  |  |  ⟨α, β, γ, δ1, δ2⟩ ⟵ ComputeWidthPlanes(C)
7  |  |  a ⟵ α
8  |  |  b ⟵ β
9  |  |  c ⟵ γ
10 |  |  μ ⟵ min(−δ1, −δ2)
11 |  |_ ω ⟵ |δ2 − δ1| + 1
12 end
```

This leads to the incremental procedure described in Algorithm 1. The function ComputeWidthtPlanes(C) at line 6 computes the width planes of C according to the method described in Section 3.1. The returned tuple contains the coefficients of these planes.

Complexity: In [9], Gärtner and Herrmann showed that the complexity of computing the function ComputeWidthPlanes(C) is $\mathcal{O}(n^2)$, where n is the number of points in $\mathcal{E}$. As the other instructions of Algorithm 1 run in constant time,

[1] Consider a plane P_f containing a face f of the convex hull. By convexity, this convex hull is completely contained in one of the closed half-spaces defined by P_f. The face f is *visible* from a point if that point is located in the open half-space on the other side of P_f.

we obtain a complexity of $\mathcal{O}(n^2)$ for our incremental procedure. We need to use this incremental procedure each time we add a point to $\mathcal{E}$. Thus, we obtain an $\mathcal{O}(n^3)$ worst case complexity for a set $\mathcal{E}$ of n points. Nevertheless, in practice, the recognition process seems rather linear.

4 Linear Programming Method

The second method relies on linear programming and permits to solve the problem by considering the norm $\|\cdot\|_\infty$. We recall in the following section the general formulation of a linear programming problem and the simplex algorithm. The problem of recognition of blurred pieces of discrete planes is then modelled in that way in Section 4.2.

4.1 The Simplex Algorithm

Formulation. We try to identify a minimum point $x^* \in \mathbb{R}^d$ of a function $f(x) : \mathbb{R}^d \to \mathbb{R}$ where $x = (x_1, \ldots, x_d)$. Moreover, x^* must satisfy a set of n constraints $G = (g_i(x) \leq b_i)_{1\leq i \leq n}$. LP is the specialization of mathematical programming to the case where both, the *objective function* f and the problem constraints G are linear. Let $A(n \times d)$ denote a matrix of n rows and d columns. Let $c(d)$, $b(n)$ and $x(d)$ denote three column vectors of size d and n. Thus, we can write our LP problem in such a way: Min $c^t.x$ subject to $A \cdot x \leq b$ and $x \geq 0$. We call the *standard form* the equivalent rewriting: Min $c'^t.x'$ subject to $A' \cdot x' = b$ and $x' \geq 0$ where $A' = [A|Identity(n \times n)]$, $c' = [c|Zero(n)]$. The n inserted variables in the standard form are called the *slack variables.*

The simplex algorithm. This method, developed by George Dantzig 1947, provides a powerful computational tool (see [12] for details). It operates on the formulation of the standard form. We have $n + d$ variables and n equalities in the system $Ax = b$, we can extract a nonsingular matrix B of rank n relative to this system of equations.

The *basis* corresponds to the indices of the columns extracted from A to create B. In the simplex method, the *nonbasic variables*, denoted by $x_N = (x_i)_{\substack{1\leq i\leq n+d \\ i\notin basis}}$ are forced to be zero. The *basic variables* $x_B = (x_i)_{i\in basis}$ are thus equal to $B^{-1}b$. A solution x associated with a basis B is called *feasible* when it verifies $x_B \geq 0$.

The simplex algorithm starts from a feasible solution. At each iteration, the program computes a new basis in such a way that the new basic solution is feasible and that the objective function has decreased or remains unchanged. To build the new basis, one nonbasic variable is reclassified as basic and vice versa. Which variable can we choose ? Let N denote the columns of A whose indices are not in the basis. From $Ax = b$, we have: $[B|N].[x_B, x_N] = b$. As B is a nonsingular matrix, we obtain: $x_B = B^{-1}.(b - N.x_N)$. The objective function can be rewritten as: $f(x) = c^t.x = {c_B}^t.x_B + {c_N}^t.x_N = ({c_N}^t - {c_B}^t B^{-1} N)x_N + c_B^t B^{-1} b$. This rewriting is not depending on the variables x_B. Thus, as the variables are positive, if there exists no negative value in the *reduced cost vector* $rc^t = c_N{}^t - c_B{}^t B^{-1} N$, we have found the minimum x^*.

If there exists a negative value, then we can decrease the current value of the objective function by increasing the corresponding variable x_l of x_N. As x_l is no more zero, at the next iteration, it will be reclassified as a basic variable. By increasing x_l, the values of the basic variables change. If they all increase, the problem is *unbounded*, it means that the minimum value for the objective function is $-\infty$. In the other case, where some basic variables decrease when x_l increases, the first basic variable x_k that reaches zero will stop the increase of x_l. Thus, x_k leaves the basis.

To determine the index k, let consider the equalities $x_B = B^{-1}.(b - N.x_N)$. Only x_l is now nonzero among x_N, so we have $x_B = B^{-1}.b - B^{-1} A_l x_l$. Let $\overline{b}$ and P denote $B^{-1}b$ and $B^{-1}A_l$. Values in $\overline{b}$ are positive, so only the indices associated with a positive value in P are of interest. The previous condition $\overline{b} - P.x_l \geq 0$ implies that for all i in the basis with $P_i > 0$, we have: $x_l \leq \overline{b_i}/P_i$. It follows that $k = index\ of\ min_{i,P_i>0}\{\overline{b_i}/P_i\}$.

function Min-Simplex(A,b,c,basis)
Repeat
1- Extract B, c_B from A // relative to the current basis
2- $\overline{b} = B^{-1}b$
3- $rc' = c^t - (c_B{}^t B^{-1}).A$ // equivalent version of rc
4- If $(rc' \geq 0)$ **return** $\overline{b}$ // optimum found ($\leq$ for a Max)
5- Choose l such that $rc'_l < 0$ // x_l enters the basis (> 0 for a Max)
6- $P = B^{-1} A_l$
7- If $P \leq 0$ **return** unbounded
8- $k = min_{i,P_i>0}\{\overline{b_i}/P_i\}$ // (same thing for a Max)
9- $basis \leftarrow basis \backslash \{k\} \cup \{l\}$

Duality theorem. Associated with each *Primal* LP problem is a companion problem called the *Dual*. The main theorem of LP proves that the Primal problem is infeasible iff the Dual problem is unbounded and vice versa. Moreover, one problem has an optimum iff the other problem has an optimum. The two optimum values are equal. Moreover, if c_B and B are the matrices associated with the optimum in the Dual, then the optimum in the Primal is equal to $c_B^t B^{-1}$.

Primal:	(i) Min $c^t.x$	$\longleftrightarrow$	**Dual:**	Max $b^t.\lambda$
Subject to:	(ii) $A.x \geq b$	$\longleftrightarrow$	Subject to:	$\lambda \geq 0$
	(iii) $A.x = b$	$\longleftrightarrow$		$\lambda \in \mathbb{R}$
	(iv) $x \geq 0$	$\longleftrightarrow$		$A^t.\lambda \leq c$
	(v) $x \in \mathbb{R}$	$\longleftrightarrow$		$A^t.\lambda = c$

4.2 Modelling the Recognition Problem

In this way, we compute the minimum vertical distance between two parallel planes whose slopes relative to the x-axis and the y-axis are between $\pm\pi/2$. Indeed, let us recall the given problem, we are looking for the characteristics

a, b, c, μ, ω of an optimal discrete plane bounding $\mathcal{P}$ for a set of n points by minimizing the vertical distance between its two leaning planes. By considering $\alpha = -\frac{a}{c}$, $\beta = -\frac{b}{c}$, $h = \frac{\mu}{c}$ and $e = \frac{\omega-1}{c}$, the problem may be reformulated as follows: for a given set of n points (x_i, y_i, z_i), we want to find two planes $P : z(x, y) = \alpha.x + \beta.y + h$ and $P' : z'(x, y) = \alpha.x + \beta.y + h + e$ such that all the points are located between P and P' and such that e is minimal. We obtain one couple of inequalities for each entered point: $\alpha.x_i + \beta.y_i + h \le z_i$ and $\alpha.x_i + \beta.y_i + h + e \ge z_i$.

Primal

Min e

$$\begin{cases} \text{-}\alpha.x_i - \beta.y_i - h \ge \text{-}z_i \\ \alpha.x_i + \beta.y_i + h + e \ge z_i \\ i = 1, \dots, n \end{cases}$$

$|\alpha| \le 1,\ |\beta| \le 1$

$\alpha, \beta, h \in \mathbb{R},\ e \ge 0$

Dual standard form

Max $[\text{-}z_1 \dots \text{-}z_n \mid z_1 \dots z_n \mid \text{-}1\ \text{-}1\ \text{-}1\ \text{-}1\ 0\ ].\lambda$

$$\begin{bmatrix} \text{-}x_1 \dots \text{-}x_n & x_1 \dots x_n & \text{-}1\ 1\ 0\ 0\ 0 \\ \text{-}y_1 \dots \text{-}y_n & y_1 \dots y_n & 0\ 0\ \text{-}1\ 1\ 0 \\ \text{-}1 \dots \text{-}1 & 1 \dots 1 & 0\ 0\ 0\ 0\ 0 \\ 0 \dots 0 & 1 \dots 1 & 0\ 0\ 0\ 0\ 1 \end{bmatrix} \begin{bmatrix} \lambda_1 \\ \dots \\ \lambda_{2n+5} \end{bmatrix} = \begin{bmatrix} 0 \\ 0 \\ 0 \\ 1 \end{bmatrix}$$

$\lambda \ge 0$

We gather the two different types of inequalities on each side of the matrix. Working in the Primal problem with the standard form forces to manage a large sparse matrix of size $(2n+4) \times (2n+8)$. The Dual allows to bypass this problem with a $4 \times (2n+5)$ matrix ((i), (ii) and (v) in 4.1). We can easily check that the basis $\{\lambda_1, \lambda_{2n+1}, \lambda_{2n+3}, \lambda_{2n+5}\}$ where $B^{-1}b = [0\ 0\ 0\ 1]^t \ge 0$ is always a feasible basis for the Dual problem.

Geometrical Interpretation of the Dual Problem

The basis of the Dual problem is associated with four inequalities in the Primal problem. So when λ_i is in the basis, the ith inequality in the Primal problem corresponds to an equality. For example, when $\lambda_i, 1 \le i \le n$ is in the basis, the ith inequality implies $\alpha.x_i + \beta.y_i + h = z_i$, this means that the point p_i belongs to the lower plane P. When $n < i \le 2n$, the point p_{i-n} belongs to the upper plane P'. In the same way, the variables $\lambda_{2n+1}, \dots, \lambda_{2n+5}$ are associated with the cases: $\alpha = 1, \alpha = -1, \beta = 1, \beta = -1$ or $e = 0$.

The vector $c_B^t B^{-1}$ in the Dual transforms the current basis into the primal variables. This follows from the previous remark. Let K denote the matrix corresponding to the equalities retained in the Primal problem. The current system verifies: $K \cdot [\ \alpha\ \beta\ h\ e\ ]^t = b_B^{Primal}$. Thus, we have: $[\ \alpha\ \beta\ h\ e\ ] = (K^{-1} \cdot b_B^{Primal})^t = (b_B^{Primal})^t \cdot (K^t)^{-1} = c^t_{B\ Dual} \cdot B^{-1}_{Dual}$

Reduced cost optimality condition. The simplex algorithm maximizes a function in the Dual. So, it stops when it finds an rc vector with negative values (line 4). We easily verify that: $rc^t = c^t - (c_B{}^t B^{-1}).A = [\ (\text{-}z_i + [\alpha.x_i + \beta.y_i + h])_{1 \le i \le n} \mid (z_i - [\alpha.x_i + \beta.y_i + h + e])_{1 \le i \le n} \mid \text{-}1{+}\alpha \mid \text{-}1{-}\alpha \mid \text{-}1{+}\beta \mid \text{-}1{-}\beta \mid \text{-}e\ ]$. As all these values are negative, this implies that the inequalities of the Primal are all verified. The Dual program stops when it finds two parallel planes that include all the points and that have valid slopes.

The **objective function** in the Dual is quite obscure. Nevertheless relative to the theorem of Duality, the dual objective function must represent the same thing than the Primal function. In fact, we have $f(\lambda) = c^t_{BDual}(B^{-1}b_{Dual}) = (c^t_{BDual}B^{-1})b_{Dual} = [\ \alpha\ \beta\ h\ e\ ] \times c_{Primal} = e$.

The core of the algorithm. Each iteration is associated with a feasible basis. We only consider in the following the two most important cases with all the basic variables λ_i such that $1 \leq i \leq 2n$. Other subcases can be processed without difficulty. The configuration $1 \leq i, j, k, l \leq n$ for the indices of the basic variables is not possible because the corresponding matrix B would be singular.

Configuration 1: $1 \leq i, j, k \leq n < l \leq 2n$. In this case, the three points $p_i(x_i, y_i), p_j(x_j, y_j), p_k(x_k, y_k)$ define the lower plane P and the parallel plane P' is supported by p_{l-n}. The matrix B is equal to $[\ -x_i\ -y_i\ -1\ 0 \mid -x_j\ -y_j\ -1\ 0 \mid -x_k\ -y_k\ -1\ 0 \mid x_{l-n}\ y_{l-n}\ 1\ 1\ ]$. Wlog, we can assume that the point p_{l-n} corresponds to the origin, this allows to simplify the writing of the matrix B to $[\ -x_i\ -y_i\ -1\ 0 \mid -x_j\ -y_j\ -1\ 0 \mid -x_k\ -y_k\ -1\ 0 \mid 0\ 0\ 1\ 1\ ]$. Let N_i denote the two-dimensional vector (x_i, y_i). The vector $B^{-1}b$ is equal to: $[\frac{N_k \wedge N_j}{det(B)}; \frac{N_i \wedge N_k}{det(B)}; \frac{N_j \wedge N_i}{det(B)}; 1]$. As the matrix B is nonsingular and $det(B) = -det([\ x_i\ y_i\ 1 \mid x_j\ y_j\ 1 \mid x_k\ y_k\ 1])$ the three points p_i, p_j, p_k must not be colinear. Suppose that the three points $N_i N_j N_k$ lie in clockwise order, so $det(B) \geq 0$. As $B^{-1}b \geq 0$, $N_k \wedge N_j, N_i \wedge N_k$ and $N_j \wedge N_i$ are positive. Such a situation can appear only when the point p_{l-n} lies inside the triangle $N_i N_j N_k$ relative to the projection into the O_{xy} plane.

Configuration 2: $1 \leq i, j \leq n < k, l \leq 2n$. The planes P (resp. P') is supported by the segment $p_i p_j$ (resp. $p_{k-n} p_{l-n}$). As they are parallel, this couple of planes is unique. Consider that the point p_{l-n} is centered on the origin, we have $B^{-1}b = [N_k \wedge N_j/\Delta; N_i \wedge N_k/\Delta; ...; ...]$ with $\Delta = (N_i - N_j) \wedge N_k$. Δ is nonzero iff the segment $p_i p_j$ and the segment $p_{k-n} p_{l-n}$ are not colinear in the Oxy plane. It follows that $N_k \wedge N_j$ and $N_k \wedge N_i$ have not the same sign. Thus, the segment $p_i p_j$ crosses the line $(p_{k-n} p_{l-n})$. When we center the origin on p_i, we symmetrically obtain the same result. Thus, this case is associated with two segments $p_i p_j$ and $p_{k-n} p_{l-n}$ that intersect each other relative to a projection into the Oxy plane.

Configuration 3: It is equivalent to the first configuration.

Variables interchanging. We traverse all the set of points. For each point, we consider its vertical distance from P when it lies under P or from P' when it lies above P'. If no points are found, our problem is solved. Otherwise, we select the point that is the vertically farthest point from P and P'. The associated variable λ_u enters the basis. In the **Configuration 1**, we have three equalities of the type: $\alpha.x + \beta.y + h = z$. When we select a variable λ_u of the same type, it means with $1 \leq u \leq n$, we can not withdraw p_{l-n}, otherwise we would obtain a basis with four equalities of the same type and this configuration is not possible. Thus the new basis will remain in configuration 1. So, the new point p_u replaces the sole point among p_i, p_j and p_k that will preserve the constraint: p_{u-n} lies

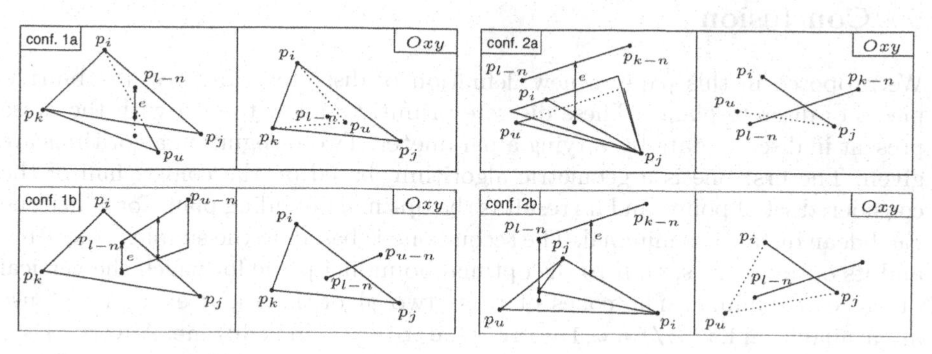

Fig. 3. Different configurations relative to the basis and the entering variable

inside the new triangle relative to the Oxy plane. As p_{u-n} is under the plane P defined by $p_i p_j p_k$, the current thickness e has also increased (see Fig. 3.1a). In the other case where $n < u \leq 2n$, two possibilities can appear. When p_{u-n} lies inside the triangle, it simply replaces its equivalent point p_l and e increases. When p_{u-n} lies outside, we cannot achieve a configuration of type 1, thus we move to a configuration of type 2. For this, the segment that supports P' is also $p_{u-n}p_{l-n}$. The other segment corresponds to the sole edge of the triangle that crosses this segment relative to the Oxy plane. p_{u-n} lies at a vertical distance greater than the one defined by the triangle and p_{l-n}. Moreover, this distance is equal to the distance between the two retained segments, so the new configuration increases the value of e (see Fig. 3.1b). In the **Configuration 2**, when a variable λ_u, $1 \leq u \leq n$ is selected, we have two possibilities. To remain in the same configuration, p_u must replace a point in such a way that the two new segments cross each other relative to the Oxy plane (see Fig. 3.2a). When this is not attainable, one of the two points p_{k-n} or p_{l-n} inevitably belongs to the triangle $p_i p_j p_u$ and we then shift to a configuration of type 1 (see Fig. 3.2b). Other interchangings can be deduced from the ones exposed in this section.

Convergence and complexity. As the Primal is feasible (choose a large value for e), the Dual is never unbounded and we can suppress the processing of this particular case. As we select a point outside of two parallel planes, we know that the vertical distance (the objective function) strictly increases at each iteration. Thus, unlike in the general case, the simplex algorithm applied to this recognition problem can not cycle. Moreover, we have at most $C^4_{2n+5} = \mathcal{O}(n^4)$ possible feasible basis. Thus, we obtain an $\mathcal{O}(k^4)$ time complexity where k represents the number of the vertices of the convex hull of the given points. In practice, this quantity is relatively small compared to the number of points.

The incremental version. When a new point is inserted, it may lie between the two planes P and P'. In this case, the previous solution remains optimal and nothing has to be done. Otherwise, two columns are added to the matrix A in the Dual. Next, using the last processed feasible basis, we launch a new sequence of iterations until the new optimum solution is found.

5 Conclusion

We proposed in this paper a new definition of discrete primitives: the blurred pieces of discrete planes. These discrete primitives allow to deal with the noise present in discrete data by varying a parameter. Two recognition algorithms are given. The first one is a geometric algorithm, based on the convex hull of the considered set of points and its result is the optimal bounding plane for which the Euclidean distance is minimal. The second one is based on the simplex algorithm and its output corresponds to the optimal bounding plane for which the vertical distance is minimal. The codes of these two algorithms and examples of use are available on `http://www.loria.fr/~debled/BlurredPlane`. A work about the comparison between these two methods is in progress. Moreover we intend to use these algorithms in the framework of the boundary segmentation of 3D noisy discrete objects. Our aim is to obtain an algorithm of polyhedrization of 3D noisy discrete objects by controlling the approximations done.

References

1. Klette, R., Rosenfeld, A.: Digital Geometry. Morgan Kaufmann publishers, Elsevier (2004)
2. Debled-Rennesson, I., Rémy, J.L., Rouyer-Degli, J.: Linear segmentation of discrete curves into fuzzy segments. Discrete Applied Math. **151** (2005) 122–137
3. Debled-Rennesson, I., Feschet, F., Rouyer-Degli, J.: Optimal blurred segments decomposition of noisy shapes in linear time. Computers & Graphics **30**(1) (2006)
4. Reveillès, J.: Géométrie discrète, calculs en nombres entiers et algorithmique. Thèse d'Etat – Université Louis Pasteur (1991)
5. Buzer, L.: An elementary algorithm for digital line recognition in the general case. In: DGCI-2005. Number 3429 in LNCS, Springer-Verlag (2005) 299–310
6. Debled-Rennesson, I., Tabbone, S., Wendling, L.: Multiorder polygonal approximation of digital curves. Electronic Letters on Computer Vision and Image Analysis **5**(2) (2005) 98–110 Special Issue on Document Analysis.
7. Andres, E.: Le plan discret. In: Colloque de géométrie discrète en imagerie: fondements et applications, Strasbourg, France (1993)
8. Houle, M., Toussaint, G.: Computing the width of a set. IEEE Trans. on Pattern Analysis and Machine Intelligence **10**(5) (1988) 761–765
9. Gärtner, B., Herrmann, T.: Computing the width of a point set in 3-space. J. Exp. Algorithmics **4** (2001) 3
10. Toussaint, G.: Solving geometric problems with the rotating calipers. In: Proceedings of IEEE MELECON'83, Athens, Greece (1983)
11. de Berg, M., van Kreveld, M., Overmars, M., Schwarzkopf, O.: Computational Geometry : Algorithms and Applications. Springer-Verlag, Heidelberg (2000)
12. Chvatal, V.: Linear Programming. Freeman, New York (1983)

The Number of Line-Convex Directed Polyominoes Having the Same Orthogonal Projections

Péter Balázs

Department of Computer Algorithms and Artificial Intelligence
University of Szeged
Árpád tér 2, H-6720 Szeged, Hungary
pbalazs@inf.u-szeged.hu

Abstract. The number of line-convex directed polyominoes with given horizontal and vertical projections is studied. It is proven that diagonally convex directed polyominoes are uniquely determined by their orthogonal projections. The proof of this result is algorithmical. As a counterpart, we show that ambiguity can be exponential if antidiagonal convexity is assumed about the polyomino. Then, the results are generalised to polyominoes having convexity property along arbitrary lines.

Keywords: Discrete tomography; line-convex directed polyomino; reconstruction from projections.

1 Introduction

The reconstruction of two-dimensional discrete sets (the finite subsets of $\mathbb{Z}^2$) from their projections is a frequently studied area of discrete tomography [1] and it has its applications in pattern recognition, image processing, electron microscopy, and radiology [2,3,4,5,6]. For practical reasons the number of projections used in the reconstruction is small (usually two or four). Thus, in certain cases the number of discrete sets having the same projections can be extremely large leading to a reconstructed set which is possibly quite dissimilar to the original one. A commonly used technique to reduce the number of solutions is to suppose having some a priori information of the set to be reconstructed such as convexity, connectedness and directedness. Some properties imposed on the set to be reconstructed completely eliminate ambiguity and make it possible to develope efficient reconstruction algorithms [7,8,9]. However, there are classes of discrete sets where ambiguity is only partially eliminated making the reconstruction of the set very difficult [10,11]. In this paper we investigate the problem of ambiguity when the set to be reconstructed must be line-convex, connected, and directed.

This contribution is structured as follows. First, the necessary definitions are introduced in Section 2. In Section 3 we study diagonally convex directed polyominoes. We give a uniqueness result for this class and derive an algorithm for

A. Kuba, L.G. Nyúl, and K. Palágyi (Eds.): DGCI 2006, LNCS 4245, pp. 77–85, 2006.

reconstructing the uniquely determined polyomino from its horizontal and vertical projections. In Section 4 we show that ambiguity can remain very high when the polyominoes to be reconstructed are antidiagonally convex. In Section 5 we consider the possibility to adapt the results of Sections 3 and 4 to polyominoes that are convex along an arbitrary set of directions. Finally, in Section 6 we conclude our results.

2 Preliminaries

The finite subsets of $\mathbb{Z}^2$ defined up to translation are called *discrete sets.* A discrete set F can be represented by a set of unitary cells or by a binary matrix $\hat{F} = (\hat{f}_{ij})_{m\times n}$ where $\hat{f}_{ij} = 1$ if and only if $(i,j) \in F$. To stay consistent we assume that the vertical axis of the two-dimensional integer lattice is directed top down (see Fig. 1). The *horizontal* and *vertical projections* of F are the vectors $\mathcal{H}(F) = H = (h_1,\ldots,h_m)$, and $\mathcal{V}(F) = V = (v_1,\ldots,v_n)$, respectively, where

$$h_i = \sum_{j=1}^{n} \hat{f}_{ij} \quad (i = 1,\ldots,m) \quad \text{and} \quad v_j = \sum_{i=1}^{m} \hat{f}_{ij} \quad (j = 1,\ldots,n) \ . \qquad (1)$$

Not any pair of vectors are the projections of some discrete set. In the following we suppose that $H \in \mathbb{N}_0^m$ and $V \in \mathbb{N}_0^n$ are *compatible* which means that they satisfy the following conditions

(i) $h_i \le n$, for $1 \le i \le m$, and $v_j \le m$, for $1 \le j \le n$;
(ii) $\sum_{i=1}^{m} h_i = \sum_{j=1}^{n} v_j$, i.e., the two vectors have the same total sums.

Two points $P = (p_1,p_2)$ and $Q = (q_1,q_2)$ in $\mathbb{Z}^2$ are said to be *4-adjacent* if $|p_1 - q_1| + |p_2 - q_2| = 1$. The sequence of distinct points $P_0,\ldots,P_k$ is a *4-path* from point P_0 to point P_k in a discrete set F if each point of the sequence is in F and P_l is 4-adjacent to P_{l-1} for each $l = 1,\ldots,k$. A discrete set F is 4-connected if for any two points in F there is a 4-path between them. Such a set is also called as *polyomino.* A 4-path in a discrete set F is a *northeast path* (or shortly, *NE-path*) from point P_0 to point P_k if each point P_l of the path is in north or east to P_{l-1} for each $l = 1,\ldots,k$. The discrete set F is *directed* if there is a NE-path in F from point $(m,1)$ to any other point of F.

Given two integers a and b such that they are coprimes and $(a,b) \neq (0,0)$ we define the kth line of the discrete rectangle $R = \{1,\ldots,m\} \times \{1,\ldots,n\}$ parallel to the vector (a,b) by

$$S_k^{(a,b)} = \{(i,j) \in R \mid bi - aj = k\} \ . \qquad (2)$$

Throughout this paper, without loss of generality, we will assume that $b \ge 0$ and $b = 0$ if and only if $a = 1$. The discrete set F is *line-convex* along the direction (a,b) if for every $k \in \mathbb{Z}$ and $(i_1,j_1),(i_2,j_2) \in S_k^{(a,b)} \cap F$ the discrete line segment with the endpoints (i_1,j_1) and (i_2,j_2) is in F, i.e., if $(i,j) \in S_k^{(a,b)}$ such that $(i,j) = (i_1 + t(i_2 - i_1), j_1 + t(j_2 - j_1))$ where $t \in [0,1]$ then $(i,j) \in F$. In particular, the discrete set is

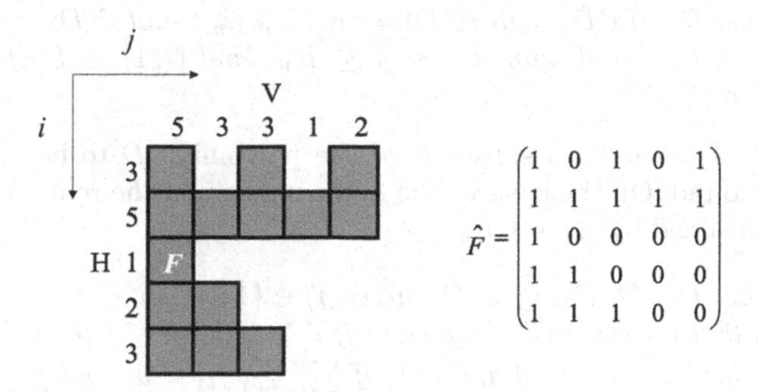

Fig. 1. A diagonally convex directed polyomino F, and the correspondig binary matrix $\hat{F}$. The projections of the polyomino are the vectors H and V.

- *horizontally convex* if it is line-convex along the direction $(0, 1)$,
- *vertically convex* if it is line-convex along the direction $(1, 0)$,
- *diagonally convex* if it is line-convex along the direction $(1, 1)$,
- *antidiagonally convex* if it is line-convex along the direction $(-1, 1)$.

For example, Fig. 1 shows a diagonally convex directed polyomino, which is vertically, horizontally, and antidiagonally non-convex.

3 Diagonally Convex Directed Polyominoes

Polyominoes are widely used in physics and chemistry for modelling and they have long been studied by mathematicians and computer scientists (see [11, 12, 13] and the references given there). Concerning polyominoes with some line-convexity properties some important results are already known. In [11] the authors studied the number of (horizontally and vertically) convex polyominoes reconstructible from their orthogonal projections and showed that in this class ambiguity can be very high. Moreover, in [12] a method is given to enumerate diagonally convex directed polyominoes according to several parameters (sources, diagonals, horizontal and vertical edges, etc.). Recently, in [14] the author stressed the importance of finding classes of polyominoes where the reconstruction from two projections can be solved uniquely in polynomial time. The only class known so far satisfying this condition was investigated in [8, 9]. The results given there can be summarized in the following.

Theorem 1. *Every horizontally or vertically convex directed polyomino can be reconstructed from its horizontal and vertical projections uniquely in $O(mn)$ time.*

In this section we show that the class of diagonally convex directed polyominoes (let us denote this class by $\mathcal{DCD}_4$) also satisfies this condition. As an inmediate consequence of the directedness and 4-connectedness we can say

Lemma 1. *Let $D \in \mathcal{DCD}_4$ with $\mathcal{H}(D) = (h_1, \ldots, h_m)$ and $\mathcal{V}(D) = (v_1, \ldots, v_n)$. Then, $(m, j) \in D$ if and only if $1 \leq j \leq h_m$, and $(i, 1) \in D$ if and only if $m - v_1 < i \leq m$.*

With the aid of Lemma 1 a subset F of the polyomino D to be reconstructed can easily be found. On the basis of the following lemma the remaining elements of D are determined by the set F.

Lemma 2. *Let $D \in \mathcal{DCD}_4$, $F \subset D$ and $(i, j) \in \{1, \ldots, m-1\} \times \{2, \ldots, n\}$ be a position such that for every $(i', j') \neq (i, j)$ if $i' \geq i$ and $j' \leq j$ then $(i', j') \in D \leftrightarrow (i', j') \in F$. Then, $(i, j) \in D$ if and only if $\sum_{t=i+1}^{n} \hat{f}_{tj} < v_j$ and $\sum_{t=1}^{j-1} \hat{f}_{it} < h_i$.*

Proof. Let (i, j) be a position satisfying the conditions of the lemma. The necessary part is trivial since $i' \geq i$, $j' \leq j$ and $(i', j') \neq (i, j)$ implies $(i', j') \in D \leftrightarrow (i', j') \in F$ and so the inequalities $\sum_{t=i+1}^{n} \hat{f}_{tj} < v_j$ and $\sum_{t=1}^{j-1} \hat{f}_{it} < h_i$ must hold. To prove the sufficient part assume that $\sum_{t=i+1}^{n} \hat{f}_{tj} < v_j$ and $\sum_{t=1}^{j-1} \hat{f}_{it} < h_i$ and contrary $(i, j) \notin D$, i.e., $\hat{d}_{ij} = 0$. If $i = 1$ then the contradiction follows from $\sum_{t=i+1}^{n} \hat{f}_{tj} < v_j$ and the fact that $(i', j) \in F \leftrightarrow (i', j) \in D$ holds for every position (i', j) if $i' > 1$. Similarly, if $j = n$ then the contradiction follows from $\sum_{t=1}^{j-1} \hat{f}_{it} < h_i$ and the fact that $(i, j') \in F \leftrightarrow (i, j') \in D$ holds for every position (i, j') if $j' < n$. In any other cases, since the conditions of the lemma hold, there exist $i'' < i$ and $j'' > j$ for which $\hat{d}_{i'',j} = \hat{d}_{i,j''} = 1$. Since D is directed there is a NE-path from $(m, 1)$ to (i'', j) such that for every point (c_1, c_2) of this path $c_2 \leq j$ holds. Therefore the diagonal $S_{i-j}^{(1,1)}$ contains at least one point of D, say (i_1, j_1) for which $j_1 < j$. Similarily, we get that there is a NE-path from $(m, 1)$ to (i, j'') and therefore the diagonal $S_{i-j}^{(1,1)}$ contains at least one point of D, say (i_2, j_2) for which $j_2 > j$. We get $\hat{d}_{i_1,j_1} = 1$, $\hat{d}_{ij} = 0$ and $\hat{d}_{i_2,j_2} = 1$ with $j_1 < j < j_2$ which contradicts the diagonal convexity (see Fig. 2). □

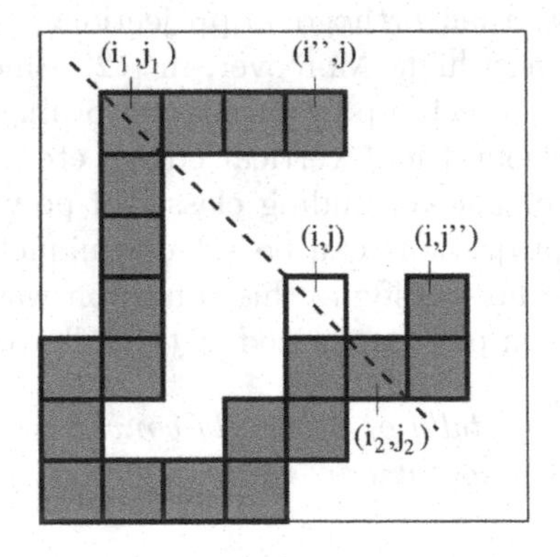

Fig. 2. Proof of Lemma 2

The following theorem states that if we assume that the directed polyomino to be reconstructed is diagonally convex then ambiguity can be completely eliminated.

Theorem 2. *Let $H \in \mathbb{N}^m$ and $V \in \mathbb{N}^n$. In the class $\mathcal{DCD}_4$ there is at most one polyomino P such that $\mathcal{H}(P) = H$ and $\mathcal{V}(P) = V$.*

Proof. On the basis of Lemma 1 the first column and the last row of P are uniquely determined by v_1 and h_m, repsectively, i.e., a subset F of the polyomino P (consisting of all the positions of the last row and first column of P) can be found. Then, for the position $(m-1, 2)$ the conditions of Lemma 2 hold. Therefore on the basis of Lemma 2 it can be decided whether the position $(m-1, 2)$ belongs to P and if so then we set $F = F \cup \{(m-1, 2)\}$. Taking each position bottom up left to right F always satisfies the conditions of Lemma 2 and so the above method can be repeated. If H and V are the projections of a diagonally convex polyomino then in the end we get $F = P$. Uniqueness follows from the construction. □

The proof of Theorem 2 is constructive, i.e., an algorithm similar to the one in [9] can also be described to reconstruct the possibly existing polyomino of the class $\mathcal{DCD}_4$ with given horizontal and vertical projections.

Algorithm DCD4
Input: Two compatible vectors, $H \in \mathbb{N}^m$ and $V \in \mathbb{N}^n$.
Output: The binary matrix $\hat{F}$ representing the uniquely determined polyomino of $\mathcal{DCD}_4$ having projections H and V (if there is such a solution).

Step 1 $\hat{F} := (0)_{m \times n}$; $H' := H$; $V' := V$;
Step 2 **for** $i := m - v_1 + 1, \ldots, m$ $\{\hat{f}_{i1} := 1;\ h'_i - -;\}$
  for $j := 1, \ldots, h_m$ $\{\hat{f}_{mj} := 1;\ v'_j - -;\}$
Step 3 **for** $i := m - 1, \ldots, 1$
  for $j := 2, \ldots, n$
   if $(h'_i > 0$ **and** $v'_j > 0)$ **then** $\{\hat{f}_{ij} := 1;\ h'_i - -;\ v'_j - -;\}$
Step 4 **if** $(\mathcal{H}(F) \neq H$ **or** $\mathcal{V}(F) \neq V$ **or** F is not a diagonally convex polyomino)
  then exit(no solution) **else return** $\hat{F}$;

This algorithm works as follows. Step 1 is for the initialization of the matrix $\hat{F}$ and the auxiliary vectors H' and V'. In Step 2 a subset F of the polyomino to be reconstructed is defined on the basis of Lemma 1. Then, in each iteration of Step 3 we check whether the conditions of Lemma 2 hold and if so then we update the matrix $\hat{F}$ and the vectors H' and V'. Due to the vectors H' and V' Step 3 runs in $O(mn)$ time. Finally, in Step 4 we check whether the reconstructed set is a diagonally convex polyomino and has the given projections H and V which can also be done in $O(mn)$ time. Summarizing this we can say

Corollary 1. *Let $H \in \mathbb{N}^m$ and $V \in \mathbb{N}^n$. If there is a polyomino $P \in \mathcal{DCD}_4$ such that $\mathcal{H}(P) = H$ and $\mathcal{V}(P) = V$ then Algorithm DCD4 reconstructs it in $O(mn)$ time.*

4 Antidiagonally Convex Directed Polyominoes

In this section we show that there is a drastic change in the number of directed polyominoes if instead of diagonal convexity it is assumed that the polyominoes to be reconstructed are antidiagonally convex. We will use the concept of *switching component* of a binary matrix $\hat{F}$ which is a submatrix of $\hat{F}$ of the form

$$\begin{pmatrix} 0 & 1 \\ 1 & 0 \end{pmatrix} \text{ or } \begin{pmatrix} 1 & 0 \\ 0 & 1 \end{pmatrix} . \tag{3}$$

Interchanging 0s and 1s in a switching component the horizontal and vertical projections of F do not change [15]. Let us denote the class of antidiagonally convex directed polyominoes by $\mathcal{ACD}_4$. The following theorem shows that assuming antidiagonal convexity on the directed polyomino having given horizontal and vertical projections does not eliminate ambiguity.

Theorem 3. *In the class $\mathcal{ACD}_4$ there can be exponentially large number of polyominoes with the same horizontal and vertical projections.*

Proof. We show that for any $k \in \mathbb{N}$ there are at least 2^k number of discrete sets of size $(6k-1) \times 3$ in the class $\mathcal{ACD}_4$ with the same horizontal and vertical projections. Let

$$\hat{B} = (1 \;\; 1 \;\; 0), \quad \hat{F}_1 = \begin{pmatrix} 1 & 1 & 0 \\ 1 & 1 & 1 \\ 1 & 1 & 0 \\ 1 & 0 & 1 \\ 1 & 1 & 1 \end{pmatrix}, \quad \text{and} \quad \hat{F}_k = \begin{pmatrix} \hat{F}_1 \\ \hat{B} \\ \hat{F}_{k-1} \end{pmatrix} \quad \text{for} \quad k > 1 . \tag{4}$$

For a given $k \in \mathbb{N}$ and for any $l \in \mathbb{N}$ $(1 \le l \le k)$ we will refer to the submatrix of $\hat{F}_k$ consisting of the rows $6(l-1)+i \quad (i = 1, \ldots, 5)$ as the *lth level of* $\hat{F}_k$ and to the submatrix $\hat{B}$ in the row $6l$ as the *lth bridge of* $\hat{F}_k$ (omitting the case $k = l$). For any l the positions $(6(l-1)+4, 2)$, $(6(l-1)+4, 3)$, $(6(l-1)+1, 2)$, and $(6(l-1)+1, 3)$ form a switching component in $\hat{F}_k$ and we will refer to it as the *lth switching component of* $\hat{F}_k$. Let $\hat{F}_1'$ be the binary matrix that we get by interchanging the 0s and 1s in the first switching component of $\hat{F}_1$, i.e.,

$$\hat{F}_1' = \begin{pmatrix} 1 & 0 & 1 \\ 1 & 1 & 1 \\ 1 & 1 & 0 \\ 1 & 1 & 0 \\ 1 & 1 & 1 \end{pmatrix} . \tag{5}$$

Clearly, $F_1, F_1' \in \mathcal{ACD}_4$ and so the theorem holds for the case $k = 1$. For the case $k > 1$ let $\hat{F}_k^S$ where $S = \{s_1, \ldots, s_n\} \subseteq \{1, \ldots, k\}$ $(n \le k)$ denote the binary matrix that we get from $\hat{F}_k$ by switching the s_1th, ..., s_nth switching components. Note, that from the viewpoint of directedness, 4-connectedness and antidiagonal convexity the lth bridge effects only on the $(l+1)$-th and lth levels and vice versa (if they exist). Then, in order to prove that $F_k^S \in \mathcal{ACD}_4$ for any $k \in \mathbb{N}$ and $S \subseteq \{1,\ldots,k\}$ it is sufficient to study the submatrices of $\hat{F}_k^S$ consisting of the lth level and the lth bridge and the lth bridge and the $(l+1)$-th level. These matrices can only be of the form $\begin{pmatrix} \hat{B} \\ \hat{F}_1 \end{pmatrix}$, $\begin{pmatrix} \hat{F}_1 \\ \hat{B} \end{pmatrix}$, $\begin{pmatrix} \hat{B} \\ \hat{F}_1' \end{pmatrix}$, or $\begin{pmatrix} \hat{F}_1' \\ \hat{B} \end{pmatrix}$. It can be shown directly that the four sets represented by these matrices are antidiagonally convex. For a given k S can be any subset of $\{1,\ldots,k\}$ therefore the number of solutions with the same projections is at least 2^k and so the theorem is proven (see Fig. 3 for the case $k = 2$). □

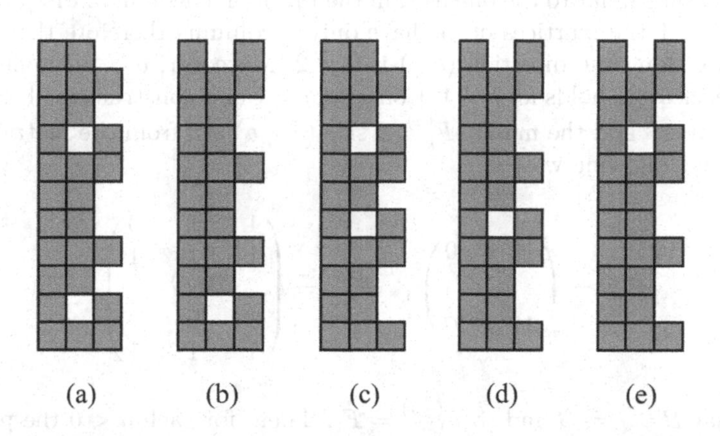

Fig. 3. Exponentially large number of discrete sets of $\mathcal{ACD}_4$ with the same horizontal and vertical projections. (a)-(d) Proof of Theorem 3 for the case $k = 2$. The sets left to right are $F_2^{\{\}}$, $F_2^{\{1\}}$, $F_2^{\{2\}}$, and $F_2^{\{1,2\}}$, respectively. (e) One more set with the same projections.

Remark 1. The bound 2^k in the proof of Theorem 3 is not tight. See, for example, the discrete set in Fig. 3e.

As a consequence of this theorem we get

Corollary 2. *If there is an algorithm that reconstructs all the discrete sets of $\mathcal{ACD}_4$ with the horizontal and vertical projections H and V, respectively, then there are some pair of vectors H and V for which the time complexity of the algorithm is not polynominal.*

5 Generalisation to Arbitrary Line-Convexity

From Sections 3 and 4 it is clear that the direction of convexity has an important role in whether or not ambiguity can be eliminated. In this section we study in more detail how the direction of convexity effects on the number of directed polyominoes. First, note that without further modification Theorem 2 can be stated in a more general way to polyominoes that are line-convex along the direction (a, b) if $a > 0$ holds. Moreover, the construction given in the proof of Theorem 3 can be adapted to polyominoes that are line-convex along any direction (a, b) for which $a < 0$ in the following way.

Theorem 4. *Let $\mathcal{C}_{(a,b)}$ be the class of polyominoes that are line-convex along the direction (a, b) such that $a < 0$. Then, there can be exponentially large number of polyominoes of $\mathcal{C}_{(a,b)}$ with the same horizontal and vertical projections.*

Proof. (Sketch.) Assume that a direction (a, b) is given with $a < 0$. We will give a construction similar to the one used in the proof of Theorem 3. First, note that for each $k \geq 1$ the martices of (4) have only 3 columns therefore they are also line-convex along the direction (a, b) if $b \geq 2$. Moreover, $a \neq 1$ implies $b \neq 0$. Thus, the theorem holds for $b \neq 1$. Consider $b = 1$ and construct the bridge $\hat{B}^{(a)}$ of size $(-a) \times 3$ and the matrix $\hat{F}_1^{(a)}$ of size $(4 - a) \times 3$ from the matrices given in (4) in the following way

$$\hat{B}^{(a)} = \begin{pmatrix} 1 & 1 & 0 \\ & \vdots & \\ 1 & 1 & 0 \end{pmatrix}, \quad \hat{F}_1^{(a)} = \begin{pmatrix} 1 & 1 & 0 \\ 1 & 1 & 1 \\ & \hat{B}^{(a)} & \\ 1 & 0 & 1 \\ 1 & 1 & 1 \end{pmatrix}. \tag{6}$$

Notice that $\hat{B}^{(-1)} = \hat{B}$ and so $\hat{F}_1^{(-1)} = \hat{F}_1$. Then, for each $a < 0$ the proof can be finished similarly as the proof of Theorem 3. □

Now, we can state the main result of our paper.

Theorem 5. *Let $\mathcal{D} = \{(a_i, b_i) \mid i = 1, \ldots, l\}$ be a finite set of directions and P be a polyomino line-convex along every directions of $\mathcal{D}$. Then, P is uniquely determined by its horizontal and vertical projections and it can be reconstructed in $O(mn)$ time if there exist a direction $(a, b) \in \mathcal{D}$ with $a \geq 0$. If $a_i < 0$ for all directions $(a_i, b_i) \in \mathcal{D}$ $(i = 1, \ldots, l)$ then there can be exponentially large number of polyominoes that are line-convex along the directions of $\mathcal{D}$ and have the same horizontal and vertical projections as P.*

Proof. It follows from Theorem 1 and the discussion of the above paragraph. The only non-trivial statement is the case if $a_i < 0$ for all directions $(a_i, b_i) \in \mathcal{D}$ $(i = 1, \ldots, l)$. In this situation we can have two cases. If there is at least one direction (a_j, b_j) such that $b_j = 1$ then we have to to apply the construction of (6) with the argument $a = \min\{a_i \mid b_i = 1, (a_i, b_i) \in \mathcal{D}\}$. Otherwise, (that is, if $b_i \neq 1$ for all $i = 1, \ldots, l$) we can simply use the construction of (4). □

Remark 2. Theorem 5 can be extended to any infinte set $\mathcal{D}$ of directions except the case if $b_i \neq 1$ for all $(a_i, b_i) \in \mathcal{D}$. In this latter case the integer $\min\{a_i \mid b_i = 1, (a_i, b_i) \in \mathcal{D}\}$ does not exist.

6 Conclusions

We have studied the problem of ambiguity in classes of directed discrete sets. We have shown that in the class of diagonally convex directed polyominoes the horizontal and vertical projections uniquely determine the polyomino and this polyomino can be reconstructed in $O(mn)$ time. However, assuming antidiagonal convexity about the polyomino the number of solutions with the same horizontal and vertical projections can be extremely large. Then, the results were generalised to polyominoes having arbitrary line-convexity. It is an open question whether in the classes where non-uniqueness holds a reconstruction algorithm can be given to find a solution in polynomial time. Another interesting question is if it is possible to generalise Theorem 5 to arbitrary infinite set of directions, too.

References

1. Herman, G.T., Kuba, A. (eds.): Discrete Tomography: Foundations, Algorithms and Applications. Birkhäuser, Boston (1999)
2. Crewe, A.V., Crewe, D.A.: Inexact Reconstruction: Some Improvements. Ultramicroscopy **16** (1985) 33–40
3. Gordon, R., Herman, G.T.: Reconstruction of Pictures from their Projections. Graphics Image Process. **14** (1971) 759–768
4. Kuba, A.: The Reconstruction of Two-Directionally Connected Binary Patterns from their Two Orthogonal Projections. Comp. Vision, Graphics, and Image Proc. **27** (1984) 249–265
5. Prause, G.M.P., Onnasch, D.G.W.: Binary Reconstruction of the Heart Chambers from Biplane Angiographic Image Sequences. IEEE Trans. Medical Imaging **MI-15** (1996) 532–546
6. Schilferstein, A.R., Chien, Y.T.: Switching Components and the Ambiguity Problem in the Reconstruction of Pictures from their Projections. Pattern Recognition **10** (1978) 327–340
7. Balázs, P., Balogh, E., Kuba, A.: Reconstruction of 8-Connected but not 4-Connected *hv*-Convex Discrete Sets. Discrete App. Math. **147** (2005) 149–168
8. Del Lungo, A.: Polyominoes Defined by Two Vectors. Theor. Comput. Sci. **127** (1994) 187–198
9. Kuba, A., Balogh, E.: Reconstruction of Convex 2D Discrete Sets in Polynomial Time. Theor. Comput. Sci. **283** (2002) 223–242
10. Barcucci, E., Del Lungo, A., Nivat, M., Pinzani, R.: Reconstructing Convex Polyominoes from Horizontal and Vertical Projections. Theor. Comput. Sci. **155** (1996) 321–347
11. Del Lungo, A., Nivat, M., Pinzani, R.: The Number of Convex Polyominoes Recostructible from their Orthogonal Projections. Discrete Math. **157** (1996) 65–78
12. Feretic, S., Svrtan, D.: Combinatorics of Diagonally Convex Directed Polyominoes, Discrete Math. **157** (1996) 147–168
13. Golomb, S.W.: Polyominoes. Charles Scriber's Sons, New York (1965)
14. Balázs, P.: A Decomposition Technique for Reconstructing Discrete Sets from Four Projections. Image and Vision Computing, submitted
15. Ryser, H.J.: Combinatorial Mathematics. The Carus Mathematical Monographs **14** (1963)

A Network Flow Algorithm for Binary Image Reconstruction from Few Projections

Kees Joost Batenburg[1,2]

[1] Leiden University, P.O. Box 9512, 2300 RA Leiden, The Netherlands
[2] CWI, P.O. Box 94079, 1090 GB Amsterdam, The Netherlands

Abstract. Tomography deals with the reconstruction of images from their projections. In this paper we focus on tomographic reconstruction of binary images (i.e., black-and-white) that do *not* have an intrinsic lattice structure from a small number of projections. We describe how the reconstruction problem from only two projections can be formulated as a network flow problem in a graph, which can be solved efficiently. When only two projections are used, the reconstruction problem is severely underdetermined and many solutions may exist. To find a reconstruction that resembles the original image, more projections must be used. We propose an iterative algorithm to solve the reconstruction problem from more than two projections. In every iteration a network flow problem is solved, corresponding to two of the available projections. Different pairs of projection angles are used for consecutive iterations. Our algorithm is capable of computing high quality reconstructions from very few projections. We evaluate its performance on simulated projection data and compare it to other reconstruction algorithms.

1 Introduction

Tomography deals with the reconstruction of images from a number of their projections [1, 2]. In many applications, such as the reconstruction of medical CT images, a large number of different pixel values may occur in the reconstruction. Typically, the number of projections that is required to obtain sufficiently accurate reconstructions is large in such cases (more than a hundred).

For certain applications, however, it is known in advance that only a few possible gray values may occur. Many objects scanned in industry for nondestructive testing or reverse engineering purposes are made of one homogeneous material, resulting in two possible gray values: the material and the surrounding air. Another example is medical digital subtraction angiography, where one obtains projections of the distribution of a contrast in the vascular system.

The field of *discrete tomography* deals with the reconstruction of images from a small number of projections, when the set of pixel values is known to have only a few discrete values [3]. By using this prior information about the set of possible values, it may be possible to dramatically reduce the amount of projection data that is required to obtain accurate reconstructions.

A. Kuba, L.G. Nyúl, and K. Palágyi (Eds.): DGCI 2006, LNCS 4245, pp. 86–97, 2006.

In [4] the author proposed an algorithm for reconstructing binary images that are defined on a lattice, using smoothness assumptions. This algorithm exploits the fact that the reconstruction problem for only two projections can be solved in polynomial time. The proposed reconstruction procedure is iterative: in each iteration a new reconstruction is computed using only two projections *and* the reconstruction from the previous iteration. The new reconstruction simultaneously resembles the image from the previous iteration and adheres to the two selected projections.

In this paper we describe a new iterative algorithm for reconstructing binary images that do *not* have an intrinsic lattice structure (i.e., subsets of the plane), which is based on ideas similar to those used in [4]. To solve the two-projection subproblems efficiently, a different pixel grid has to be used in each iteration, corresponding to the selected pair of projections. The reconstruction problem can then be solved as a special case of the *minimum cost network flow* problem in graphs, for which efficient polynomial time algorithms are available [5].

In [6] a more general version of our algorithm is described and more details are provided on the algorithmic steps. We refer the reader to that paper for further details.

We restrict ourselves to parallel beam tomography. Let $D = \{\theta_1, \ldots, \theta_d\}$ be a set of disjoint real numbers in the interval $[0, \pi)$, the *projection angles*. Let n be a positive integer. For $i = 0, \ldots, n$, put $t_i = -n/2 + i$. Let $\theta \in [0, \pi)$. For $t, t' \in \mathbb{R}$, $t < t'$, define the *strip* $S_\theta(t, t')$ as

$$S_\theta(t,t') = \left\{ (x,y) \in \mathbb{R}^2 : \begin{array}{l} x\cos\theta + y\sin\theta \geq t \quad \text{and} \\ x\cos\theta + y\sin\theta \leq t' \end{array} \right\}.$$

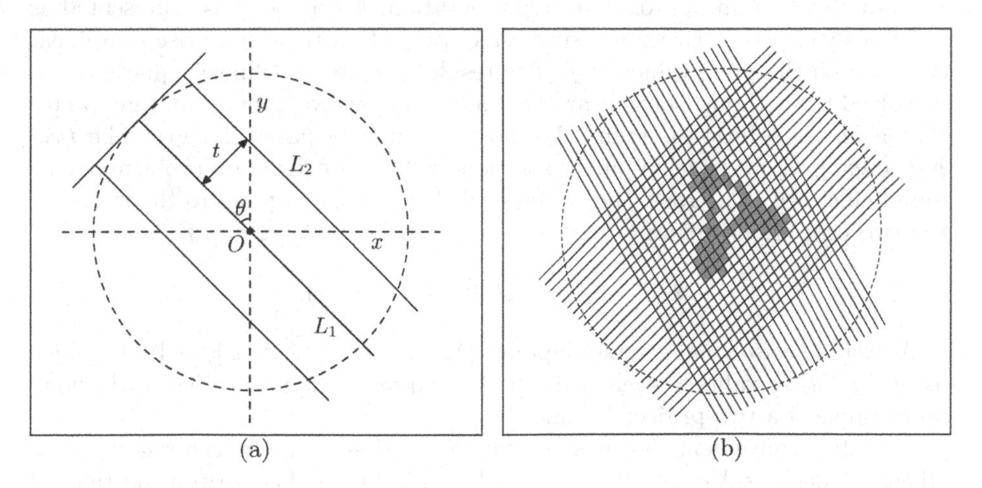

Fig. 1. (a) Left: Schematic view of the parallel beam geometry which contains the lines $L_1 : x\cos\theta + y\sin\theta = 0$ and $L_2 : x\cos\theta + y\sin\theta = t$. (b) Right: A two-projection image.

Figure 1a shows the geometric meaning of the last definition. Define the *imaging area* I as

$$I = \bigcap_{k=1}^{d} S_{\theta_k}(t_0, t_n).$$

We will now define the basic reconstruction problem of reconstructing a binary image (i.e., black-and-white) from few projections. We consider the unknown image as a mapping $f : I \to \{0,1\}$.

Problem 1. Let $p_1 = (p_{11} \dots p_{1n})^T, \dots, p_d = (p_{d1} \dots p_{dn})^T \in \mathbb{R}^n$ be vectors of nonnegative real numbers (the measured strip projections for projection angles $\theta_1, \dots, \theta_d \in D$ respectively). Construct a function $f : I \to \{0,1\}$ such that

$$\iint_{S_{\theta_k}(t_{i-1},t_i)} f(x,y)\, \mathrm{d}y\, \mathrm{d}x \quad = \quad p_{ki} \qquad \text{for } i = 1, \dots, n,\ k = 1, \dots, d.$$

We call an integral of the form $\iint_{S_{\theta_k}(t_{i-1},t_i)} f(x,y)\, \mathrm{d}y\, \mathrm{d}x$ a *strip projection.*

In the next section we deal with the reconstruction problem from only two projections, which is severely underdetermined. In Section 3 we describe our iterative algorithm for reconstruction from more than two projections. Experimental results of the algorithm from Section 3 are presented in Section 4.

2 Reconstruction from Two Projections

This section deals with the reconstruction problem from only two projections, for angles $\bar{\theta}_1$ and $\bar{\theta}_2$. To represent a mapping $f : I \to \{0,1\}$ in a computer we have to resort to an approximate representation. An image f is represented as a 2D array of pixels. Every measured strip projection then gives rise to a linear equation on the pixel values of f. The resulting system of linear equations can be solved by methods from linear algebra, but in this way one cannot guarantee that a *binary* solution is found. We now introduce a particular grid, the *two-projection grid,* for which a binary solution of the reconstruction problem can be found efficiently. The rows and columns of this grid correspond to the strips for the two projections angles $\bar{\theta}_1$ and $\bar{\theta}_2$. Define *grid cell* C_{ij} $(1 \le i, j \le n)$ as

$$C_{ij} = S_{\bar{\theta}_1}(t_{i-1}, t_i) \cap S_{\bar{\theta}_2}(t_{j-1}, t_j).$$

A *two-projection image* is a mapping $\{1, \dots, n\} \times \{1, \dots, n\} \to \{0,1\}$ which assigns a binary value to each grid cell of a two-projection grid. Figure 1b shows an example of a two-projection image.

It is often convenient to consider a two-projection image X as a *matrix* (x_{ij}), where x_{ij} denotes $X((i,j))$ (the pixel value of cell C_{ij}). The strip projections of X for the projection angles $\bar{\theta}_1$ and $\bar{\theta}_2$ can be computed directly by summation of all entries in each row of X, or column respectively, and multiplying the result by the cell area a. For $k = 1, 2$, define $P_k(X) \in \mathbb{R}^n$ as the vector of strip projections of X for angle $\bar{\theta}_k$. Define the *one-count* of X by $S(X) = \sum_{1\leq i,j\leq n} x_{ij}$. We denote the area of a single grid cell by a. Note that all grid cells have the same shape and size.

We will now define a reconstruction problem for two-projection images. As the projection data may contain noise or other errors, we don't require that the image adheres perfectly to the measured projection data. In the next section, where we consider reconstruction from more than two projections, we require a generalization of the reconstruction problem which incorporates prior knowledge. We state this general reconstruction problem here.

Problem 2. Let $\bar{\theta}_1, \bar{\theta}_2 \in [0, \pi)$ be two disjoint projection angles. Let $p_1, p_2 \in \mathbb{R}^n$ be two vectors of nonnegative real numbers (the *measured projection data*). Let $W = (w_{ij}) \in \mathbb{R}^{n\times n}$, $\bar{T} \in \mathbb{N}_{>0}$ and $\alpha \in \mathbb{R}$. Construct a matrix $X \in \{0,1\}^{n\times n}$ such that $S(X) = \bar{T}$ and

$$\alpha(|P_1(X) - p_1|_1 + |P_2(X) - p_2|_1) - \sum_{1\leq i,j\leq n} a w_{ij} x_{ij}$$

is minimal.

In any instance of Problem 2, the one-count $S(X)$ is considered to be fixed at $\bar{T}$. A good value for $\bar{T}$ can be computed from the measured projection data by taking $\bar{T} = (|p_1|_1 + |p_2|_1)/2a$. Putting $\alpha = 1$ and $w_{ij} = 0$ for $1 \leq i,j \leq n$, yields the basic reconstruction problem (without the use of prior knowledge). The matrix W is called the *weight map.* The weight map is used extensively in the algorithm for reconstructing a binary image from more than two projections that we describe in the next section. It is used to express a preference for each pixel to obtain a value of either 0 or 1 in the reconstruction.

Problem 2 can be solved efficiently by formulating it as a minimum cost flow problem in a particular graph. Efficient algorithms are available for solving the resulting network flow problem. The basic idea of using network flow methods for the reconstruction of binary images from two projections was first described by Gale in 1957 [7], in the context of reconstructing binary *matrices* from their row and column sums.

For the sake of brevity, we make one simplifying assumption: we assume that all measured strip projections are integral multiples of the pixel area a. If the strip projections do not satisfy this assumption they can simply be rounded to the nearest multiple of a. If n is large, the effect of this rounding step is neglegible. Details on how to compute an exact solution of Problem 2, without the assumption on the strip projections, are given in [6].

In the remainder of this section we assume that the reader is familiar with the basic concepts of *network flows.* The book [5] provides an excellent introduction to this subject.

With a pair of projection angles $(\bar{\theta}_1, \bar{\theta}_2)$ and their respective measured projections (p_1, p_2), we associate a *directed* graph $G = (V, E)$, where V is the set of nodes and E is the set of edges. We call G the *associated graph*, see Figure 2. The set V contains a node s (the *source*), a node t (the *sink*), one node for each

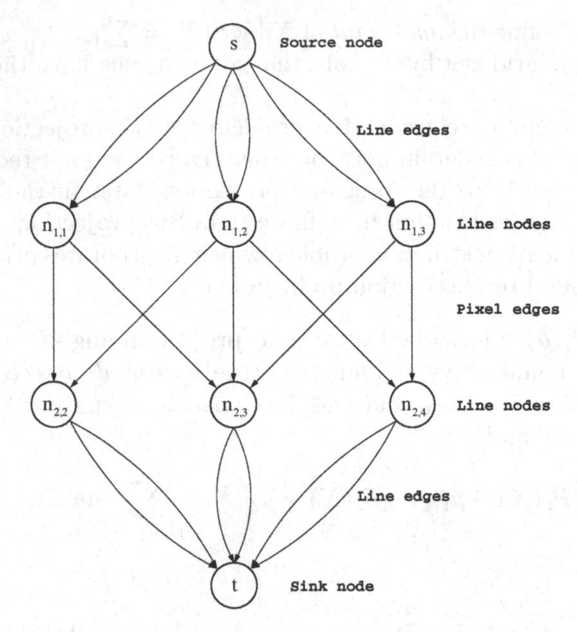

Fig. 2. Basic structure of the associated graph

strip of projection angle $\bar{\theta}_1$ and one node for each strip of projection angle $\bar{\theta}_2$. The node that corresponds to $S_{\bar{\theta}_k}(t_{i-1}, t_i)$ has label $n_{k,i}$. We call the nodes $n_{k,i}$ *line nodes.*

Each edge e of G has an associated *capacity* u_e and *cost* c_e. Every pair $(n_{1,i}, n_{2,j})$ of nodes is connected by a directed edge. We call these edges *pixel edges.* Each pixel edge $(n_{1,i}, n_{2,j})$ has capacity 1 and cost $-aw_{ij}$. For each pair $(s, n_{1,i})$ the graph G contains two parallel edges. The first edge has capacity p_{1i}/a and cost 0, the second edge has capacity $n - p_{1i}/a$ and cost $2\alpha a$. Similarly, G contains two parallel edges for every pair $(n_{2,j}, t)$ with capacities p_{2j}/a and $n - p_{2j}/a$ and costs 0 and $2\alpha a$ respectively.

An integral *flow* in G is a mapping $Y : E \to \mathbb{N}_{\geq 0}$ such that $Y(e) \leq u_e$ for all $e \in E$ and such that for all $v \in V \backslash \{s, t\}$:

$$\sum_{w:\,(w,v)\in E} Y((w,v)) = \sum_{w:\,(v,w)\in E} Y((v,w)).$$

Every integral flow Y corresponds to a two-projection image $X = x_{ij}$, defined by $x_{ij} = Y((n_{1,i}, n_{2,j}))$. Define the *total cost* of an integral flow Y by

$$C(Y) = \sum_{e\in E} c_e Y(e).$$

In [6] it is proved that any integral flow of minimal cost in G corresponds to a solution of Problem 2. Note that all edge capacities of G are integers. The integral minimum cost flow problem can be solved in polynomial time for graphs that have integral arc capacities and costs. To obtain integral costs in our case, we multiply the edge costs by a sufficiently large number and round the results. Note that scaling all costs by the same factor does not change which flow has minimal cost.

3 More Than Two Projections

Unfortunately, there is no straightforward generalization of the network flow approach to the case of more than two projections. We propose an iterative algorithm, which uses the fact that the two-projection problem can be solved efficiently. The algorithm computes a reconstruction from more than two projections by solving a series of two-projection subproblems, each using two projection angles. The algorithm uses the concept of a weight map, as introduced in Problem 2. Our algorithm aims to find an *approximate* solution of the reconstruction problem. In each iteration a new pair of projection angles is selected. An instance of Problem 2 is then solved on the two-projection grid that corresponds to those two angles. The weight map is computed using the reconstruction from the previous iteration, in such a way that solutions are preferred which resemble the reconstruction from the previous iteration. Additionaly, a preference for locally smooth regions is incorporated in the weight map. The reconstruction from the previous iteration was computed using a different pair of projections, which are thus incorporated into the new reconstruction. Repeating this argument, projections from earlier iterations are also incorporated.

Compute a start solution X^0 on the standard square pixel grid;
$A := (\sum_{k=1}^{d} |p_k|_1)/d$;
$\tau := 0$;
while (stop criterion is not met) **do**
begin

  $\tau := \tau + 1$;

  Select a new pair of projection angles $\bar{\theta}_1^\tau, \bar{\theta}_2^\tau \in D$;

  Compute the weight map $W^\tau = (w_{ij}^\tau)$ for the two-projection grid corresponding to $(\bar{\theta}_1^\tau, \bar{\theta}_2^\tau)$, using the previous reconstruction $X^{\tau-1}$;

  Compute $\bar{T} := A/a_\tau$;

  Compute a solution X^τ with $S(X^\tau) = \bar{T}$ of Problem 2 on the two-projection grid for angles $(\bar{\theta}_1^\tau, \bar{\theta}_2^\tau)$, using the weight map W^τ;

end

Return the final reconstruction X^*;

Fig. 3. Basic steps of the algorithm

Figure 3 shows the basic steps of the algorithm. First, a start solution is computed, using all projections simultaneously. The start solution should provide a good first approximation of the unknown image, while being easy to compute. The start solution can be computed on the standard $n \times n$ square pixel grid. For the experiments in Section 4 we used the SIRT (Simultaneous Iterative Reconstruction Technique, see Chapter 7 of [1]) to compute the start solution, which yields a gray value reconstruction.

Subsequently, the "total area A of the white region" (i.e., the region where the function value of the unknown image f is 1) is computed as $(\sum_{k=1}^{d} |p_k|_1)/d$.

Next, the algorithm enters the main loop. In each iteration τ of the main loop a new pair $(\bar{\theta}_1^\tau, \bar{\theta}_2^\tau)$ of projection angles is first selected, which determines the two-projection grid for iteration τ. We refer to the cell (i, j) of this grid as C_{ij}^τ. To choose the projection angles, the projections of the current image for each of the angles $\theta_1, \ldots, \theta_d$ are computed first. The new pair of angles is chosen such that the angle between $\bar{\theta}_1^\tau$ and $\bar{\theta}_2^\tau$ is at least $\pi/3$ and the total deviation of the computed projections from the prescribed projections is largest, according to the sum-norm.

Next, the number of white pixels $\bar{T} = \text{round}(A/a_\tau)$ is computed, where a_τ denotes the area of a grid cell in the current two-projection grid. Note that the grid cell area is different for each two-projection grid.

Subsequently the weight map $W^\tau = (w_{ij}^\tau)$ is computed from the previous reconstruction. We denote the grid cells of the new two-projection grid by C_{ij}^τ. Define $m_{ij} \in \mathbb{R}^2$ as the center of mass of cell C_{ij}^τ. The pixel weight w_{ij}^τ of pixel (i, j) depends directly on the average gray value of $X_{\tau-1}$ in a small neighbourhood of m_{ij}. Most common pixel neighbourhood definitions that are used in image processing, such as the 4-neighbourhood and 8-neighbourhood, are very suitable for square pixel grids. However, as our algorithm deals with many different pixel grids for which the pixel sizes and shapes may vary, we have to use a more general neighbourhood concept. Let r be a positive real number, the *neighbourhood radius*. We used $r = 1.5$ for all experiments in Section 4. Let Γ_{ij} be the average gray value inside a circle of radius r, centered in m_{ij}, in the two-projection image $X_{\tau-1}$. This value can be computed by intersecting each of the overlapping pixels of $X_{\tau-1}$ with the circular neighbourhood of m_{ij} and weighting the pixel values by the intersection area. Note that we always have $\Gamma_{ij} \in [0, 1]$.

Define $g : [-1, 1] \rightarrow \mathbb{R}$ by

$$g(v) = \begin{cases} v & \text{if} \quad |v| \neq 1 \\ 2v & \text{if} \quad |v| = 1. \end{cases}$$

We call g the *local weight function*. Together with the neighbourhood radius r, the local weight function determines the preference for locally smooth regions. The pixel weight w_{ij}^τ is computed as

$$w_{ij}^\tau = g(2(\Gamma_{ij} - \frac{1}{2}))$$

The basic idea of the local weight function is that, as the neighbourhood of a pixel becomes more white, the preference to make this pixel white in the next iteration increases. The same holds for black neighbourhoods. There is an additional increase in the (absolute value of) the pixel weight if the neighbourhood is completely homogeneous. If a pixel neighbourhood consists of 50% black and 50% white pixels, no preference is expressed as the pixel weight is zero.

The weight map W^τ, the value $\bar{T}$ and the projections for angles $(\bar{\theta}_1^\tau, \bar{\theta}_2^\tau)$ define an instance of Problem 2. Solving this problem by the network flow approach yields the new reconstruction X^τ.

To determine when the algorithm should terminate, the strip projections of X_τ are computed for all projection angles $\theta_1, \ldots, \theta_d$. Subsequently the norm of the error with respect to the measured projections is computed. The algorithm terminates if the error has not decreased for T iterations, where T is a constant. In all experiments of Section 4 we used $T = 30$.

4 Experimental Results

In this section we present reconstruction results of our algorithm for a set of four phantom images. For each of the phantoms we computed the measured projection data by simulation. In this way the noise level can be controlled, which is difficult for datasets obtained by real measurements.

We implemented the iterative network flow algorithm in C++, using the RelaxIV library [8] to solve the min cost flow problems. All experiments were performed on a 2.4GHz PentiumIV PC.

The four phantom images that we use to evaluate the reconstruction time and the reconstruction quality of our algorithm are shown in Figure 4.

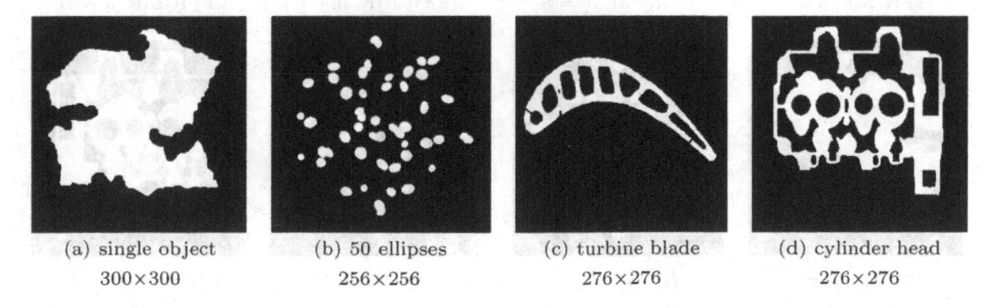

(a) single object 300×300 (b) 50 ellipses 256×256 (c) turbine blade 276×276 (d) cylinder head 276×276

Fig. 4. Four phantom images

First, we compare our reconstruction results from perfect projections to two alternative approaches. It is well known that continuous tomography algorithms such as Filtered Backprojection require a large number of projections. Algebraic reconstruction methods, such as ART and SIRT (see Chapter 7 of [1]) typically perform much better than backprojection algorithms if the number of projections is very small. Our algorithm uses the SIRT algorithm to compute a start solution. As a first comparison, we compare the final reconstruction computed by our algorithm to the continuous SIRT reconstruction. In [9,10], Weber et al. describe a linear programming approach to binary tomography which incorporates a smoothness prior. We implemented the *R-BIF* approach which is described in [10] using the ILOG CPLEX interior point solver [11]. The real-valued pixel values are rounded to binary values as a postprocessing step, which was also done in [9]. Besides the projection data, the linear program depends on a parameter α, which determines the preference for smoothness. For our set of four phantoms we found that $\alpha = 0.2$ works well. Figure 5 shows the reconstruction results for the four phantoms. Table 1 shows a quantitative comparison.

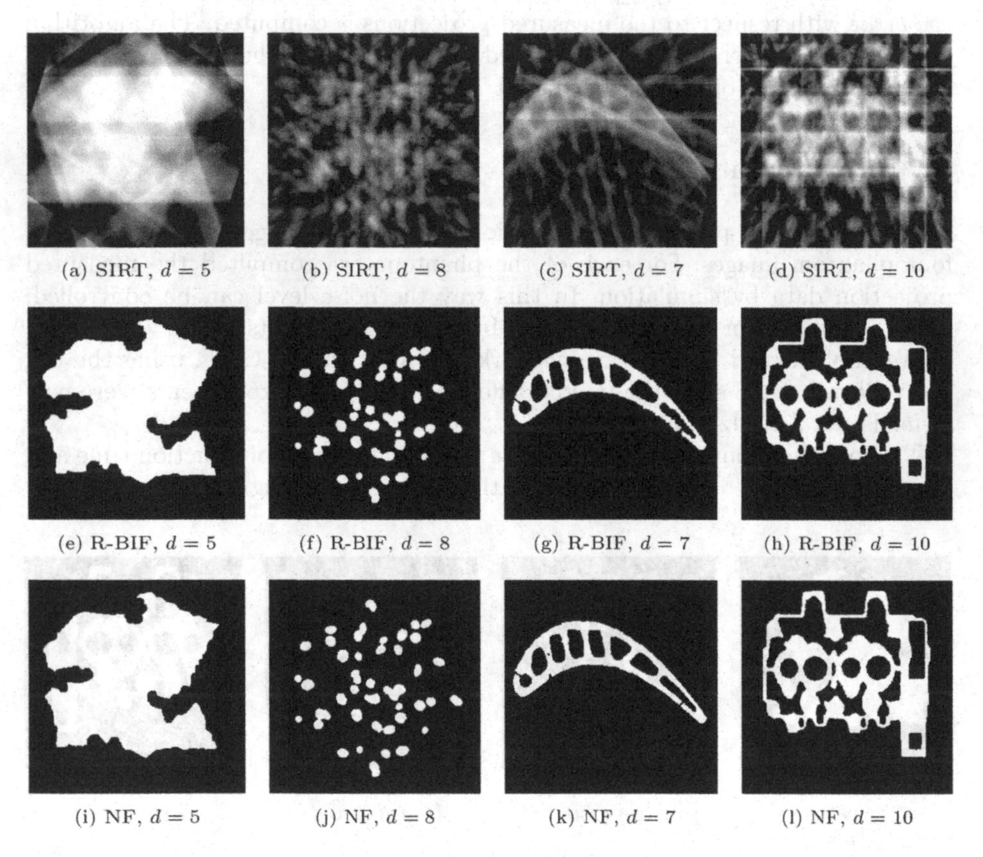

(a) SIRT, $d = 5$ (b) SIRT, $d = 8$ (c) SIRT, $d = 7$ (d) SIRT, $d = 10$

(e) R-BIF, $d = 5$ (f) R-BIF, $d = 8$ (g) R-BIF, $d = 7$ (h) R-BIF, $d = 10$

(i) NF, $d = 5$ (j) NF, $d = 8$ (k) NF, $d = 7$ (l) NF, $d = 10$

Fig. 5. Reconstruction results for the four phantoms from parallel beam projections, using SIRT (top), R-BIF from [10] (middle) and our network flow algorithm (NF, bottom). The figure captions show the number d of projections that was used.

The number d of projections is chosen for each phantom as the minimum number for which our algorithm computes an accurate reconstruction from the projection data. The projection angles are equally spaced between 0 and 180

Table 1. Quantitative comparison between the R-BIF reconstruction and the reconstruction computed by our network flow algorithm. The table shows the reconstruction time in minutes and the total number of pixel differences between the reconstruction and the original phantom.

phantom	size	#proj.	R-BIF		Network Flow	
			#errors	time(min)	#errors	time(min)
single object	300×300	5	220	9.2	58	2.0
50 ellipses	256×256	8	216	3.5	152	1.5
turbine blade	276×276	7	214	3.2	108	1.8
cylinder head	276×276	10	375	10	665	1.9

degrees. Each projection consists of n strip projections, that each have a width equal to the phantom image pixel width.

Our algorithm is capable of computing an accurate reconstruction of each of the four phantom images from a small number of projections. The reconstruction quality of the R-BIF algorithm appears to be similar, although there are some small differences between the reconstructions by both methods. Both algorithms clearly outperform the continuous SIRT algorithm.

We observed that the number of projections that is required to compute an accurate reconstruction is the same for the R-BIF algorithm as for our algorithm, for each of the four phantoms. As there are major differences between both algorithms, this suggests that this minimal number of projections is an intrinsic property of the images themselves. It may be very difficult or even impossible to do better by using a different algorithm.

The minimal number of projection that is required to reconstruct a given (unknown) image depends strongly on characteristics of the unknown image, as can be seen by the varying number of projections that is required for the four phantom images.

Our algorithm has several advantages compared to the R-BIF linear programming approach. First, the network flow algorithm is faster than general linear programming as used in [10], even though we used a high performance interior point solver for solving the linear program. It was demonstrated in [12] for the case of lattice images that the network flow approach for 2D reconstruction can be extended to a highly efficient algorithm for 3D reconstruction. Within each iteration a series of 2D reconstruction problems is solved, instead of one big 3D problem. The 2D reconstructions can be computed fast (because each subproblem is relatively small) and in parallel. A similar extension is possible for our new algorithm. Dealing with large volumes seems to be very difficult when using general linear programming, as the number of variables becomes huge for large 3D volumes.

Another advantage of our algorithm is that it can deal with noise effectively, as we will show in the next subsection. The linear programming approach from [9,10] is not capable of handling noisy data. It can be extended to the case of noisy data, as described in [13]. However, the presented algorithm for dealing with noisy data solves a *series* of linear programs, which results in far longer reconstruction times.

More recently, Schüle et al. developed a different reconstruction algorithm based on D.C. programming [14]. The results of this algorithm seem to be very promising and it does not have some of the drawbacks of the linear programming approach. We intend to perform a comparison between a wider set of reconstruction algorithms in a future publication.

4.1 Noisy Projection Data

So far all experiments were carried out with perfect projection data. We now focus on the reconstruction of images from noisy projection data, which is practically more realistic. We also assume that the noise is independent for each projected strip. To be more precise, we assume that the noise is additive and that it follows a Gaussian distribution with $\mu = 0$ and σ constant over all projected strips. The standard deviation σ is expressed as $\sigma = vy$, where y denotes the average measured strip projection over all projected strips. For example, taking $v = 0.05$ results in independently distributed additive noise for each strip projection, following the $\mathcal{N}(0, 0.05y)$ distribution.

Figure 6 shows reconstruction results for the cylinder head phantom, using 10 parallel beam projections and varying noise levels. The reconstruction quality decreases gradually as the noise level is increased. Up to $v = 0.02$, the noise hardly has any visible influence on the reconstruction quality.

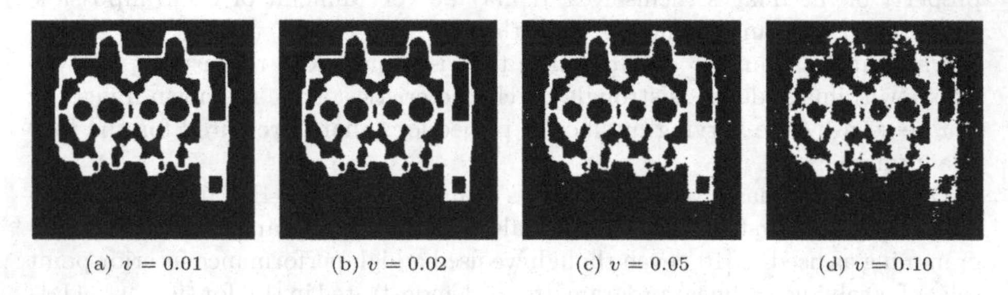

(a) $v = 0.01$ (b) $v = 0.02$ (c) $v = 0.05$ (d) $v = 0.10$

Fig. 6. Reconstruction results from 10 parallel beam projections for increasing noise levels

5 Conclusions

We have described a novel algorithm for the reconstruction of binary images from a small number of their projections. Our algorithm is iterative. In each iteration a reconstruction problem is solved that depends on two of the projections and the reconstruction from the previous iteration. The two-projection reconstruction problem is equivalent to the problem of finding a flow of minimal cost in the associated graph. This equivalence allows us to use network flow algorithms for solving the two-projection subproblems.

The reconstruction results show that the reconstruction quality of our algorithm is far better than for the SIRT algorithm from continuous tomography. A comparison with the R-BIF linear programming approach from [10], which uses a smoothness prior, shows that both algorithms yield comparable reconstruction quality. Our algorithm runs faster than the linear programming approach and can be easily extended to 3D reconstruction or reconstruction from noisy projections. These extensions are difficult to accomplish efficiently for the linear programming approach.

Our results for noisy projection data shows that the algorithm is capable of dealing with significant noise levels. The reconstruction quality decreases gradually as the noise level is increased.

In future research we intend to perform extensive comparisons between different discrete tomography algorithms. Such a comparison is often difficult, as each algorithm makes different assumptions on the class of images, the detector setting, etc. Generalization of our algorithm to 3D reconstruction is straightforward, following the same approach as in [12].

References

1. Kak, A.C., Slaney, M.: Principles of Computerized Tomographic Imaging. SIAM (2001)
2. Natterer, F.: The Mathematics of Computerized Tomography. SIAM (2001)
3. Herman, G.T., Kuba, A. (eds.): Discrete Tomography: Foundations, Algorithms and Applications. Birkhäuser Boston (1999)
4. Batenburg, K.J.: Reconstructing Binary Images from Discrete X-Rays. CWI Report PNA-E0418 (2004) (submitted for publication)
5. Ahuja, R.K., Magnanti, T.L., Orlin, J.B.: Network Flows: Theory, Algorithms, and Applications. Prentice-Hall (1993)
6. Batenburg, K.J.: A Network Flow Algorithm for Binary Tomography Without an Intrinsic Lattice. preprint: `www.math.leidenuniv.nl/~kbatenbu` (2006)
7. Gale, D.: A Theorem on Flows in Networks. Pacific J. Math. **7** (1957) 1073–1082
8. Bertsekas, D.P., Tseng, P.: RELAX-IV: A Faster Version of the RELAX Code for Solving Minimum Cost Flow Problems. LIDS Technical Report LIDS-P-2276. MIT (1994)
9. Weber, S., Schnörr, C., Hornegger, J.: A Linear Programming Relaxation for Binary Tomography with Smoothness Priors. Electron. Notes Discrete Math. **12** (2003)
10. Weber, S., Schüle, T., Schnörr, C., Hornegger, J.: A Linear Programming Approach to Limited Angle 3D Reconstruction from DSA Projections. Special Issue of Methods of Information in Medicine **4** (2004) Schattauer Verlag. 320-326
11. ILOG CPLEX: `http://www.ilog.com/products/cplex/`
12. Batenburg, K.J.: A New Algorithm for 3D Binary Tomography. Electron. Notes Discrete Math. **20** (2005) 247-261
13. Weber, S., Schüle, T., Hornegger, J., Schnörr, C.: Binary Tomography by Iterating Linear Programs from Noisy Projections. Proceedings of IWCIA 2004. Lecture Notes in Computer Science, Vol. 3322 (2004) 38-51
14. Schüle, T., Schnörr, C., Weber, S., Hornegger, J.: Discrete Tomography by Convex-Concave Regularization and D.C. Programming. Discrete Appl. Math. **151** (2005) 229-243

Fast Filling Operations Used in the Reconstruction of Convex Lattice Sets

Sara Brunetti[1], Alain Daurat[2], and Attila Kuba[3]

[1] Dipartimento di Scienze Matematiche e Informatiche, Università di Siena, Pian dei Mantellini 44, 53100, Siena, Italy
sara.brunetti@unisi.it
[2] LSIIT CNRS UMR 7005, Université Louis Pasteur (Strasbourg 1), Pôle API, Boulevard Sébastien Brant, 67400 Illkirch-Graffenstaden, France
daurat@dpt-info.u-strasbg.fr
[3] Department of Image Processing and Computer Graphics, University of Szeged, Árpád tér 2. H-6720 Szeged, Hungary
kuba@inf.u-szeged.hu

Abstract. Filling operations are procedures which are used in Discrete Tomography for the reconstruction of lattice sets having some convexity constraints. In [1], an algorithm which performs four of these filling operations has a time complexity of $O(N^2 \log N)$, where N is the size of projections, and leads to a reconstruction algorithm for convex polyominoes running in $O(N^6 \log N)$-time. In this paper we first improve the implementation of these four filling operations to a time complexity of $O(N^2)$, and additionally we provide an implementation of a fifth filling operation (introduced in [2]) in $O(N^2 \log N)$ that permits to decrease the overall time-complexity of the reconstruction algorithm to $O(N^4 \log N)$. More generally, the reconstruction of Q-convex sets and convex lattice sets (intersection of a convex polygon with $\mathbb{Z}^2$) can be done in $O(N^4 \log N)$-time.

Keywords: Discrete Tomography, Convexity, Filling Operations.

1 Introduction

One of the most intensively studied fields of discrete tomography is the reconstruction of lattice sets or, specially, binary matrices. Several algorithms have been published for reconstructing such sets. It is well-known that a binary matrix from its row and column sums can be reconstructed in polynomial time [3]. The interesting question is which sub-class of binary matrices can be reconstructed in polynomial time. In most cases some kind of (discrete) convexity is supposed on the sets. For example, Kuba published an algorithm [4] to reconstruct so-called *hv*-convex lattice sets from two projections. As it turned out later the reconstruction problem in this class is NP-complete [5]. Barcucci et al. showed [6] that a sub-class of *hv*-convex lattice sets, namely, the class of *hv*-convex polyominoes can be reconstructed in polynomial time. This result was extended also to a bigger class, that of *hv*-convex 8-connected lattice sets [2]. A new bigger class of convex sets, the so-called Q-convex sets was studied by Brunetti and

A. Kuba, L.G. Nyúl, and K. Palágyi (Eds.): DGCI 2006, LNCS 4245, pp. 98–109, 2006.

Daurat [7] and proved that even in this class the reconstruction can be solved in polynomial time.

Most of the algorithms reconstructing sets presenting some convexity properties use special procedures called filling operations. These operations can be applied in iterative procedures to approach the final solutions with two sequences of sets. The first sequence is a sequence of decreasing upper bounds and the second one is a sequence of increasing lower bounds of the solutions.

Originally in [6], four filling operations were defined. In [1], an efficient algorithm was given to apply the filling operations. In [2] a fifth filling operation was introduced to decrease the overall complexity of the reconstruction algorithm. Unfortunately, the algorithm for the filling operations of [1] cannot be generalized with this fifth operation (in [2] this point was not treated). In this paper we provide an implementation of all the five filling operations in the same complexity as the algorithm of [1]. As a result, we get an improvement in the time-complexity of the reconstruction algorithm.

The structure of this paper is the following. Section 2 contains the necessary definitions and notations. The filling operations, the new reconstruction algorithm, its analysis, and a possible generalization are described in Section 3. Section 4 shows the application of the new operation in the case of Q-convex and convex sets. Finally, in Section 5 we show some statistical results connected with the application of the filling operations in computer experiments.

2 Definitions

A lattice set is a finite subset of $\mathbb{Z}^2$. A lattice direction is given by an integer vector $\boldsymbol{p} = (p_x, p_y)$, and it can also be represented by a linear form $p(x, y) = p_y x - p_x y$. The horizontal direction (resp. vertical direction) denoted by h (resp. v) is determined by the vector $(1, 0)$ (resp. $(0, 1)$).

A lattice set is line-convex with respect to a direction p if its intersection with each line in the direction p is made of consecutive points. A set which is line-convex w.r.t. to the horizontal and vertical directions is called hv-convex.

The projection of a lattice set E along a direction p, denoted by X_pE, is the function which gives the number of points on any line of direction p, more precisely:

$$X_pE(k) = |\{M \in E \,:\, p(M) = k\}| \text{ for any } k \in \mathbb{Z}$$

where p is the linear form associated to $\boldsymbol{p}$.

In this article we are interested in the reconstruction of set E which satisfies some convexity constraints from its projections. More precisely if $\mathcal{M}$ is a class of lattice sets, and $\mathcal{D}$ is a finite set of lattice directions, the reconstruction problem for the class $\mathcal{M}$ and the directions $\mathcal{D}$ is the following.

RECONSTRUCTION($\mathcal{M}, \mathcal{D}$)
Data: A function $f : \mathcal{D} \times \mathbb{Z} \to \mathbb{Z}_+$ which gives a non-negative integer $f(p, k)$ for any line $p = k$ with $p \in \mathcal{D}$, and such that $\{(p, k) \,:\, f(p, k) > 0\}$ is finite.

Task: Reconstructing a lattice set $E \in \mathcal{M}$ such that $X_pE(k) = f(p,k)$ for any $(p,k) \in \mathcal{D} \times \mathbb{Z}$

In the whole paper $[a,b]$ denotes the discrete interval $\{k \in \mathbb{Z} \, : \, a \leq k \leq b\}$.

3 Filling Operations

3.1 Preliminaries

A filling operation is a procedure which has been used in many reconstruction algorithms [2, 4, 6, 7, 8]. Formally, a filling operation takes function f of RECONSTRUCTION$(\mathcal{M},\mathcal{D})$, and a pair of sets (α,β) such that $\alpha \subset \beta$ and returns a new pair of sets (α',β') with $\alpha \subseteq \alpha' \subseteq \beta' \subseteq \beta$.

We now present classical filling operations which can be used for any class contained in that of line-convex sets w.r.t. $\mathcal{D}$.

To simplify the description of these operations, we first describe them for the set $\mathcal{D} = \{h, v\}$ consisting of the horizontal and vertical directions. We denote $h_i = f(h,i)$ and $v_j = f(v,j)$. We also suppose without loss of generality that there exist $m, n \in \mathbb{Z}_+$ such that $h_i = 0$ for $i \notin [1,m]$ and $v_j = 0$ for $j \notin [1,n]$.

For any $i \in [1,m]$ we denote the set $\{(i,j) \, : \, j \in \mathbb{Z}, \, (i,j) \in \alpha\}$ by α_i^h. On an analogous way we define α_j^v, β_i^h, β_j^v for $i \in [1,m]$ and $j \in [1,n]$.

The following notations are used for the extremities of each α_i^h and β_i^h.

$$l(\alpha_i^h) = \min(\{j \, : \, (i,j) \in \alpha_i^h\}), \qquad r(\alpha_i^h) = \max(\{j \, : \, (i,j) \in \alpha_i^h\})$$

$$l(\beta_i^h) = \min(\{j \, : \, (i,j) \in \beta_i^h\}), \qquad r(\beta_i^h) = \max(\{j \, : \, (i,j) \in \beta_i^h\})$$

with the conventions $\min(\emptyset) = +\infty$, $\max(\emptyset) = -\infty$.

With this notation, the four filling operations of [6] on horizontal lines can be defined as:

- If $\alpha_i^h \neq \emptyset$ then $\oplus\alpha_i^h = \{(i,j) \in \beta_i^h \, : \, l(\alpha_i^h) \leq j \leq r(\alpha_i^h)\}$.
- $\otimes\alpha_i^h = \{(i,j) \in \beta_i^h \, : \, r(\beta_i^h) - h_i < j < l(\beta_i^h) + h_i\}$.
- If $\alpha_i^h \neq \emptyset$, $j' = \max(\{j \, : \, (i,j) \notin \beta_i^h \text{ and } j < l(\alpha_i^h)\}$, $j'' = \min(\{j \, : \, (i,j) \notin \beta_i^h \text{ and } j > r(\alpha_i^h)\}$ then $\ominus\beta_i^h = \{(i,j) \in \beta_i^h \, : j' < j < j''\}$.
- If $\alpha_i^h \neq \emptyset$, then $\odot\beta_i^h = \{(i,j) \in \beta_i^h \, : \, r(\alpha_i^h) - h_i < j < l(\alpha_i^h) + h_i\}$.

A fifth filling operation $\odot'$ has been introduced in [2,7,8]. It permits to reduce the overall complexity of the reconstruction algorithm: In all the reconstruction algorithms the first step of the algorithm is fixing arbitrarily some points on the border of the reconstructing sets (these points are in general called bases or feet). Without the operation $\odot'$, at least four fixed points were necessary, but with it, only two are necessary. (See [7, p.43-44] for more details.) The operation $\odot'$ simply removes the components of β_i^h (maximum sequences of consecutive elements of β_i^h) which are smaller than the corresponding projection.

To define it formally we need a notation for the extremities of each component. So the sequence $(c_k)_{1 \leq k \leq 2r} = c(\beta_i^h)$ is defined by:

$$c_k < c_{k+1} \text{ and } \{j \, : \, (i,j) \in \beta_i^h\} = \bigcup_{k=1}^{r} [c_{2k-1}, c_{2k} - 1]. \tag{1}$$

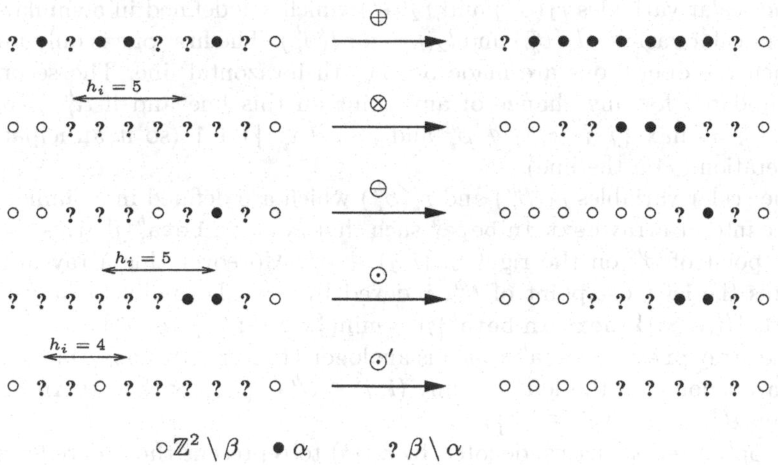

? • ? • • ? ? • ? ○ $\xrightarrow{\oplus}$? • • • • • • ? ○

$h_i = 5$

○ ○ ? ? ? ? ? ? ? ○ $\xrightarrow{\otimes}$ ○ ○ ? ? • • • ? ? ○

○ ○ ? ? ? ○ ? • ? ○ $\xrightarrow{\ominus}$ ○ ○ ○ ○ ○ ○ ? • ? ○

$h_i = 5$

? ? ? ? ? ? • • ? ○ $\xrightarrow{\odot}$ ○ ○ ○ ? ? ? • • ? ○

$h_i = 4$

○ ? ? ○ ? ? ? ? ? ○ $\xrightarrow{\odot'}$ ○ ○ ○ ○ ? ? ? ? ? ○

○ $\mathbb{Z}^2 \setminus \beta$ • α ? $\beta \setminus \alpha$

Fig. 1. The filling operations

Then the operation $\odot'$ is defined by:

$$\odot' \beta_i^h = \bigcup_{\substack{1 \le k \le r \\ c_{2k} - c_{2k-1} \ge h_i}} [c_{2k-1}, c_{2k} - 1].$$

We can also define these five operations on the vertical lines.

The reconstruction algorithms described in [2,6,7] iteratively apply the filling operations in a fixed order on all the lines of $[1, m] \times [1, n]$ (iterative step). The kth iteration gives rise to a new couple (α_k, β_k) from $(\alpha_{k-1}, \beta_{k-1})$, and the iterative process ends when an invariant couple is obtained, that is, $(\alpha_{k'}, \beta_{k'}) = (\alpha_{k'-1}, \beta_{k'-1})$. There are several methods to construct the initial couple of sets (α_0, β_0). For example in [6], β_0 is the complete rectangle $[1, m] \times [1, n]$ and α_0 consists of a set of points (called feet) located on the four edges of β. More generally, we assume that $\alpha_0 \subseteq \beta_0 \subseteq [1, m] \times [1, n]$.

In [1,6] only the first four filling operations are considered. If $N = \max(m, n)$, the whole iterative process runs in $O(N^4)$-time in [6]. In [1] the author proves that this process can be executed in $O(N^2 \log N)$-time. The best time-complexity with the five filling operations is $O(N^3)$ [7]. Now we will describe a procedure which performs the five filling operations in $O(N^2 \log N)$-time.

3.2 The New Algorithm

At first we describe the data structures we use in the algorithm.

For each horizontal line of index i, we use the following data.

- The scalar variables $l_1(\alpha_i^h)$ and $l_2(\alpha_i^h)$ for $l(\alpha_i^h)$. The first one is only updated when the operations are performed on the ith horizontal line. The second one is updated for any change of any point on this line.

- The scalar variables $r_1(\alpha_i^h)$ and $r_2(\alpha_i^h)$ which are defined in a similar way.
- The scalar variable $l_1(\beta_i^h)$ and $l_2(\beta_i^h)$ for $l(\beta_i^h)$. The first one is only updated when the operations are made on the ith horizontal line. The second one is updated for any change of any point on this line and if $\alpha_i^h \neq \emptyset$, then $l_2(\beta_i^h) = \max(\{j : (i,j) \notin \beta_i^h \text{ and } j < l(\alpha_i^h)\}) + 1$ (so it anticipates the operation $\ominus$ on the line).
- The scalar variables $r_1(\beta_i^h)$ and $r_2(\beta_i^h)$ which are defined in a similar way.
- The integer array $\texttt{next\_in\_beta}_i^h$ such that $\texttt{next\_in\_beta}_i^h[j]$ gives the nearest point of β_i^h on the right of $(i,j) \in \beta_i^h$. Moreover this array also indicates the leftmost point of β_i^h, indexed by $-\infty$. Formally for any $(i,j) \in \beta_i^h \cup \{(i,-\infty)\}$, $\texttt{next\_in\_beta}_i^h[j] = \min(\{k > j : (i,j) \in \beta_i^h\})$.
- The array $\texttt{prev\_in\_beta}_i^h$ which is analogous to $\texttt{next\_in\_beta}_i^h$ but gives the nearest point on the left: For any $(i,j) \in \beta_i^h \cup \{(i,+\infty)\}$ $\texttt{prev\_in\_beta}_i^h[j]$ is $\max(\{k < j : (i,j) \in \beta_i^h\})$.
- An optimized structure denoted by $c(\beta_i^h)$ to represent the ordered sequence of intervals $[c_{2k-1}, c_{2k} - 1]$ defined by (1).
  We suppose that we have the following operations on this structure:
  - insert($c(\beta_i^h), u, v$): insertion of a new interval $[u, v-1]$ which is disjoint with the intervals $[c_{2k-1}, c_{2k} - 1]$.
  - delete($c(\beta_i^h), u, v$): deletion of an interval $[u, v-1]$.
  - search($c(\beta_i^h), x$): returns (c_{2k-1}, c_{2k}) such that $c_{2k-1} \leq x < c_{2k}$.

  We use a structure such that these operations are made in $O(\log(r))$-time. For example, an implementation with AVL trees permits this (see [9]).
- Another ordered sequence $d(\beta_i^h)$ of intervals $[d_{2k-1}, d_{2k} - 1]$ which represents the components which are to be deleted by the operation $\odot'$. This sequence is represented by the same structure as $c(\beta_i^h)$.

There are also data not associated directly to a line.

- The sets α and β are simply implemented by a boolean two-dimensional array.
- A set `lines_to_treat` stores the lines which are to be treated by the filling operations. The only operations used on this structure are the vacuity test, the extraction of one arbitrary element and the insertion of one element (if not already present in the structure). These operations can be executed in constant time if the set is implemented as an array B of booleans coupled with an array A (implementing a stack) of the elements both indexed in $[1, m+n]$ and an integer variable cA for the cardinality. Precisely the implementation is the following.

```
isempty(lines_to_treat)
  return(cA = 0)
extract(lines_to_treat)
  x ← A[cA]; B[x] ← false; cA ← cA − 1
  if x ≤ m then
    return(h = x)
  else
    return(v = x − m)
```

```
    end if
```

```
  add_line(lines_to_treat, h = i)
```

```
    if not(B[i]) then
      B[i] ← true; cA ← cA + 1; A[cA] ← i
    end if
```

```
  add_line(lines_to_treat, v = j)
```

```
    if not(B[m + j]) then
      B[m + j] ← true; cA ← cA + 1; A[cA] ← m + j
    end if
```

Now we can describe precisely the algorithm for the filling operations.

The two first procedures put a point in α or remove a point from β. They update the data for the horizontal and vertical lines passing through the point.

```
put_in_alpha_h(i, j)
```

```
  if (i, j) ∉ β then
    exit(no solution)
  end if
  if (i, j) ∈ α then
    return                                                    // Nothing to do !
  end if
  α ← α ∪ {(i, j)}
  for (p, i', j') ∈ {(h, i, j), (v, j, i)} do
    l2(α^p_i') ← min(l2(α^p_i'), j'); r2(α^p_i') ← max(r2(α^p_i'), j')
  end for
  add_line(lines_to_treat, v = j)
```

```
remove_from_beta_h(i, j)
```

```
  if (i, j) ∈ α then
    exit(no solution)
  end if
  if (i, j) ∉ β then
    return                                                    // Nothing to do !
  end if
  β ← β \ {(i, j)}
  for (p, i', j', x) ∈ {(h, i, j, h_i), (v, j, i, v_j)} do
    if j' = l2(β^p_i') or j' < l2(α^p_i') then
      l2(β^p_i') ← next_in_beta^p_i'[j']
    end if
    if j' = r2(β^p_i') or j' > r2(α^p_i') then
      r2(β^p_i') ← prev_in_beta^p_i'[j']
    end if
    next_in_beta^p_i'[prev_in_beta^p_i'[j']] ← next_in_beta^p_i'[j']
    prev_in_beta^p_i'[next_in_beta^p_i'[j']] ← prev_in_beta^p_i'[j']
    (u, u') ← search(c(β^p_i'), j'); delete(c(β^p_i'), u, u')
    if u < j' then
      insert(c(β^p_i'), u, j')
    end if
    if j' + 1 < u' then
      insert(c(β^p_i'), j' + 1, u')
    end if
    if u' − u < x then
```

```
      delete(d(β^p_{i'}), u, u')
    end if
    if 0 < j' − u < x then
      insert(d(β^p_{i'}), u, j')
    end if
    if 0 < u' − (j' + 1) < x then
      insert(d(β^p_{i'}), j' + 1, u')
    end if
  end for
  add_line(lines_to_treat, v = j)
```

This procedure applies the filling operations on a horizontal line.

treat_line($h = i$)

```
  (d_k)_{1≤k≤2r} ← d(β^h_i)                                          // Operation ⊙'
  for all k ∈ [1, r] and j ∈ [d_{2k−1}, d_{2k} − 1] do
    remove_from_beta_h(i, j)
  end for
  if l_1(α^h_i) = +∞ then                                            // Operation ⊕
    for all j ∈ [l_2(α^h_i) + 1, r_2(α^h_i) − 1] do
      put_in_alpha_h(i, j)
    end for
  else
    for all j ∈ [l_2(α^h_i) + 1, l_1(α^h_i) − 1] ∪ [r_1(α^h_i) + 1, r_2(α^h_i) − 1] do
      put_in_alpha_h(i, j)
    end for
  end if
  for all j ∈ [l_1(β^h_i), l_2(β^h_i) − 1] ∪ [r_2(β^h_i) + 1, r_1(β^h_i)] do     // Operation ⊖
    remove_from_beta_h(i, j)
  end for
  if r_2(β^h_i) − h_i + 1 ≤ l_2(β^h_i) + h_i − 1 then                // Operation ⊗
    if l_2(α^h_i) = +∞ then
      for all j ∈ [r_2(β^h_i) − h_i + 1, l_2(β^h_i) + h_i − 1] do
        put_in_alpha_h(i, j)
      end for
    else
      for all j ∈ [r_2(β^h_i) − h_i + 1, l_2(α^h_i) − 1] ∪ [r_2(α^h_i) + 1, l_2(β^h_i) + h_i − 1] do
        put_in_alpha_h(i, j)
      end for
    end if
  end if
  if l_2(α^h_i) ≠ +∞ then                                            // Operation ⊙
    for all j ∈ [l_2(β^h_i), r_2(α^h_i) − h_i] ∪ [l_2(α^h_i) + h_i, r_2(β^h_i)] do
      remove_from_beta_h(i, j)
    end for
  end if
  l_1(α^h_i) ← l_2(α^h_i); r_1(α^h_i) ← r_2(α^h_i); l_1(β^h_i) ← l_2(β^h_i); r_1(β^h_i) ← r_2(β^h_i)
```

The procedures put_in_alpha$_v(i, j)$, remove_from_beta$_v(i, j)$, treat_line ($v = j$) are similar. This is the main procedure for the filling operations.

filling_operations(α_0, β_0)

```
  α ← α_0; β ← β_0
```

```
β ← β \ {(i, j) : h_i = 0 or v_j = 0}
for all l ∈ {h = i : 1 ≤ i ≤ m and h_i > 0} ∪ {v = j : 1 ≤ j ≤ n and v_j > 0} do
  add_line(lines_to_treat, l)
end for
initialize l1, l2, r1, r2, next_in_beta, prev_in_beta, c, d for all the lines of lines_to_treat
while not(isempty(lines_to_treat)) do
  l ← extract(lines_to_treat)
  treat_line(l)
end while
return(α, β)
```

3.3 Correctness of the Algorithm

- At the end of the executions of the procedure `put_in_alpha` and `remove_from_beta`, all the variables l_2, r_2, `next_in_beta`, `prev_in_beta`, c, d are updated according to the actual α and β.
- The modifications of α and β done by `treat_line` correspond exactly to the five filling operations $\odot'$, $\oplus$, $\ominus$, $\otimes$, $\odot$ executed in this order. In particular, if an instruction "**exit**(no solution)" is executed, the filling operations lead to a situation where $\alpha \not\subseteq \beta$.
- At the end of the procedure `treat_line`$(h = i)$, the line $h = i$ is invariant w.r.t. the five filling operations, so during the execution of the algorithm all the lines which are *not* in `lines_to_treat` are invariant w.r.t. to the filling operations: when `lines_to_treat` is empty, (α, β) is invariant w.r.t. to the filling operations.
- The algorithm stops after a finite number of steps because $(|\beta \setminus \alpha|, |$`lines_to_treat`$|)$ decreases lexicographically at each iteration of `filling_operations`.

3.4 Analysis of Complexity

Let $N = \max(\{m, n\})$.

- The procedures `put_in_alpha` and `remove_from_beta` are executed in $O(1)$ and $O(\log N)$-time respectively.
- The procedure `treat_line` has a time complexity $O(1 + p \log N)$ where p is the number of times the procedure `put_in_alpha`, or `remove_from_beta` run.
- The procedure `put_in_alpha`^{h} is never done more than once on a point and `remove_from_beta`^{h} is never done more than twice: once for the first four filling operations and a second time for the fifth operation $\odot'$. So these two procedures are executed less than $2N^2$ times.
- Similarly the procedures `put_in_alpha`^{v} and `remove_from_beta`^{v} are executed less than $2N^2$ times.
- The procedure `treat_line` is repeated less than $2N + 8N^2$ times because `lines_to_treat` is filled first with less than $2N$ lines and then a line is added to it only from `put_in_alpha` or `remove_from_beta`. So the global time-complexity of the algorithm is $O(N^2 \log N)$.

3.5 Differences with Gebala's Algorithm

The procedures performing the filling operations which are described by Gebala in [1] have the same structure than the ones presented here. However our algorithm presents several improvements:

- Gebala's algorithm does not apply the fifth filling operation.
- Gebala's algorithm uses a tree (free_0) to store the points on each line which are not in β. In our algorithm this structure is not needed because we use the arrays `next_in_beta` and `prev_in_beta`. Thanks to this, there is no loop in the procedures `put_in_alpha` and `remove_from_beta` which simplifies the analysis of the complexity of these procedures. Moreover these arrays need only $O(1)$-time operations.
- Gebala uses two trees (tree_{row} and tree_{col}) in the place of `lines_to_treat`.

In fact if we restrict our algorithm to work with the first four filling operations, the ordered sequences c and d are not necessary and so our algorithm runs in $O(N^2)$-time, that is better compared to the complexity $O(N^2 \log N)$ of Gebala's algorithm. Unfortunately, the additional fifth filling operation increases the time-complexity of our algorithm to $O(N^2 \log N)$.

3.6 Extension to Any Finite Set of Lattice Directions

Let $\mathcal{D}$ be a finite set of lattice directions, $\mathcal{M}$ be a class of lattice sets containing the line-convex sets w.r.t. $\mathcal{D}$. We suppose that f is a function as in Reconstruction$(\mathcal{M}, \mathcal{D})$. The size of f will be measured by $N = \max_{p \in \mathcal{D}}(\max(\{k : f(p,k) > 0\}) - \min(\{k : f(p,k) > 0\}) + 1)$.

The filling operations described above can be easily generalized to any set of directions:

- The procedures `put_in_alpha` and `remove_from_beta` must update the data for all the lines parallel to one of the directions of $\mathcal{D}$.
- The procedure `treat_line` is unchanged.
- The initial β_0 is always included in $\mathcal{G} = \{M \in \mathbb{Z}^2 \ : \ \forall p \in \mathcal{D} \ \min(\{k : f(p,k) > 0\}) \leq p(M) \leq \max(\{k : f(p,k) > 0\})\}$ which contains less than N^2 points.
- The time-complexity of the whole algorithm is still $O(N^2 \log N)$ as the procedures `put_in_alpha` and `remove_from_beta` are done at most two times on each point and each direction.

4 Consequence on the Reconstruction of Convex Sets

We now consider two special classes of lattice sets for which the new implementation of the filling operations improves the complexity of the algorithm solving the reconstruction problem.

4.1 Reconstruction of Q-Convex Sets

Let $p = bx - ay$ and $q = dx - cy$ define two lattice directions, and M be a point of $\mathbb{Z}^2$; the four quadrants around M are the four regions delimited by the lines of directions p and q and passing through M. More precisely;

$$\begin{aligned}
Z_0^{pq}(M) &= \{M' \in \mathbb{Z}^2 : p(M') \leq p(M) \text{ and } q(M') \leq q(M)\} , \\
Z_1^{pq}(M) &= \{M' \in \mathbb{Z}^2 : p(M') \geq p(M) \text{ and } q(M') \leq q(M)\} , \\
Z_2^{pq}(M) &= \{M' \in \mathbb{Z}^2 : p(M') \geq p(M) \text{ and } q(M') \geq q(M)\} , \\
Z_3^{pq}(M) &= \{M' \in \mathbb{Z}^2 : p(M') \leq p(M) \text{ and } q(M') \geq q(M)\} .
\end{aligned}$$

Definition 1. *A lattice set E is Q-*convex *w.r.t. $\mathcal{D} = \{p, q\}$ if $Z_k^{pq}(M) \cap E \neq \emptyset$ for all $k \in \{0, 1, 2, 3\}$ implies $M \in F$..*

Definition 2. *A lattice set is Q-convex w.r.t. a set $\mathcal{D}$ of directions if it is Q-convex w.r.t. every pair of directions included in $\mathcal{D}$.*

We denote the class of the Q-convex sets w.r.t. $\mathcal{D}$ by $\mathcal{Q}(\mathcal{D})$. In [10] it is proved that there is an algorithm for RECONSTRUCTION($\mathcal{Q}(\mathcal{D}), \mathcal{D}$) which runs in time $O(N^2(N^2 + F(N)))$, where $F(N)$ is the complexity of the filling operations. We can deduce:

Theorem 1. RECONSTRUCTION($\mathcal{Q}(\mathcal{D}), \mathcal{D}$) *can be solved in $O(N^4 \log N)$-time where $N = \max_{p \in \mathcal{D}}(\max(\{k : f(p, k) > 0\}) - \min(\{k : f(p, k) > 0\}) + 1)$.*

4.2 Reconstruction of Convex Lattice Sets

Definition 3. *A lattice set is convex if it is the intersection of a convex polygon and $\mathbb{Z}^2$. We denote the class of convex lattice sets by $\mathcal{C}$.*

If $(p_i)_{i=1...4}$ are four lattice directions determined by the vectors $(\boldsymbol{p}_i)_{i=1...4} = (a_i, b_i)_i$ and with the slopes $(\lambda_i)_i = (-b_i/a_i)_i$ then the cross-ratio of the four directions $(p_i)_{i=1...4}$ denoted by $[p_1, p_2, p_3, p_4]$ is the element of $\mathbb{R} \cup \{\infty\}$ defined by:

$$[p_1, p_2, p_3, p_4] = \frac{(\lambda_3 - \lambda_1)(\lambda_4 - \lambda_2)}{(\lambda_3 - \lambda_2)(\lambda_4 - \lambda_1)} .$$

The *ordered* cross-ratio of $(p_i)_{i=1..4}$ is $[p_{\sigma(1)}, p_{\sigma(2)}, p_{\sigma(3)}, p_{\sigma(4)}]$, where σ is the permutation such that $\lambda_{\sigma(i)} < \lambda_{\sigma(i+1)}$. The ordered cross-ratio of four lattice directions is always a rational number which is greater than 1.

It is known that if $\mathcal{D}$ is a set of directions containing four directions whose ordered cross-ratio is not in $\{4/3, 3/2, 2, 3, 4\}$, then the convex lattice sets and Q-convex lattice sets w.r.t $\mathcal{D}$ are uniquely determined by their projections along $\mathcal{D}$ (see [11,12]). From the same scheme as in [7,10] we can deduce:

Theorem 2. *If $\mathcal{D}$ is a set of directions containing four directions whose ordered cross-ratio is not in $\{4/3, 3/2, 2, 3, 4\}$, then* RECONSTRUCTION($\mathcal{C}, \mathcal{D}$) *can be solved in $O(N^4 \log N)$-time, where $N = \max_{p \in \mathcal{D}}(\max(\{k : f(p, k) > 0\}) - \min(\{k : f(p, k) > 0\}) + 1)$.*

5 The Filling Operations in Practice

In this paper we have proved that the five filling operations can be made until the invariance of α and β in $O(N^2 \log N)$ time. However if we do not apply the fifth filling operation $\odot'$ this complexity decreases to $O(N^2)$. Let us consider the following algorithm which does contain the fifth filling operation.

`filling_operations2`(α_0, β_0)

$\alpha \leftarrow \alpha_0$
$\beta \leftarrow \beta_0$
repeat
  Apply the four operations $\oplus$, $\ominus$, $\odot$, $\otimes$ to (α, β) until invariance of α and β
  Apply the operation $\odot'$ to (α, β)
until the last operation $\odot'$ leaves (α, β) invariant
return(α, β)

The time-complexity of this algorithm is $O(lN^2)$, where l is the number of iterations of the **repeat** loop. The only theoretical upper bound we have found for l is N^2. To have a better estimation of l we have conducted the following experiment:

- We have considered the set of directions $\mathcal{D} = \{h, v\}$ and the class of lattice sets $\mathcal{Q}(\mathcal{D})$.
- We have generated 10^6 sets $\mathcal{Q}(\mathcal{D})$ having a fixed sum of $m + n$ by the algorithm described in [13].
- For each set, we have computed its projections, the initial sets α_0, β_0 given by the algorithm described in [7], and then the algorithm `filling_operations2` is applied.

Table 1. The number of iterations in the algorithm `filling_operations2` applied to the reconstruction to Q-convex sets w.r.t. the horizontal and vertical directions

l \ $m+n$	10	30	50	70	90	110
1	996977	994865	996970	997909	998468	998764
2	3023	5134	3030	2091	1532	1236
3	0	1	0	0	0	0

Table 1 gives the frequencies of the number l of iterations. In this experiment we have always $l \leq 3$. So it seems reasonable to make the conjecture that l is bounded by a constant. With it, the time-complexity of `filling_operations2` is $O(N^2)$.

6 Conclusion and Perspectives

In this paper, we presented an implementation of the five filling-operations in $O(N^2 \log N)$-time, where N is the size of the projections. The new implementation permitted to reconstruct Q-convex sets in $O(N^4 \log(N))$-time from projections in the same directions as the ones used for Q-convexity. This represented an improvement of the previous fastest algorithm which run in $O(N^5)$-time.

The introduction of the fifth operation has permitted to reduce the complexity of the reconstruction because it allowed to fix two points instead of four. Additional considerations could perhaps induce a faster algorithm. In particular, the phase which fixes some points (bases) could be faster in the case of three directions and more, because in this case experiments show that these bases are very rarely needed (see [14, Annexe B]). This could lead to an algorithm with a complexity of $O(N^2)$-time, but at the moment we have only experimental hints.

Acknowledgments

This work was partially supported by the NSF Grant DMS 0306215, the OTKA Grant T048476 and the CNRS program IRMC.

References

1. Gebala, M.: The reconstruction of convex polyominoes from horizontal and vertical projections. In: Proc. of SOFSEM '98. Volume 1521 of LNCS. (1998) 350–359
2. Brunetti, S., Del Lungo, A., Del Ristoro, F., Kuba, A., Nivat, M.: Reconstruction of 4- and 8-connected convex discrete sets from row and column projections. Linear Algebra Appl. **339** (2001) 37–57
3. Ryser, H.J.: Combinatorial properties of matrices of zeroes and ones. Canad. J. Math. **9** (1957) 371–377
4. Kuba, A.: Reconstruction of two-directionally connected binary patterns from their two orthogonal projections. Comp. Vis. Graph. Image process. **27** (1984) 249–265
5. Woeginger, G.H.: The reconstruction of polyominoes from their horizontal and vertical projections. Inform. Process. Lett. **77**(5-6) (2001) 225–229
6. Barcucci, E., Del Lungo, A., Nivat, M., Pinzani, R.: Reconstructing convex polyominoes from horizontal and vertical projections. Theoret. Comput. Sci. **155** (1996) 321–347
7. Brunetti, S., Daurat, A.: An algorithm reconstructing convex lattice sets. Theoret. Comput. Sci. **304** (2003) 35–57
8. Brunetti, S., Daurat, A.: Reconstruction of discrete sets from two or more X-rays in any direction. In: Proc. of IWCIA 2000, Université de Caen (2000) 241–258
9. Knuth, D.E.: Balanced Trees (section 6.2.3). In: Sorting and Searching. Volume 3 of The Art of Computer Programming. Addison-Wesley (1998) 458–475
10. Brunetti, S., Daurat, A.: Reconstruction of Q-convex sets. In Herman, G.T., Kuba, A., eds.: Advances in Discrete Tomography and its Applications. Appl. Numer. Harmon. Anal. Birkhäuser (To Appear)
11. Gardner, R.J., Gritzmann, P.: Discrete tomography: Determination of finite sets by X-rays. Trans. Amer. Math. Soc. **349** (1997) 2271–2295
12. Daurat, A.: Determination of Q-convex sets by Xrays. Theoret. Comput. Sci. **332** (2005) 19–45
13. Brunetti, S., Daurat, A.: Random generation of Q-convex sets. Theoret. Comput. Sci. **347** (2005) 393–414
14. Daurat, A.: Convexité dans le plan discret. Application la tomographie. PhD thesis, LLAIC1, and LIAFA Université Paris 7 (2000)

Reconstruction Algorithm and Switching Graph for Two-Projection Tomography with Prohibited Subregion

Akira Kaneko[1,*] and Rina Nagahama[2]

[1] Department of Information Sciences, Ochanomizu University, 2-1-1, Otsuka, Bunkyo-ku, Tokyo, 112-8610, Japan
kanenko@is.ocha.ac.jp
http://www.is.ocha.ac.jp/~kaneko/
[2] Graduate School of Humanities and Sciences
Ochanomizu University
rinax@atom.is.ocha.ac.jp

Abstract. We consider the two-projection tomography problem, assuming a priori known prohibited region. We show that a modification of Ryser's reconstruction algorithm gives a solution. We then study the relation of the switching graph for the solution sets with and without the prohibited region. Finally, we apply our idea to get a better reconstruction figure imposing prohibited region artificially.

1 Introduction

We consider the reconstruction problem of a discrete plain figure F contained in a rectangle I from its two projections $f_y(x)$ and $f_x(y)$ along the y- and x-axis, respectively. This is equivalent to finding a binary matrix from its column- and row-sums, but we prefer the geometric notation better related with the continuous tomography. See [11] for general reference on this problem. In this report we assume that F is a priori known to have no building cell (that is, filled with 0 for the binary matrix formulation) in a subregion J of I, and consider the reconstruction problem with this constraint.

We show that under the assumption of uniqueness for J Ryser's algorithm for the reconstruction without constraint can be modified to obtain a solution of this problem. Then we study the structure of the solution set by means of a graph, extending our former work [7] for the full solution set without constraint.

Then we apply our idea to obtain a better solution of the reconstruction problem without constraint, by setting artificial constraint as a priori knowledge. This works faster than the strategy of successive improvement adopted so far. In the final section we try characterization for the prohibited region.

The reconstruction problem with rectangular constraint was considered by Brualdi and Dahl [3]. In [4] an equivalent result for unique figure J is announced

* Partially supported by Grant-in-Aid for Scientific Research No. 16540140.

A. Kuba, L.G. Nyúl, and K. Palágyi (Eds.): DGCI 2006, LNCS 4245, pp. 110–121, 2006.

without proof. We hope that our reconstruction algorithm is simple and practical. As further related works, Fulkerson [5] considered reconstruction of binary square matrices with zero diagonal, which is a typical non-unique figure. Also, Kuba [10] studied problem of reconstruction with prescribed 1's, which is intimately related with the present problem, but not equivalent (see Remark 1 (2) of §2). The work of Anstee [1] can also be understood of reconstruction with prescribed 1's. We thank the referee for informing us these references.

2 Setting of the Problem and Reconstruction Algorithm

Let I denote the rectangular region $a \le x < b$, $c \le y < d$ in the first quadrant of $\boldsymbol{R}^2$, where $a, b, c, d \in \boldsymbol{Z}$. Let F be a subregion of I which is the union of integer cells $C_{ij} := [i, i+1) \times [j, j+1)$. We shall denote its characteristic function also by F. Thus its y-projection $f_y(x)$, or x-projection $f_x(y)$ is defined as

$$f_y(x) = \int_c^d F(x,y)dy, \qquad f_x(y) = \int_a^b F(x,y)dx. \tag{1}$$

These take integer values, representing the number of cells in the respective columns or rows. The reconstruction problem from the two projections is to find F from the projection data $f = \{f_y(x), f_x(y)\}$. Define the arrangements by

$$f_{xy}(x) = \text{meas}\{y; f_x(y) \ge x\}, \quad f_{yx}(y) = \text{meas}\{x; f_y(x) \ge y\},$$
$$f_{yxy}(x) = \text{meas}\{y; f_{yx}(y) \ge x\},$$

where meas denotes the one-dimensional length. In the discrete case, this is equivalent (modulo measure 0) to the permutation of the columns or rows in decreasing order and finally view all from the x-axis. Then the consistency condition, that is, the condition for the existence of a solution, given by Lorentz, Gale and Ryser is

$$\forall x \quad \int_0^x f_{xy}(t)dt \ge \int_0^x f_{yxy}(t)dt, \qquad \text{and} \qquad \int_0^\infty f_{xy}(t)dt = \int_0^\infty f_{yxy}(t)dt,$$

The uniqueness of the solution is assured if and only if the equality holds for all x in the first inequality above. For further information about this problem see the survey article [11]. We here recall only necessary materials.

First we review Ryser's reconstruction algorithm in a form given in [6]. We shall call this hereafter the Ryser-Kaori algorithm without constraint.

1. Choose the tallest column from $f_y(x)$. In case of tie, choose the leftmost one.
2. Remove this column from $f_y(x)$ and at the same time, remove the same number of cells from $f_x(y)$ one for each row by the strategy of the longest row first.
3. Modify the graph of $f_x(y)$ crashing the removed cells toward the y-axis.
4. Return to 1 if there still remain cells in the projection data.

Now we have a subregion $J \subset I$ where we should not place any cell in reconstruction. In what follows we assume that J constitutes a unique figure, that is, there is no other figure having the same projection data. Note the following.

Lemma 1. *J is unique if and only if by a permutation of columns and rows it is brought to the form of Lorentz's renormalization of the projections, that is, to the form of union of height-decreasing adjacent subrectangles with the lower edge common to that of I (see Figure 1) :*

$$J = J_1 \cup J_2 \cup \cdots \cup J_r. \tag{2}$$

In fact, the sufficiency is obvious. For the necessity, we can obviously find a permutation bringing the y-projection $f_y(x)$ to the monotone decreasing form. Then by the assumption of uniqueness, the y-projection $f_{xy}(x)$of the x-projection $f_x(y)$ agrees with this. Since the y-projection is achieved by a permutation of rows, after these two permutations the y- and x-projection of J agree with this monotone figure. In view of the uniqueness, J itself has the same form.

Thus, we shall assume henceforth without loss of generality that J has the above form (2).

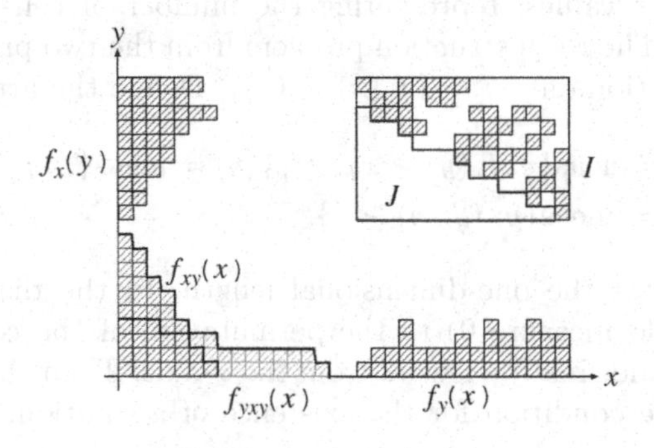

Fig. 1. Figure with prohibited region and its projections

Now we explain the reconstruction algorithm. We assume that the projection data are J-consistent, that is, there exists at least a solution with J as prohibited region.

1. For $i = 1$ to r do
2. Choose the tallest column from $f_y(x)$ among those above J_i. In case of tie, choose the leftmost one.
3. Remove this column from $f_y(x)$ and at the same time, remove the same number of cells from $f_x(y)$ one for each level by the strategy of the longest level first, but among the rows not touching J_i.
4. Modify the graph of $f_x(y)$ crashing the removed cells.
5. Return to 2 if there still remain cells above J_i.
6. End for.

Theorem 1. *The above algorithm successfully gives a reconstruction figure which does not contain any cell in J, provided that the projection data are J-consistent.*

Proof. We proceed by the induction of the total number of cells in the figure. The case of one cell is trivial. Assume that the assertion is true for any J and for any J-consistent projection data up to n cells, and consider a problem with $n+1$ cells. By the assumption of J-consistency, there exists a solution figure F which we may not know concretely. Following the above algorithm, we first choose the tallest column among those above J_1. For each cell in this column, we pick up a cell from $f_x(y)$ at the longest row not touching J_1. If the chosen cell exists in the figure F, we are correctly diminishing the data. If we chose a cell, say P, from a row where there was no cell in that column of the figure F, then there should exist a cell of F, say P', in the column of P which was not chosen by the reconstruction algorithm. On the other hand, P comes from a cell of F, say Q, in another column by x-projection, There are several candidates of such Q, but we claim that among them there is at least one such that P' and Q constitute a switching component in F, that is, the place R in Figure 2 is vacant. In fact, if all the counterparts in the row of P' are occupied by the cells of F, then the x-projection at the row of P' will have length greater than that at the row of P. This violates the rule of algorithm that we should pick up the cell from the longer rows of $f_x(y)$ first. This argument applies to all cells chosen in relation to this column. Thus after removing the column from $f_y(x)$ and the corresponding cells from $f_x(y)$ there remain projection data which come from a true figure F' obtained by several switching as mentioned above from F. This means that the remaining projection data are J-consistent, and by the induction hypothesis, we can obtain a solution of reconstruction with the constraint. By adding the first treated column to this solution, we obtain a solution for given size. □

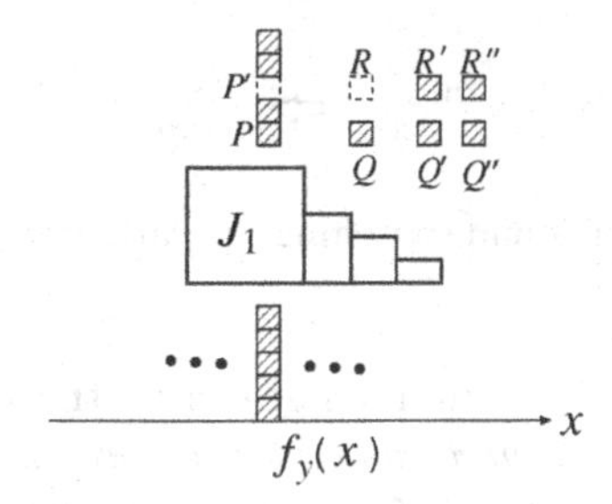

Fig. 2. Proof of justification of reconstruction algorithm

The converse is obvious: if our algorithm ends up using all the cells in the projection, we obtain a reconstruction with the given constraint J. Thus it presents a practical criterion for the J-consistency.

Remark 1. (1) The direct application of Ryser-Kaori algorithm, that is, processing from the tallest of all columns ignoring J, does not work. Figure 3 is such an example.

(2) One may think of an alternative algorithm such as filling the prohibited region by cells and applying Ryser-Kaori algorithm without constraint, then removing the cells in the prohibited region will give a desired solution. But in general it is not easy to obtain a reconstruction of which the prohibited region is filled with cells. See, however, §5 in this respect.

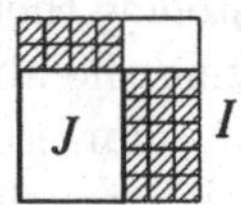

Fig. 3. Necessity of modification of reconstruction order

3 Switching Graph

It is well known that a figure is non-unique, namely, there is another figure with the same y- and x-projections, if and only if it contains a switching component. For a non-unique figure we can relate a graph to the solution set by considering each solution as a vertex, and connecting a pair of solutions by an edge if and only if they are transformed by one switching operation. This graph seems to have been first introduced by Brualdi [2] under the name of interchange graph. Later, [9] re-intruduced it and called Ryser graph. Ignoring these, we called it the switching graph and studied its properties with many examples. Further we gave a direction to each edge showing the type modification from type 2 to type 1 (see Figure 4), thus producing the switching digraph ([7]). Since we employ in the sequel permutation of columns and rows which may change the direction of engaged edges, we only consider the switching graph in this paper.

Fig. 4. Switching components and switching operation: type 2 (left) and type 1 (right)

We shall denote by G_J the switching graph for the solution set with constraint J, and simply call it the *J-constraint switching graph.* If distinction is preferable, we shall call the switching graph G of all the solutions without constraint the *full switching graph* and further add the projection data like $G[f]$ or $G_J[f]$. It is obvious from the definition that for any J (not necessarily unique) G_J becomes a full subgraph of G. Although G is known to be connected by Ryser's theorem, it is not obvious if G_J is connected, too. We shall first establish this.

Theorem 2. *Let J be a prohibited region which is a unique figure. Then the J-constraint switching graph G_J is a connected full subgraph of the full switching graph G.*

Proof. The prohibited region J is constructed step by step, by adding a cell each time from left and from bottom, so that

$$\emptyset = J_0 \subset J_1 \subset J_2 \subset \cdots \subset J_m = J,$$

where each J_k consists of k cells in the renormalized form, that is, consists of subrectangles with the bottom edge on the x-axis and with decreasing heights. The corresponding constraint switching graphs G_{J_k} constitute a decreasing sequence of full subgraphs

$$G = G_{J_0} \supset G_{J_1} \supset G_{J_2} \supset \cdots \supset G_{J_m} = G_J.$$

Assume that G_J is not connected. Since G is connected, there exists the minimal k such that G_j, $j = 0, 1, \ldots, k$ is connected but G_{k+1} is not. Note that $J_{k+1} = J_k \cup \{P\}$, for some cell P. Let A, B be two vertices of $G_{J_{k+1}}$ which are connected by a path γ in G_{J_k}, but never in $G_{J_{k+1}}$. We can assume that the other vertices of γ are not in $G_{J_{k+1}}$, and moreover, γ is the shortest among such paths. These vertices are obtained from A by several switching operations employing the cells of A and using the place or cell at P. Since the region to the left and below P are totally contained in J_k, the cells of A participating in the switching operations together with P must lie to the right and above P. Thus in observing what happens along the path γ we can restrict our consideration inside this rectangle. (There may be cells below P and to the right of J_k, or above J_k and to the left of P. But these do not mutually switch. Hence the argument below is not essentially affected by these cells.) This means that we can only consider the case where J consists of a single cell to the leftmost and the lowest place, which we shall denote by J by abbreviation of notation, just like the initial step of an induction argument.

Note also that since J is in the prohibited region, the figure A does not have a cell at J. By the same reason, the figures corresponding to the vertices of γ other than the endpoints should all have a cell at J. In fact, if there exists a vertex C in the midst, which does not contain J in its cell, then it is a due vertex of $G_{J_{k+1}}$, hence either of the subpaths $\overline{AC}$ or $\overline{CB}$ would be a path shorter than γ and connecting two vertices in G_{J_k} which are disjoint in $G_{J_{k+1}}$. This violates the choice of γ. Just by the same reason, vertices not adjacent to the endpoints of γ do not contain any cell which constitute a switching component with J.

Thus we assume hereafter that J is the left-lower corner cell of I and show a contradiction, assuming that γ is a minimal path connecting two vertices A, Z of G_1 in G. The first edge of γ corresponds to the switching of $P, Q \in A$ bringing P to the hole J, and Q to some vacant place Q' in $I \setminus J$, thus producing a new figure B, the second vertex of γ. First note that

(0) Z can never be the next vertex of B.

In fact, if so, J must switch with another cell R, as in Figure 5. But if position U is vacant in A, then we can execute this modification without using the position J, namely as a path in G_1, by the series of switchings Q-R, P-U. On the other hand, if U is occupied, then the same modification is also realized as a path in G_1 by the switchings P-U, Q-R. Thus the vertex C next to B should be in $G \setminus G_1$. The passage from B to C, or more generally, any inner edge of γ,

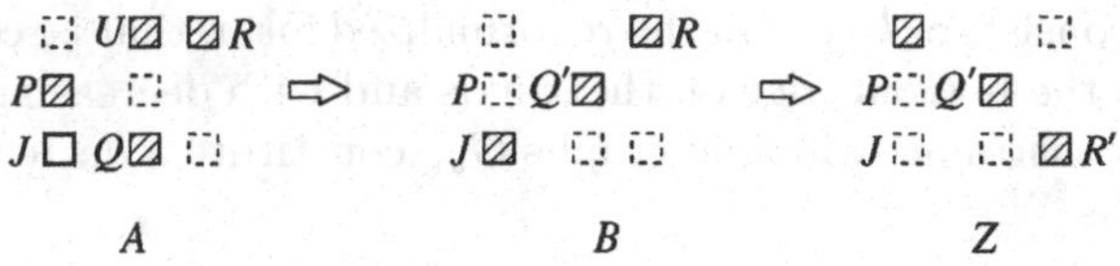

Fig. 5. Case where the length of γ is 2

never corresponds to the switching of cells independent of those touched before, because otherwise, that switching could be preprocessed before the first edge, thus shortening γ. Hence the second edge B to C corresponds to either of

(i) the switching of Q' with another cell $R \in I \setminus J$ (see Figure 6 upper),
(ii) the switching of a new pair R, S in $I \setminus J$ executed using the vacant place after P or Q moved.

Since the essence of the problem does not change by the reflection with respect to the diagonal passing through J, we can assume without loss of generality that it is the place of Q which is used in (ii), as in Figure 6 lower.

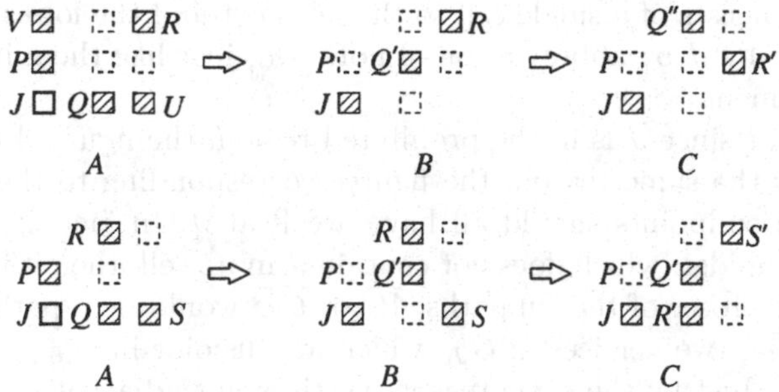

Fig. 6. Path A–B–C; upper: case (i), lower: case (ii) (only concerned cells are shown)

Let us consider case (i). Note that R can never be in the same column as P. In fact, if so, we may directly switch R with Q, obtaining the same figure C, thus shortening γ. Hence the above Figure 6 represents the general situation. Next note that the places denoted U, V in A are both occupied by the cells of A. In fact, assume e.g. that U is vacant. Then, we can preprocess the switching of Q and R, to Q'' and U. Then the above portion of γ is shortened to one edge corresponding to the switching of P and U. The same is true of V. Now we have to consider the next edge. Assume first that it ends at Z. Then a new cell S switches with J. In this last figure, however, we can see the switchable

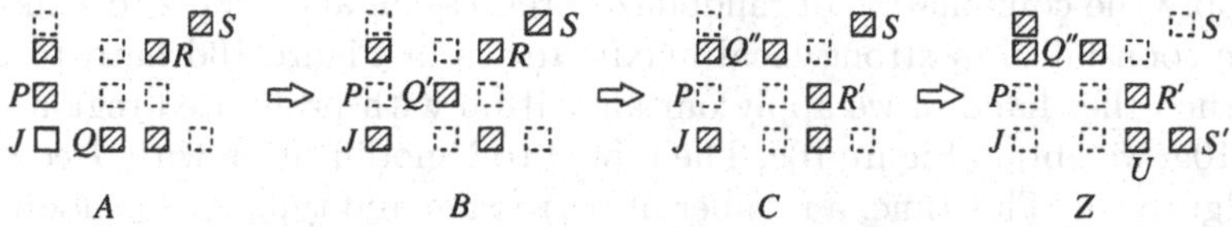

Fig. 7. Case (i) where γ = path A–B–C–Z

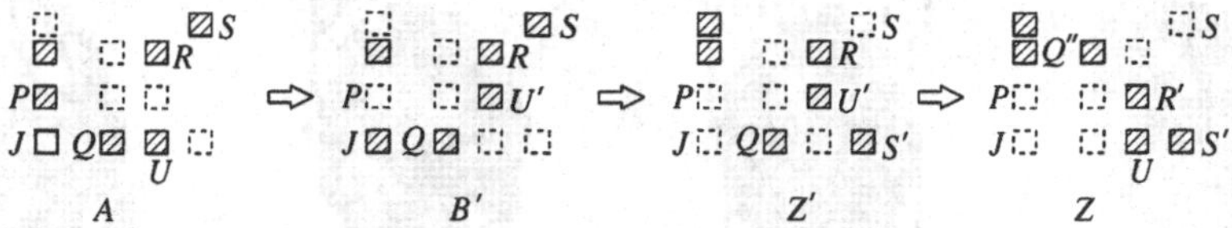

Fig. 8. Shortened path A–B'–Z'–Z

pair Q''-U. After switching this, we obtain another figure Z' contained in G_1, which can be connected with A by a shorter path as in Figure 8. This contradicts the minimality of γ. There is a case where S is in the same row as P, but the conclusion is the same.

Thus the above path continues to one more inner vertex A–B–C–D with $D \in G \setminus G_1$. We can show in this way that we can never reach the end vertex Z. See [8] for the detailed proof. The case (ii) can be discussed similarly. □

Remark 2. When the prohibited region J is not unique, the connectivity of G_J is not necessarily assured. Figure 9 presents such an example. There G_J consists of the two shadowed vertices to the right at the top and the bottom of G.

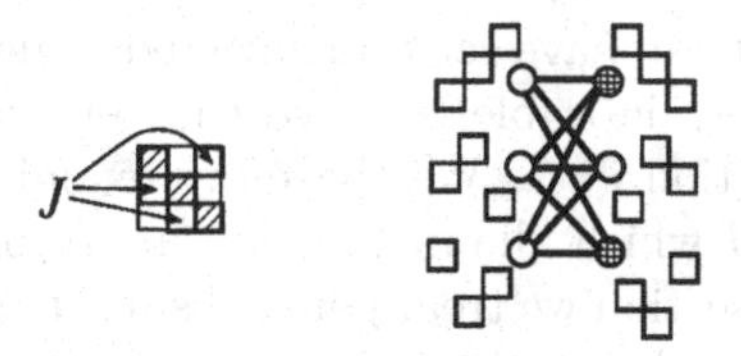

Fig. 9. Example of disconnected constraint switching graph

4 Experiments

We apply our construction to discretized slant ellipse. Figure 10a is the original figure. Figure 10b is the Ryser-Kaori reconstruction without constraint, and Figure 10c the type 2 to 1 modification. The value indicated in each figure denotes that of the standard weight function introduced in [6], which increases by the type modification:

$$w(F) := \sum_{C_{ij} \subset F} ij, \qquad \text{where } C_{ij} = [i, i+1) \times [j, j+1).$$

If we apply the combination of randomized regression and type 2 to 1 modification, we soon fall in a strong local maximum as in Figure 10d hard to get rid of. On the other hand, if we apply our algorithm with prohibited region J as in Figure 10g, we obtain Figure 10e. The type 2 to 1 modification with J-constraint gives Figure 10f. This time, a number of regression and type 2 to 1 modification with J-constraint easily regains the original figure as shown in 10g.

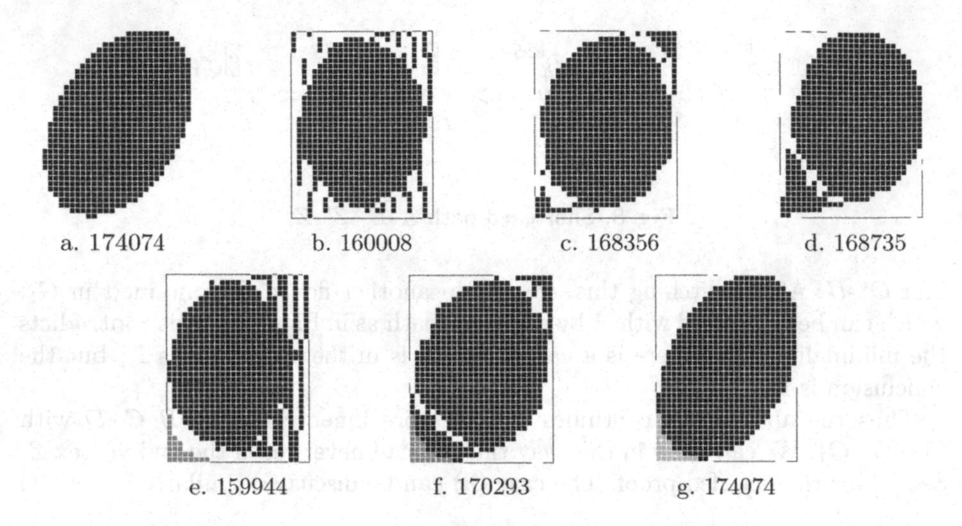

Fig. 10. Reconstruction of slant ellipse

5 Condition for Prohibited Region

In the above experiment, we have set a prohibited region from the known original figure. It is, however, desirable that we can set such a region only on the knowledge of the projection data. We therefore examine here a necessary and sufficient condition for J which allows at least one reconstruction for the given projection data. In this section we treat general subsets as prohibited region and do not necessarily require their uniqueness.

Lemma 2. *Let J be unique, and let $f = \{f_y(x), f_x(y)\}$, $f' = \{f'_y(x), f'_x(y)\}$ be two pairs of J-consistent projection data. Assume that*

$$\forall x \;\; f_y(x) \leq f'_y(x), \quad \forall y \;\; f_x(y) \leq f'_x(y), \quad \|f'_y - f_y\|_{L^1} = 1, \quad \|f'_x - f_x\|_{L^1} = 1,$$

that is, they are different only by one cell. Then there exist reconstruction F from f and F' from f', each with J-constraint, such that the Hamming distance of F, F' is equal to 1, that is, differing only by one cell.

Proof. We shall assume that the columns and rows are so arranged that J has the renormalized form. We proceed by induction on the size n of f. For $n = 0$, the assertion is trivial. Let it be true for any J and for any J-consistent projection data, where f has up to $n-1$ cells, and consider the case of n cells. We apply the modified Ryser-Kaori algorithm for reconstruction with J constraint. Let us take the tallest column from $f_y(x)$ in the range of J_1. By permutation of these columns, we can assume without loss of generality that this column is leftmost and that it is the tallest of $f'_y(x)$ among J_1, too. In fact, if this is not the case, it means that the cell was added to another column of $f_y(x)$ of the same height in the range of J_1. Then we could take this latter column from the beginning. Now process the lowest cell of this column by the modified Ryser-Kaori algorithm and stop. This should produce the same cell P outside J to the reconstruction figure. By the permutation of rows above J_1, we can assume without loss of generality that P is just the next cell above the leftmost column of J_1, hence $J \cup \{P\}$ is still a unique figure. Since this algorithm can be continued to finally produce a legitimate solution with J constraint for each projection data, the remaining projection data should be $J \cup \{P\}$-consistent, and have $n-1$, resp. n cells. Thus by the induction hypothesis, these should have at least one reconstruction F_0, F'_0 with $J \cup \{P\}$-constraint of Hamming distance $= 1$. Then $F = F_0 \cup \{P\}$, $F' = F'_0 \cup \{P\}$ will be a desired pair of reconstructions with J-constraint. □

Lemma 3. *Let $f = \{f_y(x), f_x(y)\}$ be a consistent pair of projection data. Then for a cell P, the projection data augmented by P, $f + \text{proj}(P)$, is consistent if and only if there exists a reconstruction F for f for which the place P is vacant.*

Proof. The sufficiency is obvious: if f admits a reconstruction F for which the place P is vacant, then, the augmented projection data will be those for a valid figure $F \cup \{P\}$. Let us prove the necessity. Assume that the augmented projection data is consistent. Then there are reconstructions for the augmented data. If there is one F' among them for which the place P is filled, then, $F = F' \setminus \{P\}$ will be a reconstruction for the original data with the place P vacant. Thus assume that the place P is always vacant in any reconstruction F' for the augmented projection data and nevertheless that P is always present in any reconstruction F for the original data. Then the Hamming distance of F and F' gains one at P, and at least 2 outside P, because $|F \setminus \{P\}| = n-1$ and $|F' \setminus \{P\}| = |F'| = n+1$. Thus it is ≥ 3, contradicting Lemma 2 (applied with $J = \emptyset$). □

Employing Lemma 3 repeatedly, we can add a cell at any place as well as the corresponding augmentation of the projection data keeps the consistency of Lorentz-Ryser. Thus we can finally reach a unique figure. But this does not imply that we can adopt the place of thus added cells as the constraint set. We have, however, the following criterion for prohibited region which is verifiable only from the projection data:

Theorem 3. *Let $f = \{f_y(x), f_x(y)\}$ be a consistent projection data. A region J (not necessarily unique) can be set as a prohibited region to the data f if the augmented projection data $f + \text{proj}(J)$ satisfies the consistency condition of Lorentz-Ryser, and if J is incrementally constructed cell by cell in such a way that each column is filled until it is maximal satisfying the consistency. Any (not necessarily unique) subset of a set constructed in this way is again an admissible* *prohibited region. Especially, we can construct a subregion J such that $I \setminus J$ is a reconstruction from f.*

Thus a subregion of such J consisting of rectangles in the renormalized form can serve for our discussion hitherto.

Proof. Since the order of adding cells is inessential, we can assume without loss of generality, that we add cells starting from the leftmost column and proceed upward, verifying the consistency cell by cell. In view of Lemma 3, the first cell can be added if and only if f admits a reconstruction with this cell vacant. Hence this cell certainly constitute a part of J. Assume that we proceed in the first column and successfully added $J_{0k} = \bigcup_{j=0}^{k} C_{0j}$, where $C_{0j} = [0,1) \times [j, j+1)$, having a reconstruction from $f + \mathrm{proj}(J_{0k})$ with these cells filled. The next cell $C_{0,k+1}$ can be added, again in view of Lemma 3, if there exists a reconstruction from $f + \mathrm{proj}(J_{0k})$ with this cell vacant. But this may have also vacant place among J_{0k}. We claim that nevertheless there is a reconstruction with J_{0k} filled and with $C_{0,k+1}$ vacant. This is not obvious, but can be proved by an elementary argument similar to that of Theorem 2. We omit the details. If on the other hand, $C_{0,k+1}$ cannot be added, it is filled for all reconstructions from $f + \mathrm{proj}(J_{0k})$, hence especially for those with J_{0k} filled. Thus we proceed to the next upper cell. Continuing this, we finally fill whole the first column with original and added cells. Therefore this column does not concern the problem hereafter, and the argument goes just in the same way with the next column. □

Though the cell-handling order can be arbitrary, a well arranged one is preferable to obtain a better figure. In Figure 11 we show two examples against the projection data of Figure 10a. In each figure, the black region shows J and the complement presents a reconstruction canonical in some sense.

Remark 3. We cannot omit the assumption of maximality of J in the above theorem. Actually we have a counter-example as in Figure 12. This shows in the

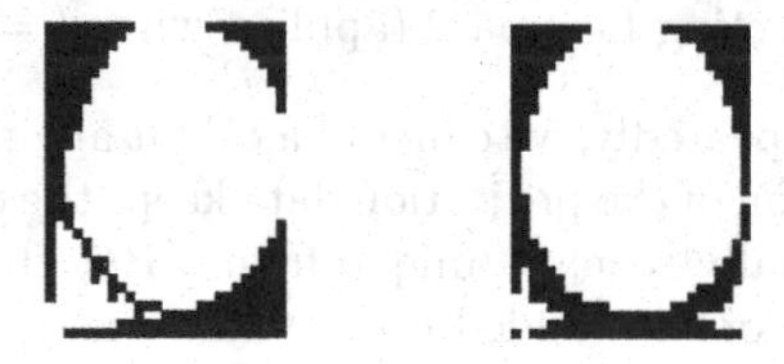

Fig. 11. Reconstructions of slant ellipse with maximal prohibited region; left: from leftmost column; right: from the highest column

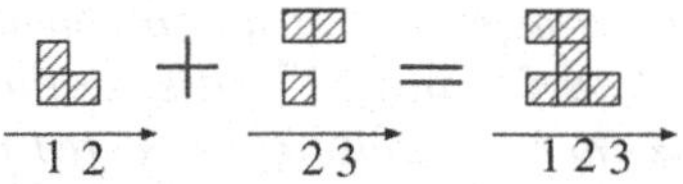

Fig. 12. Counter example without the assumption of maximality

same time that the assertion "if f, f' are two pairs of consistent projection data and $f \cup f'$ is also consistent, then there exist reconstructions F, F' of f, f' such that $F \cap F' = \emptyset$" is false even if all projections are unique.

In general, it is difficult to find the relation between switching graphs $G[f]$ and $G[f \cup \text{proj}(J)]$. We now see, however, that they are connected through the common connected full subgraph $G_J[f]$.

References

1. Anstee, R. P.: Properties of a class of $(0,1)$-matrices covering a given matrix. Canadian J. Math. **34** (1982) 438 – 453
2. Brualdi, R. A. : Matrices of zeros and ones with fixed row and column sum vectors. Linear Alg. and Appl. **33** (1980) 159 – 231
3. Brualdi, R. A., Dahl, G.:Matrices of zeros and ones with given line sums and a zero block. Linear Alg. and Appl. **371** (2003) 191 – 207
4. Dahl, G., Brualdi, R.A.: Matrices of zeros and ones with given line sums. Electronic Notes in Discrete Mathematics **20** (2005) 83–97
5. Fulkerson, D. R.: Zero-one matrices with zero trace. Pacific J. Math. **10** (1960) 831 - 836
6. Kaneko, A., Huang, L.: Reconstruction of plane figures from two projections. In: Herman, T. G., Kuba, A. (eds.): Discrete Tomography. Birkhauser (1999) 115–135
7. Kaneko A., Nagahama, R.: Switching graphs and digraphs associated with total reconstructed sets from two projection data. Nat. Sci. Report Ochanomizu Univ. **56**-2 (2005) 33–45
8. Kaneko A., Nagahama, R.: Reconstruction from Two Projections with Prohibited Subregion – Algorithm, Switching Graph and Consistency. Tech. Rep. Dep. Info. Sci. **06-2** May (2006)
9. Kong, T. Y., Herman, T. G.: Tomographic equivalence and switching operations. In: Herman, T. G., Kuba, A. (eds.): Discrete Tomography. Birkhauser (1999) 59 – 84
10. Kuba, A.: Reconstruction of unique binary matrices with prescribed elements. Acta Cybernetica **12** (1995) 57–70
11. Kuba, A., Herman, T. G.: Discrete tomography: a historical overview. In: Herman, T. G., Kuba, A. (eds.): Discrete Tomography. Birkhauser (1999) 3 – 34

A Geometry Driven Reconstruction Algorithm for the Mojette Transform

Nicolas Normand, Andrew Kingston, and Pierre Évenou

IRCCyN-IVC, École polytechnique de l'Université de Nantes,
Rue Christian Pauc, La Chantrerie,
44306 Nantes Cedex 3, France
{nicolas.normand, andrew.kingston, pierre.evenou}@univ-nantes.fr

Abstract. The Mojette transform is an entirely discrete form of the Radon transform developed in 1995. It is exactly invertible with both the forward and inverse transforms requiring only the addition operation. Over the last 10 years it has found many applications including image watermarking and encryption, tomographic reconstruction, robust data transmission and distributed data storage. This paper presents an elegant and efficient algorithm to directly apply the inverse Mojette transform. The method is derived from the inter-dependance of the "rational" projection vectors (p_i, q_i) which define the direction of projection over the parallel set of lines $b = p_i l - q_i k$. Projection values are acquired by summing the value of image pixels, $f(k, l)$, centered on these lines. The new inversion is up to 5 times faster than previously proposed methods and solves the redundancy issues of these methods.

1 Introduction

The Mojette transform is a form of Radon transform. It is an entirely discrete mapping which requires only the addition operation and is exactly invertible. It was first proposed by Guédon, Barba and Burger in 1995 [1] in the context of psychovisual image coding. It has since been applied in many aspects of image processing such as image analysis [2], image watermarking [3], image encrytion [4], image compression [5] and tomographic image reconstruction from projections [6,7]. The unique properties of the transform have also made it a useful multiple description tool with applications in robust data transmission [8] and distributed data storage [9]. A summary of the evolution and applications of the mojette transform entitled "The Mojette Transform: the First Ten Years" [10] was presented at the last DGCI conference.

Since the Mojette transform is pre-dominantly used as a tool, (e.g., for image analysis, to apply a watermark, for channel coding), the transform and inversion procedure should be as efficient as possible especially for real-time applications. This paper presents an inversion algorithm which uses a geometrical approach to streamline the reconstruction process.

Section 2 recalls the definition and some important properties of the Mojette transform as well as the methods for exact inversion utilised to date. Section 3

A. Kuba, L.G. Nyúl, and K. Palágyi (Eds.): DGCI 2006, LNCS 4245, pp. 122–133, 2006.

outlines the proposed geometry driven inversion method. Simple cases where q_i or p_i is constant for all I projections are presented in Sections 3.1 and 3.2. These results generalised for the inverse to apply to an arbitrary set of (p_i, q_i) in Sections 3.3 and 3.4. A comparison between this method and previously proposed reconstructions is presented in Section 4 followed by a conclusion in Section 5.

2 The Mojette Transform

2.1 The Forward Transform (Projection)

The linear integration of the discrete 2D function $f(k,l)$ is obtained via the Mojette transform over a set of I pre-defined rational angles, $\theta_i = \tan^{-1}(q_i/p_i)$. The pairs of integers defining the angles, (p_i, q_i) must be relatively prime, i.e., $\gcd(p_i, q_i) = 1$, and since linear integration is directionally independant, q_i is restricted to $\mathbb{Z}^+$ (except for the case $p_i = 1, q_i = 0$) to ensure $\theta_i \in [0, \pi[$. Assuming a Dirac pixel model the linear integrations become sums over the pixels centred on the lines $b = q_i k - p_i l$. The Mojette projection operator is defined as

$$\mathcal{M}\{f(k,l)\} = \mathrm{Proj}(p_i, q_i, b) = \sum_{k=0}^{P-1}\sum_{l=0}^{Q-1} f(k,l)\delta(b + p_i l - q_i k), \tag{1}$$

where $\delta(\eta)$ is the Kronecker function, i.e., $\delta(\eta) = 1$ if $\eta = 0$, otherwise $\delta(\eta) = 0$. An example of these projections is given in Fig. 1. The number of linesums called "bins" per projection, B, for a $P \times Q$ image is found as

$$B_i(P, Q, p_i, q_i) = (Q-1)|p_i| + (P-1)q_i + 1, \tag{2}$$

with $b \in [0, B_i - 1]$ for $p_i \leq 0$ and $b \in [-(Q-1)p_i, (P-1)q_i]$ otherwise. For a transform with I projections, unique inversion is possible provided the Katz criterion [11] is satisfied, i.e.,

$$P \leq \sum_{i=0}^{I-1} |p_i| \quad \text{or} \quad Q \leq \sum_{i=0}^{I-1} q_i. \tag{3}$$

This criterion was generalised by Normand, Guédon, Phillipé and Barba [12] for images of arbitrary shape. Their scheme generates the minimum sized ghost functions ,(i.e., functions that exist in the image but disappear in the projections, refer to [11] for more detail) as a sequence of 2D convolutions with all two pixel structuring elements formed from the set of projection slopes q_i/p_i by 1 at $(0,0)$ and -1 at (p_i, q_i). Any array which cannot contain the minimum ghost generated by the projection set therefore has an empty null-space and must have a unique inverse.

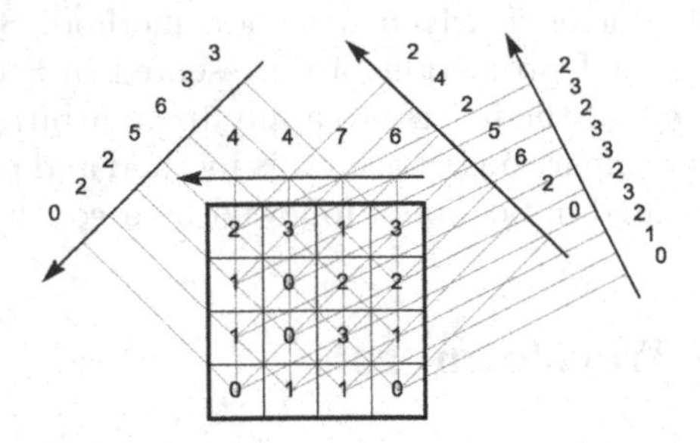

Fig. 1. Four projections of a 4×4 image $f(k,l)$, Proj$(-1,1,b)$, Proj$(0,1,b)$, Proj$(1,1,b)$ and Proj$(2,1,b)$

2.2 The Inverse Transform (Reconstruction)

When the Mojette transform was first proposed a recursive algebraic method was used for reconstruction. The following year a fast and more direct technique, requiring addition operations only was proposed by Normand, Guédon, Philippé and Barba [12]. It solves for one pixel at a time and subtracts this value from the bins that include this pixel in each of the I projections [12]. The reconstruction propagates from the image corners (where there is only one pixel value per bin) to the centre. The first step of the inversion for $f(k,l)$, as given in Fig. 1, is shown in Fig. 2a. For each of the PQ pixels there are $O(I)$ operations, so the complexity of this technique is $O(IPQ)$. If the number of projections, I, is chosen to be $\log(PQ)$, the Mojette transform has similar complexity to that of the fast Fourier transform [12].

There are two minor problems with this method. First, locating the bins in a projection which can be back projected (i.e., those bins for which only one pixel value remains unknown along its corresponding line of projection). Second, determining which one of the pixels, (k,l), in the line of projection, $b = q_i k - p_i l$, is yet to be reconstructed.

A simple method is utilised to overcome these problems. Two "comptabilité" (or accounting) images are projected with the same projection sets and reconstructed simultaneously with the unknown image. The first of these is a *unitary image*, i.e., an image where $f(k,l) = 1$ for all pixels. The second is an *index image* which labels the pixels according to a raster scan, i.e., $f(k,l) = k + lP$. The projections of these images assist with the respective problems above. This inversion technique will be referred to as the Comptabilité Mojette Inversion (CMI) method.

In recent years, two reconstruction methods involving back-projection have been proposed in the context of applying the Mojette transform to reconstruct medical images from continuous projections. The first of these is an exact method which was discovered by Servières, Normand, Guédon and Bizais [7]. Given all I possible projections in the $P \times Q$ array, back-projection ($\mathcal{M}^*$) yields $I-1$ times the original pixel value plus the sum of the image, f_{sum} (which can be found as

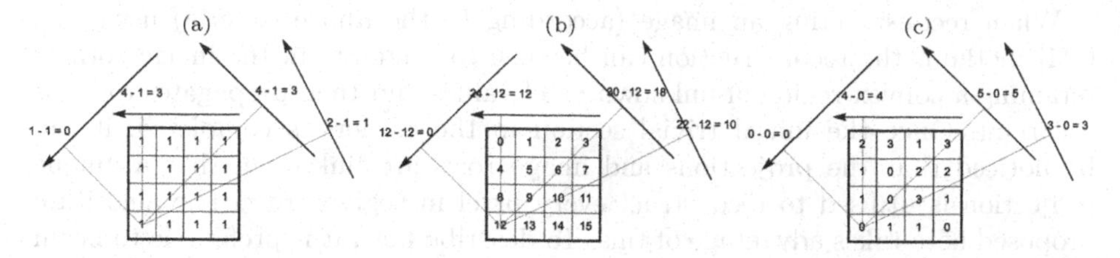

Fig. 2. Reconstructon via the CMI method. (a) A candidate bin is selected in the projections of the *unitary image*. (b) The value in the corresponding projection bin of the *index image* gives the pixel to be reconstructed. (c) The value in the corresponding projection bin of the image gives the pixel value. The projections of all 3 images are then updated simultaneously.

$\sum_b \text{Proj}(p_i, q_i, b)$ for any projection), i.e.,

$$\begin{aligned}\mathcal{M}^*\{\text{Proj}(p_i, q_i, b)\} = \tilde{f}(k', l') &= \sum_{i=0}^{I-1} \text{Proj}(p_i, q_i, q_i k - p_i l) \\ &= \sum_{i=0}^{I-1} f(k, l)\delta(q_i(k - k') - p_i(l - l')) \\ &= (I - 1)f(k', l') + f_{\text{sum}}.\end{aligned} \tag{4}$$

The set of projections can be found from (p_i, q_i) being all the points visible from the origin, i.e., Farey points, of the $P \times Q$ array and all symmetries, $(-p_i, q_i)$. Assuming a uniform density of Farey points in the plane, approximately $I = 12PQ/\pi^2$ projections are required.

The second back-projection technique uses the conjugate gradient method [13] to minimise $||\mathcal{M}^*b - \mathcal{M}^*\mathcal{M}\tilde{f}||^2$ where $\tilde{f}$ is the reconstructed image. Both of these inversions are relatively stable in the presence of noise and therefore ideal in this context. The exact back-projection however requires a very large number of projections and the conjugate gradient method, while it does give the inverse, is unnecessary in the case of reconstruction from uncorrupted discrete projections.

Since the Mojette transform is often used as a codec in data transmission, the most efficient inversion possible is required for real time applications. The following section outlines a very efficient inversion method which is similar to the CMI method but determines the inter-dependance of projections using graph theory to remove the accounting problems.

3 A Geometry Driven Reconstruction

For this method of reconstruction, it is assumed that $\sum_{i=0}^{I-1} q_i = Q$. Any redundant projections are ignored. The reconstruction is performed from left to right, (reconstruction from right to left, top to bottom and bottom to top are symmetries of this method). These two properties imply this algorithm can reconstruct images of infinite size, "on the fly", only the image height Q must be finite, P is not restricted.

When reconstructing an image (according to the above criteria) using the CMI method, the reconstruction can be seen to originate in the image corners retaining a convex region of unknown pixel values and then propagates towards the right. Once the initial trivial section in the corners is completed, it can be noticed that the projections and image rows are linked, in that a unique projection is utilised to reconstruct every pixel in a given row. The algorithm proposed here takes advantage of this. To describe how, it is preferable to begin with a simplified case where all projections have a common value for q_i of 1.

3.1 The Case Where $q_i = 1$ for $i \in \mathbb{Z}_Q$

This case commonly arises when the Mojette transform is utilised for multiple description coding in packet data transmission [8]. If the set of projections are sorted by p_i, (i.e., $p_0 < p_1 < \ldots$), and reconstruction is performed from left to right, then row r of the image, $f(k, r)$ for $k \in \mathbb{Z}_P$, is reconstructed by the Mojette projection, $\mathrm{Proj}(p_r, 1, b)$.

Proof. Assume on the contrary that projection, $\mathrm{Proj}(p_{Q-1}, 1, b)$, is used to reconstruct the pixel value $f(k, l)$ on a row other than $Q - 1$, i.e., $0 \leq l < Q - 1$. This implies that the pixel value $f(k+p_{Q-1}, l+1)$ has already been reconstructed by some projection other than $\mathrm{Proj}(p_{Q-1}, 1, b)$, say $\mathrm{Proj}(p_r, 1, b)$. Thus the pixel value $f(k + p_{Q-1} - p_r, l)$ must have been reconstructed and since p_{Q-1} is the largest in the set of p_i, then $k + p_{Q-1} - p_r > k$ and this pixel is further right than $f(k, l)$. However, it can not be known if reconstructing from left to right; A contradiction.

Therefore, only row $Q - 1$ can be reconstructed by $\mathrm{Proj}(p_{Q-1}, 1, b)$. Since reconstruction requires that only one pixel remains unknown in the line of projection, $\mathrm{Proj}(p_{Q-1}, 1, b)$ can not be used to reconstruct any other row. Therefore, this proof can be repeated to show $\mathrm{Proj}(p_{Q-2}, 1, b)$ must reconstruct row $Q - 2$ and so on, down to $\mathrm{Proj}(p_0, 1, b)$.

Intuitively this can be seen as ordering the projections by the slope (or angle) of their corresponding line of projection. This gives a convex hull to the reconstruction region that ensures the lines of back-projection can cut the hull such that only one pixel on the line lies within the hull; The condition necessary for reconstruction.

Since each projection corresponds to one row of the image, a dependancy graph can be constructed to show the relationship between the projections in reconstruction. The dependancy graph for the example image of Fig. 1 is given in Fig. 3a. Here vertices correspond to pixels and directed edges represent the dependancies of each pixel on other pixels being reconstructed in the inversion process.

Two simple paths can be found to traverse the graph (as shown in Fig. 3b-i and b-iv for the example). The reconstruction process involves beginning to the left of the image, so that only the rightmost of the vertices in the path intersect image pixels in column zero, and reconstructing pixel values according to the path. The distance the path is initially shifted is referred to as the *offset*.

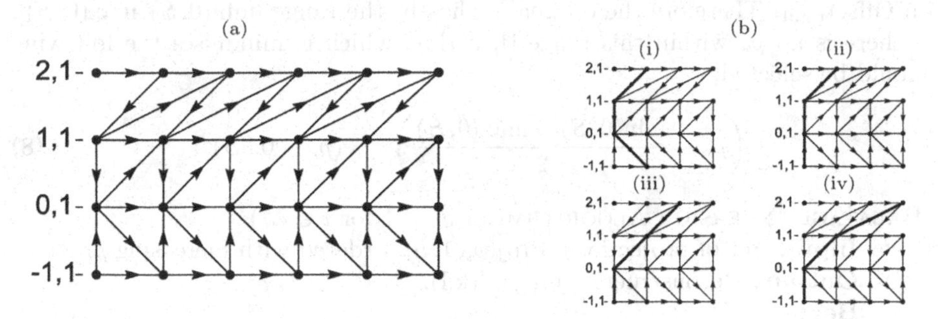

Fig. 3. (a) The dependancy graph for the example image. (b) The 4 possible reconstruction paths.

The path is then shifted right 1 pixel and the entire process repeated until the last pixel value in column $P-1$ of image is reconstructed. This process will be referred to as the Balayage (or sweeping) Mojette Inversion (BMI) method.

The path through the graph termed a *reconstruction path* does not necessarily have to traverse from one side to the other. Two seperate paths originating from opposite sides of the graph can terminate at a common vertex in the graph (some examples are shown in Fig. 3b-ii and b-iii). To optimise the reconstruction algorithm, it is desired to find the most compact path possible.

The offset for two paths in the graph terminating on row r due to all projection vectors with negative gradient is found as

$$\begin{aligned}\text{Offset}^-(r) &= \sum_{i=1}^{r} \max(0,-p_i) + \sum_{i=r}^{Q-2} \max(0,-p_i) \\ &= \max(0,-p_r) + \sum_{i=1}^{Q-2} \max(0,-p_i) = \max(0,-p_r) + S^-,\end{aligned} \tag{5}$$

where $S^- = \sum_{i=1}^{Q-2} \max(0,-p_i)$. Similarly the offset due to all projection vectors with positive gradient is found as:

$$\begin{aligned}\text{Offset}^+(r) &= \sum_{i=1}^{r} \max(0,p_i) + \sum_{i=r}^{Q-2} \max(0,p_i) \\ &= \max(0,p_r) + \sum_{i=1}^{Q-2} \max(0,p_i) = \max(0,p_r) + S^+,\end{aligned} \tag{6}$$

where $S^+ = \sum_{i=1}^{Q-2} \max(0,p_i)$. The width of the reconstructed path is determined by the maximum of these two offsets. The objective is therefore to find an r which minimises

$$\text{Offset}_{\text{total}} = \max(\text{Offset}^-(r), \text{Offset}^+(r)). \tag{7}$$

Let $S = S^- - S^+ = \sum_{i=1}^{Q-2} -p_i$. Note that if $S^- > S^+$ then any $p_r \in [0,S]$ has no effect on $\text{Offset}_{\text{total}}$. Similarly, if $S^- < S^+$ then any $p_r \in [S,0]$ has no effect on $\text{Offset}_{\text{total}}$. Therefore the optimal p_r lies in the range $[\min(0,S),\max(0,S)]$. If there is no p_r within this range then that which is minimises the following should be selected;

$$\left(p_r - \frac{\min(0,S)+\max(0,S)}{2}\right)^2 = (p_r - 0.5S)^2. \quad (8)$$

Balayage Inversion Algorithm(for $q_i = 1$ for $i \in \mathbb{Z}_I$)

▷ **Input:** Set of projections, $\text{Proj}(p_i, 1, b)$, ordered with increasing p_i
▷ **Output:** Reconstructed image, f(k,l).

```
     Begin
 2     ▷ Compute S⁻, S⁺ and S
 2     S_minus ← S_plus ← 0
 3     for i ← 1 to Q − 2 do
 4     |   S_minus ← S_minus + max(0, −p_i)
 5     |   S_plus ← S_plus + max(0, p_i)
 6     S ← S_minus − S_plus
       ▷ Determine the rendezvous row r
 7     temp ← (p_0 − 0.5S)²
 8     r ← 0
 9     for i ← 1 to Q − 1 do
10     |   if (p_i − 0.5S)² < temp then
11     |   |   temp ← (p_i − 0.5S)²
12     |   |   r ← i
       ▷ Determine the initial image column offset for each projection
13     offset(r) ← max(max(0, −p_r) + S_minus, max(0, p_r) + S_plus)
14     for i ← r + 1 to Q − 1 do
15     |   offset(i) ← offset(i − 1) + p_{i−1}
16     for i ← r − 1 downto 0 do
17     |   offset(i) ← offset(i + 1) + p_{i+1}
       ▷ Begin reconstructing image, f(k, l), at column k = −offset(r)
18     for k ← −offset(r) to P − 1 do
19     |   for l ← 0 to r − 1 do
20     |   |   f(k, l) ← Proj_l(k − p_k l)
21     |   |   for i ← 0 to Q − 1 do
22     |   |   |   Proj_i(k − p_i l) ← Proj_i(k − p_i l) − f(k, l)
23     |   for l ← Q − 1 downto r do
24     |   |   k' ← k + offset(r) − offset(l)
25     |   |   f(k', l) ← Proj_l(k' − p_k l)
26     |   |   for i ← 0 to Q − 1 do
27     |   |   |   Proj_i(k' − p_i l) ← Proj_i(k' − p_i l) − f(k', l)
28   End
```

This reconstruction procedure can be trivially generalised to the case where all $q_i = m$ for $m \in \mathbb{Z}^+$. In this instance each projection $\text{Proj}(p_i, m, b)$ reconstructs m consecutive rows of the image. The reconstruction paths are simlar to that for the above case with $q_i = 1$ but with m passes shifted down a row each time.

An example directed graph and reconstruction path for $m = 3$ is presented in Fig. 4a. Another simple case for the reconstruction occurs when p_i is constant for all projections as is discussed in the next section.

3.2 The Case Where $p_i = m$ for $i \in \mathbb{Z}_I$

This case where $m = 1$ is the most common type of angle set used for transforming images with minimal redundancy as described in [12]. Since p_i is constant, for a proof similar to that given in section 3.1 to apply, the projections must be sorted in order of decreasing q_i, (i.e., $q_0 > q_1 > \ldots$), then the rth set of q_i rows of the image, i.e., from row $R + 1$ up to row $R + q_r$ where $R = \sum_{i=0}^{r-1} q_i$, are reconstructed by projection $\text{Proj}(m, q_r, b)$.

Since all $q_i > 0$ the set of reconstruction paths all have the same total offset of $(I - 1)m$ as shown for the example in Fig. 4b. The reconstruction is similar to that for constant q_i, in that it requires multiple passes ($q_{\max}$ in this case), however the number of vertices included in each subsequent pass decreases as shown for the example.

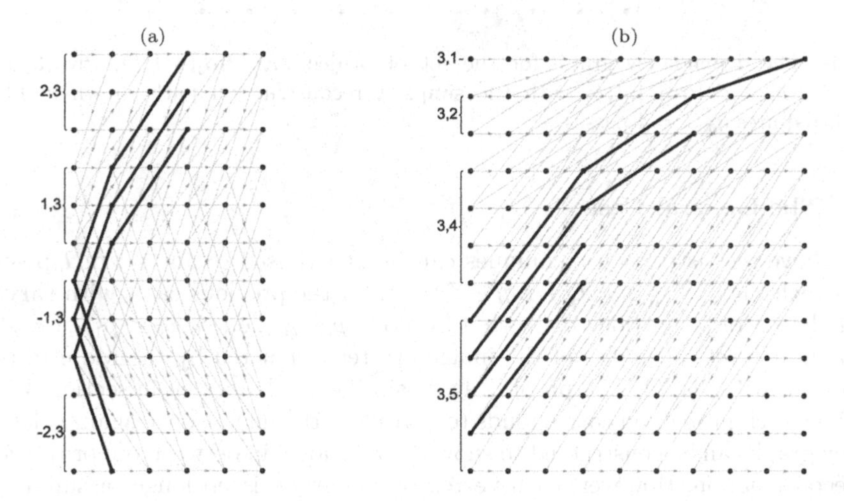

Fig. 4. The dependancy graphs (grey) and simplest reconstruction paths (black) for the set of projections (a) $\text{Proj}(\text{Proj}(-2, 3, b)$, $\text{Proj}(-1, 3, b)$, $\text{Proj}(1, 3, b)$ and $\text{Proj}(2, 3, b)$ which requires 3 passes and (b) $\text{Proj}(3, 1, b)$, $\text{Proj}(3, 2, b)$, $\text{Proj}(3, 4, b)$ and $\text{Proj}(3, 5, b)$ which requires $q_{\max} = 5$ passes

3.3 The Case Where $p_i \geq p_{i-1}$ and $q_i \leq q_{i-1}$

The constant q_i and constant p_i cases from Sections 3.1 and 3.2 can be amalgamated if, when ordered by slope, the projections have the property that p_i is increasing and q_i is decreasing, i.e., $p_i \geq p_{i-1}$ and $q_i \leq q_{i-1}$ for $(0 < i < I)$. For this case the reconstruction path is straightforward, similar to that for constant

p_i path with $q_{\max}$ passes. The paths through the region with $p_i \geq 0$ can be constructed independantly to those in the region with $p_i \leq 0$ similar to the case with constant q_i. An example has been presented in Fig. 5.

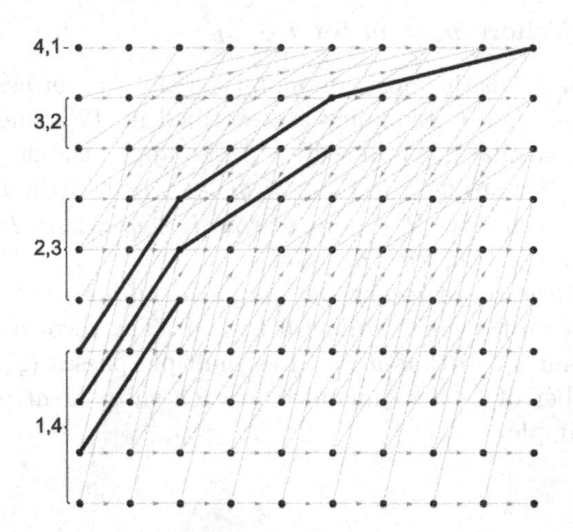

Fig. 5. The dependancy graph for the set of projections Proj$(4, 1, b)$, Proj$(3, 2, b)$, Proj$(3, 2, b)$ and Proj$(4, 1, b)$ with the simplest reconstruction path shown in black which requires $q_{\max} = 4$ passes

3.4 The General Case

The above reconstruction techniques can be generalised to any set of I projections such that $\sum_{i=0}^{I-1} q_i = Q$ for $q_i \in \mathbb{Z}^+$. As for the previous cases with varying q_i, if the set of projections are sorted by slope p_i/q_i, (i.e., $p_0/q_0 < p_1/q_0 < \ldots$), then the rth set of q_i rows of the image are reconstructed by the Mojette projection, Proj(p_r, q_r, b). The proof is again similar to that given in section 3.1.

Since each projection corresponds to q_i rows of the image, once again a dependancy graph can be constructed to show the relationship between the projections in reconstruction. However, each vertex of the graph is no longer assured of 2 originating and 2 terminating directed edges. There may only be a single terminating edge and there may be zero or many originating edges depending on the set of (p_i, q_i) used to define the projections.

To ensure a reconstruction path in this instance, only the pixels located immediately inside the edge of the convex hull created by the lines of projection ordered by slope are considered. As for the constant q_i case, the paths that terminate at the rows with minimum slope are used to generate the most compact convex hull. An example of the selection process is given in Fig. 6a with the dashed line giving the complex hull. The directed graph is then used to determine the reconstruction paths required for these selected vertices as shown in Fig 6b for the example.

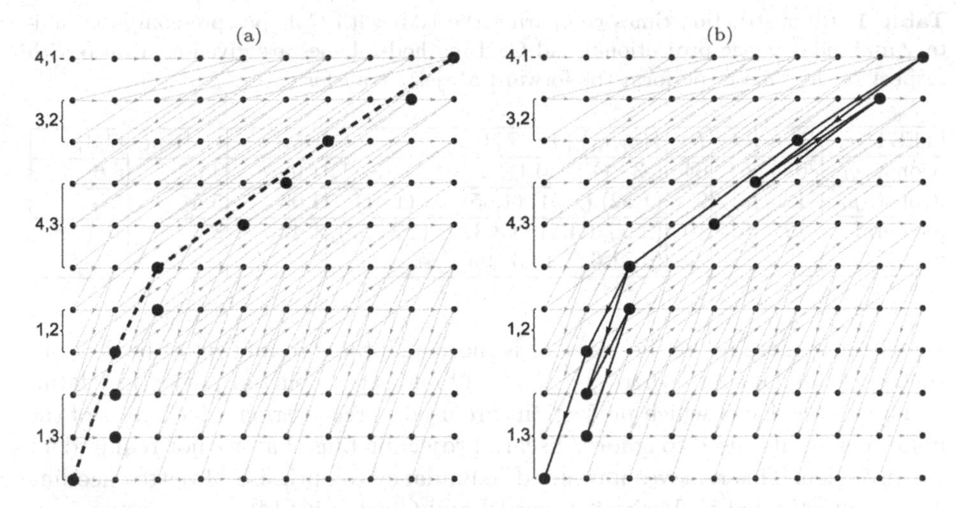

Fig. 6. Determining the vertices of the dependancy graph to be considered for the set of projections (a) Proj(Proj(4, 1, b), Proj(3, 2, b), Proj(4, 3, b),Proj(1, 2, b) and Proj(1, 3, b) (b) A reconstruction path from the directed graph to reconstruct these vertices

4 Discussion

Although very similar in nature, the balayage inversion algorithm can be shown to be up to 5 times faster than the comptabilité inversion algorithm. Assuming the pixel count and pixel label projections have not been pre-computed (which is possible in the case of image transmission where the incoming array size is known), these must both be determined in $2\times O(PQI)$ operations and during the inversion process, I projection value bins, pixel count bins and pixel label bins must be updated for each of the $P \times Q$ pixels ($3 \times O(PQI)$ operations), giving a total of approximately $5\times O(PQI)$ operations. In contrast the balayage algorithm requires updating I projection bins for each image pixel once in a single pass across the image in $O((P + \text{Offset})QI)$ operations. This has been demonstrated for three types of angle sets in Table 1 comparing the computation times for the BMI, the CMI with pre-computed unitary and index image projections (CMI-pc) and the complete CMI. In implementation the actual gain in efficiency can be up to an order of magnitude since there is a periodic pattern to the BMI process that can be exploited while this is not the case for the CMI method where the position of the next pixel to be reconstructed is not predetermined at all.

Since the proposed method is based on the inter-dependancy of each projection, the algorithm removes the need to search through the projections to find the next candidate bin that can be back-projected. As a pixel is reconstructed the predetermined dependancy graph indicates the pixel that can be reconstructed next and by what projection. This is highlighted by the relative performance of the CMI method for the General case of Table 1 where the projections have a large number of possible reconstruction bins to manage. The CMI

Table 1. Reconstruction times comparing the BMI with CMI-pc (pre-computed unitary and index image projections) and CMI methods. Times are given as a ratio with respect to the time to perform the forward Mojette transform.

TEST	P	Q	I	$\{(p_i, q_i) \mid i \in \mathbb{Z}_I\}$	BMI	CMI-pc	CMI
Const. q_i	4096	64	64	(-31,1), (-30,1), ... (31,1), (32,1)	1.08	4.6	6.6
Const. p_i	512	512	9	(1,52),(1,54),(1,55), ... (1,61)	1.05	5.5	7.5
General	512	512	10	(±95,31), (±63,32), (±31,32), (±31,64), (±31,96)	1.42	8.6	10.6

algorithm is more robust however, it is more adaptable to any set of projections such as redundant sets where $\sum_{i=0}^{I-1} q_i > Q$ and sets of partial projection data.

The knowledge of which projections are used to reconstruct which rows of the image can be also used to ignore/discard projection bins that are not required for the inversion. This removes unwanted redundancy to optimise Mojette encoding as was investigated by Verbert, Ricordel and Guédon in [14].

5 Conclusion and Future Work

A new inversion algorithm for the Mojette transform has been presented which takes advantage of the knowledge of the interdependancy of projections in reconstruction. The method is more direct and more efficient than previous methods however is less robust in terms of adaptability to any set of projections. This method of reconstruction also automatically enables optimal encoding by the Mojette transform by identifying which projection bins are required for inversion. Developing a BMI method that can be applied to reconstruct a redundant set of projections and can adapt to sets of partial projections as well as determining optimal coding incorporating redundancy are topics of future research.

Acknowledgements

AK holds a postdoctoral position at l'Université de Nantes supported by a grant from the Région Pays de la Loire, France.

References

1. Guédon, J., Barba, D., Burger, N.: Psychovisual image coding via an exact discrete Radon transform. In Wu, L., ed.: Proc. Visual Communications and Image Processing 1995 (VCIP95), Taipei, Taiwan, CORESA (1995) 562–572
2. Guédon, J., Normand, N.: The Mojette transform: applications in image analysis and coding. The International Society for Optical Engineering **3024**(2) (1997) 873–884

3. Autrusseau, F., Guédon, J.: Image watermarking for copyright protection and data hiding via the Mojette transform. In Sonka, M., Fitzpatrick, J., eds.: Proc. SPIE; Security and Watermarking of Multimedia Contents IV. Volume 4675., San Jose, CA (2002) 378–386
4. Autrusseau, F., Guédon, J., Bizais, Y.: Mojette cryptomarking scheme for medical images. In Sonka, M., Fitzpatrick, J., eds.: Proc. SPIE; Medical Imaging 2003: Image Processing. Volume 5032. (2003) 958–965
5. Autrusseau, F., Parrein, B., Servières, M.: Lossless compression based on a discrete and exact Radon transform: A preliminary study. In: Proc. IEEE International Conference on Acoustics, Speech and Signal Processing 2006 (ICASSP), Toulouse, France (2006)
6. Servières, M., Guédon, J., Normand, N.: A discrete tomography approach to PET reconstruction. In: Proc. 7th International Conference on Fully 3D Reconstruction in Radiology and Nuclear Medicine, Saint-Malo, France (2003)
7. Servières, M., Normand, N., Guédon, J., Bizais, Y.: The Mojette transform: Discrete angles for tomography. In Herman, G., Kuba, A., eds.: Proc. of Workshop on Discrete Tomography and its Applications, New York City, Elsevier Science Publishers (2005)
8. Parrein, B., Guédon, J., Normand, N.: Multimedia forward error correcting codes for wireless LAN. Annals of Telecommunications **3**(4) (2003) 448–463
9. Guédon, J., Parrein, B., Normand, N.: Internet distributed image information system. Integrated Computer-Aided Engineering **8** (2001) 205–214
10. Guédon, J., Normand, N.: The Mojette transform: the first ten years. In Andres, E., Damiand, G., Lienhardt, P., eds.: Proc. 12th International Conference on Discrete Geometry for Computer Imagery. Volume LNCS3429., Poitiers, France, Springer-Verlag (2005) 79–91
11. Katz, M.: Questions of uniqueness and resolution in reconstruction from projections. Springer Verlag (1977)
12. Normand, N., Guédon, J., Philippé, O., Barba, D.: Controlled redundancy for image coding and high-speed transmission. In Ansari, R., Smith, M., eds.: Proc. SPIE Visual Communications and Image Processing 1996. Volume 2727., SPIE (1996) 1070–81
13. Servières, M., Idier, J., Normand, N., Guédon, J.: Conjugate gradient Mojette reconstruction. In Fitzpatrick, J., Reinhardt, J., eds.: Proc. SPIE Medical Imaging 2005: Image Processing. Volume 5747. (2005) 2067–74
14. Verbert, P., Ricordel, V., Guédon, J.: Analysis of the Mojette transform projections for an efficient coding. In: Proc. International Workshop on Image Analysis for Multimedia Interactive Services (WIAMIS), Barcelona, Spain (2004)

Quantised Angular Momentum Vectors and Projection Angle Distributions for Discrete Radon Transformations

Imants Svalbe[1], Shekhar Chandra[1],
Andrew Kingston[2], and Jean-Pierre Guédon[2]

[1] School of Physics, Monash University, Australia 3800
[2] IRCCyN-IVC, École polytechnique de l'Université de Nantes, France

Abstract. A quantum mechanics based method is presented to generate sets of digital angles that may be well suited to describe projections on discrete grids. The resulting angle sets are an alternative to those derived using the Farey fractions from number theory. The Farey angles arise naturally through the definitions of the Mojette and Finite Radon Transforms. Often a subset of the Farey angles needs to be selected when reconstructing images from a limited number of views. The digital angles that result from the quantisation of angular momentum (QAM) vectors may provide an alternative way to select angle subsets. This paper seeks first to identify the important properties of digital angles sets and second to demonstrate that the QAM vectors are indeed a candidate set that fulfils these requirements. Of particular note is the rare occurrence of degeneracy in the QAM angles, particularly for the half-integral angular momenta angle sets.

Keywords: Discrete projection, tomography, digital angles, finite Radon transforms.

1 Introduction

The ultimate quality with which digital images can be reconstructed from projected views is highly sensitive to the selection of the viewing angles [1, 2, 3]. Conventional CT view angles, Figure 1(a), are constrained by the configuration of the x-ray source and detectors. In contrast, digital image angles are constrained only by pixellation of the discrete array on which the image is to be reconstructed, Figure 1(b). Simply dividing an angle interval into equal or integral steps does not provide descriptive digital angle sets. This becomes even more critical for asymmetric digital images that have one or more elongated axes. True digital angle sets should satisfy the following properties:

1. Generate a set of $O(N)$ discrete angles for a symmetric $N \times N$ array (to balance $O(N)$ view angles with $O(N)$ projected elements in each view. For an asymmetric discrete array, far fewer than N angles would be needed).
2. These angles should be constructed in a way that accommodates the integer spacing of pixels.

A. Kuba, L.G. Nyúl, and K. Palágyi (Eds.): DGCI 2006, LNCS 4245, pp. 134–145, 2006.

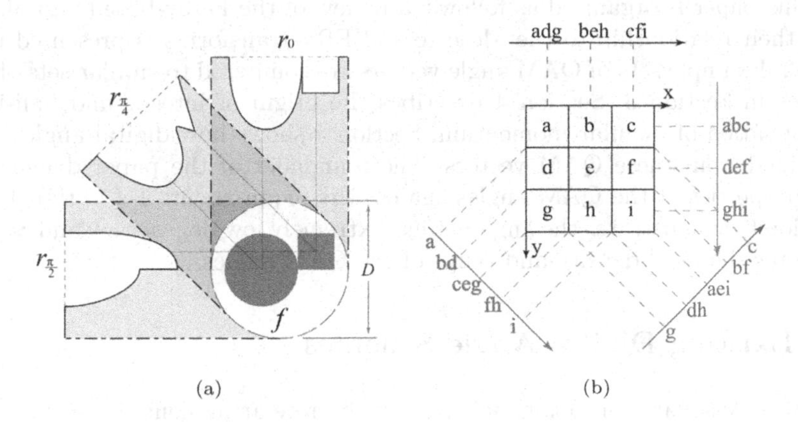

Fig. 1. (a) real space CT projections of a continuous object at three analogue angles. (b) discrete projections of a simple digital object taken at four "grid-friendly" angles.

3. A digital angle set should have the properties of being uniformly distributed over the range $0\,^\circ - 180\,^\circ$, or at least be locally uniform over a more limited angle range.
4. Be increasing; any new angle set for $N' > N$ should contain all of the previously generated angles for N, with each resulting angle remaining unique.

A digital angle set that satisfies these criteria is derived from the Farey series [4]. A different scheme to span analogue real and Fourier spaces to create digital angles was developed through the pseudo-polar Fourier transform scheme of [5]. The Farey angles are important here as they arise intrinsically in the Mojette Transform [6] and the Finite Radon Transform (FRT) [7]. The discrete angle properties of these transforms are reviewed in Section 2.

This paper examines an alternative set of angles designed for use on a discrete grid that are derived from the spatial quantisation of the 3D angular momentum in quantum physics. We will show that the digital angle set derived from the QAM vectors satisfy the above criterion. The idea of using an analogue version of the quantum vector spaces has already been utilised by [8] to represent polar colour variables.

This "naturally" occurring quantum angle set has attractive properties that seem to be relatively unexplored. The quantised angular momenta (QAM) angle distribution has the property of being increasing and appears to be locally uniform, with a slow and smooth decrease in density at larger angles. The set of all possible QAM angles is almost, but not quite, unique. Understanding the formation and distribution of the small number of redundant QAM angles is important when choosing appropriate digital projection and reconstruction angles. This alternative set of quantum-based angles may improve the attainable quality of digital image reconstructions and help optimise the number of projected views required, particularly for the projection of asymmetric digital objects.

This paper is organised as follows: a review of the Farey-based digital angles and their relationship to the Mojette and FRT transforms is presented in Section 2. Example sets of QAM angle vectors are compared to similar sets of Farey angles in Section 3. Section 4 describes the origin of integer and half-integer quantisation of angular momentum. Section 5 shows how digital angles can be obtained from these QAM vectors. The remainder of the paper demonstrates the properties of the QAM angles against the requirements 1-4 outlined above; Section 6 demonstrates the angle set has extremely low degeneracy and Section 7 explores the local density and range of the QAM angles.

2 Existing Digital Angle Schemes

An $N \times N$ square (or hexagonal) regular discrete array generates a natural set of projection angles [9, 10]. These unique angles have tangents that are based on the ratios of relatively prime integers belonging to members of the set of Farey fractions F_N ranging from $1/N$ up to $1/1$. Taking the arctangent of these fractions produces a set of angles lying between $0°$ and $45°$. An example showing $F_4 = 0, \frac{1}{4}, \frac{1}{3}, \frac{1}{2}, \frac{2}{3}, \frac{3}{4}, 1$ for $N = 4$ has been depicted in Figure 2(a). The density of Farey angles is remarkably even and exhibits no degeneracy or replication of any angles as the integer N increases to infinity. Each Farey angle is defined by a unique vector θ_{ab} that links the origin $(0, 0)$ to a co-prime pair of Cartesian coordinates (a, b). Farey angles ranging from $0°$ to $180°$ degrees are obtained using (a, b), (b, a), $(-a, b)$ and $(-b, a)$ as four-fold symmetric vectors oriented at θ_{ab}, 90-θ_{ab}, 90+θ_{ab} and 180-θ_{ab} respectively, as shown for F_{20} in Figure 2(b). Any $N \times N$ image can be reconstructed exactly if $N > 1 + \max(|a_i|, |b_i|)$ for any set of projections taken at rational angles a_i/b_i. The Katz criterion [11] ensures that ambiguous "ghost functions" do not exist in projections of the reconstructed image space (and hence that the reconstructed image is unique).

The Mojette transform developed in [6] is a generalisation of the FRT [7]. Many properties of the FRT have been investigated and applied by Kingston and Svalbe [12]. The Mojette and FRT formalisms both map between digital images and digital projections. This is done exactly and invertibly, with no interpolation (hence preserving image sharpness) by a deliberate selection of grid-dependent digital view angles and projection paths.

The FRT restricts N to be prime. This endows the projections with the property of minimal information redundancy [13, 14] and enables the use of very simple projection and reconstruction algorithms. Each 2D square array of prime size p has a pre-determined set of $p + 1$ rational slopes that define the digital projection orientations. These orientations are a subset of the Farey series for F_N, as shown in Figure 2(c). The subset of Farey fractions (a/b) that are selected at each prime size p has its own interesting behaviour, as discussed in [15]. The FRT has robust, efficient real-space and Fourier-space reconstruction algorithms based on simple addition. It automatically satisfies the Katz reconstruction criterion.

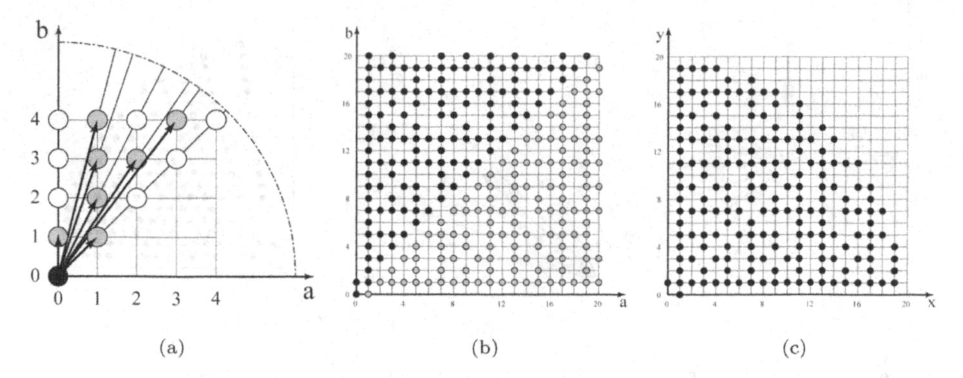

Fig. 2. (a) Farey angles (F_4) made of co-prime integer ratios a/b ($a < b$). (b) the set F_{20} depicted as co-ordinates (a, b) giving θ_{ab} (grey points), and the reflected set (b, a) giving $90° - \theta_{ab}$ (black points). (c) The first half ($0° < \theta_{ab} < 90°$) of the FRT angle set is a subset of the extended Farey set, shown here for $p = 379$ and F_{20}.

The Mojette Transform allows much more flexibility in the shape and size of the discrete array chosen to represent some discrete object. A set of Farey vectors, tailored to satisfy the Katz criterion, are selected for projection orientations based on the shape and size of the array. Some degree of information redundancy arises in this more general Mojette projection representation. However this redundancy may be exploited usefully in the design of very efficient data transmission, storage and encryption schemes [16, 17]. The algorithms to reconstruct images from Mojette projections [6] are more complex than for the FRT, largely because of the increased level of redundancy.

3 Quantised Angular Momentum Directions

The 3D angular momentum vector in quantum physics also generates a "natural" set of discrete angles. The angular momentum vector (**j**) can only take on values that are integer or half-integer multiples of the reduced Planck's constant ($\hbar$). When this vector is aligned with respect to an externally imposed reference direction (such as that of the total local magnetic field), the magnitude of the z-projection of **j** can change only in integer increments, Figure 3(a). This alignment constraint results ultimately from the quantisation of stable energy states for bound particles, and the sensitivity of the state energy to the orientation of the particle orbit.

In atoms and nuclei, the use of quantum mechanics is essential. There the observed magnitude of **j** ranges from 0 to 8 as electron or nucleon orbits are filled in the atoms and nuclei from hydrogen to uranium. The correspondence principle argues that, in the classical (large angular momentum) limit, the 3D vector **j** is equally free to take any alignment direction with respect to any z-axis. The spectra of vibrational modes which arise as allowed excitations in finite discrete lattices is another relevant physical example of where a set of

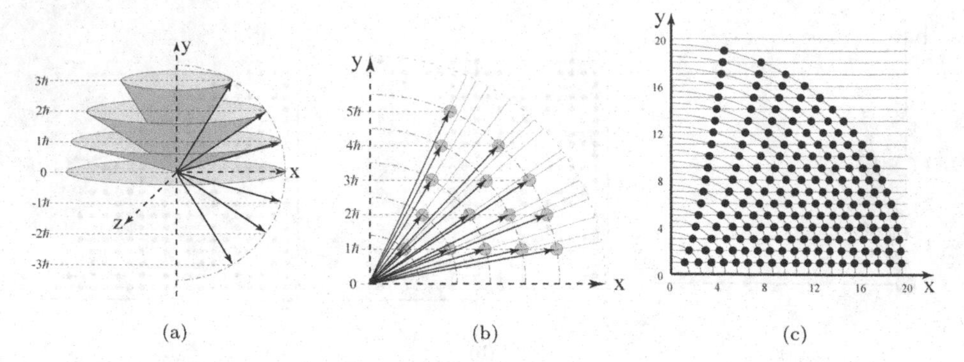

Fig. 3. (a) Depiction of the allowed quantum angular momentum (QAM) vector precession orientations for angular momentum $\mathbf{j} = 3$. (b) The quantised z-projection of the QAM vectors for $\mathbf{j} = 5$. (c) The full integer $\mathbf{j}$ QAM angle set for $|\mathbf{j}| \leq 19$.

discrete angle trajectories emerges naturally from the underlying matrix of the data structure [18].

The QAM vectors provide a locally smooth and reasonably uniform global coverage of angle space, particularly over the interval $0\,° - 45\,°$, for large angular momentum values, Figure 3(c). At angles closer to $90\,°$, the density decreases steadily and this property may find application in the limited field of view case encountered when reconstructing SPECT data using cone-beams or for PET/CT tomosynthesis.

The size of the set of QAM angle vectors increases as the magnitude of the maximum angular momentum $\mathbf{j}$ is increased, adding $2j + 1$ vectors when $j - 1$ increments to j. In the main, the added vectors are new and do not occur at angles generated previously by smaller values of j. There are, however, some values of j that do result in degenerate angles.

The occurrence of such replicated angles in the QAM set is rare, even more so for the half-integer QAM case. In this paper, these relatively rare degenerate angles will be examined with a view to being able to predict and quantify any clumping in the local smoothness of the QAM angle set. Quite localised non-uniformities in the density of angles also occur for the Farey sets [10]. The analytic work of [19] has examined and modelled the details of those variations.

4 Quantised Angular Momentum

Classical angular momentum is a measure of the "turning moment" of a moving object about some axis: it is a vector quantity of magnitude proportional to the radius $\mathbf{r}$ of the object from the axis and to the linear momentum $\mathbf{p} = m\mathbf{v}$ of the object, where $\mathbf{v}$ is the velocity of the particle which has mass m. Formally, $\mathbf{L} = \mathbf{r} \times \mathbf{p}$, with the direction of vector $\mathbf{L}$ being normal to the plane defined by the vectors $\mathbf{r}$ and $\mathbf{p}$. In classical mechanics, the direction of $\mathbf{L}$ is free to take any direction in space and $|\mathbf{L}|$ is a continuous variable, given that values for $\mathbf{r}$ and $\mathbf{p}$ are unconstrained.

In quantum mechanics, the measurable momentum and location of all objects are subject to the Heisenberg Uncertainty Principle that reflects the "graininess" of space-time and the particles that can exist within it. Momentum and position cannot be simultaneously specified to a precision below the value of the reduced Planck's constant ($\hbar$). This automatically imposes an uncertainty on the angular momentum, it being a product of position and linear momentum. This uncertainty means that the angular momentum vector can only be observed to change by amounts proportional to $\hbar$.

If we take a quantised angular momentum vector $\mathbf{L}$ in 3D (x, y, z) space with components L_x, L_y, L_z, then the uncertainty in $\mathbf{L}$ means that if the direction of L_z is fixed, the vector must "precess" (with unknown phase) around the z-axis so that L_x and L_y are uncertain, Figure 3(a). The uncertainty in $\mathbf{L}$, and the expression for its magnitude, emerges naturally in quantum mechanics after separation of the radial and angular part in the solution to the Schrödinger equation when applied to any particle constrained by a potential well.

In quantum mechanics, all fundamental particles (like electrons) have an internal angular momentum (called spin $\mathbf{S}$). This spin is also quantised, so that changes in spin must also occur in $\hbar$ steps. Spin can be either a symmetric or anti-symmetric component of the quantised particle's wavefunction. Fermion objects have half-integral spin ($\mathbf{S} = (n + 1/2)\hbar$) whilst bosons have integral spin ($\mathbf{S} = n\hbar$) or positive integer n. It appears that the fermion and boson QAM distributions, whilst being very similar, turn out to have remarkably different angle degeneracy properties. A particle with rotational and spin angular momentum has a total angular momentum $\mathbf{j}$, subject to the same uncertainty and quantisation, with $\mathbf{j} = \mathbf{L} + \mathbf{S}$.

The length of the quantised angular momentum vector is given by $|\mathbf{j}| = \sqrt{j(j+1)}\hbar$. It is the $j(j+1)$, rather than a j^2 term, that endows interesting properties to the QAM angle distribution. The direction of j is defined by the angle θ measured in cylindrical coordinates relative to the direction of the xy plane. The z-projection of j, j_z is constrained to have $|j_z| = m\hbar$, where m is integral for integer j and half-odd integral for half-integer j values.

5 QAM Vectors as Points in the Plane

If we denote the radius of the projection of the vector $\mathbf{j}$ on to the x-y plane as r_m, then $r_m = \sqrt{j(j+1) - m^2}$, see Figure 4(a). The angle θ of the vector $\mathbf{j}$ from the origin to (r_m, m) is given by the arctangent of the gradient, taken here as $g(j, m) = m/r_m$.

The equations for lines of constant $|m|$ form simple parabolas, for example, $r_m = \sqrt{j}$ for $|m| = j$. The half-integer points fall exactly in between the integer points, as both sets of allowed (r_m, m) values lie on the same quadratic curves as seen in Figure 3(c).

The r_m values will not, in general, be integers, but can be scaled and rounded to the nearest integer at any required precision. Here (r_m, m) have the same role as the (a, b) integers used in Section 2 for the Farey and FRT examples. The z-projection m is an integer or half-integer but in either case changes in integer steps. As r_m may be irrational, the exact process used to convert this value to an integer may be important in any application.

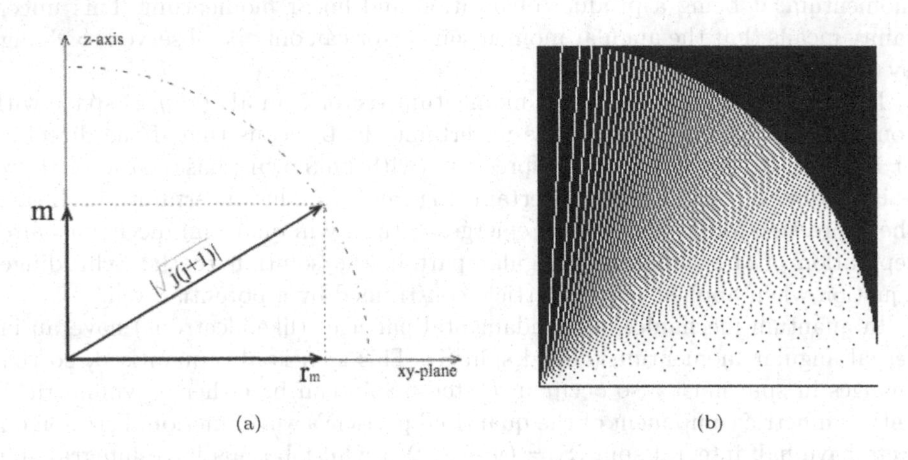

Fig. 4. (a) Definition of the QAM projections m and r_m for a given j. (b) Plot of the density of the QAM angle distribution for $j = 199/2$. Each white point represent a selected angle.

For $j \gg 1$, $\sqrt{j(j+1)}$ asymptotes to $j + \frac{1}{2}$ (and so can never be an integer, nor an exact $\frac{1}{2}$ integer). Hence the direction of any QAM vector $\mathbf{j}$ cannot ever be aligned exactly along the z-axis, as $j_z = m$ must be an integer or half-integer. The maximum angle of θ occurs when $m = j$, so that $r_m = \sqrt{j}$ and $g = tan(\theta) = j/\sqrt{j} = \sqrt{j}$. The maximum angle is 88.7° for $j = 2000$ (and reaches only 70.5° for $j = 8$, for a j typical of the atomic case).

The minimum angle for integer QAM is zero degrees, which occurs when $m = 0$ with $r_m = \sqrt{j(j+1)}$. The minimum half integer angle occurs at $|m| = \frac{1}{2}$. For large j, the minimum gradient g then approaches $1/(2j)$. This corresponds to a minimum angle of 1.8° for $j = 15/2$, the typical maximum half integer j for the atomic case, falling to 0.014° when $j = 3999/2$.

6 QAM Angle Degeneracy

The QAM angle set can adapt to the size and shape of a discrete array, as $|\mathbf{j}|$ can be matched to the array size and $|m|$ can be used to accomodate asymmetry in the array shape. The QAM digital angles then meet the design properties 1,2 and 3 that are outlined in Section 1. Are there integer j vectors that link the origin to points (r_m, m) and $(r_{m'}, m')$ that have the same angle? For this to be true, the values j and j' (and hence m' and m, or $r_{m'}$ and r_m) must scale according to

$$(m'/m)^2 = \frac{j'(j'+1)}{j(j+1)}. \tag{1}$$

The next section examines integer j values where (1) is satisfied.

6.1 Degeneracy of the QAM Angles for Integer j

For $r_m = nm$, that is a gradient of $g = m/(nm) = 1/n$, (i.e. $1 : n$), then $j(j+1) - m^2 = (nm)^2$ means that $N = \sqrt{1 + 4(n^2+1)m^2}$ needs to have integer solutions, where $N = 2j+1$. N also corresponds, incidentally, to the total number of allowed QAM vector projections for a given j.

Then $j(j+1) = (n^2+1)m^2 = (n^2+1)k^2l^2$, where $m = kl$, with k, and l integers. We choose to factor m into a product of two integers, kl, because then we can identify j and $j+1$ as separate squared quantities that enable m to be an integer. The values of k and l (and hence j and m) can be determined recursively, starting with the values $k = l = 1$. Then

$$j = (n^2+1)k^2 \text{ and } j+1 = l^2. \tag{2}$$

The ratio of k/l approximates $1/\sqrt{n^2+1}$ because

$$l^2 = (n^2+1)k^2 + 1. \tag{3}$$

The values obtained for k and l turn out to be exceptionally good integer approximations for the irrational number $\sqrt{n^2+1}$. The next integral solution at j', m' turns out to given by

$$j' = (n^2+1)k'^2 \text{ and } j'+1 = l'^2 \text{ with } m' = k'l', \text{ where} \tag{4}$$

$$k' = nk + l \text{ and } l' = (n^2+1)k + nl. \tag{5}$$

The recursive relationship (5) determines all of the redundant solutions for gradients $(1 : n)$. Table 1 shows example redundant (j, m) values for the gradients $1:1$ and $1:3$.

Finding integers k,l that give integer values of r_m has a parallel to the approximation of surds (such as $a+\sqrt{b}$) using continued fractions, where the sequence of continued fraction values is periodic. For example, the value of $\sqrt{2}-1$ (which corresponds to $\sqrt{(n^2+1)} = \sqrt{2}$ for $n = 1$) can be found as $1/(2+1/(2+1/(2+\ldots)$, which can be written recursively as $a_{r+1} = 1/(2+a_r)$ where $a_0 = 0$. The fractions k/l used to find (j, m) values that each have exactly the same gradient $1 : n$ can also be found using $a_{r+1} = 1/(2n + a_r)$ and adding n to each fraction.

The j values that replicate a given gradient have quadratic separation in j. The gap between the next (j', m') with the same slope as (j, m) grows very rapidly. The ratio or scale, s, between successive values of j is given by m'/m (or j'/j or N'/N) and can be shown to be $s = 2n^2 + 1 + 2n\sqrt{n^2+1}$.

Table 1. Examples of degenerate QAM angles for integer j and gradients of $1:n$

$g=1:1$ $\quad j(j+1)=(2)k^2l^2$				
N	j	m	k	l
3	1	1	1	1
17	8	6	2	3
99	49	35	5	7
577	288	204	12	17
3363	1681	1189	29	41
19601	9800	6930	70	99

$g=1:3$ $\quad j(j+1)=(10)k^2l^2$				
N	j	m	k	l
19	9	3	1	3
721	360	114	6	19
27379	13689	4329	37	110

The ratios for gradients $n:1$ have the same values of r_m and m in exchanged roles as for $1:n$, so these angles and their j,m values can be found from the $1:n$ result. The tabulated N and j values for $n:1$ are the same as for $1:n$, but the value of m is just n times that for $1:n$.

For $g=p:q$ (as well as $g=\sqrt{p/q}$) there is also a similar sparse redundancy in the representation of angles. Results for $g=3:5$ are given in Table 2.

Table 2. QAM angle redundancy for the gradient 3:5

$g=3:5$ $\quad j(j+1)=(34/9)k^2l^2$				
N	j	m	k	l
35	17	9	9	1
2449	1224	630	18	35

The total number of redundant angles for the integer QAM case appears to increase approximately linearly with increasing j. For $0 < j \leq 2000$, we found just 242 redundant angles in the integer QAM angle distribution out of a total number of $j(j+1)/2 = 2,001,000$ angles.

6.2 Degeneracy of the j QAM Angles for Half-Integer j

For half integer QAM, j and m are both required to be odd, so we write:

$$(2n+1)(2n+3)/4 - (2m+1)^2/4 = r_m^2, \tag{6}$$

where n and m are any positive integers. Then $r_m^2 = (n^2 - m^2) + (2n - m) + 1/2$, so that r_m can never have integer values for the half-integer QAM case. For the gradient to be degenerate, i.e. $g = m/r_m = m'/r_{m'}$, then $[j(j+2)/4 - m^2/4]m'^2 = [j'(j'+2)/4 - m'^2/4]m^2$, and

$$(m'/m)^2 = \frac{j'(j'+2)}{j(j+2)}. \tag{7}$$

Condition (7) turns out to be much harder to satisfy than (1) for the integer QAM case. Scaled solutions for j and j' that give redundant angles are only possible if $j' = \alpha^2 j$ and $j' + 2 = \beta^2(j+2)$ with $m' = (\alpha\beta)m$, and where α, β are integral, which is similar to the constraint (4) for integer QAM.

For $0 < j \leq 3999/2$, we found only 16 redundant angles in the first 2 million possible angles (as compared to 242 redundant angles for j up to 2000 for the integer case). Note the repeated occurrence of the integers 845, 2023 and 3969 in Table 3. These integers have highly composite forms, for example $3969 = 3^4.7^2$. If the half integer and integer angle distributions are pooled (this does not occur in real quantum systems), the number of redundant angles increases, at approximately double the integer rate, to reach a total of 510 redundancies for j up to 2000. The QAM angle set hence satisfies the fourth digital angle property as set out in Section 1.

Table 3. Degenerate values of j, m for half-integer QAM angles

$\mathbf{G = m/r_m}$	**G**	**j, m**	**j′, m′**
0.101015	$1/\sqrt{2.7^2}$	9/2,1/2	3969/2, 399/2
0.127001	$1/\sqrt{2.31}$	7/2, 1/2	2023/2, 255/2
0.101499	$1/\sqrt{2.17}$	5/2, 1/2	845/2, 143/2
0.267261	$1/\sqrt{2.7}$	3/2, 1/2	243/2, 63/2
0.316228	$1/\sqrt{2.5}$	9/2, 3/2	3969/2, 1197/2
0.408248	$1/\sqrt{2.3}$	7/2,3/2	2023/2, 765/2
0.581238	$5/\sqrt{2.37}$	9/2, 5/2	3969/2, 1995/2
0.588348	$3/\sqrt{2.13}$	5/2, 3/2	845/2, 429/2
0.707107	$1/\sqrt{2}$	1/2, 1/2	25/2, 15/2 and 361/2, 209/2
0.988849	$7/(5\sqrt{2})$	9/2, 7/2	3969/2, 2793/2
1.224745	$\sqrt{3/2}$	3/2, 3/2	243/2, 189/2
1.581139	$\sqrt{5/2}$	5/2, 5/2	845/2, 715/2
1.870829	$\sqrt{7/2}$	7/2, 7/2	2023/2, 1785/2
2.12132	$3/\sqrt{2}$	9/2,9/2	3969/2, 3591/2

7 Distribution of QAM Angles

The uniformity of the QAM angle distribution has been examined using the same unevenness criterion (D) as used for the Farey angle set [10]. The QAM sets are much more uneven than the Farey sets because of the smooth decrease in angle density as the z-axis projected value of j increases. For the half integer case at $j = 199/2$, $D = 65.09$ over 2203 angles (selected for $m < r_m$ and

$r_m < j/\sqrt{2} = 70.71$ here, from a total of 5050 angles). For the Farey sequence $F_{86}(1/86, 1/85, \ldots 85/86)$ which has 2273 angles, $D = 2.106$.

Figure 4(b) shows the density of m, r_m points from the QAM distribution in the 2D plane which satisfies the property 3 qiven in Section 1. Note that the points "missing" for the low angles are reflections of the positions of the real points at large angles.

8 Conclusions and Further Work

The QAM vectors produce an interesting set of digital angles with remarkably little degeneracy, especially for the half-integer angular momenta where the conditions required for integer-based solutions are harsher. In physical systems, the QAM directions are limited to either whole or half-integral values, but the above investigation can be extended to include angles for 1/3 (quark-like) or other fractional quantisation values. More work is needed on how to best round or interpolate the (often irrational) values of r_m to integers when applying these digital angle sets.

Acknowledgements

SC holds a postgraduate scholarship from the Faculty of Science and School of Physics at Monash University. AK holds a postdoctoral position supported by a grant from Pays De Loire, France.

References

1. Servières, M., Normand, N., Subirats, P., Guédon, J.: The mojette transform: Discrete angles for tomography. In Herman, G., Kuba, A., eds.: Proc. Workshop on Discrete Tomography and its Applications. Elsevier Science Publishers, New York (2005)
2. Subirats, P., Servières, M., Normand, N., Guédon, J.: Angular assessement of the mojette filtered back projection. Proceedings of SPIE - The International Society for Optical Engineering **5370 III** (2004) 1951 – 1960
3. Svalbe, I., van der Spek, D.: Reconstruction of tomographic images using analog projections and the digital radon transform. Linear Algebra and Its Applications **339** (2001) 125 – 45
4. Hardy, G., Wright, E.: An Introduction to the Theory of Numbers. 5th edn. Clarendon Press, Oxford (1979)
5. Averbuch, A., Coifman, R., Donoho, D., Israeli, M., Walden, J.: The pseudo-polar fft and its applications. Technical Report YALEU/DCS/RR, Yale University **1178** (1999)
6. Guédon, J., Normand, N.: The mojette transform: the first ten years. Discrete Geometry for Computer Imagery. 12th International Conference, DGCI 2005. Proceedings (LNCS) **3429** (2005) 79 – 91
7. Matúš, F., Flusser, J.: Image representation via a finite radon transform. IEEE Transactions on Pattern Analysis and Machine Intelligence **15**(10) (1993) 996 – 1006

8. Nölle, M., Őmer, B.: Representation of cyclic colour spaces within quantum space. Proceedings DICTA 2005. Digital Image Computing: Techniques and Applications (2006) 6
9. Svalbe, I.: Sampling properties of the discrete radon transform. Discrete Applied Mathematics **139**(1-3) (2004) 265 – 81
10. Svalbe, I., Kingston, A.: On correcting the unevenness of angle distributions arising from integer ratios lying in restricted portions of the farey plane. Combinatorial Image Analysis. 10th International Workshop, IWCIA 2004. Proceedings (LNCS) **3322** (2004) 110 – 21
11. Katz, M.: Questions of Uniqueness and Resolution in Reconstruction from Projections. Springer-Verlag (1977)
12. Kingston, A., Svalbe, I.: Projective transforms on periodic discrete image arrays. Advances in Imaging and Electron Physics **139** (2006) 75–177
13. Svalbe, I.: An image labeling mechanism using digital radon projections. Proceedings 2001 International Conference on Image Processing **3** (2001) 1015 – 18
14. Kingston, A., Svalbe, I.: Geometric effects in redundant keys used to encrypt data transformed by finite discrete radon projections. Proc. IEEE Digital Imaging Computing: techniques and applications, Cairns, Australia. (2005)
15. Svalbe, I., Kingston, A.: Intertwined digital rays in discrete radon projections pooled over adjacent prime sized arrays. In Sanniti di Baja, G., Svensson, S., eds.: 11th Discrete Geometry for Computer Imagery. Volume 2886. LNCS (2003) 485–494
16. Autrusseau, F., Guédon, J., Bizais, Y.: Mojette cryptomarking scheme for medical images. Proceedings of the SPIE - The International Society for Optical Engineering **5032** (2003) 958 – 65
17. Normand, N., Guédon, J.P., Philippe, O., Barba, D.: Controlled redundancy for image coding and high-speed transmission. Proceedings of the SPIE - The International Society for Optical Engineering **2727** (1996) 1070 – 81
18. Homolya, S., Osborne, C., Svalbe, I.: Density of states for vibrations of fractal drums. Physical Review E (Statistical, Nonlinear, and Soft Matter Physics) **67**(2) (2003) 26211 – 1
19. Boca, F., Corbeli, C., Zahrescu, A.: Distribution of lattice points visible from the origin. Commun. Math. Phys. **213** (2003) 433–470

A Benchmark Evaluation of Large-Scale Optimization Approaches to Binary Tomography

Stefan Weber[1], Antal Nagy[2], Thomas Schüle[1],
Christoph Schnörr[1], and Attila Kuba[2]

[1] Dept. Math. & Comp. Science, CVGPR-Group, University of Mannheim,
D-68131 Mannheim, Germany
{wstefan, schuele, schnoerr}@uni-mannheim.de

[2] Department of Image Processing and Computer Graphics,
University of Szeged, Árpád tér 2. H-6720 Szeged, Hungary
{nagya, kuba}@inf.u-szeged.hu

Abstract. Discrete tomography concerns the reconstruction of functions with a finite number of values from few projections. For a number of important real-world problems, this tomography problem involves thousands of variables. Applicability and performance of discrete tomography therefore largely depend on the criteria used for reconstruction and the optimization algorithm applied. From this viewpoint, we evaluate two major optimization strategies, simulated annealing and convex-concave regularization, for the case of binary-valued functions using various data sets. Extensive numerical experiments show that despite being quite different from the viewpoint of optimization, both strategies show similar reconstruction performance as well as robustness to noise.

1 Introduction

Discrete tomography (DT) is an active field of research covering a number of important problems across various application areas [1, 2]. A key aspect of DT is the reconstruction of functions under non-standard conditions, in contrast to conventional tomography. A necessary condition for making such reconstructions feasible is to restrict the range of the functions to be reconstructed to a finite set. Challenging application problems that can be naturally modeled in this way include non-destructive testing [3], electron microscopy [5], and medical imaging [4, 6].

A major problem in connection with DT concerns optimization. In fact, most applications like DT in medical imaging involve thousands of variables representing the discrete-valued function to be computed. Solving such large-scale combinatorial problems to reach global optimality is generally not possible, hence, optimization strategies providing a good compromise between the quality of sub-optimal solutions and runtime are of primary interest.

For these reasons, we study in this paper two different optimization strategies that showed promising performance in recent work:

A. Kuba, L.G. Nyúl, and K. Palágyi (Eds.): DGCI 2006, LNCS 4245, pp. 146–156, 2006.

- The first strategy based on the classical approach of simulated annealing (SA). It is a stochastical optimization strategy, a random-search technique that is based on the physical phenomenon of metal cooling [10]. The system of metal particles, here the values of the image pixels, gradually reaches the minimum energy level where the metal freezes into a crystalline structure.
- The second optimization strategy, convex-concave regularization, was proposed in [7]. It combines convex relaxations of reconstruction functionals with concave minimization to enforce discrete decisions. A local minimum is determined by solving a sequence of convex optimization problems, each of which can be solved to global optimality. The method involves a single regularization parameter only requiring an application-dependent choice.

In Section 2 we briefly describe the general reconstruction problem. Section 3 details the algorithms related to the two optimization strategies which are evaluated. The evaluation criteria (data sets, performance measures, parameter settings) are specified in section 4, and our quantitative numerical results are presented and discussed in Section 5. We conclude and outline further work in Section 7.

2 Reconstruction Problem

We consider the reconstruction problem of transmission tomography for binary objects. As explained in Fig. 1(a), the imaging process is represented by the algebraic system of equations

$$Ax = b\,, \qquad A \in \mathbb{R}^{m \times n}\,,\; x \in \{0,1\}^n\,,\; b \in \mathbb{R}^m\,, \tag{1}$$

where A and b are given, and the binary indicator vector x representing the unknown object has to be reconstructed. Though we restrict ourselves here to parallel beam geometry, Fig. 1(b), this algebraic representation is general enough to suit other geometries as well.

3 Two Optimization Strategies

This section describes two approaches capable to numerically solve large-scale instances of the general reconstruction problem (1).

3.1 Simulated Annealing

Actually, a possible way of solving (1) at least approximately is to reformulate it as an optimization problem. Formally, we should find the minimum of the following objective function

$$C(x) = ||Ax - b||^2 + \gamma \cdot \Phi(x)\,, \qquad \text{where } x \text{ is a binary-valued vector}\,. \tag{2}$$

The first term on the right hand side ensures that we have an x satisfying (1) at least approximately. The second term allows us to include *a priori* knowledge

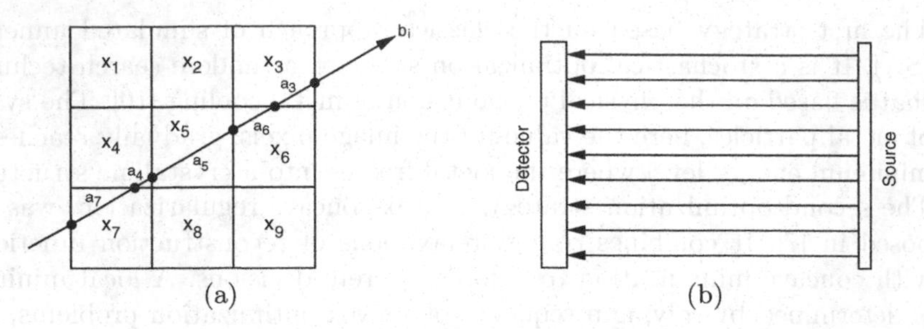

Fig. 1. (a) Discretization model for transmission tomography. The measured projection data are given in terms of a vector $b \in \mathbb{R}^m$. Each component b_i corresponds to a projection ray measuring the absorption along the ray through the volume which is discretized into cells. The absorption a_j in each cell is assumed to be proportional to the density of the unknown object. $x_1, x_2, \ldots$ are binary variables indicating whether the corresponding cells belong to the object ($x_k = 1$) or not ($x_k = 0$). Assembling all projection rays into a linear system gives $Ax = b$, $x \in \{0,1\}^n$, from which the unknown binary object, represented by x, has to be determined. **(b) Parallel beam geometry.** Multiple projections are gathered by rotating the source-detector system around a center point.

about x into the optimization if there are several good binary vector candidates that keep $||Ax - b||^2$ low. In our experiments we have used the following $\Phi(x)$ function

$$\Phi(x) = \sum_{j=0}^{n} \sum_{l \in Q_j^m} g_{l,j} \cdot |\xi_j - \xi_l| \,, \tag{3}$$

where Q_j^m is the set of the indexes of the $m \times m$ adjacent pixels of the j-th lattice pixel and $g_{l,j}$ is the corresponding element of a matrix representing a 2D $m \times m$ Gaussian matrix. The $g_{l,j}$ scalar weights the differences according to the distance of the two adjacent, l-th and j-th pixels. Using this regularization term we can force the optimization algorithm to find binary matrices with possibly compact regions of 0s and 1s.

For solving (2) the *simulated annealing* (SA) optimization method [10] was used.

3.2 Convex-Concave Regularization and DC-Programming

We also consider the one-parameter family of functionals introduced in [7]:

$$J_\mu(x) := \|Ax - b\|^2 + \frac{\alpha}{2} \sum_{j=1}^{n} \sum_{l \in Q_j^1} (x_j - x_l)^2 - \mu \frac{1}{2} \langle x, x - e \rangle \,, \quad x \in [0,1]^n \,. \tag{4}$$

The first terms in (4) and (2) coincide. The second term in (4) is similar to (3), but involves nearest neighbors only, i.e. $m = 1$, with uniform weighting. This term is controlled by the regularization parameter α. Proper values depend on the application and have to be supplied by the user. The third term in (4), together with the convex domain of definition $x \in [0,1]^n$, pertains to the second optimization strategy, to be explained below, that was used to minimize (4). It is a concave functional which gradually enforces the binary constraint $x \in \{0,1\}^n$ by increasing the value of μ (e denotes the vector with all components equal to 1).

Algorithm 1. SA Algorithm

Require: $\gamma \geq 0$ {regularization parameter supplied by the user}
Require: $T_{\text{start}} > 0$ {start temperature supplied by the user}
Require: $T_{\min} > 0$ {minimum temperature supplied by the user}
Require: $1 > T_{\text{factor}} > 0$ {The multiplicative constant for reducing the temperature supplied by the user}
Require: $1 > R_{\text{objective}} > 0$ {The ratio between the first and the current value of the objective function supplied by the user}

```
x := (0, ..., 0)^T
T := T_start
C_start := C_old := ||Ax − b||^2 + γ · Φ(x)
repeat
  for i = 0 to sizeof(x) do
    choose a random position j in the vector x
    x̃ := x
    x̃[j] := 1 − x[j] {change the value of x in the position j}
    C_new := ||Ax̃ − b||^2 + γ · Φ(x̃)
    z := random()
    ΔC := C_new − C_old
    if ΔC < 0 or exp(−ΔC/T) > z , then
      x := x̃ {accept changes}
      C_old := C_new
    end if
  end for
  T := T * T_factor
until T > T_min or C_old/C_start > R_objective
```

Functional (4) can be represented by the sum of a convex and a concave function

$$J_\mu(x) = g(x) - h_\mu(x) \, , \quad x \in [0,1]^n \, , \tag{5}$$

where

$$g(x) := \|Ax - b\|^2 + \frac{\alpha}{2} \sum_i^n \sum_{j \in Q_i^m} (x_i - x_j)^2 \, , \tag{6}$$

$$=: \|Ax - b\|^2 + \alpha \langle x, L^\top L x \rangle \, , \tag{7}$$

$$h_\mu(x) := \mu \frac{1}{2} \langle x, x - e \rangle \, . \tag{8}$$

As a consequence, (4) naturally belongs to the class of dc-programs (dc: difference of convex functions) and thus provides a basis for algorithm design. It is shown in [7] that the following algorithm converges to a *binary* local minimum of the criterion J_μ:

Algorithm 2. DC Algorithm

Require: $\alpha \geq 0$ {regularization parameter supplied by the user}
Require: $\epsilon_{in} > 0$ {termination criterion for the inner loop}
Require: $\epsilon_{out} > 0$ {termination criterion for the outer loop}
Require: $\epsilon_\mu > 0$ {determines the increment μ_Δ by eqn. (9)}
  $x := (0, ..., 0)^\top$
  $\mu = 0$
  repeat
    repeat
      $\tilde{x} := x$
      $x := \underset{x \in [0,1]^n}{\operatorname{argmin}} \{g(x) - \langle x, \nabla h_\mu(\tilde{x}) \rangle\}$
    until $\|x - \tilde{x}\|_2 < \epsilon_{in}$
    $\mu := \mu + \mu_\Delta$
  until $\max_i \{\min\{x_i, 1 - x_i\}\} < \epsilon_{out}$

We point out that each x computed in the inner loop is the global optimum of a *convex* optimization problem. Our current implementation involves [9] for this step, but many other convex optimization techniques could be applied as well.

Furthermore, while the decomposition (5) with (7) and (8) is the most natural one, a range of alternative decompositions of the functional J_μ are possible to which algorithm 1 can be applied. We refer to [8] for further details.

4 Evaluation

4.1 Data Sets

For evaluation purposes both reconstruction algorithms were tested on the same data set of binary images. The images are software phantoms consisting of discretized versions of geometrical objects like circles, ellipses, etc. – see Fig. 2.

For each phantom, the image reconstruction problem (1) was compiled by taking parallel projections from different directions. The number p of projections ranged between 2, 3, 5, and 6. For $p \in \{2, 3, 5\}$, directions were uniformly chosen within $[0°, 90°]$, and within $[0°, 150°]$ for $p = 6$. For each direction, the number of measurements was 96 for phantom 1 and 384 for phantom 2 and 3.

In addition to noiseless projection data, we also used noisy data for the evaluation. To this end, the projection data were superimposed by noise with Gaussian distribution $\mathcal{N}(0, \sigma)$, $\sigma \in \{0.5, 1.5, 5\}$. Negative values that may rarely be generated in this way, do not make sense physically and were clipped to the value zero.

Fig. 2. Phantom images of size 64×64, 256×256, and 256×256 used for the experimental evaluation

4.2 Performance Measures

Let x^* be the ground truth image and x be a solution to the reconstruction problem (1) computed by either optimization algorithm. We use the following error measures for a quantitative evaluation:

$$E_1(x) := \|Ax - b\|_2 ,$$
$$E_2(x) := \frac{1}{\sum_{i=1}^n x_i} \|x - x^*\|_1 .$$

For interpreting the corresponding numerical results in the tables below, readers should keep in mind that these two measures scale quite differently. While a single pixel error results in a change of E_1 of about 10^1, say, the order of change of E_2 will be 10^{-2} only.

4.3 Parameter Settings

To compare both approaches numerically, we used a fixed parameter set for each reconstruction algorithm. These values were used throughout all experiments.

Simulated Annealing Algorithm:

$$\begin{aligned} \gamma &= 14.0 \\ T_{\text{start}} &= 4.0 \\ T_{\text{min}} &= 10^{-14} \\ T_{\text{factor}} &= 0.97 \\ R_{\text{objective}} &= 0.00001 \end{aligned}$$

DC Algorithm:

$$\begin{aligned} \alpha &= 0.25 \\ \epsilon_{in} &= 0.1 \\ \epsilon_{out} &= 0.01 \\ \epsilon_{\mu} &= 10 \end{aligned}$$

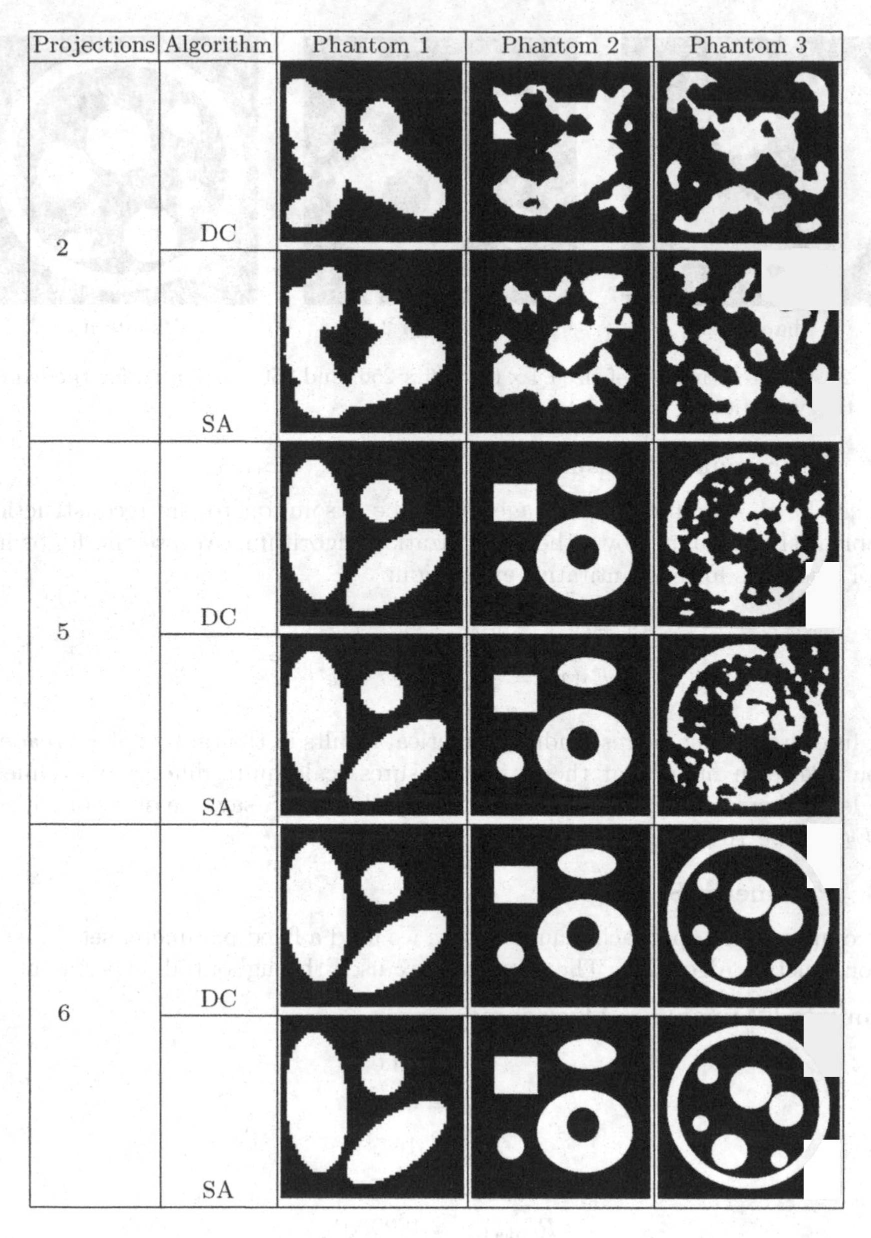

Projections
Algorithm
Phantom 1
Phantom 2
Phantom 3
2
DC
SA
5
DC
SA
6
DC
SA

Fig. 3. The phantom images reconstructed from noise free projections ($p = 2, 5, 6$)

The $\mu-$increment was computed by evaluating the following equation

$$\mu_\Delta := \frac{\epsilon_\mu n^{1/2} \lambda_{\min}(Q)}{\|x - \frac{1}{2}e\|}, \quad Q := A^\top A + \alpha L^\top L. \tag{9}$$

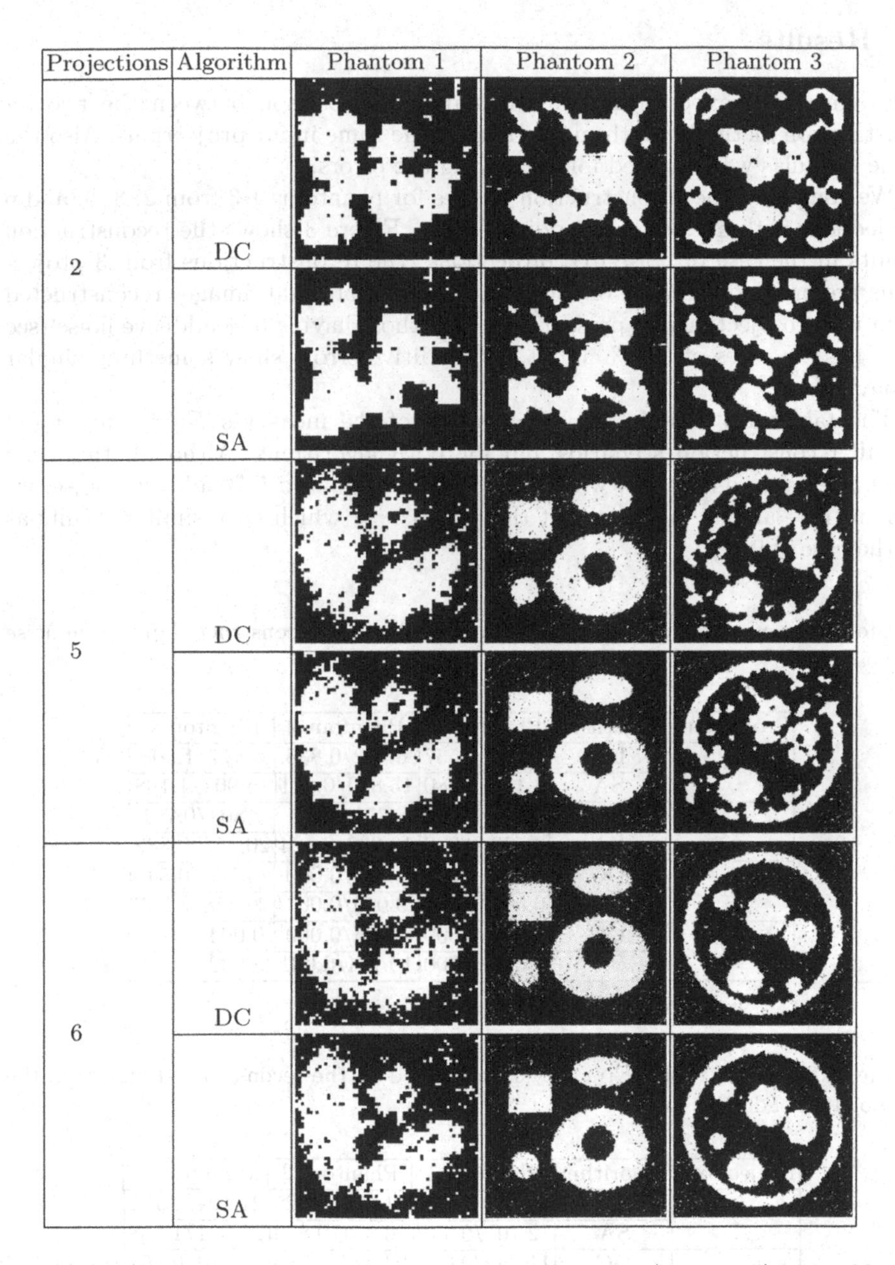

Fig. 4. The phantom images reconstructed from projections ($p = 2, 5, 6$) with additive 5 % noise

Here, x denotes the solution of the very first inner loop for $\mu = 0$, and $\lambda_{\min}$ is the smallest eigen value of the matrix Q that can be computed offline and beforehand. For details and an interpretation of (9), we refer to [8].

5 Results

The aim of the experiments was to make a comparison between the two reconstruction methods. Both methods used the same input projections. Also the same formulas were applied for measuring the errors.

We computed the reconstruction images for phantoms 1-3 from 2, 3, 5, and 6 projections with and without additive noise. Figure 3 shows the reconstruction results in the case of noise-free projections (the reconstructions from 3 projections are omitted, because of the lack of space). From all images reconstructed from noisy projections we present here only those having 5 % additive noise (see Fig. 4). The cases of 0.5 % and 1.5 % additive errors show something similar behavior).

The tables 1-3 contain the error values of the measures $E_1(x)$ and $E_2(x)$ for all reconstruction scenarios and for both algorithms. Although the same experiments were repeated with 0 %, 0.5 %, 1.5 %, and 5 % additive noise, we present here all the tables except the case 0.5 % (which gave similar results as in the case of 0 %).

Table 1. The error values $E_1(x)/E_2(x)$ measured on the reconstructed images in noise free case

Projections	Algorithm	Phantom 1	Phantom 2	Phantom 3
2	DC	3.464/0.537	6.782/0.835	5.477/1.108
	SA	8.173/0.480	15.870/0.841	16.901/1.198
3	DC	0.000/0.000	8.351/0.471	7.804/0.751
	SA	6.779/0.020	19.028/0.524	20.453/0.882
5	DC	0.000/0.000	0.005/0.000	14.761/0.545
	SA	0.000/0.000	9.040/0.001	26.478/0.537
6	DC	0.000/0.000	0.005/0.000	0.004/0.000
	SA	0.000/0.000	10.134/0.001	9.632/0.001

Table 2. The error values $E_1(x)/E_2(x)$ measured on the reconstructed images in the case of 1.5 % additive noise

Projections	Algorithm	Phantom 1	Phantom 2	Phantom 3
2	DC	9.708/0.492	21.375/0.829	21.101/1.181
	SA	12.707/0.442	26.854/0.853	26.391/1.188
3	DC	11.892/0.080	24.024/0.489	23.414/0.761
	SA	15.993/0.093	31.156/0.565	30.878/0.918
5	DC	19.020/0.080	31.135/0.026	29.182/0.551
	SA	23.323/0.059	41.052/0.021	39.057/0.536
6	DC	18.795/0.102	31.298/0.034	33.203/0.045
	SA	25.324/0.058	45.537/0.020	43.371/0.042

Table 3. The error values $E_1(x)/E_2(x)$ measured on the reconstructed images in the case of 5 % additive noise

Projections	Algorithm	Phantom 1	Phantom 2	Phantom 3
2	DC	25.599/0.479	59.701/0.857	59.270/1.160
	SA	28.633/0.519	63.004/0.839	62.437/1.162
3	DC	27.276/0.305	68.470/0.525	68.802/0.798
	SA	31.936/0.295	73.966/0.563	73.522/0.903
5	DC	44.607/0.292	82.858/0.114	80.945/0.589
	SA	48.423/0.265	91.049/0.103	87.514/0.597
6	DC	47.855/0.342	86.354/0.123	86.269/0.151
	SA	53.585/0.287	98.575/0.102	95.214/0.145

6 Discussion

Both algorithm perform very similar on the tested reconstruction problems. Consider first the noise free reconstructions. The methods were able to reconstruct Phantom 1 from 3 or more projections. Phantom 2 was more difficult, 5 or more projections are necessary for the almost perfect reconstruction. The most difficult object was Phantom 3, it needs 6 projections for a good quality reconstruction.

The DC method gives smaller errors in almost all cases in Table 1. It is interesting that the measure $E_1(x)$ was smaller for DC than SA in every cases. The reason can be explained as follows. $E_1(x)$ measures the differences between the input projections (b) and the projections of the reconstructed object (Ax). For this reason $E_1(x)$ takes into account only the projections and not the original object. (That is, $E_1(x)$ can be very small even if the object x is far from the original one.) Our results shows that the difference between the projections is not so strongly weighted in the objective function of SA (2). At the same time SA reaches similarly low values of $E_2(x)$ as DC does.

Consider now the results of noisy projections. It is clear that DC gives again better $E_1(x)$ values. The differences in the $E_2(x)$ values are small, if we have 5 or more projections then SA seems to give solutions being nearer to the original object.

7 Conclusion

Summarizing the results we can say that there is no huge difference between the qualities of the reconstructed images of the two methods.

Acknowledgments

The CVGPR-group gratefully acknowledges support by Siemens Medical Solutions, Forchheim, Germany. The fifth author was supported during his stay in Mannheim by the Alexander von Humboldt-Foundation. This work was partially supported by the NSF Grant DMS 0306215 and the OTKA Grant T048476.

References

1. Herman, G.T., Kuba, A., eds.: Discrete Tomography: Foundations, Algorithms, and Applications. Birkhäuser, Boston (1999)
2. Herman, G.T., Kuba, A., eds.: Advances in Discrete Tomography and Its Applications. Birkhäuser, Boston (2006). To appear
3. S. Krimmel, J. Baumann, Z. Kiss, A. Kuba, A. Nagy, J. Stephan: Discrete tomography for reconstruction from limited view angles in non-destructive testing. Electronic Notes in Discrete Mathematics **20** (2005) 455–474
4. G. T. Herman, A. Kuba: Discrete tomography in medical imaging. Proc. of the IEEE **91** (2003) 1612–1626
5. Liao, H.Y., Herman, G.T.: A method for reconstructing label images from a few projections, as motivated by electron microscopy. Annals of Operations Research. To appear
6. Weber, S., Schüle, T., Schnörr, C., Hornegger, J.: A linear programming approach to limited angle 3d reconstruction from dsa projections. Special Issue of Methods of Information in Medicine **4** (2004) 320–326
7. Schüle, T., Schnörr, C., Weber, S., Hornegger, J.: Discrete tomography by convex-concave regularization and D.C. programming. Discrete Applied Mathematics **151** (2005) 229–243
8. Schnörr, C., Schüle, T., Weber, S.: Variational Reconstruction with DC-Programming. In Herman, G.T., Kuba, A., eds.: Advances in Discrete Tomography and Its Applications. Birkhäuser, Boston (2006). To appear
9. Birgin, E.G., Martínez, J.M., Raydan, M.: Algorithm 813: SPG - software for convex-constrained optimization. ACM Transactions on Mathematical Software **27** (2001) 340–349
10. Metropolis, N., A. Rosenbluth, A. T. Rosenbluth, M., Teller, E.: Equation of state calculation by fast computing machines. J. Chem. Phys. **21** (1953), 1087–1092.

Construction of Switching Components

Steffen Zopf

Department of Image Processing and Computer Graphics,
University of Szeged, Árpád tér 2. H-6720 Szeged, Hungary
steffen@inf.u-szeged.hu, ZS0307@aol.com

Abstract. Switching components play an important role investigating uniqueness of problems in discrete tomography. General projections and additive projections as well as switching components w.r.t. these projections are defined. Switching components are derived by combining other switching components.

The composition of switching components into minimal ones in case of additive projections is proved. We also prove, that the product of minimal switching components is also minimal.

1 Introduction

In many scenarios of discrete tomography the so-called switching components play an important role. In the current paper we consider switching components from a more general point of view. The paper is intended to serve as a starting point for further description or even construction of the set of all switching components w.r.t. a given projection. This may be done via composing switching components into bigger ones. For projections derived from lower dimensional projections – let them be additive, as in the unabsorbed or absorbed case, or even more general – we can decide whether a switching component is minimal. This will possibly give the opportunity to describe or construct the set of all the switching components of this kind of projections.

2 Switching Components

2.1 Generalized Projections

Consider a finite or countably infinite set $\mathcal{L}$ (for example an integer lattice) and a set $\mathcal{R}$ containing the so-called rays of $\mathcal{L}$. This set may, but need not be a set of certain subsets of $\mathcal{L}$. A ray of $\mathcal{L}$ will be denoted by R, the *set of the rays* is $\mathcal{R}(\mathcal{L})$.

Given a (finite or countably infinite) set $\mathcal{L}$ and a set of rays $\mathcal{R} = \mathcal{R}(\mathcal{L})$ on $\mathcal{L}$. Furthermore, let $\mathbb{F}$ be a (commutative) ring, for example the ring of the integer, real or complex numbers.

Definition 1. *A function $\mathcal{P}$ is called a generalized ($\mathbb{F}$-valued) projection of the pair $(\mathcal{L}, \mathcal{R}(\mathcal{L}))$, if for every (finite) subset $G \subset \mathcal{L}$ and for every ray $R \in \mathcal{R}(\mathcal{L})$ $\mathcal{P}$*

A. Kuba, L.G. Nyúl, and K. Palágyi (Eds.): DGCI 2006, LNCS 4245, pp. 157–168, 2006.

returns a value $\mathcal{P}(G, R) \in \mathbb{F}$. *In order to emphasize the projection as a function on the subsets of the set* $\mathcal{L}$ *we will also write*

$$\mathcal{P}^{(R)}(G) \;:=\; \mathcal{P}(G, R)\,.$$

Definition 2. *A generalized projection* $\mathcal{P}$ *is called an additive projection, if*

$$\mathcal{P}^{(R)}(G_1 \cup G_2) \;=\; \mathcal{P}^{(R)}(G_1) \;+\; \mathcal{P}^{(R)}(G_2) \tag{1}$$

for all rays $R \in \mathcal{R}(\mathcal{L})$
for all (finite) sets $G_1, G_2 \subset \mathcal{L}$ with $G_1 \cap G_2 = \emptyset$.

Let $\mathcal{P}$ an additive projection. Considering the projection on an arbitrary ray R as above, we then get for a (finite) subset G of $\mathcal{L}$

$$\mathcal{P}^{(R)}(G) \;=\; \sum_{g\in G} \mathcal{P}^{(R)}(\{g\}) \;=\; \sum_{g\in\mathcal{L}} \chi_G(g)\cdot\omega_g^{(R)}\,. \tag{2}$$

with the weights $\omega_g^{(R)} \;:=\; \mathcal{P}^{(R)}(\{g\}) \in \mathbb{F}$.

2.2 Product of Projections

Let $(\mathcal{L}_1, \mathcal{R}_1(\mathcal{L}_1), \mathcal{P}_1)$ and $(\mathcal{L}_2, \mathcal{R}_2(\mathcal{L}_2), \mathcal{P}_2)$ be two projections. $\mathcal{L} := \mathcal{L}_1 \times \mathcal{L}_2$.

Let a ray on the base set $\mathcal{L}$ be defined either as a pair (g_1, r_2) with $g_1 \in \mathcal{L}_1$ and $r_2 \in \mathcal{R}_2(\mathcal{L}_2)$ or as a pair (r_1, g_2) with $r_1 \in \mathcal{R}_1(\mathcal{L}_1)$ and $g_2 \in \mathcal{L}_2$. Hence, the rays on $\mathcal{L}$ form the set

$$\mathcal{R}(\mathcal{L}) \;=\; (\mathcal{L}_1 \times \mathcal{R}_2(\mathcal{L}_2)) \;\dot\cup\; (\mathcal{R}_1(\mathcal{L}_1) \times \mathcal{L}_2)\,.$$

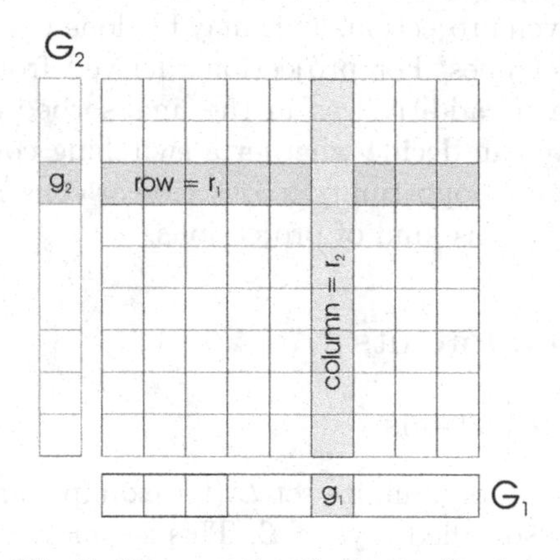

Fig. 1. Rays on the Cartesian Product

Let the transections of a subset $G \subset \mathcal{L}$ be defined by

$$T_2(G\,|\,g_1) \;:=\; \{g \in \mathcal{L}_2 \,|\, (g_1, g) \in G\}$$

and

$$T_1(G\,|\,g_2) \;:=\; \{g \in \mathcal{L}_1 \,|\, (g, g_2) \in G\}\,.$$

Definition 3. *Let*

$$\begin{array}{lll}\forall\ (g_1, r_2) \in \mathcal{L}_1 \times \mathcal{R}_2(\mathcal{L}_2) & \mathcal{P}^{(g_1,r_2)}(G) & := \mathcal{P}_2^{(r_2)}(T_2(G \,|\, g_1)) \\ \forall\ (r_1, g_2) \in \mathcal{R}_1(\mathcal{L}_1) \times \mathcal{L}_2 & \mathcal{P}^{(r_1,g_2)}(G) & := \mathcal{P}_1^{(r_1)}(T_1(G \,|\, g_2))\ . \end{array} \tag{3}$$

We call the triple $(\mathcal{L}, \mathcal{R}(\mathcal{L}), \mathcal{P})$ the product of the projections $(\mathcal{L}_1, \mathcal{R}_1(\mathcal{L}_1), \mathcal{P}_1)$ and $(\mathcal{L}_2, \mathcal{R}_2(\mathcal{L}_2), \mathcal{P}_2)$ and denote it by $\mathcal{P}_1 \times \mathcal{P}_2$.

2.3 Generalized Switching Components

Let $\mathcal{L}$ be a given set, $\mathcal{R}(\mathcal{L})$ a set of rays in $\mathcal{L}$ and $\mathcal{P}$ a given projection. Let G be a (finite) subset of $\mathcal{L}$ and c and c^S two functions on G with values in $\{0\,;1\}$.

Definition 4. *We say a function ε on $\mathcal{L}$ switchable w.r.t. the pair (G, c) if*

$$\forall\ g \in G\ :\ \varepsilon(g) = c(g)\ .$$

Definition 5. *If ε is switchable w.r.t. (G, c), then we say the function $\varepsilon^S = \varepsilon^S_{(G,\,c,\,c^S)}$ that we get by replacing the values of ε by those of c^S on the set G, the switching result of ε with respect to the triplet (G, c, c^S).*

$$\varepsilon^S(g) = \varepsilon^S_{(G,\,c,\,c^S)}(g) := \begin{cases} c^S(g) \text{ if } g \in G \\ \varepsilon(g) \ \ \text{otherwise} \end{cases}\ . \tag{4}$$

Definition 6. *The triplet (G, c, c^S) is said a switching component with respect to the projection function $\mathcal{P}$, if whenever ε is switchable w.r.t. (G, c) the projection values $\mathcal{P}\varepsilon$ and $\mathcal{P}\varepsilon^S_{(G,\,c,\,c^S)}$ are identical.*

$$\forall\ R \in \mathcal{R}(\mathcal{L})\ :\ \mathcal{P}^{(R)}(\varepsilon^S_{(G,\,c,\,c^S)}) \ = \ \mathcal{P}^{(R)}(\varepsilon)\ . \tag{5}$$

For a switching component $S = (G, c, c^S)$ we call the set G the *domain* of S and denote it by $G = dom(S)$.

Lemma 1

(i) *For every set $G \subseteq \mathcal{L}$ and every function $c : G \to \{0; 1\}$ the triplet (G, c, c) is a switching component of each projection $\mathcal{P}$.*
(ii) *If (G, c, c^S) is a switching component with respect to the projection function $\mathcal{P}$, then (G, c^S, c) is also a switching component.*
(iii) *If $(G, c, c^{(1)})$ and $(G, c^{(1)}, c^{(2)})$ are switching components with respect to the projection function $\mathcal{P}$ then $(G, c, c^{(2)})$ is also a switching component.*

Proof
The proof is evident and can be omitted. □

For a switching component $S = (G, c, c^S)$ we say the switching component (G, c^S, c) the *switched switching component* and denote it by $S^{sw} = (G, c^S, c)$.
The empty switching component is denoted by $E = (\emptyset, \emptyset, \emptyset)$.

Definition 7. *Let $\mathcal{P}$ be a projection on $\mathcal{L}$ and (G, c, c^S) a switching component w.r.t. this projection. The switching component is called a minimal switching component if whenever (G', c', c'^S) is also a switching component w.r.t. $\mathcal{P}$ and G' is a subset of G and $c|_{G'} = c'$ and $c^S|_{G'} = c'^S$ then $G' = G$ or $G' = \emptyset$.*

Lemma 2. *If $\mathcal{P}$ is an additive projection on $\mathcal{L}$ and $S = (G, c, c^S)$ is a minimal switching component w.r.t. $\mathcal{P}$, then*

$$\forall\, g \in G \;:\; c(g) \neq c^S(g) \;.$$

Proof
Let $G_e = \{g \in G \,|\, c(g) = c^S(g)\}$. Consider the set $G - G_e$ and the restrictions of c and c^S to this set. Let R be an arbitrary ray of $\mathcal{R}(\mathcal{L})$ and $\varepsilon : \mathcal{L} \to \{0\,;1\}$ a function on $\mathcal{L}$ with values in $\{0\,;1\}$. Suppose that ε is switchable w.r.t. $(G - G_e, c)$.

Since $\mathcal{P}$ is additive we can calculate the projections value $\mathcal{P}^{(R)}$ of ε on a ray $R \in \mathcal{R}$ as follows

$$\mathcal{P}^{(R)}(\varepsilon) \;=\; \mathcal{P}_{G^c}(\varepsilon) + \mathcal{P}_{G-G_e}(\varepsilon) + \mathcal{P}_{G_e}(\varepsilon) \;,$$

where $\mathcal{P}_G(f) := \mathcal{P}^{(R)}(f \cdot \chi_G)$ for a set G and a function f, and $G^c := \mathcal{L} - G$.

ε equals c on $G - G_e$, i.e.

$$\begin{aligned}
\mathcal{P}^{(R)}(\varepsilon) &= \mathcal{P}_{G^c}(\varepsilon) + \mathcal{P}_{G-G_e}(c) + \mathcal{P}_{G_e}(c) - \mathcal{P}_{G_e}(c) + \mathcal{P}_{G_e}(\varepsilon) \\
&= \mathcal{P}_{G^c}(\varepsilon) + \mathcal{P}_{G}(c) - \mathcal{P}_{G_e}(c) + \mathcal{P}_{G_e}(\varepsilon) \;.
\end{aligned}$$

Since (G, c, c^S) is a switching component we have $\mathcal{P}_G(c) = \mathcal{P}_G(c^S)$. Additionally, c and c^S are identical on G_e, thus $\mathcal{P}_{G_e}(c) = \mathcal{P}_{G_e}(c^S)$ also holds. Replacing these expressions we get

$$\begin{aligned}
\mathcal{P}^{(R)}(\varepsilon) &= \mathcal{P}_{G^c}(\varepsilon) + \mathcal{P}_{G}(c^S) - \mathcal{P}_{G_e}(c^S) + \mathcal{P}_{G_e}(\varepsilon) \\
&= \mathcal{P}_{G^c}(\varepsilon) + \mathcal{P}_{G-G_e}(c^S) + \mathcal{P}_{G_e}(\varepsilon) \;,
\end{aligned}$$

which means that replacing the values of c on $G - G_e$ by those of c^S already results in the same projection value, i.e. $(G - G_e, c|_{G-G_e}, c^S|_{G-G_e})$ is also a switching component w.r.t. $\mathcal{P}$. Since (G, c, c^S) is minimal, $G_e = \emptyset$. This is what we wanted to show. □

3 Deriving Switching Components

In the following we are going to derive switching components from other, known switching components.

3.1 Composition of Switching Components

Let $\mathcal{P}$ be a given projection, $S_1 = (G_1, c_1, c_1^S)$ and $S_2 = (G_2, c_2, c_2^S)$ two switching components with respect to $\mathcal{P}$.

Definition 8. *If the functions c_1^S and c_2 are identical on the set $G_1 \cap G_2$, i.e.*

$$\forall\, g \in G_1 \cap G_2 \ :\ c_1^S(g) \ = \ c_2(g) \ ,$$

then the two switching components are called composible.

Note, that the composible relation betwen switching components is not symmetric.

Lemma 3. *Suppose that S_1 and S_2 are two composible switching components. Defining c and c^S as*

$$c(g) \ = \ \begin{cases} c_1(g) \text{ if } g \in G_1 \\ c_2(g) \text{ if } g \in G_2 - G_1 \end{cases}$$

and

$$c^S(g) \ = \ \begin{cases} c_1^S(g) \text{ if } g \in G_1 - G_2 \\ c_2^S(g) \text{ if } g \in G_2 \end{cases} ,$$

the triplet $(G_1 \cup G_2, c, c^S)$ is a switching component with respect to the projection $\mathcal{P}$.

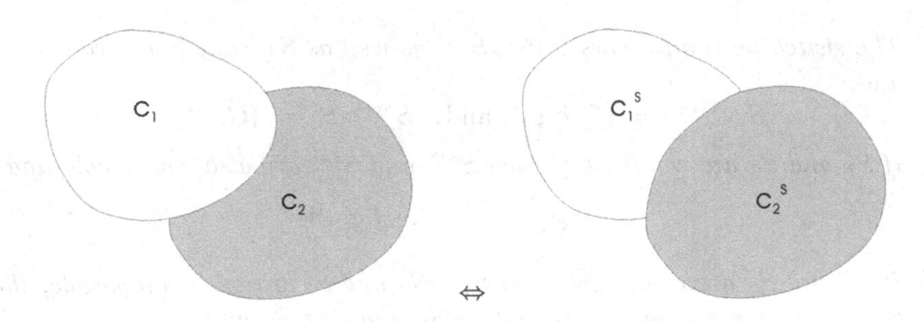

Fig. 2. The composition of two switching components

Proof
Let $R \in \mathcal{R}(\mathcal{L})$, and ε a function $\mathcal{L} \to \{0\,;1\}$ switchable w.r.t to $(G_1 \cup G_2, c)$.

Because of the definition of c, ε is identical to c_1 on G_1. Thus, the switching component (G_1, c_1, c_1^S) can be applied, and as the result we get the following

$$\mathcal{P}^{(R)}(\varepsilon) \ = \ \mathcal{P}^{(R)}(\varepsilon_1) \ ,$$

where

$$\varepsilon_1(g) \ = \ \begin{cases} c_1^S(g) \text{ if } g \in G_1 \\ \varepsilon(g) \text{ otherwise} \end{cases} .$$

Now, (G_2, c_2, c_2^S) can be applied, since c_1^S and c_2 are identical on the set $G_1 \cap G_2$ and ε and also ε_1 are equal to c_2 on $G_2 - G_1$ and we have

$$\mathcal{P}^{(R)}(\varepsilon_1) \ = \ \mathcal{P}^{(R)}(\varepsilon_2) \ ,$$

where

$$\varepsilon_2(g) = \begin{cases} c_2^S(g) & \text{if } g \in G_2 \\ \varepsilon_1(g) & \text{otherwise} \end{cases} = \begin{cases} c_2^S(g) & \text{if } g \in G_2 \\ c_1^S(g) & \text{if } g \in G_1 - G_2 \\ \varepsilon(g) & \text{otherwise} \end{cases} .$$

This is the function after applying $(G_1 \cup G_2, c, c^S)$ and, hence, the projection value is the same. □

Definition 9. *Given two composible switching components S_1 and S_2. The switching component constructed in Lemma 3 is called the composition of the switching components, and we write*

$$(G_1 \cup G_2, c, c^S) = (G_1, c_1, c_1^S) \circ (G_2, c_2, c_2^S)$$

Lemma 4. *Let $\mathcal{L}$ be a set, $\mathcal{R}(\mathcal{L})$ a set of rays of $\mathcal{L}$ and $\mathcal{P}$ a projection on $(\mathcal{L}, \mathcal{R}(\mathcal{L}))$. Let $S = (G, c, c^S)$, $S_1 = (G_1, c_1, c_1^S)$, $S_2 = (G_2, c_2, c_2^S)$ and $S_3 = (G_3, c_3, c_3^S)$ be switching components w.r.t. $(\mathcal{L}, \mathcal{R}(\mathcal{L}), \mathcal{P})$. For the composition of switching components w.r.t. $(\mathcal{L}, \mathcal{R}(\mathcal{L}), \mathcal{P})$ the following properties hold.*

(i) S and the empty switching component E are composible and

$$S \circ E = E \circ S = S .$$

(ii) The switching components S and S^{sw} as well as S^{sw} and S are composible and

$$S \circ S^{sw} = (G, c, c) \quad \text{and} \quad S^{sw} \circ S = (G, c^S, c^S)$$

(iii) If S_1 and S_2 are composible, then S_2^{sw} and S_1^{sw} are also composible and

$$S_2^{sw} \circ S_1^{sw} = (S_1 \circ S_2)^{sw} .$$

(iv) If S_1 and S_2 are composible and $S_1 \circ S_2$ and S_3 are also composible, then S_2 and S_3 are composible as well as S_1 and $S_2 \circ S_3$ and

$$(S_1 \circ S_2) \circ S_3 = S_1 \circ (S_2 \circ S_3) .$$

Proof
The proof of these properties follow immediately from the definition. □

3.2 Symmetric Composition of Switching Components

In the following, let $\mathcal{P}$ be an additive projection. $S_1 = (G_1, c_1, c_1^S)$ and $S_2 = (G_2, c_2, c_2^S)$ two switching components w.r.t. $\mathcal{P}$.

Definition 10. *The switching components S_1 and S_2 are called symmetric composible, if*

$$\forall\, g \in G_1 \cap G_2 \ : \ c_1^S(g) = c_2(g) \quad \text{and} \quad c_2^S(g) = c_1(g) .$$

Note, that the symmetric composible relation betwen switching components is a symmetric one.

Lemma 5. *Assume, that S_1 and S_2 are symmetric composible. Define c and c^S on $G_1 \triangle G_2$, the symmetric difference of G_1 and G_2, as*

$$c(g) := \begin{cases} c_1(g) & \text{if } g \in G_1 - G_2 \\ c_2(g) & \text{if } g \in G_2 - G_1 \end{cases}$$

and

$$c^S(g) := \begin{cases} c_1^S(g) & \text{if } g \in G_1 - G_2 \\ c_2^S(g) & \text{if } g \in G_2 - G_1 \end{cases}$$

Then the triplet $(G_1 \triangle G_2, c, c^S)$ is a switching component with respect to the projection $\mathcal{P}$.

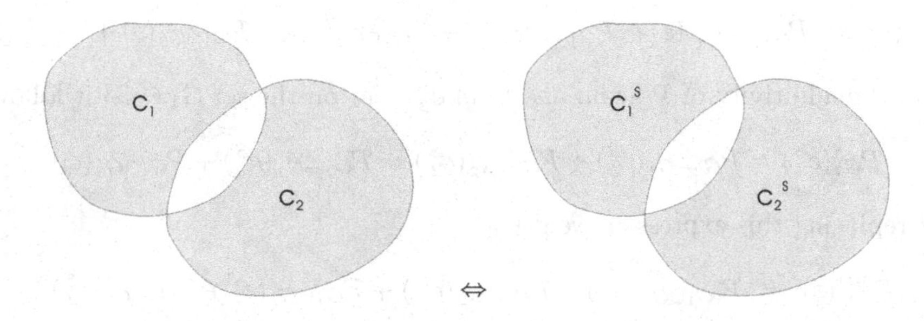

Fig. 3. The symmetric composition of two switching components

Proof
Let $R \in \mathcal{R}(\mathcal{L})$ be a ray, and $\varepsilon : \mathcal{L} \longrightarrow \{0\,;1\}$ a function switchable w.r.t. $(G_1 \triangle G_2, c)$.

For a set G and a function f, we denote

$$\mathcal{P}_G(f) := P^{(R)}(f \cdot \chi_G) .$$

G^c denotes the complement of the set G w.r.t. to the set $\mathcal{L}$, i.e. $G^c = \mathcal{L} - G$.

Since the projection $\mathcal{P}$ is additive and ε equals c_1 and c_2 on $G_1 - G_2$ and $G_2 - G_1$ respectively, for the projection of ε we then have

$$\mathcal{P}^{(R)}(\varepsilon) = \mathcal{P}_{(G_1 \cup G_2)^c}(\varepsilon) + \mathcal{P}_{G_1 - G_2}(c_1) + \mathcal{P}_{G_1 \cap G_2}(\varepsilon) + \mathcal{P}_{G_2 - G_1}(c_2) .$$

From the additivity of $\mathcal{P}$ we have

$$\mathcal{P}_{G_1 - G_2}(c_1) = \mathcal{P}_{G_1}(c_1) - \mathcal{P}_{G_1 \cap G_2}(c_1)$$

and, hence,

$$\mathcal{P}^{(R)}(\varepsilon) = \mathcal{P}_{(G_1 \cup G_2)^c}(\varepsilon) + \mathcal{P}_{G_1}(c_1) - \mathcal{P}_{G_1 \cap G_2}(c_1) + \mathcal{P}_{G_1 \cap G_2}(\varepsilon) + \mathcal{P}_{G_2 - G_1}(c_2) .$$

(G_1, c_1, c_1^S) is a switching component w.r.t. $\mathcal{R}$, that's why we can replace c_1 by c_1^S

$$\mathcal{P}^{(R)}(\varepsilon) = \mathcal{P}_{(G_1 \cup G_2)^c}(\varepsilon) + \mathcal{P}_{G_1}(c_1^S) - \mathcal{P}_{G_1 \cap G_2}(c_1) + \mathcal{P}_{G_1 \cap G_2}(\varepsilon) + \mathcal{P}_{G_2 - G_1}(c_2) .$$

Since $\mathcal{P}$ is additive and $c_1^S = c_2$ on the set $G_1 \cap G_2$

$$\mathcal{P}_{G_1}(c_1^S) = \mathcal{P}_{G_1-G_2}(c_1^S) + \mathcal{P}_{G_1\cap G_2}(c_1^S) = \mathcal{P}_{G_1-G_2}(c_1^S) + \mathcal{P}_{G_1\cap G_2}(c_2)$$

and

$$\mathcal{P}_{G_2-G_1}(c_2) + \mathcal{P}_{G_1\cap G_2}(c_2) = \mathcal{P}_{G_2}(c_2)$$

and, thus,

$$\mathcal{P}^{(R)}(\varepsilon) \;=\; \mathcal{P}_{(G_1\cup G_2)^c}(\varepsilon) + \mathcal{P}_{G_1-G_2}(c_1^S) - \mathcal{P}_{G_1\cap G_2}(c_1) + \mathcal{P}_{G_1\cap G_2}(\varepsilon) + \mathcal{P}_{G_2}(c_2)\,.$$

From (G_2, c_2, c_2^S) also being a switching component w.r.t. $\mathcal{P}$ we get

$$\mathcal{P}^{(R)}(\varepsilon) \;=\; \mathcal{P}_{(G_1\cup G_2)^c}(\varepsilon) + \mathcal{P}_{G_1-G_2}(c_1^S) - \mathcal{P}_{G_1\cap G_2}(c_1) + \mathcal{P}_{G_1\cap G_2}(\varepsilon) + \mathcal{P}_{G_2}(c_2^S)\,.$$

From the additivity of $\mathcal{P}$ again and from $c_2^S = c_1$ on the set $G_1 \cap G_2$ it follows

$$\mathcal{P}_{G_2}(c_2^S) = \mathcal{P}_{G_2-G_1}(c_2^S) + \mathcal{P}_{G_1\cap G_2}(c_2^S) = \mathcal{P}_{G_2-G_1}(c_2^S) + \mathcal{P}_{G_1\cap G_2}(c_1)$$

and replacing this expression we get

$$\mathcal{P}^{(R)}(\varepsilon) \;=\; \mathcal{P}_{(G_1\cup G_2)^c}(\varepsilon) + \mathcal{P}_{G_1-G_2}(c_1^S) + \mathcal{P}_{G_1\cap G_2}(\varepsilon) + \mathcal{P}_{G_2-G_1}(c_2^S)\,.$$

The expression on the right hand side is an expression for $\mathcal{P}^{(R)}(\varepsilon^S)$ and, hence, as the final result we have

$$\mathcal{P}^{(R)}(\varepsilon) = \mathcal{P}^{(R)}(\varepsilon^S)\,.$$

This is what we wanted to prove. □

Definition 11. *For two symmetric composible switching components S_1 and S_2, the switching component constructed in Lemma 5 is called the symmetric composition of the two switching components, and we write*

$$(G_1 \triangle G_2, c, c^S) \;=\; (G_1, c_1, c_1^S) \odot (G_2, c_2, c_2^S)\,.$$

The following properties of the symmetric composition hold.

Lemma 6. *Let $\mathcal{L}$ be a set, $\mathcal{R}(\mathcal{L})$ be a set of rays on $\mathcal{L}$ and $\mathcal{P}$ be an additive projection on these rays. Furthermore, let $S = (G, c, c^S)$, $S_1 = (G_1, c_1, c_1^S)$, $S_2 = (G_2, c_2, c_2^S)$, and $S_3 = (G_3, c_3, c_3^S)$ switching components w.r.t. $(\mathcal{L}, \mathcal{R}(\mathcal{L}), \mathcal{P})$.*

(i) S and E are always symmetric composible and

$$S \odot E \;=\; E \odot S \;=\; S\,.$$

(ii) S and S^{sw} are symmetric composible and

$$S \odot S^{sw} \;=\; S^{sw} \odot S \;=\; E\,.$$

(iii) If $S_1 \odot S_2$ and S_3 are symmetric composible then S_1 and $S_2 \odot S_3$ are also symmetric composible and

$$(S_1 \odot S_2) \odot S_3 = S_1 \odot (S_2 \odot S_3) .$$

(iv) If S_1 and S_2 are symmetric composible, then

$$S_2 \odot S_1 = S_1 \odot S_2 .$$

Proof
The proof of these properties follows immediately from the definition. □

Proposition 1. *Let S be an arbitrary switching component w.r.t. an additive projection $\mathcal{P}$. There exists a constant switching component S_0 and a sequence of* ***minimal*** *switching components $\{S_i\}_{i=1}^N$ with pairwise disjunct domains with the symmetric composition being S*

$$S = S_0 \odot S_1 \odot \cdots \odot S_N$$

with

$$c|_{S_0} \equiv c^S|_{S_0} \quad \text{and} \quad dom(S_i) \cap dom(S_j) = \emptyset \quad \text{for } i \neq j .$$

Proof
This easily follows from the definitions. □

3.3 Product of Switching Components

Let $\mathcal{P}_1$ be a generalized projection w.r.t. $(\mathcal{L}_1, \mathcal{R}_1(\mathcal{L}_1))$ and $\mathcal{P}_2$ a generalized projection w.r.t. $(\mathcal{L}_2, \mathcal{R}_2(\mathcal{L}_2))$ and $\mathcal{P} = \mathcal{P}_1 \times \mathcal{P}_2$ their product as defined in Section 2.2.

Furthermore, let $S_1 = (G_1, c_1, c_1^S)$ and $S_2 = (G_2, c_2, c_2^S)$ be a switching component w.r.t. $\mathcal{P}_1$ and $\mathcal{P}_2$, resp.

The following figure gives an idea how to create the switching component on the set $G_1 \times G_2$ or on a subset of $G_1 \times G_2$ based on the two given switching components.

Let

$$G_1 \times_{(c_1, c_2)} G_2 := G_1 \times G_2 - \{(g_1, g_2) \,|\, c_1(g_1) = c_1^S(g_1) \wedge c_2(g_2) = c_2^S(g_2)\} .$$

For $c_1(g_1) \neq c_1^S(g_1) \;\wedge\; c_2(g_2) \neq c_2^S(g_2)$ we define

$$c(g_1, g_2) = \begin{cases} 0 \text{ if } c_1(g_1) = c_2(g_2) \;\wedge\; c_1^S(g_1) = c_2^S(g_2) \\ 1 \text{ if } c_1(g_1) = c_2^S(g_2) \;\wedge\; c_1^S(g_1) = c_2(g_2) \end{cases}$$

and

$$c^S(g_1, g_2) = \begin{cases} 1 \text{ if } c_1(g_1) = c_2(g_2) \;\wedge\; c_1^S(g_1) = c_2^S(g_2) \\ 0 \text{ if } c_1(g_1) = c_2^S(g_2) \;\wedge\; c_1^S(g_1) = c_2(g_2) \end{cases} .$$

For the other cases we define

$$c(g_1, g_2) = \begin{cases} 0 & \text{if } c_1(g_1) = c_1^S(g_1) = 0 \ \vee \ c_2(g_2) = c_2^S(g_2) = 0 \\ 1 & \text{if } c_1(g_1) = c_1^S(g_1) = 1 \ \vee \ c_2(g_2) = c_2^S(g_2) = 1 \end{cases}$$

and

$$c^S(g_1, g_2) = \begin{cases} 0 & \text{if } c_1(g_1) = c_1^S(g_1) = 0 \ \vee \ c_2(g_2) = c_2^S(g_2) = 0 \\ 1 & \text{if } c_1(g_1) = c_1^S(g_1) = 1 \ \vee \ c_2(g_2) = c_2^S(g_2) = 1 \end{cases} .$$

Lemma 7. *The triple* $(G_1 \times_{(c_1, c_2)} G_2, c, c^S)$ *with the two functions* c *and* c^S *as defined above is a switching component w.r.t. the product projection* $\mathcal{P} = \mathcal{P}_1 \times \mathcal{P}_2$.

Proof

Let $\varepsilon : A \to \{0;1\}$ an arbitry function for which

$$\forall \ (g_1, g_2) \in G_1 \times_{(c_1, c_2)} G_2 \ : \ \varepsilon(g_1, g_2) \ = \ c(g_1, g_2)$$

and we investigate the function ε^S with

$$\varepsilon^S(g_1, g_2) \ := \begin{cases} c^S(g_1, g_2) & \text{if } (g_1, g_2) \in G_1 \times_{(c_1, c_2)} G_2 \\ \varepsilon(g_1, g_2) & \text{otherwise} \end{cases}$$

We define

$$\forall \ g \in G_1 \ : \ \varepsilon_1^{(g_2)}(g) \ := \ \varepsilon(g, g_2)$$

and

$$\forall \ g \in G_2 \ : \ \varepsilon_2^{(g_1)}(g) \ := \ \varepsilon(g_1, g)$$

The appropriate projections $\mathcal{P}_1$ and $\mathcal{P}_2$ can be applied to the functions $\varepsilon_1^{(g_2)}$ and $\varepsilon_2^{(g_1)}$.

First, let $(g_1, r_2) \in \mathcal{L}_1 \times \mathcal{R}_2(\mathcal{L}_2)$. For $g_1 \notin G_1$ the following equality is trivial.

$$\mathcal{P}^{(g_1, r_2)}(\varepsilon^S) \ = \ \mathcal{P}_2^{(r_2)}([\varepsilon_2^S]^{(g_1)}) \ = \ \mathcal{P}_2^{(r_2)}(\varepsilon_2^{(g_1)}) \ = \ \mathcal{P}^{(g_1, r_2)}(\varepsilon)$$

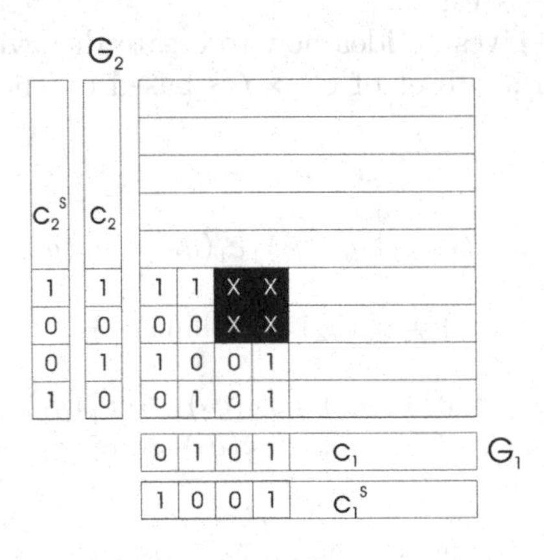

Fig. 4. Product of Switching Components

Let now $g_1 \in G_1$ and $c_1(g_1) \neq c_1^S(g_1)$. As we can see from the definition of c, in this case it holds that

$$c(g_1, g) \equiv c_2(g) \quad \text{and} \quad c^S(g_1, g) \equiv c_2^S(g) \quad \text{for } g \in G_2 ,$$

or

$$c(g_1, g) \equiv c_2^S(g) \quad \text{and} \quad c^S(g_1, g) \equiv c_2(g) \quad \text{for } g \in G_2 .$$

Since (G_2, c_2, c_2^S) is a switching component for the projection $\mathcal{P}_2$, again we have

$$\mathcal{P}^{(g_1,r_2)}(\varepsilon^S) = \mathcal{P}^{(g_1,r_2)}(\varepsilon)$$

If $g_1 \in G_1$ and $c_1(g_1) = c_1^S(g_1)$ we can see from the definition of c that $c^S(g_1, g) \equiv c(g_1, g)$ for $g \in G_2$, if $c_2(g_2) \neq c_2^S(g_2)$. Hence, $\varepsilon(g_1, g) \equiv \varepsilon^S(g_1, g)$ for these $g \in G_2$ and, again,

$$\mathcal{P}^{(g_1,r_2)}(\varepsilon^S) = \mathcal{P}^{(g_1,r_2)}(\varepsilon) .$$

The second case, namely if $(r_1, g_2) \in \mathcal{R}_1(\mathcal{L}_1) \times \mathcal{L}_2$, can be shown similarly. □

Definition 12. *We call the switching component $S = (G_1 \times_{(c_1, c_2)} G_2, c, c^S)$ the product of the switching components $S_1 = (G_1, c_1, c_1^S)$ and $S_2 = (G_2, c_2, c_2^S)$ and denote it by $S = S_1 \times S_2 = (G_1, c_1, c_1^S) \times (G_2, c_2, c_2^S)$.*

Proposition 2. *If $S_1 = (G_1, c_1, c_1^S)$ and $S_2 = (G_2, c_2, c_2^S)$ are two minimal (non-empty) switching components w.r.t. the generalized projections $\mathcal{P}_1$ and $\mathcal{P}_2$ then their product $S = S_1 \times S_2$ is also a minimal switching component w.r.t. to the product projection $\mathcal{P} = \mathcal{P}_1 \times \mathcal{P}_2$.*

Proof
Suppose that $S = S_1 \times S_2$ is not a minimal switching component, i.e. there exists a (G', c', c'^S) switching component w.r.t. $\mathcal{P} = \mathcal{P}_1 \times \mathcal{P}_2$ for which $\emptyset \subsetneq G' \subsetneq G = G_1 \times_{(c_1, c_2)} G_2$ and $c(S)|_{G'} = c'$ and $c(S)^S|_{G'} = c'^S$.

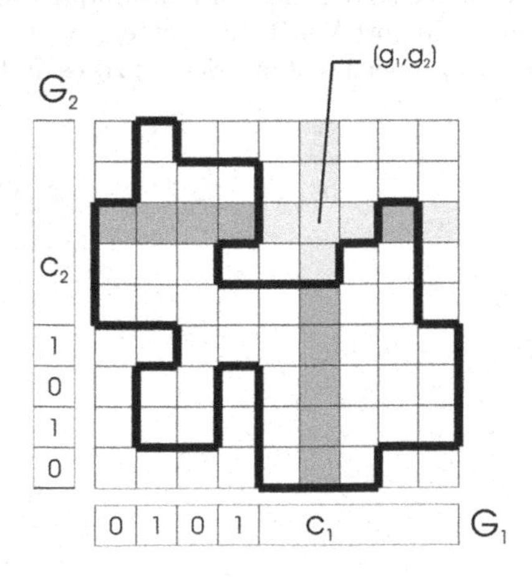

Fig. 5. Product of minimal switching Components

Let $(g_1, g_2) \in G_1 \times_{(c_1, c_2)} G_2 - G'$ and consider the sets $T_2(G' \mid g_1)$ and $T_1(G' \mid g_2)$.

Since $(g_1, g_2) \in G_1 \times_{(c_1, c_2)} G_2$ we have $c_1(g_1) \neq c_1^S(g_1)$ or $c_2(g_2) \neq c_2^S(g_2)$.

If $c_1(g_1) \neq c_1^S(g_1)$ then

$$c \mid T_2(G \mid g_1) \equiv c_2 \quad \text{or} \quad c \mid T_2(G \mid g_1) \equiv c_2^S$$

and, hence, $T_2(G' \mid g_1) = T_2(G \mid g_1)$ or $T_2(G' \mid g_1) = \emptyset$ because c_2 is minimal.

Similarly, if $c_2(g_2) \neq c_2^S(g_2)$ then

$$c \mid T_1(G \mid g_2) \equiv c_1 \quad \text{or} \quad c \mid T_1(G \mid g_2) \equiv c_1^S$$

and, hence, $T_1(G' \mid g_2) = T_1(G \mid g_2)$ or $T_1(G' \mid g_2) = \emptyset$ because c_1 is minimal.

Equality cannot hold because we supposed $(g_1, g_2) \in G_1 \times_{(c_1, c_2)} G_2 - G'$. Hence, the appropriate transection sets are all empty.

As one of the possibilties, let's suppose now, that $c_1(g_1) \neq c_1^S(g_1)$. In this case $T_2(G' \mid g_1) = \emptyset$. This implies that $\forall j \in G_2 : (g_1, j) \notin G'$ and so, whenever $j \in G_2 \wedge c_2(j) \neq c_2^S(j)$ we have $T_1(G' \mid j) = \emptyset$ also.

From this, on the other hand, it follows that whenever $i \in G_1 \wedge c_1(i) \neq c_1^S(i)$ we have $T_2(G' \mid i) = \emptyset$, and as the final consequence $G' = \emptyset$. This completes the proof. □

References

1. Balogh, E., Kuba, A., Del Lungo, A., and Nivat, M. : Reconstruction of binary matrices from absorbed projections. In: Braquelaire, A., Lachaud, J.-O., Vialard, A., eds.: Discrete Geometry for Computer Imagery, LNCS 2301. Springer-Verlag, Berlin Heidelberg New York (2002) 392–403
2. Kuba, A., Nivat, M. : Reconstruction of Discrete Sets with Absorption. Linear Algebra and Its Applications **339** (2000) 171–194
3. Kuba, A., Nivat, M. : A sufficient condition for non-uniqueness in binary tomography with absorption. Technical Report Vol. 1953. University of Szeged (2001)
4. Ryser, H. : Combinatorical properties of matrices of zeros and ones. Canad. J. Math. **9** (1957) 371–377

Minimal Non-simple and Minimal Non-cosimple Sets in Binary Images on Cell Complexes

T. Yung Kong

Department of Computer Science, Queens College, City University of New York,
65-30 Kissena Boulevard, Flushing, NY 11367, U.S.A.

Abstract. The concepts of *weak component* and *simple* 1 are generalizations, to binary images on the n-cells of n-dimensional cell complexes, of the standard concepts of "26-component" and "26-simple" 1 in binary images on the 3-cells of a 3D cubical complex; the concepts of *strong component* and *cosimple* 1 are generalizations of the concepts of "6-component" and "6-simple" 1. Over the past 20 years, the problems of determining just which sets of 1's can be minimal non-simple, just which sets can be minimal non-cosimple, and just which sets can be minimal non-simple (minimal non-cosimple) without being a weak (strong) foreground component have been solved for the 2D cubical and hexagonal, 3D cubical and face-centered-cubical, and 4D cubical complexes. This paper solves these problems in much greater generality, for a very large class of cell complexes of dimension ≤ 4.

1 Introduction

In a binary image, the n-dimensional cells of an n-dimensional cell complex (most often, the 2D or 3D cubical complex) are labeled 1 or 0. Cells labeled 1 are referred to as 1's of the image, and cells labeled 0 are referred to as 0's.

We say that a 1 of the image is *simple* if "the topology of image is preserved" (in a sense which will be made precise in Sect. 4) when that 1 is changed into a 0. We say that a 1 is *cosimple* if the topology of the image is preserved in another, complementary, sense when the 1 is changed into a 0.

In the case of the 2D cubical complex, these are two of the oldest concepts of digital topology, and date back to the 1960's. Rosenfeld's concept of an "8-deletable" pixel in [20] is mathematically equivalent to our concept of a simple 1 in a binary image on the 2D cubical complex. The concept of a "4-deletable" pixel in [20] is similarly equivalent to our concept of a cosimple 1. Today, simple and cosimple 1's in binary images on the 2D cubical complex are often called "8-simple" and "4-simple", respectively. In binary images on the 3D cubical complex, simple 1's are often called "26-simple", cosimple 1's are often called "6-simple", and a number of non-trivial characterizations of such 1's have been published (e.g., in [2,21]).

A subset of the set of 1's of a binary image is said to be *simple* (*cosimple*) if the elements of that subset can be arranged in a sequence $D_1, \ldots, D_k$ in which each element D_i is simple (cosimple) after its predecessors $D_1, \ldots, D_{i-1}$ have all

A. Kuba, L.G. Nyúl, and K. Palágyi (Eds.): DGCI 2006, LNCS 4245, pp. 169–188, 2006.

been changed to 0's. Such sequences were apparently first studied by Ronse [18] in the 1980's, in the case of binary images on the 2D cubical complex.

A subset $\mathcal{S}$ of the set of 1's is said to be *minimal non-simple* or *MNS* (*minimal non-cosimple* or *MNCS*) if $\mathcal{S}$ is non-simple (non-cosimple) but every proper subset of $\mathcal{S}$ is simple (cosimple). MNS and MNCS sets were first introduced by Ronse [19], for the 2D cubical complex. (In that context, Ronse referred to MNS sets as "8-MND sets" and referred to MNCS sets as "4-MND sets"; MND stood for "minimal non-deletable".)

The principal application of the concepts of simple and cosimple sets of 1's is to the theory of parallel thinning algorithms for binary images. Each iteration of such an algorithm deletes (i.e., changes to 0) all 1's for which the configuration of nearby 1's and 0's satisfies the algorithm's deletion condition. Thinning algorithms are expected to "preserve topology" in the sense that the set of 1's deleted by the algorithm should always be simple or always be cosimple.

The concepts of MNS and MNCS sets provide the basis for a systematic method of verifying that a proposed parallel thinning algorithm satisfies either of these conditions. In the types of cell complex which seem most likely to be used in applications, only a few kinds of set can ever be MNS or MNCS, and such sets can have only a few elements. (For example, in the case of the 2D cubical complex Ronse showed in [19] that a set of 1's can be MNS only if every pair of those 1's are 8-adjacent—which implies that no MNS set can contain more than four 1's.) If we can deduce from a given parallel thinning algorithm's deletion condition that the set of 1's which are changed to 0's at a single iteration can never include a non-simple (non-cosimple) set of one of the kinds that can be MNS (MNCS), then we will have proved that the set of 1's that are changed to 0's at any iteration of the algorithm is always a simple (cosimple) set, so that the thinning algorithm does indeed "preserve topology" in the corresponding sense.

It can happen that a certain kind of set can be MNS (MNCS), but *only* in the very special case where the set is a weak (strong) component of the 1's. (Here the concepts of weak and strong components are generalizations, to sets of n-dimensional cells of nD cell complexes, of the well known concepts of 8- and 4-components, respectively, in sets of 2-cells of the 2D cubical complex.) For example, in the case of the 2D cubical complex Ronse showed in [19] that a set of two 1's that are 8-adjacent but not 4-adjacent can be MNS only if it is an 8-component of the 1's (i.e., only if neither of the 1's is 8-adjacent to any other 1 of the image). Knowing that sets of certain kinds cannot be MNS (MNCS) unless they are weak (strong) components of the 1's can considerably simplify the application of the verification method described above.

This motivates the problem of determining just which kinds of set can be MNS, just which kinds can be MNCS, and just which kinds can be MNS (MNCS) without being a weak (strong) component of the 1's. Ronse [19] solved these problems for the 2D cubical complex. Hall [6, Sect. 4] essentially solved the problems for the 2D hexagonal complex. The problems were solved for the 3D cubical complex by Ma [15] and Kong [10]. Gau and Kong [4] solved the problems for the 3D face-centered-cubical complex (whose 3-dimensional cells are rhombic dodecahedra) and, more recently, for the 4D cubical complex [5,11].

In this paper, we solve these problems for a very general class of cell complexes of dimension ≤ 4, namely the *xel-complexes* which we define in Sect. 3. The cubical, 2D hexagonal, and 3D face-centered-cubical complexes mentioned above, and most other complexes that have been considered in digital topology (such as the 3D body-centered-cubical complex [7,14], whose 3-dimensional cells are truncated octahedra), are simple examples of xel-complexes.

2 Contractibility, Polyhedra, and Polyhedral Cells

A set S in $\mathbb{R}^n$ is said to be *contractible* if S is nonempty, and S can be continuously deformed over itself to some point p in S. More precisely, S is contractible if and only if $S \neq \emptyset$ and there is a continuous mapping $h : S \times [0,1] \to S$ such that, for every point $s \in S$ and some point p in S, $h(s,0) = s$ and $h(s,1) = p$. A contractible set is necessarily connected.

Every convex set is contractible. More generally, if $\mathcal{P}$ is any nonempty collection of convex sets such that $\bigcap \mathcal{P} \neq \emptyset$, then $\bigcup \mathcal{P}$ is contractible—because if p is any point in $\bigcap \mathcal{P}$ then the map $h : \bigcup \mathcal{P} \times [0,1] \to \bigcup \mathcal{P}$ that is defined by $h(s,t) = tp + (1-t)s$ has the above-mentioned properties.

On the other hand, it is an easy consequence of basic results of algebraic topology that the boundary of a k-simplex—i.e., the set of all points that lie on one or more of the $(k-1)$-dimensional faces of the k-simplex—is not contractible.

In this paper a set in $\mathbb{R}^n$ is called a *polyhedron* if it is expressible as a union of a finite collection of simplexes (which may possibly include simplexes of different dimensions). Note that the empty set is a polyhedron, and that a polyhedron need not be connected. Evidently, the union of two polyhedra is a polyhedron. It is also not hard to prove that the intersection of two polyhedra is a polyhedron.

There is a simple characterization of contractible polyhedra in $\mathbb{R}^3$: A polyhedron P in $\mathbb{R}^3$ is contractible if and only if P is nonempty, connected, and simply connected, and $\mathbb{R}^3 \setminus P$ is connected. This characterization follows from well known results of algebraic topology—the Alexander duality theorem, and the theorems of Whitehead and Hurewicz [16, Chs. 5, 7, and 8].

For any integer $k \geq 0$, a *polyhedral k-cell* is a polyhedron that is homeomorphic to a k-simplex. A *polyhedral cell* is a set that is a polyhedral k-cell for some integer k; the integer k (which is always uniquely determined) is the *dimension* of the polyhedral cell. The dimension of a polyhedral cell C is denoted by $\dim(C)$. Note that a polyhedral 0-cell consists of just one point. A polyhedral cell is closed and bounded, and is contractible because it is homeomorphic to a simplex (which is a contractible set because it is nonempty and convex).

If C is a polyhedral k-cell, and $h : \sigma \to C$ is a homeomorphism of a k-simplex σ onto C, then the image under h of the boundary of the simplex σ is called the *manifold boundary* or just the *boundary* of C, and is denoted by ∂C. (This set does not depend on our choice of h and σ.) If C is a polyhedral 0-cell then $\partial C = \emptyset$.

3 Xel-Complexes

A *xel-complex* is a collection $\mathbf{K}$ that satisfies the following conditions for some positive integer n, which we call the *dimension* of $\mathbf{K}$ and denote by $\dim(\mathbf{K})$:

1. Each member of $\mathbf{K}$ is a polyhedral k-cell for some $k \leq n$, and $\bigcup \mathbf{K} = \mathbb{R}^n$.
2. No bounded region of $\mathbb{R}^n$ intersects infinitely many members of $\mathbf{K}$.
3. For all distinct $X, Y \in \mathbf{K}$, either $X \subsetneq \partial Y$, or $Y \subsetneq \partial X$, or $X \cap Y = \partial X \cap \partial Y$.
4. For all $X, Y \in \mathbf{K}$, either $X \cap Y = \emptyset$ or $X \cap Y \in \mathbf{K}$.
5. For all $X, Y \in \mathbf{K}$ such that $\emptyset \neq Y \subsetneq X$, $\bigcup\{D \in \mathbf{K} \mid D \subsetneq X \text{ and } D \cap Y = \emptyset\}$ is a contractible polyhedron.
6. For all $X, Y \in \mathbf{K}$ such that $X \cap Y = \emptyset$, there exist $X', Y' \in \mathbf{K}$ such that $\dim(X') = \dim(Y') = n$, $X' \supseteq X$, $Y' \supseteq Y$, and $X' \cap Y' = \emptyset$.

The only places in this paper where we use conditions 5 and 6 are in the proofs of assertion 4 of our first main theorem (Theorem 3) and assertions 3 and 4 of our second main theorem (Theorem 4).

Each member of a xel-complex $\mathbf{K}$ will be called a *xel* of $\mathbf{K}$, and a xel X will be called a *k-xel* if $\dim(X) = k$. An *mD xel-complex* is a xel-complex $\mathbf{K}$ for which $\dim(\mathbf{K}) = m$. The above conditions imply that if X and Y are xels of $\mathbf{K}$ such that $X \subsetneq Y$, then $X \subsetneq \partial Y$; in such cases we say X is a *proper face* of Y. So if C_1 and C_2 are distinct intersecting xels of $\mathbf{K}$ neither of which is a proper face of the other, then $C_1 \cap C_2 = \partial C_1 \cap \partial C_2$ is a proper face of C_1 and of C_2.

A simple and important example of an nD xel-complex is the *nD cubical complex*, whose xels are the Cartesian products $E_1 \times \ldots \times E_n$ in which each set E_i either is a singleton set of the form $\{i + 0.5\}$ for some integer i, or is a closed unit interval $[i - 0.5, i + 0.5]$ for some integer i. Here $E_1 \times \ldots \times E_n$ is a k-xel of the xel-complex if $n - k$ of the n E's are singleton sets and the other k E's are closed unit intervals. (Thus a k-xel of this xel-complex is an upright closed k-dimensional unit (hyper)cube in $\mathbb{R}^n$ whose vertices are located at points each of whose coordinates differs from an integer by exactly 0.5.)

If X and Y are n-xels of an nD xel-complex $\mathbf{K}$, then X is said to be *weakly adjacent* to Y if $X \neq Y$ and $X \cap Y \neq \emptyset$, and X is said to be *strongly adjacent* to Y if $X \cap Y$ is an $(n-1)$-xel of $\mathbf{K}$. If $\mathcal{T}$ is any set of n-xels of $\mathbf{K}$, then each equivalence class of the reflexive transitive closure of the restriction to $\mathcal{T}$ of the "is weakly adjacent to" relation is called a *weakly connected component* of $\mathcal{T}$. Similarly, each equivalence class of the reflexive transitive closure of the restriction to $\mathcal{T}$ of the "is strongly adjacent to" relation is called a *strongly connected component* of $\mathcal{T}$. We say $\mathcal{T}$ is *weakly connected* if $\mathcal{T} = \emptyset$ or if there is just one weakly connected component of $\mathcal{T}$. Similarly, we say $\mathcal{T}$ is *strongly connected* if $\mathcal{T} = \emptyset$ or if there is just one strongly connected component of $\mathcal{T}$. (In the 2D (3D) cubical complex, a set of 2-(3-)xels is strongly connected if and only if it is 4-(6-)connected, and is weakly connected if and only if it is 8-(26-)connected.) Evidently, every strongly connected set is weakly connected.

We now state (without proof) a number of properties of xel-complexes which will be used in proving our main theorems. Readers are encouraged to at least convince themselves that the 2D and 3D cubical complexes have these properties.

Property 1. If X is a xel of a xel-complex $\mathbf{K}$, then ∂X is a union of xels of $\mathbf{K}$. □

Property 2. If X_1 and X_2 are xels of a xel-complex $\mathbf{K}$ such that $X_1 \subsetneq X_2$, then $\dim(X_1) < \dim(X_2)$. □

Property 3. If X is a xel of a xel-complex $\mathbf{K}$, and $\dim(X) > 0$, then ∂X contains at least $\dim(X) + 1$ distinct 0-xels of $\mathbf{K}$. □

Property 4. If Z is an $(n-1)$-xel of an nD xel-complex $\mathbf{K}$, then there are n-xels $X_1, X_2 \in \mathbf{K}$ such that $X_1 \cap X_2 = \partial X_1 \cap \partial X_2 = Z$, and no other xel of $\mathbf{K}$ intersects $Z \setminus \partial Z$. □

Property 5. If X and X' are distinct n-xels of an nD xel-complex $\mathbf{K}$ such that $X \cap X' \neq \emptyset$, then there exists a sequence $X_0, X_1, \ldots, X_k$ of n-xels of $\mathbf{K}$ such that $X_0 = X$, $X_k = X'$, and, for $1 \leq i \leq k$, $X_{i-1} \cap X_i$ is an $(n-1)$-xel of $\mathbf{K}$ that contains $X \cap X'$. □

Property 6. If X and C are xels of a xel-complex $\mathbf{K}$ such that $X \subsetneq \partial C$, then there is a $(\dim(C) - 1)$-xel Y of $\mathbf{K}$ such that $X \subseteq Y \subsetneq \partial C$. □

4 MNS and MNCS Sets in Binary Images

Let $\mathbf{K}$ be an nD xel-complex, for some positive integer n, and let $\mathcal{G}$ be the set of all n-xels of $\mathbf{K}$. A function $\mathbb{I} : \mathcal{G} \to \{0, 1\}$ for which either $\mathbb{I}^{-1}[\{1\}]$ is finite or $\mathbb{I}^{-1}[\{0\}]$ is finite will be called a *binary image on* $\mathbf{K}$ or, more briefly, a $\mathbf{K}$*-image*. Note that $\mathbb{I}(X)$ is only defined if $X \in \mathcal{G}$ (i.e., if X is an n-xel of $\mathbf{K}$)—$\mathbb{I}(X)$ is undefined if X is a xel of lower dimension in $\mathbf{K}$.

If $\mathbb{I}$ is a $\mathbf{K}$-image, then each n-xel in $\mathbb{I}^{-1}[\{1\}]$ is called a 1 of $\mathbb{I}$, and each n-xel in $\mathbb{I}^{-1}[\{0\}]$ is called a 0 of $\mathbb{I}$. If $\mathcal{D}$ is any subset of the set of 1's of a $\mathbf{K}$-image $\mathbb{I}$, then we write $\mathbb{I} - \mathcal{D}$ to denote the $\mathbf{K}$-image whose set of 1's is $\mathbb{I}^{-1}[\{1\}] \setminus \mathcal{D}$. Changing $\mathbb{I}$ to $\mathbb{I} - \mathcal{D}$ is referred to as *deletion* of the set $\mathcal{D}$ from $\mathbb{I}$.

We write $\mathbb{I}^{\mathbf{c}}$ to denote the $\mathbf{K}$-image defined by $\mathbb{I}^{\mathbf{c}}(X) = 1 - \mathbb{I}(X)$ for all $X \in \mathcal{G}$. Thus the set of 1's of $\mathbb{I}^{\mathbf{c}}$ is the set of 0's of $\mathbb{I}$.

Each weakly (strongly) connected component of $\mathbb{I}^{-1}[\{1\}]$ will be called a *weak foreground component* (*strong foreground component*) of $\mathbb{I}$. Each weakly (strongly) connected component of $\mathbb{I}^{-1}[\{0\}]$ will be called a *weak background component* (*strong background component*) of $\mathbb{I}$.

If $D \in \mathbb{I}^{-1}[\{1\}]$, then D is said to be *simple* in $\mathbb{I}$ if (loosely speaking) "the deletion of $\{D\}$ from $\mathbb{I}$ preserves topology". A precise definition of this concept is as follows:

Definition 1. *Let* $\mathbf{K}$ *be a xel-complex, and let* D *be a* 1 *of a* $\mathbf{K}$*-image* $\mathbb{I}$*. Then we say* D *is* simple *in* $\mathbb{I}$ *if* $\bigcup(\mathbb{I}^{-1}[\{1\}] - \{D\})$ *is a deformation retract of* $\bigcup \mathbb{I}^{-1}[\{1\}]$*.*

In other words, D is simple in $\mathbb{I}$ if and only if the union of all the 1's of $\mathbb{I}$ can be continuously deformed over itself onto the union of all the 1's of $\mathbb{I}$ other than D, in such a way that all points in the latter union remain fixed throughout the deformation process.

The idea of defining simpleness in terms of continuous deformation is an old one that dates back to the 1960's: In the case of the 2D cubical complex, the above definition is very similar to an informally stated connectivity preservation condition given by Hilditch in an early paper on thinning [8, p. 411, condition 5].

A complementary concept to that of a simple 1 is that of a *cosimple* 1:

Definition 2. *Let* $\mathbf{K}$ *be a xel-complex, and let* D *be a* 1 *of a* $\mathbf{K}$*-image* $\mathbb{I}$. *Then we say* D *is* cosimple *in* $\mathbb{I}$ *if* D *is simple in* $(\mathbb{I} - \{D\})^{\mathbf{c}}$. *Equivalently,* D *is* cosimple *in* $\mathbb{I}$ *if and only if* $\bigcup \mathbb{I}^{-1}[\{0\}]$ *is a deformation retract of* $\bigcup(\mathbb{I}^{-1}[\{0\}] \cup \{D\})$.

Let $\mathbb{I}$ be a $\mathbf{K}$-image for some xel-complex $\mathbf{K}$, and let D be any 1 of $\mathbb{I}$. Then we define two sets $\mathbf{Attach}(D, \mathbb{I})$ and $\mathbf{Coattach}(D, \mathbb{I})$ of xels in ∂D as follows:

$$\mathbf{Attach}(D, \mathbb{I}) = \{X \in \mathbf{K} \mid X \subsetneq \partial D \quad \text{and} \quad \exists Q \in \mathbb{I}^{-1}[\{1\}] \setminus \{D\} \ (X \subsetneq \partial Q)\}$$
$$\mathbf{Coattach}(D, \mathbb{I}) = \{X \in \mathbf{K} \mid X \subsetneq \partial D \quad \text{and} \quad \exists Q \in \mathbb{I}^{-1}[\{0\}] \ (X \subsetneq \partial Q)\}$$

If a xel X is in $\mathbf{Attach}(D, \mathbb{I})$ or in $\mathbf{Coattach}(D, \mathbb{I})$, then so is every proper face of X. Note also that $\mathbf{Coattach}(D, \mathbb{I}) = \mathbf{Attach}(D, (\mathbb{I} - \{D\})^{\mathbf{c}})$. Conditions 3 and 4 in the definition of a xel-complex and Property 2 imply that $\bigcup \mathbf{Attach}(D, \mathbb{I}) = D \cap \bigcup(\mathbb{I}^{-1}[\{1\}] \setminus \{D\})$ and $\bigcup \mathbf{Coattach}(D, \mathbb{I}) = D \cap \bigcup \mathbb{I}^{-1}[\{0\}]$.

We can now state essentially discrete characterizations of simple and cosimple 1's in binary images on xel-complexes of dimension ≤ 4:

Theorem 1. *Let* $\mathbf{K}$ *be an* nD *xel-complex, where* $n \leq 4$, *and let* D *be a* 1 *of a* $\mathbf{K}$*-image* $\mathbb{I}$. *Then:*

1. D *is simple in* $\mathbb{I}$ *if and only if* $\bigcup \mathbf{Attach}(D, \mathbb{I})$ *is contractible.*
2. D *is cosimple in* $\mathbb{I}$ *if and only if* $\bigcup \mathbf{Coattach}(D, \mathbb{I})$ *is contractible.* □

Note that, since D is cosimple in $\mathbb{I}$ if and only if D is simple in $(\mathbb{I} - \{D\})^{\mathbf{c}}$, and since $\mathbf{Coattach}(D, \mathbb{I}) = \mathbf{Attach}(D, (\mathbb{I} - \{D\})^{\mathbf{c}})$, the two assertions of this theorem are really equivalent. The "if" parts of the theorem can be deduced from the fact that if A and B are contractible polyhedra such that $B \subseteq A$, then B is a deformation retract of A.[1] The "only if" parts of the theorem can be proved using methods of algebraic topology.[2]

---

[1] A self-contained proof of this fact is given in [13, Sect. 4].

[2] More specifically, it follows from the excision theorem and the exact homology sequence of a pair [16, Ch. 4] that if D is simple in $\mathbb{I}$ then the polyhedron $\bigcup \mathbf{Attach}(D, \mathbb{I})$ is nonempty and its reduced homology groups are all trivial. A polyhedron in $\mathbb{R}^3$ or in the boundary of a polyhedral 4-cell is contractible if (and only if) it has these properties. This is a consequence of (1) the theorems of Whitehead and Hurewicz [16, Chs. 7, 8] and (2) the fact that a polyhedron P in $\mathbb{R}^3$ or in the boundary of a polyhedral 4-cell is simply connected if its first homology group $H_1(P)$ is trivial. In the case where P is in $\mathbb{R}^3$, a proof of (2) is given in [12]. The truth of (2) for a polyhedron P in $\mathbb{R}^3$ implies its truth for a polyhedron P in the boundary of a polyhedral 4-cell X, because if $P \subsetneq \partial X$ then, by Thm. 2 of [17, Ch. 36], there is a homeomorphism $h : \partial X \to \mathbb{R}^3 \cup \{\infty\}$ such that $h[P]$ is a polyhedron in $\mathbb{R}^3$.

For $n \leq 4$, if D is a polyhedral n-cell then a polyhedron $P \subseteq \partial D$ is contractible if and only if P is connected, $(\partial D) \setminus P$ is connected, and the Euler characteristic of P is 1.[3] An important consequence of this and Theorem 1 is that it is computationally straightforward to determine whether or not a given 1 of a binary image on a xel-complex of dimension ≤ 4 is simple or cosimple.

The concepts of simple and cosimple 1's are extended to finite sets of 1's as follows:

Definition 3. *Let* $\mathbf{K}$ *be a xel-complex, and let* $\mathcal{D}$ *be a set of* 1*'s of a* $\mathbf{K}$*-image* $\mathbb{I}$*. Then we say* $\mathcal{D}$ *is* simple (cosimple) *in* $\mathbb{I}$ *if* $\mathcal{D}$ *is a finite set and there is an enumeration* $D_1, \ldots, D_k$ *of all the elements of* $\mathcal{D}$ *such that, for* $1 \leq i \leq k$, D_i *is a simple* (*cosimple*) 1 *in the* $\mathbf{K}$*-image* $\mathbb{I} - \{D_j | 1 \leq j < i\}$.

Note that the empty set is both simple and cosimple in every $\mathbf{K}$-image. Also note that, if D is a 1 of $\mathbb{I}$, then the singleton set $\{D\}$ is simple (cosimple) in $\mathbb{I}$ if and only if D is simple (cosimple) in $\mathbb{I}$.

Important properties[4] of simple sets of 1's are that the deletion of such a set can never split a weak foreground component, can never completely eliminate a weak foreground component, can never merge different strong background components, and can never create a new strong background component. More precisely, if $\mathcal{D}$ is a set of 1's that is simple in $\mathbb{I}$, then each weak foreground component of $\mathbb{I}$ contains exactly one weak foreground component of $\mathbb{I} - \mathcal{D}$, and each strong background component of $\mathbb{I} - \mathcal{D}$ contains exactly one strong background component of $\mathbb{I}$.

Analogously, deletion of a cosimple set can never split a strong foreground component, can never completely eliminate a strong foreground component, can never merge different weak background components, and can never create a new weak background component: If $\mathcal{D}$ is a set of 1's that is cosimple in $\mathbb{I}$, then each strong foreground component of $\mathbb{I}$ contains exactly one strong foreground component of $\mathbb{I} - \mathcal{D}$, and each weak background component of $\mathbb{I} - \mathcal{D}$ contains exactly one weak background component of $\mathbb{I}$.

We are now ready to define the principal concepts of this paper, namely MNS and MNCS sets:

Definition 4. *Let* $\mathbf{K}$ *be a xel-complex, and let* $\mathcal{D}$ *be a set of* 1*'s of a* $\mathbf{K}$*-image* $\mathbb{I}$*. Then we say* $\mathcal{D}$ *is* minimal non-simple, *or* MNS (minimal non-cosimple, *or* MNCS) *in the* $\mathbf{K}$*-image* $\mathbb{I}$ *if* $\mathcal{D}$ *is non-simple* (*non-cosimple*) *in* $\mathbb{I}$*, but every proper subset of* $\mathcal{D}$ *is simple* (*cosimple*) *in* $\mathbb{I}$.

Note that, if D is any 1 of $\mathbb{I}$, then the singleton set $\{D\}$ is MNS (MNCS) in $\mathbb{I}$ if and only if D is non-simple (non-cosimple) in $\mathbb{I}$. Note also that all MNS and MNCS sets are finite, because simple and cosimple sets are, by definition, finite.

---

[3] This can be deduced from the fact stated in the second sentence of footnote 2 and the Alexander duality theorem [16, Ch. 4].

[4] These properties can be deduced from the first sentence of footnote 2 and the Alexander duality theorem, which imply that if D is a 1 of $\mathbb{I}$ that is simple in $\mathbb{I}$ then $\bigcup \mathbf{Attach}(D, \mathbb{I})$ is a nonempty connected proper subset of ∂D whose complement in ∂D is also connected.

If a finite set $\mathcal{Q}$ of 1's of a **K**-image $\mathbb{I}$ is non-simple (non-cosimple) in $\mathbb{I}$, then $\mathcal{Q}$ must contain a subset that is MNS (MNCS) in $\mathbb{I}$. Thus if $\mathcal{P}$ is a set of 1's of $\mathbb{I}$ such that no subset of $\mathcal{P}$ is MNS (MNCS) in $\mathbb{I}$, then *every* subset of $\mathcal{P}$ is simple (cosimple) in $\mathbb{I}$. We say that a set $\mathcal{P}$ of 1's of $\mathbb{I}$ is *hereditarily simple* (*hereditarily cosimple*) in $\mathbb{I}$ if $\mathcal{P}$ has this property. It can be shown that, if $\mathbb{I}$ is a binary image on the 3D cubical complex, then $\mathcal{P}$ is hereditarily simple (hereditarily cosimple) in $\mathbb{I}$ if and only if $\mathcal{P}$ is $\mathcal{P}_{26}$-simple ($\mathcal{P}_6$-simple) in the sense of Bertrand [1].

The arguments in this paper will be based on the characterizations of MNS and MNCS sets that are stated in the following theorem:

Theorem 2. *Let* **K** *be an nD xel-complex, where* $n \leq 4$, *and let* $\mathcal{D}$ *be a set of 1's of a* **K**-*image* $\mathbb{I}$. *Then:*

1. $\mathcal{D}$ *is MNS in* $\mathbb{I}$ *if and only if the following conditions hold for all* $D \in \mathcal{D}$:
   MNS0 $\mathcal{D}$ *is nonempty and finite.*
   MNS1 D *is non-simple in* $\mathbb{I} - (\mathcal{D} \setminus \{D\})$.
   MNS2 D *is simple in* $\mathbb{I} - \mathcal{D}'$ *for every* $\mathcal{D}' \subsetneq \mathcal{D} \setminus \{D\}$.
2. $\mathcal{D}$ *is MNCS in* $\mathbb{I}$ *if and only if the following conditions hold for all* $D \in \mathcal{D}$:
   MNCS0 $\mathcal{D}$ *is nonempty and finite.*
   MNCS1 D *is non-cosimple in* $\mathbb{I} - (\mathcal{D} \setminus \{D\})$.
   MNCS2 D *is cosimple in* $\mathbb{I} - \mathcal{D}'$ *for every* $\mathcal{D}' \subsetneq \mathcal{D} \setminus \{D\}$. □

Both assertions of this theorem are special cases of Prop. 6 in [9]. Explanations of why the hypotheses of that proposition are satisfied are given in [5, p. 123] (for the MNS case) and in [11, p. 326] (for the MNCS case).

We say that a set $\mathcal{S}$ of n-xels of an nD xel-complex **K** *can be MNS* (*can be MNCS*) if there exists a **K**-image $\mathbb{I}$ in which $\mathcal{S}$ is an MNS (MNCS) set of 1's. We say that $\mathcal{S}$ *can be MNS* (*MNCS*) *without being a weak* (*strong*) *foreground component* if there exists a **K**-image $\mathbb{I}$ in which $\mathcal{S}$ is an MNS (MNCS) set of 1's and $\mathcal{S}$ is not a weak (strong) foreground component of $\mathbb{I}$. The main goals of this paper are to determine, for every xel-complex **K** of dimension ≤ 4, exactly which sets of xels can be MNS, exactly which sets can be MNCS, and exactly which sets can be MNS (MNCS) without being a weak (strong) foreground component.

5 Properties of Contractible Polyhedra in $\mathbb{R}^3$ or in the Boundary of a Polyhedral 4-Cell

The proofs of our main theorems will depend on the following fact:

Property 7. Let A and B be polyhedra in $\mathbb{R}^3$ or in the boundary of a polyhedral 4-cell, such that at least two of the following three statements are true:

1. Each of A and B is contractible.
2. $A \cup B$ is contractible.
3. $A \cap B$ is contractible.

Then all three of these statements are true. □

The hypotheses of Property 7 evidently imply that none of the polyhedra A, B, and $A \cap B$ is empty. Indeed, if any of these sets is empty then $A \cap B = \emptyset$ (so that statement 3 is false), and either A and B are disjoint nonempty closed sets (in which case $A \cup B$ is disconnected and statement 2 is false) or one of A and B is empty (in which case statement 1 is false).

Property 7 is a consequence of the reduced Mayer-Vietoris sequence (see, e.g., [16, pp. 128–129]) and the fact, mentioned in footnote 2, that a polyhedron in $\mathbb{R}^3$ or in the boundary of a polyhedral 4-cell is contractible if and only if it is nonempty and its reduced homology groups are all trivial.

The following lemma and its corollary state some consequences of Property 7. Note that the hypotheses of assertions 1 and 2 of the lemma imply that each member of the collection $\mathcal{S}$ is contractible, since "every nonempty subcollection" includes subcollections that consist of just one member.

Lemma 1. *Let $\mathcal{S}$ be a nonempty finite collection of polyhedra in $\mathbb{R}^3$ or in the boundary of a polyhedral 4-cell. Then:*

1. *$\bigcap \mathcal{S}$ is contractible if every nonempty subcollection of $\mathcal{S}$ has a contractible union.*
2. *$\bigcup \mathcal{S}$ is contractible if every nonempty subcollection of $\mathcal{S}$ has a contractible intersection.*

Proof. First, we prove assertion 1. Assertion 1 is evidently true if $|\mathcal{S}| = 1$. Now assume as an induction hypothesis that, for some integer $l > 1$, assertion 1 is true whenever $|\mathcal{S}| < l$. Suppose $|\mathcal{S}| = l$, and every nonempty subcollection of $\mathcal{S}$ has a contractible union. We need to show that $\bigcap \mathcal{S}$ is contractible. Let $\mathcal{S} = \{A_i \mid 1 \leq i \leq l\}$, $\mathcal{S}' = \mathcal{S} \setminus \{A_l\}$, and $\mathcal{S}'' = \{A_l \cup A_i \mid 1 \leq i \leq l-1\}$. Since every nonempty subcollection of $\mathcal{S}$ has a contractible union, we have that:

(a) A_l is contractible.
(b) Every nonempty subcollection of $\mathcal{S}'$ has a contractible union.
(c) Every nonempty subcollection of $\mathcal{S}''$ has a contractible union.

It follows from (b), (c), and the induction hypothesis that each of the two sets $\bigcap \mathcal{S}' = \bigcap_{i=1}^{l-1} A_i$ and $\bigcap \mathcal{S}'' = A_l \cup \bigcap_{i=1}^{l-1} A_i$ is contractible. This, (a), and Property 7 imply that $A_l \cap \bigcap_{i=1}^{l-1} A_i = \bigcap \mathcal{S}$ is contractible, as required. This proves assertion 1. By a symmetrical argument, with unions in place of intersections, and vice versa, assertion 2 is also true. □

Corollary 1. *Let $\mathcal{S}$ be a nonempty finite collection of polyhedra, in $\mathbb{R}^3$ or in the boundary of a polyhedral 4-cell, that satisfies one of the following conditions:*

1. *Every nonempty proper subcollection of $\mathcal{S}$ has a contractible union.*
2. *Every nonempty proper subcollection of $\mathcal{S}$ has a contractible intersection.*

Then $\mathcal{S}$ satisfies both of these conditions. Moreover, $\bigcup \mathcal{S}$ is contractible if and only if $\bigcap \mathcal{S}$ is contractible.

Proof. If condition 1 holds, and $\mathcal{S}'$ is any nonempty proper subcollection of $\mathcal{S}$, then every nonempty subcollection of $\mathcal{S}'$ has a contractible union, and so $\bigcap \mathcal{S}'$ is contractible by assertion 1 of Lemma 1. Hence condition 2 holds if condition 1 holds. Since condition 2 holds, if $\bigcap \mathcal{S}$ is contractible then every nonempty subcollection of $\mathcal{S}$ has a contractible intersection, and so $\bigcup \mathcal{S}$ is contractible by assertion 2 of Lemma 1. By symmetrical arguments, condition 1 holds if condition 2 holds, and $\bigcap \mathcal{S}$ is contractible if $\bigcup \mathcal{S}$ is contractible. □

Another property of contractible polyhedra that we will need is:

Property 8. Let X be a polyhedral n-cell, and let $\mathcal{T}$ be a nonempty finite collection of polyhedra in ∂X such that:

1. $\bigcap \mathcal{T} = \emptyset$.
2. $\bigcap \mathcal{T}'$ is a contractible set whenever $\emptyset \neq \mathcal{T}' \subsetneq \mathcal{T}$.

Then $|\mathcal{T}| - 1 \leq n$, and $\bigcup \mathcal{T} = \partial X$ if and only if $|\mathcal{T}| - 1 = n$. □

The hypotheses of Property 8 imply that the polyhedron of the nerve complex of $\mathcal{T}$ is the boundary of a $(|\mathcal{T}|-1)$-simplex. So it follows from the nerve theorem [3, Thm. 10.6(i)] that the $|\mathcal{T}| - 2^{\text{nd}}$ Betti number of $\bigcup \mathcal{T}$ is 1 if $|\mathcal{T}| \geq 3$. Property 8 can be deduced from this and the fact that $\bigcup \mathcal{T}$ is a polyhedron in ∂X.

6 The Fundamental Lemma

We now use the results of Sect. 5 to establish some key facts (stated in the following lemma) on which the proofs of our main theorems will be based.

Lemma 2 (Fundamental Lemma). *Let X be an n-xel of a xel-complex $\mathbf{K}$, where $n \leq 4$, let $(X_i)_{1 \leq i \leq k}$ be a nonempty finite family of xels of $\mathbf{K}$ in ∂X, and let $P \subseteq \partial X$ be a union of xels of $\mathbf{K}$ for which*

$$P \cup \bigcup \{X_i \mid i \in \mathcal{M}\} \text{ is contractible whenever } \emptyset \neq \mathcal{M} \subsetneq \{1, \ldots, k\} \qquad (*)$$

Then:

1. *For all $\mathcal{S}$ such that $\emptyset \neq \mathcal{S} \subsetneq \{X_i \mid 1 \leq i \leq k\}$, $P \cap \bigcap \mathcal{S}$ is contractible if and only if P is contractible and $\bigcap \mathcal{S} \neq \emptyset$.*
2. *If $\bigcap_{i=1}^k X_i = \emptyset$, then P is contractible if and only if $P \cup \bigcup_{i=1}^k X_i$ is contractible.*
3. *If $P \cap \bigcap_{i=1}^k X_i$ is contractible, then P is contractible if and only if $P \cup \bigcup_{i=1}^k X_i$ is contractible.*
4. *If P is contractible, and there is some $\mathcal{S}$ such that $\emptyset \neq \mathcal{S} \subsetneq \{X_i \mid 1 \leq i \leq k\}$ and $\bigcap \mathcal{S} = \bigcap_{i=1}^k X_i$, then $P \cup \bigcup_{i=1}^k X_i$ is contractible.*
5. *If $\bigcap_{i=1}^k X_i \neq \emptyset$ but $P \cap \bigcap_{i=1}^k X_i = \emptyset$ and $P \cup \bigcup_{i=1}^k X_i$ is contractible, then $P = \emptyset$.*
6. *If $\bigcap_{i=1}^k X_i \neq \emptyset$ but $P \cap \bigcap_{i=1}^k X_i = \emptyset$ and P is contractible, then $k \leq n$, and $P \cup \bigcup_{i=1}^k X_i = \partial X$ if and only if $k = n$.*

Proof. We claim that it is enough to prove this lemma in the case where no two of the X_i's are equal. For assertions 1 – 5 this is because, if $(X'_i)_{1\leq i\leq k'}$ is a family of *distinct* sets such that $\{X'_i \mid 1 \leq i \leq k'\} = \{X_i \mid 1 \leq i \leq k\}$, then when we replace X_i with X'_i and k with k' it is evident that (*) still holds and none of assertions 1 – 5 changes in meaning. In the case of assertion 6, if $X_j = X_l$ for some $j \neq l$, then the case $\mathcal{M} = \{1,\ldots,k\} \setminus \{j\}$ of (*) implies that $P \cup \bigcup_{i=1}^k X_i$ is contractible, and so the hypotheses of assertion 6 are inconsistent with assertion 5. In other words, if assertion 5 is true then assertion 6 is vacuously true if $X_j = X_l$ for some $j \neq l$. This justifies our claim, and in the rest of this proof we shall assume that the X_i's are all distinct.

To prove assertion 1, let $\mathcal{S}$ satisfy $\emptyset \neq \mathcal{S} \subsetneq \{X_i \mid 1 \leq i \leq k\}$ and let $\mathcal{S}^* = \{P \cup Y \mid Y \in \mathcal{S}\}$. By (*), every nonempty subcollection of $\mathcal{S}^*$ has a contractible union. Hence, by assertion 1 of Lemma 1, $P \cup \bigcap \mathcal{S} = \bigcap \mathcal{S}^*$ is contractible. If $\bigcap \mathcal{S} = \emptyset$ then $P \cap \bigcap \mathcal{S} = \emptyset$ is not contractible, which is consistent with assertion 1. Now suppose $\bigcap \mathcal{S} \neq \emptyset$. Then $\bigcap \mathcal{S}$ is a xel of $\mathbf{K}$ (by condition 4 of the definition of a xel-complex) and is therefore contractible. Since each of $P \cup \bigcap \mathcal{S}$ and $\bigcap \mathcal{S}$ is contractible, it follows from Property 7 that P is contractible if and only if $P \cap \bigcap \mathcal{S}$ is contractible. This proves assertion 1.

Next, we prove assertion 2. Suppose $\bigcap_{i=1}^k X_i = \emptyset$ (which implies $k \geq 2$). By (*), every nonempty proper subcollection of $\{P \cup X_1 \cup X_i \mid 2 \leq i \leq k\}$ has a contractible union, and so it follows from Corollary 1 of Lemma 1 that $P \cup \bigcup_{i=1}^k X_i = \bigcup_{i=2}^k (P \cup X_1 \cup X_i)$ is contractible if and only if $(P \cup X_1) \cup \bigcap_{i=2}^k X_i = \bigcap_{i=2}^k (P \cup X_1 \cup X_i)$ is contractible.

There are now two cases: $\bigcap_{i=2}^k X_i = \emptyset$, and $\bigcap_{i=2}^k X_i \neq \emptyset$. In the first case, the set $(P \cup X_1) \cup \bigcap_{i=2}^k X_i = P \cup X_1$ is contractible (by (*)), so it follows from the equivalence established in previous paragraph that $P \cup \bigcup_{i=1}^k X_i$ is also contractible. Moreover, in this case it follows from assertion 1 of Lemma 1 that $P = P \cup \bigcap_{i=2}^k X_i = \bigcap_{i=2}^k (P \cup X_i)$ is contractible as well, because every nonempty subcollection of $\{P \cup X_i \mid 2 \leq i \leq k\}$ has a contractible union (by (*)). Thus the sets P and $P \cup \bigcup_{i=1}^k X_i$ are both contractible, which is consistent with assertion 2.

In the second case, where $\bigcap_{i=2}^k X_i \neq \emptyset$, the set $(P \cup X_1) \cup \bigcap_{i=2}^k X_i$ is the union of the set $\bigcap_{i=2}^k X_i$ (which is a xel of $\mathbf{K}$, by condition 4 in the definition of a xel-complex, and is therefore contractible) with the set $P \cup X_1$ (which is contractible because of (*)). Hence, by Property 7, we have that $(P \cup X_1) \cup \bigcap_{i=2}^k X_i$ is contractible if and only if $(P \cup X_1) \cap \bigcap_{i=2}^k X_i = P \cap \bigcap_{i=2}^k X_i$ is contractible. But, by assertion 1, $P \cap \bigcap_{i=2}^k X_i$ is contractible if and only if P is contractible. We conclude that $(P \cup X_1) \cup \bigcap_{i=2}^k X_i$ is contractible if and only if P is contractible. As we showed earlier that $(P \cup X_1) \cup \bigcap_{i=2}^k X_i$ is contractible if and only if $P \cup \bigcup_{i=1}^k X_i$ is contractible, assertion 2 is proved.

To prove assertion 3, suppose $P \cap \bigcap_{i=1}^k X_i$ is contractible. This implies that $\bigcap_{i=1}^k X_i \neq \emptyset$, and so $\bigcap_{i=1}^k X_i$ is a xel of $\mathbf{K}$ (by condition 4 in the definition of a xel-complex) and is therefore contractible. Since $P \cap \bigcap_{i=1}^k X_i$ and $\bigcap_{i=1}^k X_i$ are both contractible, it follows from Property 7 that P is contractible if and only if $P \cup \bigcap_{i=1}^k X_i$ is contractible. But, since every nonempty proper subcollection of $\{P \cup X_i \mid 1 \leq i \leq k\}$ has a contractible union (by (*)), $P \cup \bigcap_{i=1}^k X_i = \bigcap_{i=1}^k (P \cup X_i)$ is contractible if and only if $P \cup \bigcup_{i=1}^k X_i = \bigcup_{i=1}^k (P \cup X_i)$ is contractible, by Corollary 1 of Lemma 1. This proves assertion 3.

To prove assertion 4, suppose that P is contractible, and that there is some $\mathcal{S}$ such that $\emptyset \neq \mathcal{S} \subsetneq \{X_i \mid 1 \leq i \leq k\}$ and $\bigcap \mathcal{S} = \bigcap_{i=1}^k X_i$. Now if $\bigcap \mathcal{S} = \bigcap_{i=1}^k X_i = \emptyset$ then assertion 4 is true, by assertion 2. If, on the other hand, $\bigcap \mathcal{S} = \bigcap_{i=1}^k X_i \neq \emptyset$, then assertion 1 implies that $P \cap \bigcap_{i=1}^k X_i = P \cap \bigcap \mathcal{S}$ is contractible, and so assertion 4 is true, by assertion 3. This proves assertion 4.

To prove assertion 5, suppose $\bigcap_{i=1}^k X_i \neq \emptyset$ but $P \cap \bigcap_{i=1}^k X_i = \emptyset$. Let $P' = P \cup \bigcap_{i=1}^k X_i$. Then the hypotheses of the lemma still hold when we replace P with P', and $P' \cap \bigcap_{i=1}^k X_i = \bigcap_{i=1}^k X_i$ is a xel of $\mathbf{K}$ (by condition 4 in the definition of a xel-complex) and is therefore contractible. Hence assertion 3 of the lemma (with P' in place of P) implies that if $P \cup \bigcup_{i=1}^k X_i = P' \cup \bigcup_{i=1}^k X_i$ is contractible then P' is contractible. However, P' is contractible only if $P = \emptyset$ (for if $P \neq \emptyset$ then $P' = P \cup \bigcap_{i=1}^k X_i$ is disconnected, as P and $\bigcap_{i=1}^k X_i$ are disjoint nonempty closed sets).

To prove assertion 6, suppose $\bigcap_{i=1}^k X_i \neq \emptyset$ but $P \cap \bigcap_{i=1}^k X_i = \emptyset$ and P is contractible. Let $\mathcal{T} = \{X_i \mid 1 \leq i \leq k\} \cup \{P\}$. Then it follows from assertion 1 that every nonempty proper subcollection of $\mathcal{T}$ has a contractible intersection. Moreover, $\bigcap \mathcal{T} = P \cap \bigcap_{i=1}^k X_i = \emptyset$. So it follows from Property 8 that $k \leq n$, and that $P \cup \bigcup_{i=1}^k X_i = \bigcup \mathcal{T} = \partial X$ if and only if $k = n$. $\square$

7 The Main Theorems

Theorem 3 (First Main Theorem). *Let* $\mathbf{K}$ *be an* nD *xel-complex, where* $1 \leq n \leq 4$, *and let* $\mathcal{T}$ *be a nonempty finite collection of* n*-xels of* $\mathbf{K}$. *Then:*

1. *If* $\bigcap \mathcal{T} = \emptyset$, *then there is no* $\mathbf{K}$*-image* $\mathbb{I}$ *such that* $\mathcal{T}$ *is MNS in* $\mathbb{I}$.
2. *If* $\bigcap \mathcal{T} \neq \emptyset$, *and* $\mathcal{T}$ *is a weak foreground component of a* $\mathbf{K}$*-image* $\mathbb{I}$, *then* $\mathcal{T}$ *is MNS in* $\mathbb{I}$.
3. *If* $\bigcap \mathcal{T}$ *is a 0-xel of* $\mathbf{K}$, *and* $\mathcal{T}$ *is MNS in a* $\mathbf{K}$*-image* $\mathbb{I}$, *then* $\mathcal{T}$ *is a weak foreground component of* $\mathbb{I}$.
4. *If* $\bigcap \mathcal{T}$ *is an* m*-xel of* $\mathbf{K}$ *for some* $m \geq 1$, *then there is a* $\mathbf{K}$*-image* $\mathbb{I}$ *such that* $\mathcal{T}$ *is MNS in* $\mathbb{I}$ *and* $\mathcal{T}$ *is not a weak foreground component of* $\mathbb{I}$.

Proof. Let $k = |\mathcal{T}| - 1$, let $\mathcal{T} = \{X, T_1, \ldots, T_k\}$ and, for $1 \leq i \leq k$, write X_i for $X \cap T_i$.

We first prove assertions 1 and 3. For this purpose we may assume $k \neq 0$, as this is implied by the hypotheses of assertions 1 and 3. Suppose there is a $\mathbf{K}$-image $\mathbb{I}$ such that $\mathcal{T}$ is an MNS set of 1's of $\mathbb{I}$. We will deduce that $\bigcap \mathcal{T} \neq \emptyset$ (which will prove assertion 1). We will further deduce that if $\bigcap \mathcal{T}$ is a 0-xel of $\mathbf{K}$ then $\mathcal{T}$ is a weak foreground component of $\mathbb{I}$ (which will prove assertion 3).

Let $P = X \cap \bigcup(\mathbb{I}^{-1}[\{1\}] \setminus \mathcal{T})$. Thus $P = \bigcup \mathbf{Attach}(X, \mathbb{I} - \{T_i \mid 1 \leq i \leq k\})$. Then $\bigcup \mathbf{Attach}(X, \mathbb{I} - (\{T_i \mid 1 \leq i \leq k\} \setminus \mathcal{W})) = P \cup \bigcup\{X_i \mid T_i \in \mathcal{W}\}$ for any subcollection $\mathcal{W}$ of $\{T_i \mid 1 \leq i \leq k\}$.

Since $\mathcal{T} = \{X, T_1, \ldots, T_k\}$ is MNS in $\mathbb{I}$, it follows from Theorem 2 that X is simple in $\mathbb{I} - (\{T_i \mid 1 \leq i \leq k\} \setminus \mathcal{W})$ for every nonempty subcollection $\mathcal{W}$ of $\{T_i \mid 1 \leq i \leq k\}$, and that X is non-simple in $\mathbb{I} - \{T_i \mid 1 \leq i \leq k\}$. Hence, by Theorem 1, $P = \bigcup \mathbf{Attach}(X, \mathbb{I} - \{T_i \mid 1 \leq i \leq k\})$ is not contractible, but $P \cup \bigcup\{X_i \mid T_i \in \mathcal{W}\} = \bigcup \mathbf{Attach}(X, \mathbb{I} - (\{T_i \mid 1 \leq i \leq k\} \setminus \mathcal{W}))$ is contractible whenever $\emptyset \neq \mathcal{W} \subseteq \{T_i \mid 1 \leq i \leq k\}$.

The collection of sets $\{P \cup \bigcup\{X_i \mid T_i \in \mathcal{W}\} \mid \emptyset \neq \mathcal{W} \subseteq \{T_i \mid 1 \leq i \leq k\}\}$ is the same as the collection of sets $\{P \cup \bigcup \mathcal{S} \mid \emptyset \neq \mathcal{S} \subseteq \{X_i \mid 1 \leq i \leq k\}\}$. Hence $P \cup \bigcup \mathcal{S}$ is contractible whenever $\emptyset \neq \mathcal{S} \subseteq \{X_i \mid 1 \leq i \leq k\}$. Since this implies $P \cup X_i$ is contractible for $1 \leq i \leq k$, and since we saw above that P is not contractible, none of the sets X_i is empty, and so each set $X_i = X \cap T_i$ is a xel of $\mathbf{K}$. Thus we have established the following:

(a) P, X, and the family $(X_i)_{1 \leq i \leq k}$ satisfy the hypotheses of the Fundamental Lemma.
(b) P is not contractible.
(c) $P \cup \bigcup_{i=1}^{k} X_i$ is contractible.

Assertion 2 of the Fundamental Lemma now implies:

$$\bigcap \mathcal{T} = \bigcap_{i=1}^{k} X_i \neq \emptyset \qquad (\dagger)$$

This proves assertion 1.

Now suppose $\bigcap \mathcal{T} = \bigcap_{i=1}^{k} X_i$ is a 0-xel of $\mathbf{K}$. If $P \cap \bigcap_{i=1}^{k} X_i \neq \emptyset$ then $P \cap \bigcap_{i=1}^{k} X_i$ is the 0-xel $\bigcap_{i=1}^{k} X_i$ (as a 0-xel has no nonempty proper subset), and so $P \cap \bigcap_{i=1}^{k} X_i$ is contractible, which contradicts assertion 3 of the Fundamental Lemma (in view of (a), (b), and (c) above). Hence $P \cap \bigcap_{i=1}^{k} X_i = \emptyset$. In view of this, (a), (c), (†), and assertion 5 of the Fundamental Lemma, we have that $X \cap \bigcup(\mathbb{I}^{-1}[\{1\}] \setminus \mathcal{T}) = P = \emptyset$.

As X is an arbitrary element of $\mathcal{T}$, it follows that $T \cap \bigcup(\mathbb{I}^{-1}[\{1\}] \setminus \mathcal{T}) = \emptyset$ for every $T \in \mathcal{T}$. Moreover, every two elements of $\mathcal{T}$ are weakly adjacent (since $\bigcap \mathcal{T} \neq \emptyset$), and so $\mathcal{T}$ is weakly connected. Hence $\mathcal{T}$ is a weak foreground component of $\mathbb{I}$. This proves assertion 3.

To prove assertion 4, suppose $\bigcap \mathcal{T}$ is an m-xel of $\mathbf{K}$ for some $m \geq 1$. Then, by Property 3 of a xel-complex, there exist two distinct 0-xels $\{q_1\}$ and $\{q_2\}$ of $\mathbf{K}$ in $\bigcap \mathcal{T}$. By condition 6 of the definition of a xel-complex, there exist n-xels Q_1 and Q_2 of $\mathbf{K}$ such that $q_1 \in Q_1$, $q_2 \in Q_2$, and $Q_1 \cap Q_2 = \emptyset$. Let $\mathbb{I}^*$ be the $\mathbf{K}$-image whose set of 1's is $\mathcal{T} \cup \{Q_1, Q_2\}$.

We claim that $\mathcal{T}$ is MNS in $\mathbb{I}^*$. To justify this claim, let $P^* = X \cap (Q_1 \cup Q_2)$, so $\bigcup \mathbf{Attach}(X, \mathbb{I}^* - \{T_i \mid 1 \leq i \leq k\}) = P^*$. Then, for any $\mathcal{W} \subseteq \{T_i \mid 1 \leq i \leq k\}$, $\bigcup \mathbf{Attach}(X, \mathbb{I}^* - (\{T_i \mid 1 \leq i \leq k\} \setminus \mathcal{W})) = P^* \cup \bigcup\{X_i \mid T_i \in \mathcal{W}\}$. So (since X is an arbitrary element of $\mathcal{T}$) our claim that $\mathcal{T}$ is MNS in $\mathbb{I}^*$ will follow from Theorems 1 and 2 if we can show that:

(a) P^* is not contractible.
(b) $P^* \cup \bigcup \mathcal{S}$ is contractible whenever $\emptyset \neq \mathcal{S} \subseteq \{X_i \mid 1 \leq i \leq k\}$.

Here (a) is true since $P^* = X \cap (Q_1 \cup Q_2)$ is disconnected (as $Q_1 \cap Q_2 = \emptyset$). To prove (b), let $\emptyset \neq \mathcal{S} \subseteq \{X_i \mid 1 \leq i \leq k\}$, let $\mathcal{S}_1 = \mathcal{S} \cup \{X \cap Q_1\}$, and let $\mathcal{S}_2 = \mathcal{S} \cup \{X \cap Q_2\}$. Then $(\bigcup \mathcal{S}_1) \cup (\bigcup \mathcal{S}_2) = P^* \cup \bigcup \mathcal{S}$. Note that if $\mathcal{X}$ is $\mathcal{S}$, $\mathcal{S}_1$, or $\mathcal{S}_2$, then the intersection of any nonempty subcollection of $\mathcal{X}$ is nonempty, and is therefore a xel of $\mathbf{K}$. So if $\mathcal{X}$ is $\mathcal{S}$, $\mathcal{S}_1$, or $\mathcal{S}_2$ then every nonempty subcollection of $\mathcal{X}$ has a contractible intersection, which implies (by assertion 2 of Lemma 1) that $\bigcup \mathcal{X}$ is contractible. Thus each of the sets $\bigcup \mathcal{S}$, $\bigcup \mathcal{S}_1$, and $\bigcup \mathcal{S}_2$ is contractible. Since $(\bigcup \mathcal{S}_1) \cap (\bigcup \mathcal{S}_2) = (\bigcup \mathcal{S}) \cup (X \cap Q_1 \cap Q_2) = \bigcup \mathcal{S}$ is also contractible, we see from Property 7 that $P^* \cup \bigcup \mathcal{S} = (\bigcup \mathcal{S}_1) \cup (\bigcup \mathcal{S}_2)$ is contractible. This proves (b) and completes the proof of assertion 4.

To prove assertion 2, suppose $\bigcap \mathcal{T} \neq \emptyset$, and let $\mathbb{I}'$ be any $\mathbf{K}$-image of which $\mathcal{T}$ is a weak foreground component. We will show that $\mathcal{T}$ is MNS in $\mathbb{I}'$.

Now $\mathbf{Attach}(X, \mathbb{I}' - \{T_i \mid 1 \leq i \leq k\}) = \emptyset$, as $\mathcal{T}$ is a weak foreground component of $\mathbb{I}'$. We also have that $\bigcup \mathbf{Attach}(X, \mathbb{I}' - (\{T_i \mid 1 \leq i \leq k\} \setminus \mathcal{W}))$ $= \bigcup \{X_i \mid T_i \in \mathcal{W}\}$ for any subcollection $\mathcal{W}$ of $\{T_i \mid 1 \leq i \leq k\}$. So (since X is an arbitrary element of $\mathcal{T}$) our claim that $\mathcal{T}$ is MNS in $\mathbb{I}'$ will follow from Theorems 1 and 2 if we can just show that $\bigcup \mathcal{S}$ is contractible whenever $\emptyset \neq \mathcal{S} \subseteq \{X_i \mid 1 \leq i \leq k\}$. Now the intersection of any nonempty subcollection of $\{X_i \mid 1 \leq i \leq k\}$ is nonempty (as $\bigcap \mathcal{T} \neq \emptyset$) and is therefore a xel of $\mathbf{K}$. Thus if $\emptyset \neq \mathcal{S} \subseteq \{X_i \mid 1 \leq i \leq k\}$ then every nonempty subcollection of $\mathcal{S}$ has a contractible intersection, and so $\bigcup \mathcal{S}$ is contractible (by assertion 2 of Lemma 1), as required. □

The proof of our second main theorem will depend on two more lemmas, which we now establish. Note that if $\mathcal{S}$ is a finite set of 1's of a binary image $\mathbb{I}$ on a xel-complex of dimension ≤ 4, then it follows from assertion 2 of the first lemma below that $\mathcal{S}$ is cosimple if (and only if) the intersection of $\mathcal{S}$ with each strong foreground component of $\mathbb{I}$ is cosimple, and so $\mathcal{S}$ cannot be MNCS in $\mathbb{I}$ if $\mathcal{S}$ intersects more than one strong foreground component of $\mathbb{I}$.

Lemma 3. *Let $\mathbf{K}$ be an nD xel-complex, where $n \leq 4$, and let $\mathcal{T}$ be any set of n-xels of $\mathbf{K}$. Let $\mathbb{I}_1$ and $\mathbb{I}_2$ be $\mathbf{K}$-images such that $\mathbb{I}_1(X) = \mathbb{I}_2(X) = 1$ for every $X \in \mathcal{T}$, and $\mathbb{I}_1(X) = \mathbb{I}_2(X) = 0$ for every n-xel X that is not in $\mathcal{T}$ but is strongly adjacent to an n-xel in $\mathcal{T}$ (i.e., $\mathcal{T}$ is a union of strong foreground components both of $\mathbb{I}_1$ and of $\mathbb{I}_2$). Then:*

1. *For every $T \in \mathcal{T}$, $\mathbf{Coattach}(T, \mathbb{I}_1) = \mathbf{Coattach}(T, \mathbb{I}_2)$.*
2. *For every $T \in \mathcal{T}$, T is cosimple in $\mathbb{I}_1$ if and only if T is cosimple in $\mathbb{I}_2$.*
3. *For every $\mathcal{T}' \subseteq \mathcal{T}$, $\mathcal{T}'$ is MNCS in $\mathbb{I}_1$ if and only if $\mathcal{T}'$ is MNCS in $\mathbb{I}_2$.*

Proof. To prove assertion 1, let $T \in \mathcal{T}$. We now show that $\mathbf{Coattach}(T, \mathbb{I}_1) \subseteq \mathbf{Coattach}(T, \mathbb{I}_2)$. Let $Y \in \mathbf{Coattach}(T, \mathbb{I}_1)$. Then $Y \subsetneq \partial T$ and there is an n-xel $Q \in \mathbb{I}_1^{-1}[\{0\}]$ such that $Y \subsetneq \partial Q$. Thus $Y \subseteq T \cap Q$ and so, by Property 5, there exists a sequence $Q_0, Q_1, \ldots, Q_k$ of n-xels of $\mathbf{K}$ such that $Q_0 = T$, $Q_k = Q$, and, for $1 \leq i \leq k$, $Q_{i-1} \cap Q_i$ is an $(n-1)$-xel of $\mathbf{K}$ that contains Y. Now $Q_0 = T \in \mathcal{T}$ and $Q_k = Q \notin \mathcal{T}$ (since $Q \in \mathbb{I}_1^{-1}[\{0\}]$). Let Q_j be the first element of the sequence $Q_0, Q_1, \ldots, Q_k$ that does not belong to $\mathcal{T}$. Then, $Q_{j-1} \in \mathcal{T}$. Since $Q_{j-1} \cap Q_j$ is an

$(n-1)$-xel of $\mathbf{K}$, the n-xel Q_j is strongly adjacent to an n-xel in $\mathcal{T}$, and therefore $\mathbb{I}_1(Q_j) = \mathbb{I}_2(Q_j) = 0$. As $Q_j \in \mathbb{I}_2^{-1}[\{0\}]$ and $Y \subseteq Q_{j-1} \cap Q_j \subsetneq Q_j$ (which implies that $Y \subsetneq \partial Q_j$), it follows that $Y \in \mathbf{Coattach}(\mathcal{T}, \mathbb{I}_2)$. As Y is an arbitrary xel in $\mathbf{Coattach}(\mathcal{T}, \mathbb{I}_1)$, this shows that $\mathbf{Coattach}(\mathcal{T}, \mathbb{I}_1) \subseteq \mathbf{Coattach}(\mathcal{T}, \mathbb{I}_2)$. By a symmetrical argument, $\mathbf{Coattach}(\mathcal{T}, \mathbb{I}_2) \subseteq \mathbf{Coattach}(\mathcal{T}, \mathbb{I}_1)$. This proves assertion 1. Assertion 2 follows from assertion 1 and Theorem 1. Assertion 3 follows from assertion 2 and Theorem 2, because if $\mathcal{W}$ is any subset of $\mathcal{T}$ then the hypotheses of the lemma must still hold when $\mathcal{T}$, $\mathbb{I}_1$, and $\mathbb{I}_2$ are respectively replaced by $\mathcal{T} \setminus \mathcal{W}$, $\mathbb{I}_1 - \mathcal{W}$, and $\mathbb{I}_2 - \mathcal{W}$. □

Lemma 4. *Let $\mathbf{K}$ be an nD xel-complex, and let $\mathcal{T}$ be a nonempty finite collection of n-xels of $\mathbf{K}$ such that $\bigcap \mathcal{T} \neq \emptyset$ and there is no $\mathcal{T}' \subsetneq \mathcal{T}$ such that $\bigcap \mathcal{T}' = \bigcap \mathcal{T}$. Then $|\mathcal{T}| \leq n+1$. Moreover, if $|\mathcal{T}| = n+1$ then $\bigcap \mathcal{T}^*$ is an $(n+1-|\mathcal{T}^*|)$-xel of $\mathbf{K}$ whenever $\emptyset \neq \mathcal{T}^* \subseteq \mathcal{T}$.*

Proof. Let $k = |\mathcal{T}| - 1$ and let $T^0, T^1, \ldots, T^k$ be an enumeration of the elements of $\mathcal{T}$. Since $\bigcap \mathcal{T} \neq \emptyset$, $\bigcap_{i=0}^{l} T^i$ is a xel of $\mathbf{K}$ for $0 \leq l \leq k$. Hence:

$$\dim(T^0) - \dim(\bigcap_{i=0}^{k} T^i) = \sum_{l=0}^{k-1} (\dim(\bigcap_{i=0}^{l} T^i) - \dim(\bigcap_{i=0}^{l+1} T^i)) \qquad (\star)$$

But we must have $\bigcap_{i=0}^{l} T^i \supsetneq \bigcap_{i=0}^{l+1} T^i$ for $0 \leq l \leq k-1$ (for if $\bigcap_{i=0}^{l} T^i = \bigcap_{i=0}^{l+1} T^i$ then $\bigcap(\mathcal{T} \setminus \{T^{l+1}\}) = \bigcap \mathcal{T}$), and so it follows from Property 2 that $\dim(\bigcap_{i=0}^{l} T^i) - \dim(\bigcap_{i=0}^{l+1} T^i) \geq 1$ for $0 \leq l \leq k-1$. Hence the right side of $(\star)$ is $\geq k$. Since the left side of $(\star)$ is $\leq \dim(T^0) = n$, we have that $n \geq k$, and so $|\mathcal{T}| = k+1 \leq n+1$.

Now suppose $|\mathcal{T}| = n+1$. Then $k = n$ and the left side of $(\star)$ is $\leq k$, so no term on the right exceeds 1 and we have that $\dim(\bigcap_{i=0}^{l} T^i) - \dim(\bigcap_{i=0}^{l+1} T^i) = 1$ for $0 \leq l \leq k-1$. Hence $\dim(\bigcap_{i=0}^{l} T^i) = n - l = (n+1) - |\{T^i \mid 0 \leq i \leq l\}|$ for $0 \leq l \leq k$, since $\dim(\bigcap_{i=0}^{0} T^i) = \dim(T^0) = n$. As this holds for any enumeration $T^0, T^1, \ldots, T^k$ of $\mathcal{T}$, the lemma is proved. □

Theorem 4 (Second Main Theorem). *Let $\mathbf{K}$ be an nD xel-complex, where $1 \leq n \leq 4$, and let $\mathcal{T}$ be a nonempty finite collection of n-xels of $\mathbf{K}$. Then:*

1. *If $\bigcap \mathcal{T} = \emptyset$, then there is no $\mathbf{K}$-image $\mathbb{I}$ such that $\mathcal{T}$ is MNCS in $\mathbb{I}$.*
2. *If there is some $\mathcal{T}' \subsetneq \mathcal{T}$ such that $\bigcap \mathcal{T}' = \bigcap \mathcal{T}$, then there is no $\mathbf{K}$-image $\mathbb{I}$ such that $\mathcal{T}$ is MNCS in $\mathbb{I}$.*
3. *If $\bigcap \mathcal{T} \neq \emptyset$ and there is no $\mathcal{T}' \subsetneq \mathcal{T}$ such that $\bigcap \mathcal{T}' = \bigcap \mathcal{T}$, and $|\mathcal{T}| = n+1$, then $\mathcal{T}$ is MNCS in a $\mathbf{K}$-image if and only if $\mathcal{T}$ is a strong foreground component of that $\mathbf{K}$-image.*
4. *If $\bigcap \mathcal{T} \neq \emptyset$ and there is no $\mathcal{T}' \subsetneq \mathcal{T}$ such that $\bigcap \mathcal{T}' = \bigcap \mathcal{T}$, and $|\mathcal{T}| \leq n$, then there is a $\mathbf{K}$-image $\mathbb{I}$ such that $\mathcal{T}$ is MNCS in $\mathbb{I}$ and $\mathcal{T}$ is not a strong foreground component of $\mathbb{I}$.*

Proof. Let $k = |\mathcal{T}| - 1$, let $\mathcal{T} = \{X, T_1, \ldots, T_k\}$ and, for $1 \leq i \leq k$, write X_i for $X \cap T_i$.

We first prove assertions 1 and 2, and the "only if" part of assertion 3. For this purpose we may assume $k \neq 0$, as this is implied by the hypotheses of assertions 1, 2, and 3. Suppose there is a $\mathbf{K}$-image $\mathbb{I}$ such that $\mathcal{T}$ is an MNCS set of 1's of $\mathbb{I}$. We will deduce that $\bigcap \mathcal{T} \neq \emptyset$ (which will prove assertion 1). We will further deduce that there is no set $\mathcal{T}' \subsetneq \mathcal{T}$ such that $\bigcap \mathcal{T}' = \bigcap \mathcal{T}$ (which will prove assertion 2). Then we will prove the "only if" part of assertion 3 by showing that $\mathcal{T}$ must be a strong foreground component of $\mathbb{I}$ if $|\mathcal{T}| = n+1$.

Let $P = X \cap \bigcup \mathbb{I}^{-1}[\{0\}] = \bigcup \mathbf{Coattach}(X, \mathbb{I})$. Then $\bigcup \mathbf{Coattach}(X, \mathbb{I} - \mathcal{W}) = P \cup \bigcup\{X_i \mid T_i \in \mathcal{W}\}$ for any subcollection $\mathcal{W}$ of $\{T_i \mid 1 \leq i \leq k\}$.

Since $\mathcal{T} = \{X, T_1, \ldots, T_k\}$ is MNCS in $\mathbb{I}$, it follows from Theorem 2 that X is cosimple in $\mathbb{I} - \mathcal{W}$ for every proper subcollection $\mathcal{W}$ of $\{T_i \mid 1 \leq i \leq k\}$, and that X is non-cosimple in $\mathbb{I} - \{T_i \mid 1 \leq i \leq k\}$. Hence, by Theorem 1, $P \cup \bigcup_{i=1}^k X_i = \bigcup \mathbf{Coattach}(X, \mathbb{I} - \{T_i \mid 1 \leq i \leq k\})$ is not contractible, but $P \cup \bigcup\{X_i \mid T_i \in \mathcal{W}\} = \bigcup \mathbf{Coattach}(X, \mathbb{I} - \mathcal{W})$ is contractible for every collection $\mathcal{W} \subsetneq \{T_i \mid 1 \leq i \leq k\}$. As a special case of the latter fact, P is contractible.

The collection of sets $\{P \cup \bigcup\{X_i \mid T_i \in \mathcal{W}\} \mid \mathcal{W} \subsetneq \{T_i \mid 1 \leq i \leq k\}\}$ includes the collection of sets $\{P \cup \bigcup \mathcal{S} \mid \mathcal{S} \subsetneq \{X_i \mid 1 \leq i \leq k\}\}$. Hence $P \cup \bigcup \mathcal{S}$ is contractible whenever $\mathcal{S} \subsetneq \{X_i \mid 1 \leq i \leq k\}$. Since this implies $P \cup \bigcup(\{X_j \mid 1 \leq j \leq k\} \setminus \{X_i\})$ is contractible for $1 \leq i \leq k$, and since we saw above that $P \cup \bigcup_{i=1}^k X_i$ is not contractible, none of the sets X_i is empty, and so each set $X_i = X \cap T_i$ is a xel of $\mathbf{K}$. Thus we have established the following:

(a) P, X, and the family $(X_i)_{1 \leq i \leq k}$ satisfy the hypotheses of the Fundamental Lemma.
(b) P is contractible.
(c) $P \cup \bigcup_{i=1}^k X_i$ is not contractible.

Assertion 2 of the Fundamental Lemma now implies that $\bigcap \mathcal{T} = \bigcap_{i=1}^k X_i \neq \emptyset$. This proves assertion 1.

To prove assertion 2, we suppose there is a set $\mathcal{T}' \subsetneq \mathcal{T}$ such that $\bigcap \mathcal{T}' = \bigcap \mathcal{T}$, and deduce a contradiction. We may assume without loss of generality that $T_1 \in \mathcal{T} \setminus \mathcal{T}'$. Then $\bigcap_{i=2}^k X_i = X \cap \bigcap_{i=2}^k T_i \subseteq \bigcap \mathcal{T}' = \bigcap \mathcal{T} = X \cap \bigcap_{i=1}^k T_i = \bigcap_{i=1}^k X_i$, which implies $\bigcap_{i=2}^k X_i = \bigcap_{i=1}^k X_i$. This and (a) – (c) above contradict assertion 4 of the Fundamental Lemma, and so we have established assertion 2.

To prove the "only if" part of assertion 3, we continue to suppose that $\mathcal{T}$ is MNCS in the $\mathbf{K}$-image $\mathbb{I}$, but now also suppose that $|\mathcal{T}| = n+1$ (so that $k = n$). We need to deduce that $\mathcal{T}$ is a strong foreground component of $\mathbb{I}$.

By assertions 1 and 2, $\bigcap \mathcal{T} \neq \emptyset$ and there is no set $\mathcal{T}' \subsetneq \mathcal{T}$ such that $\bigcap \mathcal{T}' = \bigcap \mathcal{T}$. So Lemma 4 implies that, for any two distinct elements T and T' of $\mathcal{T}$, the intersection $T \cap T'$ is an $(n-1)$-xel of $\mathbf{K}$. Hence $\mathcal{T}$ is strongly connected. It also follows from Lemma 4 that $\bigcap_{i=1}^k X_i = \bigcap \mathcal{T}$ is a 0-xel of $\mathbf{K}$.

Now if $P \cap \bigcap_{i=1}^k X_i \neq \emptyset$ then $P \cap \bigcap_{i=1}^k X_i$ is the 0-xel $\bigcap_{i=1}^k X_i$, and so $P \cap \bigcap_{i=1}^k X_i$ is contractible, which contradicts assertion 3 of the Fundamental Lemma (in view of (a) – (c) above).

Hence we may assume $P \cap \bigcap_{i=1}^k X_i = \emptyset$. Then assertion 6 of the Fundamental Lemma implies that $\bigcup \mathbf{Coattach}(X, \mathbb{I} - \{T_i \mid 1 \leq i \leq k\}) = P \cup \bigcup_{i=1}^k X_i = \partial X$.

It follows that there is no n-xel Y of $\mathbf{K}$ such that Y is a 1 of $\mathbb{I} - \{T_i \mid 1 \le i \le k\}$ and $X \cap Y$ is an $(n-1)$-xel. (For if such an n-xel Y exists, and $Z = X \cap Y$, then (by Property 4) no point in $Z \setminus \partial Z$ lies on a 0 of $\mathbb{I} - \{T_i \mid 1 \le i \le k\}$ and all points of $Z \setminus \partial Z$ must lie in $\partial X \setminus \bigcup \mathbf{Coattach}(X, \mathbb{I} - \{T_i \mid 1 \le i \le k\})$.) Hence X is not strongly adjacent to any 1 of $\mathbb{I} - \mathcal{T}$. As X is an *arbitrary* element of $\mathcal{T}$ (and we already know $\mathcal{T}$ is strongly connected) it follows that $\mathcal{T}$ is a strong foreground component of $\mathbb{I}$. This proves the "only if" part of assertion 3.

It remains to establish the "if" part of assertion 3, and assertion 4. For this purpose we suppose that $\bigcap \mathcal{T} \neq \emptyset$, and that there is no set $\mathcal{T}' \subsetneq \mathcal{T}$ for which $\bigcap \mathcal{T}' = \bigcap \mathcal{T}$. (To begin with, we do not suppose that $|\mathcal{T}| = n+1$.) We will define a $\mathbf{K}$-image $\mathbb{I}^*$, and show that $\mathcal{T}$ is MNCS in $\mathbb{I}^*$.

Let $\mathcal{H}$ be the set of all n-xels of $\mathbf{K}$ that intersect the xel $\bigcap \mathcal{T}$, and let $\overline{\mathcal{H}}$ be the set of all n-xels of $\mathbf{K}$ that do not intersect the xel $\bigcap \mathcal{T}$. Let $\mathbb{I}^*$ be the $\mathbf{K}$-image whose set of 1's is $\mathcal{H}$ (and whose set of 0's is $\overline{\mathcal{H}}$). We will show that $\mathcal{T}$ is MNCS in $\mathbb{I}^*$.

Now $\overline{\mathcal{H}}$ consists of the xels of dimension n in the set $\{C \in \mathbf{K} \mid C \cap \bigcap \mathcal{T} = \emptyset\}$. Moreover, condition 6 in the definition of a xel-complex implies that each xel of dimension $< n$ in $\{C \in \mathbf{K} \mid C \cap \bigcap \mathcal{T} = \emptyset\}$ is contained in an n-xel in $\overline{\mathcal{H}}$. Hence $\bigcup \overline{\mathcal{H}} = \bigcup \{C \in \mathbf{K} \mid C \cap \bigcap \mathcal{T} = \emptyset\}$. Therefore $\bigcup \mathbf{Coattach}(X, \mathbb{I}^*) = X \cap \bigcup \overline{\mathcal{H}} = \bigcup \{X \cap C \mid C \in \mathbf{K} \text{ and } C \cap \bigcap \mathcal{T} = \emptyset\} \bigcup \{D \in \mathbf{K} \mid D \subsetneq X \text{ and } D \cap \bigcap \mathcal{T} = \emptyset\}$.

Let $P = \bigcup \mathbf{Coattach}(X, \mathbb{I}^*)$. Then, for any $\mathcal{W} \subseteq \{T_i \mid 1 \le i \le k\}$, we have that $\bigcup \mathbf{Coattach}(X, \mathbb{I}^* - \mathcal{W}) = P \cup \bigcup \{X_i \mid T_i \in \mathcal{W}\}$. Now we observe that $\{X_i \mid T_i \in \mathcal{W}\}$ is a proper subset of $\{X_i \mid 1 \le i \le k\}$ whenever $\mathcal{W}$ is a proper subset of $\{T_i \mid 1 \le i \le k\}$. (This is because there cannot exist $j \neq j'$ for which $X_j = X_{j'}$. For if such j and j' existed then $X \cap T_j \cap T_{j'} = X_j \cap X_{j'} = X_j = X \cap T_j$, which would imply that $\bigcap \mathcal{T} = \bigcap (\mathcal{T} \setminus \{T_{j'}\})$, contrary to our hypothesis that there is no set $\mathcal{T}' \subsetneq \mathcal{T}$ for which $\bigcap \mathcal{T}' = \bigcap \mathcal{T}$.) In view of this, and since X is an arbitrary element of $\mathcal{T}$, if we can show that the following statements (i) and (ii) are both true, then Theorems 1 and 2 will imply that $\mathcal{T}$ is indeed MNCS in $\mathbb{I}^*$:

(i) $P \cup \bigcup \mathcal{S}$ is contractible whenever $\mathcal{S} \subsetneq \{X_i \mid 1 \le i \le k\}$.
(ii) $P \cup \bigcup_{i=1}^{k} X_i$ is not contractible.

Recall that $P = \bigcup \{D \in \mathbf{K} \mid D \subsetneq X \text{ and } D \cap \bigcap \mathcal{T} = \emptyset\}$. If $k = 0$, then $\bigcap \mathcal{T} = X$, $P = \emptyset$, (i) is vacuously true, and (ii) is true.

Now suppose $k \neq 0$. Then condition 5 of the definition of a xel-complex implies that P is contractible, since $\bigcap \mathcal{T}$ is a nonempty proper subset of X. The intersection of any nonempty subcollection of $\{X_i \mid 1 \le i \le k\}$ is contractible, as it is nonempty (since $\bigcap \mathcal{T} \neq \emptyset$) and is therefore a xel of $\mathbf{K}$. Now let $\mathcal{S}'$ be any nonempty proper subcollection of $\{X_i \mid 1 \le i \le k\}$. Then $\bigcap \mathcal{T}$ is a nonempty proper subset of $\bigcap \mathcal{S}'$, since there is no set $\mathcal{T}' \subsetneq \mathcal{T}$ such that $\bigcap \mathcal{T}' = \bigcap \mathcal{T}$. Hence $P \cap \bigcap \mathcal{S}' \bigcup \{E \in \mathbf{K} \mid E \subsetneq \bigcap \mathcal{S}' \text{ and } E \cap \bigcap \mathcal{T} = \emptyset\}$ is contractible, by condition 5 of the definition of a xel-complex.

The observations in the preceding paragraph imply that, if $k \neq 0$, then the intersection of any nonempty proper subcollection of $\{X_i \mid 1 \le i \le k\} \cup \{P\}$ is contractible. It follows, by Corollary 1 of Lemma 1, that the union of any nonempty proper subcollection of $\{X_i \mid 1 \le i \le k\} \cup \{P\}$ is contractible. This proves (i). Corollary 1 also tells us that $P \cup \bigcup_{i=1}^{k} X_i = \bigcup(\{X_i \mid 1 \le i \le k\} \cup \{P\})$ is contractible if and only if $\bigcap(\{X_i \mid 1 \le i \le k\} \cup \{P\}) = P \cap \bigcap_{i=1}^{k} X_i$ is contractible. But $P \cap \bigcap_{i=1}^{k} X_i = P \cap \bigcap \mathcal{T} = \emptyset$ is not contractible, and so we have proved (ii). Thus we have shown that $\mathcal{T}$ is MNCS in $\mathbb{I}^*$.

Now suppose, again, that $|\mathcal{T}| = n + 1$. Since $\mathcal{T}$ is MNCS in $\mathbb{I}^*$, the "only if" part of assertion 3 implies that $\mathcal{T}$ must be a strong foreground component of $\mathbb{I}^*$, and so it follows from assertion 3 of Lemma 3 that $\mathcal{T}$ is MNCS in any **K**-image of which $\mathcal{T}$ is a strong foreground component. This establishes the "if" part of assertion 3.

Finally, we suppose, instead, that $|\mathcal{T}| \le n$ (so that $k \le n-1$), and complete the proof of assertion 4 by deducing that $\mathcal{T}$ is not a strong foreground component of $\mathbb{I}^*$. First of all, we claim that $\bigcup \mathbf{Coattach}(X, \mathbb{I}^* - \{T_i \mid 1 \le i \le k\}) = P \cup \bigcup_{i=1}^{k} X_i \subsetneq \partial X$. Recall that $P = \bigcup\{D \in \mathbf{K} \mid D \subsetneq X \text{ and } D \cap \bigcap \mathcal{T} = \emptyset\}$. If $k = 0$ then $\bigcap \mathcal{T} = X$ and $P \cup \bigcup_{i=1}^{k} X_i = P = \emptyset$, so that our claim is valid. If $k \ne 0$ then, since P is contractible (as we observed earlier), and since $\bigcap \mathcal{T} \ne \emptyset$, $P \cap \bigcap \mathcal{T} = \emptyset$, and $k \le n - 1$, the validity of our claim follows from assertion 6 of the Fundamental Lemma.

Let p be any point in $\partial X \setminus \bigcup \mathbf{Coattach}(X, \mathbb{I}^* - \{T_i \mid 1 \le i \le k\})$. Then (by Properties 1 and 6) there must exist an $(n-1)$-xel $Z \subsetneq \partial X$ such that $p \in Z$. Since $p \notin \bigcup \mathbf{Coattach}(X, \mathbb{I}^* - \{T_i \mid 1 \le i \le k\})$, we also have that $Z \not\subseteq \bigcup \mathbf{Coattach}(X, \mathbb{I}^* - \{T_i \mid 1 \le i \le k\})$, and so the n-xel Y of **K** such that $X \cap Y = \partial X \cap \partial Y = Z$ (which must exist, by Property 4) is a 1 of $\mathbb{I}^* - \{T_i \mid 1 \le i \le k\}$. Hence X is strongly adjacent to a 1 of $\mathbb{I}^* - \{T_i \mid 1 \le i \le k\}$ and $\mathcal{T}$ is not a strong foreground component of $\mathbb{I}^*$. This completes the proof. □

Note that, in view of Lemma 4, every nonempty finite collection $\mathcal{T}$ of n-xels of **K** must satisfy the hypotheses of one of the four assertions of Theorem 4.

8 Concluding Remarks

We say that a set $\mathcal{T}$ of n-dimensional xels of an n-dimensional xel-complex *can be minimal non-simple* (*can be minimal non-cosimple*) if there exists a binary image in which $\mathcal{T}$ is a minimal non-simple (minimal non-cosimple) set of 1's. We say that $\mathcal{T}$ *can be minimal non-simple* (*minimal non-cosimple*) *without being a weak* (*strong*) *foreground component* if there exists a binary image in which $\mathcal{T}$ is a proper subset of a weak (strong) foreground component and $\mathcal{T}$ is a minimal non-simple (minimal non-cosimple) set.

This paper has determined just which sets of xels can be minimal non-simple, just which sets can be minimal non-cosimple, and just which sets can be minimal non-simple (minimal non-cosimple) without being a weak (strong) foreground component, in arbitrary xel-complexes of dimension ≤ 4. A number of earlier papers [4,5,6,10,11,15,19] have solved these problems for particular xel-complexes—specifically, the 2D, 3D, and 4D cubical, 2D hexagonal, and 3D face-centered-cubical complexes. This paper generalizes that earlier work.

We have established that, for $n \le 4$, a nonempty finite collection $\mathcal{T}$ of n-dimensional xels of an n-dimensional xel-complex can be minimal non-simple if and only if $\bigcap \mathcal{T} \neq \emptyset$. We have shown, too, that $\mathcal{T}$ can be minimal non-simple without being a weak foreground component if and only if $\bigcap \mathcal{T}$ is an m-dimensional xel for some $m \geq 1$.

We have further established that $\mathcal{T}$ can be minimal non-cosimple if and only if $\bigcap \mathcal{T} \neq \emptyset$ and there is no nonempty proper subcollection $\mathcal{T}'$ of $\mathcal{T}$ such that $\bigcap \mathcal{T}' = \bigcap \mathcal{T}$, and we have shown that $\mathcal{T}$ can be minimal non-cosimple without being a strong foreground component if and only if, in addition, $|\mathcal{T}| \leq n$.

References

1. Bertrand, G.: On P-simple points. C. R. Acad. Sci. Paris, Série I **321** (1995) 1077–1084
2. Bertrand, G.: A Boolean characterization of three-dimensional simple points. Pattern Recognition Letters **17** (1996) 115–124
3. Björner, A.: Topological methods. In: R.L. Graham, M. Grötschel, and L. Lovász, editors, Handbook of Combinatorics, Vol. II, pages 1819–1872, MIT Press (1995)
4. Gau, C.J., Kong, T.Y.: Minimal nonsimple sets of voxels in binary images on a face-centered cubic grid. International Journal of Pattern Recognition and Artificial Intelligence **13** (1999) 485–502
5. Gau, C.J., Kong, T.Y.: Minimal nonsimple sets in 4D binary images. Graphical Models **65** (2003) 112–130
6. Hall, R.W.: Tests for connectivity preservation for parallel reduction operators. Topology and Its Applications **46** (1992) 199–217
7. Herman, G.T.: Geometry of Digital Spaces. Birkhäuser (1998)
8. Hilditch, C.J.: Linear skeletons from square cupboards. In: B. Meltzer and D. Michie, editors, Machine Intelligence IV, pages 403–420, Edinburgh University Press (1969)
9. Kong, T.Y.: On the problem of determining whether a parallel reduction operator for n-dimensional binary images always preserves topology. In: R.A. Melter and A.Y. Wu, editors, Vision Geometry II (Boston, September 1993), Proceedings, pages 69–77. Proc. SPIE 2060 (1993)
10. Kong, T.Y.: On topology preservation in 2D and 3D thinning. International Journal of Pattern Recognition and Artificial Intelligence **9** (1995) 813–844
11. Kong, T.Y., Gau, C.J.: Minimal nonsimple sets in 4-dimensional binary images with (8,80)-adjacency. In: R. Klette and J. Žunić, editors, Combinatorial Image Analysis: 10th International Workshop (IWCIA 2004, Auckland, New Zealand, December 2004), Proceedings, pages 318–333. Springer (2004)
12. Kong, T.Y., Roscoe, A.W.: Characterizations of simply-connected finite polyhedra in 3-space. Bulletin of the London Mathematical Society **17** (1985) 575–578
13. Kong, T.Y., Saha, P.K., Rosenfeld, A.: Strongly normal sets of contractible tiles in N dimensions. Pattern Recognition, in press
14. Kovalevsky, V.A.: Discrete topology and contour definition. Pattern Recognition Letters **2** (1984) 281–288
15. Ma, C.M.: On topology preservation in 3D thinning. CVGIP: Image Understanding **59** (1994) 328–339
16. Maunder, C.R.F.: Algebraic Topology. Dover Publications (1996)
17. Moise, E.E.: Geometric Topology in Dimensions 2 and 3. Springer (1977)

18. Ronse, C.: A topological characterization of thinning. Theoretical Computer Science **43** (1986) 31–41
19. Ronse, C.: Minimal test patterns for connectivity preservation in parallel thinning algorithms for binary digital images. Discrete Applied Mathematics **21** (1988) 67–79
20. Rosenfeld, A.: Connectivity in digital pictures. Journal of the Association for Computing Machinery **17** (1970) 146–160
21. Saha, P.K., Chaudhuri, B.B.: Detection of 3D simple points for topology preserving transformation with applications to thinning. IEEE Transactions on Pattern Analysis and Machine Intelligence **16** (1994) 1028–1032

Combinatorial Relations for Digital Pictures

Valentin E. Brimkov[1], Davide Moroni[2], and Reneta Barneva[3]

[1] Mathematics Department, SUNY Buffalo State College, Buffalo, NY 14222, USA
brimkove@buffalostate.edu

[2] Laboratory of Signals and Images, Institute of Science and Information Technologies, Area della Ricerca CNR di Pisa, 56124 Pisa, Italy
davide.moroni@isti.cnr.it

[3] Department of Computer Science, SUNY Fredonia, NY 14063, USA
barneva@cs.fredonia.edu

Abstract. In this paper we define the notion of gap in an arbitrary digital picture S in a digital space of arbitrary dimension. As a main result, we obtain an explicit formula for the number of gaps in S of maximal dimension. We also derive a combinatorial relation for a digital curve.

Keywords: Digital geometry, digital picture, gap, brim.

1 Introduction

A gap is a location in a digital picture (that is any finite set of pixels/voxels in 2D/3D) through which a "discrete path" can pass. Gaps are considered in rendering pixelized/voxelized scenes, which is done by casting digital rays from the image to the scene [1, 2]. Therefore, it is useful to know whether a digital picture has gaps of certain type or is gap-free. This is particularly interesting when dealing with digital curves or surfaces. It is also helpful to have an estimation for the number of gaps (if any) in a considered digital object, possibly as a function of other object parameters. Such kind of information may help better understand the topological structure of a binary picture and is of potential interest in property-based image analysis. Of special interest are the gaps of maximal dimension (to be defined later) since they can be penetrated by a digital ray of any connectivity. Moreover, estimations of the number of such kind of gaps may be useful for evaluating the performance of some polyhedra decomposition algorithms (see comments in Section 4). Moreover, digital picture gap-freeness appears to be equivalent to the notion of well-composedness of a set of pixels proposed by Latecki, Eckhardt, and Rosenfeld [3]. This last paper demonstrates the advantages of using well-composed (gap-free) sets in image analysis.

Theoretical studies of this sort are related to combinatorial topology, but are also of interest in several other disciplines, such as digital geometry, combinatorial image analysis, and theory of computer graphics. A classical result is the famous Descartes-Euler formula $v - e + f = 2$ that relates the number of vertices (v), edges (e), and facets (f) of a polytope. For various applications of this last formula and other similar results to image analysis and digital geometry, see Chapters 4 and 6 of [4].

A. Kuba, L.G. Nyúl, and K. Palágyi (Eds.): DGCI 2006, LNCS 4245, pp. 189–198, 2006.

Conditions for existence of gaps in digital lines and planes are available, e.g., in [5,6,7]. The notion of gap has been used in higher dimensions, too [8]. However, a rigorous definition that applies to arbitrary digital pictures is still missing. Approaches to estimating the number of gaps have been, overall, unclear.

A recent work [9] provided the formula

$$g = v - 2(p + c - h) + b, \tag{1}$$

where g is the number of gaps, v the number of vertices, p the number of pixels, h the number of holes, c the number of connected components, and b the number of 2×2 grid squares in a digital picture. For another similar result we refer to [10].

In the present paper we define the notion of gap in arbitrary dimension and obtain a formula for the number of gaps of maximal dimension n. We also derive a combinatorial relation for an n-dimensional digital curve.

In the next section we introduce some basic notions and notations of digital topology. In Section 3 we present our main results. In Section 3.4 we comment on a computer program that was developed to facilitate our theoretical research. We conclude with some remarks in Section 4.

2 Preliminaries

In this sections we introduce some basic notions of digital geometry to be used in the sequel. We conform to terminology used in [4] (see also [11]).

All considerations take place in the *grid cell model* that consists of the grid cells of $\mathbb{Z}^n$, together with the related topology. In the grid cell model we represent n-cells as hyper-cubes, called *hyper-voxels*, or *voxels*, for short. Their edges and vertices are 1-*cells* and 0-*cells*, respectively. For every $i = 0, 1, \ldots, n$, the set of all cells of dimension i (or i-cells) is denoted by $\mathbb{C}_n^{(i)}$. Further, we define the space $\mathbb{C}_n = \bigcup_{k=0}^{n} \mathbb{C}_n^{(i)}$. We say that two n-cells e, e' are k-adjacent for $0 \leq k \leq n-1$ if they share a k-cell. Two n-cells are *strictly* k-adjacent if they are k-adjacent but not $(k+1)$-adjacent.

A digital object $S \subset \mathbb{C}_n$ is a finite set of n-cells. A k-*path* $(0 \leq k \leq n-1)$ in S is a sequence of voxels from S such that every two consecutive voxels on the path are k-adjacent. Two voxels of a digital object S are k-*connected* (in S) iff there is a k-path in S between them. A subset G of S is k-*connected* iff there is a k-path connecting any two pixels of G. The maximal (by inclusion) k-connected subsets of a digital object S are called k-*components* of S. Components are nonempty, and distinct k-components are disjoint.

The grid cell model can be considered as an *abstract cell complex* $(\mathbb{C}_n, <, dim)$ (see [12]), where $<$ is a *bounding relation*, that is antisymmetric, irreflexive, and transitive, and such that for every $e, e' \in \mathbb{C}_n$, $e < e'$ if and only if eIe' and $dim(e) < dim(e')$. The relation $<$ is a partial order on $\mathbb{C}_n$. The corresponding order topology $\tau(<)$ is called the *grid cell topology*.[1] In the rest of the paper,

[1] In that topology the open sets are precisely the sets $U \subseteq \mathbb{C}_n$, such that, for every $u \in U$ and every $v \in \mathbb{C}_n$ with $u < v$, we have $v \in U$.

we will assume that the abstract cell complex $(\mathbb{C}_n, <, dim)$ is equipped with the topology $\tau(<)$. Then, for any subset A of $\mathbb{C}_n$, its *boundary* ∂A is defined as the set of all points x of $\mathbb{C}_n$ such that every open neighborhood of x meets A and $\mathbb{C}_n \setminus A$, while its *interior* $int(A)$ is the set of all points x of $\mathbb{C}_n$ such that there exists some open neighborhood of x contained in A. The points of $int(A)$ will be called *internal points* of A.

Given a digital object S, note that its closure $\bar{S}$ is naturally a subcomplex of $\mathbb{C}_n$. In the sequel, we will denote by S_k the set of k-cells of $\bar{S}$, i.e., $S_k = \bar{S} \cap \mathbb{C}_n^{(k)}$. In particular, we have $S_n = \bar{S} \cap C_n^{(n)} = S$.

3 Combinatorial Relations

In this section we first introduce the notions of tandem, gap, and brim of arbitrary dimension. Then we obtain a formula for the number of gaps of maximal dimension and a combinatorial relation for digital curves.

3.1 Tandems, Gaps, and Brims

A $\underbrace{2 \times \cdots \times 2}_{k} \times \underbrace{1 \times \cdots \times 1}_{n-k}$ grid parallelepiped in $\mathbb{C}_n$ will be called $2^k 1^{n-k}$-*block* $(0 \leq k \leq n)$. In particular, any voxel is a 1^n-block. See Figure 1a for illustrations.

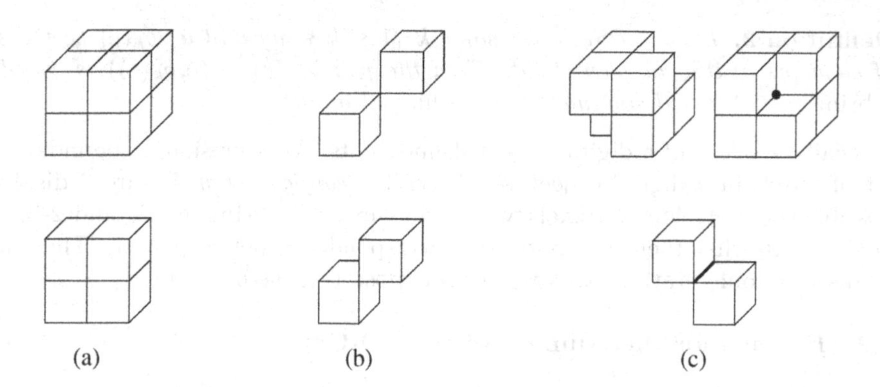

Fig. 1. Illustration to some notions in 3D. (a) *Top:* 2^3-block; *Bottom:* $2^2 1^1$-block. (b) *Top:* 0-tandem; *Bottom:* 1-tandem. (c) *Top:* Configuration exposing a 0-gap (in two different orientations); *Bottom:* Configuration exposing a 1-gap.

Now we are able to give the following definition.

Definition 1. *A pair $t_k = (v_1, v_2)$ of two strictly k-adjacent voxels v_1 and v_2, for $0 \leq k \leq n-1$, is called a k-*tandem. *Then the complement of t_k w.r.t. a $2^{n-k}1^k$-block, for $0 \leq k \leq n-2$, determines a k-*gap *of S.*

Remark 1. Technically, the complement of an $(n-1)$-tandem to a $2^1 1^{n-1}$-block can be considered as a $(n-1)$-gap. These are similar to "tunnels" known in classic combinatorial topology, see [4]. Since tunnels are well-studied object of essentially diverse type, we will not consider them here.

There are $n-1$ types of gaps: $0, 1, 2, \ldots,$ and $(n-2)$-gaps. For a given digital object S, the number of its tandems and gaps will be denoted by $b_0, b_1, \ldots, b_{n-1}$ and $g_0, g_1, \ldots, g_{n-2}$, respectively. Figure 1b,c illustrates tandems and gaps in dimension three.

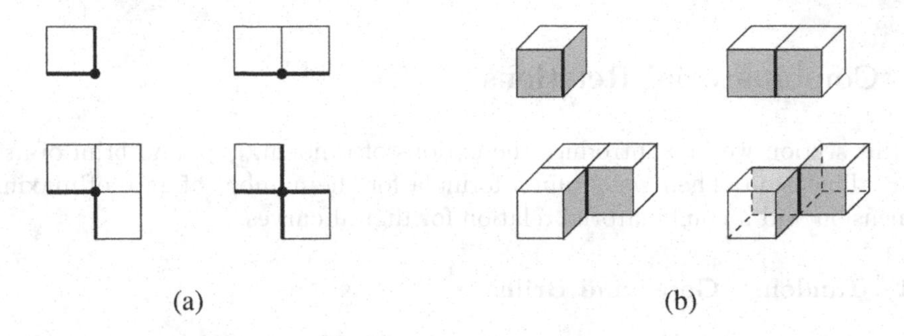

Fig. 2. (a) Possible 1-brims in 2D. (b) Possible 2-brims in 3D.

In the sequel we will also use the following technical notion.

Definition 2. *Let $c \in \partial S_{k-1}$ for some k $(1 \le k \le n)$ and let $b_k(c)$ be the set of elements of ∂S_k incident to it. Then the pair $br_k(c) = (c, b_k(c))$ is called a* k-brim *of S. We will say that $br_k(c)$ is* hinged *on c.*

Basically, k-brims of a digital object delineate its "k-dimensional" boundary. A set of voxels in a digital object will be called *configuration.* Figure 2 displays possible configurations of pixels/voxels that expose 1-brims in $\mathbb{C}_2$ and 2-brims in $\mathbb{C}_3$. (Note that there is one-to-one correspondence between both. There are 19 distinct configurations of voxels that expose 1-brims in $\mathbb{C}_3$.)

3.2 Formula for the Number of $(n-2)$-Gaps

For a given digital object $S \subset \mathbb{C}_n$, let $s_i = |S_i|$, $0 \le k \le n$. In this section we prove the following theorem.

Theorem 1. *For a given digital object $S \subset \mathbb{C}_n$,*

$$g_{n-2} = -2n(n-1)s_n + 2(n-1)s_{n-1} - s_{n-2} + b, \tag{2}$$

where b is the number of $2^2 1^{n-2}$-blocks of S.

Proof. For any $c \in S_{k-1}$, $1 \le k \le n-1$, we define

$$I_k(c) = \{c' \in S_k : c \text{ is incident with } c'\}.$$

We also define

$$\begin{aligned} intS_{k-1} =& \{c \in S_{k-1} : c \in intS\} \\ \partial S_{k-1} =& \{c \in S_{k-1} : c \in \partial S\} \\ \partial S_k =& \{c \in S_k : c \in \partial S\} \end{aligned}$$

It is easy to see that a $(k-1)$-cell belongs to $intS_{k-1}$ iff it is incident with $2^{n-(k-1)}$ n-cells of S. Otherwise, it belongs to the boundary of S.

For $c \in S_{n-1}$ we can consider $I_n(c) = \{c' \in S_n : c \text{ is incident with } c'\}$. The possible values for $|I_n(c)|$ are 1 and 2. More precisely, we have

$$\begin{aligned} intS_{n-1} =& \{c \in S_{n-1} : I_{n-1}(c) = 2\} \\ \partial S_{n-1} =& \{c \in S_{n-1} : I_{n-1}(c) = 1\} \\ S_{n-1} =& intS_{n-1} \cup \partial S_{n-1} \end{aligned}$$

Let us denote $s^{int}_{n-1} = |intS_{n-1}|$, and $s^{\partial}_{n-1} = |\partial S_{n-1}|$. Then $s_{n-1} = s^{int}_{n-1} + s^{\partial}_{n-1}$. Since every n-cell of S is incident with $2n$ $(n-1)$-cells from S_{n-1}, we obtain

$$2n|S| = s^{\partial}_{n-1} + 2s^{int}_{n-1}.$$

From here we get

$$s^{int}_{n-1} = ns_n - \frac{s^{\partial}_{n-1}}{2}.$$

Next we consider incidence relations between elements of ∂S_{n-1} and S_{n-2}. For any $c \in S_{n-2}$ we consider the brim hinged on c:

$$br_{n-1}(c) = \{c' \in \partial S_{n-1} : c \text{ is incident with } c'\}.$$

The possible values for $|br_{n-1}(c)|$ are 0, 2, and 4. This partitions S_{n-2} as follows:

$$S_{n-2} = S^0_{n-2} \cup S^2_{n-2} \cup S^4_{n-2}, \tag{3}$$

where $S^i_{n-2} = \{c \in S_{n-2} : |br_{n-1}(c)| = i\}$, for $i = 0,2,4$. If denote $\bar{s}^i_{n-2} = |S^i_{n-2}|$, $i = 0,2,4$, we get $s_{n-2} = \bar{s}^0_{n-2} + \bar{s}^2_{n-2} + \bar{s}^4_{n-2}$. From here, we obtain $\bar{s}^2_{n-2} = s_{n-2} - \bar{s}^0_{n-2} - \bar{s}^4_{n-2}$.

Every cell $x \in S^{\partial}_{n-1}$ is incident with $2(n-1)$ cells $y \in S_{n-2}$. Then it follows that

$$\begin{aligned} 2(n-1)s^{\partial}_{n-1} =& 4\bar{s}^4_{n-2} + 2\bar{s}^2_{n-2} = 4\bar{s}^4_{n-2} + 2(s_{n-2} - \bar{s}^0_{n-2} - \bar{s}^4_{n-2}) = \\ =& 2\bar{s}^4_{n-2} + 2s_{n-2} - 2\bar{s}^0_{n-2} \end{aligned}$$

from where we obtain

$$s^{\partial}_{n-1} = \frac{\bar{s}^4_{n-2} + s_{n-2} - \bar{s}^0_{n-2}}{n-1}.$$

Then

$$s_{n-1} = s^{int}_{n-1} + s^{\partial}_{n-1} = ns_n - \frac{s^{\partial}_{n-1}}{2} + s^{\partial}_{n-1} = ns_n + \frac{s^{\partial}_{n-1}}{2},$$

i.e.,

$$s_{n-1} = ns_n + \frac{\bar{s}^4_{n-2} + s_{n-2} - \bar{s}^0_{n-2}}{2(n-1)}. \tag{4}$$

Thus

$$2(n-1)s_{n-1} = 2n(n-1)s_n + \bar{s}^4_{n-2} + s_{n-2} - \bar{s}^0_{n-2},$$

and

$$\bar{s}^4_{n-2} = -2n(n-1)s_n + 2(n-1)s_{n-1} - s_{n-2} + \bar{s}^0_{n-2}.$$

We also have the following fact.

Fact 1. *For any $n \geq 2$, the sets of $(n-2)$-gaps and $(n-2)$-tandems are determined by the same configurations.*

Then it is enough to observe that $\bar{s}^4_{n-2} = g_{n-2}$ is the number of $(n-2)$-gaps (that are also $(n-2)$-tandems) and $\bar{s}^0_{n-2} = b$ the number of $2^2 1^{n-2}$-blocks of S, and we obtain the result stated. □

Note that for $n = 2$ the only gaps in S are the 0-gaps. For this case equality (4) has the form $s_1 = 2s_2 + \frac{1}{2}(g_0 + s_0 - b)$, where b is the number of (2×2)-blocks in S. Now, by Euler-Poincaré characteristic we have $s_0 - s_1 + s_2 = \beta_0 - \beta_1 + \beta_2$, where $\beta_0, \beta_1, \beta_2$ are the Betti numbers [4]. From here we get $s_2 - (2s_2 + \frac{1}{2}(s_0 - b + g_0)) + s_0 = \beta_2 - \beta_1 + \beta_0$.

Since S is homotopic to a 1D CW-complex, we have $\beta_2 = 0$. Moreover, β_0 is the number of connected components of S, while β_1 is the number of its holes. From here we immediately obtain formula (1).

3.3 Relations for Digital Curves

A digital curve admits various equivalent definitions [13]. One of them is the following. A *simple digital k-curve* is a set $\Gamma = \{c_1, c_2, \dots, c_l\}$ of voxels that satisfy the following two axioms: (A1) c_i is k-adjacent to c_j iff $i = j \pm 1(\text{modulo } l)$, and (A2) ρ is one-dimensional with respect to k-adjacency. To get acquainted with the classical definition of dimension of a digital object the reader is referred to [14]. For further developments and various results see [13, 4] and the bibliography therein. For example, we have the following:

Fact 2. *Let M be a finite set of pixels which is one-dimensional with respect to 0-adjacency. Then M does not contain any $2^2 1^{n-2}$-block.*

Figure 3 illustrates curves in $\mathbb{C}_2$ and $\mathbb{C}_3$.

Theorem 2. *Let $\Gamma \subset \mathbb{C}_n$ be a digital 0-curve. Then:*

$$g_{n-2} = -2n(n-1)s_n + 2(n-1)s_{n-1} - s_{n-2}.$$

Moreover, letting $b_0, \dots b_{n-1}$ be the number of its k-tandems, for $0 \leq k \leq n-1$ we have the relation

$$s_k = 2^{n-k}\binom{n}{k}s_n - \sum_{i=0}^{n-k-1} 2^i \binom{k+i}{k} b_{i+k} \tag{5}$$

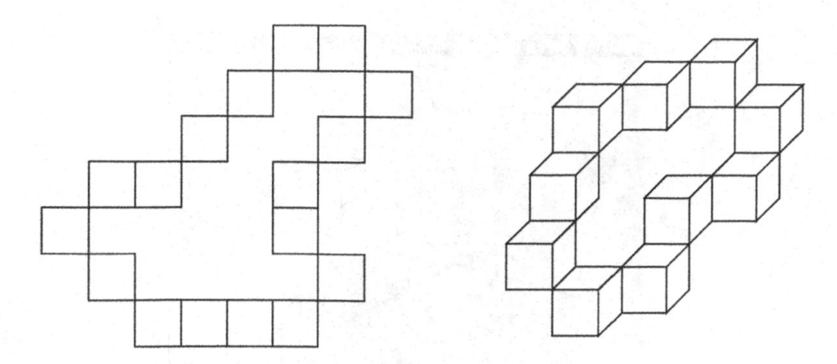

Fig. 3. Simple closed curves in C_2 (left) and C_3 (right)

Proof. Let $\Gamma = \langle c_1, c_2, \ldots, c_m \rangle$ be a closed digital 0-curve, i.e., it satisfies conditions (A1) and (A2) and Fact 2 applies, as well. None that Γ consists of consecutive tandems of the form $(c_1, c_2), (c_2, c_3), \ldots, (c_{m-1}, c_m), (c_m, c_1)$.

The first assertion follows immediately from Theorem 1 and Fact 2.

For the second assertion, let c be a k-cell for $k \neq n$.

We say that c is a *totally boundary cell* if c is incident with exactly one n-cell. If c is not totally boundary, then c belongs to the closure of the shared face of a tandem t_j in dimension $j \geq k$; we then say that *c is involved in t_j*.

Since Γ is a 0-curve, every k-cell is incident with at most two n-cells and, thus, every non totally boundary cell is involved in exactly one tandem. Now the number of k-cells involved in a j-dimensional tandem t_j is easily seen to be $2^{j-k}\binom{j}{k}$. Therefore the number of non totally boundary cells s_k^{ntb} is:

$$s_k^{\mathrm{ntb}} = b_k + 2\binom{k+1}{k} b_{k+1} + \ldots + 2^{n-1-k}\binom{n-1}{k} b_{n-1}, \tag{6}$$

whereas the number of totally boundary k-cells is given by $s_k^{\mathrm{tb}} = s_k - s_k^{\mathrm{ntb}}$. Since every n-cell is incident with $2^{n-k}\binom{n}{k}$ k-cells, we have:

$$\begin{aligned} 2^{n-k}\binom{n}{k} s_n &= 1 \cdot s_k^{\mathrm{tb}} + 2 \cdot s_k^{\mathrm{ntb}} \\ &= s_k + s_k^{\mathrm{ntb}} \end{aligned} \tag{7}$$

The second assertion now follows straightforwardly from eq. (6) and eq. (7). □

Remark 2. Note that $(n-2)$-gaps are the only gaps a digital curve Γ may have. Note also that if Γ is a digital $(n-2)$-curve,[2] then the number of $(n-2)$-gaps of Γ matches the number of "linear segments" into which Γ can be decomposed.

Remark 3. Since Γ is a closed curve, its Euler-Poincaré characteristic $\chi(\Gamma)$ is zero. We then have:

$$0 = \chi(\Gamma) = \sum_{k=0}^{n} (-1)^k s_k$$

[2] That is, any two consecutive voxels of Γ are $(n-2)$-adjacent.

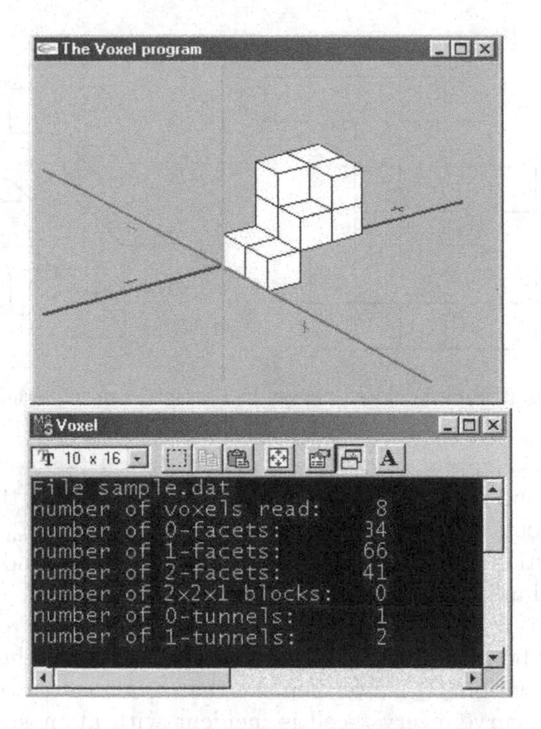

Fig. 4. Sample output of the computer program

Using the expression for the s_i found in eq. (5), we recover, after elementary manipulations, the not-surprising relation:

$$s_n = b_0 + b_1 + \ldots + b_{n-1}$$

The used approach allows to obtain similar (although more complex and thus less compact and elegant) relations for k-curves with $k \neq 0$, as well as for arbitrary digital object.

3.4 Experimental Software

The theoretical results described in the previous sections have been supported and verified by an experimental computer program.

Given a digital picture S represented by the coordinates of its voxels, our program takes as an input a file with the list of the voxel coordinates. It outputs the number of the 0-, 1-, and 2-facets, $2^2 1^1$-blocks, and 0- and 1-gaps of S. Computation of the number of the 0-/1-gaps is performed by appropriate scanning of S by $2 \times 2 \times 2$-cubes/$2 \times 2 \times 1$- blocks and counting the distinct gaps. The number of 1-gaps can alternatively be found by using formula (2).

The program is written in Visual Studio C++ 6.0 and uses OpenGL. It runs under Windows 98 or higher. It allows to visualize the digital picture S and to interactively rotate it along the Ox-, Oy-, and Oz- axes so that the object can be seen from different viewpoints. In Figure 4, a snapshot of the running program is displayed.

4 Concluding Remarks

In this paper we provided a rigorous definition of gaps in a digital picture and derived a formula for the number of $(n-2)$-gaps, as well as certain combinatorial relations for digital curves. A supporting computer program has been developed as well.

Knowledge of the number of gaps of maximal dimension can be useful in several aspects. Among these we would like to mention an application to the well-known polyhedron decomposition problem [15, 16], that is to partition a given non-convex polyhedron into as small as possible number of convex polytopes. Specifically, let P be the rectilinear polyhedron defined as a union of a set of voxels of $\mathbb{C}_3$. It is not hard to see that the number of gaps in the discrete surface constituted by the boundary voxels of P is an upper bound for the number r of "notches" of P, that are locations causing non-convexity.[3] The fact is that all bounds on the number of convex polytopes obtained by decomposition algorithms are in terms of that parameter r. A more careful study of this aspect is seen as a further task. Another one is seen in seeking approaches that would allow to obtain more compact characterizations of lower dimensional gaps in digital pictures.

Acknowledgements

The authors thank the two anonymous referees for a number of useful remarks and suggestions.

References

1. Cohen-Or, D., Kaufman, A.: 3D Line Voxelization and Connectivity Control. IEEE Computer Graphics and Applications **17** (6) (1997) 80–87
2. Kaufman, A., Cohen, D., Yagel, R.: Volume Graphics. IEEE Computer **26** (7) (1993) 51–64
3. Latecki, L., Eckhardt, U., Rosenfeld, A.: Well-Composed Sets. Computer Vision and Vision Understanding **61** (1995) 70–83
4. Klette, R., Rosenfeld, A.: Digital Geometry – Geometric Methods for Digital Picture Analysis. Morgan Kaufmann, San Francisco (2004)
5. Brimkov, V.E., Coeurjolly, D., Klette, R.: Digital Planarity – a Review. CITR-TR 142, Auckland (2004)

[3] Notch (or reflex edge) is an edge of a polyhedron where the inner dihedral angle subtended by two incident facets is greater than 180 degrees.

6. Andres, E., Nehlig, Ph., Françon, J.: Tunnel-Free Supercover 3D Polygons and Polyhedra. In: Fellner D., Szirmay-Kalos L. (guest eds.): EUROGRAPHICS'97 (1997) C3–C13
7. Brimkov, V.E., Andres, E., Barneva, R.P.: Object Discretizations in Higher Dimensions. Pattern Recognition Letters **23** (2002) 623–636
8. Andres, E., Acharya, R., Sibata, C.: Discrete Analytical Hyperplanes. Graphical Models and Image Processing **59** (1997) 302–309
9. Brimkov, V.E., Maimone, A., Nordo, G., Barneva, R.P., Klette, R.: The Number of Gaps in Binary Pictures. In: Bebis et al. (eds.): Advances in Visual Computing. Lecture Notes in Computer Science, Vol. 3804. Springer-Verlag, Berlin Heidelberg New York (2005) 35–42
10. Brimkov, V.E., Maimone, A., Nordo, G.: Counting Gaps in Binary Pictures. In: 11th International Workshop on Combinatorial Image Analysis. Lecture Notes in Computer Science, Vol. 4040. Springer-Verlag, Berlin Heidelberg New York (2006) 16–24
11. Kong, T.Y.: Digital Topology. In: Davis, L.S. (ed.): Foundations of Image Understanding. Kluwer, Boston Massachusetts (2001) 33–71
12. Kovalevsky, V.A.: Finite Topology as Applied to Image analysis. Computer Vision, Graphics and Image Processing **46**(2) (1989) 141–161
13. Brimkov, V.E., Klette, R.: Curves, Hypersurfaces, and Good Pairs of Adjacency Relations. In: Klette, R., Zunic, J.K. (eds.): 10th International Workshop on Combinatorial Image Analysis. Lecture Notes in Computer Science, Vol. 3322. Springer-Verlag, Berlin Heidelberg New York (2004) 270–284
14. Mylopoulos, J.P., Pavlidis, T.: On the Topological Properties of Quantized Spaces. I. The Notion of Dimension. J. ACM **18** (1971) 239–246
15. Dielissen, V.J., Kaldewaij, A.: Rectangular Partition is Polynomial in Two Dimensions but NP-complete in Three. Information Processing Letters **38** (1991) 1–6
16. Hershberger, J.E., Snoeyink, J.S.: Erased Arrangements of Lines and Convex Decompositions of Polyhedra. Computational Geometry: Theory and Applications **9** (1998) 129–143

Reusing Integer Homology Information of Binary Digital Images*

Rocío González-Díaz, Belén Medrano**, Javier Sánchez-Peláez, and Pedro Real

Departamento de Matemática Aplicada I
Universidad de Sevilla, Seville, Spain
{rogodi, belenmg, fjsp, real}@us.es
http://www.us.es/gtocoma

Abstract. In this paper, algorithms for computing integer (co)homology of a simplicial complex of any dimension are designed, extending the work done in [1,2,3]. For doing this, the homology of the object is encoded in an algebraic-topological format (that we call AM-model). Moreover, in the case of 3D binary digital images, having as input AM-models for the images I and J, we design fast algorithms for computing the integer homology of $I \cup J$, $I \cap J$ and $I \setminus J$.

1 Introduction

Efficient algorithms for computing topological information are powerful tools in the fields of Data Mining, Pattern Recognition, Geometric Modeling and nD Digital Image Processing. Nevertheless, topological notions (such as the cup product on cohomology, cohomology operations, fundamental group, homotopy groups, etc) are hard to adapt into an n-dimensional discrete framework; and the number of available computational tools are limited. It is a fact that the cup product on cohomology is a topological invariant which contains more information than homology groups when we deal with an object of dimension greater than or equal to 3. Since cohomology is essentially an algebraic notion, it seems reasonable to encode it using a classical algebraic-topological cover: chain homotopy equivalences. In the setting of Simplicial Topology, we use here this extra algebraic-topological information (that we will define as an AM-model for a simplicial complex) to compute the cup product on integer cohomology as well as cohomological numbers derived from it, extending the work developed in [1,2]. Our computational approach follows the philosophy of the Effective Homology Theory developed by F. Sergeraert in [4,5]. In particular, we prove that all the algorithms for computing integer homology based in the matrix reduction method to Smith normal form (for example [6,7,8,9]) can be translated to our setting with no extra computational cost in time. Finally, we successfully apply this computational algebraic topological approach to 3D binary digital images

* Partially supported by the PAICYT research project FQM–296 "Computational Topology and Applied Math" from Junta de Andalucía.

** Fellow associated to University of Seville under a Junta de Andalucia research grant.

A. Kuba, L.G. Nyúl, and K. Palágyi (Eds.): DGCI 2006, LNCS 4245, pp. 199–210, 2006.

and we prove that a suitable extended notion of AM-model for binary 3D digital images can be reused under voxel-set operations (union, intersection and difference).

2 Integer Homology, Chain Contractions and AM-Models

In [9], an algorithm improving the efficiency of the classical integer reduction homology algorithm is described. Their technique is mainly based on the results of [8], in which a matrix reduction to integer Smith normal form is determined in an efficient way. There is no problem for translating this method to our framework since it consists in constructing a chain homotopy equivalence from the previously calculated Smith normal form, without additional computational cost. Moreover, our strategy of saving more algebraic information outperforms the previous algorithms for computing integer homology in several points such as: 1) cohomological features can be computed; 2) we can efficiently control the topological changes after addition or deletion of simplices.

First, we give a brief summary of concepts and notations. The terminology follows Munkres book [6]. We will consider that the ground ring is $\mathbf{Z}$.

Simplicial Complexes. Considering an ordering on a vertex set V, a q–simplex with $q+1$ affinely independent vertices $v_0 < \cdots < v_q$ of V is the convex hull of these points, denoted by $\langle v_0, \dots, v_q \rangle$. If $i < q$, an *i–face* of σ is an i–simplex whose vertices are in the set $\{v_0, \dots, v_q\}$. A simplex is *maximal* if it does not belong to any higher dimensional simplex. A *simplicial complex* K is a collection of simplices such that every face of a simplex of K is in K and the intersection of any two simplices of K is a face of each of them or empty. The set of all the q–simplices of K is denoted by $K^{(q)}$. The *dimension of* K is the dimension of the highest dimensional simplex in K.

Chains and Homology. Let K be a simplicial complex. A *q–chain* a is a formal sum of simplices of $K^{(q)}$. The q–chains form the *qth chain group* of K, denoted by $C_q(K)$. The *boundary* of a q–simplex $\sigma = \langle v_0, \dots, v_q \rangle$ is the $(q-1)$–chain: $\partial_q(\sigma) = \sum_{i=0}^{q} (-1)^i \langle v_0, \dots, \hat{v}_i, \dots, v_q \rangle$, where the hat means that v_i is omitted. By linearity, ∂_q can be extended to q–chains. The collection of boundary operators connect the chain groups $C_q(K)$ into the *chain complex* $C(K)$: $\cdots \overset{\partial_2}{\to} C_1(K) \overset{\partial_1}{\to} C_0(K) \overset{\partial_0}{\to} 0$. An essential property is that $\partial_q \partial_{q+1} = 0$. In a more general setting, a *chain complex* $\mathcal{C}$ is a sequence $\cdots \xrightarrow{d_2} C_1 \xrightarrow{d_1} C_0 \xrightarrow{d_0} 0$ of abelian groups C_q and homomorphisms d_q, such that for all q, $d_q d_{q+1} = 0$. The set of all the homomorphisms d_q is called the *differential* of $\mathcal{C}$. A q–chain $a \in C_q$ is called a *q–cycle* if $d_q(a) = 0$. If $a = d_{q+1}(a')$ for some $a' \in C_{q+1}$ then a is called a *q–boundary.* Denote the groups of q–cycles and q–boundaries by Z_q and B_q respectively. Define the *qth homology group* to be the quotient group Z_q/B_q, denoted by $H_q(\mathcal{C})$. We say that a is a *representative q–cycle* of a homology generator α if $\alpha = a + B_q$. We denote $\alpha = [a]$. The qth *betti number* β_q is the rank of the free part of $H_q(\mathcal{C})$. Intuitively, β_0 is the number of components of connected pieces, β_1 is the number of independent "holes" and β_2 is the number of "cavities".

Chain Contractions. A *chain contraction* [10] of a chain complex $\mathcal{C}$ to a chain complex $\mathcal{C}'$ is a set of three homomorphisms (f, g, ϕ) such that: $f : \mathcal{C} \to \mathcal{C}'$ and $g : \mathcal{C}' \to \mathcal{C}$ are chain maps; fg is the identity map of $\mathcal{C}'$; and $\phi : \mathcal{C} \to \mathcal{C}$ is a chain homotopy of the identity map id of $\mathcal{C}$ to gf, that is, $\phi\partial + \partial\phi = id - gf$. In this case, $\mathcal{C}$ and $\mathcal{C}'$ have isomorphic homology groups [6, p. 73].

AM-Models. An *AM-model* for a simplicial complex K is the set (C, M, f, g, ϕ) where C is a basis of $C(K)$, M is a subset of generators of $C(K)$ and (f, g, ϕ) is a chain contraction from $C(K)$ to $M(K)$ where $M(K)$ is the chain complex generated by M with differential $\partial|_{M(K)}$ such that in each dimension q, the matrix A of the differential $\partial_q|_{M(K)}$ coincides with its Smith normal form and satisfies that any non-null entry of A is greater than 1. Moreover, if the homology is free or the ground ring is a field, then $M(K)$ is isomorphic to the homology of K. It is necessary to emphasize that given a simplicial complex K, it is possible to define different AM-models for K since the chain complex $M(K)$ and the morphisms f, g and ϕ can admit different formulae.

A translation of the integer reduction homology algorithm in terms of chain contractions has been made in [11]. Here we rewrite this work using a more algorithmic language. This algorithm consists in reducing the matrix A_q of the boundary operator in each dimension q, to its Smith normal form A'_q, relative to some basis $\{a_1, \ldots a_r\}$ of $C_q(K)$ and $\{e_1, \ldots, e_s\}$ of $C_{q-1}(K)$ such that $\{a_{\ell+1}, \ldots, a_r\}$ is a basis of $Z_q(K)$, and $\{\lambda_1 e_1, \ldots, \lambda_\ell e_\ell\}$ is a basis of $B_{q-1}(K)$ [6, pp. 56-61].

Algorithm 1. *Computing an AM-model for a Finite Simplicial Complex.*

INPUT: A simplicial complex K of dimension d.
Initially: $C_q := K^{(q)}$, $M_q := K^{(q)}$ and $C'_q = \{\ \}$ for $0 \le q \le d$,
$f(\sigma) := \sigma$, $g(\sigma) := \sigma$, $\phi(\sigma) := 0$ for every $\sigma \in K$.
For $q = 1$ to $q = d$ do
Reduce the matrix A_q of the boundary operator ∂_q relative to the basis C_q and M_{q-1} to its Smith normal form A'_q relative to some basis $\{a_1, \ldots, a_r\}$ of C_q and $\{e_1, \ldots, e_s\}$ of M_{q-1} where:
$\partial_q(a_i) = e_i$, for $1 \le i \le t \le$ min (r, s);
$\partial_q(a_i) = \lambda_i e_i$, $\lambda_i \in \mathbf{R}$, for $t < i \le \ell \le$ min (r, s);
and $\partial_q(a_i) = 0$ for $\ell < i \le r$.
Define $C_{q-1} := C'_{q-1} \cup \{e_1, \ldots, e_s\}$, $M_{q-1} := \{e_{t+1}, \ldots, e_s\}$,
$C_q := \{a_1, \ldots, a_r\}$, $C'_q := \{a_1, \ldots, a_t\}$, $M_q := \{a_{t+1}, \ldots, a_r\}$,
$f(a_i) := 0$, $f(e_i) := 0$ and $\phi(e_i) := a_i$ for $1 \le i \le t$.
OUTPUT: The set $(C_0 \cup \cdots \cup C_d, M_0 \cup \cdots \cup M_d, f, g, \phi)$.

The following result shows that although $M(K)$ is not isomorphic to the homology of K, we can directly obtain the integer homology from it.

Theorem 2. *Let K be a finite simplicial complex of any dimension. The set $(C_0 \cup \cdots \cup C_d, M_0 \cup \cdots \cup M_d, f, g, \phi)$ defines an AM-model for K, being* $C =$

$C_0 \cup \cdots \cup C_d$ a basis of $C(K)$, $M(K)$ the chain complex generated by $M = M_0 \cup \cdots \cup M_d$ and with differential $\partial|_{M(K)}$, and (f, g, ϕ) a chain contraction from $C(K)$ to $M(K)$. Moreover, the integer homology of K and integer homology generators can be directly obtained from M and $\partial|_{M(K)}$.

If K has m simplices, an AM-model for K can be computed in time and storage $\mathcal{O}(m^3)$.

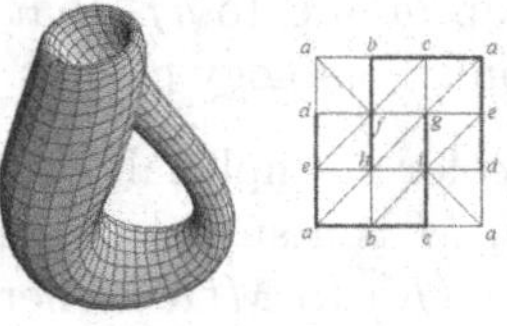

Fig. 1. The Klein bottle and a triangulation of it

Example 1. Consider the simplicial complex K in Figure 1 whose underlying space is the Klein bottle [6, p. 283]. Running the algorithm above, we obtain that the vertex $\langle a \rangle$ belongs to C_0 and $M_0(K)$, the cycles $\alpha_1 := \langle a, b\rangle + \langle b, c\rangle - \langle a, c\rangle$ and $\alpha_2 := \langle a, d\rangle + \langle d, e\rangle - \langle a, e\rangle$ belong to C_1 and M_1; and the 2-chain consisting in the sum of all the triangles in K, $\beta := -\langle a, b, f\rangle - \langle b, c, f\rangle + \langle a, c, g\rangle - \langle a, e, g\rangle + \langle e, g, i\rangle - \langle e, d, i\rangle + \langle c, d, i\rangle - \langle a, c, d\rangle + \langle b, c, i\rangle + \langle a, b, h\rangle - \langle a, e, h\rangle - \langle e, d, f\rangle + \langle a, d, f\rangle + \langle c, f, g\rangle - \langle f, g, h\rangle - \langle h, g, i\rangle - \langle b, h, i\rangle$, is an element of C_2 and M_2. The rest of the elements of C_0 are the boundaries of the 1-simplices marked in blue in Figure 1. These 1-simplices are also elements of C_1. Denote by x one of these 1-simplices. The rest of the elements of C_1 are the boundaries of all the 2-simplices except for $\langle f, g, h\rangle$. These 2-simplices belong to C_2. Denote by y one of these 2-simplices. The images of the maps (f, g, ϕ) on the generators of $C(K)$ and $M(K)$ are described in the table below:

C	M	f	g	ϕ
$\langle a\rangle$	$\langle a\rangle$	$\langle a\rangle$	$\langle a\rangle$	0
α_1	α_1	α_1	α_1	0
α_2	α_2	α_2	α_2	0
β	β	β	β	0
x		0		0
∂x		0		x
y		0		0
∂y		0		y

Summing up, $M_0 = \{\langle a\rangle\}$, $M_1 = \{\alpha_1, \alpha_2\}$ and $M_2 = \{\beta\}$. Moreover, $\partial(\langle a\rangle) = 0$, $\partial(\alpha_1) = 0$, $\partial(\alpha_2) = 0$ and $\partial(\beta) = 2\alpha_2$. Therefore we obtain that $H_0(K) \simeq \mathbf{Z}$, $H_1(K) \simeq \mathbf{Z} \oplus \mathbf{Z}/\mathbf{Z}2$ and representative cycles of the homology generators are $\langle a\rangle$ for $H_0(K)$, α_1 for the free part of $H_1(K)$ and α_2 for the torsion part.

3 Cohomology Computations with Integer Coefficients

In this section, we extend the work done in [11, 1, 2] (with coefficients in a field) for computing cohomology features over the coefficient domain $\mathbf{Z}$. The interest for computing cohomology (the dual notion of homology) is that cohomology has an additional multiplicative structure, the cup product, from which we can derive finer invariants than homology. Observe that working with coefficients in a field, homology groups are free and isomorphic to cohomology groups. Nevertheless, working with coefficients in $\mathbf{Z}$, homology and cohomology of simplicial complexes can have torsion part and, in this case, they are not isomorphic.

Cochains and Cohomology. Let $\mathcal{C}$ be a chain complex. The *cochain complex* $\mathcal{C}^*$ in each dimension q is the group of *q–cochains* with coefficients in $\mathbf{Z}$, $C^q = \{c : C_q \to \mathbf{Z}$ such that c is a homomorphism$\}$. If $\{a_1, \dots, a_n\}$ is a basis of C_q then a basis of C^q is $\{a_1^*, \dots, a_n^*\}$, where $a_i^* : C_q \to \mathbf{Z}$ is given by $a_i^*(a_i) = 1$ and $a_i^*(a_j) = 0$ for $1 \leq i, j \leq n$ and $j \neq i$. For each q, the differential d_{q+1} on C_{q+1} induces the *codifferential* $\delta_q : C^q \to C^{q+1}$ via $\delta_q(c) = cd_{q+1}$, so that δ_q raises dimension by one. Define Z^q to be the kernel of δ_q and B^{q+1} to be its image. These groups are called the group of *q–cocycles* and *q–coboundaries*, respectively. Define the qth *cohomology group*, $H^q(\mathcal{C}) = Z^q/B^q$ for $q \geq 0$.

The following result shows that we can directly obtain the integer cohomology of K from an AM-model for it. This assertion is not given in [11,1,2].

Theorem 3. *Let K be a finite simplicial complex of any dimension. Given an AM-model (C, M, f, g, ϕ) for K, the integer cohomology of K and integer cohomology generators can be directly obtained from M and $\partial|_{M(K)}$.*

Example 2. Consider the AM-model (C, M, f, g, ϕ) obtained in Example 1 for the simplicial complex K whose underlying space is the Klein bottle. Starting from the chain complex $M(K)$ whose basis is $\{\langle a\rangle, \alpha_1, \alpha_2, \beta\}$ and differential $\partial|_{M(K)}$, we construct in an straightforward way the cochain complex $M^*(K)$ whose basis is $\{\langle a\rangle^*, \alpha_1^*, \alpha_2^*, \beta^*\}$ and codifferential δ given by: $\delta(\langle a\rangle^*) = \langle a\rangle^*\partial|_{M(K)} = 0$, $\delta(\alpha_1^*) = \alpha_1^*\partial|_{M(K)} = 0$, $\delta(\alpha_1^*) = \alpha_2^*\partial|_{M(K)} = 2\beta^*$, $\delta(\beta^*) = \beta^*\partial|_{M(K)} = 0$. Therefore we obtain that $H^0(K) \simeq \mathbf{Z}$, $H^1(K) \simeq \mathbf{Z}$ and $H^2(K) \simeq \mathbf{Z}/\mathbf{Z}2$; and the generators are: $\langle a\rangle^*$ for $H^0(K)$, α_1^* for $H^1(K)$ and β^* for $H^2(K)$.

Cup Product. The cochain complex $C^*(K)$ is a ring with the *cup product* $\smile$: $C^p(K) \times C^q(K) \to C^{p+q}(K)$ given by: $(c \smile c')(\langle v_0, \dots, v_{p+q}\rangle) = c(\langle v_0, \dots, v_p\rangle) \cdot c'(\langle v_p, \dots, v_{p+q}\rangle)$. It induces an operation $\smile$: $H^p(K) \times H^q(K) \to H^{p+q}(K)$, via $[c] \smile [c'] = [c \smile c']$, that is bilinear, associative, commutative up to a sign, independent of the ordering of the vertices of K and homotopy-type invariant [6, p. 289].

Working with coefficients in $\mathbf{Z}/\mathbf{Z}2$, a new cohomology invariant called HB1 is obtained in [1,2]. The idea is to put into a matrix form the multiplication table of the cup product of cohomology generators of dimension 1. The following algorithm compute HB1 working with integer coefficients. Assuming that K has m simplices, the complexity of this algorithm is $\mathcal{O}(m^6)$. This algorithm is an straightforward extension of that given in [1,2].

Algorithm 4. *Algorithm for computing HB1 with integer coefficients.*

```
INPUT: An AM-model (C, M, f, g, φ) for a simplicial complex K.
Let {α_1,...,α_p} and {β_1,...,β_m} be the set of 1 and 2-cycles in M.
For i = 1 to p do
    For j = i to p do
        For k = 1 to m do
            b_((i,j),k) := (α_i^* f ⌣ α_j^* f)(g(β_k)).
HB1:= the rank of the 2D matrix of integers B_(p(p+1)/2)×m = (b_((i,j),k)).
OUTPUT: The integer HB1.
```

The implementation of the algorithms described above working with coefficients in **Z/Z2** has been made by J. Sánchez-Peláez and P. Real. We have tested it on several 3D objects. We give here an example of the computation of the cohomology, cohomology generators and the invariant HB1.

Example 3. Consider the simplicial complex T whose underlying space is showed in Figure 2 (on the left). It consists in 11847 simplices. The running time for computing an AM-model for T and the homology of T using a Pentium 4, 3.2 GHz, 1Gb RAM was 2 seconds. We obtain that $\beta_0 = 1$, $\beta_1 = 4$ and $\beta_2 = 3$. The running time for computing the cup product was 1.5 seconds. In Figure 2 (on the center), the 1 and 2-simplices on which the representative cocycles are non-null are drawing. The table on the right of Figure 2 shows the results of the cup product of any two cohomology generators of dimension 1. Finally, HB1= 2.

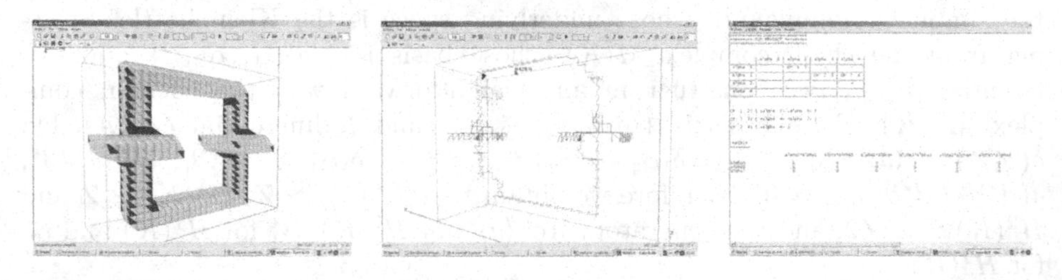

Fig. 2. The simplicial complex T, representative cocycles of the generators of $H^1(T)$ and $H^2(T)$ and the multiplication table of the cup product

4 AM-Models for 3D Digital Images

Three dimensional digital images are usually captured into the cubic grid or computed from 2D projections. There are, however, capturing techniques such as CT or MRI to produce images into other grids, such as the face-centered cubic (fcc) and the body-centered cubic (bcc) grids [12]. An important issue in Digital Volume Processing is to design efficient algorithms for analysis and processing in these grids, since it is very easy to obtain data structures for the fcc and bcc grids. On the other hand, the only Voronoi adjacency relation on the bcc grid is the 14–adjacency. Using this adjacency, it is straightforward to associate to a digital image I, a unique simplicial complex $K(I)$ (up to isomorphism) with the same topological information as I. The i–simplices of $K(I)$ ($i \in \{0,1,2,3\}$) are constituted by the different sets of i mutually 14–neighbor black points in I.

Definition 1. *Let I be a 3D binary digital image. An* AM-model *for I is defined as an AM-model for its simplicial representation $K(I)$.*

Since simplicial complexes considered in this section are embedded in $\mathbf{R}^3$, their homology groups vanishes for dimensions greater than 3 and they are torsion–free for dimensions 0, 1 and 2 (see [13, ch.10]). Therefore the chain complex $M(K(I))$ is isomorphic to the homology of I.

In the following table we present the running time for computing integer homology generators of the 3D digital images showed in figure 3. We have to say that these images have been created in a cubic grid. For these reason, we consider a special 14-adjacency in the cubic grid in the way that it is isomorphic to the bcc grid.

Image I	Number of voxels in I	Time for computing	β_0	β_1	β_2
A	26308	50 seconds	2	9	3
B	31012	38 seconds	138	419	13
C	18842	27 seconds	1	277	5

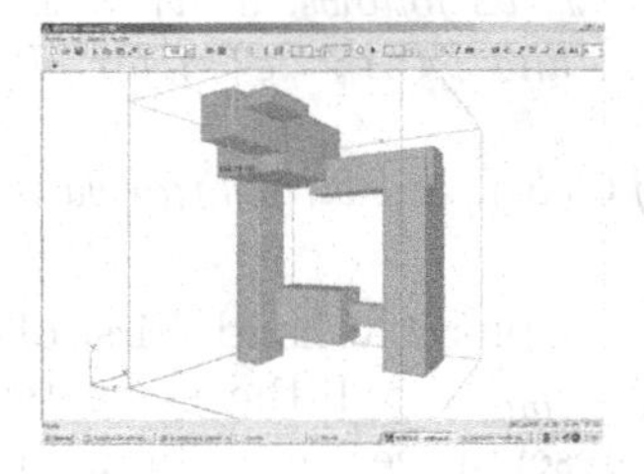
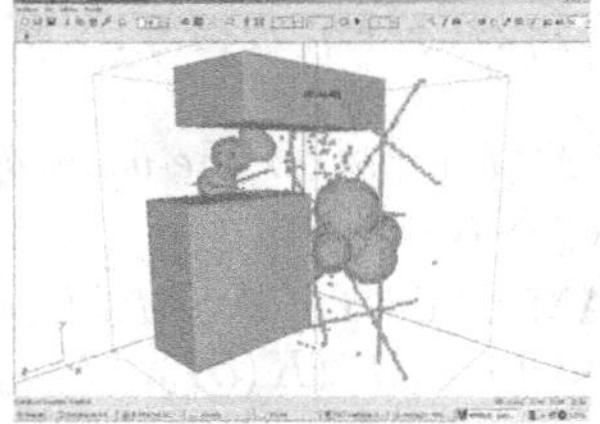
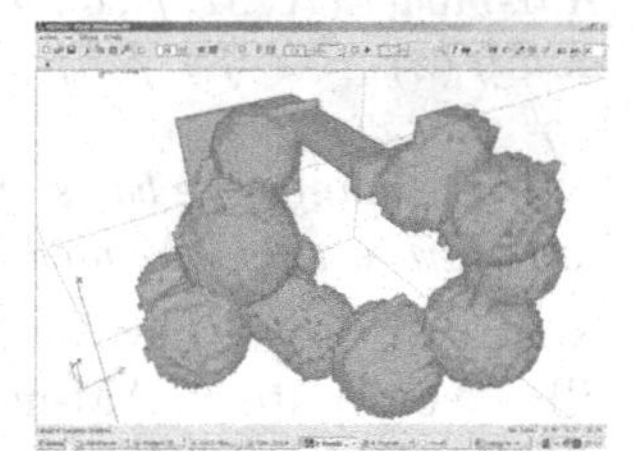

Fig. 3. The 3D digital images A, B and C

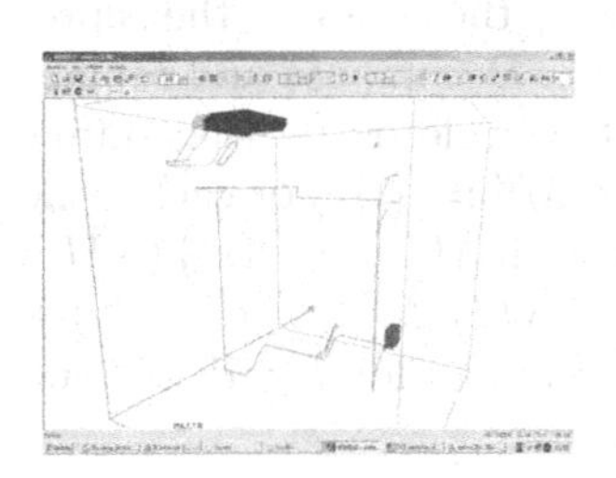
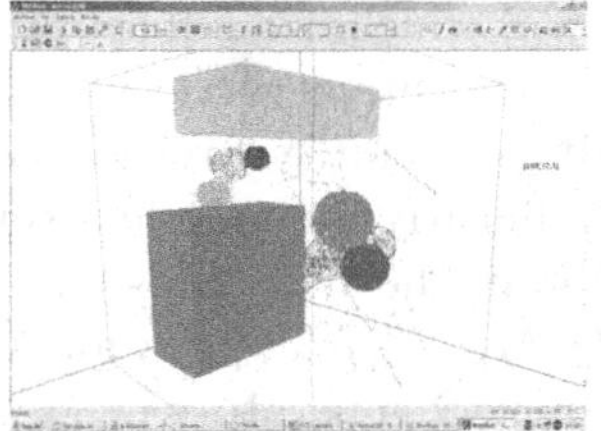
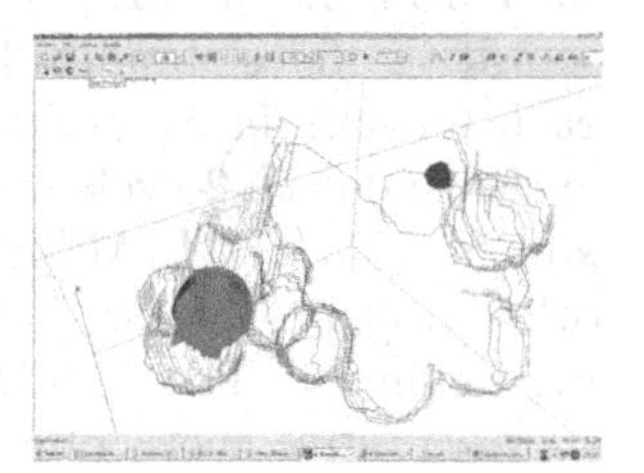

Fig. 4. Representative cycles of the homology generators of the images A, B and C

4.1 Computing "Good" Homology Generators

In [14], algorithms for obtaining "optimal" generators of the first homology group are developed using Dijkstra's shortest path algorithm for any oriented 2-manifolds. Here, in the context of digital volumes we sketch some techniques for drawing "good" representative cycles of homology generators.

Given an AM-model $(K(I), M_I, f_I, g_I, \phi_I)$ for a 3D digital image I, we say that x is the representative cycle of the generator $h \in M_I$ obtained by g_I if $g_I(h) = x$. Our interest now is to get a new AM-model $(K(I), M'_I, f'_I, g'_I, \phi'_I)$ with "good" representative cycles of homology generators obtained by g'_I. This means that each representative cycle obtained by g'_I belongs to the boundary of the image ∂I (it is constituted by the set of black voxels in I with a 14-neighbor white voxel). Moreover, it is required that in dimension 0, it must be a vertex; in dimension 1, an *elementary cycle* (it is connected, each vertex is shared by exactly two edges and two consecutive edges can not belong to the same triangle in $K(I)$) and in dimension 2, an *elementary cavity* (it is a connected 2-cycle with exactly one white connected component inside and three triangles can not belong to the same tetrahedra).

Now, for obtaining good representative cycles of homology generators we can use the following new result.

Lemma 1. *Let (K, M, f, g, ϕ) be an AM-model for a simplicial complex K. Let h be a generator of M and x a chain in $C(K)$ such that $x = g(h)$. Let x' be a chain in $C(K)$ such that $\partial(x') = 0$ and $f(x') = h$. Then, it is possible to define a new AM-model $(K, M, f', g', \phi'))$ for K such that $g'(h) := x'$ as follows: $g'(h) := x'$ and $g'(z) := z$ if $z \neq h$; $\phi'(x) := \phi(x')$, $\phi'(x') := \phi(x)$ and $\phi'(z) := \phi(z)$ for all $z \neq x', x$.*

If we change the basis of $M(K)$ and/or the basis of $C(K)$, it is straightforward to obtain a new AM-model for K.

Now, suppose we have an AM-model $(K(I), M_I, f_I, g_I, \phi_I)$ for I at hand. First of all, we compute an AM-model for ∂I, $(K(\partial I), M_{\partial I}, f_{\partial I}, g_{\partial I}, \phi_{\partial I})$. If the elements of $M_{\partial I}$ are denoted by $\{\alpha_1, \dots, \alpha_n\}$ then the set of representative cycles obtained by $g_{\partial I}$ is $S_{\partial I} = \{g_{\partial I}(\alpha_1), \dots, g_{\partial I}(\alpha_n)\}$. Since all the homology generators of I are homology generators of ∂I, find a subset M'_I of $\{f_I g_{\partial I}(\alpha_1), \dots, f_I g_{\partial I}(\alpha_n)\}$, which is a basis of $M(K(I))$. Obtain the new AM-model $(K(I), M'_I, f'_I, g'_I, \phi'_I)$ for I using Lemma 1. Now, denote by $S_I = \{c_1, \dots, c_m\}$ the set of all the representative cycles obtained by g'_I which is a subset of $S_{\partial I}$. Decompose and replace each 0-cycles in S_I by its constitutive vertices, each 1-cycle by its elementary cycles and each 2-cycle by its elementary cavities. Let $M''_I := \{\ \}$. For each cycle s in S_I, if $\{f'_I(s)\} \bigcup M''_I$ is a linearly-independent set then $M''_I := \{f_I(s)\} \bigcup M''_I$; otherwise, $M''_I := M''_I$. Obtain the new AM-model $(K(I), M''_I, f''_I, g''_I, \phi''_I)$ using Lemma 1. Then, $(K(I), M''_I, f''_I, g''_I, \phi''_I)$ is an AM-model for I with "good" representative cycles of homology generators of I.

4.2 AM-Models After Adding or Deleting a Voxel

Now, we study the problem of topologically controlling a digital image using AM-models when it suffers local changes (addition or deletion of one voxel).

More concretely, we show how to compute an AM-model for a digital image when a voxel is added or deleted using the algebraic-topological information computed before. Assuming that I has m voxels, the complexity of the algorithms in this subsection is $\mathcal{O}(m^2)$. The key idea for both algorithms is that when a q-simplex is added to or deleted from an AM-model of K, we only have to put into a Smith normal form the matrix of ∂_q for obtaining the new AM-model. Moreover, take into account that adding or deleting a voxel v of I means to add or delete a set of simplices of $K(I \cup \{v\})$ having v as a vertex. Since we work with simplicial complexes representing 3D digital images considering the 14-adjacency, the maximum number of simplices having v as a vertex is 74.

Let (K, M, f, g, ϕ) be an AM-model for a simplicial complex K. The differential of $M(K)$ is null since the homology is torsion free. Moreover, the value of all the possible non-null entries of the Smith normal form of the matrix of the differential of $C(K)$ in each dimension only can only be 1.

AM-Models After Adding a Voxel. As we have mentioned before, the addition of a voxel v to I means the addition to $K(I)$ of all the simplices of $K(I \cup \{v\}) \setminus K(I)$. In each step of the process, one simplex is added. Assuming that I has m voxels, the following algorithm computes an AM-model for the image $I \cup \{v\}$ with integer coefficients in $\mathcal{O}(m^2)$.

Algorithm 5. *Incremental Algorithm for computing an AM-model for a 3D Binary Digital Image.*

INPUT: An AM-model $AMM_I = (K(I), M, f, g, \phi)$ for I and a voxel $v \notin I$. Let $\{\sigma_1, \dots, \sigma_n\}$ $(n \leq 74)$ be the ordered-by-increasing-dimension set of all the simplices of $K(I \cup \{v\}) \setminus K(I)$.
$K_0 := K(I)$.
For $i = 1$ to $i = n$ do:
  Let q be the dimension of σ_i; let $C_q = \{a_1, \dots, a_r\}$,
  $M_q = \{a_{t+1}, \dots, a_r\}$, $C_{q-1} = \{e_1, \dots, e_s\}$ and $M_{q-1} = \{e_{t+1}, \dots, e_{s+1}\}$;
  let $\partial_q(a_j) = e_j$ for $1 \leq j \leq t$ and $\partial_q(a_j) = 0$ for $t < j \leq \min(r, s)$;
  let $\partial_q(\sigma_i) = \sum_{\ell=1}^{s} \lambda_\ell e_\ell$ where $\lambda_\ell \in \mathbf{R}$.
  Define $a := \sigma_i - \sum_{\ell=1}^{t} \lambda_\ell a_\ell$ and $C_q := \{a_1, \dots, a_r, a\}$.
  If $\lambda_\ell = 0$ for $\ell > t$ then
    $f(a) := a$, $g(a) := a$, $\phi(a) := 0$ and $M_q := \{a_{t+1}, \dots, a_r, a\}$.
  Else obtain the Smith normal form of the matrix of ∂_q
    relative to some base $\{e_1, \dots, e_t, e'_{t+1}, \dots, e'_s\}$ of $C_{q-1}(K)$.
    Define $f(a) := 0$, $\phi(a) := 0$, $\phi(e'_{t+1}) := a$, $f(e'_{t+1}) := 0$,
    $C_{q-1} := \{e_1, \dots, e_t, e'_{t+1}, \dots, e'_s\}$ and $M_{q-1} := \{e'_{t+2}, \dots, e'_s\}$.
  $K_i := K_{i-1} \cup \{\sigma_i\}$.
OUTPUT: An AM-model (K_n, M, f, g, ϕ) for $I \cup \{v\}$.

AM-Models After Deleting a Voxel from a 3D Digital Image. The deletion of a voxel v from I means the deletion from $K(I)$ of all the simplices having v as a vertex. In each step of the process one simplex is deleted. Suppose that an AM-model for a digital image I with m voxeles has been computed and after this a voxel is deleted. The following algorithm computes an AM-model for the image $I \setminus \{v\}$ with integer coefficients in $\mathcal{O}(m^2)$.

Algorithm 6. *Decremental Algorithm for computing an AM-model for a 3D Digital Image I.*

INPUT: An AM-model $(K(I), M, f, g, \phi)$ for I and a voxel $v \in I$.
Let $\{\mu_1, \dots, \mu_n\}$ $(n \leq 72)$ be the ordered-by-decreasing-dimension set of all the simplices of $K(I)$ having v as a vertex.
$K_0 := K(I)$.
For $i = 1$ to $i = n$ do
  Let q be the dimension of σ_i; let $C_q = \{a_1, \dots, a_r\}$,
  $M_q = \{a_{t+1}, \dots, a_r\}$, $C_{q-1} = \{e_1, \dots, e_s\}$ and $M_{q-1} = \{e_{t+1}, \dots, e_{\ell+1}\}$;
  let $\partial_q(a_j) = e_j$ for $1 \leq j \leq t$ and $\partial_q(a_j) = 0$ for $t < j \leq$ min (r, s)
  Find the element $a_k \in C_q$ such that
  $C_q := \{a_1, \dots, \hat{a_k}, \dots, a_r\}$ is a base of $C_q(K_{i-1} \setminus \{\sigma_i\})$.
  If $1 \leq k \leq t$ then
    $M_{q-1} := \{e_k, e_{t+1}, \dots, e_{\ell+1}\}$, $f(e_k) := e_k$, $g(e_k) := e_k$ and $\phi(e_k) := 0$.
  Else $M_q := \{a_{t+1}, \dots, \hat{a_k}, \dots, a_r\}$.
  $K_i := K_{i-1} \setminus \{\sigma_i\}$.
OUTPUT: An AM-model (K_m, M, f, g, ϕ) for $I \setminus \{v\}$.

4.3 AM-Models Under Voxel-Set Operations on 3D Digital Images

In this subsection, we efficiently reuse the AM-model information for digital images under voxel-set operations (union, intersection and difference).

Let I and J be two digital images. We will not consider these trivial cases: $I = \emptyset$, $J = \emptyset$, $I \cap J = \emptyset$, $I \subseteq J$ and $J \subseteq I$. Let $AMM_I := (K(I), M_I, f_I, g_I, \phi_I)$ and $AMM_J := (K(J), M_J, f_J, g_J, \phi_J)$ be AM-models for I and J, respectively.

We give now the pseudocode of the algorithms we have developed for computing AM-models for $I \cup J$, $I \cap J$ and $I \setminus J$ starting from AM-models for I and J. Denote by $Fr_I(J) = \{v_1, \dots, v_m\}$ the set of all the voxels of $I \setminus J$ that are 14-neighbors of a voxel of J. Algorithm 7 is a common preprocessing to the three voxel-set operations treated here. In this algorithm, an AM-model for the image $I \setminus Fr_I(J)$ is calculated.

Algorithm 7. *Preprocessing.*

INPUT: The AM-model AMM_I and the set $Fr_I(J) = \{v_1, \dots, v_m\}$.
$I_{m+1} := I$.
For $i = m$ to $i = 1$ do
  apply Algorithm 6 to v_i and the AM-model $(K(I_{i+1}), M_I, f_I, g_I, \phi_I)$.
  $I_i := I_{i+1} \setminus \{v_i\}$.
OUTPUT: An AM-model $(K(I_1), M_I, f_I, g_I, \phi_I)$ for $I \setminus Fr_I(J)$.

For computing an AM-model for $I \cup J$, we first compute an AM-model for $(I \cup J) \setminus Fr_I(J)$ using Algorithm 7 and after that we add the voxels of $Fr_I(J)$ using Algorithm 5.

Algorithm 8. *Computing an AM-model for* $I \cup J$.

INPUT: The AM-models AMM_I for I and AMM_J for J and the set $Fr_I(J)$.
Apply Algorithm 7 to AMM_I and $Fr_I(J) = \{v_1, \ldots, v_m\}$.
Define $I_0 := (I \cup J) \setminus Fr_I(J)$;
$f(\mu) := f_J(\mu)$, $\phi(\mu) := \phi_J(\mu)$ if $\mu \in K(J)$;
$f(\mu) := f_I(\mu)$, $\phi(\mu) := \phi_I(\mu)$ if $\mu \in K_0 \setminus K(J)$; $M := Imf$;
$g(\alpha) := g_J(\alpha)$ if $\alpha \in M_J$ and $g(\alpha) := g_I(\alpha)$ if $\alpha \in M \setminus M_J$.
For $i = 1$ to $i = m$ do
    apply Algorithm 5 to v_i and the AM-model $(K(I_{i-1}), M, f, g, \phi)$.
    $I_i := I_{i-1} \cup \{v_i\}$.
OUTPUT: an AM-model $(K(I_m), M, f, g, \phi)$ for $I \cup J$.

Algorithm 7 is also the essential step for computing an AM-model for $I \cap J$.

Algorithm 9. *Computing an AM-model for* $I \cap J$.

INPUT: The AM-model AMM_I for I and the set $Fr_I(J)$.
Apply Algorithm 7 to AM_I and $Fr_I(J)$.
Define $f(\mu) := f_I(\mu)$ and $\phi(\mu) := \phi_I(\mu)$ if $\mu \in K(I \cap J)$;
    $M := Imf$ and $g(\alpha) := g_I(\alpha)$ if $\alpha \in M$.
OUTPUT: an AM-model $(K(I \cap J), M, f, g, \phi)$ for $I \cap J$.

For computing an AM-model for $I \setminus J$, we first apply Algorithm 7. Second, we consider the voxels that are in $I \setminus (J \cup Fr_I(J)$. Finally, we add the voxels of $Fr_I(J)$ using Algorithm 5.

Algorithm 10. *Computing an AM-model for* $I \setminus J$.

INPUT: The AM-model AMM_I for I and the set $Fr_I(J)$.
Apply Algorithm 7 to AMM_I and $Fr_I(J) = \{v_1, \ldots, v_m\}$.
Define $I_0 := I \setminus (J \cup Fr_I(J)$; $f(\mu) := f_I(\mu)$ and $\phi(\mu) := \phi_I(\mu)$ if $\mu \in K(I_0)$
    $M := Imf$; $g(\alpha) := g_I(\alpha)$ if $\alpha \in M$.
For $i = 1$ to $i = m$ do
    apply Algorithm 5 to v_i and the AM-model $(K(I_{i-1}), M, f, g, \phi)$.
    $I_i := I_{i-1} \cup \{v_i\}$.
OUTPUT: an AM-model $(K(I_m), M, f, g, \phi)$ for $I \setminus J$.

5 Comments

The algebraic-topological representation of simplicial complexes of any dimension showed here, allows us to compute topological invariants derived from the integer cohomology ring. Moreover, we give a positive answer to the problem of efficiently reusing AM-models for determining homological information of new 3D binary digital images constructed from the previous ones using voxel-set operations.

There is considerable scope for further research: 1) To compute cohomology operations or homotopy groups of simplicial complexes using AM-models.

2) To suitably extend our method to nD binary digital images in any grid using simplicial analogous techniques [15, 16, 17, 18].

Potential applications of our particular method in computer vision and digital image processing involving not only 3D object but also higher dimensional structures can be encountered in Medical Imaging and Object Modeling. Our method seems to be especially well adapted to segmentation under topological constraints and elimination of small topological noise.

References

1. González–Díaz R., Real P.: Towards Digital Cohomology. DGCI 2003, LNCS, Springer **2886** (2003) 92–101
2. González–Díaz R., Real P.: On the Cohomology of $3D$ Digital Images. Discrete Applied Math **147** (2005) 245–263
3. Gonzalez-Diaz R., Medrano B., Real P., Sánchez-Peláez J.: Algebraic Topological Analysis of Time-Sequence of Digital Images. Lecture Notes in Computer Science **139** (2005) 208–219
4. Sergeraert F.: Homologie effective. I, II. C. R. Acad. Sci. Paris Sér. I Math. 304 (1987), no. 11, 279–282, no. 12, 319–321
5. Sergeraert F.: The computability problem in algebraic topology. Adv. Math. 104 (1994), no. 1, 1–29
6. Munkres J.R.: Elements of Algebraic Topology. Addison–Wesley Co. 1984
7. Agoston M.K.: Algebraic Topology, a first course. Marcel Dekker Ed., 1976
8. Dumas J.G., Saunders B.D., Villad G.: On efficient sparse integer matrix Smith normal form computations. Journal of Symbolic Computation **32** (2001) 71–99
9. Peltier S., Alayrangues S., Fuchs L., Lachaud J.: Computation of homology groups and Generators. LNCS, **3429** (2005) 195-205
10. MacLane S.: Homology. Classic in Math., Springer–Verlag, 1995
11. González–Díaz R., Real P.: Computation of Cohomology Operations on Finite Simplicial Complexes. Homology, Homotopy and Applications **5 (2)** (2003) 83–93
12. Herman, G.T.: Geometry of Digital Spaces. Birkhauser, Boston (1998).
13. Alexandroff P., Hopf H.: Topologie I. Springer, Berlin 1935
14. Erickson J., Whittlesey K.: Greedy Optimal Homotopy and Homology Generators. Proc. of the sixteenth annual ACM-SIAM symposium on Discrete algorithms (2005) 1038 - 1046
15. Kong T.Y.: A Digital Fundamental Group. Computer Graphics **13** (1989) 159–166
16. Kong T.Y., Roscoe A.W., Rosenfeld A.: Concepts of Digital Topology. Topology and its Applications **46** (1992) 219–262
17. Khalimsky E.D., Kopperman R.D., Meyer P.R.: Computer Graphics and Connected Topologies on Finite Ordered Sets. Topology and Appl. **36** (1990) 1–17
18. Ayala R., Domínguez E., Francés A.R., Quintero A.: Homotopy in Digital Spaces. DGCI 2000. LNCS, Springer–Verlag **1953** (2000) 3–14

On the Lattice Structure of Subsets of Octagonal Neighborhood Sequences in $\mathbb{Z}^n$

András Hajdu[1] and Lajos Hajdu[2,*]

[1] Faculty of Informatics, University of Debrecen,
P.O. Box 12, H-4010 Debrecen, Hungary
hajdua@inf.unideb.hu

[2] Institute of Mathematics, University of Debrecen,
and the Number Theory Research Group of the Hungarian Academy of Sciences,
P.O. Box 12, H-4010 Debrecen, Hungary
hajdul@math.klte.hu

Abstract. In this paper we investigate the lattice properties of several special, but important subsets of S_n, the set of nD octagonal neighborhood sequences in $\mathbb{Z}^n$, with respect to two ordering relations $\sqsupseteq^*$ and $\sqsupseteq$. Both orderings have some natural meaning, especially $\sqsupseteq^*$ compares the "speed" how neighborhood sequences spread in $\mathbb{Z}^n$. We summarize our and the previous related results in a table. In particular, our theorems can be considered as extensions of some results from [1,2,3].

1 Introduction

Motions in the digital space $\mathbb{Z}^n$ play an important role in several parts of discrete mathematics, including discrete geometry and digital image processing. The most important and most well-known motions in $\mathbb{Z}^2$ are the so-called city-block (or von Neumann) and the chessboard (or Moore) motions. They are based upon the classical 4-neighborhood and 8-neighborhood relations, respectively. If we combine these relations in a strictly alternating way, then we get the so-called octagonal distance. These motions and the induced distance functions were (partly) introduced and extensively studied in the pioneer paper of Rosenfeld and Pfaltz [4]. Das, Chakrabarti and Chatterji [5] investigated periodic octagonal neighborhood sequences (i.e. arbitrary periodic combinations of the 4- and 8 neighborhood relations). They performed similar investigations also in $\mathbb{Z}^n$. Such sequences are dealt with in many papers, see e.g. [1,6,5,7,8] and the references given there. Fazekas, Hajdu and Hajdu [2] extended the theory to the case of arbitrary octagonal neighborhood sequences, where any (not necessarily periodic) sequences are considered. Such sequences are much more appropriate for certain purposes. For example, the authors in [9] constructed digital metrics on $\mathbb{Z}^2$ based upon such sequences, which provide the best approximation to the Euclidean distance in a certain sense. Using periodic sequences, only some finite parts of such sequences can be obtained, see e.g. [6,10].

* Research supported in part by the János Bolyai Research Fellowship of the Hungarian Academy of Sciences, and by the OTKA grants F043090, F034981 and T042985.

A. Kuba, L.G. Nyúl, and K. Palágyi (Eds.): DGCI 2006, LNCS 4245, pp. 211–222, 2006.

In [2] Fazekas, Hajdu and Hajdu described the lattice structure of S_n, the set of (general) octagonal neighborhood sequences on $\mathbb{Z}^n$, with respect to two ordering relations $\sqsupseteq^*$ and $\sqsupseteq$. The structure of P_n, the subset of S_n consisting of the periodic octagonal neighborhood sequences was also investigated here. Further, the same authors in [3] (among other results) described the lattice structure of M_n, i.e. the set of neighborhood sequences inside S_n which generate metrics.

In this paper we investigate the lattice structures of some other subsets of S_n, as well. Our main motivation is to tie up some loose ends in this field, and also to extend existing results to other subsets, not examined so far. We start with U_n, the set of ultimately periodic octagonal neighborhood sequences. The study of such sequences is motivated by their importance, as they yield a natural extension of the periodic sequences. We also refer to the paper of Hajdu, Hajdu and Tijdeman [11] where such sequences are closely investigated, under more general circumstances. We also consider the so-called Lyndon sequences L_n. Such sequences play a central role in many problems of word theory, see e.g. the book [12]. Finally, as a natural generalization of the sets P_n, U_n, M_n we consider the set D_n consisting of sequences A such that all symbols in A have densities. Beside these results we provide some theorems for P_n and M_n, as well. The extension of the investigated families is meaningful in applications, e.g. for the U_n sequences that are generated by finite data. The main interest of comparing the sequences lies in deciding about their relative spreading speed in $\mathbb{Z}^n$ (see also [13]).

The structure of the paper is as follows. In section 2 we introduce some concepts and notation. Section 3 contains our main results summarized in a table, as well. We give auxiliary results and prove our theorems in section 4.

2 Basic Concepts and Notation

In this section we introduce the necessary notation. First we recall some standard notions from the theory of words; for more details see e.g. [12,14]. Let Γ be a finite alphabet. As usual, finite sequences consisting of symbols from Γ are called Γ-words, or simply words. The concatenation uv of two words u and v is understood in the well-known way, and by $|u|$ we denote the number of symbols in u. If u is a word and $n \in \mathbb{N}$, then u^n means the n-fold concatenation $uu \dots u$. If W is a (one-sided) infinite sequence of symbols from Γ, then we call W a Γ-sequence, or simply a sequence. If u is a word and W is a sequence then the concatenation uW is defined in the classical way, as well as the prefixes and suffixes of words and sequences. Finally, if u is a word then $\overline{u}$ denotes the sequence $uuu \dots$.

Now we introduce some standard notation concerning neighborhood sequences, see e.g. [5, 2]. Let $m, n \in \mathbb{N}$ with $m \leq n$. The points $p = (p_1, \dots, p_n)$, $q = (q_1, \dots, q_n)$ in $\mathbb{Z}^n$ are m-neighbors, if the following conditions hold:

- $|p_i - q_i| \leq 1$, for all $1 \leq i \leq n$,
- $\sum_{i=1}^{n} |p_i - q_i| \leq m$.

The sequence $A = (A(i))_{i=1}^{\infty}$, where $A(i) \in \{1, \dots, n\}$ for all $i \in \mathbb{N}$, is called an n-dimensional (shortly nD) neighborhood sequence. The set of such sequences is denoted by S_n. Obviously, S_n can be considered as the set of Γ-sequences with $\Gamma = \{1, \dots, n\}$. For $A \in S_n$ and $k \in \{1, \dots, n\}$, $i, j \in \mathbb{N}$ with $i \leq j$ put

$$\mathbf{k}(A, i) = \#\{A(l) \mid A(l) = k,\ l = 1, \dots, i\}$$

and let

$$\mathbf{k}(A, i, j) = \mathbf{k}(A, j) - \mathbf{k}(A, i).$$

The density s_k of the symbol k in A is defined as

$$s_k(A) = \lim_{i \to \infty} \frac{\mathbf{k}(A, i)}{i}, \quad \text{for } k \in \{1, \dots, n\},$$

if the limit exists. We say that $A \in S_n$ have density, if $s_k(A)$ exists for every $k = 1, \dots, n$. In this case $s(A) = (s_1(A), \dots, s_n(A))$ is called the density tuple of A.

If for an $A \in S_n$ we have

$$A = A(1)A(2)\dots A(h)\overline{A(h+1)A(h+2)\dots A(h+l)}$$

for some $h, l \in \mathbb{N}$ then A is called an ultimately periodic neighborhood sequence. In case of $h = 0$, i.e. when $A = \overline{A(1)A(2)\dots A(l)}$, A is called periodic with period l. The set of periodic and ultimately periodic sequences is denoted by P_n and U_n, respectively.

Let $p, q \in \mathbb{Z}^n$ and $A \in S_n$. The point sequence $p = p_0, p_1, \dots, p_t = q$, where p_{i-1} and p_i are $A(i)$-neighbors in $\mathbb{Z}^n$ $(1 \leq i \leq t)$, is called an A-path from p to q of length t. The A-distance $d(p, q; A)$ of p and q is defined as the length of the shortest A-path(s) between them. As a brief notation, we also use $d(A)$ for the A-distance. If $d(A)$ is a metric on $\mathbb{Z}^n$, then A is called a metrical neighborhood sequence. The set of such sequences is denoted by M_n.

A neighborhood sequence $A \in S_n$ is called a Lyndon sequence if for any (infinite) suffix B of A, either $A = B$, or A lexicographically precedes B. The set of such sequences is denoted by L_n. As it turns out, we have $M_n \subseteq L_n$. The literature of (finite) Lyndon words is very extensive; see e.g. [12]. For some basic results and properties of Lyndon sequences see e.g. [15,16].

In our structural investigations we examine some subsets of S_n with respect to two partial orderings, $\sqsupseteq^*$ and $\sqsupseteq$ [1,2]. These orderings are defined in the following way. For $A, B \in S_n$ write $A \sqsupseteq^* B$ if and only if $d(p, q; A) \leq d(p, q; B)$ for every $p, q \in \mathbb{Z}^n$, and set $A \sqsupseteq B$ if and only if $A(i) \geq B(i)$ for every $i \in \mathbb{N}$. Obviously, $\sqsupseteq^*$ is a refinement of $\sqsupseteq$, that is $A \sqsupseteq B$ implies $A \sqsupseteq^* B$ for every $A, B \in S_n$. Further, from [2] we know that if $A, B \in S_n$ then $A \sqsupseteq^* B$ if and only if

$$\sum_{i=1}^{j} \min\{k, A(i)\} \geq \sum_{i=1}^{j} \min\{k, B(i)\}, \quad \text{for all } j \in \mathbb{N},\ k = 1, \dots, n.$$

We will use this property throughout the paper, without any further reference.

We also note that it is obvious that $d(A)$ is not a metric on $\mathbb{Z}^n$ for every $A \in S_n$. Nagy [17,18] proved that $A \in M_n$ if and only if $B \sqsupseteq^* A$ for any (infinite) suffix B of A. We will use this assertion without any further mentioning.

Now we recall a few basic concepts and facts from lattice theory. They also will be used throughout the paper without any further notice. Let H be a partially ordered set. We say that H is a lattice, if for any $A, B \in H$ the greatest lower bound $A \wedge B$ and the least upper bound $A \vee B$ of these elements exist. If for any $S \subseteq H$ the greatest lower bound $\bigwedge S$ and the least upper bound $\bigvee S$ of S also exist, then the lattice H is called complete. It is well-known that if $\bigwedge S$ exists for all subset S of H, then $\bigvee S$ also exists for any subset, and vice versa. The lattice H is distributive, if for any $A, B, C \in H$ we have

$$(A \wedge B) \vee C = (A \vee C) \wedge (B \vee C) \quad \text{and} \quad (A \vee B) \wedge C = (A \wedge C) \vee (B \wedge C).$$

In our investigations we consider greatest lower bounds and least upper bounds both with respect to $\sqsupseteq^*$ and $\sqsupseteq$. To make a distinction, the simple notation $\wedge$ and $\vee$ will refer to $\sqsupseteq$, while in case of $\sqsupseteq^*$ we use $\wedge_*$ and $\vee_*$. In particular, a theorem from [2] guarantees that for any $A, B \in S_2$ we have

$$\mathbf{2}(A \wedge_* B, i) = \min\{\mathbf{2}(A, i), \mathbf{2}(B, i)\}$$

and

$$\mathbf{2}(A \vee_* B, i) = \max\{\mathbf{2}(A, i), \mathbf{2}(B, i)\}$$

for all $i \in \mathbb{N}$. A very important remark is that though we will work in several subsets of S_n, this notation will always refer to the corresponding upper or lower bounds in S_n.

3 The Lattice Structure of Subsets of S_n

In this section we give our results about the lattice structure of certain special, but important subsets of S_n. We investigate the lattice structures under both orderings $\sqsupseteq^*$ and $\sqsupseteq$. We start with noting that the lattice structure of S_n under these orderings were clarified in [2]; see also Table 1. In particular, $(S_2, \sqsupseteq^*)$ and $(S_n, \sqsupseteq)$ for $n \geq 2$ are complete distributive lattices. However, $(S_n, \sqsupseteq^*)$ is not a lattice for $n \geq 3$.

3.1 Periodic Neighborhood Sequences — P_n

We start with the investigation of the simplest and most studied subset of S_n, i.e. with the set P_n of periodic sequences. For the survey of the related literature we refer to [2].

As it is known, $(P_n, \sqsupseteq^*)$ is not a lattice for any $n \geq 2$, while $(P_n, \sqsupseteq)$ is a lattice for every $n \geq 2$ (cf. also Table 1). However, it is possible to prove something "positive" for $\sqsupseteq^*$ also in this case. Namely, we have

Theorem 1. *Let* $0 \leq \alpha \leq 1$, *and put* $H = \{A \in P_2 \mid s_2(A) = \alpha\}$. *Then* $(H, \sqsupseteq^*)$ *is a distributive lattice.*

3.2 Ultimately Periodic Neighborhood Sequences — U_n

As a natural extension of periodic sequences, now we consider ultimately periodic ones. Our first result shows that while $(P_2, \sqsupseteq^*)$ is not a lattice, the larger set $(U_2, \sqsupseteq^*)$ has this property.

Theorem 2. *$(U_2, \sqsupseteq^*)$ is a non-complete distributive lattice.*

However, it turns out that for larger values of n the structure of $(U_n, \sqsupseteq^*)$ is not so nice, and we get

Theorem 3. *$(U_n, \sqsupseteq^*)$ is not a lattice for $n \geq 3$.*

The following result shows that the structural behavior of U_n with respect to $\sqsupseteq$ is nice, just as in case of P_n.

Theorem 4. *$(U_n, \sqsupseteq)$ is a non-complete distributive lattice for $n \geq 2$.*

3.3 Metrical Neighborhood Sequences — M_n

From [3] we now that $(M_2, \sqsupseteq^*)$ forms a complete lattice, which is not distributive. However, this is the only "positive" result, as neither $(M_n, \sqsupseteq^*)$ with $n \geq 3$, nor $(M_n, \sqsupseteq)$ with $n \geq 2$ are lattices. Here we provide the following related theorem. Note that by a result from [9], every $A \in M_2$ has density.

Theorem 5. *Let $0 \leq \alpha \leq 1$, and set $H = \{A \in M_2 \mid s_2(A) = \alpha\}$. Then $(H, \sqsupseteq^*)$ is a complete lattice.*

Fazekas, Hajdu and Hajdu in [3] made the observation that for any $A, B \in M_2$ we have $A \wedge_* B \in M_2$, however, $A \vee_* B \in M_2$ does not always hold. As $(M_2, \sqsupseteq^*)$ is a lattice, this implies that for any $A \in S_2$ there exists a uniquely determined $B \in M_2$ such that for any $C \in M_2$ with $C \sqsupseteq^* A$ we have $C \sqsupseteq^* B$. This B is called the metrical closure of A, and is denoted by $MC(A) = B$. The authors in [3] derived several properties of the metrical closure, e.g. they provided an algorithm which generates $MC(A)$ for a fixed A. Now we prove the following result.

Theorem 6. *For any $A \in P_2$ we have $MC(A) \in U_2$.*

Note that by a simple example one can check that $A \in P_2$ does not necessarily implies $MC(A) \in P_2$. Further, unfortunately we are not able to prove (or disprove) the analogue of the above theorem for ultimately periodic sequences. So we formulate the following open problem:

Problem 1. Is it true that for any $A \in U_2$ we have $MC(A) \in U_2$?

3.4 Lyndon Sequences — L_n

In this section we investigate the lattice properties of Lyndon sequences. This investigation is motivated by the simple but important observation that metrical neighborhood sequences are Lyndon sequences. Note that the opposite statement is false, as the Lyndon sequence $1112211211\overline{2}$ is not a metrical neighborhood sequence. Unfortunately, it turns out that the set of Lyndon sequences does not form a nice structure neither under $\sqsupseteq^*$, nor with respect to $\sqsupseteq$.

Theorem 7. *$(L_n, \sqsupseteq^*)$ and $(L_n, \sqsupseteq)$ are not lattices for any $n \geq 2$.*

3.5 Neighborhood Sequences Having Density — D_n

It is obvious that sequences from P_n and U_n have densities. It is also known from [3] that the same holds for sequences from M_n. Hence it is natural to investigate general neighborhood sequences having density. In this section we give our results about the lattice structure of the set D_n of such sequences.

Theorem 8. *$(D_2, \sqsupseteq^*)$ is a non-complete distributive lattice.*

Similarly to all the other investigated subsets of S_n, neither D_n behaves nicely for higher dimensions with respect to $\sqsupseteq^*$.

Theorem 9. *$(D_n, \sqsupseteq^*)$ is not a lattice for $n \geq 3$.*

The following statement implies that D_2 has worse structural properties with respect to $\sqsupseteq$ than to $\sqsupseteq^*$.

Theorem 10. *$(D_n, \sqsupseteq)$ is not a lattice for $n \geq 2$.*

3.6 Summarizing the Lattice Properties

We summarize the lattice properties of the investigated subsets of S_n under the orderings $\sqsupseteq^*$ and $\sqsupseteq$ in Table 1. Though $\sqsupseteq^*$ shows a rather negative behavior from 3D on, this ordering describes better the relative spreading speed. Moreover $\sqsupseteq^*$ supersedes $\sqsupseteq$ in case of metrical expectations. The 3D space is also important from practical point of view, and this domain can be addressed by restricting the investigated subsets for a more successful comparison. Introducing a new ordering can be an alternative to overcome the negative results, as well. In this paper we focused only on the relations investigated in the literature so far.

4 Auxiliary Results and Proofs of the Theorems

In this section we give the proofs of our theorems. For this purpose we need some auxiliary results. Our first statement shows that if $A, B \in S_2$ have the same density tuples, and they belong to some special subset of S_2, then the same is true for $A \wedge_* B$ and $A \vee_* B$. More precisely, we have

Table 1. Lattice properties of subsets of S_n with respect to $\sqsupseteq^*$ and $\sqsupseteq$

	$\sqsupseteq^*$ $(n = 2)$			$\sqsupseteq^*$ $(n \geq 3)$	$\sqsupseteq$ $(n \geq 2)$		
	Lattice	Complete	Distributive	Lattice	Lattice	Complete	Distributive
S_n	+	+	+	−	+	+	+
P_n	−	−	−	−	+	−	+
U_n	+	−	+	−	+	−	+
M_n	+	+	−	−	−	−	−
L_n	−	−	−	−	−	−	−
D_n	+	−	+	−	−	−	−

Lemma 1. *Let $H \in \{P_2, U_2, D_2\}$ and let $A, B \in H$ with $s_2(A) = s_2(B)$. Then $A \wedge_* B, A \vee_* B \in H$.*

Proof. Assume first that $H = P_2$, i.e. $A, B \in P_2$. Without loss of generality we may assume that l is a common period length for both A and B. As $s_2(A) = s_2(B)$ we have $\mathbf{2}(A, l) = \mathbf{2}(B, l)$. Moreover,

$$\mathbf{2}(A \wedge_* B, l) = \mathbf{2}(A \vee_* B, l) = \mathbf{2}(A, l) = \mathbf{2}(B, l)$$

also hold. Hence for all $i \in \{1, \ldots, l\}$ we get that

$$(A \wedge_* B)(i) = (A \wedge_* B)(i + l) \text{ and } (A \vee_* B)(i) = (A \vee_* B)(i + l).$$

Thus $A \wedge_* B, A \vee_* B \in P_2$.

Let now $H = U_2$, that is $A, B \in U_2$. Without loss of generality we may assume that $A = A_1\overline{A_2}$ and $B = B_1\overline{B_2}$ with $c := |A_1| = |B_1|$. Let l be the least common multiple of $|A_2|$ and $|B_2|$. Then by $s_2(A) = s_2(B)$ we have $T := \mathbf{2}(A, c, c + l) = \mathbf{2}(B, c, c + l)$, and thus also $\mathbf{2}(A, c, c + kl) = \mathbf{2}(B, c, c + kl) = kT$ for any $k \in \mathbb{N}$. Hence we have

$$\mathbf{2}(A \vee_* B, c + kl + i) = \max\{\mathbf{2}(A, c + kl + i), \mathbf{2}(B, c + kl + i)\} =$$

$$= \max\{\mathbf{2}(A, c + i), \mathbf{2}(B, c + i)\} + kT = \mathbf{2}(A \vee_* B, c + i) + kT$$

for any $i = 1, \ldots, l$, $k \in \mathbb{N}$. Consequently, $A \vee_* B \in U_2$. As a similar argument shows that $A \wedge_* B \in U_2$, the lemma is proved when $H = U_2$.

Finally, let $H = D_2$, so $A, B \in D_2$. As $s_2(A) = s_2(B)$ and

$$\min\{\mathbf{2}(A, i), \mathbf{2}(B, i)\} = \mathbf{2}(A \wedge_* B, i) \leq \mathbf{2}(A \vee_* B, i) = \max\{\mathbf{2}(A, i), \mathbf{2}(B, i)\}$$

for any $i \in \mathbb{N}$, the statement follows in this case, as well. □

Proof (of Theorem 1). By Lemma 1 for any $A, B \in P_2$ we have $A \vee_* B, A \wedge_* B \in P_2$. As $(S_2, \sqsupseteq^*)$ is a distributive lattice, the theorem follows. □

Lemma 2. *Let $A, B \in D_2$ with $s_2(A) > s_2(B)$. Then there exists a $K \in \mathbb{N}$ such that $(A \wedge_* B)(i) = B(i)$ and $(A \vee_* B)(i) = A(i)$ for $i > K$.*

Proof. As $s_2(A) > s_2(B)$, there exists a $K \in \mathbb{N}$ such that for $i > K$ we have $\mathbf{2}(A,i) > \mathbf{2}(B,i)$. Hence the statement immediately follows. □

Proof (of Theorem 2). Let $A, B \in U_2$. If $s_2(A) = s_2(B)$ then by Lemma 1, and if $s_2(A) \neq s_2(B)$ then by Lemma 2 we obtain that $A \wedge_* B, A \vee_* B \in U_2$. Hence as $(S_2, \sqsupseteq^*)$ is a distributive lattice, the same is true for $(U_2, \sqsupseteq^*)$. To prove that $(U_2, \sqsupseteq^*)$ is non-complete, put $A_1 = \overline{12}$ and for $i \in \mathbb{N}$ let $A_{i+1} = \overline{A_1 \dots A_i 1^{i+1} 2}$. Writing $\mathcal{A} = \{A_i \mid i \in \mathbb{N}\}$, it is obvious that $A := \bigwedge_* \mathcal{A} = 121^2 21^3 21^4 2\dots$ does not belong to U_2. Suppose that there exists the greatest lower bound B of $\mathcal{A}$ in $(U_2, \sqsupseteq^*)$. Then $A \sqsupseteq^* B$, hence from $s_2(A) = 0$ we have $s_2(B) = 0$ which implies that B is of the form $B = B_1\overline{1}$. Put $k = |B_1|$, and consider the sequence $C = C_1\overline{1} \in U_2$ with $|C_1| = 2k+1$ and $C(j) = A(j)$ for $j = 1, \dots, 2k+1$. Then $A \sqsupseteq^* C$ and from $\sum_{i=1}^{2k+1} B(i) < \sum_{i=1}^{2k+1} C(i)$ we derive the contradiction $B \not\sqsupseteq^* C$. Hence the lattice $(U_2, \sqsupseteq^*)$ is not complete. □

Proof (of Theorem 3). Fix $n \geq 3$ and put $A_1 = \overline{31}$, $A_2 = \overline{2}$, $B_1 = 2\overline{13}$ and $B_2 = 13\overline{1}$. Now for any $B_3 \in U_n$ with $B_3 \neq B_1$, $A_1 \sqsupseteq^* B_3$ and $A_2 \sqsupseteq^* B_3$ imply $B_3 \not\sqsupseteq^* B_1$. This means that if the greatest lower bound of A_1 and A_2 exists in U_n then it must be B_1. However, $A_1 \sqsupseteq^* B_2$ and $A_2 \sqsupseteq^* B_2$ but $B_1 \not\sqsupseteq^* B_2$. This together with the fact that A_1, A_2, B_1, $B_2 \in U_n$ imply that A_1 and A_2 have no greatest lower bound in U_n with respect to $\sqsupseteq^*$ and the statement follows. □

Proof (of Theorem 4). Let $A, B \in U_n$. Then we may write $A = A_1\overline{A_2}$ and $B = B_1\overline{B_2}$ with $|A_1| = |B_1|$ and $|A_2| = |B_2|$. Hence it is obvious that $A \vee B, A \wedge B \in U_n$, and we get that $(U_n, \sqsupseteq)$ is a distributive lattice. To show that the lattice is non-complete, let $B_i = 12^i$ $(i \in \mathbb{N})$ and put $A_k = B_1 \dots B_k\overline{1}$ $(k \in \mathbb{N})$ and $\mathcal{A} = \{A_k \mid k \in \mathbb{N}\}$. Suppose that $\mathcal{A}$ has a least upper bound in U_n with respect to $\sqsupseteq$, which is given by $C = C_1\overline{C_2}$. Put $B = B_1B_2\dots$, and observe that $B \in S_n \setminus U_n$. Hence, as obviously $\bigvee \mathcal{A} = B$, we have $C \sqsupseteq B$. Let j be minimal with $C(j) = 2$ and $B(j) = 1$, and define $D \in U_2$ by

$$D(l) = \begin{cases} B(l), & \text{for } 1 \leq l \leq j, \\ 2, & \text{for } l > j. \end{cases}$$

Then clearly, $D \sqsupseteq A_k$ for all $k \in \mathbb{N}$, however, $D \sqsupseteq C$ does not hold. This shows that the lattice $(U_n, \sqsupseteq)$ is not complete. □

Proof (of Theorem 5). Let L be an arbitrary subset of M_2 such that $s_2(A) = s_2(B) = \alpha$ for all $A, B \in L$. Theorem 7 of [3] implies that $\bigwedge_* L \in M_2$. Define the sequence $A \in S_2$ in the following way. If $\alpha = 1$ then put $A(1) = 2$, otherwise let $A(1) = 1$. Then if $A(i)$ is already defined, set $A(i+1) = 2$ if $(i+1)\alpha \leq \mathbf{2}(A,i)+1$ and let $A(i+1) = 1$ otherwise. By Lemma 2 from [9] we get that $A \in M_2$. Further, it is clear that $s_2(A) = \alpha$ and that $A \sqsupseteq^* B$ holds for any $B \in M_2$ with $s_2(B) = \alpha$. Let C be the least upper bound of L in M_2 with respect to $\sqsupseteq^*$. Then we have $A \sqsupseteq^* C$, which immediately gives $s_2(C) = \alpha$, and the theorem follows. □

To prove Theorem 6 we need two lemmas. The first one gives a characterization of the metrical property of neighborhood sequences.

Lemma 3. *Let $A \in S_2$. Then $A \in M_2$ if and only if for all $t \in \mathbb{N}$, $j \leq t$ and $k_1, \ldots, k_j \in \mathbb{N}$ with $\sum_{i=1}^{j} k_i = t$ we have $\mathbf{2}(A,t) \geq \sum_{i=1}^{j} \mathbf{2}(A,k_i)$.*

Proof. First we show necessity. Let $A \in M_2$. For $t = 1$ the statement is obvious. Let $T \in \mathbb{N}$, and assume that the condition is valid for all t with $1 \leq t < T$. Further, let $k_1, \ldots, k_j \in \mathbb{N}$ with $\sum_{i=1}^{j} k_i = T$. Then as $A \in M_2$ we have

$$\mathbf{2}(A,T) \geq \mathbf{2}(A,k_1) + \mathbf{2}(A,T-k_1) \geq \sum_{i=1}^{j} \mathbf{2}(A,k_i)$$

and the necessity follows by induction. To prove sufficiency, take an $A \in S_2$ having the appropriate condition. Then for arbitrary $k_1, k_2 \in \mathbb{N}$ with $k_1 < k_2$ we have

$$\mathbf{2}(A,k_1+k_2) \geq \mathbf{2}(A,k_1) + \mathbf{2}(A,k_2)$$

which yields $\mathbf{2}(A,k_1,k_1+k_2) \geq \mathbf{2}(A,k_2)$. Hence $A \in M_2$, and the lemma follows. □

The following lemma shows that when creating the metrical closure of a periodic sequence A, after a certain index A "does not matter" any more, it is sufficient to take care of the metricity condition only.

Lemma 4. *Let $A \in P_2$ having period length l and $B \in M_2$. If $\mathbf{2}(A,t) \leq \mathbf{2}(B,t)$ holds for all $t = 1, \ldots, l$ then $B \sqsupseteq^* A$.*

Proof. Obviously, $\mathbf{2}(A,k) \leq \mathbf{2}(B,k)$ for $k = 1, \ldots, l$. Assume that $k > l$ and write $k = ql + r$ with $q, r \in \mathbb{N}$, where $0 \leq r < l$. Now using our conditions and Lemma 3 we have

$$\mathbf{2}(A,k) = q\mathbf{2}(A,l) + \mathbf{2}(A,r) \leq q\mathbf{2}(B,l) + \mathbf{2}(B,r) \leq \mathbf{2}(B,k),$$

and the statement follows. □

Proof (of Theorem 6). Let $A \in P_2$ with period length l. Put $B = MC(A)$ and denote by r the minimal index with $1 \leq r \leq l$ for which $\frac{\mathbf{2}(B,r)}{r} = \max_{1 \leq i \leq l} \left\{ \frac{\mathbf{2}(B,i)}{i} \right\}$.

Now we show that for an arbitrary $t \in \mathbb{N}$ with $t > l$ there exist non-negative integers $a_1, \ldots, a_l$ with $\sum_{i=1}^{l} a_i i = t$ such that $\mathbf{2}(B,t) = \sum_{i=1}^{l} a_i \mathbf{2}(B,i)$. Note that by Lemma 3 and Theorem 8 of [3] we get that $B(t) = 2$ if and only if for some $k \in \mathbb{N}$ with $1 \leq k < t$ we have $\mathbf{2}(B,k) = \mathbf{2}(B,t-k,t-1)+1$. Hence $\mathbf{2}(B,l+1) = \mathbf{2}(B,l) + \mathbf{2}(B,1)$ if $B(l+1) = 1$, and $\mathbf{2}(B,l+1) = \mathbf{2}(B,k) + \mathbf{2}(B,l+1-k)$ with the appropriate k if $B(l+1) = 2$. That is, our claim holds for $t = l+1$.

Assume now that $T > l+1$, and that the property holds for every t with $l < t < T$. Again, we have $\mathbf{2}(B,T) = \mathbf{2}(B,T-1) + \mathbf{2}(B,1)$ if $B(T) = 1$, and $\mathbf{2}(B,T) = \mathbf{2}(B,k) + \mathbf{2}(B,T-k)$ with the appropriate k if $B(T) = 2$. In both cases we get by the induction hypothesis that the property is valid for T, as well. Hence our claim follows. We call $a_1, \ldots, a_l$ a minimal combination system for t.

Let $u \in \mathbb{N}$ such that $u \geq u_0 := \sum_{i=1}^{l} ri$ and take a minimal combination system $a_1, \ldots, a_l$ for u. Then we have $a_i \geq r$ for some $i \in \{1, \ldots, l\}$. However, as $r\mathbf{2}(B,i) \leq i\mathbf{2}(B,r)$, we may assume that $a_r > 0$. Let now $b_1, \ldots, b_l$ be a minimal combination system for $u+r$. Similarly as above, we may assume that $b_r > 0$. Hence as

$$\mathbf{2}(B,u) + \mathbf{2}(B,r) \leq \mathbf{2}(B,u+r) = \sum_{i=1}^{l} b_i \mathbf{2}(B,i)$$

and by Lemma 3

$$\mathbf{2}(B,u) \geq \sum_{i=1}^{l} b_i \mathbf{2}(B,i) - \mathbf{2}(B,r)$$

holds, we get

$$\mathbf{2}(B,u) + \mathbf{2}(B,r) = \mathbf{2}(B,u+r).$$

Note that the latter relation is valid for any $u \geq u_0$, in particular also if we write $u+1$ in place of u. This implies

$$\mathbf{2}(B,r) = \mathbf{2}(B,u,u+r) = \mathbf{2}(B,u+1,u+r) + \mathbf{2}(B,u,u+1),$$

$$\mathbf{2}(B,r) = \mathbf{2}(B,u+1,u+r+1) = \mathbf{2}(B,u+1,u+r) + \mathbf{2}(B,u+r,u+r+1).$$

Thus $\mathbf{2}(B,u,u+1) = \mathbf{2}(B,u+r,u+r+1)$, or in other words $B(u+1) = B(u+r+1)$. This proves that $B \in U_2$, and the theorem follows. □

Proof (of Theorem 7). To prove the statement in case of $\sqsupseteq^*$, we give a counterexample. Let

$$A_1 = 1122211222222112221212 11\overline{2}, \quad A_2 = 112221122221211222112222\overline{2}.$$

Then $A_1, A_2 \in L_n$, and we have

$$C = A_1 \wedge_* A_2 = 1122211222\boxed{21}2112221122212\overline{2} \notin L_n.$$

Moreover, let

$$A = 1122211222\boxed{12}2112221122212\overline{2}.$$

Then $A \in L_n$. If $B \in L_n$ with $C \sqsupseteq^* B \sqsupseteq^* A$ then as only the eleventh and twelveth elements (boxed above) of A and C are different, we get that $A = B$. Thus we deduce that if the greatest lower bound of A_1 and A_2 in L_n exists, then it must be A. However, for $D = 111122222221211222112221\overline{2} \in L_n$ we have $A_1 \sqsupseteq^* D$ and $A_2 \sqsupseteq^* D$, but $A \not\sqsupseteq^* D$. This shows that A cannot be the greatest lower bound of A_1 and A_2 in L_2. Thus $(L_2, \sqsupseteq^*)$ is not a lattice.

Now we consider the ordering $\sqsupseteq$. Let $A = 1222221\overline{2}$, $B = 1222122\overline{2}$. As it can be easily seen, $A, B \in L_n$. We show that A and B do not have a greatest lower bound in L_n. Let $C = 1212121\overline{2}$ and $D = 1122121\overline{2}$. Clearly, $C, D \in L_n$, $A \sqsupseteq C$, $B \sqsupseteq C$, $A \sqsupseteq D$ and $B \sqsupseteq D$. Moreover, neither C nor D can be the greatest lower bound of A and B in L_n, since C and D cannot be compared. Looking at the first few elements of A, B, C and D we obtain that if the greatest lower bound E of A and B in L_n exists, then we must have $E = 1222121\ldots$. However, such a sequence cannot belong to L_n and the statement follows. □

Proof (of Theorem 8). Let $A, B \in D_2$. By Lemmas 1 and 2 it is obvious that $A \wedge_* B, A \vee_* B \in D_2$. Thus, as S_2 is a distributive lattice, D_2 is also a distributive lattice with respect to $\sqsupseteq^*$.

To prove that the lattice is not complete, let $A \in S_2 \setminus D_2$ arbitrary with $\liminf\limits_{k\to\infty} \frac{\mathbf{2}(A,k)}{k} = 0$. For $i \in \mathbb{N}$ put

$$A_i(j) = \begin{cases} A(j), & \text{for } j \leq i, \\ 2, & \text{for } j > i, \end{cases}$$

and $P = \bigcup\limits_{i=1}^{\infty} A_i$. Then as $A_i \in D_2$ with $s_2(A_i) = 1$ for every $i \in \mathbb{N}$, we have $P \subseteq D_2$ and $\bigwedge_* P = A$.

Assume that $B \in D_2$ is the least upper bound of P in D_2. Then as $A \sqsupseteq^* B$, $\frac{\mathbf{2}(A,k)}{k} \geq \frac{\mathbf{2}(B,k)}{k}$ must hold for all $k \in \mathbb{N}$ which by taking lim inf yields $s_2(B) = 0$.

Consider now the minimal index $t \in \mathbb{N}$ for which $A(t) \neq B(t)$. Then $A \sqsupseteq^* B$ implies $A(t) = 2$ and $B(t) = 1$. Put

$$C(i) = \begin{cases} B(i), & \text{for } i < t, \\ 2, & \text{for } i = t, \\ 1, & \text{for } i > t. \end{cases}$$

Then $C \in D_2$ with $s_2(C) = 0$, and we have $A \sqsupseteq^* C$. However, as $\mathbf{2}(C,t) > \mathbf{2}(B,t)$ we get $B \not\sqsupseteq^* C$ which contradicts the assumption that B is the greatest lower bound of P in D_2. □

Proof (of Theorem 9). The validity of the statement can be easily checked by the same example as in the proof of Theorem 3. We omit the details. □

Proof (of Theorem 10). Let $A = \overline{12}$ and let $B = V_1 W_1 V_2 W_2 V_3 W_3 \ldots$, where

$$V_k = (12)^{2^{k-1}} \text{ and } W_k = (21)^{2^{k-1}}$$

for $k \in \mathbb{N}$. Note that $A, B \in D_2$ with $s_2(A) = s_2(B) = \frac{1}{2}$. For $C = A \vee B$ we have

$$C = B_1 1^2 B_2 1^4 B_3 1^8 B_4 1^{16} \ldots,$$

where $B_k = (12)^{2^{k-1}}$ for $k \in \mathbb{N}$. It is easy to check that

$$\liminf_{k\to\infty} \frac{\mathbf{2}(C,k)}{k} = \frac{1}{4} \text{ and } \limsup_{k\to\infty} \frac{\mathbf{2}(C,k)}{k} = \frac{1}{3},$$

whence $C \notin D_2$. Suppose that $D \in D_2$ is the greatest lower bound of A and B in D_2. Let $t \in \mathbb{N}$ be the minimal index such that $D(t) \neq C(t)$. Now $C \sqsupseteq D$ implies $D(t) = 1$ and $C(t) = 2$. For $i \in \mathbb{N}$ put

$$E(i) = \begin{cases} C(i), & \text{if } i \leq t, \\ 1, & \text{if } i > t. \end{cases}$$

Then we have $E \in D_2$ with $s_2(E) = 0$, and $C \sqsupseteq E$. However, since $E(t) > D(t)$ we obtain $E \not\sqsupseteq D$ which contradicts the assumption that D is the greatest lower bound of A and B in D_2. □

References

1. Das, P.P.: Lattice of octagonal distances in digital geometry. Pattern Recognition Lett. **11** (1990) 663–667
2. Fazekas, A., Hajdu, A., Hajdu, L.: Lattice of generalized neighbourhood sequences in nD and ∞D. Publ. Math. Debrecen **60** (2002) 405–427
3. Fazekas, A., Hajdu, A., Hajdu, L.: Metrical neighborhood sequences in $\mathbb{Z}^n$. Pattern Recognition Letters **26** (2005) 2022–2032
4. Rosenfeld, A., Pfaltz, J.L.: Distance functions on digital pictures. Pattern Recognition **1** (1968) 33–61
5. Das, P.P., Chakrabarti, P.P., Chatterji, B.N.: Distance functions in digital geometry. Inform. Sci. **42** (1987) 113–136
6. Das, P.P.: Best simple octagonal distances in digital geometry. J. Approx. Theory **68** (1992) 155–174
7. Das, P.P., Chakrabarti, P.P., Chatterji, B.N.: Generalised distances in digital geometry. Inform. Sci. **42** (1987) 51–67
8. Fazekas, A.: Lattice of distances based on 3D-neighbourhood sequences. Acta Math. Acad. Paedagog. Nyházi. (N.S.) **15** (1999) 55–60
9. Hajdu, A., Hajdu, L.: Approximating the Euclidean distance by digital metrics. Discrete Math. **283** (2004) 101–111
10. Mukherjee, J., Das, P.P., Aswatha Kumar, M., Chatterji, B.N.: On approximating Euclidean metrics by digital distances in 2D and 3D. Pattern Recognition Lett. **21** (2000) 573–582
11. Hajdu, A., Hajdu, L., Tijdeman, R.: General neighbourhood sequences in $\mathbb{Z}^n$. Disc. Appl. Math. (to appear)
12. Lothaire, M.: Combinatorics on Words. Addison-Wesley, Reading, MA (1983) xix+238 pp.
13. Hajdu, A., Hajdu, L.: Velocity and distance of neighbourhood sequences. Acta Cybernet. **16** (2003) 133–145
14. Nivat, M.: Infinite words, infinite trees, infinite computations. In: Math Centre Tracts **109** (1979) 1–52
15. Siromoney, R., Mathew, R., Dare, V.R., Subramanian, K.G.: Infinite Lyndon words. Inf. Proc. Letters **50** (1994) 101–104
16. Yucas, J.L.: Counting special sets of binary Lyndon words. Ars Combinatoria **31** (1991) 21–29
17. Nagy, B.: Distance functions based on generalised neighbourhood sequences in finite and infinite dimensional spaces. In: E. Kovács, Z. Winkler (eds.),Fifth International Conference on Applied Informatics, Eger, Hungary, 2001, pp. 183–190
18. Nagy, B.: Distance functions based on neighbourhood sequences. Publ. Math. Debrecen **63** (2003) 483–493

On the Connectedness of Rational Arithmetic Discrete Hyperplanes

Damien Jamet[1] and Jean-Luc Toutant[2]

[1] LORIA - INRIA Lorraine
615 rue du Jardin Botanique - BP 101
F-54602 Villers-lès-Nancy, France
jamet@loria.fr
[2] LIRMM - CNRS UMR 5506 - Université Montpellier II
161 rue Ada - 34392 Montpellier Cedex 5, France
toutant@lirmm.fr

Abstract. While connected arithmetic discrete lines are entirely characterized by their arithmetic thickness, only partial results exist for arithmetic discrete hyperplanes in any dimension. In the present paper, we focus on 0-connected rational arithmetic discrete planes in $\mathbb{Z}^3$. Thanks to an arithmetic reduction on a given integer vector $\mathbf{n}$, we provide an algorithm which computes the thickness of the thinnest 0-connected arithmetic plane with normal vector $\mathbf{n}$.

1 Introduction

In [1], J.-P. Reveillès initiated a new approach of linear discrete objets and introduced arithmetic discrete lines as sets of pairs of integers satisfying a double Diophantine inequality : the *arithmetic discrete line* with *normal vector* $\mathbf{n} \in \mathbb{R}^2$, *translation parameter* $\mu \in \mathbb{R}$ and *thickness* $w \in \mathbb{R}$ is the set $\mathbf{D}(\mathbf{n}, \mu, w) = \{\mathbf{x} \in \mathbb{Z}^2,\ 0 \leq \mathbf{n} \cdot \mathbf{x} + \mu < w\}$, where $\mathbf{n} \cdot \mathbf{x} = n_1 x_1 + n_2 x_2$ is the usual Euclidean scalar product of $\mathbf{n}$ and $\mathbf{x}$. Geometrically, an arithmetic discrete line can be viewed as a set of integer points of the plane $\mathbb{R}^2$ included in a band delimited by two parallel Euclidean lines (see Fig. 1). The thickness parameter w plays a key role in the topology of the arithmetic discrete lines: given $\mathbf{n} \in \mathbb{R}^2$ and $\mu \in \mathbb{R}$, the thinnest 0-connected (resp. 1-connected) arithmetic discrete line among the ones with normal vector $\mathbf{n}$ and translation parameter μ is the arithmetic discrete line $\mathbf{D}(\mathbf{n}, \mu, \|\mathbf{n}\|_\infty)$ (resp. $\mathbf{D}(\mathbf{n}, \mu, \|\mathbf{n}\|_1)$) (see Section 2 for the definition of the 0-connectedness and 1-connectedness) [1].

The definition of arithmetic discrete lines extends naturally in dimension 3 to the *arithmetic discrete planes* and in any dimension $d \geq 2$ to the *arithmetic discrete hyperplanes* [2]. It is thus natural to try to exhibit a similar relation between the κ-connectedness of an arithmetic discrete hyperplane and its thickness. In fact, the 2-dimensional case is somewhat confusing since a 0-connected (resp. 1-connected) arithmetic discrete line is also 1-separating (resp. 1-separating) in $\mathbb{Z}^2$ (see Section 2).

A. Kuba, L.G. Nyúl, and K. Palágyi (Eds.): DGCI 2006, LNCS 4245, pp. 223–234, 2006.

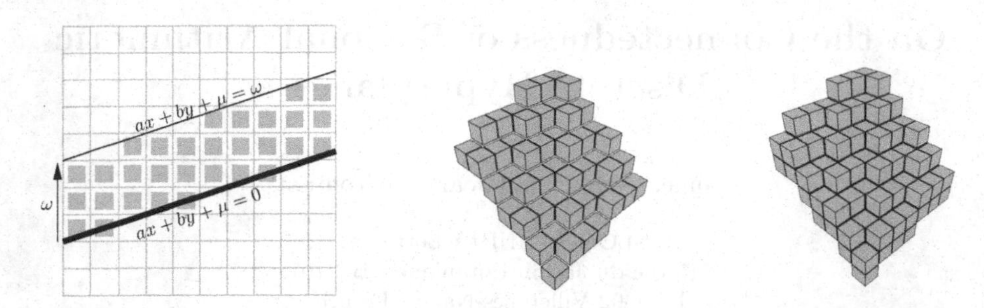

Fig. 1. From left to right: an arithmetic discrete line - a naive discrete plane - a standard discrete plane

In the particular case of rational arithmetic discrete hyperplanes (rememeber that an arithmetic discrete hyperplane is *rational* if its normal vector $\mathbf{n} \in \mathbb{R}^d$ is colinear to an integer vector, or equivalently, if the $\mathbb{Q}$-vector space spanned by $\{n_1, \ldots, n_d\}$ is of dimension 1), several approaches have been attempted [2,3,4] although none of them provides an explicit formula to compute the thickness of the thinnest 0-connected rational arithmetic discrete hyperplane with any given normal vector.

In [4], V. Brimkov and R. Barneva partially solved this request for rational arithmetic discrete planes whose the normal vector $\mathbf{n} \in \mathbb{Z}^2$ satisfies particular conditions (for instance when $|n_1| + 2|n_2| \leq |n_3|$) and provided an algorithm for the entire problem. Unfortunately, their algorithm seems to incorrect and does not generally return the right thickness (see Section 4).

In [3], Y. Gérard investigated a problem close to the one we are interested in in the present paper: given an arithmetic discrete hyperplane $\mathbf{P}(\mathbf{n}, \mu, w)$ and $\kappa \in \{0, \ldots, d-1\}$, is $\mathbf{P}(\mathbf{n}, \mu, w)$ κ-connected ? In other words, given the graph $\mathbf{G}(\mathbf{n}, \mu, w)$ whose vertices are the points of $\mathbf{P}(\mathbf{n}, \mu, w)$ and whose edges are the pairs $\{\mathbf{x}, \mathbf{y}\}$ of κ-adjacent points of $\mathbf{P}(\mathbf{n}, \mu, w)$, does $\mathbf{G}(\mathbf{n}, \mu, w)$ admit a unique connected component ? The main difficulty of this problem is the possibly infiniteness of $\mathbf{G}(\mathbf{n}, \mu, w)$. Assuming $\dim_{\mathbb{Q}}\{n_1, \ldots, n_d\} = 1$, one reduces $\mathbf{G}(\mathbf{n}, \mu, w)$ to a finite graph by quotienting $\mathbf{G}(\mathbf{n}, \mu, w)$ iteratively by a subgroup of rank 1 of the lattice of periods of $\mathbf{P}(\mathbf{n}, \mu, w)$. Since $\mathbf{G}(\mathbf{n}, \mu, w)$ is injectively projectable in $\mathbb{Z}^d$, then, with at most d such quotienting processes, one reduces $\mathbf{G}(\mathbf{n}, \mu, w)$ to a finite graph with the same connectedness as $\mathbf{G}(\mathbf{n}, \mu, w)$.

In the present paper, we deal with the determination of the thickness of the thinnest 0-connected rational arithmetic discrete plane with a given normal vector $\mathbf{n}$. For this purpose, we give a short and elementary algorithm which takes a vector $\mathbf{n} \in \mathbb{Z}^3$ as entry and returns the thickness w of the thinnest 0-connected arithmetic discrete plane with normal vector $\mathbf{n}$. While Y. Gérard, V. Brimkov and R. Barneva's approaches need to determine a connected component, our algorithm is *entirely* arithmetic and does not need to consider any connectivity graph.

Here is the sketch of the present paper. Section 2 is devoted to the basic notions useful for the remaining. In Section 3, we investigate the notions of κ-connectedness and κ-separatingness and state a first comparison between their characterization in the case of rational arithmetic discrete lines and rational arithmetic discrete hyperplanes. In Section 4, we focus on V. Brimkov and R. Barneva's investigation [4]. After having recalled some of their results, we exhibit a counter example of the algorithm they proposed. In Section 5, we introduce an arithmetic reduction on the integer vectors preserving the 0-connectedness of arithmetic discrete planes. We end this section by designing an elementary and quite short algorithm which computes the minimal thickness by iterating this arithmetic reduction.

2 Basic Notions

The aim of this section is to introduce the basic notions and definitions we use throughout the present paper.

Let d be an integer equal or greater than 2 and let $\{\mathbf{e_1}, \ldots, \mathbf{e_d}\}$ denote the canonical basis of the Euclidean vector space $\mathbb{R}^d$. Let us call a *discrete set* any subset of the *discrete space* $\mathbb{Z}^d$. In the following, for the sake of clarity, we denote by $(x_1, \ldots, x_d)$ the point (resp. vector) $\mathbf{x} = \sum_{i=1}^{d} x_i \mathbf{e_i} \in \mathbb{R}^d$. An integer point $\mathbf{x} \in \mathbb{Z}^d$ is called a *voxel* (resp. a *pixel* if $d = 2$). A subset of $\mathbb{Z}^d$ is called a *discrete set*.

Let $\kappa \in \{0, \ldots, d-1\}$. Two voxels $\mathbf{x} \in \mathbb{Z}^d$ and $\mathbf{x}' \in \mathbb{Z}^d$ are said to be *κ-adjacent* if $\|\mathbf{x} - \mathbf{x}'\|_\infty = 1$ and $\|\mathbf{x} - \mathbf{x}'\|_1 \leq d - \kappa$. In other words, $\mathbf{x} \in \mathbb{Z}^d$ and $\mathbf{x}' \in \mathbb{Z}^d$ are *κ-adjacent* if they are distinct, the differences of their coordinates are at most 1 and $\mathbf{x}$ and $\mathbf{x}'$ have at most $d - \kappa$ different coordinates (resp. at least κ identical components). A *κ-path* is a (finite or infinite) sequence of consecutive κ-adjacent voxels. If $(\gamma_i)_{1 \leq i \leq n}$ is a finite κ-path, then we say that γ *links* the voxel γ_1 to the voxel γ_n. A subset $E \subseteq \mathbb{Z}^d$ is said *κ-connected* if, for each pair of voxels $(\mathbf{x}, \mathbf{x}') \in E^2$, there exists a κ-path in E linking $\mathbf{x}$ to $\mathbf{x}'$. Given a discrete set $E \subseteq \mathbb{Z}^d$ and given $\kappa \in \{0, \ldots, d-1\}$, one says that E is *κ-separating* in $\mathbb{Z}^d$ if its complement in $\mathbb{Z}^d$ has (at least) two κ-connected components.

In [1], J.-P. Reveillès introduced the arithmetic discrete line as a set of integer points satisfying a double Diophantine inequality. This definition extends in a natural way to higher dimensions:

Definition 1 (Arithmetic discrete hyperplane [1,2]). *The* arithmetic discrete hyperplane *with* normal vector $\mathbf{n} \in \mathbb{Z}^d$, translation parameter $\mu \in \mathbb{Z}$ *and* thickness $w \in \mathbb{Z}$ *is the discrete set* $\mathbf{P}(\mathbf{n}, \mu, w)$ *defined by:*

$$\mathbf{P}(\mathbf{n}, \mu, w) = \left\{\mathbf{x} \in \mathbb{Z}^d,\ 0 \leq \mathbf{n} \cdot \mathbf{x} + \mu < w\right\}, \tag{1}$$

where $\mathbf{n} \cdot \mathbf{x}$ denotes the usual Euclidean scalar product in $\mathbb{R}^d$. If $w = \|\mathbf{n}\|_\infty$ (resp. $w = \|\mathbf{n}\|_1$) then $\mathbf{P}(\mathbf{n}, \mu, w)$ is said naive *(resp.* standard*). If $d = 2$ the arithmetic discrete hyperplane $\mathbf{P}(\mathbf{n}, \mu, w)$ is called an* arithmetic discrete line *and is denoted by $\mathbf{D}(\mathbf{n}, \mu, w)$. If $d = 3$ the arithmetic discrete hyperplane $\mathbf{P}(\mathbf{n}, \mu, w)$ is called an* arithmetic discrete plane.

Remark 1. Throughout the present paper, when $\mathbf{P}(\mathbf{n}, \mu, w)$ is a rational arithmetic hyperplane, we assume, with no loss of generality, that $\gcd\{n_1, \ldots, n_d\} = 1$, $\mu \in \mathbb{Z}$ and $w \in \mathbb{Z}$. Moreover, since the isometry group of the unit cube $[-0.5, 0.5]^d$ acts on the set of arithmetic discrete hyperplanes and since any isometry of $[-0.5, 0.5]^d$ preserves the κ-connectedness of any arithmetic discrete hyperplane, whatever $\kappa \in \{0, \ldots, d-1\}$, then in the following, except when explicitly mentioned, we suppose the normal vector $\mathbf{n} \in \mathbb{Z}^d$ to satisfy $0 \leq n_1 \leq \cdots \leq n_d$.

In Section 3, we recall some partial results on the connectedness of arithmetic discrete lines and give a first extension of them to arithmetic discrete hyperplanes.

3 κ-Connected Arithmetic Discrete Lines *vs.* κ-Separating Arithmetic Discrete Hyperplanes

Let us first deal with the case $d = 2$. In [1], J.-P. Reveillès showed how the κ-connectedness of an arithmetic discrete line depends only on its normal vector and its thickness:

Theorem 1 ([1]). *Let $\mathbf{D}(\mathbf{n}, \mu, w)$ be the arithmetic discrete line with normal vector $\mathbf{n} \in \mathbb{Z}^2$, translation parameter $\mu \in \mathbb{Z}$ and thickness $w \in \mathbb{Z}$. Then $\mathbf{D}(\mathbf{n}, \mu, w)$ is 0-connected (resp. 1-connected) if and only if $w \geq \|\mathbf{n}\|_\infty$ (resp. $w \geq \|\mathbf{n}\|_1$).*

It becomes natural to try to extend Theorem 1 to higher dimensions, that is, given $\mathbf{n} \in \mathbb{Z}^d$, $\mu \in \mathbb{Z}$ and $\kappa \in \{0, \ldots, d-1\}$, to try to characterize the thickness of the thinnest κ-connected arithmetic discrete hyperplane with normal vector $\mathbf{n}$ and translation parameter μ.

Let us give a helpful reduction of our problem: if $\mu \in \mathbb{Z}$ and $\mathbf{n} \in \mathbb{Z}^d$, then the κ-connectedness (resp. κ-separatingness in $\mathbb{Z}^d$) of $\mathbf{P}(\mathbf{n}, \mu, w)$, whatever $d \geq 2$ and $\kappa \in \{0, \ldots, d-1\}$, does not depend on the translation parameter μ. Indeed, it is a direct consequence of the following lemma:

Lemma 1. *Let $\mathbf{P}(\mathbf{n}, \mu, w)$ be an arithmetic discrete hyperplane with $d \geq 2$, $\mu \in \mathbb{Z}$ and $\mathbf{n} \in \mathbb{Z}^d$. For all $\mu' \in \mathbb{Z}$, there exists a vector $\boldsymbol{\alpha} \in \mathbb{Z}^d$ such that $\mathbf{P}(\mathbf{n}, \mu, w) = \mathbf{P}(\mathbf{n}, \mu', w) + \boldsymbol{\alpha}$.*

Proof. It obviously follows form Bezout's Lemma applied on the coordinates of $\mathbf{n}$. □

From now on, we consider only rational arithmetic discrete hyperplanes with a null translation parameter. Thanks to Lemma 1, in the determination of the thickness of the thinnest arithmetical discrete hyperplane with a given rational normal vector, this assumption is not restrictive. From now on, in order to simplify the notation, we denote by $\mathbf{P}(\mathbf{n}, w)$ the arithmetic discrete hyperplane with normal vector $\mathbf{n}$, translation parameter 0 and thickness w.

Definition 2 (**κ-Connecting thickness**). *Let $\mathbf{n} \in \mathbb{Z}^d$ and $\kappa \in \{0, \dots, d-1\}$. The thickness w_κ of the thinnest κ-connected arithmetic discrete hyperplane with normal vector $\mathbf{n}$ is called the* κ-connecting thickness *of* $\mathbf{n}$.

Let us now investigate the κ-connectedness of arithmetic discrete planes ($d = 3$). It is not difficult to exhibit a 0-connected arithmetic discrete plane $\mathbf{P}(\mathbf{n}, w)$ thinner than the naive one, that is, satisfying $w < \|\mathbf{n}\|_\infty$ (see Fig. 2). Similarly, one easily finds a 2-connected arithmetic discrete plane $\mathbf{P}(\mathbf{n}, w)$ thinner than the standard one, that is, with $w < \|\mathbf{n}\|_1$.

(a) A 0-connected arithmetic discrete plane thinner than the naive one

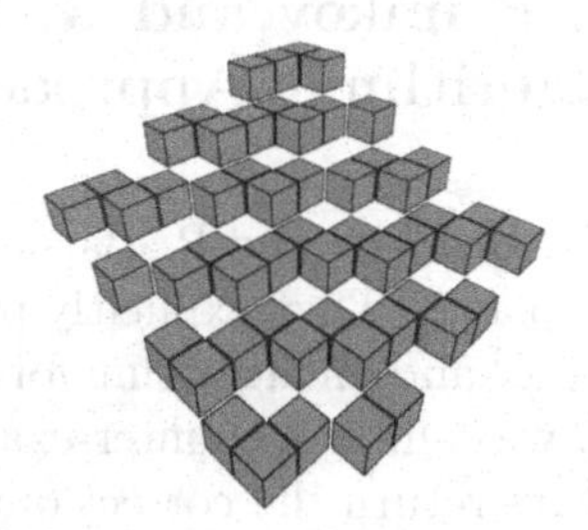

(b) A 1-connected arithmetic discrete plane thinner than the standard one

Fig. 2. Connected arithmetic discrete planes

Nevertheless, although Theorem 1 does not seem to extend naturally to higher dimensions, it admits a quite nice generalization of it concerning the κ-separating arithmetic discrete hyperplane. For the sake of clarity, we introduce the following notation, providing a norm on $\mathbb{R}^d$:

NOTATION. — Let $\mathbf{x} \in \mathbb{R}^d$ and let σ be a permutation over the set $\{1, \dots, d\}$ such that, for all $i \in \{1, \dots, d-1\}$, $|x_{\sigma(i)}| \leq |x_{\sigma(i+1)}|$. For all $\kappa \in \{0, \dots, d-1\}$, we denote by $]\mathbf{x}[_\kappa$ the following number:

$$]\mathbf{x}[_\kappa = \sum_{i=d-\kappa}^{d} |x_{\sigma(i)}|.$$

In other words, $]\mathbf{x}[_\kappa$ is equal to the sum of the $(\kappa + 1)$ greatest absolute values of the coordinates of $\mathbf{x}$.

One checks that, for each $\kappa \in \{0, \dots, d-1\}$, the map $] \cdot [_\kappa : \mathbb{R}^d \longrightarrow \mathbb{R}^d$ is a norm on $\mathbb{R}^d$. Moreover, one has $] \cdot [_0 = \| \cdot \|_\infty$ and $] \cdot [_{d-1} = \| \cdot \|_1$.

In the particular case of $d = 2$, for $\kappa \in \{0, 1\}$, the κ-connected arithmetic discrete lines are exactly the $(2 - (\kappa + 1))$-separating ones in $\mathbb{Z}^2$ and Theorem 1 is reformulated as follows:

Theorem 2 ([1]). *Let $\mathbf{D}(\mathbf{n}, w)$ be the arithmetic discrete line with normal vector $\mathbf{n} \in \mathbb{Z}^2$ and thickness $w \in \mathbb{Z}$. Let $\kappa \in \{0, 1\}$. Then, $\mathbf{D}(\mathbf{n}, w)$ is $(1 - \kappa)$-separating in $\mathbb{Z}^2$ if and only if $w \geq]\mathbf{n}[_\kappa$.*

In fact, as previously mentioned, the κ-separatingness of an arithmetic discrete hyperplane $\mathbf{P}(\mathbf{n}, w)$, whatever the dimension d, is entirely characterized by $]\mathbf{n}[_\kappa$. Indeed, Theorem 2 extends in the most natural way to every dimension:

Theorem 3 ([2]). *Let $\mathbf{P}(\mathbf{n}, w)$ be the arithmetic discrete hyperplane with normal vector $\mathbf{n} \in \mathbb{Z}^d$ and thickness $w \in \mathbb{Z}$. Let $\kappa \in \{0, \ldots, d-1\}$. The arithmetic discrete hyperplane $\mathbf{P}(\mathbf{n}, w)$ is $(d-\kappa-1)$-separating in $\mathbb{Z}^d$ if and only if $w \geq]\mathbf{n}[_\kappa$.*

4 V. Brimkov and R. Barneva's Investigation: An Algorithmic Approach [4]

In [4], V. Brimkov and R. Barneva investigated 0-connected rational arithmetic discrete planes. They explicitly provided the 0-connecting thickness of some vectors $\mathbf{n} \in \mathbb{Z}^3$ and an algorithm for computing it in the general case. In the present section, we exhibit a counter-example to this algorithm and deduce that it does not always return the correct output.

Let $\mathbf{P}(\mathbf{n}, w)$ be a rational arithmetic discrete plane. It is well known that if $w \geq \|\mathbf{n}\|_\infty$ then $\mathbf{P}(\mathbf{n}, w)$ is 0-connected (see [2] Cor. 10 p. 307). Hence, if w_0 is the 0-connecting thickness of $\mathbf{n}$, then $w_0 \leq \|\mathbf{n}\|_\infty$. In [4], V. Brimkov and R. Barneva reduced the determination of w_0 to the determination of the 0-connectedness of a subset of $\mathbb{Z}^2$ as follows:

Theorem 4 ([4]). *Let $\mathbf{P}(\mathbf{n}, w)$ be a rational arithmetic discrete plane with $\|\mathbf{n}\|_\infty = |v_3|$ and $w \leq \|\mathbf{n}\|_\infty$. The arithmetic dscrete plane $\mathbf{P}(\mathbf{n}, w)$ is 0-connected in $\mathbb{Z}^3$ if and only if the set $\{\mathbf{x} \in \mathbb{Z}^2,\ v_1x_1 + v_2x_2 \mod v_3 \in [0, w[\}$ is 0-connected in $\mathbb{Z}^2$.*

Remark 2. Let us remember that, thanks to Remark 1, the condition $\|\mathbf{n}\|_\infty = |v_3|$ in Theorem 4 is not restrictive. Up to an isometry, one can similarly treat the cases $\|\mathbf{n}\|_\infty = |v_1|$ and $\|\mathbf{n}\|_\infty = |v_2|$.

For the sake of clarity, we introduce the following notation:

NOTATION. — Let $\mathbf{P}(\mathbf{n}, w)$ be an arithmetic discrete plane with $\|\mathbf{n}\|_\infty = |v_3|$ and $w \leq \|\mathbf{n}\|_\infty$. We denote by $\mathbf{\Pi}(\mathbf{n}, w)$ the set $\{\mathbf{x} \in \mathbb{Z}^2,\ v_1x_1 + v_2x_2 \mod v_3 \in [0, w[\}$. In what follows, since $\mathbf{\Pi}(\mathbf{n}, w)$ can be indexed by (a subset of) $\mathbb{Z}^2$, we call $\mathbf{\Pi}(\mathbf{n}, w)$ the *array of remainders* of $\mathbf{P}(\mathbf{n}, w)$. For $\mathbf{x} \in \mathbb{Z}^2$, the number $v_1x_1 + v_2x_2 \mod v_3$ is called the *remainder* of $\mathbf{x}$. Let us notice that this denomination is not exactly the one used in [4,5], but is equivalent in the way we use it.

With this notation, from Theorem 4, it follows:

Corollary 1. *Let $\mathbf{P}(\mathbf{n}, w)$ be a rational arithmetic discrete plane with $\|\mathbf{n}\|_\infty = |v_3|$ and $w \leq \|\mathbf{n}\|_\infty$. The arithmetic dscrete plane $\mathbf{P}(\mathbf{n}, w)$ is 0-connected in $\mathbb{Z}^3$ if and only if the set $\mathbb{Z}^2 \setminus \mathbf{\Pi}(\mathbf{n}, w)$ is not 0-separating in $\mathbb{Z}^2$.*

Before describing V. Brimkov and R. Barneva's algorithm, let us introduce a notation:

Notation. — Let $\mathbf{n} \in \mathbb{Z}^3$ such that $0 \leq n_1 \leq n_2 \leq n_3$ and $\gcd\{n_1, n_2, n_3\} = 1$. We denote by $\mathbf{\Gamma}(\mathbf{n})$ the set of 1-paths in $\mathbf{\Pi}(\mathbf{n}, \|\mathbf{n}\|_\infty)$ linking two points of maximal remainder, that is, $n_3 - 1$. For a 1-path $\gamma \in \mathbf{\Gamma}(\mathbf{n})$, we denote by

$$\min(\gamma) = \min\{n_1 i_1 + n_2 i_2 \mod n_3, (i_1, i_2) \in \gamma\}.$$

In other words, $\min(\gamma)$ is the smallest remainder reached in γ.

In [4], V. Brimkov and R. Barneva stated:

Theorem 5 ([4]). *Let $\mathbf{n} \in \mathbb{Z}^3$ such that $0 \leq n_1 \leq n_2 \leq n_3$ and $\gcd\{n_1, n_2, n_3\} = 1$. Let $w_0 \in \mathbb{Z}$ be the 0-connecting thickness of $\mathbf{n}$. Then $w_0 = \max\{\min(\gamma) \in \Gamma(\mathbf{n})\} + 1$.*

Given a vector $\mathbf{n} \in \mathbb{Z}^3$ satisfying $0 \leq n_1 \leq n_2 \leq n_3$ and $\gcd\{n_1, n_2, n_3\} = 1$, the problem of determining w_0 can thus be reduced to the following one: how to compute $\max\{\min(\gamma) \in \Gamma(\mathbf{n})\}$ in a reasonable time ? V. Brimkov and R. Barneva assumed that only exclusively down-right or up-right searches (with additional conditions) in $\mathbf{\Pi}(\mathbf{n}, \|\mathbf{n}\|_\infty)$ are necessary to compute w_0 (see [4]). This assertion is false and here is a counter-example:

Example 1. Let $\mathbf{n} = (4, 7, 16)$. Let w_0 be the 0-connecting thickness of $\mathbf{n}$. In Figure 3(a), both light paths are computed by V. Brimkov and R. Barneva's algorithm. Minimal remainders of each one are respectively 3 and 5, and the algorithm returns $w_0 = \max\{3, 5\} + 1 = 6$. In Figure 3(b), one sees that $\mathbf{\Pi}(\mathbf{n}, 6)$ is not 0-connected, and by Theorem 4, so is $\mathbf{P}(\mathbf{n}, 6)$. In fact, the correct 0-connecting thickness for the vector $\mathbf{n}$ is 7 as shown in Figure 3(c). This value is obtained with the dark grey path in Figure 3(a), which cannot be computed using exclusively up-right or down-right searches.

5 Arithmetic Reduction of an Arithmetic Discrete Plane

We have seen in Section 4, that V. Brimkov and R. Barneva's algorithm needs a graph traversal for computing the 0-connecting thickness of a given integer vector. Similarly, Y. Gérard proposed an algorithm, based on a graph traversal too, testing whether a given rational arithmetic discrete hyperplane is κ-connected. In the present section, we propose a reduction acting on the normal vector and the arithmetic thickness of an arithmetic discrete plane $\mathbf{P}(\mathbf{n}, w)$ which returns an arithmetic discrete plane $\mathbf{P}(\mathbf{n}', w')$ with the same 0-connectedness as $\mathbf{P}(\mathbf{n}, w)$ and such that $|n_1'| < |n_1|$. By iterating this reduction, we obtain in a finite time an arithmetic discrete plane $\mathbf{P}(\mathbf{n}', w')$ with a zero coordinate. The 0-connecting thickness (see Definition 2) of such a vector is easy to determine:

Lemma 2. *Let $\mathbf{P}(\mathbf{n}, w)$ be a rational arithmetic discrete plane. Let us suppose there exists $i \in \{1, 2, 3\}$ such that $n_i = 0$. Then, $\mathbf{P}(\mathbf{n}, w)$ is 0-connected if and only if $w \geq \|\mathbf{n}\|_\infty$. In other words, the 0-connecting thickness of $\mathbf{n}$ is $\|\mathbf{n}\|_\infty$.*

(a) 1-connected paths in the 2-dimensional representation

(b) Array of remainders $\mathbf{\Pi}(\mathbf{n}, 6)$

(c) Array of remainders $\mathbf{\Pi}(\mathbf{n}, 7)$

Fig. 3. Computation of V. Brimkov and R. Barneva's algorithm [4] on the vector $\mathbf{n} = (4, 7, 16)$

Proof. It is well known that, if $w \geq \|\mathbf{n}\|_\infty$ then $\mathbf{P}(\mathbf{n}, w)$ is 0-connected [2]. Conversely, let us suppose, with no loss of generality, that $n_1 = 0$ and $0 \leq n_2 \leq n_3$. Let $\mathbf{x} \in \mathbb{Z}^2$ such that $n_1x_1 + n_2x_2 = n_2x_2 \equiv n_3 - 1 \bmod n_3$ (remember we assume $\gcd\{n_1, n_2, n_3\} = 1$). Then, for all $k \in \mathbb{Z}$, $(x_1 + k)n_1 + x_2n_2 = x_1n_1 + x_2n_2 \equiv n_3 - 1 \bmod n_3$. Hence, for all $k \in \mathbb{Z}$, $(x_1 + k, x_2) \in \mathbf{\Pi}(\mathbf{n}, w)$ and $\mathbf{\Pi}(\mathbf{n}, w)$ is not 0-connected. The result follows from Theorem 4. □

Remember that, thanks to Theorem 4, one can reduce the determination of the 0-connectedness of the arithmetic discrete plane $\mathbf{P}(\mathbf{n}, w)$ to the one of $\mathbf{\Pi}(\mathbf{n}, w) = \{\mathbf{x} \in \mathbb{Z}^2,\ n_1x_1 + n_2x_2 \bmod n_3 \in [0, w[\}$ with $\mathbf{n} \in \mathbb{Z}^3$ and $n_3 = \|\mathbf{n}\|_\infty$. Moreover, a direct consequence of Theorem 4 is:

Lemma 3 (Symmetry Lemma [4]). *Let $\Omega : \mathbb{N}^3 \to \mathbb{N}$ be the function mapping each vector of $\mathbb{N}^3$ to its 0-connecting thickness. For all $\mathbf{n} \in \mathbb{Z}^3$, if $0 \leq n_1, n_2 \leq n_3$, then $\Omega(n_1, n_2, n_3) = \Omega(n_3 - n_1, n_2, n_3) = \Omega(n_1, n_3 - n_2, n_3) = \Omega(n_3 - n_1, n_3 - n_2, n_3)$.*

> Given a vector $\mathbf{n} \in \mathbb{Z}^3$, thanks to Lemma 3 and to the action of the isometry group of the cube on the set of arithmetic discrete planes, one suppose **with no loss of generality** and in order to compute the 0-connecting thickness of $\mathbf{n}$ that $0 \leq 2n_1 \leq 2n_2 \leq n_3$.

Let us now state the main theorem of the present section:

Theorem 6 (Arithmetic reduction). *Let $\mathbf{n} \in \mathbb{Z}^3$ such that $0 \leq 2n_1 \leq 2n_2 \leq n_3$ and let $w \in \mathbb{Z}$. Let $(q, r) \in \mathbb{N}^2$ be the unique pair of integers such that $n_2 = qn_1 + r$ and $r \in [0, n_1[$. Let $\mathbf{n}' = M \cdot \mathbf{n}$ with*

$$M = \begin{pmatrix} 1 & 0 & 0 \\ -q & 1 & 0 \\ 1-q & -1 & 1 \end{pmatrix},$$

and let $w' = w - (n_2 - n_1)$. *Then, the arithmetic discrete plane* $\mathbf{P}(\mathbf{n}, w)$ *is* 0*-connected if and only if so is the arithmetic discrete plane* $\mathbf{P}(\mathbf{n}', w')$.

In order to prove Theorem 6, let us introduce in some sense the *dual* notion of the κ-connecting thickness of a vector:

Definition 3 (κ-separating thickness). *Let* $\mathbf{n} \in \mathbb{Z}^3$ *and let* $\kappa \in \{0,1\}$. *The* κ-separating thickness $\overline{w}_\kappa$ *of* $\mathbf{n}$ *is the thickness of the thinnest* κ*-separating* $\mathbf{\Pi}(\mathbf{n}, w)$, *with* $w \in \mathbb{Z}$.

An easy computation directly gives:

Lemma 4. *Let* $\mathbf{n} \in \mathbb{Z}^3$ *such that* $0 \leq n_1, n_2 \leq n_3$ *and* $\gcd\{n_1, n_2, n_3\} = 1$. *Let* w_0 *(resp.* $\overline{w}_0$*) be the* 0*-connecting thickness (resp.* 0*-separating thickness) of* $\mathbf{n}$. *Then* $w_0 + \overline{w}_0 = n_3 + 1$.

Proof. Let $w \in \mathbb{N}$. Then

$$\begin{aligned} \mathbb{Z} \setminus \mathbf{\Pi}(\mathbf{n}, w) &= \{(x_1, x_2) \in \mathbb{Z}^2, n_1 x_1 + n_2 x_2 \bmod n_3 \in [w, n_3[\} \\ &= \{(x_1, x_2) \in \mathbb{Z}^2, n_1 x_1 + n_2 x_2 - w \bmod n_3 \in [0, n_3 - w[\} \end{aligned}$$

Let $(\alpha_1, \alpha_2) \in \mathbb{Z}^2$ such that $n_1\alpha_1 + n_2\alpha_2 \equiv -w \bmod n_3$. Thus, $\mathbb{Z} \setminus \mathbf{\Pi}(\mathbf{n}, w) + (\alpha_1, \alpha_2) = \mathbf{\Pi}(\mathbf{n}, n_3 - w)$ and $\mathbf{\Pi}(\mathbf{n}, w)$ is 0-connected if and only if $\mathbf{\Pi}(\mathbf{n}, n_3 - w)$ is not 0-separating. Since $\mathbf{\Pi}(\mathbf{n}, w_0)$ (resp. $\mathbf{\Pi}(\mathbf{n}, w_0 - 1)$) is 0-connected (resp. is not 0-connected), then $\mathbf{\Pi}(\mathbf{n}, n_3 - w_0)$ (resp. $\mathbf{\Pi}(\mathbf{n}, n_3 - w_0 + 1)$) is not 0-separating (resp. is 0-separating). Hence $\overline{w}_0 = n_3 - w_0 + 1$ and the result follows. □

Since the κ-connectedness and the κ-separatingness of a rational arithmetic discrete plane do not depend on the translation parameter, an easy computation gives the equivalent reformulation of Theorem 6:

Theorem 7 (Arithmetic reduction). *Let* $\mathbf{n} \in \mathbb{Z}^3$ *such that* $0 \leq 2n_1 \leq 2n_2 \leq n_3$ *and let* $w \in \mathbb{Z}$. *Let* $(q, r) \in \mathbb{N}^2$ *be the unique pair of integers such that* $n_2 = qn_1 + r$ *and* $r \in [0, n_1[$. *Let* $\mathbf{n}' = M \cdot \mathbf{n}$ *with*

$$M = \begin{pmatrix} 1 & 0 & 0 \\ -q & 1 & 0 \\ 1-q & -1 & 1 \end{pmatrix},$$

and let $w' = w - qn_1$. *Then,* $\mathbf{\Pi}(\mathbf{n}, w)$ *is* 0*-separating if and only if so is* $\mathbf{\Pi}(\mathbf{n}', w')$.

Proof (sketch). For clarity, let us first introduce a quite natural notation. One naturally represents a 1-path γ in $\Pi(\mathbf{n}, w)$ as a triple (A, u, B) with:

i) $A \in [0, w[$ (resp. $B \in [0, w[$) is the starting (resp. the ending) remainder of the 1-path γ.

ii) $u \in \{\pm n_1, \pm n_2, \pm(n_1 - n_3), \pm(n_2 - n_3)\}^k$ is a finite sequence of *movements*

4	8	12	0	4	8	12	0	4
13	1	5	9	13	1	5	9	13
6	10	14	2	6	10	14	2	6
15	3	7	11	15	3	7	11	15
8	12	0	4	8	12	0	4	8
1	5	9	13	1	5	9	13	1

Fig. 4. A 1-path corresponding to the triple $(1, [-n_2, n_1, n_1 - n_3, -(n_2 - n_3), -n_2, n_1, n_1, n_1 - n_3, n_1], 4)$

between A and B (see Figure 4) (the integer $k \in \mathbb{N}$ is called the *length* of u). Let us notice that the *movements* $\pm(n_1 - n_3)$ and $\pm(n_2 - n_3)$ corresponds to horizontal (resp. vertical) movements in $\Pi(\mathbf{n}, w)$ with a change of height in $\mathbf{P}(\mathbf{n}, w)$. Such a change is represented by a thick line in the array of remainders (see Figure 4).

Conversely, let (A, u, B) be a triple with $(A, B) \in [0, w[^2$ and $u \in \{\pm n_1, \pm n_2, \pm(n_1 - n_3), \pm(n_2 - n_3)\}^k$, with $k \in \mathbb{N}$, then (A, u, B) is a 1-path in $\Pi(\mathbf{n}, w)$ if and only if, for all $j \in \{0, \ldots, k\}$, $A + \sum_{i=1}^{j} u_k \in [0, w[$.

The aim of this proof is to show that $\Pi(\mathbf{n}, w)$ admits an infinite 1-path if and only if so does $\Pi(\mathbf{n'}, w')$.

Let us first prove that each pair of two 1-adjacent pixels in $\Pi(\mathbf{n'}, w')$ can be *expanded* into a 1-path in $\Pi(\mathbf{n}, w)$.

i) Let (A, n_1', B) represent a pair of two 1-adjacent pixels in $\Pi(\mathbf{n'}, w')$. Then $0 \leq A < w' = w - qn_1 \leq w$, $0 \leq B < w' = w - qn_1 \leq w$ and $(A, n_1, B) = (A, n_1', B)$ is a 1-path in $\Pi(\mathbf{n}, w)$.

ii) Let $(A, n_1' - n_3', B)$ represent a pair of two 1-adjacent pixels in $\Pi(\mathbf{n'}, w')$. Since $n_1' - n_3' = qn_1 + n_2 - n_3$ and $A < w - qn_1$, then $0 \leq A + qn_1 < w$ and $(A, \underbrace{n_1, \ldots, n_1}_{q}, n_2 - n_3, B)$ is a 1-path in Pi$(\mathbf{n}, w)$.

The other cases, namely (A, n_2', B) and $(A, n_2' - n_3', B)$, are obtained in the same way. For summarize, see Figure 5 for a correspondance between a 1-path in $\Pi(\mathbf{n'}, w')$ and a 1-path in $\Pi(\mathbf{n}, w)$. Conversely, if $\Pi(\mathbf{n}, w)$ admits an infinite 1-path, then by a similar recoding of it, one obtains an infinite 1-path in $\Pi(\mathbf{n'}, w')$. The complete proof if this theorem will appear in a forthcoming long version of the present paper. □

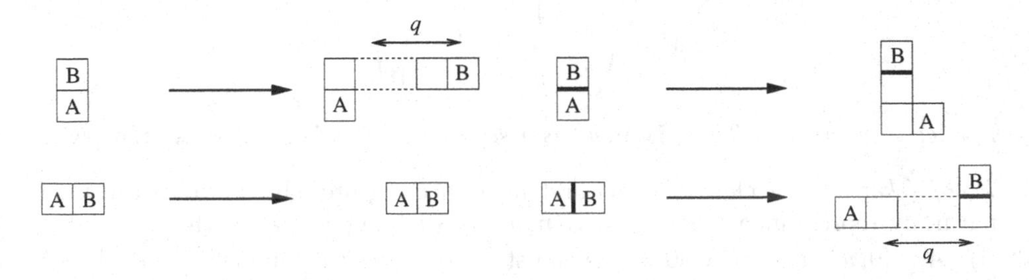

Fig. 5. Transformation of 1-paths in $\Pi(\mathbf{n'}, w')$ into 1-paths in $\Pi(\mathbf{n}, w)$

6 Algorithm

In the present section, we design an algorithm which computes the 0-connecting thickness of a given integer vector $\mathbf{n} \in \mathbb{Z}^3$. It iterates the reduction introduced in Theorem 6 until 0-connecting thickness becomes easy to determine.

The arithmetic reduction mentioned above only preserves 0-connectedness between the arithmetic discrete plane with normal vector $\mathbf{n}$ and its image under some conditions on $\mathbf{n}$. Nevertheless changing the components of a vector according to the symmetry lemma 3 or sorting them do not change the associated 0-connecting thickness. It is then possible to find from any vector $\mathbf{n}$ a vector $\mathbf{n}'$ with the same 0-connecting thickness meeting the requirement of Theorem 6. A step consisting of application of symmetry lemma, sorting, and the arithmetic reduction can be repeated and turns the vector $\mathbf{n}$ into another vector $\mathbf{n}'$ such that $n_1 \leq n'_1$, $n_2 < n'_2$ and $n_3 < n'_3$. Consequently, after a finite number of iteration, we always obtain a vector with a zero component for which the 0-connecting thickness is easy to determine.

Algorithm 1. follows from those considerations. It always terminates since the stopping condition, that is a vector with a zero component, is always reached in a finite number of iteration.

Algorithm 1. Determination of the 0-connecting thickness.

Input : $\mathbf{n} \in \mathbb{N}^3$.
Output : $w_0 \in \mathbb{Z}$, the 0-connecting thickness of $\mathbf{n}$.

$\omega \leftarrow 0$
while $n_2 \neq 0$ **do**
  {Symmetry and ordering}
  $n_1 \leftarrow \min(n_1, n_3 - n_1)$
  $n_2 \leftarrow \min(n_2, n_3 - n_2)$
  $t \leftarrow \min(n_1, n_2)$
  $n_2 \leftarrow \max(n_1, n_2)$
  $n_1 \leftarrow t$
  {Reduction}
  $q \leftarrow \lfloor n_2/n_1 \rfloor$
  $\omega \leftarrow \omega + (n_2 - n_1)$
  $n_3 \leftarrow n_3 - (n_2 + (q-1)n_1)$
  $n_2 \leftarrow n_2 - qn_1$
end while
return $\omega + n_3$

---

7 Conclusion and Perspectives

In the present paper, we presented an algorithm computing the 0-connecting thickness of any integer vector. The main difference between this algorithm and the ones already known [4,3] is that it does not need a graph traversal and only computes basic reductions on an integer vector.

In a forthcoming work, we plan to investigate the case of non-rational arithmetic planes. Since the reduction of Theorem 6 does not depend on the nature of the input vector (integer or not), we hope to extend this approach to any vector $\mathbf{n} \in \mathbb{R}^3$.

Other interesting investigations should be, on the one hand, the computation of κ-connected thickness for $\kappa \in \{1, 2\}$ and, on the other, the extension of this work to arithmetic discrete hyperplanes in any dimension.

Acknowledgements

We would like to thank E. Andres for having pointed out the original problem to the authors. We also thank P. Arnoux, V. Berthe, V. Brimkov, C. Fiorio, Y. Gérard and D. Vergnaud for many interesting discussions on the subject.

References

1. Reveillès, J.P.: Géométrie discrète, Calcul en Nombres Entiers et Algorithmique. Thèse d'Etat, Université Louis Pasteur, Strasbourg (1991)
2. Andres, E., Acharya, R., Sibata, C.: Discrete analytical hyperplanes. CVGIP: Graphical Models and Image Processing **59**(5) (1997) 302–309
3. Gérard, Y.: Periodic graphs and connectivity of the rational digital hyperplanes. Theoritical Computer Science **283**(1) (2002) 171–182
4. Brimkov, V., Barneva, R.: Connectivity of discrete planes. Theoritical Computer Science **319**(1-3) (2004) 203–227
5. Debled-Rennesson, I.: Etude et reconnaissance des droites et plans discrets. Thèse de Doctorat, Université Louis Pasteur, Strasbourg. (1995)

Homology of Simploidal Set

Samuel Peltier, Laurent Fuchs, and Pascal Lienhardt

SIC, Université de Poitiers, Boulevard Marie et Pierre Curie
86962 Futuroscope Chasseneuil Cedex, France
{peltier, fuchs, lienhardt}@sic.univ-poitiers.fr

Abstract. In this article the homology of simploidal sets is studied. Simploidal sets generalize both simplicial complexes and cubical complexes, more precisely cells of simplicial sets are cartesian products of simplices. We define one homology for simploidal sets and we prove that this homology is equivalent to the homology usually defined on simplicial complexes.

1 Introduction

The aim of this paper is to define and to study the homology of simploidal sets. Simploidal sets (see Fig. 1(a)) can be considered as a special case of cellular complexes, where cells are simploids [1,2], i.e. products of simplices. Simploidal sets include simplicial complexes and cubical complexes as particular cases, so they can be used for representing the topology of digital images. They can also be used for representing *hybrid* grids coming from finite elements methods. The notion of simploid was introduced by DAHMEN and MICCHELLI [1] to study multivariate splines.

Topological invariants provide information about the structure of an object. *Homology* is a powerful one[1] which can be computed for any dimension. Homology groups describe dimensional "holes" of a combinatorial object (connected components for dimension 0, holes for dimension 1, cavities for dimension 2,...). Homology information can be represented on combinatorial structures by computing *homology groups generators.* For example, Fig. 1(c) represents the two 1−dimensional holes of the torus (b).

For digital image analysis, topological invariants are useful for classification, indexation, or shape description [3]. Homology groups are classically computed for simplicial combinatorial structures such as abstract simplicial complexes [4] or semi-simplicial sets [5]. In this paper, we show that it is always possible to convert a simploidal set into a simplicial structure (a semi-simplicial set). So, homology groups of a simploidal set can be computed from the corresponding semi-simplicial set. Since many simplices correspond to a single simploid, this conversion of data structures can be space and time consuming. Similar arguments as those developed for cubical complexes [6] can be taken into account.

[1] Homology groups contain other classical topological invariant as Euler characteristic, Betti numbers, and orientability of a closed surface.

A. Kuba, L.G. Nyúl, and K. Palágyi (Eds.): DGCI 2006, LNCS 4245, pp. 235–246, 2006.

We propose a direct definition of homology groups for simploidal sets, i.e. we define *boundary homomorphisms* from which we can construct a free chain complex[2]. It is well-known from algebraic topology that homology defined on a triangulable cell complex (in our case : a simploidal set), is equivalent to homology defined on the associated triangulated space (in our case : the associated semi-simplicial set). We study this equivalence in a combinatorial and constructive way in order to compute simplicial generators from simploidal generators and conversely.

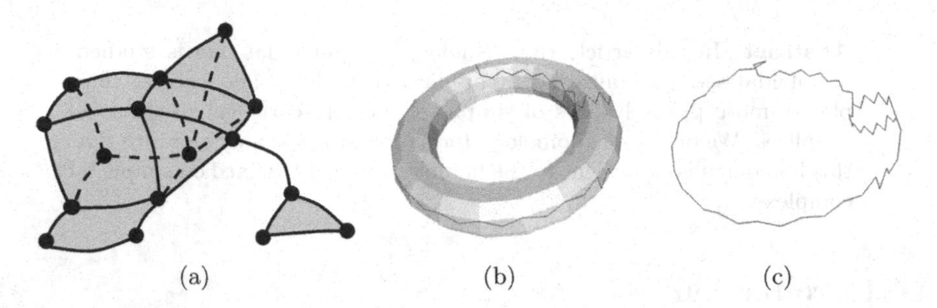

Fig. 1. (a) : a simploidal object. $(b)-(c)$: a geometric representation of the two $1-$holes of the torus.

In section 2 we recall the definition of semi-simplicial sets and basic concepts of chain, cycle, boundary, free chain complex and homology groups. In section 3, the simploidal set definition is recalled. Boundary homomorphisms for this structure is defined, and thus homology groups of these sets can be defined by constructing a free chain complex. In section 4, we study the conversion of a simploidal set into a corresponding semi-simplicial set. Then, We define a morphism between simplicial and simploidal chain groups, which associates to each simploidal chain an equivalent simplicial chain in the corresponding semi-simplicial set. After, we describe algorithms for constructing a simplicial homology generator from a simploidal one, and conversely. This construction provides a combinatorial and constructive proof of the equivalence between simploidal and simplicial homologies.

2 Homology of Semi-simplicial Sets

In this section all notions needed to define the homology groups over a combinatorial structure are introduced. Semi-simplicial sets [5,7] are used to illustrate these notions. Since our goal is the computation of homology groups of objects explicitly represented within a computer, all sets are finite.

---

[2] A free chain complex is an algebraic structure from which homology groups are defined (cf. section 2).

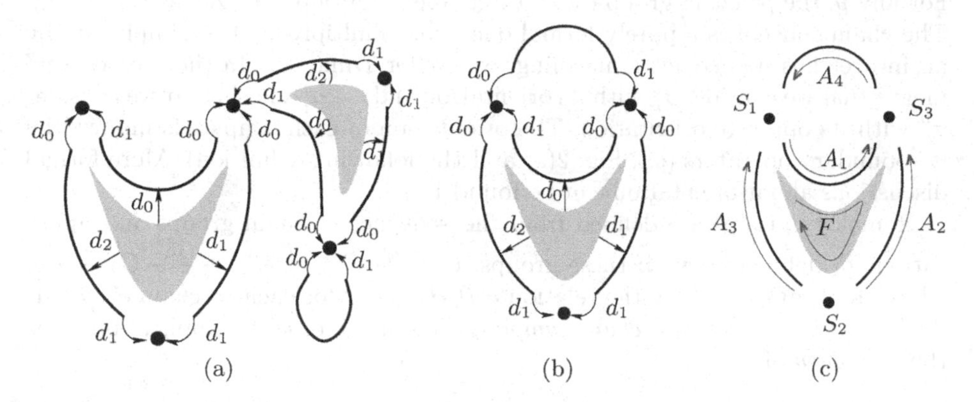

Fig. 2. $(a) - (b)$: Examples of semi-simplicial sets. (c) : positive orientation of simplices of (b).

2.1 Semi-simplicial Sets

Definition 1. *[5] Let $n \in \mathbb{N}$. A $n-$dimensional* semi-simplicial set $S = (K, (d_i^p))$ *is a family of sets $K = (K^p)_{p \in [0..n]}$ together with maps $d_i^p : K^p \to K^{p-1}$ for $i = 0, \ldots, p$, which satisfy the following property[3] :*

$$\forall p, 2 \leq p \leq n, \forall i, j, 0 \leq j < i \leq n, \forall \sigma \in K^p, \sigma d_i^p d_j^{p-1} = \sigma d_j^p d_{i-1}^{p-1}$$

The elements of K^p are *$p-$simplices*, the d_i^p are *boundary operators* (the superscripts p will be usually dropped). The relations between the boundary operators ensure that simplices are coherently assembled. Without these relations, a $p-$simplex could have more than $p+1$ distinct vertices in its boundary, i.e. $\sigma d_2 d_1$, $\sigma d_2 d_0$, $\sigma d_1 d_1$, $\sigma d_1 d_0$, $\sigma d_0 d_1$,$\sigma d_0 d_0$: relations between boundary operators ensure that $\sigma d_2 d_1 = \sigma d_1 d_1$, $\sigma d_2 d_0 = \sigma d_0 d_1$, $\sigma d_1 d_0 = \sigma d_0 d_0$. The notion of semi-simplicial set generalizes the classical notion of abstract simplicial complexes [4] in the following way: a semi-simplicial set can be associated to any abstract simplicial complex, but the converse is not true. For example, it is not possible to associate an abstract simplicial complex with the semi-simplicial set of Fig 2(a), since it contains a self-loop.

2.2 Chain, Boundary Homomorphism and Free Chain Complex

Let n_p be the number of $p-$simplices of K^p, and $K^p = \{\sigma_1^p, \cdots, \sigma_{n_p}^p\}$. A *$p-$chain* c is a combination of $p-$simplices together with integer coefficients : $c = \sum_{i=1}^{n_p} \alpha_i^p \sigma_i^p$. For example on Fig. 2$(c)$: A_1, $-A_2$ and $3A_2 - A_4$ are $1-$chains. The addition of $p-$chains consists in the addition of the corresponding simplex coefficients. The neutral element is the empty chain denoted 0 for each dimension.

[3] It could be noted that the notation xf is used instead of the classical notation $f(x)$ as it is more convenient when handling boundary operators.

For any p, the $p-$chain group C_p is a free abelian group with K^p as a basis [4]. The chain concept is a purely formal one, since multiplying a $p-$simplex σ_i^p by an integer has no geometric meaning except for 1 and -1. In these cases $1.\sigma_i^p$ means that we consider σ_i^p with its orientation and $-1.\sigma_i^p$ means that we consider σ_i^p with its opposite orientation. The orientation of each simplex is induced by its boundary operators (c.f. Fig. 2(*c*) and the following definition). More formal discussions about orientation can be found in [4].

Homology groups are defined from the sequence of chain groups and applications ∂_i defined between these groups, $C_n \xrightarrow{\partial_n} C_{n-1} \xrightarrow{\partial_{n-1}} \cdots \xrightarrow{\partial_1} C_0 \xrightarrow{\partial_0} 0$. These applications satisfy the relation $c^p\partial_p\partial_{p-1} = 0$ for each $p-$chain $c^p, p \geq 1$. Such a sequence is a *free chain complex*. The image of c^p by application ∂_p is the *boundary* of c^p.

Definition 2. *For any* $p, 1 \leq p \leq n$, *the* boundary *of* $p-$*simplex* σ^p *is the* $(p-1)-$*chain* $\sigma^p\partial_p = \sum_{i=0}^{p}(-1)^i\sigma d_i$. *The boundary of a* $0-$*simplex is the null chain.*

The definition of a $p-$chain boundary is directly deduced by linearity from the definition of a $p-$simplex boundary and the boundary applications ∂_p are homomorphisms. Usually the subscript and superscript p will be dropped. For example on Fig. 2(*c*): $F\partial = A_1 - A_2 + A_3$ and $(4A_2 - 3A_4)\partial = 4(A_2)\partial - 3(A_4)\partial = 7S_3 - 4S_2 - 3S_1$ are chain boundaries. Note that $F\partial\partial = (A_1 - A_2 + A_3)\partial = 0$.

In order to verify that applications ∂_p are actually boundary applications, we only have to check that $c\partial\partial = 0$ for each chain c composed by one simplex. This can directly be done using definition 2 and the property that for any $p-$simplex σ, $\sigma\partial\partial = 0$.

2.3 Cycles, Boundaries, Homology Groups

In order to define homology groups, we first define particular chains. A $p-$chain which boundary is null is a $p-$*cycle*. The set of $p-$cycles equipped with the addition is a $p-$chain subgroup, denoted Z_p. For example on Fig. 2(*c*): $1-$chains $A_1 - A_2 + A_3$ and $A_1 + A_4$ are $1-$cycles: $(A_1 - A_2 + A_3)\partial = (S_3 - S_1) - (S_3 - S_2) + (S_1 - S_2) = 0$ and $(A_1 + A_4)\partial = (S_3 - S_1) + (S_1 - S_3) = 0$.

A $p-$chain which is the boundary of a $(p+1)-$chain is a $p-$*boundary*. The set of $p-$boundaries equipped with the addition is also a $p-$chain subgroup, denoted B_p. Moreover, each $p-$boundary is a $p-$cycle (since $\forall c \in C^{p+1}, c\partial\partial = 0$) hence B_p is a subgroup of Z_p. For example on Fig. 2(*c*): $1-$chain $A_1 - A_2 + A_3$ is the boundary of $2-$chain F.

A $p-$dimensional *hole* is a $p-$cycle which is not a $p-$boundary. For example, on Fig. 2(*c*), $1-$cycle $A_1 + A_4$ is not a boundary.

Now an equivalence relation is defined as follow: two $p-$cycles μ_1 and μ_2 are equivalent if their difference is a boundary, i.e. $\mu_1 = \mu_2 + c\partial_{p+1}$: μ_1 and μ_2 are *homologous*[4]. Homology group H_p is the quotient of cycle group Z_p by the equivalence relation (i.e. $H_p = Z_p/B_p$). Hence two cycles belong to the same

[4] As a special case, if $\mu = c\partial_{p+1}$ then μ is homologous to 0.

equivalence class if they surround the same hole. For example on Fig. 2(c): cycles $z_2 = A_2 - A_3 + A_4$ and $A_1 + A_4$ are homologous, since $z_1 = z_2 + F\partial$.

For any p, H_p is finitely generated, i.e. there is a finite number of elements from which all others can be deduced. Hence, following the finitely generated group theorem, any group H_p is isomorphic to a direct sum [4]:

$$\underbrace{\mathbb{Z} \oplus ... \oplus \mathbb{Z}}_{\beta_p} \oplus \mathbb{Z}/t_1\mathbb{Z} \oplus ... \oplus \mathbb{Z}/t_n\mathbb{Z}.$$

Each $\mathbb{Z}$ corresponds to an equivalence class of infinite order cycles[5]. The number β_p is the pth *Betti number*. Each $\mathbb{Z}/t_i\mathbb{Z}$ corresponds to an equivalence class of cycles of finite order t_i[6]. Integers t_i are the *torsion coefficients*. A cycle of finite order is a *weak* boundary.

3 Homology of Simploidal Sets

In this section, simploidal sets are introduced. We extend classical notions of chains, cycles and boundaries and we define boundary homomorphisms for this structure. Thus, we provide a direct homology definition for simploidal sets.

3.1 Simploidal Sets

A simploid can be defined as the product of polytopes, which are "geometric" simplices [2]. We recall here the combinatorial structure of simploidal sets, which is based upon the notion of semi-simplicial set (see section 4). In a simploidal set, a simploid is defined by a $k-$tuple $(a_1, \ldots, a_k)$ of strictly positive integers, which is its *type*, k is the *length* of the simploid, $\sum_{l=1}^{k} a_l$ is its *dimension* (intuitively, a simploid is the product of simplices of respective dimensions $a_1, \cdots a_k$). Some examples of simploids are shown on Fig. 3. It should be noted that a $p-$simplex is a simploid of type (p) and that a $p-$cube is a simploid of type $(1, \ldots, 1)$ with length p.

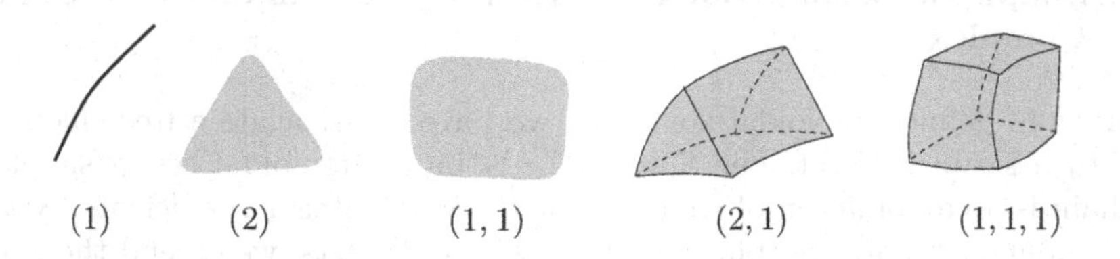

Fig. 3. Examples of simploids

[5] For any p, $h \in H_p$ is an infinite order cycle if and only if, for any α, $\alpha h \notin B_p$.

[6] For any p, $h \in H_p$ is a cycle of order t_i if and only if, for any $\alpha \in [1..t_i - 1], \alpha h \notin B$ and $t_i h \in B$.

Definition 3. *[8] A simploidal set $S = (K, (\epsilon^i_j))$ of dimension n is the union $\bigcup_{p=0}^{n} K^p$ of sets of $p-$dimensional simploids, $0 \leq p \leq n$ equipped with* border operators ϵ^i_j *such that:*

$$(\ldots, a_i, \ldots)\epsilon^i_j : \longrightarrow \begin{cases} (\ldots, a_i - 1, \ldots) \text{ if } a_i > 1 \\ (\ldots, \widehat{a}_i, \ldots) \quad \text{otherwise } (\widehat{a}_i \text{ means } a_i \text{ is removed}) \end{cases} \quad (1)$$

$$(\ldots, a_i, \ldots)\epsilon^i_k\epsilon^i_l = (\ldots, a_i, \ldots)\epsilon^i_l\epsilon^i_{k-1} \quad \text{with } k > l \text{ and } a_i > 1 \quad (2)$$

$$\begin{matrix}(\ldots, a_i, \ldots, a_j, \ldots)\epsilon^j_k\epsilon^i_l \\ \text{with } i < j\end{matrix} = \begin{cases} (\ldots, a_i, \ldots, a_j, \ldots)\epsilon^i_l\epsilon^j_k & \text{if } a_i > 1 \\ (\ldots, a_i, \ldots, a_j, \ldots)\epsilon^i_l\epsilon^{j-1}_k & \text{otherwise.} \end{cases} \quad (3)$$

Figure 4(a) shows an example of simploidal set. In the previous definition, first equation (1) denotes the action of a border operator on the simploid type. The cartesian product of a simploid s by a simploid of type (0) (i.e. a vertex) is the identity. Hence, if zero appears in the type of a simploid by the application of a boundary operator, it is removed from the type. With equation (2), the commutation relation for semi-simplicial set boundary operators is retrieved. Finally, equation (3) is the commutation relation when two boundary operators are successively applied to two different simplices. The second part of this equation allows us to take into account the shifts that are produced by suppressed zeros.

For example, if we apply the sequence of boundary operators $\epsilon^3_0\epsilon^2_1$ to a simploid of type $(2, 1, 1)$, we obtain first a simploid $(2, 1)$, due to the application of ϵ^3_0 and after a simploid (2) by the application of ϵ^2_1. In an other way, if we start by the application of ϵ^2_1, a simploid $(2, 1)$ is obtain since the zero that appears in the middle of the type is removed. Hence, we cannot apply operator ϵ^3_0. The applied operator is ϵ^2_0, so we get $(2, 1, 1)\epsilon^3_0\epsilon^2_1 = (2, 1, 1)\epsilon^2_1\epsilon^2_0$.

3.2 Simploidal Chain, Boundary Homomorphism and Free Chain Complex

In order to define simploidal homology, we have to associate a free chain complex to a simploidal set. Let $S = (K, (\epsilon^i_j))$ be a simploidal set: a simploidal $p-$chain is a combination of simploids of K^p with integer coefficients. Now, to define boundary homomorphisms $\partial^\square$ for simploidal sets, we extend the general boundary formula of a cell-product: $(a \times b)\partial = a\partial \times b + (-1)^{dim(a)} a \times b\partial$.

Definition 4. *Let s be a simploid of type $(a_1, \cdots, a_k)$.*

$$s\partial^\square = \begin{cases} 0 & \text{if } s = () \\ \sum_{i=1}^{k} \sum_{j=0}^{a_i} (-1)^{j + \sum_{l=1}^{i-1} a_l} s\epsilon^i_j & \text{otherwise} \end{cases}$$

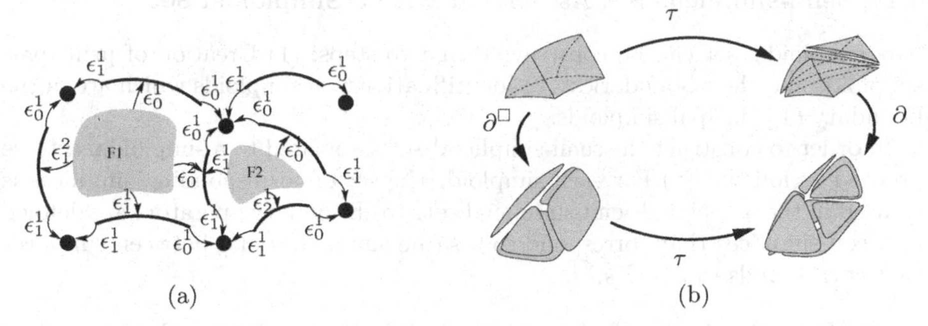

Fig. 4. (*a*) an example of simploidal set of dimension 2. (*b*) a commutative diagram which illustrates the property $c\tau\partial = c\partial^{\square}\tau$.

For example on Fig. 4(a): $F_1\partial^{\square} = (F_1\epsilon_0^1 - F_1\epsilon_1^1) - (F_1\epsilon_0^2 - F_1\epsilon_1^2)$; $F_2\partial^{\square} = F_2\epsilon_0^1 - F_2\epsilon_1^1 + F_2\epsilon_2^1$. Definition 4 is extended by linearity for simploidal chains.

To prove that $c\partial^{\square}\partial^{\square} = 0$ for any simploidal chain c (i.e. $\partial^{\square}$ are boundary homomorphisms), we prove this property for a simploidal chain containing a single simploid s (linearity ensure extension for a general chain). Then, we prove that definition 4 satisfies the general boundary formula of a cell-product[7] and conclusion follows. So, we can associate a free chain complex to a simploidal set. Now we are able to compute the homology groups for such a set, using for example the SMITH normal form transformation for incidence matrices [4].

4 Conversion Between Simploidal and Semi-simplicial Sets

As we will see in this section, it is always possible to associate a semi-simplicial set with a simploidal set. And, it is well known in algebraic topology that the homology of a triangulable space does not depend on its triangulation [4]. So we can directly conclude that simploidal homology as defined in section 3 is equivalent to simplicial homology.

In this section, we study conversions between semi-simplicial and simploidal sets. We define operator $\mathfrak{T}$ which associates to each simploid a set of simplices in the associated semi-simplicial set. We also define operator τ, which associates to each $p-$simploidal chain a $p-$simplicial chain in the associated semi-simplicial set. Operator τ preserves the boundary i.e. for any simploidal chain c, $c\tau\partial = c\partial^{\square}\tau$ (see Fig. 4(b)).

Finally, we provide algorithms for converting a simploidal chain into a simplicial chain and conversely. So we can associate a simplicial generator with each simploidal homology generator and conversely. Incidentally, we get a direct and constructive proof of the equivalence between simploidal and simplicial homology.

[7] This can be directly be proved using a recursion over the length of a simploid.

4.1 Semi-simplicial Set Associated with a Simploidal Set

Any simploidal set can be constructed in two steps: (1) Creation of principal[8] simploids and their boundaries. (2) Identifications[9] of simploids which are in the boundary of principal simploids.

In order to construct the semi-simplicial set associated to a simploidal set, we proceed as follow : (a) For each simploid, the set of corresponding simplices is created in the associated semi-simplicial set. (b) Boundary operators are defined between simplices that correspond to a same simploid s and between simplices that corresponds to s and $s\partial^{\Box}$.

Cartesian Product of Semi-simplicial Sets; Triangulation of a Simploid. We recall some notions related to the cartesian product. The definition is not provided, since it is rather long and it will not be used here. Actually, the cartesian product operation is defined on simplicial sets, which extend semi-simplicial sets by adding a second class of operators (degeneracy operators), which induces a second type of simplices (degenerate simplices, see [5]). The general definition of simplicial sets makes possible to define cartesian product in a very simple way. From which an equivalent definition of cartesian product which acts directly upon semi-simplicial sets [9] is deduced. The basic principle is the following : the cartesian product of two simplices is made of simplices (maybe having different dimensions), which can be identified by integer sequences (these integer sequences correspond to sequences of degeneracy operators). Boundary operators can also be deduced from these integer sequences and relations with boundary operators of the initial simplices.

In practice, the product of $l-$simplex σ and $m-$simplex μ, such that $l \geq m$, is a set of simplices of dimensions l to $l+m$, where the set of $r-$dimensional simplices ($l \leq r \leq l+m$) corresponds to the set of simplices denoted $(\sigma I, \mu J)$, for all disjoint sequences I and J such that :

- $I = (i_1 \cdots i_{r-l})$, $J = (j_1 \cdots j_{r-m})$
- $0 \leq i_1 < \ldots < i_{r-l} \leq r-1$, $0 \leq j_1 < \ldots < j_{r-m} \leq r-1$

For instance Fig. 5(c) illustrates the cartesian product of the two semi-simplicial sets (a).

Then it is possible to define the set of simplices associated with a simploid of length 2. We can extend this definition for any simploid $s = \sigma_1 \times \cdots \sigma_n$. The set of associated simplices is denoted $s\mathfrak{T}$. From $s\mathfrak{T}$ the set of simplices of dimension $d = a_1 + \cdots + a_n$, is $sT = \{((\cdots((\sigma_1 I_1, \sigma_2 \overline{I}_1)I_2, \sigma_3 \overline{I}_2)\cdots)I_{n-1}, \sigma_n \overline{I}_{n-1})\}$, where:

$I_i \in E^{a_{i+1}, a_1 + \cdots + a_{i+1}}, 1 \leq i \leq n-1$,
$E^{p,n}$ is the set of strictly increasing integer sequences of p integers range 0 to $n-1$,
If I is an element of $E^{p,n}$, we denote $\overline{I}$ the sequence of $E^{n-p,n}$ such that $I \cap \overline{I} = \emptyset$.

[8] A simploid is a main simploid if it is not in the boundary of another simploid.
[9] Intuitively, identifying two simploids consists in merging them.

For instance, let σ_1 (reps. σ_2, σ_3) be a 1−simplex (resp. 3−simplex, 2−simplex). The set of 6−simplices of triangulation of $\sigma_1 \times \sigma_2 \times \sigma_3$ is $\{((\sigma_1 I_1, \sigma_2 \overline{I}_1)I_2, \sigma_3 \overline{I}_2)\}$ where :

- $I_1 \in \{012, 013, 023, 123\}$, $\overline{I}_1 \in \{3, 2, 1, 0\}$,
- $I_2 \in \{01, 02, 03, 04, 05, 12, 13, 14, 15, 23, 24, 25, 34, 35, 45\}$,
- $\overline{I}_2 \in \{2345, 1345, 1245, 1235, 1234, 0345, 0245, 0235, 0234, 0145, 0135, 0134, 0125, 0124, 0123\}$,

4.2 Morphism τ Between Simploidal Chain Groups and Simplicial Chain Groups

In this section, we define a morphism τ between simploid chains and simplicial chains of the associated semi-simplicial set. We prove that τ commutes with boundary homomorphisms.

Definition of Morphism τ. The simplicial chain associated to a simploid s by τ is composed of simplices from sT taking into account their orientation such that for any simploidal chain $c^{\square}$: $c^{\square}\tau\partial = c^{\square}\partial^{\square}\tau$ (cf. Fig. 4(b)). The set of p−simplices associated to a p−simploid s is sT. To define τ, we assign a sign for each simplex of $s\tau$ such that its boundary; (1) does not contain "internal" simplices. (2) has an orientation which corresponds to the orientation of simploids of the boundary of s.

For example on Fig. 5, semi-simplicial set (c) is equivalent to simploidal set (b). The boundary of $s = \sigma \times \mu$ is $a_1 + a_4 - a_2 - a_3$. In the associated semi-simplicial set, we know that the chain corresponding to s is composed of 2−simplices $(\sigma 0, \mu 1)$ and $(\sigma 1, \mu 0)$. The unique chain composed of these two simplices that does not contain the internal edge a_5 in its boundary and such that its boundary corresponds to the boundary of s is $(\sigma 1, \mu 0) - (\sigma 0, \mu 1)$.

More generally, for a p−simploid s corresponding to the product of two simplices σ_1 and σ_2, we know that internal $(p-1)$−simplices of $s\mathfrak{T}$ are in the

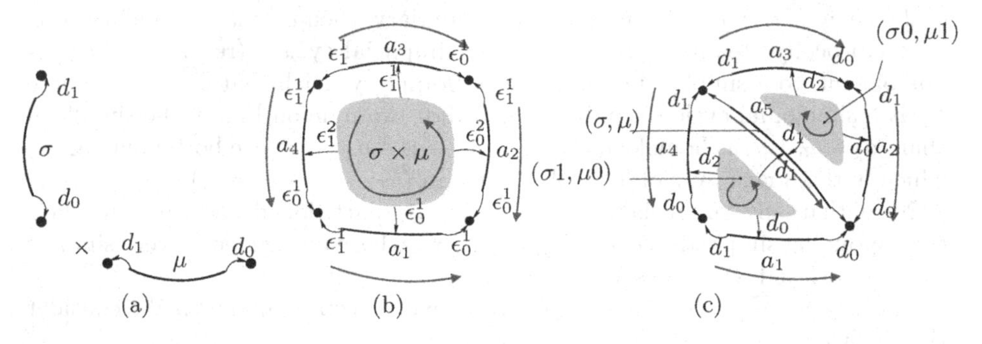

Fig. 5. (a) two simploidal sets. (b) simploidal cartesian product of (a). (c) simplicial cartesian product of (a).

boundary of two simplices $(\sigma_1 I, \sigma_2 \overline{I})$ and $(\sigma_1 I', \sigma_2 \overline{I'})$ such that there exists $i \in I, i+1 \in \overline{I}$. $(I', \overline{I'})$ is obtained from $(I, \overline{I})$ by switching i and $i+1$ (cf. [9]). Thus, we have: $(\sigma_1 I, \sigma_2 \overline{I})d_{i+1} = (\sigma_1 I', \sigma_2 \overline{I'})d_{i+1}$. For example, on Fig. 5, the two 2−simplexes of (*b*), $(\sigma 0, \mu 1)$ and $(\sigma 1, \mu 0)$ have the common internal face : $(\sigma 0, \mu 1)d_1 = (\sigma 1, \mu 0)d_1 = (\sigma, \mu) = a_5$.

As each internal simplex must vanish in the boundary of a chain $s\tau$, two simplices that have a common internal face must have opposite signs. Each sign can be deduced from the parity of integer sequences that defined simplices [10], and the sign of all simploids can be deduced from simplex,

$$((\cdots((\sigma_1 \overline{J}_{a_1}, \sigma_2 J_{a_1})\overline{J}_{a_1+a_2}, \sigma_3 J_{a_1+a_2})\cdots)\overline{J}_{a_1+\cdots+a_{k-1}}, \sigma_k J_{a_1+\cdots+a_{k-1}})$$

where $J_m = 0 \cdots m-1$. So we get the following definition :

Definition 5. *Let $s = \sigma_1 \times \cdots \times \sigma_k$ a simploid of length k and dimension $a_1 + \cdots + a_k$.*

$$s\tau = \sum_{\overline{I}_1 \cdots \overline{I}_{k-1}} (-1)^{A(\overline{I}_1, \cdots, \overline{I}_{k-1})}((\cdots((\sigma_1 I_1, \sigma_2 \overline{I}_1)I_2, \sigma_3 \overline{I}_2)\cdots)I_{k-1}, \sigma_k \overline{I}_{k-1})$$

where:

- $A(\overline{I}_1, \cdots, \overline{I}_{k-1}) = p(J_{a_1}) + p(\overline{I}_1)) + \cdots + p(J_{a_1+\cdots+a_{k-1}}) + p(\overline{I}_{k-1})$
- *for any integer sequence I, $p(I)$ is the parity of the sum of elements of I.*

So we get the commutation property such that for a simploid s, $s\tau\partial = s\partial^{\square}\tau$. The proof is not provided here since it is direct and long (see [10]).

Conversion Between Simploidal and Simplicial Generators. We introduce the following notations: Let $S^{\square}$ be a simploidal set and let S be its associated semi-simplicial set. $C^{\square}, Z^{\square}, B^{\square}$ et $H^{\square}$ (resp. C, Z, B, H) denote chain group, cycle group, boundary group and homology group of $S^{\square}$ (resp. S). The previous commutation property ensures that τ preserves cycles and boundaries, i.e. we use this property to prove that for any simploidal chain $c^{\square}$, if $c^{\square}$ is a cycle, then $c^{\square}\tau$ is a cycle and if $c^{\square}$ is a boundary, then $c^{\square}\tau$ is a boundary too.

Reciprocally, it can be proved that any simplicial cycle z (resp. boundary) is homologous to a simplicial cycle z' (resp. boundary) such that $z^{\square}\tau = z'$, where $z^{\square}$ is a simploidal cycle (resp. boundary). For example on Fig. 5, the simplicial chain $a_1 - a_5 + a_4$ is homologous to $a_1 - a_2 - a_3 + a_4$ (they are both boundaries) which is the image by τ of a simploidal chain $(a_1 - a_2 - a_3 + a_4)$.

Note that any simplicial p−chain c can be partitioned according to their corresponding simploids, i.e. $c = \sum_i \sum_j \alpha_{ij}\sigma_{ij}$ where for a given i, every simplex σ_{ij} is associated with the same simploid s_i.

Let $z = \sum_i \sum_j \alpha_{ij}\sigma_{ij}$ be a simplicial p−cycle (resp. boundary). We consider the following two cases :

• **Case (1):** For all i, s_i is a p−simploid. In this case, we can directly prove that there exists a simploidal cycle (resp. boundary) $z^{\square} = \sum_i \gamma_i s_i$ such that $z^{\square}\tau = z$; else z contains simplices which are internal to a simploid, and z is not a cycle (resp. boundary).
• **Case (2):** There exists k such that s_k is a n−simploid, $n > p$. Let c_k be the subchain of z corresponding to simploid s_k. The boundary of c_k must be in the boundary of s_k (since we consider only cycles). In this case, we propose an algorithm that constructs a simplicial p−chain c'_k homologous to c_k having the same boundary, such that each simplex of c'_k comes from a m− simploid, $m < n$. As c'_k is homologous and have the same boundary as c_k, by replacing c_k by c'_k in the expression of z we don't change the homology class of z. This operation is repeated until all simplices belong to p−simploids (corresponding to case (1)).

We do not provide here completely this algorithm as it is rather technical. The principle is to use an ordering of simplices [10] for replacing each p−simplex (in $s_k\mathfrak{T}$) of the current chain c_k by its complementary in the boundary of a $(p+1)$−simplex of $s_k\mathfrak{T}$.

For example Fig. 6(a) represents a subchain c_k which is a part of a 2−cycle. The two triangles of c_k come from the triangulation of the cube, which is a 3−simploid. (b) illustrates a chain c'_k, homologous to c_k and with same boundary. Each triangle of c'_k comes from the triangulations of simploids on the boundary of the cube.

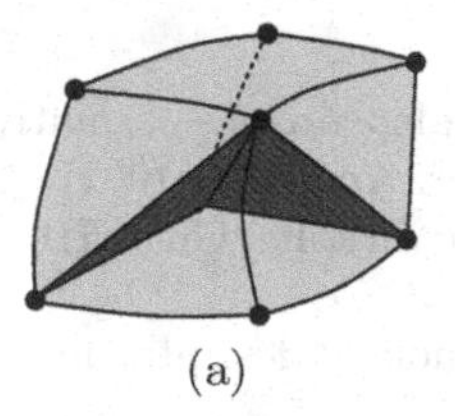

(a)

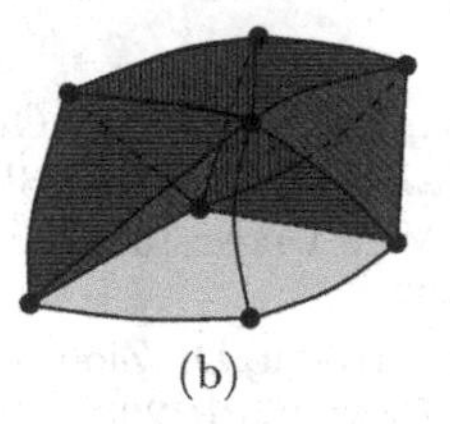

(b)

Fig. 6. (a) a 2−chain which comes from the cube (of dimension 3). (b) an homologous 2−chain which has same boundary as. Each resulting simplex of comes from simploids of the boundary of the cube.

As a conclusion, we can associate to each simploidal cycle (resp. boundary) a simplicial cycle (resp. boundary) using τ. Reciprocally, any simplicial cycle z (resp. boundary) can be transformed into an homologous cycle (resp. boundary) z' such that z' is the image of a simploidal cycle (resp. boundary). So we are able to convert any generator of a simploidal set into a generator on the associated semi-simplicial set and conversely. Incidentally, this provide a purely combinatorial and constructive proof of the equivalence between simploidal and simplicial homology.

---

[10] This ordering is based on the properties of integer sequences that define simplices on a semi-simplicial set associated to a simploidal set.

5 Conclusion

In this paper, notions of chains, cycles and boundaries have been extended to simploids. We defined boundary homomorphisms $\partial^\square$ for simploidal sets (these homomorphisms are directly defined using boundary operators ϵ_j^i): we also proved that $c\partial^\square\partial^\square = 0$ for any simploidal chain c. So we show how to associate a free chain complex to a simploidal set and thus we defined an homology for simploidal sets.

We provided algorithms for converting a simploidal set into an equivalent semi-simplicial set. Then we provided algorithms for converting simploidal homology generators into simplicial ones and reciprocally. We thus provide a purely combinatorial and constructive proof of the equivalence between simplicial and simploidal homology.

Now we want to study the adaptation to simploidal sets of existing algorithms initially defined for computing homology of simplicial structures [7]. We hope that this will lead to interesting results in terms of memory occupation and efficiency. To do so, we need to develop comparisons with, as far as the authors know, the lone studies for non simplicial complexes [6].

From a practical point of view, we are also interested to experiment these algorithms for images of dimension greater than or equal to 3 (voxel images, sequences of 3D images).

References

1. Dahmen, W., Micchelli, C.A.: On the linear independence of multivariate b-splines I. Triangulation of simploids. SIAM J. Numer. Anal. **19** (1982)
2. Moore, D.: V.10 Understanding simploids. In: Graphic Gems III. Academic Press (1992) 250–255
3. Allili, M., Corriveau, D., Ziou, D.: Morse homology desriptor for shape characterization. In: Proc. ICPR 2004. (2004)
4. Munkres, J.R.: Elements of algebraic topology. Perseus Books (1984)
5. Lang, V., Lienhardt, P.: Geometric modeling with simplicial sets. In: Proc. of Pacific Graphics'95, Seoul, Korea (1995)
6. Kaczynski, T., Mischaikow, K., Mrozek, M.: Computational Homology. Springer (2004)
7. Peltier, S., Alayrangues, S., Fuchs, L., Lachaud, J.O.: Computation of homology groups and generators. In: Proceedings of 12th International Conference of Discrete Geometry for Computer Imagery. Volume 3429. (2005) 195–205
8. Fuchs, L., Lienhardt, P.: Topological structures and free-form spaces. In: Journes franco-espagnoles de gomtrie algorithmique. (1997) 35–46
9. Lienhardt, P., Skapin, X., Bergey, A.: Cartesian product of simplicial and cellular structures. Int. Journal of Computational Geometry and Applications **14**(3) (2004) 115–159
10. Peltier, S.: Calcul de groupes d'homologie sur des structures simpliciales, simploidales et cellulaires. PhD thesis, Université de Poitiers (2006) To appear.

Measuring Intrinsic Volumes in Digital 3d Images

Katja Schladitz[1], Joachim Ohser[2], and Werner Nagel[3]

[1] Fraunhofer-Institut für Techno- und Wirtschaftsmathematik
Fraunhofer-Platz 1
67663 Kaiserslautern, Germany
katja.schladitz@itwm.fraunhofer.de

[2] Fachhochschule Darmstadt
Schöfferstraße 3
64295 Darmstadt, Germany
ohser@fh-darmstadt.de

[3] Friedrich-Schiller-Universität
Fakultät für Mathematik und Informatik
07740 Jena, Germany
nagel@minet.uni-jena.de

Abstract. The intrinsic volumes – in 3d up to constants volume, surface area, integral of mean curvature, and Euler number – are a very useful set of geometric characteristics. Combining integral and digital geometry we develop a method for efficient simultanous calculation of the intrinsic volumes of sets observed in binary images. In order to achieve consistency in the derived intrinsic volumes for both foreground and background, suitable pairs of discrete connectivities have to be used. To make this rigorous, the concepts discretization w.r.t. an adjacency system and complementarity of adjacency systems are introduced.

1 Introduction

With the fast development of new materials like foams or fiber reinforced composites there is a growing need for non-destructive testing and structure characterization. In particular, computer tomography is now able to produce high quality 3d images of very fine structures, yielding the demand for subsequent quantitative analysis.

In many applications, geometric characteristics of the whole structure have to be measured from the given image. A very attractive set of geometric characteristics are the intrinsic volumes (or quermassintegrals or Minkowski functionals). In 3d, they are, up to constants, volume, surface area, integral of mean curvature, and Euler number. For fibrous structures, the integral of mean curvature yields the total fiber length without need to segment individual fibers.

The Crofton formulae boil down computing the intrinsic volumes to computing Euler numbers in lower dimensional intersections. Discretization of these formulae (see Section 4) combined with an efficient calculation of the Euler numbers in the intersections yield a fast algorithm for simultaneously determining

A. Kuba, L.G. Nyúl, and K. Palágyi (Eds.): DGCI 2006, LNCS 4245, pp. 247–258, 2006.

the intrinsic volumes based on observations in a digital image. The backbone of the Euler number calculation are thorough investigations of digital connectivity and consistency from [1,2,3], summarized in Section 3.

The consistency results for the Euler number of foreground and background established in [2] carry over to all intrinsic volumes. A remarkable observation in this context is that the 18-neighborhood can not be used for consistent estimation of the intrinsic volumes. This counteracts the fact that (18,6) is considered to be a 'good pair' in digital topology, see e.g. [4, Chapter 7].

In 3d, the presented method's algorithmic core consists in a convolution of the binary image with a $2\times2\times2$ mask, resulting in an 8 bit gray value image [5]. All further steps are based solely on the gray value histogram whose size does not depend on image size or content. Thus the advantage over other methods for computing the intrinsic volumes [6,7] are simplicity and speed of the algorithm. The surface area is measured directly from the binary volume image without need to approximate the surface. Similar methods are [8] and [9], see Section 6 for a comparison.

2 Section Lattices and Translative Complements

In this paper, we restrict ourselves to the three-dimensional cubic primitive lattice $\mathbb{L}^3 = a\mathbb{Z}^3$, $a > 0$, where $\mathbb{Z}$ is the set of integers. For a more general approach see [10]. Let u_1, u_2, u_3 denote the standard unit vector basis of the three-dimensional Euclidean space $\mathbb{R}^3$. The closed unit cell of $\mathbb{L}^3$ is the Minkowski sum $C = [0, au_1] + [0, au_2] + [0, au_3]$. Its volume is $\operatorname{vol} C = a^3$. We denote by $\mathcal{F}^0$ the set of vertices of a polyhedron, in particular $\mathcal{F}^0(C) = \{0, a\}^3$. The set of all lattice cells covers $\mathbb{R}^3 = \bigcup_{x\in\mathbb{L}^3}(C + x)$.

The Crofton formulae for computing the intrinsic volumes of a set $X \subset \mathbb{R}^3$ use section profiles of X on affine subspaces of $\mathbb{R}^3$. In order to obtain a digitized version, we introduce section lattices of $\mathbb{L}^3$ and their translative complements in analogy to linear subspaces and their orthogonal complements:

Definition 1. *A pair* $\left(\mathbb{L}^k, {}^T\mathbb{L}^{3-k}\right)$, $k = 1, 2$, *is called a* k-dimensional section lattice $\mathbb{L}^k$ *equipped with the* translative complement ${}^T\mathbb{L}^{3-k}$, *if there exists a basis* v_1, v_2, v_3 *of* $\mathbb{L}^3$ *with*

(i) $\mathbb{L}^k = (v_1, v_k)\mathbb{Z}^k$,
(ii) ${}^T\mathbb{L}^{3-k} = (v_{k+1}, v_3)\mathbb{Z}^{3-k}$,
(iii) *there is an* $x \in \mathcal{F}^0(\check{C})$ *with* $\{v_1, v_k\} \subset \mathcal{F}^0(C + x)$ *where* $\check{C}$ *is the reflection of* C *at the origin,* $\check{C} = -C$.

Condition (iii) ensures that integration over 'local knowledge' on the image data is possible as needed later. The translative complement ${}^T\mathbb{L}^{3-k}$ has properties similar to those of the orthogonal complement of a linear subspace:

1. $\mathbb{L}^k \cap {}^T\mathbb{L}^{3-k} = 0$,
2. If $x_1, x_2 \in {}^T\mathbb{L}^{3-k}$, $x_1 \neq x_2$, then $(\mathbb{L}^k + x_1) \cap (\mathbb{L}^k + x_2) = \emptyset$,
3. $\mathbb{L}^3 = \bigcup_{x\in{}^T\mathbb{L}^{3-k}}(\mathbb{L}^k + x)$.

However, the translative complement is not necessarily uniquely determined. Nevertheless, choosing one of the translative complements arbitrarily turns out to work for all considerations presented in the following. For $\mathbb{L}^3$, there are 13 section lattices $\mathbb{L}_i^k$ for both $k = 1$ and $k = 2$. This restriction is due to condition (iii).

Table 1. The bases of the 13 section lattices $\mathbb{L}_i^k$ of $\mathbb{L}^3 = \mathbb{Z}^3$ and a possible translative complement ${}^T\mathbb{L}_i^k$ for $k = 1$ (left), and $k = 2$ (right)

i	basis of $\mathbb{L}_i^1$	basis of ${}^T\mathbb{L}_i^2$
1	$\{u_1\}$	$\{u_2, u_3\}$
2	$\{u_2\}$	$\{u_1, u_3\}$
3	$\{u_3\}$	$\{u_1, u_2\}$
4	$\{u_1 + u_2\}$	$\{u_1, u_3\}$
5	$\{-u_1 + u_2\}$	$\{u_1, u_3\}$
6	$\{u_1 + u_3\}$	$\{u_1, u_2\}$
7	$\{-u_1 + u_3\}$	$\{u_1, u_2\}$
8	$\{u_2 + u_3\}$	$\{u_1, u_3\}$
9	$\{-u_2 + u_3\}$	$\{u_1, u_3\}$
10	$\{-u_1 + u_2 + u_3\}$	$\{u_1, u_2\}$
11	$\{-u_1 + u_2 + u_3\}$	$\{u_1, u_2\}$
12	$\{u_1 - u_2 + u_3\}$	$\{u_1, u_2\}$
13	$\{u_1 + u_2 - u_3\}$	$\{u_1, u_2\}$

i	basis of $\mathbb{L}_i^2$	basis of ${}^T\mathbb{L}_i^1$
1	$\{u_1, u_2\}$	$\{u_3\}$
2	$\{u_1, u_3\}$	$\{u_2\}$
3	$\{u_2, u_3\}$	$\{u_1\}$
4	$\{u_1, u_2 + u_3\}$	$\{u_3\}$
5	$\{u_1, -u_2 + u_3\}$	$\{u_3\}$
6	$\{u_2, u_1 + u_3\}$	$\{u_3\}$
7	$\{u_2, -u_1 + u_3\}$	$\{u_3\}$
8	$\{u_3, u_1 + u_2\}$	$\{u_1\}$
9	$\{u_3, -u_1 + u_2\}$	$\{u_1\}$
10	$\{u_1 + u_3, u_2 + u_3\}$	$\{u_3\}$
11	$\{-u_1 + u_3, u_2 + u_3\}$	$\{u_3\}$
12	$\{-u_1 + u_3, -u_2 + u_3\}$	$\{u_3\}$
13	$\{u_1 + u_3, -u_2 + u_3\}$	$\{u_3\}$

3 Adjacency and Euler Number

The Crofton formulae reduce the measurement of the intrinsic volumes V_{3-k} of a poly-convex set X to measuring the Euler number χ of section profiles of X in k-dimensional sections, $k = 1, 2$. Thus it is essential to know how the Euler number $\chi(X \cap (L + y))$ can be measured when instead of $X \cap (L + y)$ only the observation $X \cap (\mathbb{L}^k + y)$ on a translated section lattice is available with $L = \operatorname{span} \mathbb{L}^k$ and $y \in {}^T\mathbb{L}^{3-k}$.

The problem of measuring the Euler number based on images was considered by several authors [11, 12, 6, 7]. Here we apply the concept of adjacency systems from [1, 2, 3]. Good pairs of adjacencies for foreground and background allow consistent calculation of the Euler number, see 3.3. It turns out that this condition differs from the usually demanded Jordan surface theorem as the 18-neighborhood can not be used for consistent calculation of the Euler number. Note that for dimensions three and higher, a complete description of good adjacencies is not yet known.

3.1 Discretization with Respect to an Adjacency System

Let $\mathbb{L}^k$ be a (section) lattice with the basis v_1, v_k and the unit cell C^k. The vertices $x_j = \sum_{i=1}^k \lambda_i v_i$ of C^k are indexed with $j = \sum_{i=1}^k 2^{i-1}\lambda_i$, $\lambda_i \in \{0, 1\}$. Clearly, the unit cell C^k has 2^k vertices $x_j \in \mathcal{F}^0(C^k)$, $j = 0, \ldots, 2^k - 1$. Analogously, we introduce the index of a subset $\xi \subseteq \mathcal{F}^0(C^k)$. Let $\mathbf{1}$ denote the indicator function of a set, i. e. $\mathbf{1}(x \in \xi) = 1$ if $x \in \xi$ and $\mathbf{1}(x \in \xi) = 0$ otherwise. An index ℓ is assigned to ξ, and we write ξ_ℓ if $\ell = \sum_{j=0}^{2^k-1} 2^j \mathbf{1}(x_j \in \xi)$, i. e. $\ell \in \{0, \ldots, \nu\}$ with $\nu = 2^{2^k} - 1$. Note that $\xi_0 = \emptyset$, $\xi_\nu = \mathcal{F}^0(C^k)$, and $\xi_{\nu-\ell} = \xi_\nu \setminus \xi_\ell$. The ξ_ℓ can be considered as a local pixel configuration of the foreground of a k-dimensional binary image. Finally, we introduce the convex hulls $F_\ell = \operatorname{conv} \xi_\ell$ forming convex polytopes with $F_\ell \subseteq C^k$ and $\mathcal{F}^0(F_\ell) \subseteq \mathcal{F}^0(C^k)$, $\ell = 1, \ldots, \nu$. Let $\mathcal{F}^j(F)$ denote the set of all j-dimensional faces of a convex polytope F. For a set $\mathbb{F}$ of convex polytopes write $\mathcal{F}^j(\mathbb{F}) = \cup\{\mathcal{F}^j(F) : F \in \mathbb{F}\}$ for the set of all j-faces. Now we are able to equip the lattice $\mathbb{L}^k$ with a (homogeneous) adjacency system defining the neighborhood of the lattice points.

Definition 2. *Let $\mathbb{F}_0 \subseteq \{F_0, \ldots, F_\nu\}$ be a set of convex polytopes $F_\ell = \operatorname{conv} \xi_\ell$ with the properties*

(i) $\emptyset \in \mathbb{F}_0$, $C \in \mathbb{F}_0$,
(ii) *if $F \in \mathbb{F}_0$ then $\mathcal{F}^i(F) \subset \mathbb{F}_0$ for $i = 0, \ldots, \dim F$,*
(iii) *if $F_i, F_j \in \mathbb{F}_0$ and $F_i \cup F_j$ is convex then $F_i \cup F_j \in \mathbb{F}_0$.*

Then the system $\mathbb{F}_0$ is a local adjacency system *and $\mathbb{F} = \bigcup_{x \in \mathbb{L}^k} \mathbb{F}_0 + x$ is called an* adjacency system *of the lattice $\mathbb{L}^k$.*

From condition (i) it follows immediately that $\mathcal{F}^0(\mathbb{F}) = \mathbb{L}^k$. The pair $\Gamma = (\mathcal{F}^0(\mathbb{F}), \mathcal{F}^1(\mathbb{F}))$ consisting of the set $\mathcal{F}^0(\mathbb{F})$ of nodes and the set $\mathcal{F}^1(\mathbb{F})$ of edges is said to be the neighborhood graph of $\mathbb{F}$. Due to homogeneity ($\Gamma + x = \Gamma$, $x \in \mathbb{L}^k$), all nodes have the same valence – the connectivity of $\mathbb{L}^k$. Note that for $n > 2$ there can be two or more adjacency systems having the same neighborhood graph. In other words, an adjacency system $\mathbb{F}$ is not uniquely determined by Γ.

Examples of Adjacency Systems for $k = 3$:

The 6-adjacency is generated from the unit cell C^3, $\mathbb{F}_0 = \cup_{j=0}^{3} \mathcal{F}^j(C^3)$.

The 14.1-adjacency is generated from the tessellation of C^3 into the 6 tetrahedra F_{139}, F_{141}, F_{163}, F_{177}, F_{197}, F_{209} which are the convex hulls of the configurations

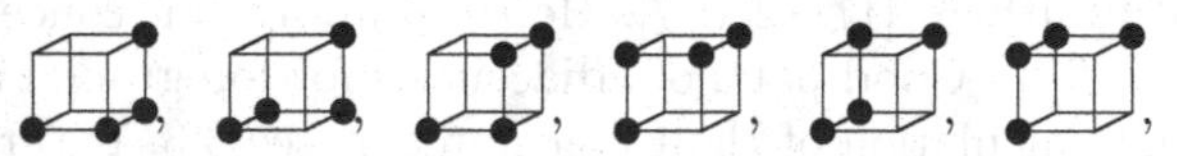

i. e. $\mathbb{F}_0$ consists of all j-faces of the tetrahedra, $j = 0, \ldots, 3$, and their convex unions. The edges of the corresponding neighborhood graph Γ are the edges of C^3, the face diagonals of C^3 containing the origin 0, the space diagonal of C^3 containing 0, and all their lattice translations. The degree of Γ is 14.

The 14.2-adjacency is generated from the tetrahedra F_{43}, F_{141}, F_{147}, F_{169}, F_{177}, and F_{212} which are the convex hulls of

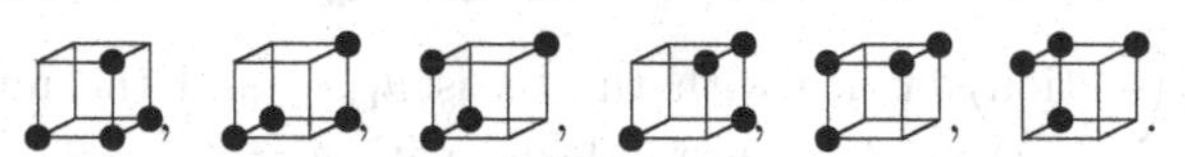

The corresponding neighborhood graph Γ differs from that one for 14.1 in the choice of one face diagonal of C^3 not containing 0.

The 26-adjacency is given by $\mathbb{F}_0 = \{F_0, \ldots, F_{255}\}$.

Definition 3. *The* discretization $X \sqcap \mathbb{F}$ *of a compact set* $X \subset \mathbb{R}^3$ *with respect to a given adjacency system* $\mathbb{F}$ *is defined as the union of all* j*-faces of the elements of* $\mathbb{F}$ *for which all the vertices hit* X, *i. e.*

$$X \sqcap \mathbb{F} = \cup\{F \in \mathbb{F} : \mathcal{F}^0(F) \subseteq X\}. \tag{1}$$

This means that a 'brick' $F \in \mathbb{F}$ is a subset of the discretization of X if and only if all vertices of F belong to X. The discretization is obtained from the observation of the set on the lattice, i.e. $X \sqcap \mathbb{F} = (X \cap \mathbb{L}^3) \sqcap \mathbb{F}$.

3.2 Euler Number

Since $X \sqcap \mathbb{F}$ forms a (not necessarily convex) polyhedron on span $\mathbb{L}^k$, the number of elements of $\mathcal{F}^j(X \sqcap \mathbb{F})$ is finite. Therefore, the Euler number $\chi(X \sqcap \mathbb{F})$ can be computed via the Euler-Poincaré formula,

$$\chi(X \sqcap \mathbb{F}) = \sum_{j=0}^{k} (-1)^j \, \#\mathcal{F}^j(X \sqcap \mathbb{F}).$$

In order to apply a 'local method' for measuring the intrinsic volumes we deduce now a local version. The discretization of a local configuration $\xi_\ell = X \cap C^k \cap \mathbb{L}^k$ of $X \cap \mathbb{L}^k$ is $\xi_\ell \sqcap \mathbb{F} = (X \sqcap \mathbb{F}) \cap C^k = X \sqcap \mathbb{F}_0$. We compute weights for the edge correction using $\kappa F = \min\{j :$ there is a $G \in \mathcal{F}^j(C^k)$ with $F \subseteq G\}$. Now define the edge-corrected localization χ_0 of χ as

$$\chi_0(\xi_\ell \sqcap \mathbb{F}) := \sum_{j=0}^{k} (-1)^j \sum_{F \in \mathcal{F}^j(\xi_\ell \sqcap \mathbb{F})} 2^{\kappa F - k}, \quad \ell = 0, \ldots, \nu. \tag{2}$$

Then, additivity and translation invariance of the Euler number and the fact that $X \sqcap \mathbb{F} = (X \cap \mathbb{L}^k) \sqcap \mathbb{F}$ yield

$$\begin{aligned} \chi(X \sqcap \mathbb{F}) &= \sum_{x \in \mathbb{L}^k} \chi_0(C^k \cap ((X \sqcap \mathbb{F}) - x)) \\ &= \sum_{x \in \mathbb{L}^k} \sum_{\ell=0}^{\nu} \chi_0(\xi_\ell \sqcap \mathbb{F}) \mathbf{1}(\xi_\ell + x \subseteq X) \mathbf{1}(\xi_{\nu-\ell} + x \subseteq X^c) \\ &= \sum_{\ell=0}^{\nu} \chi_0(\xi_\ell \sqcap \mathbb{F}) \underbrace{\sum_{x \in \mathbb{L}^k} \mathbf{1}(\xi_\ell + x \subseteq X) \mathbf{1}(\xi_{\nu-\ell} + x \subseteq X^c)}_{=: h_\ell}. \end{aligned} \tag{3}$$

Thus the Euler number can be written as a scalar product, $\chi(X \sqcap \mathbb{F}) = wh$, where the components $w_\ell = \chi_0(\xi_\ell \sqcap \mathbb{F})$ of the vector $w = (w_\ell)$ depend on $\mathbb{F}$, but they are independent of X. On the other hand, the vector $h = (h_\ell)$ is independent of $\mathbb{F}$; its components h_ℓ can be computed very efficiently from 'local information' about $X \cap \mathbb{L}^3$.

3.3 Complementarity

It is well-known that choosing an adjacency system $\mathbb{F}$ for the discretization of X implies an $\mathbb{F}_c$ for the discretization of the complementary set X^c. In other words, if the 'foreground' $X \cap \mathbb{L}^k$ is connected with respect to $\mathbb{F}$ then the 'background' $\overline{X^c} \cap \mathbb{L}^k$ must be connected with respect to a complementary adjacency $\mathbb{F}_c$. The usual criterion for complementarity is the Jordan surface theorem (Jordan-Brouwer theorem), see e. g. [13]. However, when aiming at computation of the intrinsic volumes, another criterion seems to be more appropriate: In the continuous case the consistency relation

$$\chi(X) = (-1)^{k+1} \chi\left(\overline{X^c}\right)$$

holds for all compact, poly-convex and topologically regular sets $X \subset \operatorname{span} \mathbb{L}^k$, see [2]. In the discrete case this leads to:

Definition 4. *The pair* $(\mathbb{F}, \mathbb{F}_c)$ *is called a pair of* complementary adjacency systems *if*

$$\chi(X \sqcap \mathbb{F}) = (-1)^{k+1} \chi(X^c \sqcap \mathbb{F}_c)$$

for all compact $X \subset \mathbb{R}^3$. *An adjacency system* $\mathbb{F}$ *is called* self-complementary *if* $\chi(X \sqcap \mathbb{F}) = (-1)^{k+1} \chi(X^c \sqcap \mathbb{F})$ *for all compact* X.

For a given adjacency system $\mathbb{F}$, existence of a complementary adjacency system $\mathbb{F}_c$ is not guaranteed. Even worse, until now, there is no constructive way to find the complementary system $\mathbb{F}_c$. However, most known 'good' pairs of adjacencies w.r.t. Jordan curve or surface theorems are complementary in the sense of our definition, too. Complementarity of adjacency systems can be checked the following way: Let h be defined as in (3). Then for $h^c = (h^c_\ell)$ with $h^c_\ell = \sum_{x \in \mathbb{L}^k} \mathbf{1}(\xi_\ell + x \subseteq X^c) \mathbf{1}(\xi_{\nu-\ell} + x \subseteq X)$ we obtain the relationship $h_\ell = h^c_{\nu-\ell}$, $\ell = 0, \ldots, \nu$. Using (3) and Definition 4 one can easily prove

Lemma 1. *Let* $\mathbb{F}$ *and* $\mathbb{F}_c$ *be two adjacency systems and let* w *and* w^c *be the vectors with the coefficients* $w_\ell = \chi_0(\xi_\ell \sqcap \mathbb{F})$ *resp.* $w^c_\ell = \chi_0(\xi_\ell \sqcap \mathbb{F}_c)$, $\ell = 0, \ldots, \nu$. *Then* $(\mathbb{F}, \mathbb{F}_c)$ *is a pair of complementary adjacency systems if and only if*

$$w_\ell = (-1)^{k+1} w^c_{\nu-\ell}, \quad \ell = 0, \ldots, \nu. \tag{4}$$

Examples for 3d: The 6-adjacency is complementary to the 26-adjacency. The 14.1- and the 14.2-adjacencies are constructed to be self-complementary, see [2,3]. However, the 18-adjacency is not complementary to the 6-adjacency although they are 'Jordan-Brouwer-complementary', see e.g. [13]. In order to see this, define $\mathbb{F}_{18}$ to be one of the 18-adjacencies with the neighborhood graph $\Gamma' = (\mathbb{L}^3, \mathcal{F}^1)$ where $\mathcal{F}^1$ consists of all edges and face diagonals of the cells of $\mathbb{L}^3$. Independent of the choice of $\mathbb{F}_{18}$ we get e. g. $\chi_0(\boxdot \sqcap \mathbb{F}_{18}) = \frac{1}{4}$ since the space diagonals of the unit cell do not belong to $\mathbb{F}_{18}$. On the other hand we get for the complementary configuration $\chi_0(\boxdot \sqcap \mathbb{F}_6) = -\frac{3}{4}$ where $\mathbb{F}_6$ is defined as in Section 3.1. Thus, for the pair $(\mathbb{F}_{18}, \mathbb{F}_6)$, the necessary condition for complementarity (4) is violated.

Remark: The example $(\mathbb{F}_{18}, \mathbb{F}_6)$ shows that complementarity in the sense of Definition 4 clearly differs from 'Jordan-Brouwer-complementarity'. So far, there is no general result about the relationship of the two conditions known. Jordan surface theorems for the two 14-adjacencies are expected to hold and subject of further research.

4 Intrinsic Volumes of Poly-Convex Sets

Consider a poly-convex set $X \subset \mathbb{R}^3$. The intrinsic volumes V_j, $j = 0, \ldots, 3$ are – up to constant factors – the volume $V(X)$, the surface area $S(X) = 2V_2(X)$, the integral of mean curvature $M(X) = \pi V_1(X)$, [14, p. 210] and the Euler number $\chi(X) = V_0(X)$. By means of the Crofton formula, the intrinsic volumes can be written as

$$\frac{1}{2} V_{3-k}(X) = \int_{\mathcal{L}^k} \underbrace{\int_{{}^\perp L} \chi(X \cap (L + y))\, d\lambda_{{}^\perp L}(y)}_{p_X^k(L)}\, d\mu(L), \quad k = 1, 2, \tag{5}$$

where $\mathcal{L}^k$ is the set of all k-dimensional linear subspaces of $\mathbb{R}^3$, ${}^\perp L$ denotes the orthogonal complement of $L \in \mathcal{L}^k$, $\lambda_{{}^\perp L}$ is the $3-k$-dimensional Lebesgue measure on ${}^\perp L$, μ denotes the rotation invariant probability measure on $\mathcal{L}^k$, $\mu(\mathcal{L}^k) = 1$.

Remark: Note that the Crofton formula is also the base of stereological formulae for the intrinsic volumes, see e.g. [15].

Here, the set X is observed in an image (a finite subset of the lattice $\mathbb{L}^3$) only. This implies that the integrand in the Crofton formulae (5) is known for only a finite number of elements of $\mathcal{L}^k$, and the translation $L + y$ is possible for discrete values of y, only. That is, both integrals in (5) are approximated by sums. Furthermore, the intersection $(X - y) \cap L$ must be replaced by its discretization $(X - y) \sqcap \mathbb{F}^k$ with respect to an adjacency system $\mathbb{F}^k$ in $\mathbb{L}^k$ where $L = \operatorname{span} \mathbb{L}^k$, and the translations y are from ${}^T\mathbb{L}^{3-k}$ instead of ${}^\perp L$, where ${}^T\mathbb{L}^{3-k}$ is a translative complement according to Definition 1.

4.1 Discretization of the Translative Integral

Let C^k and ${}^TC^{3-k}$ be the unit cells of $\mathbb{L}^k$ and ${}^T\mathbb{L}^{3-k}$, respectively. Denote by $\operatorname{proj} {}^TC^{3-k}$ the orthogonal projection of ${}^TC^{3-k}$ onto ${}^\perp L$. Its volume is $\operatorname{vol} \operatorname{proj} {}^TC^{3-k} = \operatorname{vol} C / \operatorname{vol} C^k$.

Then – in analogy to the rectangular quadrature rule – the inner integral in the Crofton formula (5) can be approximated by

$$\begin{aligned} p_X^k(L) &\approx \frac{\operatorname{vol} C}{\operatorname{vol} C^k} \sum_{y \in {}^T\mathbb{L}^{3-k}} \chi((X - y) \cap L)) \approx \frac{\operatorname{vol} C}{\operatorname{vol} C^k} \sum_{y \in {}^T\mathbb{L}^{3-k}} \chi((X - y) \sqcap \mathbb{F}^k)) \\ &= \frac{\operatorname{vol} C}{\operatorname{vol} C^k} \sum_{x \in \mathbb{L}^3} \chi_0(C^k \cap ((X - x) \sqcap \mathbb{F}^k)) := \tilde{p}_X^k(L). \end{aligned} \tag{6}$$

The volume of proj ${}^T C^{3-k}$ and thus $\tilde{p}_X^k(L)$ do not depend on the particular choice of ${}^T\mathbb{L}^{3-k}$. From the local Euler-Poincaré-Formula (2) it follows that

$$\sum_{x\in\mathbb{L}^3} \chi_0(C^k \cap ((X-x) \sqcap \mathbb{F}^k)) = wh \tag{7}$$

where the vector w corresponds to the adjacency system $\mathbb{F}^k$ and h can be computed via

$$h_\ell = \sum_{x\in\mathbb{L}^3} \mathbf{1}(\xi_\ell + x \subset X)\mathbf{1}(\xi_{\nu-\ell} + x \subset X^c), \quad \ell = 0,\dots,\nu. \tag{8}$$

Remark: For an effective algorithmic implementation it is useful to use one vector $\bar{h}$ with (7) which has to be determined just once for given set X and lattice $\mathbb{L}^3$. In order to use it for the dimensions $k = 1, 2$ and the different directions of sections as well as for the computing the Euler number in 3d according to (3), the vectors $\bar{w}$ of weights can be adapted appropriately [5,10], such that

$$\sum_{x\in\mathbb{L}^3} \chi_0(C \cap ((X-x) \sqcap \mathbb{F}^k)) = \bar{w}\bar{h}. \tag{9}$$

$\bar{h}$ can be interpreted as the gray value histogram of the gray value image obtained by convolution of the binary image with a suitable $2 \times 2 \times 2$ mask.

4.2 Discretization of the Integral over All Subspaces

As a consequence of the observation of X on $\mathbb{L}^3$ and condition (iii) in Definition 1 an approximation of p_X^k is known for only finitely many subspaces $L_i = \operatorname{span} \mathbb{L}_i^k$, $i = 1,\dots,13$, see Table 1. Hence, an approximation of $\int_{\mathcal{L}^k} p_X^k(L)\, d\mu(L)$ is needed. Applying a simple quadrature we get

$$\int_{\mathcal{L}^k} p_X^k(L)\, d\mu(L) \approx \int_{\mathcal{L}^k} \tilde{p}_X^k(L)\, d\mu(L) \approx \sum_{i=1}^{13} \gamma_i^{(k)}\, \tilde{p}_X^k(L_i) \tag{10}$$

where $\gamma_i^{(k)}$ are the weights corresponding to the applied quadrature rule. The choice of these weights is not trivial since the $L_1,\dots,L_{13}$ are not uniformly scattered in $\mathcal{L}^k$ and moreover, the measurement values $\tilde{p}_X^k(L_i)$ for the $p_X^k(L_i)$ are not of the same precision for different subspaces.

The weights $\gamma_i^{(k)}$ can be chosen as follows, see [5]: For $k = 1$ the unit sphere S^2 is divided into Voronoï cells with respect to the point field $L_1 \cap S^2,\dots,L_{13} \cap S^2$ containing 26 points. Then the weight $\gamma_i^{(1)}$ is the sum of the areas (Hausdorff measure) of the two Voronoï cells corresponding to the two points of $L_i \cap S^2$ divided by the surface area of S^2. For $k = 2$, the same is done with ${}^\perp L_i$ instead of L_i. The numerical values are $\gamma_i^{(k)} = 0.045\,778$ for $i = 1, 2, 3$; $\gamma_i^{(k)} = 0.036\,981$ for $i = 4,\dots,9$; $\gamma_i^{(k)} = 0.035\,196$ for $i = 10,\dots,13$.

Summarizing formulas (5), (6), and (10) we obtain the approximation $\hat{V}_{3-k}(X)$ of $V_{3-k}(X)$ as

$$\frac{1}{2}\hat{V}_{3-k}(X) = \sum_{i=1}^{13} \gamma_i^{(k)} \frac{\operatorname{vol} C}{\operatorname{vol} C_i^k} \sum_{x \in \mathbb{L}^3} \chi_0(C_i^k \cap ((X-x) \sqcap \mathbb{F}_i^k)), \quad k = 1,2 \tag{11}$$

where $\mathbb{F}_i^k$ is an adjacency system on the respective section lattice $\mathbb{L}_i^k$ and C^k its unit cell. Using this approximation, consistent measurement of the intrinsic volumes for both foreground and background is possible:

Theorem 1. *Let $(\mathbb{F}_i^k, \mathbb{F}_{c,i}^k)$ be pairs of complementary adjacency systems on the lattices $\mathbb{L}_i^k$, $i = 1, \ldots, 13$, where X and X^c are discretized w.r.t. the $\mathbb{F}_i^k$ and the $\mathbb{F}_{c,i}^k$, respectively. If X is compact then the following consistency relation holds:*

$$\hat{V}_{n-k}(X) = (-1)^{k+1}\hat{V}_{n-k}(X^c), \quad k = 1,2. \tag{12}$$

This follows directly from Definition 4.

From (6), the remark in Section 4.1, (7), (8), and (10) it can be seen that the two intrinsic volumes V_1 and V_2, and thus the integral of the mean curvature M and the surface area S, can be computed by scalar products, $\hat{S}(X) = v^{(1)}\bar{h}$ and $\hat{M}(X) = v^{(2)}\bar{h}$, respectively, where

$$v^{(1)} = 4\sum_{i=1}^{13} \frac{\operatorname{vol} C}{\operatorname{vol} C_i^1}\gamma_i^{(1)}\bar{w}_i^{(1)}, \quad v^{(2)} = 2\pi \sum_{i=1}^{13} \frac{\operatorname{vol} C}{\operatorname{vol} C_i^2}\gamma_i^{(2)}\bar{w}_i^{(2)},$$

and $\bar{w}_i^{(1)}$, $\bar{w}_i^{(2)}$ are the vectors corresponding to $\bar{w}$ (as in (9)) for $\mathbb{L}_i^1$ and $\mathbb{L}_i^2$, see also [5].

5 Congruence Classes of Configurations

In this section, we derive another version of (11) using congruence classes of the local pixel configurations ξ_ℓ w.r.t. rigid motions and counting $\mathbf{1}(\xi_\ell + x \subseteq X)$ instead of $\mathbf{1}(\xi_\ell + x \subseteq X)\mathbf{1}(\xi_\ell^c + x \subseteq X^c)$, comparable to [8,9].

To this end, chose a pair of complementary adjacency systems $(\mathbb{F}, \mathbb{F}_c)$ for the lattice $\mathbb{L}^3$. As in Section 4.2 the section lattices $\mathbb{L}_i^k$ of $\mathbb{L}^3$ are equipped with pairs $(\mathbb{F}_i^k, \mathbb{F}_{c,i}^k)$ of complementary adjacency systems, $i = 1, \ldots, 13$, $k = 1, 2$. It is assumed that the section lattices $\mathbb{L}_i^k$ and the adjacency systems $\mathbb{F}_i^k$, $\mathbb{F}_{c,i}^k$ are chosen such that for each element F of $\mathbb{F}_i^k \cup \mathbb{F}_{c,i}^k$ there exists a translation $x \in \mathbb{L}^3$ such that $F + x \subset C$. (Note that not necessarily $C^k \subset C$.) Then it is sufficient to consider local configurations $\xi_\ell \subseteq \mathcal{F}^0(C)$, $\ell = 0, \ldots, 255$.

Consider first replacing $\mathbf{1}(\xi_\ell \subset X)\mathbf{1}(\xi_\ell^c \subset X^c)$ in (3) and (8), respectively, by $\mathbf{1}(\xi_\ell \subset X)$. For each set $\xi \subseteq \mathcal{F}^0(C)$, $\xi^c = \mathcal{F}^0(C) \setminus \xi$ and a point $y \in \xi^c$ we have

$$\mathbf{1}(\xi \subset X, \xi^c \subset X^c) = \mathbf{1}(\xi \subset X, \xi^c \setminus \{y\} \subset X^c) - \mathbf{1}(\xi \cup \{y\} \subset X, \xi^c \setminus \{y\} \subset X^c).$$

Table 2. The coefficients g_{3j} of the 21 congruence classes of the local pixel configurations in 3d-images. The columns of g_{0j} correspond to $(\mathbb{F}_{26}, \mathbb{F}_6)$, $(\mathbb{F}_{14.1}, \mathbb{F}_{14.1})$, $(\mathbb{F}_{14.2}, \mathbb{F}_{14.2})$, and $(\mathbb{F}_6, \mathbb{F}_{26})$, in this order. $g_{1j} = g_{2j} = 0$ for the congruence classes 11-21. Note that the weights are those in (11). That is, the configurations are counted if the black dots are foreground and the other vertices are background.

j	η_j		g_{0j}	g_{1j}	g_{2j}	j	η_j		g_{0j}
0	ξ_0	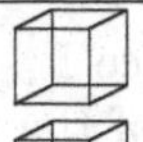	0 0 0 0	0	0	11	ξ_{195}	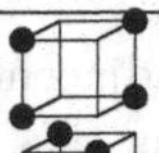	0 0 0 −6
1	ξ_1	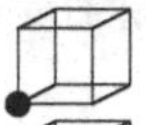	1 1 1 1	0.751	0.751	12	ξ_{105}	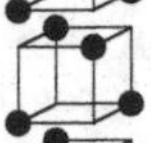	0 0 0 −2
2	ξ_3	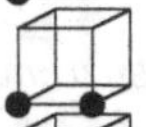	−3 −3 −3 −3	−0.861	−0.275	13	ξ_{99}	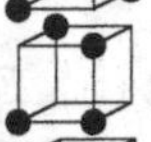	0 0 −2 −24
3	ξ_9	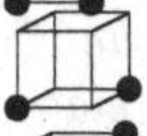	0 −3 −3 −6	−1.076	−0.314	14	ξ_{31}	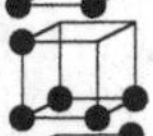	0 0 0 24
4	ξ_{129}	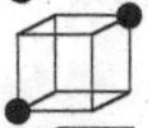	0 −1 −1 −4	−0.314	−0.163	15	ξ_{151}	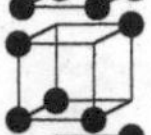	0 0 0 8
5	ξ_{11}	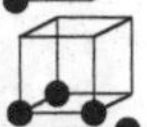	0 6 6 12	0.549	0	16	ξ_{167}	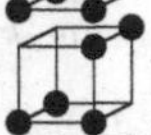	0 0 0 24
6	ξ_{131}	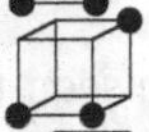	0 6 4 24	0.628	0	17	ξ_{63}	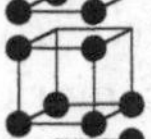	0 0 0 −12
7	ξ_{41}	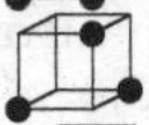	0 0 2 8	0.325	0	18	ξ_{159}	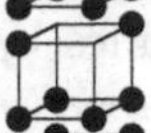	0 0 0 −12
8	ξ_{15}		3 0 0 −3	0	0	19	ξ_{231}	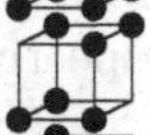	0 0 0 −4
9	ξ_{43}	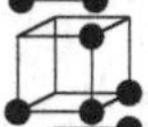	0 0 −2 −8	0	0	20	ξ_{127}	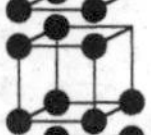	0 0 0 8
10	ξ_{139}	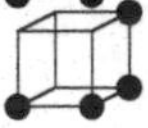	0 −6 −2 −24	0	0	21	ξ_{255}	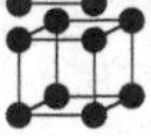	−1 0 0 −1

Recursion and translation by x yield

$$h_\ell = \sum_{x \in \mathbb{L}^3} \mathbf{1}(\xi_\ell \subset X - x, \xi_{255-\ell} \subset X - x) = \sum_{m=0}^{255} q_{m\ell} \sum_{x \in \mathbb{L}^3} \mathbf{1}(\xi_m \subset X - x),$$

$\ell = 0, \ldots, 255$, where the $q_{m\ell}$ are integers with $q_{\ell\ell} = 1$ for $\ell = 0, \ldots, 255$ and $q_{m\ell} = 0$ for $m < \ell$. Further, $\sum_{m=0}^{255} q_{m\ell} = 0$ for $\ell = 0, \ldots, 254$, as follows from the case $X = C$.

Finally, we average the approximations of the intrinsic volumes w.r.t. rotations and inversions that leave the lattice $\mathbb{L}^3$ invariant, i.e. w.r.t. the symmetry group $\{\theta_1, \ldots, \theta_{48}\}$ of the octahedron. Let $D_0, \ldots, D_{21}$ be the congruence classes of $\{\xi_0, \ldots, \xi_{255}\}$ w.r.t. translations and the octahedral group and let $\{\eta_0, \ldots, \eta_{21}\}$ be a system of representatives, $\eta_\ell \in D_\ell$. Now, using the coefficients

$$g_{3-k,j} = \frac{2}{a^{3-k}} \sum_{i=1}^{13} \gamma_i^{(k)} \frac{\operatorname{vol} C}{\operatorname{vol} C_i^k} \sum_{\ell=0}^{255} \bar{q}_{j\ell}\, \chi_0(\xi_\ell \sqcap \mathbb{F}_i^k),\ k = 1, 2,$$

and

$$g_{0j} = \sum_{\ell=0}^{255} \bar{q}_{j\ell}\, \chi_0(\xi_\ell \sqcap \mathbb{F})$$

with $\bar{q}_{j\ell} = \sum_{m=0}^{255} q_{m\ell} \mathbf{1}(\xi_m \in D_j)$ the approximations of the intrinsic volumes (11) and (3) can be rewritten as

$$\tilde{V}_{3-k}(X) = a^{3-k} \sum_{j=0}^{21} \frac{1}{48} \sum_{i=1}^{48} \sum_{x \in \mathbb{L}^3} \mathbf{1}(\theta_i^{-1}\eta_j \subset X - x)\, g_{3-k,j}, \quad k = 1, 2, 3. \quad (13)$$

The $g_{3-k,j}$ are normalized such that they are independent of the lattice distance a and $\tilde{V}_{3-k}(X)$ is the mean of $\hat{V}_{3-k}(X)$ w.r.t. the octahedral group.

For particular cases, the coefficients $g_{3-k,j}$ can easily be computed and be presented in tables. Table 2 contains the g_{3j} for the pairs $(\mathbb{F}, \mathbb{F}_c)$ of complementary adjacency systems from Section 3.1. The coefficients g_{1j} and g_{2j} are computed for the section lattices listed in Table 1. The 2-adjacency is applied for $k = 1$ and the 6-adjacency is applied for $k = 2$. The weights $\gamma_i^{(k)}$ are chosen as described in 4.2.

6 Discussion

We introduce a new method for measuring the intrinsic volumes based on weighted local $2 \times 2 \times 2$ configurations. Due to the restriction to these small configurations, an efficient and simple algorithm can be derived. Given a proper choice of connectivities for foreground and background, consistency of the results for foreground and background can be ensured. This is remarkable in particular for the Euler number, as many other algorithms ignore this.

3d imaging techniques like computed tomography and nano-tomography using transmission electron micorscopy combined with focused ion beam sample preparation often produce anisotropic lattices. The results presented in this paper carry over to these cases as well as higher dimensions, see [10]. Note that for anisotropic lattices, the sets of weights $\gamma_i^{(k)}$ for $k = 1, 2$ from 4.2 do not coincide anymore.

There are various methods of surface estimation based on weights for local pixel configurations, see e. g. [8, 9]. In Table 3 the weights $b_j^{(\mathrm{L})}$ published in [9] are compared with the weights b_j computed from the coefficients g_{2j} given in Table 2. The surface area weights for complementary representatives η_j^c are the same as for η_j.

Clearly, the weights differ, for some of the directions considerably. In [10], multi-grid convergence of the surface area approximation as given by (13) is shown for an important class of random closed sets (Boolean models). A necessary condition for this is $\sum_{j=0}^{21} g_{2j} b_j = 0$, obviously violated by the weights $b_j^{(\mathrm{L})}$.

Table 3. Weights for the surface area depending on local configurations: top: the weights from [9], bottom: the weights computed from the coefficients g_{2j} given in Table 2. Note that the weights here are to be used with (13). That is, the black dots have to be foreground while there is no condition on the others.

j	0	1	2	3	4	5	6	7	8	9	10	11	12	13
η_j														
$b_j^{(L)}$	0	0.636	0.669	1.272	1.272	0.554	1.305	1.908	0.927	0.421	1.573	1.338	2.544	1.190
b_j	0	0.376	0.659	0.646	0.588	0.839	0.768	0.813	0.927	0.914	0.856	0.785	0.874	0.845

References

1. Nagel, W., Ohser, J., Pischang, K.: An integral-geometric approach for the Euler-Poincaré characteristic of spatial images. J. Microsc. **198** (2000) 54–62
2. Ohser, J., Nagel, W., Schladitz, K.: The Euler numer of discretized sets – on the choice of adjacency in homogeneous lattices. In Mecke, K.R., Stoyan, D., eds.: Morphology of Condensed Matter, Berlin, Springer-Verlag (2002) 275–298
3. Ohser, J., Nagel, W., Schladitz, K.: The Euler number of discretised sets – surprising results in three dimensions. Image Anal. Stereol. **22** (2003) 11–19
4. Klette, R., Rosenfeld, A.: Digital Geometry. Morgan Kaufmann, San Francisco (2004)
5. Lang, C., Ohser, J., Hilfer, R.: On the analysis of spatial binary images. Journal of Microscopy **203** (2001) 303–313
6. Blasquez, I., Poiraudeau, J.F.: Efficient processing of Minkowski functionals on a 3d binary image using binary decision diagrams. In: Journal of WSCG. Volume 11., Plzen, Czech Republic, WSCG, UNION Agency-Science Press (2003)
7. Schmidt, V., Spodarev, E.: Joint estimators for the specific intrinsic volumes of stationary random sets. Stochastic Process. Appl. **115** (2005) 959–981
8. Windreich, G., Kiryati, N., Lohmann, G.: Surface area estimation in practice. In Nyström, I., di Baja, G.S., Svensson, S., eds.: Discrete Geometry for Computer Imagery. Volume 2886 of LNCS., Berlin, Heidelberg, New York, DGCI, Naples, Italy, Springer (2003) 358–367
9. Lindblad, J.: Surface area estimation of digitized 3D objects using weighted local computations. Image Vision Comp **23** (2005) 111–122
10. Ohser, J., Nagel, W., Schladitz, K.: Miles formulae for Boolean models observed on lattices. In preparation. (2006)
11. Serra, J.: Image Analysis and Mathematical Morphology, volume 1. Academic Press, London (1982)
12. Jernot, J.P., Jouannot-Chesney, P., Lantuéjoul, C.: Local contributions to the Euler-Poincaré characteristic of a set. J. Microsc. **215** (2004) 40–49
13. Lachaud, J.O., Montanvert, A.: Continuous analogs of digital boundaries: A topological approach to iso-surfaces. Graphical Models **62** (2000) 129–164
14. Schneider, R.: Convex Bodies. The Brunn–Minkowski Theory. Cambridge University Press, Cambridge (1993)
15. Weibel, E.R.: Stereological Methods: Theoretical Foundations. Volume 2. Academic Press, London (1980)

An Objective Comparison Between Gray Weighted Distance Transforms and Weighted Distance Transforms on Curved Spaces

Céline Fouard and Magnus Gedda

Centre for Image Analysis, Uppsala University,
Lägerhyddsvägen 3, SE-75237 Uppsala, Sweden
{celine, magnusg}@cb.uu.se

Abstract. In this paper, we compare two different definitions of distance transform for gray level images: the Gray Weighted Distance Transform (GWDT), and the Weighted Distance Transform On Curved Space (WDTOCS). We show through theoretical and experimental comparisons the differences, the strengths and the weaknesses of these two distances.

1 Introduction

Automatic image analysis processes are generally performed on binary images. However, when images are acquired, gray level values have specific meanings. In some images, they can represent a third (fourth) dimension for 2D (3D) images, or they can represent blurry boundaries of objects, or the object density distribution and many other features. In all the cases, the binarization process, although often mandatory to perform further automated image analysis, results in a loss of information.

To overcome this problem, more and more methods are proposed to perform image analysis directly on gray level images [1, 2, 3]. This is also the case of distance transforms which are widely used on binary images to extract shape and size information [4].

Rutovitz first proposed in [5] a Gray Weighted Distance Transform (GWDT) which uses a pixel gray value as a cost to traverse this pixel. The Gray Weighted Distance (GWD) between two pixels is then defined as the smallest weighted sum of gray level values along the discrete path between these two points. Levi and Montanari also proposed in [6] a distance transform on gray level images where the length of each path is weighted by the gray values of the pixels along the path. In their definition, the length of a path is defined as the discretization of the integral of the pixel values along the path. Saha et al. [7] proposed a theoretical framework and a dynamic programming method for the n-dimensional computation of the Gray Weighted Distance Transform. Verbeek and Verwer [8] and Kimmel et al. [9] used this Gray Weighted Distance Transform to solve the eikonal equation.

Another way of computing distance transforms on gray level images was proposed by Toivanen in [10]. The path between two points is then defined as a

A. Kuba, L.G. Nyúl, and K. Palágyi (Eds.): DGCI 2006, LNCS 4245, pp. 259–270, 2006.

$n + 1$ dimensional path constraint to lie on the hyper-surface defined by the gray level values (here considered as heights on the n dimensional image). This distance transform is thus called Weighted Distance Transform On Curve Space (WDTOCS).

Other distances have been defined on gray level images seen as supports for fuzzy sets. For example, Bloch detailed in [11] several distances between fuzzy sets. She also proposed a new geodesic distance for fuzzy sets [12]. Soille [13] also defined a geodesic measure for fuzzy sets inspired by Levi and Montanari's definition [6].

In this paper, we focus only on distance transforms, so the later distance definitions are beyond the scope of this study. As we want the path defined on gray levels to be able to reach the background, we will not consider geodesic distances either. Our aim is to understand how the two different distance transform definitions ; Gray Weighted Distance (GWD) and Weighted Distance On Curve Space (WDCOS), behave on gray level images where the gray level values have different meanings. We first propose a theoretical comparison based only on their definition and mathematical properties in Section 2. We then compare the results obtained by using these definitions to compute the radius of a fuzzy disk in Section 3. The comparison is then performed on density maps (Section 4) and height maps (Section 5).

2 Comparison from Definitions

In this paper, we consider gray level images $\mathcal{I} : \mathbb{Z}^n \longrightarrow \mathbb{R}$ as functions from the discrete points of the n-dimensional space $\mathbb{Z}^n$ to the space of real numbers $\mathbb{R}$. The gray level values correspond either to heights or to fuzzy membership functions. The notion of height comes with 2D images where the third dimension (elevation) is coded with gray level values in the image. A fuzzy membership function is defined as a mapping to the interval of real numbers $[0, 1]$. The notions of background and foreground can be extended to gray level images as follows: $B = \{p \in \mathbb{Z}^n | \mathcal{I}(p) = 0\}$, and $F = \{p \in \mathbb{Z}^n | \mathcal{I}(p) > 0\}$. In the case of a fuzzy image, the several gray levels of the foreground pixels can be seen as their *belonging degree* of the object. In the case of height maps, the gray level values correspond to the altitude of the ground.

2.1 Gray Level Distance Maps

A distance map is generally defined on a crisp image but can be extended to gray level images:

Definition 1 (Distance map). *Given a gray level image $\mathcal{I}$, the distance map of $\mathcal{I}$: $\mathcal{D}_\mathcal{I}$ is a gray level image where the value of each point of the foreground corresponds to its shortest distance to the background.*

Given a distance definition d, a point $p \in F$ of the foreground and a point $q \in B$ corresponding to its nearest background point, $\mathcal{D}_\mathcal{I}(p) = d(p, q)$. In the case of a crisp image, $d(p,q)$ only depends on the length of the path $\mathcal{P}_{pq}$ between p and q. This path can be either continuous as in Euclidean Distance Transforms [14] or discrete, as in the chamfer algorithm [15]. In the case of gray level distance maps, $\mathcal{P}_{pq}$ is a path between p and q lying on the hyper-plane defined by the gray level values of $\mathcal{I}$. In the following, we consider the continuous function $\pi : [0,1] \longrightarrow \mathbb{R}^n$ following $\mathcal{P}_{pq}$ such that $\pi(0) = \mathcal{I}(p)$ and $\pi(1) = \mathcal{I}(q)$. The length of $\mathcal{P}_{pq}$ depends not only on the spacial distance between p and q, but also on the gray level values along the path. As digital gray level images are defined on discrete grid, the gray level values along a continuous path may not bee known. This is why, even if they are theoretically defined in the continuous space $\mathbb{R}^n$, both GWDT and WDTOCS are practically computed on discrete paths. In this case, each step $[t_i, t_{i+1}]$ of the discrete path with $t_0 = p$ and $t_m = q$ is attributed a cost value w_i depending on the length $||t_i - t_{i+1}||$ of the step i, and of the gray level values $\mathcal{I}(t_i)$ and $\mathcal{I}(t_{i+1})$. The global cost of the path is the sum of all the costs of the local steps: $\mathcal{W}(\mathcal{P}_{pq}) = \sum_i w_i$, and the final distance between p and q is the minimum of the costs of all the paths: $d_{pq} = \min\{\mathcal{W}(\mathcal{P}_{pq})\}$.

2.2 GWD and WDOCS Definitions

Continuous Case. The Gray Weighted Distance is defined in the continuous case by [6] and [7] as follows: $\mathcal{D}_{GWDT} = \int_0^1 |\pi(t)|dt$. It corresponds to the surface area estimation under the curve path $\mathcal{P}_{pq}$. The Weighted Distance On Curved Space is defined in [10] as the length of the shortest geodesic path $\mathcal{P}_{pq}$ between p and q. It is expressed as follows in the continuous case: $\mathcal{D}_{WDTOCS} = \int_0^1 \left|\frac{d\pi}{dt}(t)\right| dt$. Figures of the first line of table 1 illustrate these two definitions.

Discrete Case. In the discrete case, the cost of each step for GWD and WDOCS are respectively:

$$\text{GWD:} \quad w_{GWD_i} = \frac{1}{2}\left(\mathcal{I}(t_i) + \mathcal{I}(t_{i+1})\right) \times ||t_i - t_{i+1}||$$

$$\text{WDOCS:} \quad w_{WDOCS_i} = \sqrt{\left(\mathcal{I}(t_{i+1}) - \mathcal{I}(t_i)\right)^2 + ||t_i - t_{i+1}||^2}$$

In both cases, the spatial distance between two steps $||t_i - t_{i+1}||$ can be either the Euclidean distance: $||t_i - t_{i+1}|| = \begin{cases} 1 \text{ if } t_i \text{ and } t_{i+1} \text{ are 4-neighbors} \\ \sqrt{2} \text{ if } t_i \text{ and } t_{i+1} \text{ are strict 8-neighbors} \end{cases}$

or Borgefors [15] optimal propagating weights for a binary 3×3 mask, i.e.

$$||t_i - t_{i+1}|| = \begin{cases} 0.95509 \text{ if } t_i \text{ and } t_{i+1} \text{ are 4-neighbors} \\ 1.36930 \text{ if } t_i \text{ and } t_{i+1} \text{ are strict 8-neighbors} \end{cases}$$

Table 1 summarizes the different mathematical properties of these two definitions. The last line of this table considers the metric properties (definitivity, positivity, triangular inequality and positive homogeneity) of the two distances.

2.3 About Unit Consistency

By considering gray level values as a $n+1$ image dimension, WDTOCS mixes spatial and intensity units. This may raise several problems. A practical one for

Table 1. Theoretical comparison of GWD and WDOCS

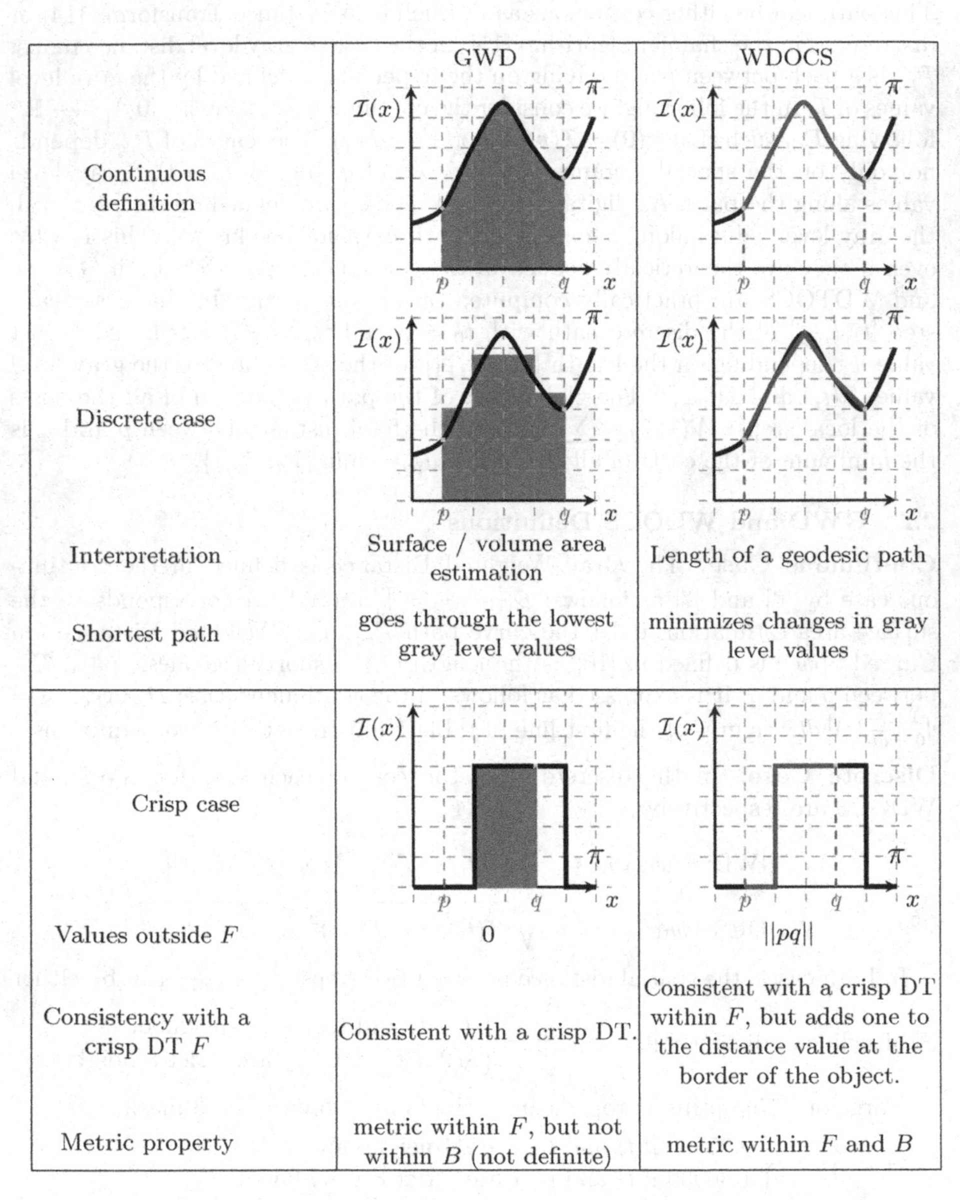

	GWD	WDOCS
Continuous definition		
Discrete case		
Interpretation	Surface / volume area estimation	Length of a geodesic path
Shortest path	goes through the lowest gray level values	minimizes changes in gray level values
Crisp case		
Values outside F	0	$\|\|pq\|\|$
Consistency with a crisp DT F	Consistent with a crisp DT.	Consistent with a crisp DT within F, but adds one to the distance value at the border of the object.
Metric property	metric within F, but not within B (not definite)	metric within F and B

example occurs when the gray level values are small with respect to the image spatial dimensions: the WDTOCS makes no difference with the binary DT as illustrated Figure 1. The experiments used in this paper to illustrate the different behaviors of WDTOCS and GWDT are performed on 2D images to allow full display of the results. Let denote the spatial dimensions of the images along x

and y directions $\mathcal{I}_x$ and $\mathcal{I}_y$ respectively. To overcome this problem in Sections 4 and 5, we scale the gray level values between 0 and $\mathcal{I}_{max} = \frac{\mathcal{I}_x+\mathcal{I}_y}{2}$ to make them comparable to the spatial dimensions (Figure 1). This unit issue does not affect GWDT as it does not mix spatial and gray level units. On the other hand, it considers the integration of distance values, which is more often associated with surface area estimation than with distances.

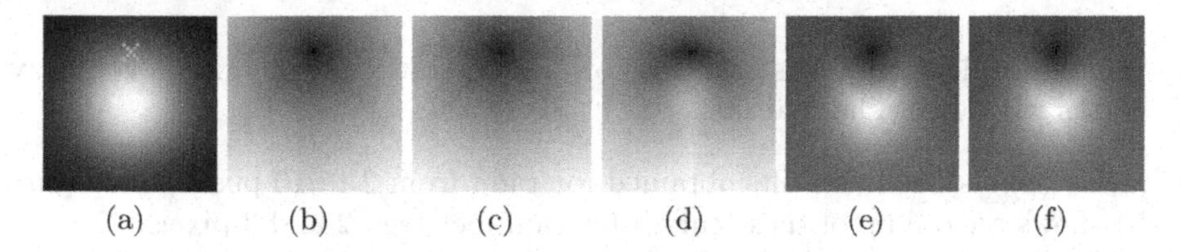

(a) (b) (c) (d) (e) (f)

Fig. 1. (a) Original Gauss image with transform start point marked. (b) Chamfer distance transform from the marked point. (c) WDTOCS using $\mathcal{I}_{max} = 1$. (d) WDTOCS using $\mathcal{I}_{max} = \frac{\mathcal{I}_x+\mathcal{I}_y}{2}$. (e) GWDT using $\mathcal{I}_{max} = 1$. (f) GWDT using $\mathcal{I}_{max} = \frac{\mathcal{I}_x+\mathcal{I}_y}{2}$.

2.4 Implementation

Both WDTOCS and GWDT are discrete path based distance transforms. They are computed with the principle of the chamfer algorithm. In [10] Toivanen proposed to compute WDTOCS by iterating Rosenfeld's raster scans [16] until stability. Saha et al. [7] and Ikonen [17] proposed a wave-front propagation implemented through a pixel-queue algorithm which starts from the border points and propagates local distances to the center of the object.

We implemented both methods (Rosenfeld's raster scans until stability and wave-front propagation algorithms) for both GWDT and WDTOCS and the numerical and visual results for the two different implementation methods are exactly the same. On small images, the calculation time is almost the same, but for larger images, the wave-propagation algorithm is more efficient.

3 Measurement of Continuous Disks Radii

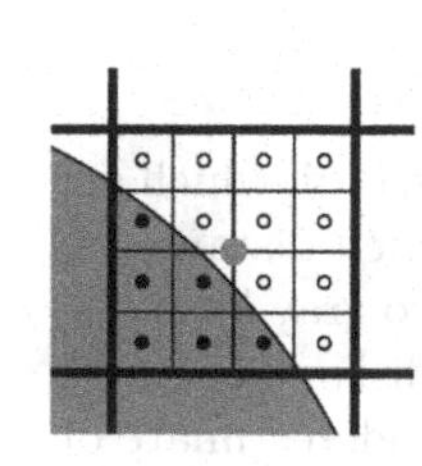

Fig. 2.

The radius of a discrete disk is obtained by taking the highest value of the distance map computed inside the disk. We produce digital fuzzy disks which are discretization of continuous disks and compare the results obtained by WDTOCS and GWDT with Euclidean distance transform and chamfer distance transform on a binarization of these fuzzy disks. The value of pixels within a fuzzy disk is 1, and 0 for a pixel outside. For border pixels, the value is calculated by subsampling the considered pixel as suggestion in [18]. Figure 2 shows an example of a pixel subsampled in 16 pixels whose value is $\frac{6}{4^2}$.

We produce such fuzzy disks with several real radii and compute:

- WDTOCS on the gray level fuzzy disks. As we saw in previous section, WDTOCS crates a *step* of value one between inside and outside pixels. To compute fuzzy disks radii, we thus remove one to the final radius value.
- GWDT on the gray level fuzzy disks
- chamfer distance transform on a binarization of the gray level fuzzy disks with a threshold at 0.5
- Euclidean distance transform [14][1] on a binarization of the gray level fuzzy disks with a threshold at 0.5

Figure 3 (a) shows the results obtained for radii from 2 to 10 pixels, and figure 3 (b) shows a close up of these curves for radii between 2 and 3 pixels.

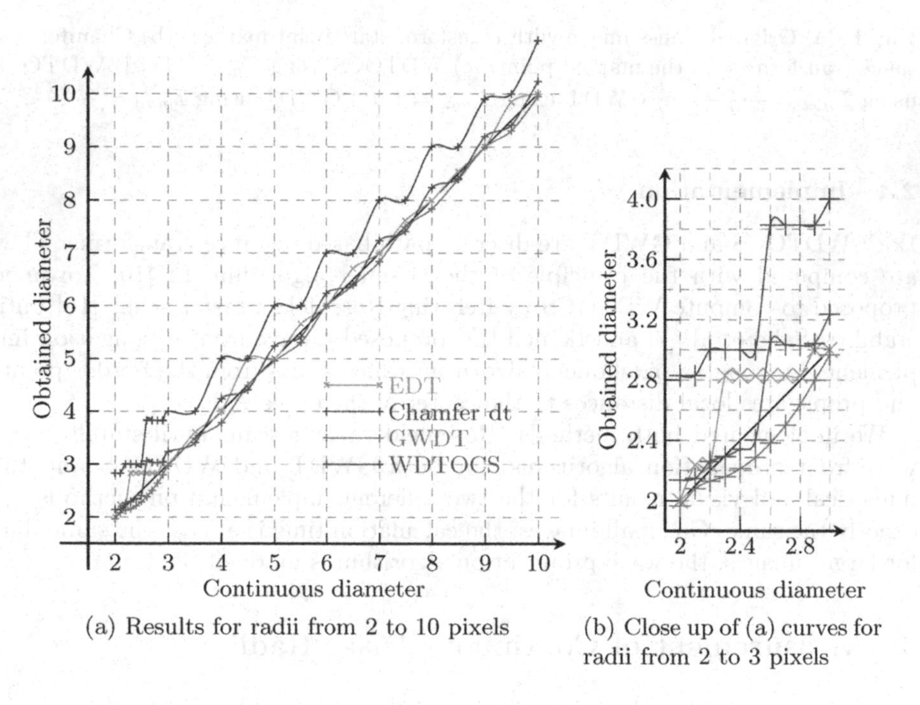

(a) Results for radii from 2 to 10 pixels

(b) Close up of (a) curves for radii from 2 to 3 pixels

Fig. 3. Radii of digitized fuzzy disks obtained with Euclidean DT (orange), Chamfer DT (purple), WDTOCS (blue) and GWDT (red)

We can see in Figure 3 (a) that WDTOCS and GWDT produce radii which are close to the real ones ; at least closer to those given by the weighted distance on the threshold disk. Figure 3 (b) shows that for small radii, the two gray level distances give better results than the Euclidean DT on threshold disks. Generally, GWDT tends to underestimate radii as WDTOCS sometimes underestimate or overestimates radii.

[1] EDT code courtesy of David Coeurjolly `http://www.cb.uu.se/~{}tc18/`

4 Comparison on Density Maps

In many areas imaging devices produce density maps, e.g. medical imaging. In these cases gray level distance transforms can be used when calculating density based distances. An example application is content based clustering of local maxima in electron tomography images of proteins [19]. In this example application, it is desirable to get high distance values when measuring distances between points in different parts of the proteins (i.e. in different high density blobs), and low distance values when measuring distances between points in the same part.

In the following, we compute point-to-point distance by computing distance transforms (GWDT or WDTOCS) from a starting point and back-tracking the path to the second point. To compare GWDT and WDTOCS for point-to-point distance measures in density maps, we use a synthetic image of two gray level blobs and a real image slice of a protein obtained by electron tomography. In both case we compute

- a distance d_{within} between two points of the same blob
- a distance d_{trans} between two points taken in two different blobs
- the ratio $r = \frac{d_{trans}}{d_{within}}$ which gives an indication of the effectiveness measure of the delineation of two density blobs (i.e. the larger r is, the better we can differentiate the two different blobs.)

Synthetic Image. The synthetic image is a 40×40 8-bit image which consists of two high-density objects. It is then inverted to map high densities to low gray level values. In Fig. 4 the gray level distance paths, and the respective distance transforms, are shown. The values of d_{within}, d_{trans}, and r are listed in Table 2.

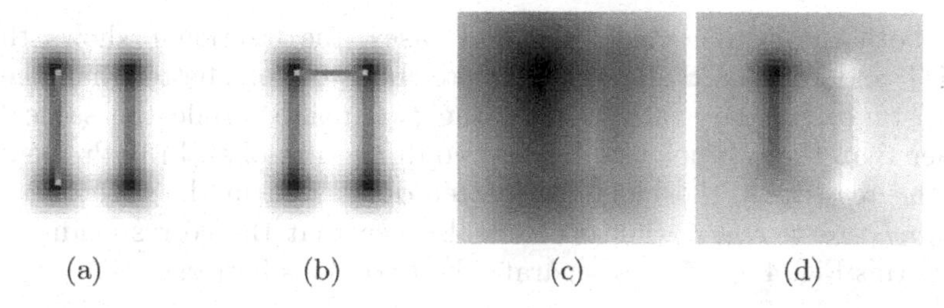

Fig. 4. (a) Inverted synthetic image with the GWDT path (red) and WDTOCS path path (blue) for the measure of d_{within} (paths are overlapping). (b) Paths corresponding to the measure of d_{trans} (c) Corresponding GWDT. (d) Corresponding WDTOCS.

Real Image. The real image is a 36×36 8-bit image, with a pixel size of 5.24 Ångström, and is taken from a slice of a $51 \times 51 \times 51$ protein density volume imaged using electron tomography. The slice shows the inertials of two blobs interconnected with lower gray levels. In Fig. 5 the gray level distance paths, and the respective distance transforms, are shown. In Table 3 the different distance values are listed.

Table 2. Distance measures in synthetic image

Measure	d_{within}	d_{trans}	$r = d_{trans}/d_{within}$
GWDT	187.72	225.38	1.20
WDTOCS	37.54	68.27	1.82
Euclidean DT	19	12	0.63

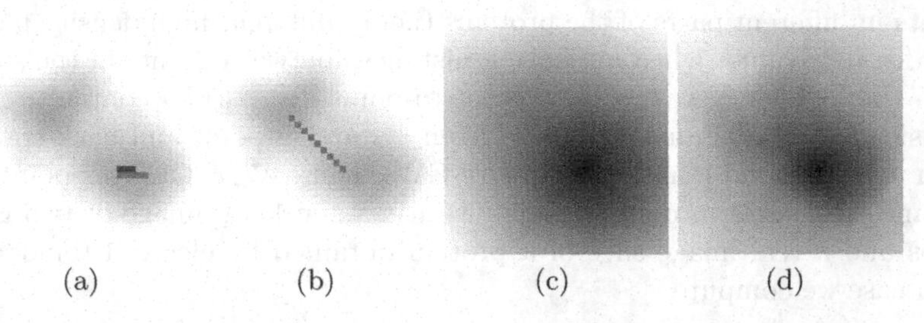

Fig. 5. (a) Inverted protein slice with the GWDT path (red) and WDTOCS path path (blue) for the measure of d_{within}. (b) Paths for the measure of d_{trans} (paths are overlapping). (c) GWDT. (d) WDTOCS.

Table 3. Distance measures in protein slice image

Measure	d_{within}	d_{trans}	$r = d_{trans}/d_{within}$
GWDT	125.13	303.40	2.42
WDTOCS	6.45	17.38	2.69
Euclidean DT	6.08	13.45	2.21

In both the synthetic and the real cases, the fraction r shows that both GWDT and WDTOCS allow to separate the two blobs better than the Euclidean distance (in the synthetic case the two points inside the same blob are further than the two points taken in two different blobs). In both the synthetic and the real cases, the blobs are better delineated in the case of WDTOCS as $r_{WDTOCS} > r_{GWDT}$ and we can also see that the corresponding distance transforms Fig. 4 and 5 also separate the two blobs better.

5 Point-to-Point Distances in Height Maps

In areas where height maps are common, e.g. remote sensing, fuzzy distance can be a valuable tool for calculating content-based distances in images. One application is shortest path-finding in terrain images[2].

[2] Remark: In the case of heights maps, all altitude of the maps correspond to altitudes of the ground. Here we consider the 0 level as absolute. Thus, if the distances of two points of the same altitude are taken at different altitudes, this can lead to a shift in the GWD value.

To compare GWDT and WDTOCS for point-to-point distance measures in height maps, we use a synthetic height map image and a height map image taken of the Grand Canyon.

Synthetic Image. The synthetic image is a 136×136 8-bit image which consists of a large central ridge. The paths between two points on the ridge calculated with GWDT and WDTOCS, along with the corresponding distance transforms are shown in Fig. 6.

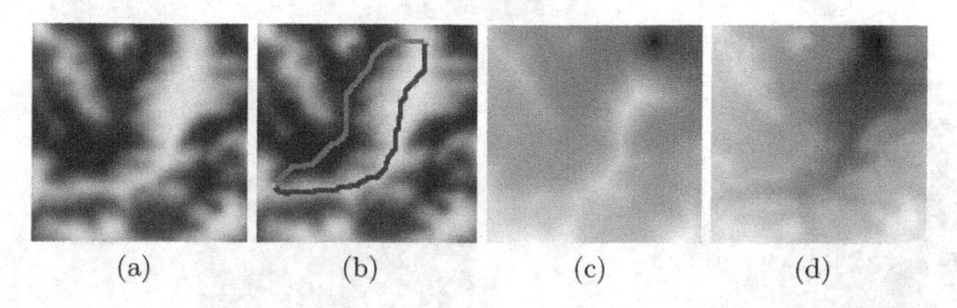

Fig. 6. (a) Original height map. (b) GWDT path (red) and WDTOCS path (blue) between two points on the ridge. (c) GWDT. (d) WDTOCS.

Fig. 6 shows the two different behaviors GWDT and WDTOCS: GWDT path goes through the lowest possible gray level values, and thus goes down from the ridge, and then up to reach the end point, while WDTOCS path minimizes the number of changes in gray level values, and thus remains on the ridge.

Real Image. The real image is a 2400×1600 8-bit image, with a pixel size of 60 meters, and is taken from a a 4097×2047 height map of the Grand Canyon area. The image shows the canyon stretching from a lake in the left part of the image, and continuing as a fissure to the far right of the image. Each pixel unit (0 to 255) corresponds to 10.004 meters, and the pixel value 0 corresponds to a base elevation of 284 meters. In Fig. 7 a surface rendered representation, along with the fuzzy distance paths and the fuzzy distance transforms, are shown.

In this case, GWDT path follows the minimum gray level values as expected and remains in the bottom of the Grand Canyon. However, the WDTOCS path surprisingly goes out the canyon, and follows a high altitude plateau before going down to the canyon again. On the corresponding WDTOCS Fig. 7 (d), we can see, that the WDOCS values are low within the canyon until the middle of the image, and then begin to be higher. To better understand what happens in this case, we compute paths between points within this region as shown in Fig. 8.

In this case, the first part of the WDTCOS path also goes away from the canyon center. This is due to the fact that the ground of this part is highly irregular as shown in Fig. 8 (c) and that WDTCOS path looks for places where the gray values vary more slowly. When the bottom ground values becomes regular again, the WDTCOS path follows the canyon.

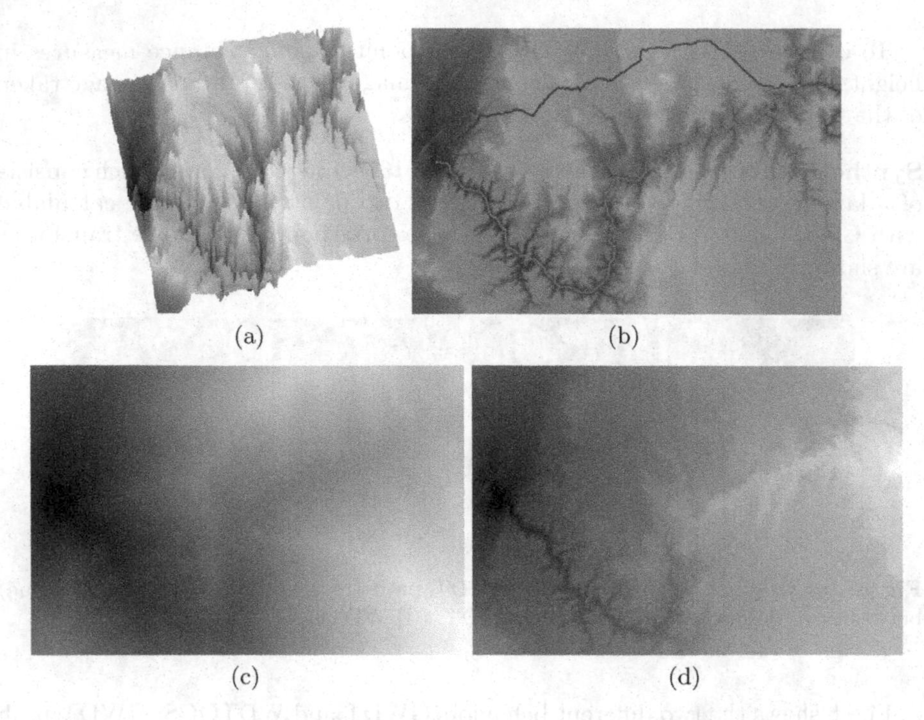

Fig. 7. (a) Grand Canyon height map surface rendered. (b) Height map with GWDT path (red) and WDTOCS path (blue) overlaid. (c) GWDT. (d) WDTOCS.

6 Discussion and Conclusion

In this paper, we compare two distance transform defined on gray level images. The Gray Weighted Distance (GWD) defines the length of a path as the spatial length of this path weighted by gray level values along this path. It can also be seen as the computation of the surface area delimited by the path as shown in Table 1. The Weighted Distance On Curved Space (WDOCS) on the other hand, defines the distance between two points as the length of the geodesic path lying on the hyper-surface defined by the gray level values. A GWD path will thus follow low gray level values whereas WDOCS paths will minimize the changes in gray level values.

The different experiments of this paper shows that the WDOCS depends highly on the scale of the gray level values, contrary to the GWD. In a general way, GWDT tends to smooth the values in the original gray level image, contrary to WDTOCS which tends to sharpen the differences between gray level values. This makes WDOCS more sensitive to the noise, whereas the GWD is more robust to local changes of gray level values. However, this also makes WDTOCS more accurate to delineate different gray valued objects in the same image.

The choice between these two distance transform definitions is highly application dependent. It depends on the aim of the application: either highlighting gray level differences or smoothing the values to obtain average distances.

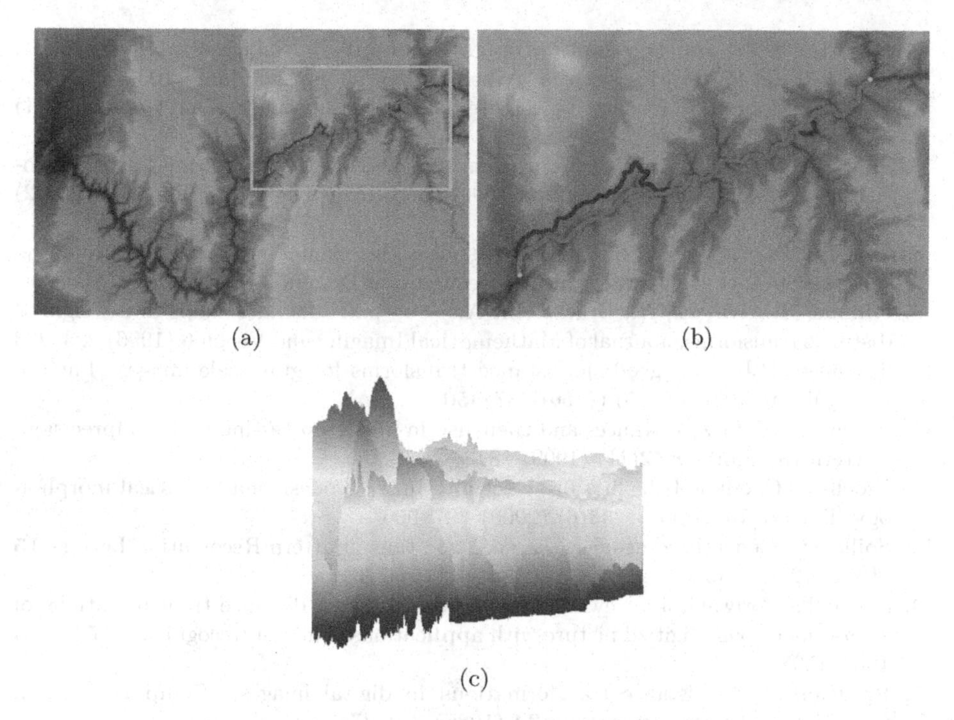

Fig. 8. (a) Height map with GWDT path (red) and WDTOCS path (blue) overlaid. (b) Close-up on paths showing WDTCOS path outside the canyon in the left part of the image. (c) The close up surface rendered and viewed sideways (from south) showing the jagged bottom of the left part of the canyon.

Acknowledgments

The authors would like to thank Dr. Natasa Sladoje and Dr Joakim Lindblad for their scientific support. The Grand Canyon data is obtained from The United States Geological Survey (USGS), with processing by Chad McCabe of the Microsoft Geography Product Unit.

References

1. Bloch, I.: Fuzzy spatial relationships for image processing and interpretation: a review. Image and Vision Computing **23**(2) (2005) 89–110
2. Sladoje, N., Nyström, I., Saha, P.K.: Mesurements of digitized objects with fuzzy borders in 2D and 3D. Image and Vision Computing **23**(2) (2005) 123–132
3. Arcelli, C., Luca, S.: Skeletoinization of labeled gray-tone images. Image and Vision Computing **23**(2) (2005) 159–167
4. Borgefors, G.: Applications using distance transforms. In Arcelli, C., Cordella, L.P., Sanniti di Baja, G., eds.: Aspects of Visual Form Processing. World Scientific, Singapore (1994) 83–108

5. Rutovitz, D.: Data structures for operations on digital images. In Cheng, G.C., ed.: Pictorial Pattern Recognition, Washington, Thompson (1968) 105–133
6. Levi, G., Montanari, U.: A gray-weighted skeleton. Information and Control **17**(1) (1970) 62–91
7. Saha, P., Wehrli, F.W., Gomberg, B.R.: Fuzzy distance transform: Theory, algorithms, and applications. Computer Vision and Image Understanding **86**(3) (2002) 171–190
8. Verbeek, P., Verwer, B.: Shading from shape, the eikonal equation solved by grey-weighted sitance transform. Pattern Recognition Letters **11** (1990) 681–690
9. Kimmel, R., Kiryati, N., Bruckstein, A.: Sub-pixel distance maps and weighted distance transforms. Journal of Mathematical Imaging and Vision **6** (1996) 223–233
10. Toivanen, P.J.: New geodesic distance transforms for gray-scale images. Pattern Recognition Letters **17**(5) (1996) 437–450
11. Bloch, I.: On fuzzy distances and their use in image processing under imprecision. Pattern Recognition **32**(11) (1999) 1873
12. Bloch, I.: Geodesic balls in a fuzzy set and fuzzy geodesic mathematical morphology. Pattern Recognition **33**(6) (2000) 897–906
13. Soille, P.: Generalized geodesy via geodesic time. Pattern Recognition Letters **15** (1994) 1235–1240
14. Saito, T., Toriwaki, J.: New algorithms for euclidean distance transformations of an n-dimensional digitizd picture with applications. Pattern Recognition **27** (1994) 1551–1565
15. Borgefors, G.: Distance transformations in digital images. Computer Vision, Graphics, and Image Processing **34** (1986) 344–371
16. Rosenfeld, A., Pfaltz, J.L.: Sequential operations in digital picture processing. Jornal of the ACM **13**(4) (1966) 471–494
17. Ikonen, L.: Pixel queue algorithm for geodesic distance transforms. In et al., E.A., ed.: Proc. of 12th Int. Conf. on dgci. Volume 3429., Springer-Verlag (2005) 228–239
18. Sladoje, N., Lindblad, J.: Estimation of moments of digitized objects with fuzzy borders. In Springer-Verlag, ed.: Proc. 13th iciap. Volume 3617. (2005) 188–195
19. Gedda, M., Svensson, S.: Separation of blob-like structures using fuzzy distance based hierarchical clustering. In Georgsson, F., Börlin, N., eds.: Symposium on Image Analysis, SSBA 2006, Umeå, Sweden, March 16-17, 2006, Proceedings. (2006) 73–76

Chordal Axis on Weighted Distance Transforms

Jérôme Hulin and Edouard Thiel

Laboratoire d'Informatique Fondamentale de Marseille (LIF, UMR 6166), France
{Jerome.Hulin, Edouard.Thiel}@lif.univ-mrs.fr
http://www.lif.univ-mrs.fr/~thiel

Abstract. Chordal Axis (CA) is a new representation of planar shapes introduced by Prasad in [1], useful for skeleton computation, shape analysis, characterization and recognition. The CA is a subset of chord and center of discs tangent to the contour of a shape, derivated from Medial Axis (MA). Originally presented in a computational geometry approach, the CA was extracted on a constrained Delaunay triangulation of a discretely sampled contour of a shape. Since discrete distance transformations allow to efficiently compute the center of distance balls and detect discrete MA, we propose in this paper to redefine the CA in the discrete space, to extract on distance transforms in the case of chamfer norms, for which the geometry of balls is well-known, and to compare with MA.

Keywords: image analysis, shape description, chordal axis, medial axis, discrete geometry, chamfer or weighted distances.

1 Introduction

Shape description consists in extracting features from a binary image, like area, width, number of holes, etc. To this purpose, Blum first proposed the notion of Medial Axis (MA) of a shape $\mathcal{S}$ in [2]. Then Pfaltz and Rosenfeld defined it in [3] as the set of centers of maximal discs in $\mathcal{S}$, a disc being maximal in $\mathcal{S}$ if it is not completely overlapped by any other disc included in $\mathcal{S}$. MA has become an important tool in image analysis and shape description, because it is a reversible coding and a global representation, centred in the shape, which allows to simplify, compress, or compute a skeleton of a shape.

Among several approaches in image analysis, we distinguish : the *continuous* approach with analytical pieces of curves, which is in general case difficult to tackle; the *semi-continuous* approach, which consists in sampling the contour of a shape, and then deals with computational geometry in Euclidean space, as Voronoï diagram, convex hull, etc; the *discrete* approach, which keeps the shape bitmap (or sample a continuous shape on the rectilinear grid) in discrete space $\mathbb{Z}^n$ and makes use of discrete geometry, often needing to redefine continuous properties in discrete ones.

MA has been studied in this three frameworks. A continuous MA is obtained in [4] with pieces of lines and arcs. The semi-continuous MA is an approximation of continuous MA, extracted from the Voronoï Diagram of the sampled contour

A. Kuba, L.G. Nyúl, and K. Palágyi (Eds.): DGCI 2006, LNCS 4245, pp. 271–282, 2006.

(e.g., see [5]). The discrete MA is generally extracted by local tests on a Distance Transform (DT), which is an image where each shape point is labelled with its distance to the background.

Working on semi-continuous framework, Prasad proposed in [1] a new representation of shapes, called Chordal Axis (CA). The aim was to correct sensitiveness to noise in sampled contour for semi-continuous MA. The definition is close to MA, but has different properties. A Maximal Chord of Tangency (MCT) of a shape $\mathcal{S}$ is a chord of a maximal disc D of $\mathcal{S}$, which separates the boundary of D into two arcs such that one at least is not tangent to the boundary of $\mathcal{S}$; the CA is the set of midpoints of the MCTs, plus the set of centers of maximal discs having at least three MCTs. Prasad then extracted a semi-continuous CA from a Delaunay triangulation of sampled contour. In order to fix some zigzags in the result, Prasad introduced in [6] a valuation of certain internal edges, leading to a shape decomposition process.

We have recently proposed an algorithm for extracting discrete MA from a DT for any chamfer norm in [7] and for Squared Euclidean Distance in [8]. The algorithm computes a test neighbourhood and Look-Up Tables and then extracts the centers of maximal balls for the given distance by local tests on DT. In this paper we naturally propose to adapt the CA in discrete framework, in the case of 2D chamfer norms. A chamfer (or weighted) distance is an integer distance defined by a mask [9]; such a distance allows very fast computation of DT with a sequential algorithm in two raster scans given in [10,11]. We focus on the masks inducing a norm, for which the geometry of ball is established (a discrete convex polyhedron, see [12]).

We recall in Section 2 some basic notions and definitions. We present and justify our method in Section 3. Results are given in Section 4, and finally we conclude in Section 5.

2 Definitions

2.1 Chordal Axis

Consider a shape $\mathcal{S}$ in $\mathbb{R}^2$ and B a maximal ball in $\mathcal{S}$. Following Prasad in [1], a chord of B is called a Maximal Chord of Tangency (MCT) if at least one of the arcs subtended by the chord is free of points of tangency with the boundary of $\mathcal{S}$. Fig. 1 gives some examples of MCTs.

The Chordal Axis (CA) of a shape is the set of all pairs (p, δ), where p and δ are either the midpoint and half the length of a maximal chord of tangency, or the center and radius of a maximal ball which has at least three maximal chords of tangency. We call *α-points* the midpoints of MCTs and *β-points* the centers of maximal balls having at least three MCTs.

In the continuous domain (see Fig. 1), the CA is generally non-connected, contrary to the MA. Indeed, the connectedness is broken as soon as a maximal disc inside a shape contains at least three maximal chords of tangency (MCTs). However, connectedness may easily be obtained if, in such maximal discs, we draw a segment from the center of the disc to the middle of each MCT.

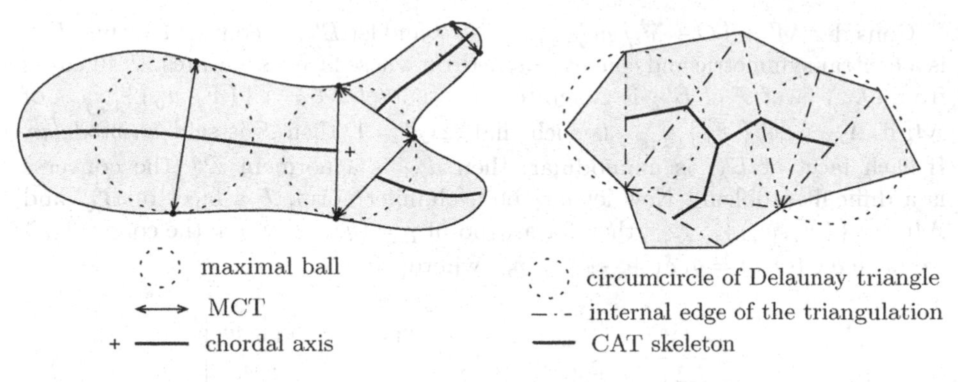

Fig. 1. Three maximal balls inside a shape and their MCTs; chordal axis

Fig. 2. Delaunay triangulation of sampled contour and CAT skeleton

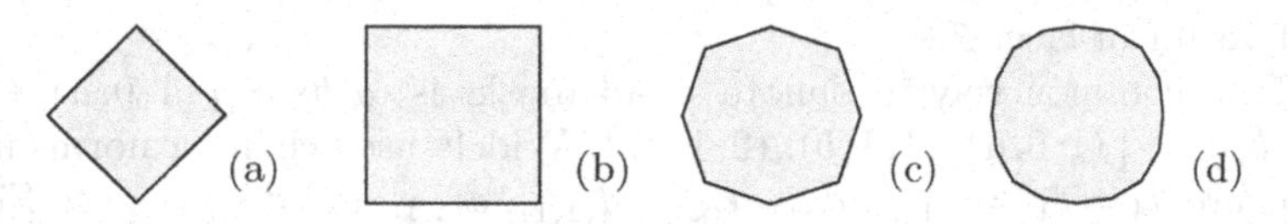

Fig. 3. Shape of chamfer balls for d_4 (a), d_8 (b), $d_{\langle 3,4 \rangle}$ (c), $d_{\langle 5,7,11 \rangle}$ (d)

In [1], Prasad defines the Chordal Axis Transform (CAT) as a semi-continuous method to extract the CA: he starts with a discrete sample of the boundary of a given shape $\mathcal{S}$, then calculates the Delaunay triangulation of these points, inside $\mathcal{S}$. The MCTs of $\mathcal{S}$ are approximated by the internal edges of the triangulation (the edges which are not on the boundary of $\mathcal{S}$). Then a skeleton based on the semi-continuous CA is constructed by connecting, inside each triangle, the midpoints of two or three internal edges, depending on the number of edges lying on the boundary of $\mathcal{S}$. The resulting skeleton (see Fig. 2) is sensitive to the irregularities of the samples of the contour of $\mathcal{S}$, giving to it angularities. In [6], a valuation on the edges of the triangulation is proposed in order to delete edges considered as weak and to smooth the skeleton; or to detect strong edges which split the object into significant parts, to achieve shape decomposition.

2.2 Chamfer Distances and Norms

Here we recall some results from [12]. A *chamfer mask* $\mathcal{M}$ in $\mathbb{Z}^n$ is a central-symmetric set $\mathcal{M} = \{(\overrightarrow{v}_i, w_i) \in \mathbb{Z}^n \times \mathbb{Z}_{+*}\}_{1 \leqslant i \leqslant m}$ containing at least a basis of $\mathbb{Z}^n$, where $(\overrightarrow{v}_i, w_i)$ are called *weightings*, $\overrightarrow{v}_i$ *vectors* and w_i *weights*. The *chamfer distance* $d_\mathcal{M}$ between two points $p, q \in \mathbb{Z}^n$ is

$$d_\mathcal{M}(p,q) = \min\left\{ \sum \lambda_i w_i \; : \; \sum \lambda_i \overrightarrow{v}_i = \overrightarrow{pq}, \; 1 \leqslant i \leqslant m, \; \lambda_i \in \mathbb{Z}_+ \right\} \; ; \qquad (1)$$

it is shown that $d_\mathcal{M}$ is always a metric. A *chamfer ball* B of center $p \in \mathbb{Z}^n$ and radius $R \in \mathbb{Z}_+$ is $B_\mathcal{M}(p,R) = \{ q \in \mathbb{Z}^n \; : \; d_\mathcal{M}(p,q) \leqslant R \}$.

Consider $\mathcal{M}' = \{ O + \overrightarrow{v_i}/w_i \}_{1 \leqslant i \leqslant m} \in \mathbb{R}^n$ and let $B'_\mathcal{M} = \text{conv}(\mathcal{M}')$, then $B'_\mathcal{M}$ is a central-symmetric and convex polyhedron whose facets separates $\mathbb{R}^n$ in cones from O. A facet $\mathcal{F}$ of $B'_\mathcal{M}$ is generated by a subset $\mathcal{M}|_\mathcal{F} = \{ (\overrightarrow{v_j}, w_j) \}_{1 \leqslant j \leqslant n}$ of $\mathcal{M}$; if $\Delta_\mathcal{F} = \det \{ \overrightarrow{v_j} \}_{1 \leqslant j \leqslant n}$ is such that $|\Delta_\mathcal{F}| = 1$, then $\mathcal{F}$ is said *unimodular.* If each facet of $B'_\mathcal{M}$ is unimodular, then $d_\mathcal{M}$ is a norm in $\mathbb{Z}^n$ (the converse is a difficult problem). Now let $d_\mathcal{M}$ be a chamfer norm, $\mathcal{F}$ a facet of $B'_\mathcal{M}$ and $\mathcal{M}|_\mathcal{F} = \{ (\overrightarrow{v_j}, w_j) \}_{1 \leqslant j \leqslant n}$; then for any point $p = (y_1, \ldots, y_n)$ in the cone $(O, \mathcal{F})$ we have $d_\mathcal{M}(O,p) = y_1\,\delta_1 + \cdots + y_n\,\delta_n$, where

$$\delta_k = \frac{(-1)^{n+k}}{\Delta_\mathcal{F}} \cdot \begin{vmatrix} v_{1,1} & \cdots & v_{1,k-1} & v_{1,k+1} & \cdots & v_{1,n} & w_1 \\ \vdots & & \vdots & \vdots & & \vdots & \\ v_{n,1} & \cdots & v_{n,k-1} & v_{n,k+1} & \cdots & v_{n,n} & w_n \end{vmatrix}^T \tag{2}$$

is the *elementary displacement* for coordinate y_k. Moreover, the chamfer ball B has the same geometry as B' (up to a scale factor), so $\overrightarrow{\delta_\mathcal{F}} = (\delta_1, \ldots, \delta_n)$ is a normal vector of facet $\mathcal{F}$.

In $\mathbb{Z}^2$, a common way to denote small masks is $\langle a, b\rangle = \{ (1,0,a),\ (1,1,b) \}$ and $\langle a,b,c\rangle = \{ (1,0,a),\ (1,1,b),\ (2,1,c) \}$. Widely used chamfer norms in image analysis are $d_4 = \ell_1 = d_{\langle 1,2\rangle}$, $d_8 = \ell_\infty = d_{\langle 1,1\rangle}$, $d_{\langle 3,4\rangle}$ and $d_{\langle 5,7,11\rangle}$, see Fig. 3.

3 Discretization of Chordal Axis

In the following we work in $\mathbb{Z}^2$ with a given chamfer norm $d_\mathcal{M}$. Let $\mathcal{S}$ be a shape, a point p of $\mathcal{S}$ is called a *boundary point* if its distance to the complement of $\mathcal{S}$ equals the smallest weight in $\mathcal{M}$. Consider B a ball of $d_\mathcal{M}$ included in $\mathcal{S}$. A point p is called a *tangency point* between B and $\mathcal{S}$ if p belongs to B and to the boundary of $\mathcal{S}$. A *tangency zone* is a maximal 8-connected set of tangency points. We exhibit three properties about tangency of discrete balls, then present our algorithms to generate the discrete CA. Our aim is to generate MCTs which are meaningful with respect to description and shape analysis.

3.1 Point Threshold

In the discrete domain (see Fig. 4.b), the local intersection between a maximal ball included in an object and the boundary of this object is seldom a unique point. Let A and B be two tangency points between a maximal ball and the boundary of a shape $\mathcal{S}$. The chord [AB] is maximal if one of the two arcs $\widehat{AB}$ is free of points of tangency with the boundary of $\mathcal{S}$. In the continuous case (Fig. 4.a), we have a unique MCT $[AB]$. In the discrete case (Fig. 4.b), two MCTs ($[A_1B_2]$ and $[A_2B_1]$) appear, because the intersections between the ball and the shape's boundary are not single points anymore. Nevertheless only one chord should characterize this ball, the extra chord being an artefact of the discretization. For a given tangency zone, we need to decide whether it is considered as a point or not. To this end we introduce a *point threshold*, denoted PTH (in pixels). We measure the farthest Euclidean distance d between the tangency zone

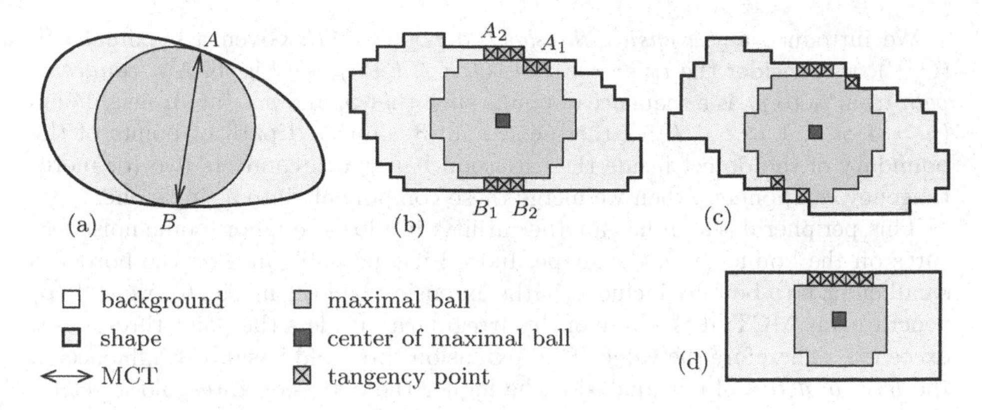

Fig. 4. Maximal Ball in continuous (a) and discrete (b),(c),(d) space

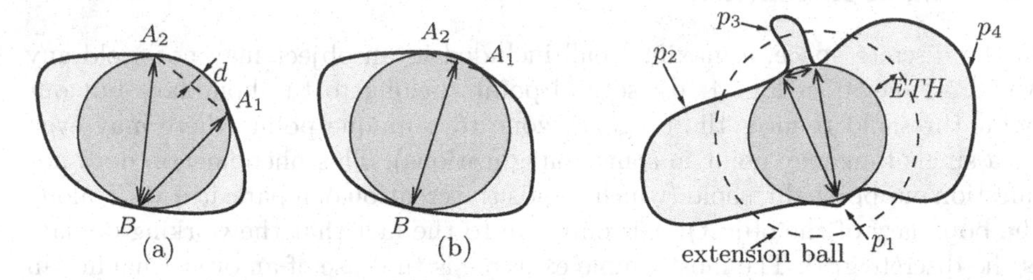

Fig. 5. Application of the point threshold

Fig. 6. Extension ball and protuberances on the boundary: neglected (p_1) and non neglected (p_2, p_3 and p_4); resulting MCTs

$\widehat{A_1A_2}$ and the chord $[A_1A_2]$ induced by the extremities of this zone (see Fig. 5). If d is less then PTH, we contract the tangency zone to a single pixel, located in the middle of the zone (case (b): one MCT). Otherwise each extremity of the tangency zone becomes the endpoint of a MCT (case (a): two MCTs).

3.2 Extension Threshold

The local intersection between a maximal ball and the boundary of the shape may be composed of several non-connected components. E.g. in Fig. 4 (c), we observe that the discretization of the image splits each tangency zone into two connected components. The immediate construction of MCTs would lead to the appearance of two parasitic chords along the boundary of the object. Before proceeding with the construction of the MCTs, we must determine for each connected component, which tangency zone it should belong to. An appropriate solution consists in exploring the peripheral domain of the maximal ball, looking for a path in the boundary of the shape which connects some tangency components.

We introduce an *extension threshold*, denoted *ETH*. Given a maximal ball (O, R), we consider the *extension ball* $(O, R + ETH)$, see Fig. 6. A *n-connected path* from p_0 to p_k is a sequence of points such that p_i and p_{i-1} are n-neighbours ($n = 4$ or 8, $1 \leqslant i \leqslant k$); if there exists an 8-connected path of points of the boundary of the object inside the extension ball which connects two (or more) tangency components, then we merge these components into a single one.

This peripheral search has another utility: it allows to ignore some noise features on the boundary of the shape. Indeed if a protuberance on the border is small enough to be fully included in the extension ball (p_1 in Fig. 6), we will not generate any MCT at the base of this irregularity (unless the point threshold is exceeded). Therefore the value of the extension threshold has direct influence on the *level of detail* of our analysis. The higher the extension threshold is set to, the less precise this analysis near the boundary of the shape.

3.3 Radius Increment

In the discrete space, a maximal ball included in an object may not yield any MCT, as shown in Fig. 4 (d): several points belong to the boundary but our point threshold reduces the tangency zone to a unique point (there may even be a single tangency point in some configurations). This phenomenon does not question our point threshold (which avoids the creation of a parasitic MCT along the boundary of the object); it is only due to the fact that the working domain is the discrete grid. The most simple example is the case of an object having an horizontal or vertical branch with an even width. In this case the maximal balls inside this branch will only be tangent to one side of the shape (the chamfer balls are central-symmetrical so their diameter, measured horizontally in pixels, is odd). No MCT would then appear in such a branch.

We propose to increment the radius of maximal balls with a certain value *RI* (Radius Increment), before proceeding with the exploration of their boundaries. Let a be the smallest weight of the chamfer mask. Then a equals the distance between two 4-neighbours, and it is sufficient to take $RI = a$ to make certain that all maximal balls have at least two tangency zones with the boundary of the shape. However, in most cases, we observe that an increment of value 1 is enough to ensure that almost all balls have two tangency zones.

3.4 Algorithms to Generate the Discrete Chordal Axis

The three proposed thresholds are independant and in practice, they enable to generate almost all useful MCTs, and avoid the creation of parasitic chords (lying along the boundary of the shape or describing noise features of the border). These thresholds are computed sequentially for each maximal ball, as follows:

1. calculate the new radius of the ball, by adding *RI*,
2. look for tangency zones in the peripheral region, using *ETH*,
3. decide for each tangency zone, if it should be contracted to a single point, using *PTH*.

We recall (see [7]) that on DT, the value $DT(p)$ for any shape point p is the radius of the greatest set $\{ q \in \mathbb{Z}^2 : DT(p) - d_{\mathcal{M}}(p,q) > 0 \}$ so the greatest ball centred in p inside the shape is $\{ q \in \mathbb{Z}^2 : d_{\mathcal{M}}(p,q) \leqslant DT(p) - 1 \}$ which has radius $DT(p) - 1$ (utilized line *4* in the procedure Gen_CA).

The boundaries of maximal balls are inspected thanks to the equations given in Section 2.2, or by consulting an image containing the distance values from any point to the top-left point of the image (computed once), then using symmetries.

Input *IN* Shape image, *DT* Distance Transform, *MA* Medial Axis
Output *CA* Chordal Axis

Procedure Gen_CA (*IN*, *DT*, *MA*, *CA*, *PTH*, *ETH*, *RI*)
1 Initialize *CA* to $\emptyset$
2 **For all** points p **in** *IN* **do**
3 **If** $p \in MA$ **then**
4 $R = DT(p) - 1$
5 **If** $R \geqslant a$ **then** Search_MCTs (p, R, *DT*, *CA*, $\{Thresholds\}$)

Procedure Search_MCTs (p, R, *DT*, *CA*, $\{Thresholds\}$)
1 IMAGE *tmp* //*image storing tangency points, initialized to* 0
2 INT $nb_{tang} = 0$ //*number of tangency zones*
3 Generate $\mathcal{C}$, the boundary of the ball of center p and radius $R + RI$
4 **For all** points q **in** $\mathcal{C}$, counterclockwise, **do**
5 **If** $tmp(q) = 0$ **and** $DT(q) = a$ **then** //*new tangency zone*
6 $nb_{tang} = nb_{tang} + 1$
7 Label recursively with the value nb_{tang} in *tmp* the 8-neighbours n_j
8 of q such that $d(p, n_j) \leqslant R + ETH$ and $DT(n_j) = a$
9 **For each** tangency zone $\widehat{AB}$ **do**
10 Compute the maximal distance d between $\widehat{AB}$ and the MCT $[AB]$
11 **If** $d \leqslant PTH$ **then** contract the tangency zone to a single point, in the middle of the zone
12 Compute the midpoints of the segments bounded by the extremities of the tangency zones, counterclockwise //*may be empty in the case of a single contracted zone*
13 Insert these points as α-points in *CA*
14 **If** $nb_{tang} \geqslant 3$ **then** insert p as β-point in *CA* //*at least three MCTs*

4 Results and Discussion

This section deals with the analysis of the chordal axis (CA) produced by our algorithm. We present the CA of different objects, and give some results in terms of shape description. We analyze the influence of the chosen chamfer norm, as well as the threshold values, on the geometry of the CA. An application to shape decomposition is presented; then a connection with the medial axis (MA) is proposed. Finally we have a look at the complexity of the algorithms.

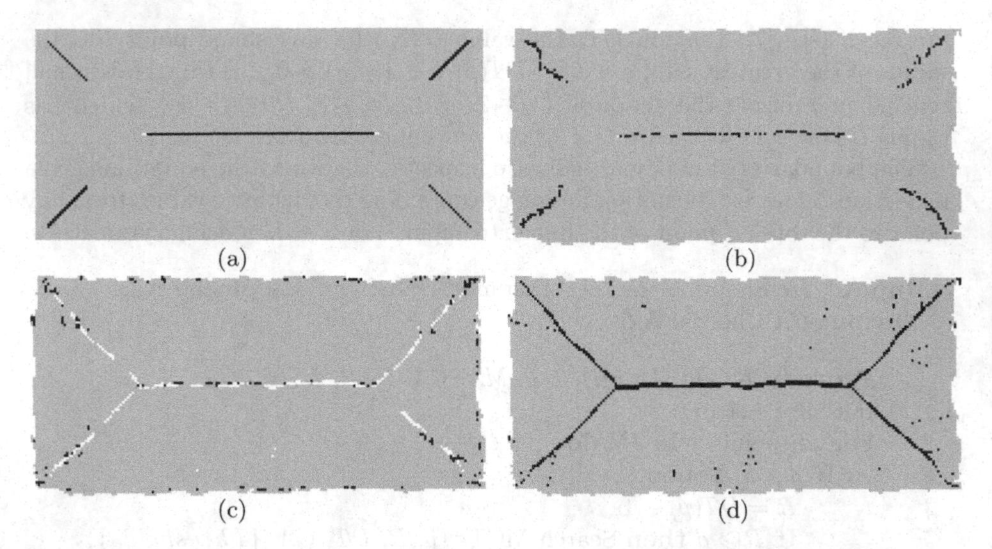

Fig. 7. CA with thresholds on a rectangle (a) and a noisy rectangle (b); CA without thresholds (c); MA (d). In (a), (b), (c) : α-points in black and β-points in white.

4.1 Characteristics of the Chordal Axis

Fig. 7 shows in (a), (b) and (c) the CA of rectangular shapes produced by our algorithm, together with the MA (d), using the chamfer norm $d_{\langle 5,7,11\rangle}$. The input rectangle (in grey) in (b), (c) and (d) has a noisy boundary. In (a), (b), (c), black points are α-points, i.e. midpoints of maximal chords of tangency, while white points are β-points, i.e. centers of maximal ball having at least three MCTs (the CA is composed of both α and β points). The CA in (b) is computed using the thresholds method described at Section 3, while (c) is generated without any threshold. The CA in (a) is coherent with the definition of the CA in the continuous plane (see Fig. 1). The differences between (b) and (c) point out the importance of the thresholds. On (b) the extension threshold ETH (set to 10 for this example) erases parasitic α-points near the border of the shape; the point threshold PTH (here at 4 pixels) avoids the apparition of superfluous β-points; the increment threshold RI (set to 1) ensures the presence of α-points in the middle of the object. Compared to the MA (d), the CA contains less points, is free of parasitic points, and is less connected. We also observe a slight deviation between the MA and the CA in the tips of branches of the shape, because α-points may be quite far from centers of maximal balls.

The influence of the chamfer norm on the CA is illustrated in Fig. 8. When using d_4 (a) or d_8 (b), we observe three annoying phenomenons with respect to shape description:

- There are too many β-points. This is because the distances d_4 and d_8 badly approximate the Euclidean distance (their balls are squares).
- In some places of the middle of the branches of the shape, there is a lack of points in the CA. This is also due to the shape of the balls of d_4 and

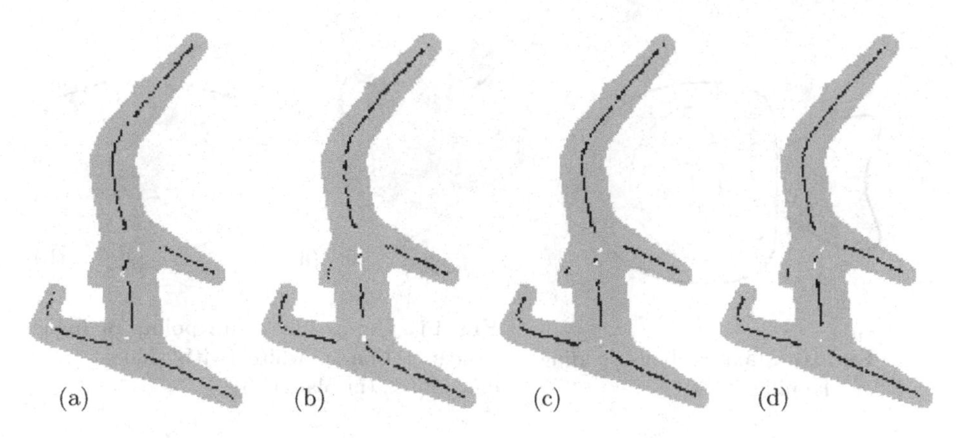

Fig. 8. CA with d_4 (a), d_8 (b), $d_{\langle 3,4 \rangle}$ (c), $d_{\langle 5,7,11 \rangle}$ (d). α-points appear in black, β-points in white.

Optimal value	If less than optimal	If greater than optimal
$1 \leqslant RI \leqslant a$	incomplete CA (see Section 3.3)	increase of the number of tangency zones $\Rightarrow$ too many MCTs
$a \leqslant ETH \leqslant 3a$	parasitic MCTs along the boundary of the shape (see Section 3.2)	rough detail level, lack of MCTs in thick branches
$3 \leqslant PTH \leqslant 6$	parasitic MCTs along the boundary of the shape (see Section 3.1)	lack of MCTs in maximal balls having a single tangency zone

Fig. 9. Influence of threshold values (a is the weight of the first vector of the chamfer mask)

d_8, which tends to create MCTs whose extremities are often located in the vertices of the squares. In certain local areas, many midpoints of chords may overlap.
- There is a strong anisotropy of the CA.

These features considerably attenuate when choosing $d_{\langle 3,4 \rangle}$ (c) and $d_{\langle 5,7,11 \rangle}$ (d); with $d_{\langle 5,7,11 \rangle}$ we obtain the best results (the average approximation error compared to d_E is only about 2%).

An important characteristic of the CA concerns the localization of its two different kinds of points:

- the α-points are located in the branches of the object, and are equidistant from each side of their branch;
- the β-points are located at the center of branching zones of the shape.

The choice of optimal values of the thresholds, for which the CA describes the shape as well as possible, has been experimentally determined by tests on different objects (with various branching zones and widths). Recommanded values and problems resulting from bad values are listed in Fig. 9.

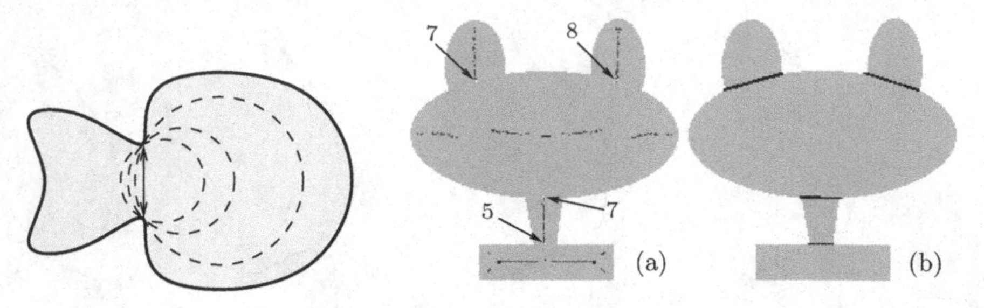

Fig. 10. Maximal balls yielding overlapping MCTs

Fig. 11. CA in black and points of high concentration in white (with values) (a); cut chords (in black) (b)

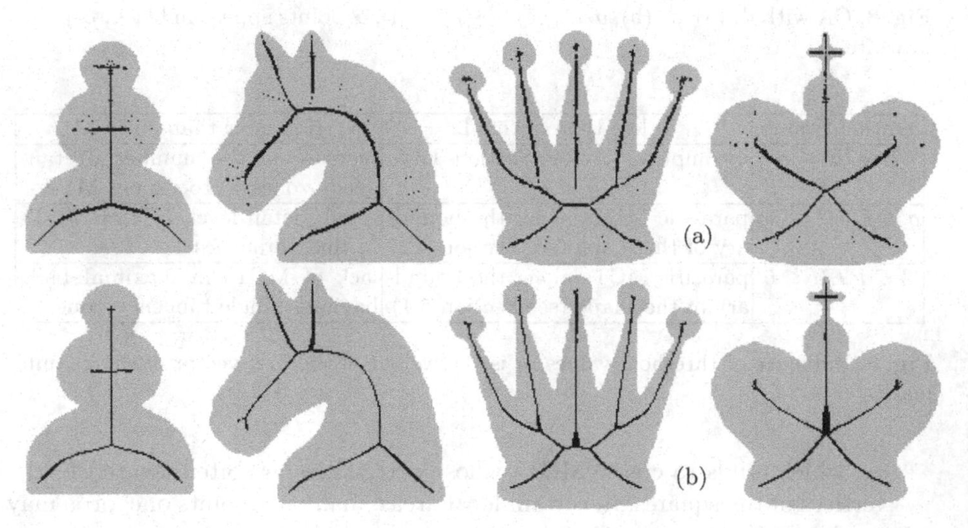

Fig. 12. MA with $d_{\langle 5,7,11 \rangle}$ (a); $\mathcal{C}$-CA with $d_{\langle 5,7,11 \rangle}$, $RI = 1, ETH = 10, PTH = 4$ (b)

4.2 Application to Shape Decomposition

The CA has an interesting characteristic: points of the axis may be superposed, as shown in Fig. 10. The three maximal balls which are drawn have maximal chords of tangency which overlap each other in the discrete space. We notice that the local concentration of the α-points of the CA is high in the zones where the object presents narrowings.

An interesting idea consists in counting, for each point of the image, the number of points of the CA (during computation). We are then able to study the concentration of points of the axis. Points of high concentration are represented in white in Fig. 11 (a). If for a given point of the image, the number of overlapping points of the CA is greater than a certain value, denoted *cut threshold*, then we consider the corresponding MCT as a *cut chord* for shape decomposition. Each cut chord splits the shape into two distinct parts. E.g. in Fig. 11 (b), the cut threshold is set to 5.

Classic methods of shape decomposition are based upon the reckoning of a skeleton, which must be thin and connected, and the study of gradient along this skeleton. The main advantage of our method is that it does not previously require the creation of a skeleton. The values of local concentrations of points of the CA are directly calculated during the CA extraction.

4.3 Connections Between the Chordal Axis and the Medial Axis

We call $\mathcal{C}$-*Chordal Axis* (or $\mathcal{C}$-*CA*) of a shape $\mathcal{S}$, the set of centers of all balls included in $\mathcal{S}$ which admit at least a maximal chord of tangency. In $\mathbb{R}^2$ the $\mathcal{C}$-CA is exactly the MA, because a ball is maximal in $\mathcal{S}$ if and only if it has at least two points of tangency with the boundary of $\mathcal{S}$ (see [2]). This property is not true in the discrete plane, however our threshold techniques allow the $\mathcal{C}$-CA to approximate the MA while filtering it. We adapt our algorithm of generation of the CA to compute the set of centers of balls inside $\mathcal{S}$ which have at least one MCT. Fig. 12 shows the MA and $\mathcal{C}$-CA of different shapes, using $d_{\langle 5,7,11 \rangle}$ and standard thresholds. Notice that both axis have a majority of common points, nevertheless there are significant differences. The $\mathcal{C}$-CA is much more connected than the MA (note that MA can be filtered by post-processing to achieve connectedness, see for instance [13]). Furthermore the $\mathcal{C}$-CA contains much less isolated parasitic points, thanks to the sequence of thresholds which play the role of noise filter when constructing the $\mathcal{C}$-CA. The $\mathcal{C}$-CA is not yet an ideal filtering of the MA: it is not guarantied to be connected, and may also be relatively thick in some configurations. A further study of the influence of the threshold values should be realized.

4.4 Complexity of the Algorithms

Two preliminary algorithms operate before the very generation of the CA. The first one is the distance transformation algorithm, which computes the DT of the object in time $O(m.L^2)$ for an image of side length L and a chamfer mask having m weightings. The second one is the MA extraction algorithm, whose complexity is $O(k.L^2)$ (k being a constant close to m, see [7]). The CA is then produced from the MA: for each point of the MA, the boundary of a maximal ball is checked. A ball of radius R pixels has about $2\Pi.R$ boundary points, the examination of the boundary of the ball takes $O(R)$, R being smaller than $L/2$. Therefore the overall complexity is $O(m.L^2 + n_{MA}.L)$, with n_{MA} being the number of points of the MA. However these points represent only a little portion of the points of the shape; in practice we observe that the time complexity is linear in the number of points of the image.

5 Conclusion

In this paper, we have adapted Prasad's continuous definition of the chordal axis (CA) [1] in the discrete plane $\mathbb{Z}^2$. We have proposed to extract it on a distance transform (DT) in the case of chamfer norms, which allows fast extraction of medial axis (MA) [7]. We have introduced three quantities, namely point threshold, extension threshold and radius increment, in the purpose of discretizing the tangency of discrete balls to the boundary of a shape, and properly characterizing maximal chords of tangency (MCT). Then our algorithm extracts the MCTs in near linear time. Changing thresholds enables to filter the noise in shapes, and concentration of CA points allows shape decomposition. Comparison between discrete CA and MA is achieved, which shows a set of new properties for shape description. Future work concerns local detection on DT, study of the reversibility degree, and extension in 3D.

References

1. Prasad, L.: Morphological analysis of shapes. CNLS Newsletter **139** (1997)
2. Blum, H.: A transformation for extracting new descriptors of shape. In Wathen-Dunn, W., ed.: Models for the Perception of Speech and Visual Form, Cambridge, MIT Press (1967) 362–380
3. Pfaltz, J., Rosenfeld, A.: Computer representation of planar regions by their skeletons. Comm. of ACM **10** (1967) 119–125
4. Montanari, U.: Continuous skeletons from digitized images. Journal of the ACM **16**(4) (1969) 534–549
5. Attali, D., Montanvert, A.: Semicontinuous skeletons of 2d and 3d shapes. In: Aspects of Visual Form Processing, World Scientific, Singapore (1994) 32–41
6. Prasad, L.: Rectification of the chordal axis transform and a new criterion for shape decomposition. In: 11th DGCI, Poitiers (2005)
7. Remy, E., Thiel, E.: Medial Axis for Chamfer Distances: computing LUT and Neighbourhoods in 2D or 3D. Pattern Recognition Letters **23**(6) (2002) 649–661
8. Remy, E., Thiel, E.: Exact Medial Axis with Euclidean Distance. Image and Vision Computing **23**(2) (2005) 167–175
9. Borgefors, G.: Distance transformations in arbitrary dimensions. Computer Vision, Graphics and Image Processing **27** (1984) 321–345
10. Rosenfeld, A., Pfaltz, J.: Sequential operations in digital picture processing. Journal of ACM **13**(4) (1966) 471–494
11. Borgefors, G.: Distance transformations in digital images. Computer Vision, Graphics and Image Processing **34** (1986) 344–371
12. Thiel, E.: Géométrie des distances de chanfrein. HDR, Univ. de la Méditerranée, Aix-Marseille 2 (2001) `http://www.lif-sud.univ-mrs.fr/~thiel/hdr`.
13. Attali, D., Sanniti di Baja, G., Thiel, E.: Skeleton simplification through non significant branch removal. Image Processing and Communications **3**(3-4) (1997) 63–72

Attention-Based Mesh Simplification Using Distance Transforms

Susana Mata[1], Luis Pastor[1], and Angel Rodríguez[2]

[1] Dpto. de Arquitectura y Tecnología de Computadores, Ciencias de la Computación e Inteligencia Artificial, U. Rey Juan Carlos (URJC)
C. Tulipán, s/n., 28933 Móstoles, Madrid, Spain
{susana.mata, luis.pastor}@urjc.es
[2] Dept. de Tecnología Fotónica, U. Politécnica de Madrid (UPM)
Campus de Montegancedo s/n, 28660 Boadilla del Monte, Spain
arodri@dtf.fi.upm.es

Abstract. Although widely used for image processing, Distance Transforms have only recently started to be used in computer graphics. This paper proposes a new mesh simplification technique based on Distance Transforms that allows taking into account the proximity of a mesh element to the focus of attention for adapting the approximation error which will be tolerated during the simplification process to the relative importance of that mesh element. Experimental results show the feasibility of this approach.

1 Introduction

Thanks to the advances in Computer Graphics realistic renderings of complex scenes can be achieved. Multiresolution modelling is one of the underlying techniques that allow to process complex scenes at interactive rates. Basic principles of this approach were set by James Clark [1] and comprehensive surveys can be found at [2, 3, 4, 5, 6].

The generation of models at different resolutions can be performed by means of simplification methods, providing reduced models that use less computational resources but look similar to the original one when rendered under certain conditions [7,8]. Within this context, the response of the human visual system plays a crucial role in determining which details become more obvious to the final observer. High visual acuity is known to be limited to a certain visual angle, perceiving a quality degradation in regions that fall outside the focus of attention [9]. Therefore, it may be useless to render a region of a model at high resolution when its quality will be degraded by our visual system.

The goal of this paper is to propose a new technique that allows taking into account the proximity of a mesh element to the focus of attention for adapting the approximation error which will be tolerated during the simplification process to the relative importance of that mesh element. More specifically, the contributions of this work can be briefly summarized as follows:

A. Kuba, L.G. Nyúl, and K. Palágyi (Eds.): DGCI 2006, LNCS 4245, pp. 283–294, 2006.

- Analyzing the applicability of Distance Transforms for detecting the proximity of mesh elements to a focus of attention defined by a point on the screen.
- Proposing a new mesh simplification criterion based on an eccentricity measure, obtained by means of a Distance Transform.
- Presenting a simplification technique using a Distance Transform for selecting the allowed approximation error depending on the eccentricity from the focus point.

The rest of the paper is organized as follows: Section 2 presents a short overview background of mesh simplification algorithms, digital Distance Transforms and the Multi-Tessellation. Section 3 describes the proposed approach, while Section 4 shows some experimental results. Finally the conclusions and future work are presented in Section 5.

2 Background

2.1 Mesh Simplification

Many mesh simplification techniques have been proposed during the last years. Among the methods based on objective metrics, work has been done in order to incorporate other attributes besides geometry like color, texture or normals [10, 11]. Perceptual metrics have also been developed during this time [12, 13]. Lindstrom and Turk use an image metric to guide the simplification process [14]. Reddy introduced a perceptive model to guide the selection of the appropriate level of detail (LOD) [15]. Luebke defined a contrast sensitivity function that predicts the perception of visual stimuli [16].

According to human ocular physiology, high visual acuity is limited by the dimension of the fovea to approximately a visual angle of 2°. Outside this area, the quality of the image perceived by our visual system declines with eccentricity [17].

Some approaches have been proposed in order to deal with the phenomena of foveal vision, either by adapting the image resolution to the eccentricity [17, 18], or by decreasing the mesh resolution in regions far away from the focus point [19, 20]. For example, Baudisch *et al.* present a selection of techniques for the design of attentive displays that take into account the distinction between foveal vision and peripheral vision [17], and Murphy and Buchowski evaluate the visual angle in world coordinates and use it for the resolution degradation in their presented gaze contingent level of detail [19].

2.2 Digital Distance Transforms

Measuring the distance between image elements may be of interest for further processing in many image analysis applications. Basics concepts regarding digital distances can be found in [21, 22].

The application of a Distance Transform to a binary image produces as output a distance image, where each element of this distance image is assigned a distance label. For any element belonging to the object, its label stores a value indicating its closest distance to the complement of the object.

A distance transform can be computed in two steps by propagating local distances over the image, regardless of its dimensions: 2D, 3D or higher dimensions [21, 23]. Initially, the elements belonging to the object are set to infinity and the elements belonging to the background are set to 0. In the case of a 2D image, during the first step the image is analyzed from top to bottom and from left to right. During the second step, the image elements are visited in reverse order, from right to left and from bottom to top. Each element is updated with the minimum value between its current value and the values of the already visited neighbours incremented by their connectivity weight. The neighbourhood of a pixel and the distance assigned to each of the neighbours define the final Distance Transform.

Distance Transforms and some variations of them in combination with other image processing techniques can be applied for representing and analyzing 3D objects in multiple applications [24, 25, 26, 27]. Distance information computed in the object space has also been applied in computer graphics environments for collision detection [28]. However, the authors are not aware of any previous work applying Distance Transforms from a focus point in the image space in order to guide a simplification process.

2.3 Multi-Tessellation

The Multi-Tessellation method, originally called Multi-Triangulation, was introduced by De Floriani *et al.* [29]. It provides a general multiresolution framework for polygonal meshes offering several attractive features like selective refinement, locality or dynamic update [30].

Multi-Tessellation, MT for short, is a hierarchical model that can be generated during an off-line simplification process, provided that the simplification method has been adapted in order to build an MT. The MT stores the mesh updates together with the approximation error.

Once the MT has been built, it can be queried at run time for extracting a simplified mesh fulfilling some defined restrictions. Some useful restrictions are already implemented in the distributed package [31], while the implementation of new ones can be easily done.

3 Method Description

The approach followed here classifies the mesh faces or vertices attending to their proximity to a focus of attention for a particular point of view. This focus of attention is set in image coordinates and its position may be obtained by several means such as eye trackers or pointing devices.

The classification process uses a Distance Transform, computed over the 2D image; this transform provides for each pixel of the image its distance to the attention point. A labelled mesh is obtained by backprojecting the distance value of a pixel to the mesh elements that project into it. These tags may be assigned either to mesh vertices or faces, and they represent the distance of the mesh element to the attention point, measured in the image space. The tags assigned to the polygonal mesh elements can then be used in different ways to guide the simplification process, providing a criterion for modifying locally the approximation error allowed in areas close to the attention point.

The projection of the mesh over the 2D visualization plane and the backprojection of distance labels for every pixel of the image is an expensive process in terms of computation time. Nevertheless, it has to be pointed out that the mesh projection process can be performed off-line in order to obtain a discrete set of edited models that can be further processed by any other multiresolution methods.

To reduce the method's computational requirements, some auxiliary information is collected previously to the simplification and rendering stage. In particular, a 2D grid is created on the visualization plane, which needs a resolution level that has to be selected by the user according to the input mesh level of detail and to the desired output mesh's level of detail.

Every cell in the 2D grid has an associated list containing the indexes of the vertices that project into it. In a straightforward way, every vertex of the 3D model can also be labelled with the indexes of the 2D grid cell where it projects. This information is very useful to speed up the backprojection of distance labels to the mesh vertices, since the computation of the Distance Transform will be carried out in an image with the same resolution as this 2D grid.

The same process can be applied for backprojecting the distance values to the mesh's faces instead of to the vertices.

The use of a Distance Transform from a focus of attention requires the modification of the technique presented in [32] by computing the Distance Transform on-line during the simplification process. Thus, the grid computation can be used for any focus of attention given a certain point of view

It must be highlighted that the computation of the 2D grid for a particular point of view is performed in a preprocessing stage, producing a mapping between the mesh vertices and the 2D grid cells that will be used later on during the simplification stage. The MT package has been used in this work to present the results obtained by integrating the distance to the focus of attention into the extraction criteria. Figure 1 shows a scheme of the proposed mesh simplification process. The following Sections describe each of the method's stages.

3.1 Mesh Mapping: 2D Grid Computation

The focus of attention is set in the screen space. Therefore the distance of the mesh elements to the attention point must be computed for a certain point of view. Given a visualization plane, the 3D mesh is projected onto it by applying the proper projection matrix to each vertex coordinates. The visualization plane is then partitioned into cells forming a grid which can be seen as a 2D digital image. Every vertex belonging to the projected polygonal mesh is tested to find

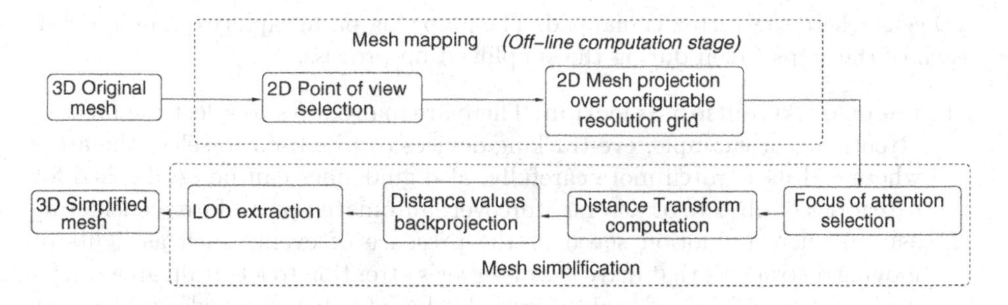

Fig. 1. Overall scheme of the mesh simplification process

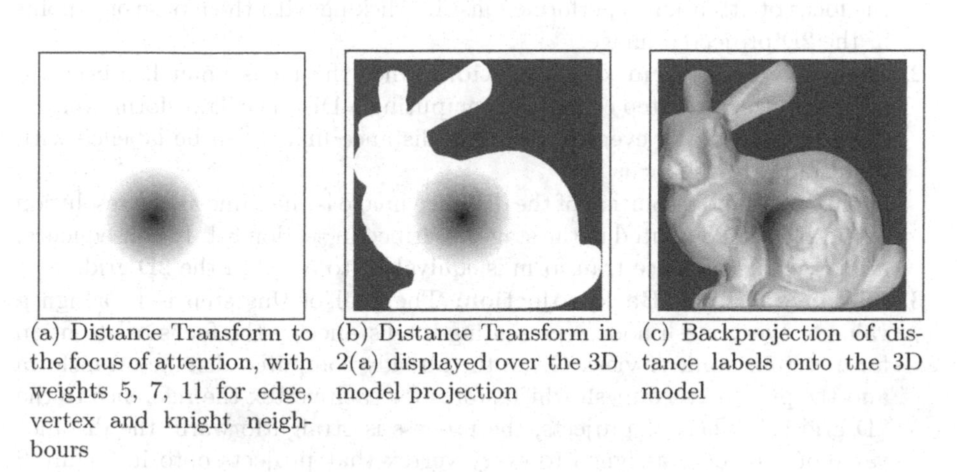

(a) Distance Transform to the focus of attention, with weights 5, 7, 11 for edge, vertex and knight neighbours

(b) Distance Transform in 2(a) displayed over the 3D model projection

(c) Backprojection of distance labels onto the 3D model

Fig. 2. Distance Transform backprojection

the cells of the 2D grid where the vertex is projected. This way, every vertex v_k stores information about the indices (i, j) of the grid cell associated to it.

Two remarks can be pointed out:

- It is possible to perform the whole process using different resolution levels for the 2D image onto which the mesh is projected. This affects in turn the resolution with which the 3D mesh analysis process is carried out.
- As mentioned above, the procedure is computationally expensive, although it is affordable as a mesh preprocessing stage.

3.2 Mesh Simplification

This stage represents the final goal of the proposed method, which is to obtain a simplified mesh where the region close to a focus of attention remains at higher resolution, while the approximation error increases with eccentricity.

Before the mesh simplification, the preprocessing stage described in the previous section has to be carried out, storing for each vertex the indexes of the

2D cells where the vertex is mapped. The following paragraphs explain in detail each of the steps taken during the simplification process.

1. **Focus of Attention Selection:** There are many ways to select the focus of attention. For example, eye-tracking devices can be used to select the areas where end user watch more carefully; also guidelines can be established for categories of objects or actors. Moreover, automated procedures considering aspects such as motion speed or the presence of events such as lights or moving distractors that drive the observer's attention to a certain area can be devised. Last, "focus of neglect" can also be introduced in order to decrease the resolution in specific areas of the object model. In this work, selecting the focus of attention is performed just by clicking with the mouse on a point in the 2D projected image.
2. **Distance Transform Computation:** Once the focus point has been established, the next step consists in computing a Distance Transform over the 2D image. This way every pixel in the distance image will be labelled with its distance to the focus.
In this work, the resolution of the distance image is the same as the resolution of the 2D grid computed in the stage described in section 3.1. In consequence, a pixel in the distance transform is equivalent to a cell in the 2D grid.
3. **Distance Values Backprojection:** The goal of this step is to assign a value to every mesh node representing its distance to the focus of attention for a certain point of view. Given the already computed distance transform and the preprocessed mesh which stores for each vertex the reference to the 2D grid cell where it projects, the process is straightforward: the distance value of a pixel is assigned to every vertex that projects onto it. Figure 2 shows this process graphically.
4. **Level Of Detail Extraction:** It is in this last stage where the extracted distance values are used for mesh simplification purposes.
Evidently, how distance labels are used depends on the selected simplification technique. The work presented here has been based on the Jade approach, a vertex decimation technique based on the global error [33]. The distance information is computed for the vertices of the original mesh. Since the vertices belonging to a simplified model are a subset of the original mesh, the precomputed distance labels are valid for any level of detail [1].
The proximity of every facet to the focus of attention is taken into account during this stage. This means that for a given error threshold, the error allowed in regions close to the focus point is reduced according to a predefined law. The implemented solution, requires the definition of two parameters:
   - Distance interval: range of distance labels which identify the region where a more accurate approximation is desired.
   - Error factor: the purpose of this parameter is to define an error threshold for the portion of the mesh within the region of interest. This threshold,

[1] Multi-Tessellations obtained through the application of the Jade method are freely distributed with the MT-Package.

different from the global error threshold, is defined as a function of the global error threshold.

The size of the region of interest can be simply modified by changing the range of distance labels that define it. In our case, the range is defined by setting a threshold over the minimum distance of the vertices belonging to a face. Other solutions can be easily devised.

The error factor allows to refine the quality of the approximation in the region of attention taking into account the threshold error fixed for the rest of the model. This way, the allowed error in the region of interest is $f \cdot e$, where e is the error permitted in the rest of the model and f is the error factor. Again, other error functions are also feasible.

It has to be noted that this approach uses only two approximation errors. The proposed technique can be easily extended in order to produce a variable approximation error which is dependent on the distance to the focus of attention. Nevertheless, the fact that the Multi-Tesselation method produces smooth transitions between regions with different levels of detail makes this approach less necessary.

4 Results

The experimental results presented in this section were obtained by applying the technique previously described to the Multi-Tessellation models distributed together with the MT-Package.

Figure 3(a) shows the full resolution mesh of the Stanford bunny rendered with smooth shading. As it was explained in paragraph 3.2.1 the focus of attention is established by clicking with the mouse on a pixel of the screen. In this first example, a pixel in the bunny's head is chosen as the focus of attention (pixel (34,89) in an image of dimensions 250x248). Figure 3(b) shows the result of computing a Distance Transform from the selected point over the image. The distance of a pixel is represented by a grey level, restricted to a maximum value of 255 only for displaying purposes.

Figure 3(e) shows a simplified model where the mesh has been refined taking into account the attention area. It can be observed how the bunny's head shows a high resolution, while the rest of the model has been strongly simplified. In these example, all the vertices with a distance label less than 178 have been preserved by setting an error factor of 0, obtaining a mesh with no simplification error in this region. Figure 3(c) shows the mesh of Figure 3(e) rendered in wireframe. The extremely high density of faces in the bunny's head can be appreciated. This restrictive error threshold can be relaxed, while still producing a simplified model where the bunny's head keeps a good resolution, as can be seen in Figure 3(d). In this case, the error factor has been set to 0.05; this means that the allowed error in the region of attention is 0.05 times the error allowed in the rest of the model.

Figure 4 shows the full resolution mesh of a shell rendered with smooth shading. Besides the quality of the simplification within the region of interest, the

(a) Original mesh of the Stanford bunny.

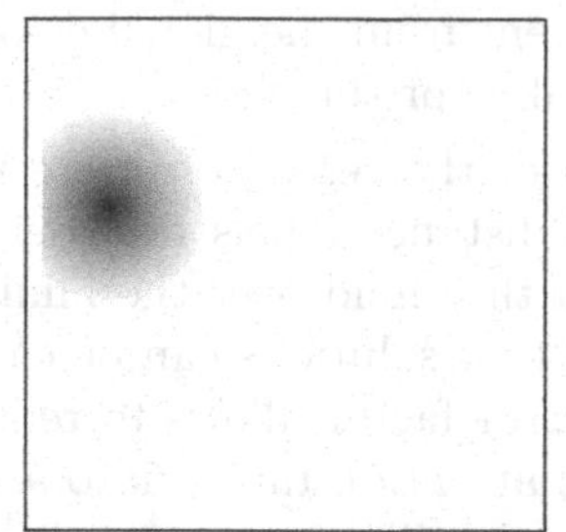

(b) Distance transform to the focus of attention computed for a given point of view.

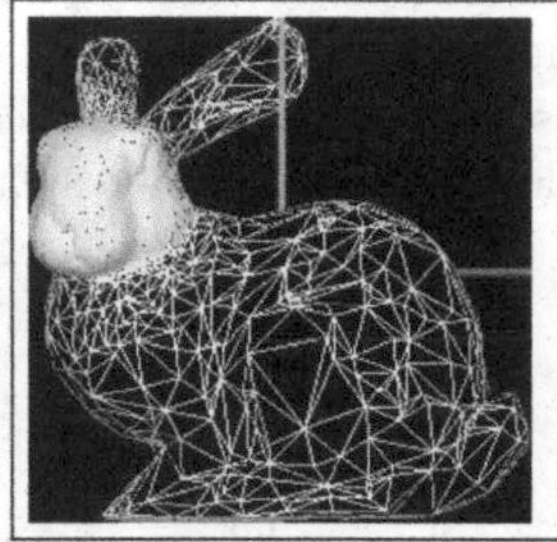

(c) Mesh with no simplification within the distance thresholded region.

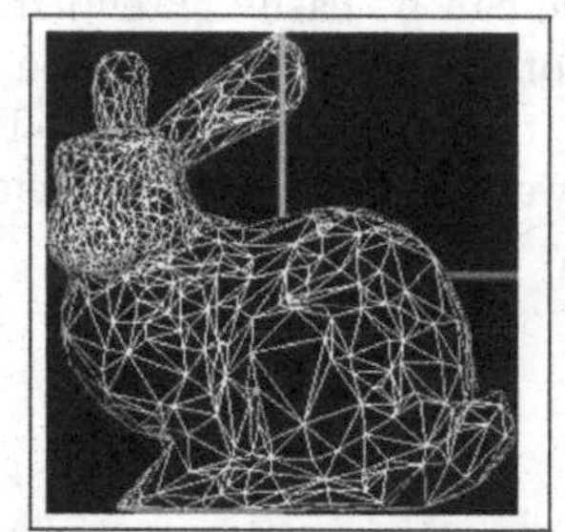

(d) Mesh with mild simplification within the distance thresholded region.

(e) Mesh in figure 3(d) rendered with smooth shading

Fig. 3. Distance Transform based simplification of a 3D mesh

(a) Smooth shading

(b) Flat shading

(c) Wireframe

Fig. 4. Original mesh of the 3D model of a shell

presented method allows to parameterize the extension of this region. It is known that the size of the focus of attention depends on some factors, like the task the observer is performing at a given time [17]. Figure 5 shows the effect of modifying the diameter of the attention region, by setting different thresholds over the distance labels which will define the region of interest. It can be observed how the high resolution area extends itself as the distance threshold is increased.

Fig. 5. Simplifications of a 3D mesh varying the distance threshold, rendered with smooth shading (5(a), 5(d), 5(g), 5(j)), flat shading (5(b), 5(e), 5(h), 5(k)) and in wireframe (5(c), 5(f), 5(i), 5(l))

Regarding computational issues, the cost in terms of memory requirements is just one extra variable per vertex. In case of tagging the mesh faces or mesh edges, an additional value per tagged element would be required. With respect to computational cost, it has to be noted that all the heavy computation is performed at preprocessing time. The most expensive step is the mesh mapping over the 2D grid, in order to collect the information needed for backprojecting the distance values. Efficient implementations for these operations using spatial data partitioning could be considered. As it was explained above, the computation of the Distance Transform can be performed involving only two passes over the 2D image.

5 Conclusions and Future Work

Simplification algorithms are usually guided by some criteria in order to select which elements of the mesh shall be removed or replaced. Introducing distance labels into the guiding metrics is a straightforward process, opening a new way to design a range of techniques which are useful for including perceptually motivated criteria in mesh simplification algorithms.

The results presented here suggest that the use of distance information is a promising approach for attention-based mesh simplification techniques, since adding distance labels to mesh elements provides a natural way to model the degradation of the perception's resolution with eccentricity.

The fact that some distance information can be assigned to any element of the mesh (vertices, edges or faces) facilitates adapting these techniques to a wide range of simplification methods. The nature of the basic underlying operator (vertex removal, edge collapse, etc) does not impose additional limitations. Furthermore, the applicability of distance labels goes from off-line simplification processing to run-time selective refinement.

The work presented here computes the mesh elements' distance to a screen point of attention for a fixed point of view. Future work includes:

- Extending the method for covering all possible points of view in a way which is both performant and computationally efficient.
- Integrating distance to the attention focus into other mesh simplification methods besides the Jade method.
- Extending the method in order to deal with multiple focuses of attention.
- Including "focus of neglect", in order to explicitly select areas where low resolution is preferred.

Acknowledgments

This work has been partially funded by the Spanish Ministry of Education and Science (grant TIC2003-08933-C02) and Government of the Community of Madrid (grant S-0505/DPI/0235). The authors would like to thank Stanford 3D Scanning Repository for making the data sets used in this research publicly available. The authors also thank to the Geometric Modelling and Computer Graphics Research Group for distributing the MT-Package. Special thanks to the reviewers for their valuable comments.

References

1. Clark, J.H.: Hierarchical geometric models for visible surface algorithms. Communications of the ACM **19** (1976) 547–554
2. Puppo, E., Scopigno, R.: Simplification, LOD and multiresolution principles and applications. In Fellner, D., Szirmay-Kalos, L., eds.: EUROGRAPHICS'97. Volume 16. (1997) Tutorial Notes PS97 TN4.
3. Garland, M.: Multiresolution modeling: Survey & future opportunities. In: EUROGRAPHICS'99. (1999)
4. Luebke, D.P.: A developer's survey of polygonal simplification algorithms. IEEE Computer Graphics and Applications **21**(3) (2001) 24–35
5. Luebke, D., Reddy, M., Cohen, J.D., Varshney, A., Watson, B., Huebner, R.: Level of Detail for 3D Graphics. Morgan Kauffmann (2003)
6. De Floriani, L., Kobbelt, L., Puppo, E.: A survey on data structures for level-of-detail models. In N.Dodgson, M.Floater, M.Sabin, eds.: Advances in Multiresolution for Geometric Modelling. Series in Mathematics and Visualization. Springer Verlag (2004) 49–74
7. Xia, J.C., Varshney, A.: Dynamic view-dependent simplification for polygonal models. In Yagel, R., Nielson, G.M., eds.: IEEE Visualization '96. (1996) 335–344
8. Hoppe, H.: View-dependent refinement of progressive meshes. In: SIGGRAPH'97, Computer Graphics Annual Conference Series, ACM (1997) 189–198
9. Ferwerda, J.A.: Elements of early vision for computer graphics (tutorial). IEEE Computer Graphics and Applications **21**(5) (2001) 22–33
10. Garland, M., Heckbert, P.: Simplifying surfaces with color and texture using quadric error metrics. In: Proceedings of IEEE Visualization'98. (1998) 263–270
11. Cohen, J., Olano, M., Manocha, D.: Appearance-preserving simplification. In: Proceedings of SIGGRAPH 98. (1998) 115–122
12. O'Sullivan, C., Howlett, S., Morvan, Y., McDonnell, R., O'Conor, K.: Perceptually adaptive graphics. In Schlick, C., Purgathofer, W., eds.: Eurographics State-of-the-Art Report (EG-STAR). Volume 6., Eurographics Association (2004) 141–164
13. Cheng, I., Boulanger, P.: A 3D perceptual metric using just-noticeable-difference. In: Proc. of Eurographics 2005, Eurographics (2005) 97–100
14. Lindstrom, P., Turk, G.: Image-driven simplification. In: ACM Transactions on Graphics. Volume 19 of 3. (2000) 204–241
15. Reddy, M.: Perceptually Modulated Level of Detail for Virtual Environments. Ph. D. dissertation, University of Edinburgh (1997)
16. Luebke, D.P.: View-dependent simplification of arbitrary polygonal environments. Ph. D. dissertation, University of North Carolina (1998)
17. Baudisch, P., DeCarlo, D., Duchowski, A.T., Geisler, W.S.: Focusing on the essential: considering attention in display design. Communications of the ACM **46**(3) (2003) 60–66 SPECIAL ISSUE: Attentive user interfaces.
18. Kleinfelder, S.: Foveated imaging on a smart focal plane. Web (1999) Retrieved june 26, 2006, from source. http://scien.stanford.edu/class/psych221/projects/99/stuartk/fovis.html
19. Murphy, H., Duchowski, A.: Gaze-contingent level of detail rendering. In: Proceedings of EuroGraphics 2001. (2001) 219–228
20. Levoy, M., Whitaker, R.: Gaze-directed volume rendering. In: SI3D '90: Proceedings of the 1990 symposium on Interactive 3D graphics, New York, NY, USA, ACM Press (1990) 217–223

21. Rosenfeld, A., Pfaltz, J.: Sequential operations in digital picture processing. Journal of the Association for Computing Machinery **13**(4) (1966) 471–491
22. Rosenfeld, A., Pfaltz, J.: Distance functions on digital pictures. Pattern Recognition **1** (1968) 33–61
23. Borgefors, G.: Distance transformations in digital images. Computer Vision, Graphics and Image Processing **34** (1986) 344–371
24. Perry, R.N., Frisken, S.F.: Kizamu: A system for sculpting digital characters. In Jain, L.C., ed.: Proc. of ACM SIGGRAPH, ACM Press (2001) 47–56 ISBN: 1-58113-374-X.
25. Svensson, S.: Representing and Analyzing 3D Digital Shape Using Distance Information. Ph. D. dissertation, Swedish University of Agricultural Sciences (2001)
26. Šrámek, M.: Distance fields in visualization and graphics. In Viola, I., Theussl, T., eds.: Proc. of CESCG'2002, Budmerice, Slovak Republic (2002) 13–16
27. Jones, M.W., Bærentzen, A., Šrámek, M.: Discrete 3D distance fields: Techniques and applications. IEEE Transactions on Visualization and Computer Graphics **12**(4) (2006) 581–599
28. Teschner, M., Kimmerle, S., Zachmann, G., Heidelberger, B., Raghupathi, L., Fuhrmann, A., Cani, M.P., Faure, F., Magnetat-Thalmann, N., Strasser, W.: Collision detection for deformable objects. In: Eurographics State-of-the-Art Report (EG-STAR), Eurographics Association, Eurographics Association (2004) 119–139
29. De Floriani, L., Magillo, P., Puppo, E.: Building and traversing a surface at variable resolution. In: Proceedings of IEEE Visualization 97. (1997) 103–110
30. De Floriani, L., Magillo, P., Puppo, E.: Efficient implementation of multi-triangulations. In: Proceedings of IEEE Visualization 98. (1998) 43–50
31. Geometric Modeling and Computer Graphics Research Group. DISI - Dipartimento di Informatica e Scienze dell'Informazione University of Genova: The MT (Multi-Tesselation) Package. Web (2005) Retrieved june 26, 2006, from source. http://gmcg.disi.unige.it/
32. Mata, S., Pastor, L., Rodrguez, A.: Silhouette detection for adaptive polygonal mesh simplification using distance transforms. In Braz, J., Jorge, J., Dias, M., Marcos, A., eds.: Proc. International Conference on Computer Graphics Theory and Applications, GRAPP'06, Setbal, Portugal, INSTICC (2006) 43–50
33. Ciampalini, A., Cignoni, P., Montani, C., Scopigno, R.: Multiresolution decimation based on global error. In: The Visual Computer. Volume 13 of 5. (1997) 228–246

Generating Distance Maps with Neighbourhood Sequences

Robin Strand[1], Benedek Nagy[3,*], Céline Fouard[1], and Gunilla Borgefors[2]

[1] Centre for Image Analysis, Uppsala University, Lägerhyddsvägen 3, SE-75237 Uppsala, Sweden
{robin, celine}@cb.uu.se

[2] Centre for Image Analysis, Swedish University of Agricultural Sciences, Lägerhyddsvägen 3, SE-75237 Uppsala, Sweden
gunilla@cb.uu.se

[3] Department of Computer Science, Faculty of Informatics, University of Debrecen, PO Box 12, 4010, Debrecen, Hungary and
Research Group on Mathematical Linguistics, Rovira i Virgili University, Tarragona, Spain
nbenedek@inf.unideb.hu

Abstract. A sequential algorithm for computing the distance map using distances based on neighbourhood sequences (of any length) in the 2D square grid; and 3D cubic, face-centered cubic, and body-centered cubic grids is presented. Conditions for the algorithm to produce correct results are derived using a path-based approach. Previous sequential algorithms for this task have been based on algorithms that compute the digital Euclidean distance transform. It is shown that the latter approach is not well-suited for distances based on neighbourhood sequences.

1 Introduction

In [1], a sequential algorithm for computing distance transforms (DTs, where each object grid point is assigned the distance value to the closest background grid point) was introduced. The authors considered only the simple L_1 (city-block) and L_∞ (chessboard) metrics and they proved that a two-scan algorithm will produce a correct distance map. This is due to the fact that the distances are path-based with fixed adjacency, i.e., the distance between two points is the length of the shortest path between the points in a graph structure. For these distances, unit distance between adjacent grid points (weights) is used. The DTs obtained from L_1 and L_∞ are very rotation-dependent. Basically, two alternative ways to decrease the rotational dependency have been introduced – weighted distances and distances based on neighbourhood sequences (n.s.-distances or octagonal distances, first defined in [2]). With weighted distances (each local step is assigned a weight), the weights are allowed to have different values than one. The literature on weighted distances is rich, see for example the

* The research is partly supported by grants OTKA F043090 and T049409.

A. Kuba, L.G. Nyúl, and K. Palágyi (Eds.): DGCI 2006, LNCS 4245, pp. 295–307, 2006.

early paper [3]. Because of the fixed adjacency, the two scan algorithm applies for weighted distances on *any* point-lattice, [4]. With weighted distances, the rotational dependency is low also for short distances. This is not the case for n.s.-distances, where the adjacency relation is allowed to vary along the path. On the other hand, all distance values in each shortest path of length n between two points consist of all integer values $1, \dots, n$. This makes the n.s.-distances well suited for morphological operations such as dilation and erosion where the object should be divided into layers.

N.s.-distances have been considered by many authors and in most papers, the theoretical properties of n.s.-distances are examined. The theory on n.s.-distances is developed in, e.g., [5,6]. Distances based on neighbourhood sequences are also of value in applications and has been used for e.g. skeletonization [7] and shading of three-dimensional objects [8]. For these applications to be efficient, a fast algorithm for computing the distance transform is of great value.

The situation for n.s.-distances is a bit more complex than for weighted distances – allowing the adjacency relation to vary implies that a two-scan algorithm is not sufficient. In previous algorithms for computing the DT for n.s.-distances [9,3,10], the scanning procedure designed for computing the Euclidean DT [9, 3, 11] were used. We will see that this approach is not appropriate for n.s.-distances.

Non-standard grids such as the face-centered cubic (fcc) grid and the body-centered cubic (bcc) grid has gained more and more attention in the last decade. One reason is that less samples can be used with the same representation/reconstruction quality compared to the cubic grid [12,13]. For example, image acquisition techniques [12,13], image processing algorithms [14,15,16] and visualization techniques [17] have been developed for these grids.

In this paper, conditions for sequential algorithms in the square, cubic, face-centered cubic, and the body-centered cubic grids to produce correct results are derived, independently from the algorithm designed for Euclidean DT, using a path-based approach.

2 Preliminaries

In this paper, we will consider the square grid $\mathbb{Z}^2$, the cubic grid $\mathbb{Z}^3$, the fcc grid $\mathbb{F}$, and the bcc grid $\mathbb{B}$. When handled in parallel, $\mathbb{G}$ is used to denote all of the four grids.

Two grid points $\mathbf{x} = (x^1, x^2, \dots, x^n), \mathbf{y} = (y^1, y^2, \dots, y^n) \in \mathbb{Z}^n$ $(n \in \mathbb{Z}_+)$ are ρ-neighbours, $1 \le \rho \le n$, if

$$\sum_{i=1}^{n} |x^i - y^i| \le \rho \text{ and } \max_{i \in \{1,2,\dots,n\}} |x^i - y^i| = 1.$$

The face-centered cubic grid $\mathbb{F}$ and the body-centered cubic grid $\mathbb{B}$ are defined as follows:

$$\mathbb{F} = \{(x^1, x^2, x^3) \in \mathbb{Z}^3 : x^1 + x^2 + x^3 \equiv 0 \pmod 2\}, \tag{1}$$

$$\mathbb{B} = \{(x^1, x^2, x^3) \in \mathbb{Z}^3 : x^1 \equiv x^2 \equiv x^3 \pmod 2\}. \tag{2}$$

Two grid points $\mathbf{x}, \mathbf{y} \in \mathbb{F}$ or $\mathbb{B}$ are ρ-neighbours, $1 \leq \rho \leq 2$ if

$$\sum_{i=1}^{3} |x^i - y^i| \leq 3 \text{ and } \max_{i \in \{1,2,3\}} |x^i - y^i| \leq \rho.$$

The neighbourhood relations in our four grids are visualized in Figure 1 by showing the Voronoi regions (the pixels (2D) and voxels (3D)) corresponding to some grid points.

The points $\mathbf{x}, \mathbf{y} \in \mathbb{G}$ are *adjacent* if $\mathbf{x}$ and $\mathbf{y}$ are ρ-neighbours for some ρ. The ρ-neighbours which are not $(\rho - 1)$-neighbours are called *strict* ρ-neighbours. A neighbourhood sequence B in $\mathbb{G}$ is a sequence $B = (b(i))_{i=1}^{\infty}$, where each $b(i)$ denotes a neighbourhood relation in $\mathbb{G}$. If B is periodic, i.e., if for some fixed $l \in \mathbb{Z}_+$, $b(i) = b(i+l)$ is valid for all $i \in \mathbb{Z}_+$, then we write $B = (b(1), b(2), \ldots, b(l))$.

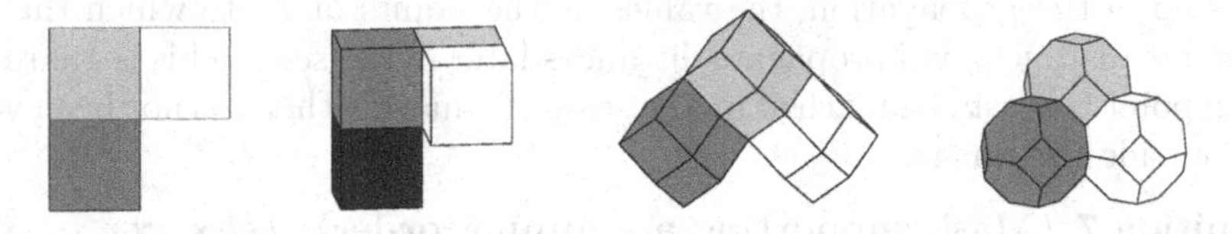

Fig. 1. The grid points corresponding to the light grey, dark grey, and black pixels/voxels are 1-, (strict) 2-, and (strict) 3-neighbours to the grid point corresponding to the white pixel/voxel, respectively. From left to right: $\mathbb{Z}^2$, $\mathbb{Z}^3$, $\mathbb{F}$, and $\mathbb{B}$.

A *path*, denoted $\mathcal{P}$, in a grid is a sequence $\mathbf{x} = \mathbf{p}_0, \mathbf{p}_1, \ldots, \mathbf{p}_n = \mathbf{y}$ of adjacent grid points. A path is a *B-path* of length n if, for all $i \in \{1, 2, \ldots, n\}$, $\mathbf{p}_{i-1}$ and $\mathbf{p}_i$ are $b(i)$-neighbours.

Definition 1. *The B-distance $d(\mathbf{x}, \mathbf{y}; B)$ between the points $\mathbf{x}$ and $\mathbf{y}$ is the length of (one of) the shortest B-path(s) between the points.*

Given a path of length n, the following notation is used: $\omega_i = \mathbf{p}_i - \mathbf{p}_{i-1}$.

A *prime vector* is a vector between a grid point and an adjacent grid point. Let $\Gamma = \{\vec{p}_1, \vec{p}_2, \ldots, \vec{p}_m\}$. The set $\{\mathbf{x} \in \mathbb{G} : \mathbf{x} = \sum \alpha_i \vec{p}_i \text{ for any } \alpha_i \in \mathbb{R}_+\}$ is called the *Γ-sector.*

We consider finite subsets of $\mathbb{G}$ for the algorithm:

Definition 2 (Image). *The finite subset $\mathcal{I}_{\mathbb{G}}$ of $\mathbb{G}$ is denoted the image domain. We call the function $f : \mathcal{I}_{\mathbb{G}} \longrightarrow \mathbb{N}$ an image.*

Definition 3 (Foreground and background). *We denote the image foreground X and the background $\overline{X}$. These sets have the following properties:*

1. $X \subset \mathcal{I}_{\mathbb{G}}$ *and* $\overline{X} \subset \mathcal{I}_{\mathbb{G}}$
2. $X \cap \overline{X} = \emptyset$
3. $X \cup \overline{X} = \mathcal{I}_{\mathbb{G}}$.

3 The Sequential Algorithm

Definition 4 (Distance map). *Given a neighbourhood sequence B, the* distance map, DM_X, *of f is a grey level image, where the value of each point of the foreground corresponds to its shortest distance to the background, i.e.*

$$DM_X : \begin{cases} \mathcal{I}_{\mathbb{G}} \longrightarrow \mathbb{N} \\ \mathbf{x} \longmapsto d(\mathbf{x}, \overline{X}; B) = \inf_{\mathbf{y} \in \overline{X}} d(\mathbf{x}, \mathbf{y}; B) \end{cases}$$

Definition 5 (Scanning mask). *A* scanning mask $\mathcal{M}$ *is the set of vectors from the origin to some grid points adjacent to* $\mathbf{0}$.

Definition 6 (Scanning order). *A* scanning order *(so) is an ordering of the $M = \text{card}(\mathcal{I}_{\mathbb{G}})$ distinct points in $\mathcal{I}_{\mathbb{G}}$, denoted $\mathbf{x}_1, \mathbf{x}_2, \ldots, \mathbf{x}_M$.*

For a scanning mask to propagate distances correctly, it is important that, in each step of the propagation, the values at the points in $\mathcal{I}_{\mathbb{G}}$ to which the mask propagate distances will propagate distances later in the scan. This is guaranteed if each point that can be reached by the scanning mask either has not been visited or is outside the image.

Definition 7 (Mask supporting a scanning order). *Let $\mathbf{x}_1, \mathbf{x}_2, \ldots, \mathbf{x}_M$ be a scanning order and $\mathcal{M}_l$ a scanning mask. The scanning mask $\mathcal{M}_l$* supports the scanning order *if*

$$\forall \mathbf{x}_i, \forall \vec{v}_j \in \mathcal{M}_l, ((\exists i' > i : \mathbf{x}_{i'} = \mathbf{x}_i + \vec{v}_j) \text{ or } (\mathbf{x}_i + \vec{v}_j \notin \mathcal{I}_{\mathbb{G}})).$$

Remark 1. If $\vec{v} \in \mathcal{M}_k$ for some k, then $-\vec{v} \notin \mathcal{M}_k$. If both $\vec{v}, -\vec{v} \in \mathcal{M}_k$, then by Definition 7, there is an i' such that $\mathbf{x}_{i'} = (\mathbf{x}_i + \vec{v}) + (-\vec{v}) = \mathbf{x}_i$. This is not possible since each grid point occurs only once in $\mathcal{I}_{\mathbb{G}}$.

Algorithm 1. *Initially, $f(\mathbf{x}) = \infty$ if $\mathbf{x} \in X$ and $f(\mathbf{x}) = 0$ if $\mathbf{x} \in \overline{X}$. The image domain $\mathcal{I}_{\mathbb{G}}$ is scanned L times using scanning orders such that the scanning masks $\mathcal{M}_i$, $1 \le i \le L$ support the scanning orders so_i.*

```
for i = 1 : L
  for all x ∈ I_G following so_i
    if f(x) < ∞
      for all v ∈ M_i
        if x and x + v are b(f(x) + 1)-neighbours
          f(x + v) ← min (f(x) + 1, f(x + v))
```

Example 1. Consider the image in $\mathbb{Z}^2$ shown in Figure 2(a) (the grid points are visualized by their pixels). The masks $\mathcal{M}_1$ and $\mathcal{M}_2$ are such that they propagate distances in the directions shown in Figure 2(b) by dashed and solid lines, respectively. The correct distance maps for $B = (1)$ (city block), $B = (2)$ (chessboard), and $B = (1, 2)$ are shown in Figure 2(c), Figure 2(d), and Figure 2(e), respectively. For $B = (1)$ and $B = (2)$, a two-scan algorithm is sufficient to propagate the distance between the two pixels in grey. Two scans are, however, not enough for $B = (1, 2)$.

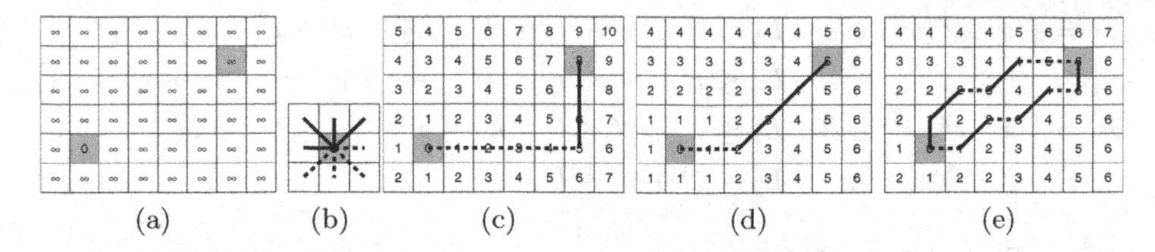

Fig. 2. Distance maps of (a) for $B = (1)$, $B = (2)$, and $B = (1, 2)$ are shown in (c), (d), and (e), respectively. Examples of shortest B-paths are shown in (c)–(e) over the distance maps. Directions supported by the masks $\mathcal{M}_1$ and $\mathcal{M}_2$ are shown as dashed and solid lines, respectively. The directions supported by the masks are shown in (b).

The distance that are propagated depends on previous propagations. Thus, if local steps from $\mathcal{M}_2$ are needed before local steps from $\mathcal{M}_1$ (as in Figure 2(e)), then two scans are not enough.

Since Algorithm 1 only propagates distances from mask i in scan i, there must be a shortest path satisfying the condition in Proposition 1 below for each pair of grid points in $\mathcal{I}_{\mathbb{G}}$ for the propagation of distances to be sufficient.

Proposition 1. *If for each neighbourhood sequence B and each $\mathbf{x}, \mathbf{y} \in \mathcal{I}_{\mathbb{G}}$ there is a shortest B-path in $\mathcal{I}_{\mathbb{G}}$ between $\mathbf{x}$ and $\mathbf{y}$ of length n and integers $T_0, T_1, \ldots, T_L$ s.t.*

$$\omega_i \in \begin{cases} \mathcal{M}_1 \text{ if } 0 = T_0 < i \leq T_1 \\ \mathcal{M}_2 \text{ if } \quad T_1 < i \leq T_2 \\ \quad \vdots \\ \mathcal{M}_L \text{ if } \quad T_{L-1} < i \leq T_L = n \end{cases}$$

then Algorithm 1 returns a distance map DM_X.

The proof of this proposition is obvious: the first scan propagates the T_1 first steps of the path, the second scan propagates the steps $T_1 + 1, \ldots, T_2$ and so on until the last scans propagates the steps $T_{L-1} + 1, \ldots, T_L$.

In the next section, conditions for the masks $\mathcal{M}_i$ to fulfill the condition in Proposition 1 are derived.

4 Theoretic Results

4.1 Conditions for a Path to Be in $\mathcal{I}_{\mathbb{G}}$

Definition 8 (Sector-preserving image domain). *The image domain is called* sector-preserving *if there are integers L_i and U_i s.t. $L_i \leq 0 < U_i$ and*

$$\mathcal{I}_{\mathbb{Z}^2} = \left\{(x^1, x^2) \in \mathbb{Z}^2 \mid L_i \leq x^i \leq U_i\right\}$$
$$\mathcal{I}_{\mathbb{Z}^3} = \left\{(x^1, x^2, x^3) \in \mathbb{Z}^3 \mid L_i \leq x^i \leq U_i\right\}$$

$$\mathcal{I}_{\mathbb{F}} = \{(x^1, x^2, x^3) \in \mathbb{F} \mid L_1 \leq x^1 + x^2 + x^3 \leq U_1,\ L_2 \leq x^1 + x^2 - x^3 \leq U_2,\\ L_3 \leq x^1 - x^2 - x^3 \leq U_3, \textit{ and } L_4 \leq x^1 - x^2 + x^3 \leq U_4\}$$
$$\mathcal{I}_{\mathbb{B}} = \{(x^1, x^2, x^3) \in \mathbb{B} \mid L_1 \leq x^1 + x^2 \leq U_1,\ L_2 \leq x^2 + x^3 \leq U_2, \textit{ and}\\ L_3 \leq x^1 + x^3 \leq U_3\}.$$

Let
$$\Gamma^1_{\mathbb{Z}^2} = \{(1,0),(0,1),(1,1)\}$$
$$\Gamma^1_{\mathbb{Z}^3} = \{(1,0,0),(0,1,0),(0,0,1),(1,1,0),(1,0,1),(0,1,1),(1,1,1)\}$$
$$\Gamma^1_{\mathbb{F}} = \{(1,1,0),(1,0,1),(0,1,1)\}$$
$$\Gamma^2_{\mathbb{F}} = \{(1,1,0),(1,0,1),(1,0,-1),(2,0,0)\}$$
$$\Gamma^1_{\mathbb{B}} = \{(1,1,1),(1,-1,1),(1,1,-1),(2,0,0)\}.$$

Definition 9 (Path with $\Gamma_{\mathbb{G}}$-sector steps). *A path that contains only steps from $\Gamma_{\mathbb{G}}$ is called a path with $\Gamma_{\mathbb{G}}$-*sector steps.

We will see that for any fixed ξ and any grid point $\mathbf{x} \in \mathcal{D}^\xi_{\mathbb{G}}$ defined below, the distance (i.e. the shortest path) between $\mathbf{0}$ and $\mathbf{x}$ is defined by a path with $\Gamma^\xi_{\mathbb{G}}$-sector steps.

$$\mathcal{D}^1_{\mathbb{Z}^2} = \{\mathbf{x} \in \mathbb{Z}^2 : x^i \geq 0\}$$
$$\mathcal{D}^1_{\mathbb{Z}^3} = \{\mathbf{x} \in \mathbb{Z}^3 : x^i \geq 0\}$$
$$\mathcal{D}^1_{\mathbb{F}} = \{\mathbf{x} \in \mathbb{F} : x^1 \geq x^2 \geq x^3 \geq 0 \text{ and } x^1 \leq x^2 + x^3\}$$
$$\mathcal{D}^2_{\mathbb{F}} = \{\mathbf{x} \in \mathbb{F} : x^1 \geq x^2 \geq x^3 \geq 0 \text{ and } x^1 > x^2 + x^3\}$$
$$\mathcal{D}^1_{\mathbb{B}} = \{\mathbf{x} \in \mathbb{B} : x^1 \geq x^2 \geq x^3 \geq 0\}$$

We also will use the following notation: for any $\mathbf{y} \in \mathbb{G}$, $\mathcal{D}_{\mathbb{G}}(\mathbf{y}) = \{\mathbf{y} + \mathbf{x} : \mathbf{x} \in \mathcal{D}_{\mathbb{G}}\}$. The following proposition follows directly from Theorem 3.5 in [18] ($\mathbb{Z}^2$ and $\mathbb{Z}^3$) and from the proofs of Theorem 2 and 5 in [15] for $\mathbb{F}$ and $\mathbb{B}$ respectively.

Proposition 2. *Let the neighbourhood sequence B and the ξ be given. For any point $\mathbf{y} \in \mathcal{D}^\xi_{\mathbb{G}}(\mathbf{x})$ such that $d(\mathbf{x}, \mathbf{y}; B) = n$, there is a shortest B-path between $\mathbf{x}$ and $\mathbf{y}$ with $\Gamma^\xi_{\mathbb{G}}$-sector steps.*

For the square/cubic grid it is obvious that we need vectors having a value 1 and one/two zero(s) to connect the points by only 1-steps. If the neighbourhood sequence contains values 2 and there are coordinate differences in at least 2 coordinates, then vectors changing 2 values simultaneously can be used in a shortest path. In the cubic grid $(1,1,1)$ is used if the neighbourhood sequence contains element 3 and there are differences in all the 3 coordinates.

In the fcc grid if $x^1 \leq x^2 + x^3$ then vectors $(1,1,0)$, $(1,0,1)$, $(0,1,1)$ can produce a shortest path independently of the used neighbourhood sequence (and in case $B = (1)$ all shortest paths built by them).

In case of $x^1 > x^2 + x^3$ then the steps $(0,1,1)$ are not needed, but we need (additional) steps to go in direction $(x^1,0,0)$. If there are values 2 in the neighbourhood sequence, then vector $(2,0,0)$ is used in a shortest path. Having not (enough) values 2 in the neighbourhood sequence a step by $(2,0,0)$ can be substituted by a step $(1,0,1)$ and a step $(1,0,-1)$.

In the bcc grid one can construct a shortest path by vectors $(1,1,1)$ and $(1,1,-1)$ if $x^1 = x^2 \geq x^3 \geq 0$. If $x^1 > x^2$ then step(s) by vector $(2,0,0)$ can also be used having values 2 in the neighbourhood sequence. Without (enough) values 2 in the neighbourhood sequence steps by vectors $(1,1,-1)$ and $(1,-1,1)$ are also used in a shortest path.

Definition 10 (Image border point). *All grid points* $\mathbf{x} \in \mathcal{I}_{\mathbb{G}}$ *such that there is a prime vector* $\vec{v}$ *such that* $\mathbf{x} + \vec{v} \notin \mathcal{I}_{\mathbb{G}}$ *are called* image border points.

Lemma 1. *Let* $\mathbf{y} \in \overline{X}$ *and* $\mathbf{x} \in X$ *be such that* $\mathbf{x} \in \mathcal{D}^{\xi}_{\mathbb{G}}(\mathbf{y})$ *for some* ξ *and there is a shortest* B*-path* $\mathcal{P}$ *between* $\mathbf{y}$ *and* $\mathbf{x}$ *defining* $d(\mathbf{x}, \overline{X}; B)$*. If either*

(A) *all image border points are in the background* or
(B) *the image domain is sector-preserving.*

then $\mathcal{P}$ *is in* $\mathcal{I}_{\mathbb{G}}$*.*

Proof. When condition (A) is fulfilled: For some $\mathbf{y} \in \overline{X}$, let $\mathbf{x} = \mathbf{p}_0, \mathbf{p}_1, \ldots, \mathbf{p}_n = \mathbf{y}$ be the B-path defining $d(\mathbf{x}, \overline{X}; B) = n$. Let $\mathbf{y} = \mathbf{q}_0, \mathbf{q}_1, \ldots, \mathbf{q}_n = \mathbf{x}$ be the B-path defining $d(\mathbf{y}, \mathbf{x}; B) = n$ and assume that $\mathbf{q}_i \notin \mathcal{I}_{\mathbb{G}}$ for some i, $0 < i < n$. Let $\mathcal{S}_{\mathbb{G},B}(\mathbf{x}, k) = \{\mathbf{y} : d(\mathbf{x}, \mathbf{y}; B) = k\}$. Now, $\mathbf{q}_i \in \mathcal{S}_{\mathbb{G},B}(\mathbf{x}, k)$ for some $k \leq n$. Therefore, by Definition 10, there is a background grid point $\mathbf{z} \in \mathcal{S}_{\mathbb{G},B}(\mathbf{x}, k-1)$ such that $\mathbf{q}_i$ and $\mathbf{z}$ are adjacent. We thus have $n = d(\mathbf{x}, \overline{X}; B) \leq d(\mathbf{x}, \mathbf{z}; B) = k-1 < n$, which is a contradiction. □

When condition (B) is fulfilled: Let $\mathbf{y} = \mathbf{p}_0, \ldots, \mathbf{p}_n = \mathbf{x}$ be a shortest path with $\Gamma^{\xi}_{\mathbb{G}}$-sector steps between $\mathbf{y}$ and $\mathbf{x}$. Assume that there is an i $(0 \leq i \leq n)$ such that $\mathbf{p}_i \notin \mathcal{I}_{\mathbb{G}}$. Since $\mathbf{y} \in \mathcal{I}_{\mathbb{G}}$, this implies

- $[\mathbb{Z}^2, \mathbb{Z}^3]$ $p_i^j > U_j$ for some j.
- $[\mathbb{F}, \mathcal{D}^1]$ $p_i^1 + p_i^2 + p_i^3 > U_1$, $p_i^1 + p_i^2 - p_i^3 > U_2$, $p_i^1 - p_i^2 - p_i^3 < L_3$, or $p_i^1 - p_i^2 + p_i^3 > U_4$.
- $[\mathbb{F}, \mathcal{D}^2]$ $p_i^1 + p_i^2 + p_i^3 > U_1$, $p_i^1 + p_i^2 - p_i^3 > U_2$, $p_i^1 - p_i^2 - p_i^3 > U_3$, or $p_i^1 - p_i^2 + p_i^3 > U_4$.
- $[\mathbb{B}]$ $p_i^1 + p_i^2 > U_1$, $p_i^2 + p_i^3 > U_2$, or $p_i^1 + p_i^3 > U_3$.

Now, for all local steps in $\Gamma^{\xi}_{\mathbb{G}}$, these inequalities are valid also for $\mathbf{p}_{i+1}$. Thus $\mathbf{p}_j \notin \mathcal{I}_{\mathbb{G}}$ for all $j \geq i$, which contradicts $\mathbf{x} \in \mathcal{I}_{\mathbb{G}}$. □

4.2 Rules to *Swap* Steps in a Path

Lemma 2. *Let the neighbourhood sequence* B *and the image* $\mathcal{I}_{\mathbb{G}}$ *be such that* either *the image domain is sector-preserving* or *all image border points are in the background. Let* $\mathcal{I}_{\mathbb{G}}$ *be such that* $\mathbf{y} \in \overline{X}$ *and let* $\mathbf{x} \in X \cap \mathcal{D}^{\xi}_{\mathbb{G}}(\mathbf{y})$ *for some* ξ *be such that the* B*-path* $\mathcal{P}$ *with* $\Gamma^{\xi}_{\mathbb{G}}$*-sector steps of length* n *between* $\mathbf{x}$ *and* $\mathbf{y}$ *is a* *shortest path defining* $d(\mathbf{x}, \overline{X}; B)$*. For any* k $(1 \leq k \leq n)$*, let* ρ_k *be such that* ω_k *corresponds to a strict* ρ_k*-neighbour. Let also* i, j *be two fixed integers such that* $1 \leq i, j \leq n$*.*

Then for any ω'_i *and* ω'_j *such that*

(A1) $\omega'_i, \omega'_j \in \Gamma^{\xi}_{\mathbb{G}}$,
(A2) ω'_i *and* ω'_j *correspond to strict* ρ_i*- and* ρ_j*-neighbours respectively,*
(A3) $\omega_i + \omega_j = \omega'_i + \omega'_j$

there is a shortest B*-path* $\mathcal{P}'$ *in* $\mathcal{I}_{\mathbb{G}}$ *of length* n *between* $\mathbf{0}$ *and* $\mathbf{x}$ *such that* $\omega'_k = \omega_k$ *if* $k \neq i, j$ *and* $\omega'_i = \omega_j$ *and* $\omega'_j = \omega_i$*.*

Proof. Let $\mathcal{P}$ be the path $\mathbf{x} = \mathbf{p}_0, \mathbf{p}_1, \ldots, \mathbf{p}_{i-1}, \mathbf{p}_i, \ldots, \mathbf{p}_{j-1}, \mathbf{p}_j, \ldots, \mathbf{p}_n = \mathbf{y}$ and let ω'_i and ω'_j satisfy (A1)–(A3). We have

$$\omega_i = \mathbf{p}_i - \mathbf{p}_{i-1}\,,\ \omega'_i = \mathbf{p}'_i - \mathbf{p}_{i-1} \tag{3}$$

$$\omega_j = \mathbf{p}_j - \mathbf{p}_{j-1}\,,\ \omega'_j = \mathbf{p}_j - \mathbf{p}'_{j-1} \tag{4}$$

$$\omega_i + \omega_j = \omega'_i + \omega'_j \quad \text{(by (A3))}. \tag{5}$$

It follows from Equations (3)-(5) that $\mathbf{p}_i - \mathbf{p}'_i = \mathbf{p}_{j-1} - \mathbf{p}'_{j-1}$, i.e., that $\mathbf{p}'_i = \mathbf{p}_i - \vec{v}$ and $\mathbf{p}'_{j-1} = \mathbf{p}_{j-1} - \vec{v}$ for some $\vec{v}$.

We will now see that the path $\mathcal{P}'$,
$\mathbf{x} = \mathbf{p}_0, \mathbf{p}_1, \ldots, \mathbf{p}_{i-1}, \mathbf{p}'_i, \ldots, \mathbf{p}'_{j-1}, \mathbf{p}_j, \ldots, \mathbf{p}_n = \mathbf{y}$ defined as

$$\begin{aligned} \mathbf{p}'_k &= \mathbf{p}_k - \vec{v} \quad \text{for any} \quad i \leq k < j \\ \mathbf{p}'_k &= \mathbf{p}_k \quad \text{otherwise.} \end{aligned} \tag{6}$$

is a shortest B-path in $\mathcal{I}_{\mathbb{G}}$ satisfying the Lemma.

By (A1), $\mathcal{P}'$ is a path with $\Gamma^{\xi}_{\mathbb{G}}$-sector steps. Thus, by Lemma 1, $\mathcal{P}'$ is in the image $\mathcal{I}_{\mathbb{G}}$.

By the definition of $\mathcal{P}'$, it is a shortest path (since it is of the same length as $\mathcal{P}$). Moreover, $\omega_k = \mathbf{p}_k - \mathbf{p}_{k-1} = \mathbf{p}'_k - \mathbf{p}'_{k-1} = \omega'_k$ for any $i < k < j$, so $\mathbf{p}_k$ and $\mathbf{p}_{k-1}$ are strict ρ-neighbours if and only if $\mathbf{p}'_k$ and $\mathbf{p}'_{k-1}$ are. The cases $k = i$ and $k = j$ follow from (A2) and when $k < i$ or $j < k$, $\mathbf{p}'_k = \mathbf{p}_k$ by definition, so $\mathcal{P}'$ is a B-path.

We can now conclude that $\mathcal{P}'$ is a shortest B-path in $\mathcal{I}_{\mathbb{G}}$ between $\mathbf{x}$ and $\mathbf{y}$. □

Example 2. Consider $\mathbb{F}$, the neighbourhood sequence $B = (1, 2, 1, 1, 2)$ and the grid point $\mathbf{x} = (6, 2, 0)$. A shortest B-path between $\mathbf{0}$ and $\mathbf{x}$ is

$$\mathbf{0} = (0,0,0), (1,1,0), (3,1,0), (4,1,-1), (5,1,0), (6,2,0) = \mathbf{x}.$$

We have

$$\omega_1 = (1,1,0),\ \omega_2 = (2,0,0),\ \omega_3 = (1,0,-1),\ \omega_4 = (1,0,1),\ \omega_5 = (1,1,0).$$

By Lemma 2, the B-paths between $\mathbf{0}$ and $\mathbf{x}$ with the following local steps are also shortest paths.

$$\begin{aligned} &\omega_1 = (1,1,0),\ \omega_2 = (2,0,0),\ \omega_3 = (1,1,0)\ \ \omega_4 = (1,0,-1),\ \omega_5 = (1,0,1) \\ &\omega_1 = (1,0,1),\ \omega_2 = (2,0,0),\ \omega_3 = (1,0,-1),\ \omega_4 = (1,1,0),\ \omega_5 = (1,1,0) \end{aligned}$$

Thus, any order of the local steps corresponding to 1-neighbours results in shortest B-paths.

Now we define sets $\mathbf{B}_i^{\mathbb{G}}$. We will see that if each of these $\mathbf{B}_i^{\mathbb{G}}$:s are in at least one mask (supporting the scan orders) in Algorithm 1 then Algorithm 1 propagate correct distance to any $\mathbf{x} \in \mathcal{D}_{\mathbb{G}}^{\xi}$.

$$\begin{array}{l|l|l}
\mathbf{B}_1^{\mathbb{Z}^2} = \{(1,0),\ (1,1)\} & \mathbf{B}_1^{\mathbb{F}} = \{(1,1,0),\ (2,0,0)\} & \mathbf{B}_1^{\mathbb{B}} = \{(1,1,1),\ (2,0,0)\} \\
\mathbf{B}_2^{\mathbb{Z}^2} = \{(0,1),\ (1,1)\} & \mathbf{B}_2^{\mathbb{F}} = \{(1,0,1),\ (2,0,0)\} & \mathbf{B}_2^{\mathbb{B}} = \{(1,-1,1),\ (2,0,0)\} \\
 & \mathbf{B}_3^{\mathbb{F}} = \{(0,1,1),\ (2,0,0)\} & \mathbf{B}_3^{\mathbb{B}} = \{(1,1,-1),\ (2,0,0)\} \\
 & \mathbf{B}_3^{\mathbb{F}} = \{(1,0,-1),\ (2,0,0)\} &
\end{array}$$

$$\begin{array}{l|l}
\mathbf{B}_1^{\mathbb{Z}^3} = \{(1,0,0),(1,1,0),(1,1,1)\} & \mathbf{B}_4^{\mathbb{Z}^3} = \{(0,1,0),(0,1,1),(1,1,1)\} \\
\mathbf{B}_2^{\mathbb{Z}^3} = \{(1,0,0),(1,0,1),(1,1,1)\} & \mathbf{B}_5^{\mathbb{Z}^3} = \{(0,0,1),(0,1,1),(1,1,1)\} \\
\mathbf{B}_3^{\mathbb{Z}^3} = \{(0,1,0),(1,1,0),(1,1,1)\} & \mathbf{B}_6^{\mathbb{Z}^3} = \{(0,0,1),(1,0,1),(1,1,1)\}
\end{array}$$

4.3 Minimal Configuration of Scanning Masks

Theorem 1. *Let the set α_i, $1 \leq i \leq L$ such that $\alpha_i \neq \alpha_k$ if $i \neq k$ and $\alpha_i \in \{1, 2, \ldots, L\}$ for all i be given. For any neighbourhood sequence B, any image $\mathcal{I}_{\mathbb{G}}$ such that all image border points are in the background, and any points $\mathbf{y} \in \overline{X}$ and $\mathbf{x} \in X \cap \mathcal{D}_{\mathbb{G}}^{\xi}(\mathbf{y})$ such that $d(\mathbf{x}, \mathbf{y}) = d(\mathbf{x}, \overline{X}; B)$ there is a shortest path between $\mathbf{x}$ and $\mathbf{y}$ such that*

$$\omega_j \in \mathbf{B}_{\alpha_1}^{\mathbb{G}} \quad \text{if } 0 = k_0 < j \leq k_1 \tag{7}$$

$$\omega_j \in \mathbf{B}_{\alpha_2}^{\mathbb{G}} \quad \text{if } k_1 < j \leq k_2 \tag{8}$$

$$\vdots$$

$$\omega_j \in \mathbf{B}_{\alpha_L}^{\mathbb{G}} \quad \text{if } k_{L-1} < j \leq k_L \tag{9}$$

for some k_i:s.

Proof. The theorem follows directly from Lemma 2 for $\mathbb{Z}^2$, $\mathbb{F}$, and $\mathbb{B}$. Since the only 2-neighbour ($(1,1)$, $(2,0,0)$, and $(2,0,0)$, respectively) is in all the $\mathbf{B}_i^{\mathbb{G}}$:s, it is enough to order the 1-neighbours such that (7)-(9) are fulfilled. Compare with Example 2.

For $\mathbb{Z}^3$, things are a bit more complicated. We argue by contradiction. Let $\mathcal{P}$ be any shortest B-path (of length n) with $\Gamma_{\mathbb{G}}$-sector steps between $\mathbf{x}$ and $\mathbf{y}$. Construct the new path $\mathcal{P}'$ as follows:

There is obviously a maximal value of k_1 such that $\omega_j \in \mathbf{B}_{\alpha_1}^{\mathbb{Z}^3}$ for all $0 < j \leq k_1$. Lemma 2 is used to find the ω_j:s. In the same way, maximal values of $k_2, \ldots, k_6$ and the ω_j:s for $j \leq k_6$ are found.

Assume that $k_6 \neq n$ (i.e. that $k_6 < n$). Since $(1,1,1)$ is in all $\mathbf{B}_i^{\mathbb{Z}^3}$:s, ω_{k_6+1} corresponds either to a 1-neighbour or a 2-neighbour.

`Case i` ω_{k_6+1} corresponds to a 1-neighbour.

We consider $\omega_{k_6+1} = (1,0,0)$ – the proofs for $(0,1,0)$ and $(0,0,1)$ are similar.

Let a and b be the values such that $\omega_j \in \mathbf{B}_1^{\mathbb{Z}^3}$ if $k_{a-1} < j \leq k_a$ and $\omega_j \in \mathbf{B}_2^{\mathbb{Z}^3}$ if $k_{b-1} < j \leq k_b$. We assume $b > a$ (the proof for $a > b$ is similar).

Since $(1,1,1)$ is in all $\mathbf{B}_i^{\mathbb{Z}^3}$:s, neither ω_{k_a+1} nor ω_{k_b+1} corresponds to 3-neighbours.

By Lemma 2, neither ω_{k_a+1} nor ω_{k_b+1} corresponds to 1-neighbours: If, say ω_{k_a+1} corresponds to a 1-neighbour, then we could use Lemma 2 to swap ω_{k_a+1} and ω_{k_6+1} contradicting that k_a is maximal.

Thus both ω_{k_a+1} and ω_{k_b+1} correspond to 2-neighbours. It follows that $\omega_j \neq (1,1,0)$ for $j > k_a$ and $\omega_j \neq (1,0,1)$ for $j > k_b$. (Otherwise we could use Lemma 2 to swap any such occurence of $(1,1,0)$ or $(1,0,1)$ with ω_{k_a+1} or ω_{k_b+1} contradicting that k_a and k_b are maximal.)

Thus, $\omega_{k_b+1} = (0,1,1)$. But then we could use Lemma 2 to set ω_{k_b+1} to $(1,0,1)$ and ω_{k_6+1} to $(0,1,0)$ contradicting that k_b is maximal.

`Case ii` ω_{k_6+1} corresponds to a 2-neighbour.

We consider $\omega_{k_6+1} = (1,1,0)$ (the proofs for $(1,0,1)$ and $(0,1,1)$ are similar).

Let now a and b be the values such that $\omega_j \in \mathbf{B}_1^{\mathbb{Z}^3}$ if $k_{a-1} < j \leq k_a$ and $\omega_j \in \mathbf{B}_3^{\mathbb{Z}^3}$ if $k_{b-1} < j \leq k_b$. We assume $b > a$ – the proof for $a > b$ is similar.

Since $(1,1,1)$ is in all $\mathbf{B}_i^{\mathbb{Z}^3}$:s, neither ω_{k_a+1} nor ω_{k_b+1} corresponds to 3-neighbours.

By Lemma 2, neither ω_{k_a+1} nor ω_{k_b+1} corresponds to 2-neighbours: If, say ω_{k_a+1} corresponds to a 2-neighbour, then we could use Lemma 2 to swap ω_{k_a+1} and ω_{k_6+1} contradicting that k_a is maximal.

Thus both ω_{k_a+1} and ω_{k_b+1} correspond to 1-neighbours. It follows that $\omega_j \neq (1,0,0)$ for $j > k_a$ and $\omega_j \neq (0,1,0)$ for $j > k_b$. (Otherwise we could use Lemma 2 to swap any such occurence of $(1,0,0)$ or $(0,1,0)$ with ω_{k_a+1} or ω_{k_b+1} contradicting that k_a and k_b are maximal.)

Thus, $\omega_{k_b+1} = (0,0,1)$. But then we could use Lemma 2 to set ω_{k_b+1} to $(0,1,0)$ and ω_{k_6+1} to $(1,0,1)$ contradicting that k_b is maximal.

Now we can conclude that, since all possible ω_{k_6+1}:s lead to contradictions, $k_6 = n$ and the proof is finished. □

We can now conclude that since the order of the $\mathbf{B}_i^{\mathbb{G}}$:s is arbitrary, Algorithm 1 will propagate the correct distance value from $\mathbf{y} \in \overline{X}$ to each point in $\mathcal{D}_{\mathbb{G}}(\mathbf{y})$ if each $\mathbf{B}_i^{\mathbb{G}}$:s is included in at least one mask in the algorithm. By symmetry, it follows that if all configurations symmetric to the $\mathbf{B}_i^{\mathbb{G}}$:s are included in at least one mask in the algorithm, distance values will be propagated from $\mathbf{y}$ to any object grid point in $\mathcal{I}_{\mathbb{G}}$. We get the following condition:

Condition 1. *If each configuration symmetric to the configurations shown in Figure 1 is included in at least one mask supporting the scan orders used in Algorithm 1, then Algorithm 1 will produce correct distance maps.*

Sets of masks in the different grids fulfilling Condition 1 are shown in Figure 3.

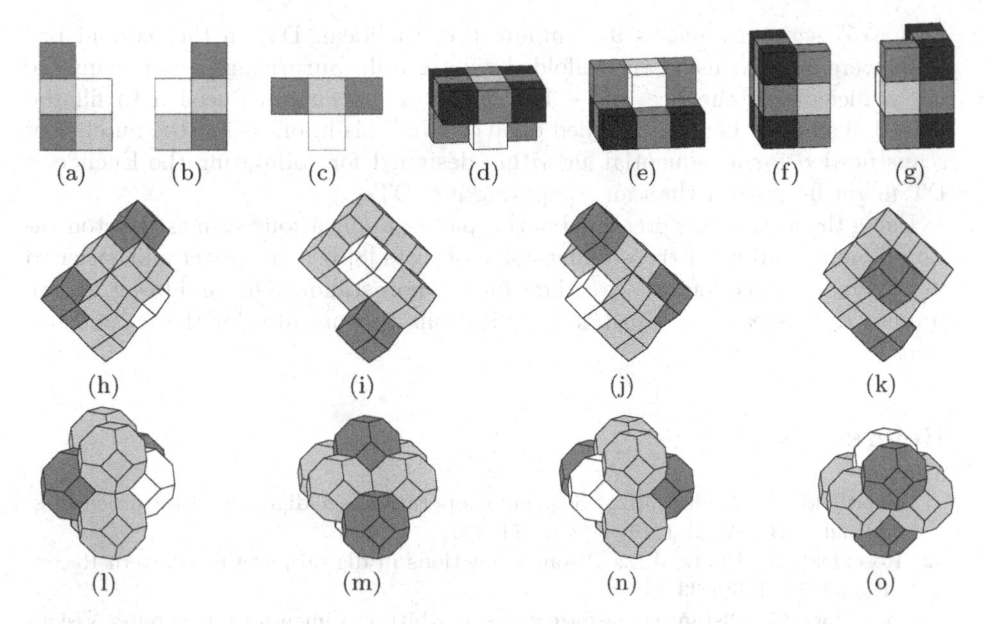

Fig. 3. The masks above (colour-coded as in Figure 1) can be used to get correct results from Algorithm 1. (a)–(c) $\mathbb{Z}^2$, (d)–(g) $\mathbb{Z}^3$, (h)–(k) $\mathbb{F}$, and (l)–(o) $\mathbb{B}$. Note that, for $\mathbb{Z}^3$, $\mathbb{F}$, and $\mathbb{B}$, any mask can be obtained by rotating any other mask in the same grid.

5 Discussion and Conclusion

The unfolded cube graph was introduced in [11] to guarantee that local distances are allowed to propagate in all possible directions. It was designed for Euclidean DTs and shows the directions supported by a mask. The unfolded cube graphs for a set of masks must fill the whole cube (direction space) to produce correct Euclidean DTs. It is easy to produce a set of masks such that the unfolded cube graph is covered but which does not produce correct distance maps for n.s.-distances as shown in Figure 4. We have shown that distance propagation from all directions is not sufficient for algorithms for n.s.-distances. The condition derived in this paper is: if each configuration symmetric to the configuration in Figure 1 is contained in at least one mask, then the algorithm produces correct DT:s. We can notice that these configurations contain one and only one vector of each neighbourhood kind.

Fig. 4. Using these masks and the 7+5 symmetric masks for $\mathbb{Z}^3$ in Algorithm 1 will result in insufficient propagation of distances, e.g., for neighbourhood sequences with $B = (1, 3)$

In [15], scanning masks to compute the Euclidean DT on the fcc and bcc grids were derived using the unfolded cube graph. Surprisingly, four scans are not sufficient for the bcc grid – five masks, i.e. scans, are needed to fill the direction space using the unfolded cube graph. This implies that the number of scans needed for a sequential algorithm designed for computing the Euclidean DT might be greater than for computing n.s.-DTs.

Using the algorithms presented in this paper, applications such as skeletonization [7] and shading of three-dimensional objects [8] become faster and easier to use. Also, since sequential algorithms for the non-standard fcc and bcc grids are presented, it is easy to adjust the applications to work also for the fcc and bcc grids.

References

1. Rosenfeld, A., Pfaltz, J.L.: Sequential operations in digital picture processing. Journal of the ACM **13**(4) (1966) 471–494
2. Rosenfeld, A., Pfaltz, J.L.: Distance functions in digital pictures. Pattern Recognition **1**(1) (1968) 33–61
3. Borgefors, G.: Distance transformations in arbitrary dimensions. Computer Vision, Graphics, and Image Processing **27** (1984) 321–345
4. Fouard, C., Strand, R., Borgefors, G.: Weighted distance transforms generalized to modules and their computation on point lattices. Technical report, Centre for Image Analysis, Uppsala University, Uppsala, Sweden (2006) Internal report 38.
5. Yamashita, M., Honda, N.: Distance functions defined by variable neighbourhood sequences. Pattern Recognition **17**(5) (1984) 509–513
6. Das, P.P., Chakrabarti, P.P., Chatterji, B.N.: Distance functions in digital geometry. Information Sciences **42** (1987) 113–136
7. Kumar, M.A., Chatterji, B.N., Mukherjee, J., Das, P.P.: Representation of 2D and 3D binary images using medial circles and spheres. International Journal of Pattern Recognition and Artificial Intelligence **10**(4) (1996) 365–387
8. Mukherjee, J., Kumar, M.A., Chatterji, B.N., Das, P.P.: Discrete shading of three-dimensional objects from medial axis transform. Pattern Recognition Letters **20**(14) (1999) 1533–1544
9. Danielsson, P.E.: Euclidean distance mapping. Computer Graphics and Image Processing **14** (1980) 227–248
10. Danielsson, P.E.: Minimal error octagonal metric in two and three dimensions. Internal report LiTH-ISY-1-1382, Linköping University, Linköping, Sweden (1992)
11. Ragnemalm, I.: The Euclidean distance transform in arbitrary dimensions. Pattern Recognition Letters **14**(11) (1993) 883–888
12. Matej, S., Lewitt, R.M.: Efficient 3D grids for image reconstruction using spherically-symmetric volume elements. IEEE Transactions on Nuclear Science **42**(4) (1995) 1361–1370
13. Garduno, E., Herman, G.T.: Optimization of basis functions for both reconstruction and visualization. Electronic Notes in Theoretical Computer Science **46** (2001) 1–17
14. Strand, R., Borgefors, G.: Distance transforms for three-dimensional grids with non-cubic voxels. Computer Vision and Image Understanding **100**(3) (2005) 294–311

15. Strand, R.: The Euclidean distance transform applied to the fcc and bcc grids. In Marques, J.S., de la Blanca, N.P., Pina, P., eds.: IbPRIA 2005, Estoril, Portugal, June 7-9, 2005, Proceedings, Volume 3522 of LNCS., Springer (2005) 243–250
16. Carvalho, B.M., Garduno, E., Herman, G.T.: Multiseeded fuzzy segmentation on the face centered cubic grid. In Ltd., S.V., ed.: Proceedings ICAPR 2001, Pattern Analysis and Applications journal (2001)
17. Theussl, T., Möller, T., Gröller, M.E.: Optimal regular volume sampling. In: VIS '01: Proceedings of the conference on Visualization '01, Washington, DC, USA, IEEE Computer Society (2001) 91–98
18. Nagy, B.: Distance functions based on neighbourhood sequences. Publicationes Mathematicae Debrecen **63**(3) (2003) 483–493
19. Strand, R., Nagy, B.: Some properties for distances based on neighbourhood sequences in the face-centered cubic grid and the body-centered cubic grid. Technical report, Centre for Image Analysis, Uppsala University, Uppsala, Sweden (2006) Internal Report 39.

Hierarchical Chamfer Matching Based on Propagation of Gradient Strengths

Stina Svensson[1] and Ida-Maria Sintorn[2]

[1] Centre for Image Analysis, Swedish University of Agricultural Sciences, Lägerhyddsvägen 3, SE-75237 Uppsala, Sweden
stina@cb.uu.se

[2] CSIRO Mathematical and Information Sciences
Locked Bag 17, North Ryde, NSW 1670, Australia
Ida-Maria.Sintorn@csiro.au

Abstract. A modification of the hierarchical chamfer matching algorithm (HCMA) with the effect that no binarisation of the edge information is performed is investigated. HCMA is a template matching algorithm used in many applications. A distance transform (DT) from binarised edges in the search image is used to guide the template to good positions. Local minima of a function using the distance values hit by the template correspond to potential matches. We propose to use distance weighted propagation of gradient magnitude information as a cost image instead of a DT from the edges. By this we keep as much information as possible until later in the matching process and, hence, do not risk to discard good matches in the edge detection and binarisation process.

1 Introduction

Chamfer matching is a template matching method based on geometric image features and can be used for both 2D and 3D images. It finds good fits between the template and edges in a search image. A generalized cost function between the edges in the search image and the template, a list of coordinate pairs corresponding to the searched pattern, is minimized. To guide the template to good positions, a distance transform (DT) is calculated from edges in the search image and the sum of the distance values hit by the superimposed template constitute the cost function. Translation, scaling, rotation, and perspective changes are for 2D images and translation, scaling, and rotation for 3D images.

In its original form, proposed by Barrow et al. in 1977 [1] as described above, chamfer matching is a fine matching algorithm which requires good start positions. In [2], Borgefors showed that using the sum of squared distances instead of the sum of distances as the cost function results in significantly fewer false minima. In [3], she embedded the chamfer matching in a resolution hierarchy. This, the hierarchical chamfer matching algorithm (HCMA), turned the original fine-matching algorithm to a fast, general and robust matching algorithm.

HCMA's usefulness is clear not the least from the number of citations (almost 200, Science Citation Index March 2006). Although introduced almost 20

A. Kuba, L.G. Nyúl, and K. Palágyi (Eds.): DGCI 2006, LNCS 4245, pp. 308–319, 2006.

years ago, it is still used in many applications today. It is described in the recent surveys of volume image registration by template matching [4] and image registration [5]. In [6], a 3D model based registration approach is taken for fusion of information from positron emission tomography (PET), magnetic resonance imaging (MRI) and magnetocardiography (MCG). For the PET-MRI registration, HCMA is used. HCMA is also used in object recognition, e.g., hand pose recognition, [7], and pedestrian detection from moving vehicles, [8].

A few suggestions of improvements to HCMA have been described. In [9] and [6], HCMA is used with an oriented DT, where not only the position of the edges in the search image, but also their orientation is taken into account. The oriented DT is used to guide the search and decreases the chance of getting stuck in false local minima. In [9], it is also suggested to use a hierarchical clustering of the template from different views to speed up the matching process. These two modifications could be incorporated in the method suggested here. Three different modifications of the DT taking into account gradient magnitude information (*salience DTs*) are presented in [10]. The use of salience DTs in various applications, including chamfer matching without the hierarchical structure and for 2D images only, is discussed. One of the salience DTs presented in [10] is similar to the distance weighted propagation of gradient magnitude information suggested here. We investigate the use of distance weighted propagation of gradient magnitude information in HCMA further. We discuss how the choice of local propagation steps influences the outcome of the matching process and give some ideas on how the matching measure need to be modified to suite the cost image. Moreover, we generalise the concept to 3D images.

HCMA and our gradient magnitude based HCMA (GM-HCMA) are described for 2D images in Section 2 and Section 3, respectively. Section 3 also contains the generalisation to 3D images. The performance of GM-HCMA is illustrated on a 2D leaf-matching example in Section 4 and on a cryo-electron tomographic (cryo-ET) 3D protein identification example in Section 5. The protein identification example was inspired by [11], where the ordinary HCMA was shown to give promising results for identifying proteins in cryo-ET images. Here, we show that similar results can be achieved using GM-HCMA.

2 HCMA

HCMA improves the chamfer matching algorithm by embedding it in a resolution pyramid, [3], [1]. Chamfer matching is performed in the image with the lowest resolution. Relevant positions, i.e., positions giving a match measure greater than a user defined *rejection factor* (RF), are used as start positions for chamfer matching in the image at the next higher resolution level. Chamfer matching is then performed level after level until the highest resolution level, i.e., the original image, is reached. We start by describing the chamfer matching performed for each resolution level and then how the image resolution pyramid is created.

Two images are needed to perform chamfer matching: a template representing the contours of the searched pattern and a search image depicting the same type of information, i.e., edge information, as the template. There are many possible ways to create a template. This is not the focus of this paper so we simply assume that a template is given. There are likewise several different ways to detect edges in an image. Here, we will detect edges as the gradient magnitudes derived as the sum of the responses from a set of Sobel operators, e.g., [12]. For 2D images, the operators are of size 3×3 pixels and horizontal, vertical, and diagonal directions are taken into account. For 3D images, the operators are of size $3 \times 3 \times 3$ voxels and directions in the x-, y-, and z-planes are considered.

An edge image is created from the gradient magnitude image by, e.g., thresholding, and a DT is calculated from the edges. In the DT, each pixel is assigned a value corresponding to its distance to the nearest edge pixel. The DT serves as a cost image which guides the template to positions giving low match measures.

When calculating a DT, the edge pixels are initially set to 0 and other pixels to ∞. Local distances from the edge pixels are propagated over the image in one forward and one backward scan. The local distances (weights) used here are 3 for an edge neighbour and 4 for a vertex neighbour, following the suggestion in [3]. Each pixel is assigned the minimum of its value and the values of its, in the current scan, already visited neighbours, each increased with their respective local distance. This method of calculating a DT is known as chamfering, which originates the name chamfer matching.

The template is represented by a set of coordinate pairs $T = \{(X_i, Y_i) \mid X_i, Y_i \in \mathbf{Z}, i = 1, \ldots, n\}$. The search for local minimal match measures starts from a set of start positions. Each position contains information on the transformation parameters translation (c_X, c_Y), rotation (θ), and scaling S. (We leave out perspective.) The template T is transformed to its current position by $T^S = T \cdot S$, and $T'(X_i, Y_i) = (c_X + X_i^S \cdot \cos\theta - Y_i^S \cdot \sin\theta, c_Y + Y_i^S \cdot \sin\theta + Y_i^S \cdot \cos\theta)$. The (X_i', Y_i') coordinates are usually not integers and are rounded to the nearest integer value. After this, T' is superimposed on the DT image. The *match measure* used, i.e., the measure on how well T' fits edges in the underlying image, is the root mean square average (RMS) of the DT values hit by the template:

$$RMS = \frac{1}{3}\sqrt{\frac{1}{n}\sum_{i=1}^{n} d(X_i', Y_i')^2},$$

where $d(X_i', Y_i')$ is the value in the DT at position (X_i', Y_i'), and the factor $\frac{1}{3}$ comes from the use of $\langle 3, 4\rangle$ DT. RMS is preferable to the arithmetic mean as match measure in order to decrease the number of false minima, [2].

To reach a local minimum of the match measure from the current position, a steepest descent and a line search method is used [13, chapter 10.5, 11.2]. Iteratively, a step is taken in the steepest descent direction if it produces a smaller match measure. If not, the step length is halved until it does, or until the step length is so small that a local minimum can be considered reached. For each step, the steepest descent direction is calculated from all four parameters (rotation, scaling, and $2\times$ translation). The gradient for each parameter is approximated by centred differences, where a translation step is the side length of a pixel, a rotation step is the angle which makes a point on the edge of the template move one pixel side in length, and a scale step is a 2% change in scale.

In HCMA, chamfer matching as described above is embedded in a resolution pyramid. The resolution pyramid is built from the binarised edges detected in the search image, which serves as the base of the pyramid (L_0). Each block of 2×2 pixels at level L_i is represented by a single pixel at level L_{i+1}. This means that for each higher level the size of the image is reduced by $\frac{3}{4}$, and the resolution is halved. The "colour" of the pixel depends on the colour of the four pixels it is assigned from. For HCMA, the highest value is chosen, i.e., an OR-pyramid is used, guaranteeing that no edges disappear between successive levels.

A DT is calculated for all levels and starting positions are then distributed at the top of the pyramid, i.e., at the lowest resolution level. The positions that after the matching at level L_i produce match measures better than RF (depending on the size of the template and the resolution level) are kept and used as starting positions at level L_{i-1}. This means that at level L_0 only fine matching from relatively good starting positions is performed. If no a priori knowledge is available, starting positions for the top level of the pyramid should be evenly distributed, i.e., as a regular grid of points in the search space.

3 GM-HCMA

Instead of using a DT from the binarised edge image as the cost image, we suggest decreasing propagation of the gradient magnitudes themselves and using the result as the cost image. By this, we keep as much information as possible until later in the process and do not risk to loose important edges by choosing a threshold that is too high.

As the gradient magnitudes are high at edges, a good match does not correspond to a low match measure but a high one. This means that we search for local maxima in the matching process. Moreover, there is no longer a match measure corresponding to a perfect match, as 0 is in HCMA. A certain match measure can be the result of a good fit at edges with low gradient magnitudes or a rather bad fit but still hitting some very high gradient magnitudes. The same phenomenon occurs for HCMA but there only needs to be considered if very occluded objects are desirable to be found. With the new cost image, this effect will be more common and therefore additional RFs need to be considered.

To explain the proposed GM-HCMA, we start by recalling the concept of the reverse distance transform (RDT) [14]. In RDT, decreasing local distance information is propagated over the image in one forward and one backward scan starting from a set of distance labelled seed points. Each pixel is assigned the maximum of its value and the values of its, in the current scan, already visited neighbours, each decreased by their respective local distance. As for the DT described in Section 2, good local distances are 3 for edge neighbours and 4 for vertex neighbours. Calculating a RDT can be seen as growing a ball for each seed until it has a size (radius) corresponding to the distance value of the seed.

Our aim is to spread decreasing gradient magnitude information in a shape preserving way in order to create a cost image where the values depend on the gradient magnitude as well as the relative distance, i.e., the shape of the objects in the image. We propagate the information in one forward and one backward scan in the same way as for RDT and use application dependent local "distances" (*weights*). To obtain a propagation stable under rotation, the edge and vertex neighbour weights are chosen so that the ratio is $\frac{3}{4}$ following the $\langle 3, 4 \rangle$ DT. The result can be thought of as smoothing the image while creating a cost image, as pixels with low gradient magnitude close to pixels with higher values will be overshadowed and not contribute to the cost image. Note that the propagation is done in full scale, which means that depending on the weights and the gradient magnitudes, the resulting cost image can have values below 0. The process to create the cost image is the same in 3D. We use face, edge, and vertex neighbour weights with the relation 3,4,5, as suggested in [15].

Calculating a resolution pyramid from the gradient magnitude image is done in the same way as in the binary case. Each pixel at a higher level is assigned the maximum value of the four pixels at the lower level it is built from. For 3D images, the resolution pyramid is created by letting each block of $2 \times 2 \times 2$ voxels in level L_i represent a single voxel in L_{i+1}, i.e, the resolution is halved and the number of voxels is reduced by $\frac{7}{8}$ between successive levels. For 3D images, the search space increases as translation and rotation are described by three parameters each, instead of two and one, respectively, as in the 2D case.

How the weights should be chosen depends on the application, image acquisition situation, and how much smoothing the user wants to build into the propagation. If the gradient magnitude information of the searched object is likely to be low compared to other parts of the image, high weights, resulting in a low smoothing factor, should be chosen. Using the logarithm of the gradient magnitude image, before the resolution pyramid is calculated, can also be considered, as this leads to a compression of the range of the gradient magnitudes. If the gradient magnitude information of the searched object is likely to be high compared to other parts, low weights should be chosen so that points with low gradient magnitude are overshadowed and noise is smoothed. This decreases the chances of getting stuck in false local maxima during the matching process. A 1D curve illustrating the effect of using higher and lower weights is shown in Figure 1. A higher weight (left) has the effect that the point with lower gradient magnitude prevents the propagation from the point with higher gradient magnitude. A lower weight (right) makes the point with higher gradient magnitude overshadow the point with lower gradient magnitude. If no a priori information is available, it is safer to use high weights as the influence of a point with high gradient magnitude then drops quickly, even though this means that the chance of getting stuck in false local maxima increases.

For GM-HCMA, good positions give high match measures (instead of low as for HCMA). However, it is not possible to give absolute values for how high good match measures should be as it depends on the gradient magnitude information available. Instead additional criteria can be used to help evaluate possible match

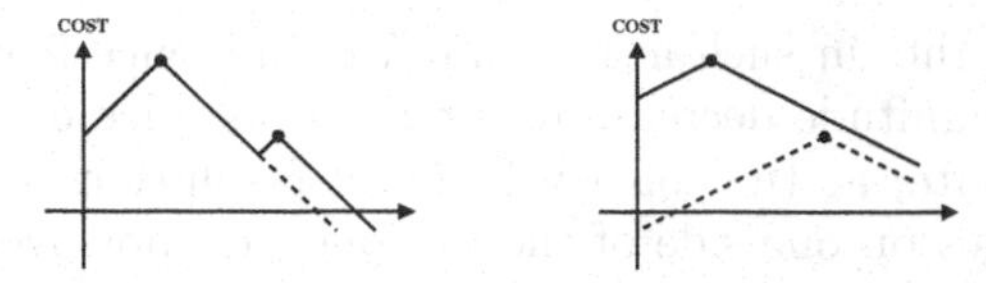

Fig. 1. 1D curves illustrating the effect of using higher (left) and lower (right) weights

positions. If the background is even and, hence, the gradient magnitudes of the searched object are likely to be of similar strength, the variance of the pixel values in the cost image contributing to the match measure gives a good indication on whether a position is relevant or not. In [16], a method for differentiating between positions giving low match measures when HCMA is applied to noisy images is presented. In such images, a low match measure does not necessarily mean that the position is correct and a rather high match measure does not automatically mean that the matching has failed. In the first case, it could be that the object matches parts of other objects and in the second, that the searched object is partially occluded or simply that the scene is noisy. In such cases, the variation of the cost along the template for a found position can be studied.

4 Comparison of GM-HCMA and HCMA for 2D Images

We illustrate GM-HCMA and compare it to HCMA using the two images, *leaf A* and *leaf B*, in Fig. 2. The images, both of size 896×592 pixels, belong to a data set used for experiments by the Computational Vision Group at California Institute of Technology (`http://www.vision.caltech.edu/archive.html`). The templates used, consisting of 2494 and 1922 pixels, respectively, are also shown in Fig. 2. These have been created manually in such a way that a "perfect" position is possible for leaf B while for leaf A the template differs slightly in shape compared to the leaf in the search image. We have chosen leaf A and B to illustrate GM-HCMA as they show GM-HCMA's advantages to HCMA as well as point out situations when GM-HCMA have difficulties (and HCMA fails).

In this example, we have used the logarithm of the gradient magnitude image to compress the range of gradient magnitudes. Assuming that we have no a priori knowledge of the gradient magnitude information for the searched object,

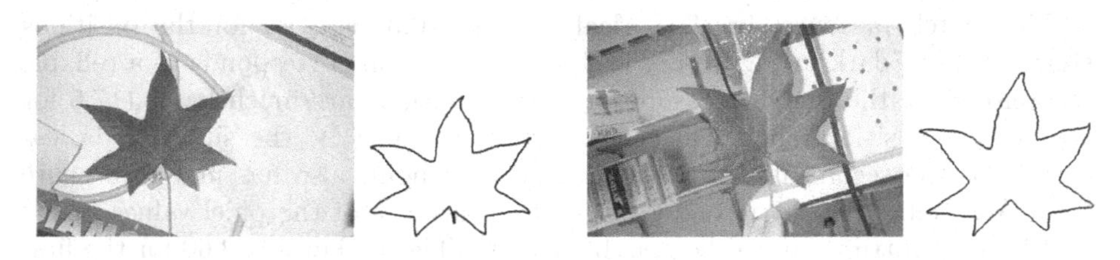

Fig. 2. Leaf A (left) and leaf B (right) and their respective templates

we have chosen weights in such a way that the influence from pixels with the highest gradient magnitude decrease to 0 over a distance corresponding to the width of the template, at the top level of the resolution pyramid. This is to make sure that edges on one side of the template do not overshadow edges on the opposite side. Hence, the weight for a step in the edge direction w_1 is set to the largest value in the gradient magnitude image divided by the minimum width of the template. The weight for a step in the vertex direction w_2 is $\frac{4}{3}w_1$. We have used five levels in the resolution pyramid and the weights $\langle 0.32, 0.43\rangle$ and $\langle 0.28, 0.37\rangle$, respectively, on all levels. The low weights are due to the use of the logarithm of the gradient magnitude instead of the gradient magnitude itself. We have implemented GM-HCMA using MATLAB and pixel type "double" as speed is not the main issue here. Implementing GM-HCMA, using a rescaling of the logarithmic values, with 8- or 16-bit images will give approximately the same computational complexity as HCMA. For translation, rotation, and scaling step lengths and optimization process, we have followed what is described in Section 2. In Fig. 3, the logarithm of the gradient magnitude image, the cost image, and the positions giving the two highest match measures are shown for level L_0. The cost image is linearly stretched (and displayed in colour) to emphasize differences.

Fig. 3. Top row: leaf A. Bottom row: leaf B. Logarithm of the gradient magnitude image, the cost image, and the positions giving the two highest match measures using GM-HCMA (blue/dark for the highest and yellow/light for the second highest).

The match measures are 20.9 (leaf A) and 20.3 (leaf B) for the positions shown in blue/dark in Fig. 3. For leaf A, this position corresponds to a reliable response since the match measure for other positions are much lower (17.5 for the second position, shown in yellow/light). For leaf B, the situation is less certain as the second position, shown in yellow/light, also has match measure 20.3. Additional information is given by the variance of the pixel values in the cost image contributing to the match measure. The variance is 2.60 for the first position (and 11.0 for the second, incorrect, position) for leaf A and 7.31 and 9.46 for the first, correct, and second, incorrect, positions with match measure

20.3 for leaf B. The low variance for the first position for leaf A indicates that it is likely to be a correct match because a low variance corresponds to a good fit for *the whole* template. Higher variances (as for leaf B) can result from partial occlusion (as is the case for the first position), i.e., most of the template fits well but parts of it does not, or from a less good fit for most of the template but where parts of it hit very high gradient magnitudes. What is the cause for the higher variance can be established by studying the graph of cost variation along the template at the match position [16]. Another more automatic possibility is to use a percentile of the values hit by the template in the RMS (and/or variance) calculation, i.e., only take into account the 75% of the pixels (those with the highest costs) hit by the template. In this example, the lower variance for the first position for leaf B still give preference to that position, without performing any additional investigations. Note that the difficulties for leaf B in finding a more clear first, and correct, position is due to the fact that there are very high gradient magnitudes on one side of the searched pattern (right) and low on the other side (left). This will be even more evident when HCMA is ran for the same example, see below. We remark that a direct comparison of match measures and variances between images should not be performed unless the images are acquired identically.

For comparison, we have also ran HCMA on leaf A and leaf B. Thresholding of the gradient magnitude image was performed to generate edge images. For leaf A, the gradient magnitudes range from 0 to 2577 and for leaf B from 0 to 2603. To illustrate the effect of the binarisation step, we ran HCMA starting from two different edge images, using the thresholds 400 and 1000, for both leaf A and leaf B. For the rest, the same settings as for GM-HCMA were used. In Fig. 4, the edge images, the DTs, and the best HCMA positions are shown for resolution level L_0. The match measures for leaf A were 11.5 and 6.3, the latter corresponding to a correct position. For leaf B, the match measures were 7.6 and 26.9, the former corresponding to a correct position. This shows that to succeed, HCMA is dependent on a careful choice of threshold for the gradient magnitude image. For leaf B, we remark, to make the link to what was pointed out for the results from GM-HCMA, that the very high gradients on one side of the searched pattern (right) and low on the other side (left) makes this thresholding difficult.

5 GM-HCMA for Identification of Proteins

The initial goal of this work was to investigate whether it was possible to use GM-HCMA to achieve better and more stable matching results than by HCMA for a specific application. In [11], HCMA was used for shape based identification of proteins, or more specifically the Immunoglobulin G (IgG) antibody, in cryo-ET images. The analysis of cryo-ET images is of interest as the technique allows imaging of individual proteins, something which is not possible with other imaging systems. By this it is possible to study protein functional dynamics. For more information on cryo-ET experiments of the IgG antibody and the structural information possible to extract from this type of images, we refer to [17].

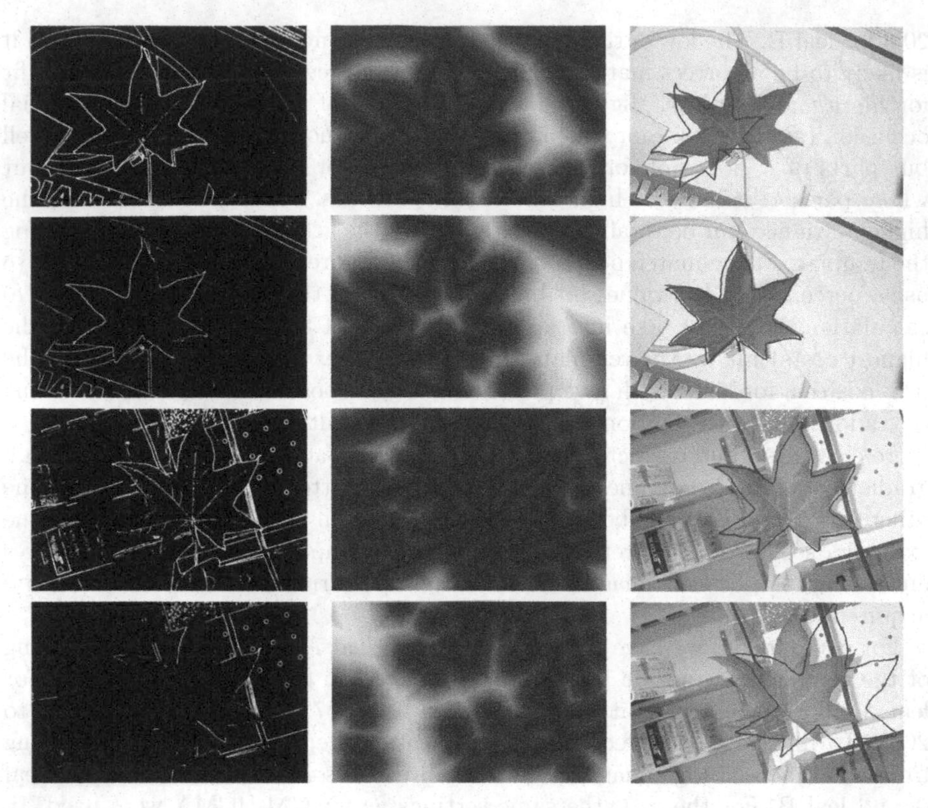

Fig. 4. Top rows: leaf A. Bottom rows: leaf B. Edge images created using two different thresholds (400 and 1000), the DTs, and the position giving the lowest match measure using HCMA.

The images resulting from cryo-ET have low contrast as the micrographs used to reconstruct the tomographic images need to be acquired using a low electron dose to prevent radiation damages. The images contain not only the proteins of interest but also other structures, such as ice crystals, artifacts from the reconstruction, and other molecules, which complicate the matching. To the left in Fig. 5, a cross section of a cryo-ET image is shown. The "Y" shaped object almost centred in the cross section is an IgG antibody. This IgG antibody is shown volume rendered in Fig. 5 (middle). The cryo-ET image is of size $150 \times 150 \times 150$ voxels with voxel size 5.24Å and a resolution of approximately 2nm.

In [11], a template created from the protein databank, [18], (PDB identification 1igt) was used in the search for IgG antibodies. This template is shown in Fig. 5 (right). The experiments in [11] showed promising results. However, as the gradient magnitude information is low in cryo-ET images and, hence, the edge detection step required for HCMA is non trivial, GM-HCMA is of interest to achieve a more robust protein identification process.

We have used the same settings for the experiments as in [11], but applied GM-HCMA instead of HCMA. The size of the template was 1220 voxels, which

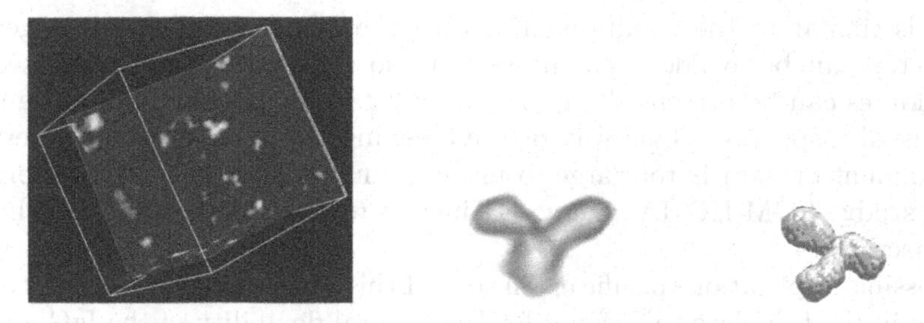

Fig. 5. Left: A cross section of a cryo-ET image with an IgG antibody shown almost in the middle. Middle: A volume rendering of the IgG antibody in the cross section. Right: The template used in GM-HCMA.

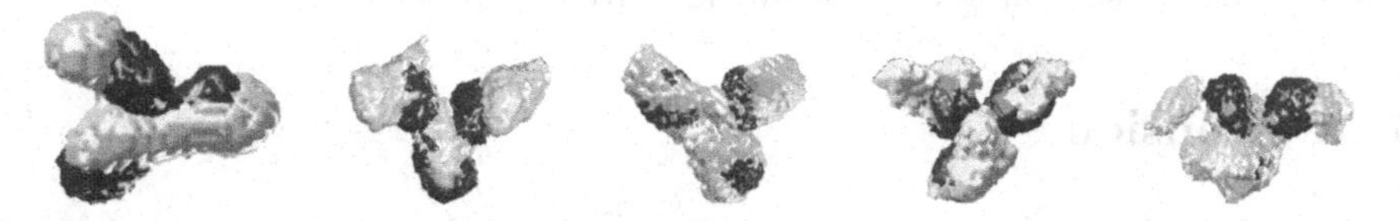

Fig. 6. The positions found by GM-HCMA (found object in yellow/light, template in blue/dark). All positions correspond to IgG antibodies except for the rightmost.

corresponds to the surface of an 1igt antibody object of size 3600 voxels. Only two resolution levels were allowed as the objects almost disappear at lower resolution than that. Moreover, we follow the suggestion in Section 3 and use the logarithm of the gradient magnitudes. The IgG antibody consists of three roundish parts, which are approximately equal in size. We have chosen the weights in such a way that the gradient magnitudes decrease to 0 over a distance corresponding to size of one of these parts, at the top level of the pyramid. GM-HCMA was applied to four cryo-ET images, of which three are the same as the three used in [11], each containing one IgG antibody. The weights were $\langle 2.20, 2.94, 3.67 \rangle$, $\langle 2.42, 3.23, 4.04 \rangle$, $\langle 2.42, 3.23, 4.03 \rangle$, and $\langle 2.41, 3.21, 4.02 \rangle$, respectively.

For three of the images, the positions found by GM-HCMA, giving the highest match measure, corresponded to an IgG antibody, with match measures 2.34, 2.24, and 2.26, respectively, and variance 0.02, 0.03, and 0.03, respectively. For the fourth image, another structure was found, with match measure 2.17 and variance 0.04. The IgG antibody was listed as the second best match, with match measure 2.16 and variance 0.04. In Fig. 6, matches are shown with the template (yellow/light) translated to the positions found by GM-HCMA and superimposed on the corresponding IgG antibody (blue/dark). The rightmost subfigure shows the structure found which does not correspond to an IgG antibody. The positions are not perfect mainly due to the fact that the template does not perfectly correspond to the searched antibody.

The initial experiments shown here indicate that GM-HCMA is a robust method for finding IgG antibody candidates in cryo-ET images. The IgG antibodies are found at the position producing the highest match measure in three of the four available volumes, and at the second highest position in the fourth.

This is similar to the results in [11], with the advantage that the edge detection step can be avoided. To ensure that no IgG antibodies are missed good candidates can be presented for the molecular biologists involved in the project for visual inspection. This way of prescreening the images is of interest since the amount of data is too large to make manual search feasible. A more thorough study of GM-HCMA on cryo-ET images will be the issue for a forthcoming manuscript.

Possible application specific extensions of this work include incorporating flexibility in the template to account for the natural flexibility of the IgG antibody. However, this implies a (much) larger search space and would thereby increase the complexity of the algorithm. Another possibility is to follow the suggestion in [9] and embed hierarchical grouping of multiple templates corresponding to different molecular configurations in the matching process.

6 Conclusion

We have described a modification of the hierarchical chamfer matching algorithm (HCMA) [3], with the advantage that no binarisation of the gradient magnitude image is necessary. Instead, distance weighted propagation of the gradient magnitude information is used to create a cost image, resulting in a gradient magnitude based HCMA (GM-HCMA). GM-HCMA needs to be tuned for the application at hand with respect to the weights used for propagating the gradient magnitudes and the evaluation of positions found by the matching process. We have shown that GM-HCMA gives comparable or improved results to HCMA, without increasing the complexity of the algorithm or its computational cost. In Section 5, we used GM-HCMA in a real application. We could achieve the same results as in [11], but without the need of a sophisticated edge binarisation method as in [11], where HCMA was used.

Acknowledgements

The cryo-ET images have been provided by Dr Sara Sandin, Department of Cell and Molecular Biology, Karolinska Institutet, Stockholm, Sweden (currently Division of Structural Studies, MRC Laboratory of Molecular Biology, Cambridge, United Kingdom). Stina Svensson is financially supported by Swedish Research Council (project 621-2005-5540).

References

1. Barrow, H.G., Tenenbaum, J.M., Bolles, R.C., Wolf, H.C.: Parametric correspondence and chamfer matching: two new techniques for image matching. In: Proc. 5th International Joint Conf. on Artificial Intelligens, Cambridge, USA (1977) 659–663
2. Borgefors, G.: An improved version of the chamfer matching algorithm. In: 7th International Conference of Pattern Recognition, Montreal (1984) 1175–1177

3. Borgefors, G.: Hierarchical chamfer matching: A parametric edge matching algorithm. IEEE Transactions on Pattern Analysis and Machine Intelligence **10**(6) (1988) 849–865
4. Ding, L., Goshtasby, A., Satter, M.: Volume image registration by template matching. Image and Vision Computing **19** (2001) 821–832
5. Zitová, B., Flusser, J.: Image registration methods: a survey. Image and Vision Computing **21** (2003) 977–1000
6. Mäkelä, T., Pham, Q.C., Clarysse, P., Nenonen, J., Lötjönen, J., Sipilä, O., Hänninen, H., Lauerma, K., Kirsi, L., Knuuti, J., Katila, T., Magnin, I.E.: A 3-D model-based registration approach for the PET, MR and MCG cardiac data fusion. Medical Image Analysis **7** (2003) 377–389
7. Thayananthan, A., Stenger, B., Torr, P.H.S., Cipolla, R.: Shape context and chamfer matching in cluttered scenes. In: 2003 IEEE Computer Soc. Conf. on Computer Vision and Pattern Recognition. Vol. 1., IEEE Computer Society (2003) 127–133
8. Gavrila, D.M.: Pedestrian detection from moving vehicle. In Vernon, D., ed.: Computer Vision - ECCV 2000: 6th European Conference on Computer Vision. Vol. 1843 of Lecture Notes in Computer Sciences. (2000) 37–49 Part II.
9. Olson, C.F., Huttenlocher, D.P.: Automatic target recognition by matching oriented edge pixels. IEEE Transactions on Image Processing **6**(1) (1997) 103–113
10. Rosin, P.R., West, G.A.W.: Salience distance transforms. Graphical Models and Image Processing **57**(6) (1995) 483–521
11. Sintorn, I.M., Borgefors, G.: Shape based identification of proteins in volume images. In Kalviainen, H., Parkkinen, J., Kaarna, A., eds.: Image Analysis: 14th Scandinavian Conference, SCIA 2005. Vol. 3540 of Lecture Notes in Computer Science., Springer-Verlag (2005) 253–262
12. Gonzalez, R.C., Woods, R.E.: Digital Image Processing. 2 edn. Prentice-Hall, Inc. (2002)
13. Nash, S.G., Sofer, A.: Linear and Nonlinear Programming. Industrial Engineering Series. McGraw-Hill International Editions (1996)
14. Rosenfeld, A., Pfaltz, J.L.: Sequential operations in digital picture processing. Journal of the Association for Computing Machinery **13**(4) (1966) 471–494
15. Borgefors, G.: On digital distance transforms in three dimensions. Computer Vision and Image Understanding **64**(3) (1996) 368–376
16. Borgefors, G., Olsson, H.Å.: Localizing and identifying objects: A method for distinguishing noise, occlusion, and other disturbances. In Pietikäinen, M., Seppänen, T., eds.: Proceedings of (2nd) Nordic Workshop on Industrial Machine Vision, Oulu, Finland (1992) ISBN 951-42-3316-6. Also published as FOA (Swedish Defence Research Establishment) Reprint B 30284-3.4 in Sep. 1992.
17. Sandin, S., Öfverstedt, L.G., Wikström, A.C., Wrange, O., Skoglund, U.: Structure and flexibility of individual immunoglobulin G molecules in solution. Structure **12** (2004) 409–415
18. Berman, H.M., Westbrook, J., Feng, Z., Gilliland, G., Bhat, T.N., Weissig, H., Shindyalov, I.N., Bourne, P.E.: The protein data bank. Nucleic Acids Research **28**(1) (2000) 235–242

Elliptical Distance Transforms and Applications

Hugues Talbot

A2SI-ESIEE / IGM
2 boulevard Blaise-Pascal
93192 Noisy-le-Grand, France
h.talbot@esiee.fr

Abstract. Discrete Euclidean distance transforms, both exact and approximate, have been studied for some time, in particular by the Discrete Geometry community.

In this paper we extend the notion of Euclidean distance transform (EDT) to elliptical distance transform (LDT). The LDT takes an additional two fixed parameters (eccentricity and orientation) in 2-D and an additional four in 3-D (two ratios and two angles) in 3-D, instead of 1 for the EDT in all cases . We study first how the LDT can be computed efficiently with good approximation in the case where all parameters are constant.

We provide an application to binary object segmentation as motivation for this work.

1 Introduction

The discrete, non-Euclidean distance transform (DT) (e.g. underlying the 4-, 6- or 8-connected grid) has been under study for quite some time, see for example [1], and more recently in arbitrary dimensions [2]. Various efficient algorithms for computing the DT, linear in the number of pixels N, have been known since about the same time [3]. This efficiency has led the DT to be used in a number of situations, for example to compute the skeleton [4], among others [5].

1.1 Euclidean Distance Transform

If one defines the DT of a discrete binary set as the function which associate each pixel to its shortest distance to the exterior of the set, one can also define a discrete Euclidean Distance Transform (EDT), simply substituting the Euclidean distance for the grid-based one. The EDT has several advantages, mainly isotropy, but is more costly to compute. There exists an obvious but unusable (quadratic in N) algorithm for computing this transform. Therefore practitioners have sought various reasonable and efficient approximation to the EDT [6]. In particular Danielsson [7] proposed a linear algorithm with excellent approximation.

Exact, linear algorithms for the EDT have recently been proposed in the literature, for example in 6-connectivity via chain propagation [8], by ordered propagation [9], using Voronoïdiagrams [10], and using dilation by paraboloid [11], among others.

A. Kuba, L.G. Nyúl, and K. Palágyi (Eds.): DGCI 2006, LNCS 4245, pp. 320–330, 2006.

1.2 Elliptical Distance Transform

The Elliptical Distance Transform (LDT[1]) was defined in [12] can be understood as an extension of the Euclidean Distance Transform (EDT). We substitute the Euclidean distance ED between two points p and q (in 2-D)

$$\mathrm{ED}(p,q) = \sqrt{(x_p - x_q)^2 + (y_p - y_q)^2}\,, \tag{1}$$

by a new function LD, with $\rho = \mathrm{ED}(p,q)$, $\rho\cos(\theta) = x_p - x_q$, $\rho\sin(\theta) = y_p - y_q$, α an arbitrary angle and $\sigma = a/b$ with a and b pair of arbitrary real numbers:

$$\mathrm{LD}(p,q) = \rho\sqrt{\frac{(b\cos(\theta-\alpha))^2 + (a\sin(\theta-\alpha))^2}{\sigma}}\,. \tag{2}$$

We have two extra parameters, it is easy to see that $\sigma = a/b$, is the axis ratio of an ellipse and α its orientation. The function LD has all the attributes of a distance (positivity, symmetry, $LD(a,b) = 0 \Leftrightarrow a = b$, triangular inequality), and its level lines are all ellipses of orientation α and axis ratio σ. By convention $\sigma > 0$, as the converse is equivalent to the convention with an additional rotation of $\pi/2$ in α. In the remainder of the paper, we assume that parameters α and σ are arbitrary, but fixed for a given computation.

1.3 Grey-Weighted and Geodesic Distance Transform

In the usual DT, whether Euclidean or not, space is considered of constant metric, i.e. the transform is invariant by translation. An extension to usual DT is to consider the grey-weighted distance transform (GWTD) [13], for which the cost of going through a pixel is equal to the grey-level value of this pixel. This concept is clear in the discrete case as it reduces to computing min-cost paths on graphs [14]. However, this idea also extends to the Euclidean DT, by considering a discrete image as as sampled continuous space with scalar metric [15]. In this case one is then numerically solving the Eikonal equation $|\nabla u| = g$, where g is the sampled metric.

The classical way to solve this equation, and thus obtain a grey-weighted Euclidean distance transform is a discrete algorithm called the Fast Marching Method (FMM) [16]. We note that by setting $g \equiv 1$ everywhere, with proper initial conditions, we compute the EDT – with two caveat. First the FMM is not linear in complexity, but only $O(N \log N)$ in the worst case (in practice very close to linear), with N the number of pixels. However with arbitrary real values for g, and not simply limited discrete integer values this is in fact optimal, as a sorting pass is inevitable. Second the FMM is not exact but only first or second-order accurate.

Recently this work was extended to Riemannian metrics – i.e metrics that are positive definite on arbitrary manifolds, with the Ordered Upwind Method

[1] Given that 'L' is pronounced as "ell" this acronym sounds adequate even if unorthodox.

(OUM) [17]. In other words, this allowed to compute the equivalent of a discrete positive definite tensor-weighted LDT. Perhaps more clearly, in DG terms this is a WDT for which at each point the weight is defined by a ratio and an orientation just like the LDT, and this weighting can vary from pixel to pixel. For a constant tensor, this obviously reduces to the LDT, with the same caveat as for the EDT vs. the FMM.

In the literature one finds mentions of the geodesic DT, which is a DT computed inside a binary set [18], for which a Euclidean version exists [19], however this reduces to the GWDT or FMM with infinite cost outside the set and constant positive cost within.

1.4 Rest of the Paper

In the following we propose an efficient algorithm for computing the LDT. This algorithm is accurate to within a fraction of a pixel.

2 Computing the LDT

The computation of the LDT following [20] is relatively simple as it is a straightforward adaptation of a known ordered propagation algorithm [21]. This algorithm carries all the approximations of the Danielsson approach, but this is not usually a problem in most applications. More importantly, this adaptation is only valid for σ ratios close to 1, which limits its usefulness.

2.1 Motivating Example

An important problem in image analysis is the separation of touching binary particles for counting purposes. A series of methods have been proposed in various cases, but a recurring theme is when the shapes of the touching particles are close to disks. This is useful when counting blood cells or fibres in cross sections, for example.

The classical approach to this problem with mathematical morphology is to compute the watershed line on the complement of the distance transform [22]. To deal with border noise, one can use suitable h-maxima on the distance transform as markers [23]. Other approaches involving the analysis of the DT along the skeleton [24], using the conditional bisector [25] and later the bisector function [26,27]. The latter approach is able to separate even the most deeply fused particles [28].

However when the particles are no longer disks but ellipses this approach breaks down, yet separating elliptical fused particles is also a common problem. For example eukariot cell nuclei are often elliptical in shape. Cylindric fibres in cross-section are not necessarily perpendicular to the cutting plane, and therefore their shape can be elliptical as well. In [20] the authors proposed to use a grid search of LDTs to find the centre of fused ellipses accurately. This is illustrated on Fig. 1.

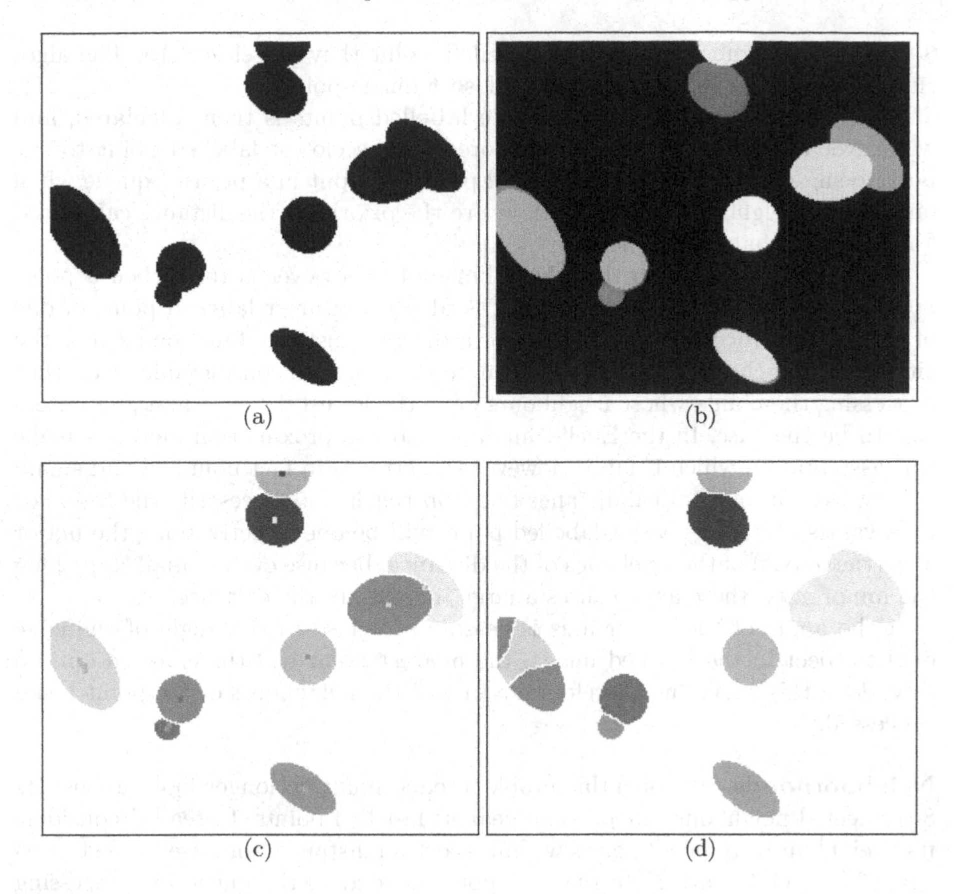

Fig. 1. Separation of fused ellipses: (a) binary input image ; (b) Fitted ellipses from the LDT search ; (c) segmentation result – each ellipse is drawn with a distinctive colour with their found centre. For comparison, (d) is the watershed segmentation based on the EDT

2.2 Algorithm for Computing the LDT

We only provide an explanation for the implementation in 2-D in the square grid, however the principle remains the same for higher dimensions and different grids. The basic idea is along the same vein as Dijkstra's algorithm and Fast Marching Methods in that each point is visited only once, and the neighbours of the point with the smallest distance are visited first.

Here the elliptical distance is defined as

$$\sqrt{u^2 + (\sigma * v)^2}\ , \tag{3}$$

where u and v are the distances along the principal axes of the ellipse and σ is fixed.

Previous Algorithm. The specific algorithm is based on the algorithm described in [20]. In that algorithm, when each point is processed, one looks at its

8-connected neighbours for which labelled point they are closest to. The algorithm uses this as the set of possible closest image points.

The distance to each of the possible labelled points is then calculated, and whichever is the closest of those is stored as the closest labelled point to the point being considered. The unlabelled points are put in a priority queue when one of their neighbours is processed, where the priority is the distance calculated for the point that added them.

The crucial assumption this algorithm makes however is that when a point is processed, its nearest labelled point is also the nearest labelled point of one of its already processed neighbours. For a discrete distance function such as the city-block and chessboard metrics it can be proven, with considerable work, that processing the points whose neighbours have the lowest distance first guarantees this to be the case. In the Euclidean case, most approximation methods make this assumption, which is false. However the error is in fact bounded and small.

However for an elliptical distance function this is not neccesarily the case. For high values of σ, the nearest labelled point will be one directly along the major axis (the u axis) of the level lines of the distance, because even a small step along the minor axis (the v axis) causes a large increase in the distance.

If the angle of the major axis is reasonably close to the angle of the edge of the object in the labelled image, the nearest point will therefore be quite a way along this edge, and therefore not one of the neighbours of the point being processed.

Neighborhoods. To solve this problem, each point no longer looks at just its 8-connected neighbours for possible nearest labelled points. Instead it considers its "neighbours" to be all points within a certain distance from itself ("distance" again being elliptical). Note that the points added to the queue for processing are still the 8-connected neighbours of the newly processed point, it is only when processing the points that the definition of 'neighbours' changes.

Clearly if this distance was big enough to include the entire image, then one of these "neighbours" always would have the same (or more specifically would be) the closest labelled point, however the algorithm would then become quadratic in complexity, and this is clearly not desirable. The question of what distance is the minimum to guarantee the accuracy of the algorithm (as the smaller the distance the faster it runs) has not yet been conclusively solved.

The value used in this function is just a little bit more than σ, designed to be approximately the smallest distance that still makes sure the 8 connected neighbours are a subset of the neighbour set. Also it should be noted that while the choice of defining the neighbours by the elliptical distance function being used has not been mathematically proven to be the best, it does intuitively seem a reasonable choice and, more importantly, gives good results.

Elliptical Distance. Another problem which had to be overcome was the fact that the distance between grid points could become significant for high values of σ. This is because a half pixel error in the distance along the minor axis is magnified by a factor of σ and for pixels close to the labelled pixels can easily become bigger than the distance itself. To fix this, the distance between two grid points is re-defined as being the minimum distance between the first grid point and a 1 pixel wide square (i.e. half a pixel to each side).

To find the elliptical distance in terms of the known x and y co-ordinates, u and v are found in terms of x and y and substituted into the elliptical distance equation 3. As the u and v axes are just a rotation by the angle α in Eq. 2:

$$u = x\cos(\alpha) + y\sin(\alpha) \;, \tag{4}$$

$$v = y\cos(\alpha) - x\sin(\alpha) \;. \tag{5}$$

We actually compute the square of the LDT, as this allows all calculations and bookkeeping to be done in integer arithmetic. The square of the elliptical distance is then

$$u^2 + (\sigma v)^2 = ax^2 + bxy + cy^2 \;,$$

where a, b and c depend only on σ and α.

To find the re-defined distance between two grid points, the distance to the closest point on the surrounding square of the second grid point needs to be found. The closest point must lie on either the horizontal side of the square visible to the the grid point or the vertical side visible (if neither horizontal side is visible the left is arbitrarily chosen to be looked at and similarly the bottom if neither vertical side is visible). If the closest point is not one of the corners it must be a local minimum of the distance function along that line, and more importantly, because there is only one local minimum along any line it must be the global minimum. For a horizontal line y is fixed, and the (squared) distance function is just a quadratic in x : $ax^2 + (by)x + cy^2$. Elementary theory on quadratics gives the global minimum at $x = -by/(2a)$, and the minimum value as $y^2(c - (b^2)/(4a))$

Similarly for vertical lines the minimum is at $y = -bx/(2c)$ and has value $x^2(a - (b^2)/(4a))$ The values $-b/2a$ and $-b/2c$ are precalculated and compared with x/y or y/x when needed. The distance is found by working out which two lines are visible, checking if the minimum of the distance function along those lines lies inside the square (if it does it is easy to show it will be the closest point on the square), and if it is not then calculating the distances to the three corners and taking the minimum.

2.3 Pseudo-code

In the pseudo code in Fig. 2, P is identical to the LD function defined in Eq. 2.

2.4 Results

Accuracy. Fig 3 shows two level lines of the LDT of a set where the background constitutes a single point in the centre of the image. The dotted lines shows the LDT computed by the algorithm from [20], the solid line shows the proposed algorithm. The parameters were $\sigma = 5$ and $\alpha = 45°$.

```
 1 - Input binary image
 2   Fill X image with 0
 3   Fill Y image with 0
 4   Fill output image O with 1
 5   Empty priority queue
 -
 6 - Scan the binary image in raster order, enqueue the points belonging to the sets
 7   which are 8-connected to the background with priority 1.
 8   For each enqueued location, set the X and Y image to the respective
 9   components of the direction to the nearest pixel in the background.
 -
10 - While the priority queue is not empty; do
11       - Dequeue lowest priority pixel C
12       - if O(C) (the output image value at this pixel) is 1, then
13           - Compute P(C), the priority of this pixel, from X(C) and Y(C),
14             i.e P(C) = P(X(C), Y(C))
15           - Set O(C) (the output image value at this position) to P(C).
16           - For each point dC in the extended elliptical neighbourhood of this pixel; do
17               - If O(C+dC) is 1 ; then
18                   - if X(C+dC) or Y(C+dC) is not 0, compute P1 = P(X(C+dC),Y(C+dC))
19                     else set P1 to +infinity
20                   - compute P2 = P(C+dC) where '+' denotes the 2-D vector addition.
21                   - If P2 < P1 then
22                         - enqueue dC with priority P2
23                         - set X(C+dC) to X(C) + x(dC), where x(A) is the X-component of A
24                         - set Y(C+dC) to Y(C) + y(dC), where y(A) is the Y-component of A
25                         - set O(C+dC) to 1
26                   + end if
27               + end if
28           + end for
29       + end if
30   + end while
```

Fig. 2. Pseudo-code for the LDT. See text for the formulation of the priority function P.

The maximum distance between the largest level lines for both algorithms is approximately one pixel. The average distance is about 0.6 pixel. However, a comparison with the current algorithm and a parametric (drawn) ellipse with the same parameters shows no difference. On the negative side, the newer version of the algorithm takes approximately twice as long to compute, with these parameters, as it needs to inspect a larger neighbourhood during the propagation.

We did not include the result of a comparison with the OUM as it is quite inaccurate for this purpose. Being first-order accurate meaning that the error increase linearly with the distance. The OUM is also quite slow (approximately an order of magnitude) compared to these special-purpose LDT.

Application. Figure 5 shows a sample of an image of overlapping cell nuclei, together with the segmentation achieved using an ellipse fitting method using the current LDF algorithm described in [20] and reproduced in algorithm in Fig. 4 :

Here, the space of possible α and σ parameters is sampled in a grid search. More sophisticated optimisation methods can be envisaged, but this is not the purpose of this article.

For display purposes it is convenient to compute a measure which is high for a good fit. The measure that we use is $s = 100/m$. We make sure that s is never 0 even for a perfect fit. We call s the *score* of a ellipse fit. To display the

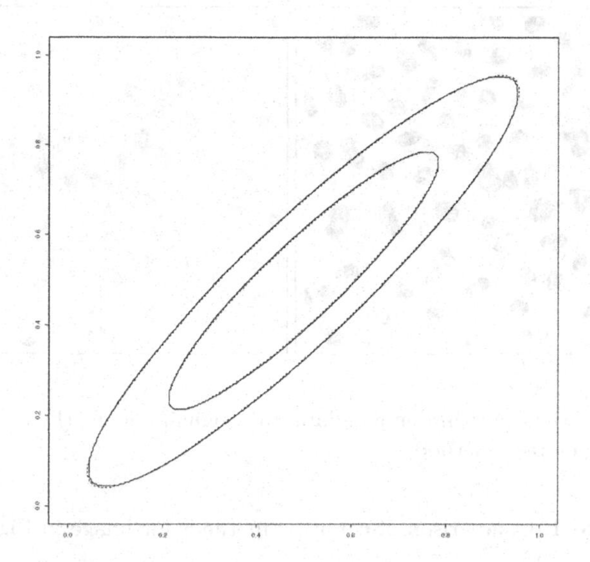

Fig. 3. Two level lines of the LDT computed by previous (dotted line) and current algorithm (solid line)

1. The EDM from the contour of the binary image is computed into image C.
2. A number of LDMs of the binary image are computed, with different values of α and σ.
3. For each LDM, special points are considered as candidate centres of ellipses. These need not include all the points of the LDM skeleton. In our application we only considered discrete regional maxima of the LDM.
4. For each special point, a candidate ellipse is generated with the same parameters (position, axis lengths and orientation) as the underlying LDM. Each pixel of the circumference of the generated ellipse is associated with its distance from the boundary of the binary image. This is provided by a simple interpolated lookup in image C. The pixel distances are sorted and a given percentile is taken as the measure m of goodness-of-fit (for example 50% yields the median distance).

Fig. 4. Algorithm for ellipse fitting

result of the fit, we draw each candidate ellipse from lowest score to highest in the grey-level of their score. As an illustration, Fig 1(b) shows the score of the fitted ellipse on the motivating example. We see that well-fitted ellipses tend to overwrite poorly-fitted ones.

On the real cell example, we sample the space with 31 different LDTs with $\delta\alpha = 30°$ and $\delta\sigma = 0.5, 1 \leq \sigma \leq 3$. The result of the ellipse fit appears reasonable, and table 1 confirms this impression. In this table we have in the first column the result of a careful manual count, in the second the result of the segmentation achieved by the proposed algorithm, the result of the segmentation achieved by the previous algorithm of [20], the result of the segmentation by watershed on an EDT, both unfiltered and filtered by h-maxima.

We see that on this sample the error rate with the new method is slightly improved compared with the older method, and much improved compared to the watershed method – it is 3 times lower.

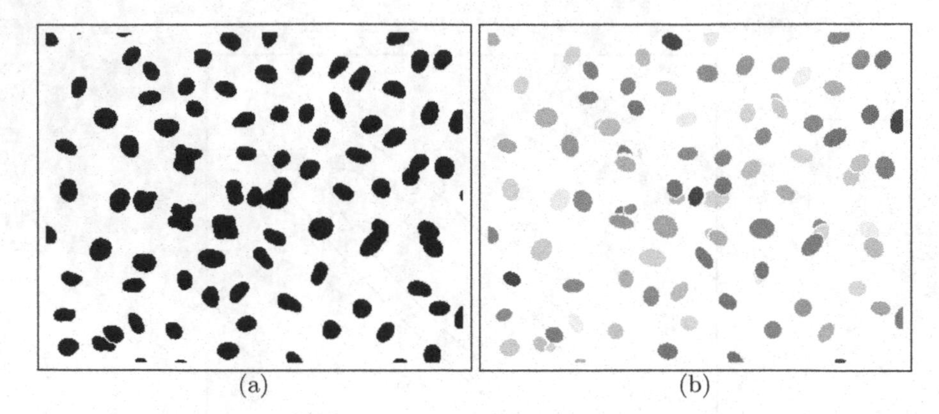

Fig. 5. Result of the algorithm on real data. (a) original image ; (b) segmentation using the overlapping ellipses method.

Table 1. Comparison of counting methods for image in Fig. 5(a)

Count method	Manual	Ellipse high acc.	Ellipse Low acc.	Unfilt. wshed	Filt wshed
Nb of cells	104	100	98	93	86
Total errors	0	6	8	15	18
Over-segmentation errors	0	1	2	2	0
Under-segmentation errors	0	5	6	13	18

On the given image low-resolution image, computing all the 30 LDTs and doing the fits takes 5 seconds on a P-IV 3GHz PC. On the full data – a 1280×1024 12-bit image, the complete computation took about one minute.

3 Conclusion and Future Work

In this paper we have presented an algorithm for computing the LDT, the elliptical distance transform and we have provided an illustration of its usefulness. While not exact, it is more accurate than a previously published version, at the cost of increased computational cost. This extra accuracy can be helpful in some applications making use of LDTs.

This algorithm is still open to improvement. In particular an exact version would certainly be desirable. The author is also interested in improving both the precision and the speed of the OUM.

Acknowledgements

The author wishes to thank Stephen Irrgang for his initial implementation of the LDT.

References

1. Rosenfeld, A., Pfaltz, J.: Distance functions on digital pictures. Pattern Recognition **1** (1968) 33–61
2. Borgefors, G.: Distance transformations in arbitrary dimensions. Computer Vision, Graphics, and Image Processing **27** (1984) 321–345
3. Rosenfeld, A., Pfaltz, J.: Sequential operations in digital picture processing. J. Assoc. Comp. Mach. **13**(4) (1966) 471–494
4. Calabi, L., Hartnett, W.: Shape recognition prairie fires, convex deficiencies and skeletons. American Mathematical Monthly **75**(4) (1968)
5. Borgefors, G.: Distance transformations in digital images. Computer Vision, Graphics, and Image Processing **34** (1986) 344–371
6. Montanari, U.: A method for obtaining skeletons using a quasi-euclidean distance. J. Assoc. Comp. Mach. **15** (1968) 600–624
7. Danielsson, P.E.: Euclidean distance mapping. Computer Graphics and Image Processing **14** (1980) 227–248
8. Vincent, L.: Exact euclidean distance function by chain propagations. In: CVPR, Maui, Hawaii, IEEE (1991) 520–525
9. Cuisenaire, O., Macq, B.: Fast euclidean distance transform by propagation using multiple neighborhoods. Computer Vision and Image Understanding **76**(2) (1994) 163–172
10. Breu, H., Gil, J., Kirkpatrick, D., Werman, M.: Linear time distance transform algorithms. IEEE Transactions on Pattern Analysis and Machine Intelligence **17**(5) (1995) 529–533
11. Mehnert, A., Jackway, P.: On computing the exact euclidean distance transform on rectangular and hexagonal grids. Journal of Mathematical Imaging and Vision **11**(3) (1999) 223–230
12. Talbot, H., Appleton, B.: Elliptical distance transforms and the object splitting problem. In Talbot, H., Beare, R., eds.: Mathematical Morphology, Proceedings of the 6th international symposium, Sydney, Australia, CSIRO Publishing (2002) 229–240
13. Rutovitz, D.: Data structures for operations on digital images. In Cheng, G., Ledley, R., Pollok, D., Rosenfel, A., eds.: Pictorial Pattern Recognition, WA, Thomson Book (1968) 105–133
14. Dijkstra, E.: A note on two problems in connexion with graph. Numerische Mathematik **1** (1959) 269–271
15. Beardon, A.: The Geometry of Discrete Groups. Graduate texts in Mathematics. Springer (1983)
16. Sethian, J.: Fast marching methods. SIAM review **41**(2) (1999) 199–235
17. Sethian, J.A., Vladimirsky, A.: Ordered upwind methods for static Hamilton-Jacobi equations. Proc. Nat. Acad. Sciences **98**(20) (2002) 11069–11074
18. Serra, J.: Image analysis and mathematical morphology. Academic Press (1982)
19. Soille, P.: Spatial distributions from contour lines: an efficient methodology based on distance transformations. Journal of Visual Communication and Image Representation **2**(2) (1991) 138–150
20. Talbot, H., Appleton, B.C.: Elliptical distance transforms and object splitting. In Talbot, H., Beare, R., eds.: Mathematical Morphology: Proceedings of the VIth International Symposium, CSIRO publishing (2002) 229–240
21. Ragnemalm, I.: The euclidean distance transform in arbitrary dimensions. Pattern Recognition Letters **14**(11) (1993) 883 – 888

22. Lantuéjoul, C.: Skeletonization in quantitative metallography. In Haralick, R.M., Simon, J.C., eds.: Issues of Digital Image Procesing, NATO, Sijthoff and Noordhoff (1980)
23. Beucher, S., Vincent, L.: Introduction aux outils morphologiques de segmentation. In: Traitement d'image en microscopie à balayage et en microanalyse par sonde électronique, Paris, ANRT, groupement microanalyse et MEB (1990) F1–F43
24. Talbot, H., Terol Villalobos, I.: Binary image segmentation using weighted skeletons. In: Image algebra and morphological image processing III. Volume 1769., San Diego, CA, SPIE (1992) 147–155
25. Meyer, F.: Cytologie quantitative et morphologie mathématique. PhD thesis, Ecole des Mines de Paris (1979)
26. Talbot, H., Vincent, L.: Euclidean skeleton and conditional bisectors. In: Visual Communications and Image Processing'92. Volume 1818., Boston, SPIE (1992) 862–873
27. Zrour, R., Couprie, M.: Discrete bisector function and euclidean skeleton. In Andres, E., Damiand G., Lienhardt, P., eds.: Discrete Geometry for Computer Imagery: 12th International Conference, DGCI 2005. Volume 3429 of LNCS., Poitiers, France, Springer (2005) 216–227
28. Talbot, H.: Analyse morphologique de fibres minérales d'isolation. PhD thesis, Ecole des Mines de Paris (1993)

A Composite and Quasi Linear Time Method for Digital Plane Recognition

Lilian Buzer

Laboratory CNRS-UMLV-ESIEE, UMR 8049
ESIEE, 2, boulevard Blaise Pascal, Cité Descartes, BP 99
93162 Noisy le Grand Cedex, France
buzerl@esiee.fr

Abstract. This paper introduces a new method for the naive digital plane recognition problem. As efficient as existing alternatives, it is the only method known to the author that also guarantees a quasi linear time complexity in the worst case. The approach presented can be used to determine if a set of n points is a naive digital hyperplane in $\mathbb{Z}^d$ in $O(n \log^2 D)$ worst case time where D represents the size of a bounding box that encloses the points. In addition, the approach succeeds in reducing the naive digital plane recognition problem to a two-dimensional convex optimization program. Thus, the solution space is planar and only simple two-dimensional geometrical methods need to be applied during the recognition process. The algorithm is a composite of simple techniques based on one-dimensional optimization: Megiddo Oracle for linear programming and two-dimensional discrete geometry.

1 Introduction

1.1 History

Naive digital plane recognition is a deeply studied problem in digital geometry (see a review in [1]). It consists of determining wether or not a given set of points is a piece of a naive digital plane. This paper presents a new technique with quasi linear worst case complexity that is very efficient in practice. The result can be extended without difficulty to the recognition of digital planes of fixed thickness relative to the infinite norm.

Currently, the approaches used in recognition are based on linear programming [2,3,4], convex hull and geometrical methodologies [5,6,7,8], combinatorial optimization [5,6,9,10,11,8] or the evenness property [12]. The methods that exploit linear programming techniques can be separated into two groups. The first group [2,4] relies on the optimal result obtained by Megiddo [3]. However, even if the approaches in this group achieve optimal linear time complexity, the resulting algorithms are too complex to be used in practice. The second group is based on efficient linear programming techniques like the simplex algorithm but their worst case complexity is often too high in practice. Methods that partially traverse the convex hull or the chords' space of the given set of points suffer from

A. Kuba, L.G. Nyúl, and K. Palágyi (Eds.): DGCI 2006, LNCS 4245, pp. 331–342, 2006.

the same problem. For example, the chord's algorithm [10] processes 10^6 voxels in about ten traversals of the point set. Nevertheless, this technique exhibits an $O(n^7)$ time complexity. Other algorithms that partially traverse the vertices and the edges of the convex hull obtain an $O(n \log n)$ time complexity, but they are less efficient. All of the previous methods always balance between efficiency in practice and a low worst case complexity. Thus, this paper presents a simple algorithm that has a quasi linear time complexity in the worst case and that is efficient in practice. Moreover, it does not require the piece of the digital plane to be rectangular as in [12, 6, 11].

In addition, the approach is different from previous approaches because it requires only $d-1$ rational parameters to recognize a valid naive digital hyperplane in $\mathbb{Z}^d$, unlike the previous algorithms that require the determination of d rational variables. Thus, for the three-dimensional recognition problem, we need only to apply planar geometrical techniques to know wether or not the two-dimensional set of solutions is empty.

In the next section, we present how our recognition problem can be transformed into a two-dimensional convex optimization program. In the next section, we present the method core of our approach. Then, we sketch the algorithm and compute its complexity in section 4. Finally, in the last section we describe how to implement some enhancements in order to improve the efficiency of our method in practice.

1.2 Definition

Arithmetic geometry provides a uniform approach to study *digital hyperplanes* in any dimension. In this paper, we only consider the case where the digital hyperplanes in $\mathbb{Z}^d$ are a function from $(x_1, \ldots, x_{d-1})$ into $\mathbb{Z}^d$. Other cases can be simply deduced by symmetry. An arithmetic plane is defined by $P_{N,\mu,\omega} = \{x \in \mathbb{Z}^d | \mu \leq N \cdot x < \mu + \omega\}$ with N the normal vector and ω the *arithmetic thickness.* Recall that when $\omega = ||N||_\infty = \max_{1 \leq i \leq d}\{|N_i|\}$ we obtain a *naive plane* (see Fig. 3 for an example). When $\omega = \sum_{i=1}^{d} |N_i|$, we obtain a *standard plane.*

2 Convex Optimization

2.1 Introduction

Property 1. Let $S = (p_j)_{1 \leq j \leq n} = (x_1^j, \ldots, x_d^j)$ denote a set of points in $\mathbb{Z}^d$. If all the points $(p_j)_{1 \leq j \leq n}$ satisfy $\gamma \leq N \cdot p_j < \gamma + 1$ with $\gamma \in \mathbb{R}$ and with N in $\mathbb{Q}^d$ then S is a subset of a naive digital plane.

Proof. Let $\frac{D_i}{N_i}$ denote the i-th component of the normal vector N. By multiplying by $\Pi_{i=1}^d N_i$, we obtain for any point p_j: $\beta \leq \sum_{i=1}^d (D_i \Pi_{k \neq i} N_k) x_i^j < \beta + \Pi_{i=1}^d N_i$ with $\beta \in \mathbb{Z}$. We can simplify this expression by g, the *gcd* of $(D_i \Pi_{k \neq i} N_k)_{1 \leq i \leq d}$. Let δ denote the ceiling of $\frac{\beta}{g}$. As $[\sum_{i=1}^d (D_i \Pi_{k \neq i} N_k) x_i^j]/g$ is an integer value, we finally obtain : $\delta \leq \sum_{i=1}^d N_i' x_i^j < \delta + ||N'||_\infty$. As δ and the components of N' are integer values, this double inequality corresponds to a naive digital hyperplane.

Definition 1. *Consider a set $S = (p_i)_{1 \leq i \leq n}$ of n points in an Euclidean space $\mathbb{R}^d$. We define a function $h_S(x_1, \ldots, x_{d-1}) : \mathbb{R}^{d-1} \to \mathbb{R}^+$ as the distance relative to the d-th axis between the two supporting hyperplanes of normal vector $(x_1, \ldots, x_{d-1}, 1)$ that enclose all the points (see [1]).*

Property 2. $h_S(x)$ is a convex function.

Proof. Let $N_x = (x_1, \ldots, x_{d-1}, 1)$ denote the current normal vector associated with the current value $x \in R^{d-1}$ that is being processed. By definition of the function $h_S(x)$, we have: $h_S(x) = Max_{p \in S}(N_x \cdot p) - Min_{p \in S}(N_x \cdot p)$. Consider the function $g_i(x) = N_x \cdot p_i$. This function is an affine function and it is also convex. Thus, the maximum defined by all the functions $(g_i)_{1 \leq i \leq n}$ is convex too. The function $h_S(x)$ can be rewritten as $Max_{p \in S}(N_x \cdot p) + Max_{p \in S}(-N_x \cdot p)$. By the same logic, the right side of this expression is also a convex function. Since the sum of two convex functions is convex, we conclude that $h_S(x)$ is convex.

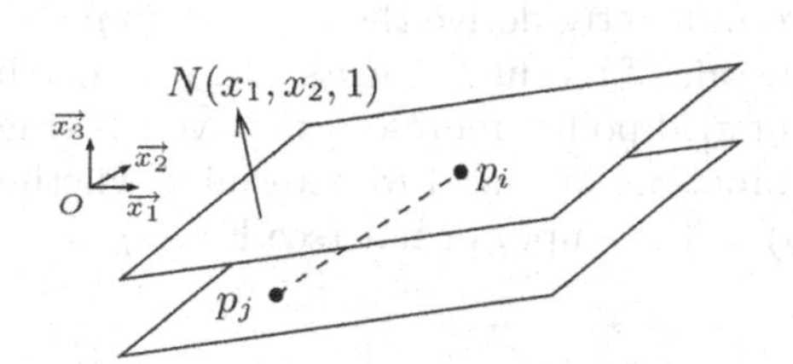

Fig. 1. Definition of $h_S(x_1, x_2)$

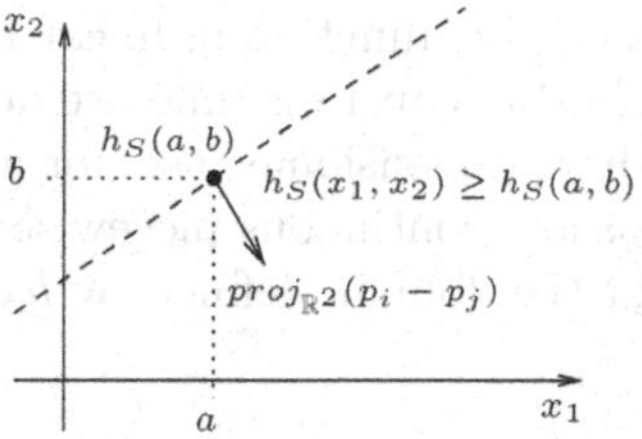

Fig. 2. A subgradient of h_S

2.2 Subgradient Computation

For a given value $x \in \mathbb{R}^{d-1}$, we only have to traverse the list of the points S in order to compute the value of $h_S(x)$. This implies that the computation of $h_S(x)$ has linear time complexity. Relative to the definition of $h_S(x)$, we know that:

$$h_S(x) = Max_{p \in S}(N_x \cdot p) - Min_{p \in S}(N_x \cdot p)$$

For a given value x, there exists two points p_i and p_j associated with the max and the min expressions (see Fig. 1). Therefore, $h_S(x) = N_x \cdot p_i - N_x \cdot p_j = N_x \cdot (p_i - p_j)$. As $T = \{p_i, p_j\}$ is included in S and we have:

$$\forall y \in \mathbb{R}^{d-1}, h_T(y) \leq h_S(y)$$

By definition, $h_T(y)$ is equal to:

$$h_T(y) = |N_y \cdot p_i - N_y \cdot p_j| = |N_y \cdot (p_i - p_j)|$$

and it follows:

$$\begin{aligned} h_T(y) &= |N_{y-x+x} \cdot (p_i - p_j)| \\ &= |N_x \cdot (p_i - p_j) + (y - x) \cdot proj_{\mathbb{R}^{d-1}}(p_i - p_j)| \\ &= |h_T(x) + (y - x) \cdot proj_{\mathbb{R}^{d-1}}(p_i - p_j)| \end{aligned}$$

Since $h_T(x)$ is positive, we have:

$$\begin{aligned} \forall y \in \mathbb{R}^{d-1}, h_S(y) &\geq h_T(y) \\ &\geq h_T(x) + (y - x) \cdot proj_{\mathbb{R}^{d-1}}(p_i - p_j) \end{aligned}$$

We recall that $h_S(x) = h_T(x)$and it follows:

$$\forall y \in \mathbb{R}^{d-1}, h_S(y) \geq h_S(x) + (y - x) \cdot proj_{\mathbb{R}^{d-1}}(p_i - p_j) \quad (1)$$

We can now conclude that the expression $proj_{\mathbb{R}^{d-1}}(p_i - p_j)$ is a subgradient of $h_S(x)$ (see Fig. 2). This value corresponds to the projection of the vector $p_i p_j$ in the $\mathbb{R}^{d-1}$ space. This means that the first $d-1$ components of the vector $p_i p_j$ are sufficient to locally determine the variation of the function h_S. When we evaluate this function in linear time, we indirectly derive the two points p_i and p_j. Thus, in constant time, we determine one of its subgradients. For example in the three-dimensional case, for a set S of grid points represented as voxels (Fig. 3), h_S is a continuous piecewise affine function. We need to determine whether or not the domain defined by $h_S(x_1, x_2) < 1$ is empty or not (see Fig. 4).

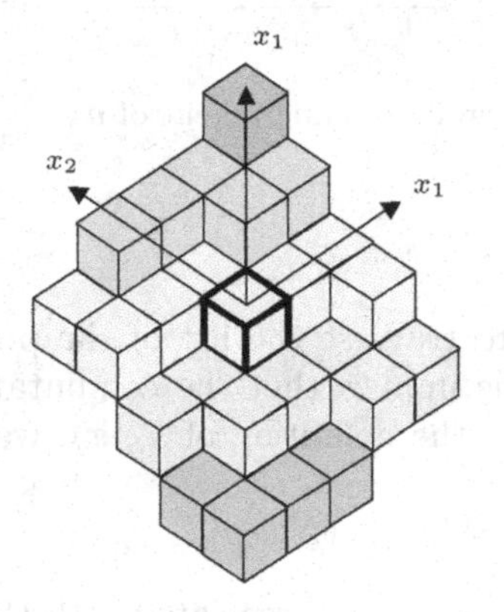

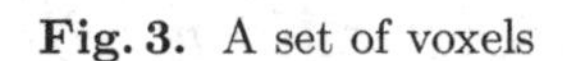

Fig. 3. A set of voxels

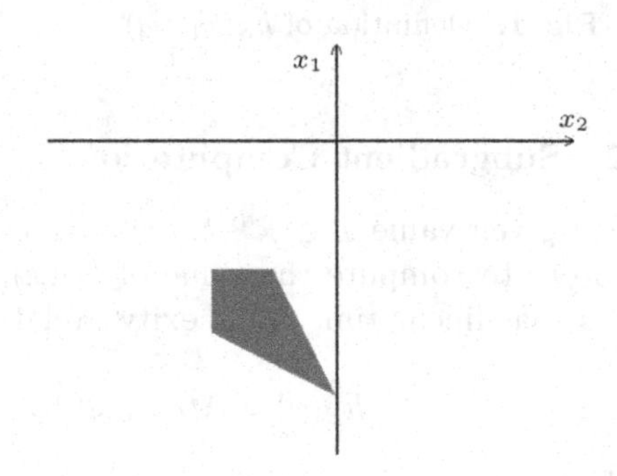

Fig. 4. Domain where $h_S(x_1, x_2) < 1$

3 The Basic Principles of Our Method

3.1 Studying the Solution Space

In the following, we focus our attention on the three-dimensional recognition problem. The core of our approach is based on the existence of the strict inequality in the double diophantine inequality. Analyzing $\mu \leq ax + by + cz <$

$\mu + ||(a,b,c)||_\infty$, which always produces integer values, the following inequality holds:

$$\mu \leq ax + by + cz \leq \mu + ||(a,b,c)||_\infty - 1$$

Most of the time, the voxels of an image are connected and so the diameter of the set of voxels is not very large. Suppose that the voxels we study all lie in a bounding box of size D. Let $||.||_\infty$ denote $||(x,y,z)||_\infty = max\{|w|,|y|,|z|\}$. Thus, any point p in this image satisfies $||p||_\infty \leq D$. In the same way, any vector k whose endpoints are supported by some voxels satisfies:

$$||k||_\infty \leq 2D$$

We can restrict our attention to the case where the naive digital planes are a mapping from (x,y) into z. This means that c is nonzero and that $||(a,b,c)||_\infty = |c|$. Wlog, we can force c to be positive. This produces $\mu' \leq a'x_i + b'y_i + z_i \leq \mu' + 1 - \frac{1}{c}$, where $a' = \frac{a}{c}$ and $b' = \frac{b}{c}$ represent the rational slopes of the digital plane. Relative to our previous assumption, we know that $|a'| \leq 1$ and that $|b'| \leq 1$. We can now determine the influence of a small variation on the normal vector $N(a',b',1)$. Let $N_\Delta(\alpha = \frac{r}{t}, \beta = \frac{s}{t}, 1)$ denote another normal vector in the neighborhood of N such that $|\alpha - a'| \leq \Delta$ and $|\beta - b'| \leq \Delta$. Thus, we have:

$$\mu' - 2D\Delta \leq \alpha x_i + \beta y_i + z_i \leq \mu' + 1 - \frac{1}{c} + 2D\Delta \tag{2}$$

Considering the convex hull of the set of points, we know that there exists two parallel planes that enclose all of the points and that have a minimal distance relative to the z-axis. These two planes are supported by four vertices of the convex hull [5]. The minimum of h_S is reached at a value associated with a normal vector whose coefficients are the vector product of two segments supported by the given voxels [7]. Therefore, there exists k and k' in $\mathbb{Z}^3$ such that the solution $(a,b,c) = k \wedge k'$. As $||k||_\infty \leq 2D$ and $||k'||_\infty \leq 2D$, we obtain:

$$1 \leq c \leq 8D^2 \tag{3}$$

Combining property 1 and inequality (2), we know that when S is a subset of a naive digital plane, any normal vector N_Δ in the neighborhood of N is valid iff: $4D\Delta < \frac{1}{c}$. It follows that when the set of solutions $h_S() < 1$ is not empty, it contains a *critical square* whose side length Δ is less than:

$$\Delta < \frac{1}{4Dc} \tag{4}$$

For example, from (3), we know that $\Delta < 1/(32 \cdot D^3)$. In this paper, we denote by Γ the value $\frac{1}{64D^3}$. In conclusion, we only have to study the function h_S relative to a uniform grid in the domain $[-1,1] \times [-1,1]$ whose resolution is given by Γ. If none of the sampled values represent a valid normal vector then the space of solutions is empty. Thus the solution space of our optimization problem reduces to a two-dimensional grid.

4 Algorithm Design

To attack our optimization problem, we could use standard algorithms from mathematical programming, including subgradient descent methods. However, we can not easily apply such algorithms. For instance, if we want to retain our philosophy of exact numerical computation with rationals, these methods will increase the size of the numerator and denominator of the normal vector at each iteration and slow down the calculation significantly. That is why the approach we present mixes methods developed in one-dimensional binary minimization, Megiddo Oracle technique and subgradient optimization. As we only pick simple techniques, we obtain a very simple approach.

4.1 Megiddo Cut

When Megiddo in [3] describes his optimization method, he cuts the current search domain by a hyperplane. If the optimum solution belongs to the cut, the problem is solved. If not, we create an Oracle (presented in section 4.3) that determines on which side of the cut the optimal solution lies. The search domain is then reduced to the solution space on the side of the cut in which the optimal solution lies. For example, suppose that the current search domain is $\{(a,b)|\{a,b\} \in [-1,1]\times[-1,1]\}$. We can choose to cut by the line $a = 0$. Next, we compute the minimum of h_S relative to this one-dimensional domain. After that step, if the optimum is less than 1, the problem is solved. Elsewhere, we call the Oracle and reduce the search domain to $[-1,0]\times[-1,1]$ or $[0,1]\times[-1,1]$. The next cut will be the line $b = 0$ and thus the resulting domains will be a square again. We perform the same operations at each iteration. When the search domain is smaller than the size of the critical square, we know that the set of solutions is empty. Using one iteration, we divide by two the side length of our search domain. Thus, in at most $log_2(2/\Gamma)$ iterations the program terminates.

4.2 One-Dimensional Binary Optimization

We know that h_S is a convex function. So, when we reduce its domain to a one-dimensional space $x_1 = e$, the resulting function h_S^e is always convex. Thus, we can apply usual optimization methods from one-dimensional convex optimization. We choose binary optimization, one of the simplest and most practical methods. For example, suppose that our two-dimensional search domain is cut by a segment $[(e,f),(e,g)]$. We choose the middle point: (e,m) with $m = \frac{f+g}{2}$ and compute a two-dimensional subgradient of h_S in $O(n)$ time. By projecting this subgradient into the one-dimensional domain, we obtain a subgradient for h_S^e at the point $\frac{f+g}{2}$ (see Fig. 5). So, we can determine which interval between $[(e,f),(e,m)]$ and $[(e,m),(e,g)]$ will be kept for the next iteration. As we work on a uniform grid, the optimization stops when the interval is bounded by two adjacent points of the grid. Thus, in $\log_2(|f-g|/\Gamma)$ iterations, we solve this subproblem.

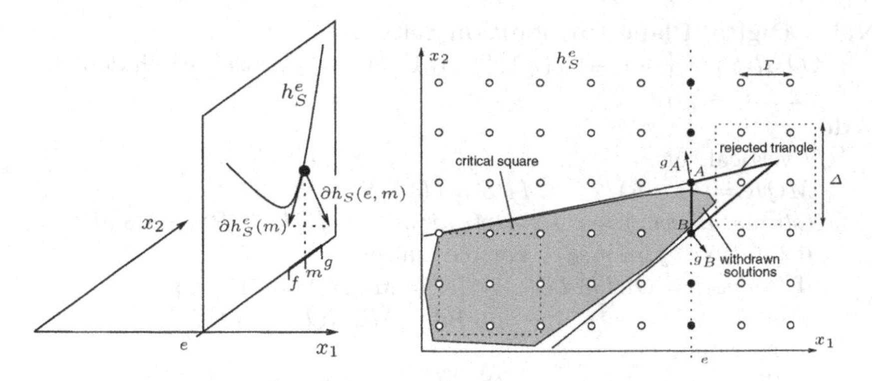

Fig. 5. Subgradient of $h_S^e : \partial h_S^e$ **Fig. 6.** Generalization of the Oracle

4.3 Subgradient and the Oracle Function

We want to determine on which side of the cut, the space of solutions lies. Megiddo Oracle computes the steepest descent from the minimum reached by h_S^e. This direction indicates on which side of the cut some solutions can lie. Nevertheless, as we work on a uniform grid, we can not determine the minimum of h_S^e. After the one-dimensional optimization step, we only know the two minimum values among the samples of the grid. Suppose that we cut by a vertical segment and that these two values are associated to two points A and B such that A lies over B (see Fig. 6). Let g_A and g_b denote the two subgradients of h_S computed at these points. When the angle (g_A, g_B) is equal to π the two subgradients are parallel and the search domain is restricted to a strip of vertical thickness equal to Γ. Thus, it can not contain a critical square and the problem has no solution. In the case where $g_A \wedge g_B < 0$, we keep the left side. However, even if the values of h_S^e at the points A and B are greater than 1, there may exist smaller values and, thus, valid solutions in the interval $]A, B[$. Selecting one of the two sides, we may withdraw some solutions that lie on the cut and on the rejected side. Nevertheless, if they exist, they lie in a triangle not large enough to contain a square of side length Δ. Thus, only the solutions lying in a critical square are of interest. Others can be withdrawn without perturbing the final result of the algorithm.

4.4 Program and Complexity

The code for the main function, **Naive Digital Plane Recognition()**, and for the one-dimensional support function, **1D-Optimization()**, used to solve the one-dimensional optimization problem follows. Each comparison processes all of the different cases.

The search domain is initialized to a square of side length 2. The program stops when the side length of the square currently being processed is less than Γ. At the ith iteration, the side length is equal to $2/2^{i-1}$. Let n_i denote the number of loops performed by the main function. Thus, n_i is equal to $2 - log_2\Gamma$.

```
Naive Digital Plane Recognition(Set,n,D)
  {P,Q,R,S} = {(-1,-1),(-1,1),(1,1),(1,-1)} // the search domain
  δ = 2   Γ = 1/(64D^3)
  do
  { // vertical cut
    MQR = (Q + R)/2      MPS = (P + S)/2
    (status, answer) = 1D-V-Optimization(MQR, MPS, Set, n, Γ)
    if (status = finished)  return  answer
    if (anwser = OnTheLeft)   {R = MQR; S = MPS;}
    else                      {Q = MQR; P = MPS;}
    // horizontal cut
    MPQ = (P + Q)/2      MRS = (R + S)/2
    (status, answer) = 1D-H-Optimization(MPQ, MRS, Set, n, Γ)
    if (status = finished)  return  answer
    if (anwser = Above)   {Q = MPQ; R = MRS;}
    else                  {P = MPQ; S = MPS;}
    δ = δ/2   //length of the search domain at the next iteration
  } while δ ≥ Γ
  return NotAPlane

1D-V-Optimization (A, B, Set, n, Γ)
  (value_A, g_A) = Evaluation(A, Set, n)
  if (value_A < 1) return (finished, A)
  (value_B, g_B) = Evaluation(B, Set, n)
  if (value_B < 1) return (finished, B)
  do
  {
    M = (A+B)/2
    (value_M, g_M) = Evaluation(M, Set, n)
    if (value_M < 1) return (finished, M)
    if (g_M.Y = 0)
      when (g_M.X < 0) return (continue, OnTheRight)
           (g_M.X > 0) return (continue, OnTheLeft)
           (g_M.X = 0) return (finished, NotAPlane)
    else
      if     (g_M.Y < 0)   (B, g_B) ← (M, g_M)
      else                 (A, g_A) ← (M, g_M)
  } while ( A.Y - B.Y ≥ Γ )
  when (g_A ∧ g_B < 0) return (continue, OnTheLeft)
       (g_A ∧ g_B > 0) return (continue, OnTheRight)
       (g_A ∧ g_B = 0) return (finished, NotAPlane)
```

Each iteration processes two instances of the one-dimensional subproblem. We can show that this subfunction performs $3 - log_2\Gamma - i$ evaluations of h_S during the vertical cut and $2 - log_2\Gamma - i$ evaluations during the horizontal cut of the ith iteration. Thus, the overall number of evaluations is equal to:

$$\sum_{i=1}^{n_i}(5-2\log_2 \Gamma - 2i) = (2-\log_2 \Gamma)^2 = (8+3\log_2 D)^2 \sim 9\log_2^2 D \quad (5)$$

4.5 Calculation

The main idea is based on the following observation. Assume that we have just completed iteration i where we processed the two normal vectors $N_1(a,b,1)$ and $N_2(a,b,1)$. In the successive iteration, $i+1$, of our program, we need to process the vector defined by $N=(N_1+N_2)/2$. Instead of computing all scalar products of N with all of the points, we can build on the calculations of the previous step. Let P denote a point and P_N, P_{N_1} and P_{N_2} denote the scalar products between this point and the different normal vectors. Then we have $2P_N = P_{N_1} + P_{N_2}$. This allows us to write a complete version of our program without using a single multiplication. This property may be interesting in some models of computation.

5 Improving Performance

For $D = 512$ and $\Delta = \frac{1}{64D^3}$, the algorithm presented in section 4, processes about one thousand evaluations of h_S. One of the fastest known algorithm in practice [10] performs about ten traversals of the point set. The method presented in section 4.4 was a concise presentation of our main algorithm. In this section, we explore some modifications that will improve its performance in practice.

5.1 Initial Step

Starting with a search space equal to $[-1,1]\times[-1,1]$ is often awkward. If we assume that the set of points is contained in a bounding box of size D, then there exists some points that lie on the borders. Suppose we determine two points $P_1(-D,y,z_1)$ and $P_2(D,y,z_2)$. They provide one constraint on the space of solutions. As they belong to a naive digital plane of normal $N(\alpha,\beta,1)$, we have: $\gamma' \leq N(-D,y_1,z_1) < \gamma'+1$ and $\gamma' \leq N(D,y_2,z_2) < \gamma'+1$. From this, it follows: $|N\cdot(2D,0,z_2-z_1)| < 1$. Thus, we can conclude:

$$\frac{z_1-z_2}{2D} - \frac{1}{2D} < \alpha < \frac{z_1-z_2}{2D} + \frac{1}{2D}$$

So, the side length of the search domain reduces by a factor $\frac{1}{D}$. This stage only requires a simple traversal of the points in order to find the extreme coordinates. The number of evaluations required for h_S reduces to $4\log_2^2 D$.

5.2 Subgradient Extension

We previously demonstrated that the knowledge of the gradient allows us to reject one part of the search domain relative to the point P. Nevertheless, we were not using all of the available information. In fact, when we know the value of $h_S(P)$, either it is less than one and the problem is finished or it reaches a value greater than one. Recall that $\partial h_S(P)$ denotes the subgradient of h_S at the point P. From (1), we keep the points P' in the search domain that verify:

$$(P' - P) \cdot \partial h_S(P) < 0 \tag{6}$$

Any point P' is associated with a value of h_S less than $h_S(P)$. As we are only interested in the values less than 1, we can reformulate (1) as follows:

$$(P' - P) \cdot \partial h_S(P) < 1 - h_S(P)$$

In the search domain, this shifts the line defined by 6 towards the solution space. The new delineations are closer to it (see Fig. 7).

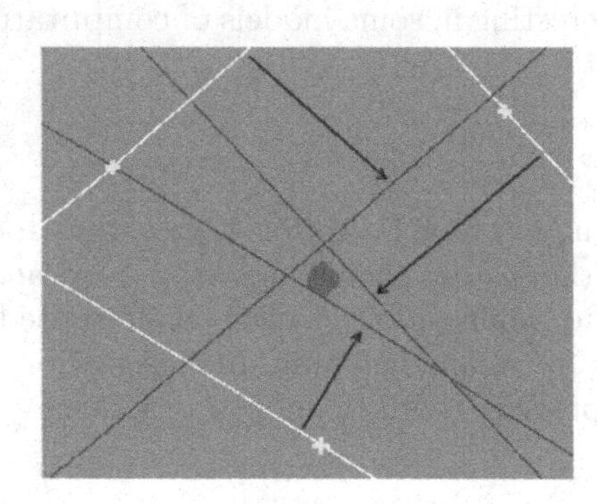

Fig. 7. Enhanced subgradient

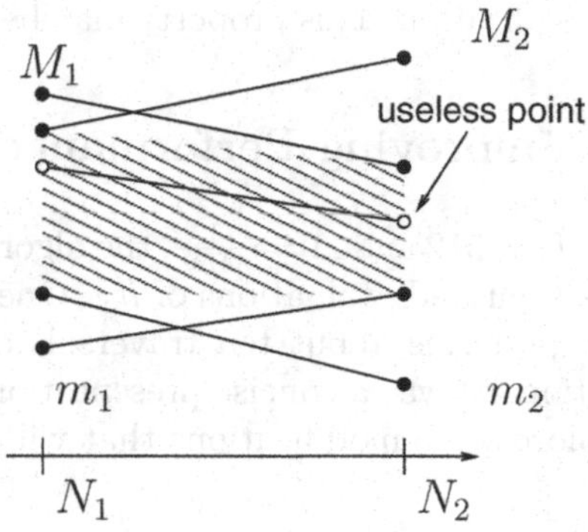

Fig. 8. The reduction criterion

5.3 History Memory

In lieu of using the information from the last subgradient computation and throwing it away after its first use, we prefer to retain the results of all of the previous subgradient computations in memory. Each time a subgradient is computed, we store the result in a list. When we want to test a point of the search domain, we do not immediately call an evaluation of h_S. Instead, we ask a more specific function if the tested point is compatible with all the results from the previous tests. This query is easy to solve because we only have to check the location of a planar point relative to a set of lines. When a previous subgradient g allows for the determination that the value of h_S is greater than 1, we do not need to perform the evaluation. The returned subgradient is equal to g. Suppose we create this function named **IsCompatible**. It could easily manage this processing and the conditions originating from the initial step. Moreover, the quantity of information we store is insignificant relative the number of voxels we process. Thus, this slight improvement will greatly increase the program efficiency in practice.

5.4 Enhancing Γ

In the previous section, we used an overestimated value for Γ in order to simplify our presentation. Nevertheless, we are not always forced to choose this extremely small value of $\frac{1}{64D^3}$. For example, we know that if there exists a valid normal vector (a,b,c) then the size Δ of the critical square is equal to $\frac{1}{4Dc}$, see (4). When we cut the search domain by a vertical line $x = \frac{r}{s}$, the one-dimensional problem would reach its exact minimum at a vector $N(\frac{r}{s}, \frac{u}{v}, 1)$. We determine the two points p_i and p_j that define the two supporting parallel planes. As the point N is a local minimum, there must exist a third point p_k that supports the two planes. Suppose that p_k is lying on the same plane as p_i. Then N is equal to $(p_i - p_k) \wedge (s, 0, r)$ and we have $||N||_\infty \leq s \cdot 2D$. So, when we perform a cut at the position $\frac{r}{s}$, it is sufficient to choose Γ equal to:

$$\Gamma_{\frac{r}{s}} = \frac{1}{16Ds}$$

In the worst case, where $s = 4D^2$, we find the previous definition of Γ. This improvement reduces the number of iterations. We must only modify the expression of Γ in the program, which is a simple operation.

5.5 Reduction

Consider a set of normal vectors that are linked to a segment in the search domain. Let N_1 and N_2 denote its two endpoints. When we determine $h_S(N_1)$, we obtain two points of S: P_1 and M_1 that define this value: $h_S(N_1) = N_1 \cdot (P_1 - M_1)$. Symmetrically, we use the same notation for N_2. Studying the other possible vectors lying on the segment, we notice that some points of S are useless (see Fig. 8). In fact when a point F of S satisfies this condition:

$$N_1 m_2 \leq N_1 F \leq N_1 M_2 \text{ and } N_2 m_1 \leq N_2 F \leq N_2 M_1$$

It can be rejected by the following precessing because it can not define the value of h_S. This condition generalizes to our search domain defined by four vertices. Although this criterion for reduction is not optimal, it is very simple to estimate, so we prefer to use it over other options. This approach allows for the suppression of a fraction of the input points and for a reduction in the cost of the subsequent evaluation of h_S. However, it must be used carefully because we do not know the ratio of the points we reject.

6 Conclusion

We describe a different approach for the naive digital plane recognition problem. We show how to transform the recognition process into a two-dimensional convex optimization program. The convex function we use corresponds to the minimal distance between two parallel planes that enclose the points and whose slopes relative to the axis are the parameters of the function. We show how to evaluate this function and how to obtain one of its subgradients in linear time. As the search domain is planar, we apply simple and well known geometric techniques in order to reduce its size until we find a solution. We exhibit a stopping criterion that only depends on the size D of a bounding box that encloses all the points. This version of our algorithm achieves an $O(n \log^2 n)$ time complexity in the worst case. We present different modifications used to speed up the optimization stage. Development of this enhanced method is in progress. The more difficult part is to use the same integer size (like 32 bits integer for example) for this algorithm and the other methods in order to equitably compare them. Moreover, it remains to estimate the yield of the reduction criterion in order to properly use it. This algorithm has been designed to be efficient in practice and especially when the set of points is dense in the bounding box.

The author thanks the reviewers for their helpful comments.

References

1. Brimkov, V., Coeurjolly, D., Klette, R.: Digital Planarity - A Review. Discrete Applied Mathematics, accepted for publication (2006)
2. Buzer., L.: A linear incremental algorithm for naive and standard digital lines and planes recognition. Graphical Models, **65**(1-3) (2003) 61-76
3. Megiddo, N.: Linear programming in linear time when the dimension is fixed. J. ACM, **31** (1984) 114-127
4. Stojmenovic, I., Tosic., R.: Digitization schemes and the recognition of digital straight lines, hyperplanes and flats in arbitrary dimensions. Vision Geometry, Contemporary Mathematics Series **119** (1991) 197-212
5. Debled-Rennesson, I.: Etude et reconnaissance des droites et plans discrets. PhD Thesis, Université Louis Pasteur, Strasbourg (1995)
6. Debled-Rennesson, I., Reveillès, J.-P.: A new approach to digital planes. Proc. Vision Geometry III, SPIE 2356 (1994) 12-21
7. Preparata, F. P., Shamos., M. I.: Computational Geometry: An Introduction. Springer, New York (1985)
8. Vittone, J., Chassery, J.-M.: Recognition of digital naive planes and polyhedrization. In Proc. Discrete Geometry for Computer Imagery, Springer, Berlin, LNCS 1953 (2000) 296-307
9. Francon, J., Schramm, J. M., Tajine, M.: Recognizing arithmetic straight lines and planes. In Proc. Discrete Geometry for Computer Imagery, LNCS 1176, Springer, Berlin (1996) 141-150
10. Gerard, Y., Debled-Rennesson, I., Zimmermann, P.: An elementary digital plane recognition algorithm. **151**(1-3) (2005) 169-183
11. Reveillès, J.-P.: Combinatorial pieces in digital lines and planes. In Proc. Vision Geometry IV, SPIE 2573 (1995) 23-34
12. Veelaert., P.: Digital planarity of rectangular surface segments. IEEE Trans. Pattern Analysis Machine Intelligence, **16** (1994) 647-652

Fusion Graphs, Region Merging and Watersheds

Jean Cousty, Gilles Bertrand, Michel Couprie, and Laurent Najman

Institut Gaspard-Monge
Laboratoire A2SI, Groupe ESIEE
Cité Descartes, BP99 93162 Noisy-le-Grand Cedex France
{j.cousty, g.bertrand, m.couprie, l.najman}@esiee.fr

Abstract. Region merging methods consist of improving an initial segmentation by merging some pairs of neighboring regions. We consider a segmentation as a set of connected regions, separated by a frontier. If the frontier set cannot be reduced without merging some regions then we call it a watershed. In a general graph framework, merging two regions is not straightforward. We define four classes of graphs for which we prove that some of the difficulties for defining merging procedures are avoided. Our main result is that one of these classes is the class of graphs in which any watershed is thin. None of the usual adjacency relations on $\mathbb{Z}^2$ and $\mathbb{Z}^3$ allows a satisfying definition of merging. We introduce the perfect fusion grid on $\mathbb{Z}^n$, a regular graph in which merging two neighboring regions can always be performed by removing from the frontier set all the points adjacent to both regions.

Introduction

Region merging methods [1, 2] consist of improving an initial segmentation by progressively merging pairs of neighboring regions until a certain criterion is satisfied. From a grayscale image, the watershed transform [3,4,5,6,7,8,9] produces a set of connected regions separated by a divide. Therefore it has long been used as an entry point for region merging methods [10]. In a general graph framework, a watershed may be thought of as a "separating set" of vertices which cannot be reduced without merging some connected components of its complementary set.

A first question arises when dealing with watersheds on a graph. Given a subset of $\mathbb{Z}^2$ equipped with the 4-adjacency relation, we observe that a watershed may contain some "interior points", *i.e.*, points which are not adjacent to any point outside the watershed (see for example the points w and z on Fig. 1c). On the other hand, such interior points do not seem to appear in any watershed on 8-connected graphs. Are the watersheds on these graphs always thin? We will prove that it is indeed true. More interestingly, we provide in this paper a framework to study the property of thinness of watersheds in any kind of graph, and we identify the class of graphs in which any watershed is necessarily thin.

A. Kuba, L.G. Nyúl, and K. Palágyi (Eds.): DGCI 2006, LNCS 4245, pp. 343–354, 2006.

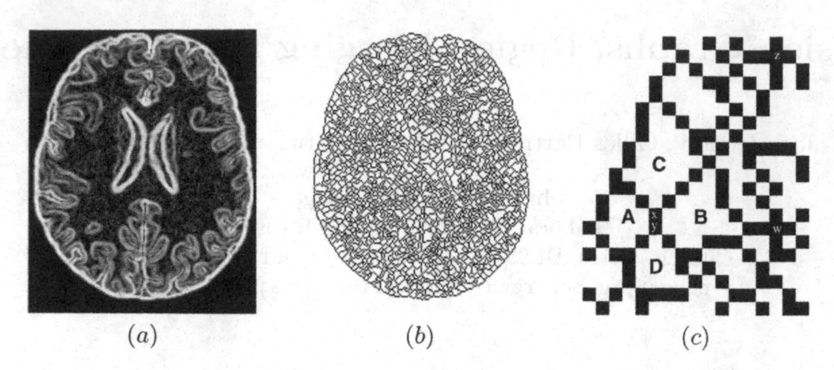

(*a*) (*b*) (*c*)

Fig. 1. (*a*): Original image (cross-section of a brain, after applying a gradient operator). (*b*): Watershed of (*a*) with the 4-adjacency (in black). (*c*): A zoom on a part of (*b*). The points z and w are interior points.

Let us now turn back to the region merging problem. What happens if we want to merge a couple of neighboring regions A and B, and if each pixel adjacent to these two regions is also adjacent to a third one, which is not wanted in the merging? Fig. 1c illustrates such a situation, where x is adjacent to regions A, B, C and y to A, B, D. This problem has been identified in particular by T. Pavlidis (see [2], section 5.6: "When three regions meet"), and has been dealt with in some practical ways, but until now a systematic study of properties related to merging in graphs has not been done. A major contribution of this article is the definition and the study of four classes of graphs, with respect to the possibility of "getting stuck" in a merging process (Sec. 2, Sec. 3). In particular, we say that a graph is a *fusion graph* if any region A in this graph can always be merged with another region B, without problems with other regions. The most striking outcome of this study is that the class of fusion graphs is precisely the class of graphs in which any watershed is thin (Th. 3). We also provide some local characterizations for two of these four classes of graphs, and prove that the two other ones cannot be locally characterized (Sec. 4).

In one of the classes of graphs introduced in Sec. 3, that we call the class of *perfect fusion graphs*, any pair of neighboring regions can always be merged, without problems with other regions, by removing all pixels adjacent to both regions. Using our framework, we analyze the status of the graphs which are the most widely used for image analysis, namely the graphs corresponding to the 4- and the 8-adjacency in $\mathbb{Z}^2$ and to the 6- and the 26-adjacency in $\mathbb{Z}^3$ (Sec. 5). We show that none of these classical graphs is a perfect fusion graph. Last, but not least, in Sec. 6 we introduce a graph on $\mathbb{Z}^n$ (for any n) that we call the perfect fusion grid, which is indeed a perfect fusion graph, and which is "between" the direct adjacency graph (which generalizes the 4-adjacency to $\mathbb{Z}^n$) and the indirect adjacency graph (which generalizes the 8-adjacency). Furthermore, in a forthcoming paper, we intend to prove that this n-dimensional grid is the unique grid (up to a translation) that possesses those two properties.

The proofs of the properties presented in this paper are given in an extended version [11].

1 Graphs and Watershed

Let E be a set, we write $X \subseteq E$ if X is a subset of E, we write $X \subset E$ if X is a proper subset of E, *i.e.*, if X is a subset of E and $X \neq E$. We denote by $\overline{X}$ the complementary set of X in E, *i.e.*, $\overline{X} = E \setminus X$. If E is a finite set, we denote by $|E|$ the number of elements of E and by 2^E the set composed of all the subsets of E.

We define a graph as a pair (E, Γ) where E is a finite set and Γ is a binary relation on E (*i.e.*, $\Gamma \subseteq E \times E$), which is reflexive (for all x in E, $(x, x) \in \Gamma$) and symmetric (for all x, y in E, $(y, x) \in \Gamma$ whenever $(x, y) \in \Gamma$). Each element of E is called a *vertex* or a *point*. We will also denote by Γ the map from E to 2^E such that, for all $x \in E$, $\Gamma(x) = \{y \in E \mid (x, y) \in \Gamma\}$. If $y \in \Gamma(x)$, we say that y is *adjacent to* x. We define also the map Γ^* such that for all $x \in E$, $\Gamma^*(x) = \Gamma(x) \setminus \{x\}$. Let $X \subseteq E$, we define $\Gamma(X) = \cup_{x \in X} \Gamma(x)$, and $\Gamma^*(X) = \Gamma(X) \setminus X$. If $y \in \Gamma(X)$, we say that y *is adjacent to* X. If $X, Y \subseteq E$ and $\Gamma(X) \cap Y \neq \emptyset$, we say that Y *is adjacent to* X.

Let $G = (E, \Gamma)$ be a graph and let $X \subseteq E$, we define the *subgraph of* G *induced by* X as the graph $G_X = (X, \Gamma \cap [X \times X])$. In this case, we also say that G_X *is a subgraph of* G. Let $G' = (E', \Gamma')$ be a graph, we say that G *and* G' *are isomorphic* if there exists a bijection f from E to E' such that, for all $x, y \in E$, y belongs to $\Gamma(x)$ if and only if $f(y)$ belongs to $\Gamma'(f(x))$.

Let (E, Γ) be a graph, let $X \subseteq E$, a *path in* X is a sequence $\pi = \langle x_0, ..., x_l \rangle$ such that $x_i \in X$, $i \in [0, l]$, and $x_i \in \Gamma(x_{i-1})$, $i \in [1, \ldots, l]$. We also say that π is a *path from* x_0 *to* x_l *in* X and that x_0 and y_l are *linked for* X. We say that X *is connected* if any x and y in X are linked for X.

Let $Y \subseteq X$. We say that Y *is a connected component of* X, or simply *a component of* X, if Y is connected and if Y is maximal for this property, *i.e.*, if $Z = Y$ whenever $Y \subseteq Z \subseteq X$ and Z connected.

We denote by $\mathcal{C}(X)$ the set of all the connected components of X.

In this paper, we study some thinness properties of watersheds in graphs.

Definition 1. *Let* (E, Γ) *be a graph. Let* $X \subseteq E$, *the* interior of X *is the set* $int(X) = \{x \in X \mid \Gamma(x) \subseteq X\}$. *We say that the set* X is thin *if* $int(X) = \emptyset$.

Let us recall the definition of line graphs ([12]). This class of graphs allows to link the framework developed in this paper and the approaches of watershed and region merging based on edges rather than vertices, *i.e.* when regions are separated by a set of edges.

Definition 2. *Let* (E, Γ) *be a graph. The* line graph of (E, Γ) *is the graph* (E', Γ') *such that* $E' = \Gamma^\star$ *and* (u, v) *belongs to* Γ' *whenever* $u \in \Gamma^\star$, $v \in \Gamma^\star$, *and* u, v *share a vertex of* E.

We say that a graph (E', Γ') is a line graph *if there exists a graph* (E, Γ) *such that* (E', Γ') *is isomorphic to the line graph of* (E, Γ).

In Fig. 2, we show a graph and its line graph. All graphs are not line graphs, in other words, there exist some graphs which are not the line graphs of any graph. The following theorem allows to characterize line graphs.

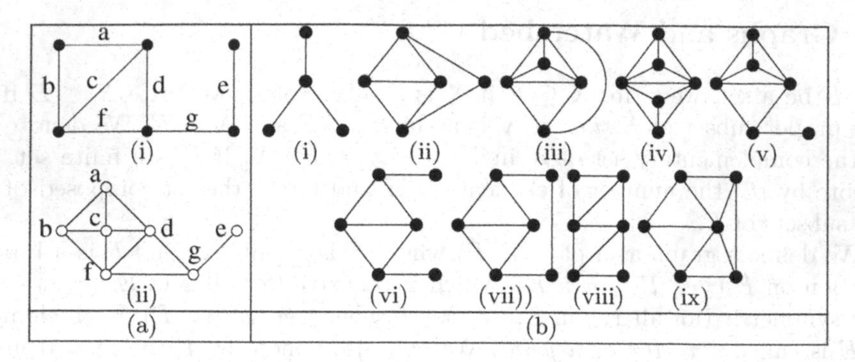

Fig. 2. (a): A graph (i) and its line graph (ii); (b): graphs for characterization of line graphs

Theorem 1 ([12]). *A graph G is a line graph if and only if none of the graphs of Fig. 2b is a subgraph of G.*

Important Remark. *From now, when speaking about a graph (E, Γ), we will assume for simplicity that E is non-empty and connected.*

Notice that, nevertheless, the subsequent definitions and properties may be easily extended to non-connected graphs.

We now introduce notions for watersheds in graphs.

Definition 3. *Let (E, Γ) be a graph. Let $X \subseteq E$, and let $p \in X$. We say that:*

- p is an inner point (for X) *if $p \in int(X)$.*
- p is W-simple (for X) *if p is adjacent to exactly one component of $\overline{X}$.*

In this definition and the following ones, the prefix "W-" stands for watershed. In Fig. 3a, x is a W-simple point for the set X constituted by the black vertices, and y is an inner point.

Definition 4. *Let $G = (E, \Gamma)$ be a graph. A set $X \subseteq E$ is a* watershed (in G) *if there is no point W-simple for X.*

A watershed X is non-trivial *if $X \neq \emptyset$ and $X \neq E$.*

Figs. 3b,c,d,e illustrate the notion of watershed. It can be seen that a watershed X is non-trivial if and only if $|\mathcal{C}(\overline{X})| \geq 2$. A watershed is a set which contains no W-simple point, but some of the examples given below show that such a set is not always thin (in the sense of Def. 1). Fig. 3b is an example of watersheds which is thin: the set of black points has no W-simple point and no inner point. The sets of black and gray points in Figs. 3c,d,e are three examples of non-thin watersheds.

2 Merging

Consider the graph (E, Γ) depicted in Fig. 4a, where a subset X of E (black vertices) separates its complementary set $\overline{X}$ into four connected components. If

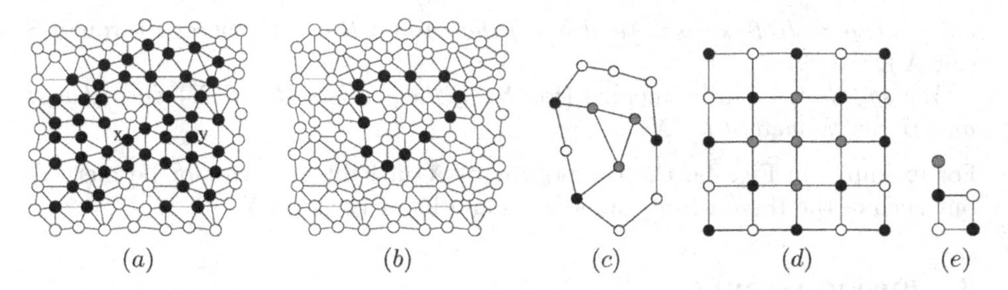

Fig. 3. Illustration of watershed. (a) A graph (E, Γ) and a subset X of E (in black). (b) A thin watershed (black points). $(c-e)$: The subset X represented by black and gray points is a watershed which is not thin: $\text{int}(X)$ is depicted by the gray points.

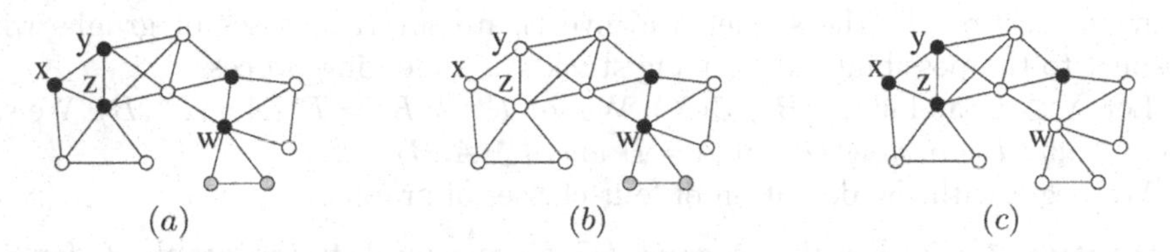

Fig. 4. Illustration of merging. (a): A graph (E, Γ) and a subset X of E (black points). (b): The black points represent $X \setminus S$ with $S = \{x, y, z\}$. (c): The black points represent $X \setminus S'$ with $S' = \{w\}$.

we replace the set X by, for instance, the set $X \setminus S$ where $S = \{x, y, z\}$, we obtain a set which separates its complementary set into three components, see Fig. 4b: we can also say that we "merged two components of $\overline{X}$ through S". This operation may be seen as an "elementary merging" in the sense that only two components of $\overline{X}$ were merged. On the opposite, replacing the set X by the set $X \setminus S'$ where $S' = \{w\}$, see Fig. 4c, would merge three components of $\overline{X}$. We also see that the component of $\overline{X}$ which is below w (in light gray) cannot be merged by an "elementary merging" since any attempt to merge it must involve the point w, and thus also the three components of $\overline{X}$ adjacent to this point. In this section, we introduce definitions and basic properties related to such merging operations in graphs.

Definition 5. *Let (E, Γ) be a graph , $X \subset E$ and $S \subseteq X$. We say that S* is F-simple (for X) *if S is adjacent to exactly two components $A, B \in \mathcal{C}(\overline{X})$ such that $A \cup B \cup S$ is connected.*

Let $p \in X$. We say that p is F-simple (for X) *if $\{p\}$ is F-simple for X.*

In this definition, the prefix "F-" stands for fusion. For example, in Fig. 4a, the point z is F-simple while x, y, w are not. Also, the sets $\{z\}$, $\{x, y\}$, $\{x, z\}$, $\{y, z\}$, $\{x, y, z\}$ are F-simple, but the sets $\{x\}$, $\{y\}$ and $\{w\}$ are not.

Definition 6. *Let (E, Γ) be a graph and $X \subset E$. Let A and $B \in \mathcal{C}(\overline{X})$, with $A \neq B$. We say that A* and B can be merged (for X) *if there exists $S \subseteq X$ such that S is F-simple for X, and A and B are precisely the two components of $\overline{X}$*

adjacent to S. In this case, we also say that A and *B* can be merged through *S* (for *X*).

We say that A can be merged (for *X*) *if there exists $B \in \mathcal{C}(\overline{X})$ such that A and B can be merged for X.*

For example, in Fig. 4a, the component of $\overline{X}$ in light gray cannot be merged, but each of the three white components can be merged for X.

3 Fusion Graphs

The preceding section and the present one constitute a theoretical basis for the study of region merging methods. The problems pointed out in the introduction can be avoided by using exclusively the notion of merging introduced in the previous section. In the sequel, we investigate several classes of graphs with respect to the possibility of "getting stuck" in a merging process.

Let $X \subset E$, and let $A, B \in \mathcal{C}(\overline{X})$. We set $\Gamma^*(A,B) = \Gamma^*(A) \cap \Gamma^*(B)$. We say that *$A$ and B are neighbors* if $A \neq B$ and $\Gamma^*(A,B) \neq \emptyset$.

We begin with the definition of four classes of graphs.

Definition 7. *We say that a graph (E,Γ) is a* weak fusion graph *if, for any $X \subset E$ such that $|\mathcal{C}(\overline{X})| \geq 2$, there exist $A, B \in \mathcal{C}(\overline{X})$ which can be merged.*

We say that a graph (E,Γ) is a fusion graph *if, for any $X \subset E$ such that $|\mathcal{C}(\overline{X})| \geq 2$, each $A \in \mathcal{C}(\overline{X})$ can be merged for X.*

We say that the graph (E,Γ) is a strong fusion graph *if, for any $X \subset E$, any A and $B \in \mathcal{C}(\overline{X})$ which are neighbors can be merged.*

We say that the graph (E,Γ) is a perfect fusion graph *if, for any $X \subset E$, any A and $B \in \mathcal{C}(\overline{X})$ which are neighbors can be merged through $\Gamma^*(A,B)$.*

Basic examples and counter-examples of weak fusion, fusion, strong fusion, perfect fusion graphs and line graphs are given in Fig. 5.

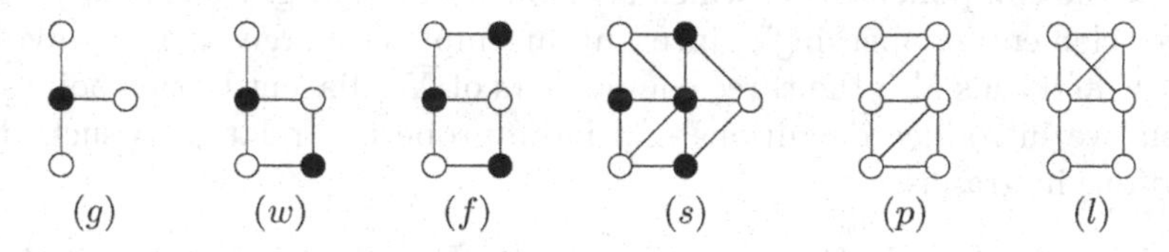

Fig. 5. Examples and counter-examples for different classes of graphs. (g): A graph which is not a weak fusion graph, (w): a weak fusion graph which is not a fusion graph, (f): a fusion graph which is not a strong fusion graph, (s): a strong fusion graph which is not a perfect fusion graph, (p): a perfect fusion graph which is not a line graph, and (l): a line graph. In the graphs (g, w, f, s), the black vertices constitute a set X which serves to prove that the graph does not belong to the pre-cited class.

We denote by $\mathcal{G}$ (resp. $\mathcal{G}_L$, $\mathcal{G}_P$, $\mathcal{G}_S$, $\mathcal{G}_F$, and $\mathcal{G}_W$) the set of all graphs (resp. line graphs, perfect fusion graphs, strong fusion graphs, fusion graphs, and weak fusion graphs).

Property 2. *We have the following strict inclusion relations:*
$\mathcal{G} \supset \mathcal{G}_W \supset \mathcal{G}_F \supset \mathcal{G}_S \supset \mathcal{G}_P \supset \mathcal{G}_L$.

Now, we present the main theorem of this section, which establishes that the class of graphs for which any watershed is thin is precisely the class of fusion graphs. As an immediate consequence of this theorem and Prop. 2, we see that all watersheds in strong fusion graphs, perfect fusion graphs and line graphs are also thin.

Theorem 3. *A graph G is a fusion graph if and only if any non-trivial watershed in G is thin.*

Observe that the graphs of Figs. 3c,d,e are not fusion graphs; we see that they may indeed contain a non-thin watershed.

We conclude this section with a nice property of perfect fusion graphs, which can be useful to design hierarchical segmentation methods based on regions splitting. Consider the example of Fig. 6a, where a watershed X (black points) in the graph G separates $\overline{X}$ into two components. Consider now the set Y (gray points) which is a watershed in the subgraph of G induced by one of these components. We can see that the union of the watersheds, $X \cup Y$, is not a watershed, since the point x is W-simple for $X \cup Y$. Prop. 4 shows that this problem cannot occur in any perfect fusion graph.

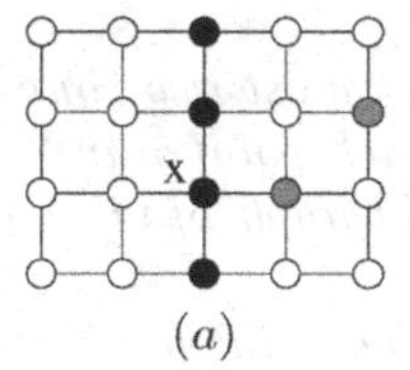

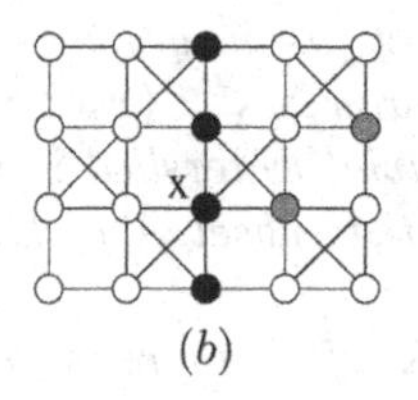

Fig. 6. Illustrations for Prop. 4. (a): The graph is not a perfect fusion graph (see Sec. 5, Prop. 8), and the union of the watersheds is not a watershed. (b): The graph is a perfect fusion graph (see Sec. 6, Prop. 10), the property holds.

Property 4. *Let $G = (E, \Gamma)$ be a graph. If G is a perfect fusion graph, then for any watershed $X \subset E$ in G and for any watershed $Y \subset A$ in G_A, where $A \in \mathcal{C}(\overline{X})$ and G_A is the subgraph of G induced by A, the set $X \cup Y$ is a watershed in G.*

4 Local Characterizations

Weak fusion, fusion, strong fusion and perfect fusion graphs are defined by conditions that must be verified for all the subsets of the vertex set. Thus, using the straightforward method based on the definition to check whether a graph belongs to one of these classes costs an exponential time with respect to the number of vertices. On the other hand, line graphs may be recognized thanks to a condition which can be checked independently in a limited neighborhood (informally speaking) of each vertex. Do such characterizations exist for the four classes of fusion graphs?

Let (E, Γ) be a graph, let $x \in E$ and $k \in \mathbb{N}$, we denote by $\Gamma^k(x)$ the kth *order neighborhood of* x, that is, $\Gamma^k(x) = \Gamma(\Gamma^{k-1}(x))$, with $\Gamma^0(x) = \{x\}$. We say that there is a local characterization of a class of graphs if there exists k an arbitrary positive integer and $\mathcal{P}$ a property on graphs such that a graph $G = (E, \Gamma)$ is in this class if and only if for all $x \in E$, $\mathcal{P}[G(x,k)]$ is true, $G(x,k)$ being the subgraph of G induced by $\Gamma^k(x)$.

Property 5. *i) There is no local characterization of weak fusion graphs.*
ii) There is no local characterization of fusion graphs.

Let x and y be two points, we say that x *and* y *are 2-adjacent* if $y \notin \Gamma(x)$ and $\Gamma^*(x) \cap \Gamma^*(y) \neq \emptyset$.

Theorem 6. *Let $G = (E, \Gamma)$ be a graph. The graph G is a strong fusion graph if and only if, for any two points $x, y \in E$ which are 2-adjacent, there exists $a \in \Gamma^*(x)$ and $b \in \Gamma^*(y)$ such that $b \in \Gamma(a)$ and $\Gamma(\{a,b\}) \subseteq [\Gamma(x) \cup \Gamma(y)]$.*

Remind that in perfect fusion graphs, any two components A, B of $\mathcal{C}(\overline{X})$ which are neighbors can be merged through $\Gamma^*(A) \cap \Gamma^*(B)$. Thus, perfect fusion graphs constitute an ideal framework for region merging methods. In the sequel, we will use the symbol $G^{\blacktriangle}$ to denote the graph (i) in Fig. 2b.

Theorem 7. *Let (E, Γ) be a graph.*
The eight following statements are equivalent:
i) (E, Γ) is a perfect fusion graph;
ii) for any $x \in E$, any $X \subseteq \Gamma(x)$ contains at most two connected components;
iii) for any non-trivial watershed Y in E, each point x in Y is F-simple;
iv) for any connected subset A of E, the subgraph of (E, Γ) induced by A is a fusion graph;
v) for any subset X of E, there is no multiple point for X;
vi) the graph $G^{\blacktriangle}$ is not a subgraph of G;
vii) any vertices x, y, z which are mutually non-adjacent are such that $\Gamma(x) \cap \Gamma(y) \cap \Gamma(z) = \emptyset$;
viii) for any $x, y \in E$ which are 2-adjacent, for any $z \in \Gamma^*(x) \cap \Gamma^*(y)$, we have $\Gamma(z) \subseteq [\Gamma(x) \cup \Gamma(y)]$.

Notice that statement *viii* bears a resemblance with the local characterization of strong fusion graphs (Th. 6). Remind that any line graph is a perfect fusion graph (Prop. 2). We can see that, thanks to Th. 7 (condition *vi*), perfect fusion graphs can be characterized in a way similar to Th. 1 which characterizes line graphs, but with a much simpler condition. Remark, for example, that all the graphs of Fig. 2b except $G^{\blacktriangle}$ are perfect fusion graphs, since none of these graphs contains $G^{\blacktriangle}$ as a subgraph.

5 Usual Grids

The aim of this section is to classify (with respect to the classes of fusion graphs) the grids which are the most commonly used in image processing. We prove that none of the grids commonly used in 2-dimensional and 3-dimensional image processing is a perfect fusion graph; several are not even fusion graphs. Thus, the most natural merging operation, which consists in merging two regions through their common boundary, is not a safe operation in these grids.

In this section, we will assume that n is a strictly positive integer.

Let E be a set and let E^n be the Cartesian product of n copies of E. An element x of E^n may be seen as a map from $\{1, ..., n\}$ to E, for each $i \in \{1, ..., n\}$, x_i is *the i-th coordinate of x.*

Let $\mathbb{Z}$ be the set of integers. We consider the families of sets H_0^1, H_1^1 such that $H_0^1 = \{\{a\} \mid a \in \mathbb{Z}\}$, $H_1^1 = \{\{a, a+1\} \mid a \in \mathbb{Z}\}$. A subset S of $\mathbb{Z}^n$ which is the Cartesian product of exactly $m \leq n$ elements of H_1^1 and $(n-m)$ elements of H_0^1 is called an *m-cube.*

In order to recover a graph structure for digital images, adjacency relations are defined on $\mathbb{Z}^n$. The following definition allows to retrieve the most frequently used adjacency relations.

Let $m \leq n$, we say that x and y in $\mathbb{Z}^n$ are *m-adjacent* if there exists an m-cube that contains both x and y. We define Γ_m^n as the binary relation on $\mathbb{Z}^n$ such that for any pair x, y in $\mathbb{Z}^n$, $(x, y) \in \Gamma_m^n$ if and only if x and y are m-adjacent.

In order to deal with graphs that can be arbitrarily large we define a *grid* as a pair (E, Γ) where E is an infinite set and Γ is a binary relation on E. Let $X \subseteq E$ we define the restriction of (E, Γ) to X as the pair (X, Γ_X) where $\Gamma_X = \Gamma \cap (X \times X)$. If X is a finite set then (X, Γ_X) is a graph. In the sequel, to simplify the notations, we will write Γ as a shortcut for Γ_X. We first examine the case of 2-dimensional usual grids.

Property 8. *Let w, h be two integers such that $w > 2$ and $h > 2$. Let $E = \{x \in \mathbb{Z}^2 \mid 0 \leq x_1 < w$ and $0 \leq x_2 < h\}$.*
i) If $\{w, h\} \neq \{3, 4\}$, (E, Γ_1^2) is not a weak fusion graph. If $\{w, h\} = \{3, 4\}$ then (E, Γ_1^2) is a weak fusion graph but not a fusion graph.
ii) The graph (E, Γ_2^2) is a fusion graph but is not a strong fusion graph.

Notice that in the literature, the graph (E, Γ_1^2) (resp. (E, Γ_2^2)) corresponds to the 4 (resp. 8)-adjacency.

Now we examine the case of 3-dimensional usual grids.

Property 9. *Let w, h and d be three integers strictly greater than 1. Let $E = \{x \in \mathbb{Z}^3 \mid 0 \leq x_1 < w, 0 \leq x_2 < h$ and $0 \leq x_3 < d\}$.*
i) The graph (E, Γ_1^3) is not a weak fusion graph.
ii) If $w \geq 5$, $h \geq 5$, $d \geq 5$, the graph (E, Γ_3^3) is not a fusion graph.

Notice that in the literature, the graph (E, Γ_1^3) (resp. (E, Γ_3^3)) corresponds to the 6 (resp. 26)-adjacency.

6 Perfect Fusion Grids

We now introduce a grid for structuring n-dimensional digital images and prove that it is a perfect fusion graph, whatever the dimension n.

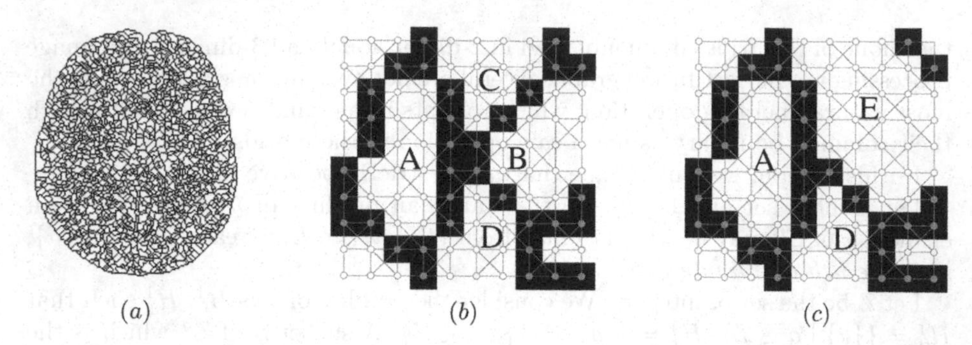

Fig. 7. (*a*): A watershed of Fig. 1a obtained on the perfect fusion grid; (*b*): a crop of (*a*) where the regions A, B, C and D correspond to the regions shown in Fig. 1c; in gray, the corresponding perfect fusion grid is superimposed; (*c*): same as (*b*) after having merged B and C to form a new region, called E.

It does thus constitute a structure on which neighboring regions can always be merged through their common neighborhood without problem with other regions. Fig. 8b gives an intuitive idea of this grid. Fig. 7a shows a watershed of Fig. 1a obtained on this grid. Remark that the problems pointed out in the introduction do not exist in this case. The watershed does not contain any inner point. Any pair of neighboring regions can be merged by simply removing from the watershed the points which are adjacent to both regions (Figs. 7b,c). Furthermore, the resulting set is still a watershed. Observe that this grid is "between" the usual grids. In a forthcoming paper, we intend to prove that this is the unique such graph.

Let C^n be the set of all n-cubes of $\mathbb{Z}^n$, we define the map B from C^n to $\mathbb{Z}^n$, such that for any $c \in C^n$, $B(c)_i = \min\{x_i \mid x \in c\}$, where $B(c)_i$ is the i-th coordinate of $B(c)$. It may be seen that c is equal to the Cartesian product: $\{B(c)_1, B(c)_1 + 1\} \times ... \times \{B(c)_n, B(c)_n + 1\}$. Thus clearly B is a bijection.

We set $\mathbb{B} = \{0, 1\}$. We set $\overline{0} = 1$ and $\overline{1} = 0$. A *binary word of length n* is an element of $\mathbb{B}^n$. If u is in $\mathbb{B}^n$, we define *the complement of* u as the binary word $\overline{u}$ such that for any $i \in \{1, ..., n\}$, $(\overline{u})_i = (\overline{u_i})$.

Definition 8. *Let f be the map from C^n to $\mathbb{B}^n$ such that for any $c \in C^n$, $f(c)_i$ is equal to $B(c)_i$ mod 2, that is the remainder in the integer division of $B(c)_i$ by 2.*

Let u be an element of $\mathbb{B}^n$, we set $C^n_u = \{c \in C^n \mid f(c) = u\}$ and $C^n_{u/\overline{u}} = C^n_u \cup C^n_{\overline{u}}$.

We define the binary relation $\Gamma^n_{u/\overline{u}} \subseteq \mathbb{Z}^n \times \mathbb{Z}^n$ as the set of pairs $(x, y) \in \mathbb{Z}^n \times \mathbb{Z}^n$ such that there exists $c \in C^n_{u/\overline{u}}$ that contains both x and y.

We define $\mathcal{P}^n$, the family of perfect fusion grids over $\mathbb{Z}^n$, *as the set $\mathcal{P}^n = \{(\mathbb{Z}^n, \Gamma^n_{u/\overline{u}}) \mid u \in \mathbb{B}^n\}$.*

Fig. 8 illustrates the above definitions for the two-dimensional case. Fig. 9 shows a watershed on a 3-dimensional perfect fusion grid. To clarify the figure, we use the following convention: any two points belonging to a same cube marked by a gray stripe are adjacent to each other.

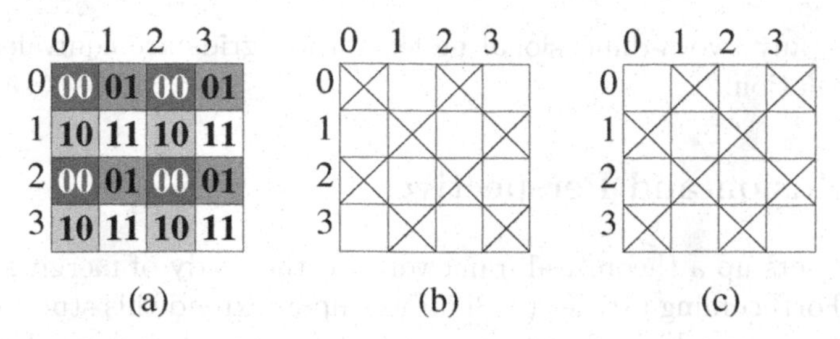

Fig. 8. Illustration of the two perfect fusions grids over $\mathbb{Z}^2$ (restricted to subsets of $\mathbb{Z}^2$). (*a*): The map f; (*b*): $(\mathbb{Z}^2, \Gamma^2_{11/00})$; (*c*): $(\mathbb{Z}^2, \Gamma^2_{10/01})$.

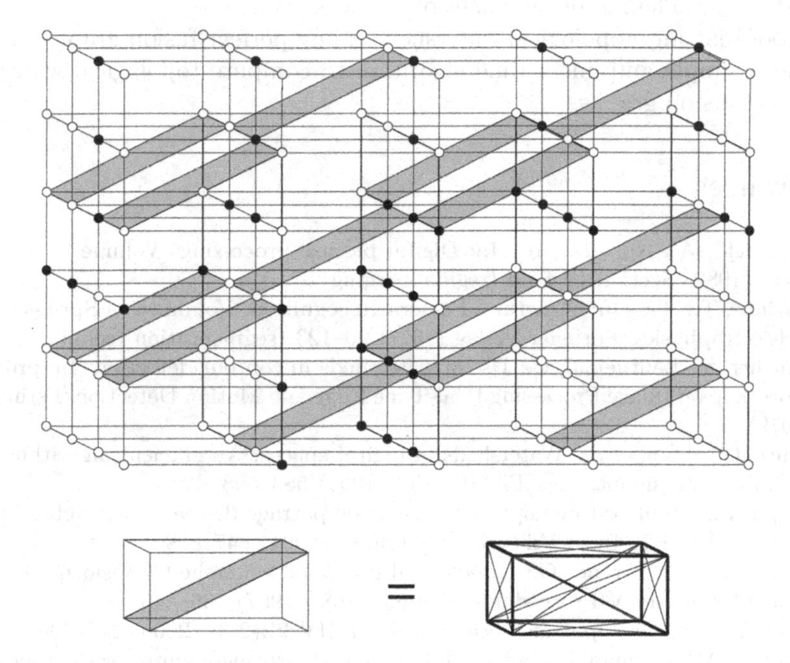

Fig. 9. A 3-dimensional perfect fusion grid. Black points constitute a set which is a watershed.

Theorem 10. *Let $u \in \mathbb{B}^n$ and let X be a finite subset of $\mathbb{Z}^n$ such that $(X, \Gamma^n_{u/\overline{u}})$ is connected. Then $(X, \Gamma^n_{u/\overline{u}})$ is a perfect fusion graph such that $\Gamma^n_1 \subseteq \Gamma^n_{u/\overline{u}} \subseteq \Gamma^n_n$. Furthermore it is a line graph.*

It may be seen that the family $\mathcal{P}^n$ contains 2^{n-1} distinct perfect fusion grids.

Let $X \subseteq \mathbb{Z}^n$ and let $t \in \mathbb{B}^n$. We define $X + t = \{x + t \mid x \in X\}$, we say that $X + t$ is a *binary translation of* X. Let m be a positive integer such that $m \leq n$. Remark that if X is an m-cube then $X + t$ is also an m-cube.

Let u and v in $\mathbb{B}^n$. Let $t \in \mathbb{B}^n$ such that for any $i \in \{1, ..., n\}$, if $u_i = \overline{v_i}$ then $t_i = 1$, otherwise $t_i = 0$. Then for any $(x, y) \in \mathbb{Z}^n \times \mathbb{Z}^n$, $(x, y) \in \Gamma^n_{u/\overline{u}}$ if and only if $(x + t, y + t) \in \Gamma^n_{v/\overline{v}}$.

In words, any two n-dimensional perfect fusion grids are equivalent up to a binary translation.

7 Conclusion and Perspectives

This article sets up a theoretical framework for the study of merging properties in graphs. Forthcoming articles ([13] for example) extend this study to the case of weighted graphs, which constitute a model for grayscale images. The notion of topological watershed [7, 6] extends the notion of watershed to weighted graphs, and possesses interesting properties which are not guaranteed by most watershed algorithms [9]. The major outcomes of [13] are:
i) a proof that any topological watershed on any perfect fusion graph is thin;
ii) a new, simple and linear-time algorithm to compute topological watersheds on perfect fusion graphs.

References

1. Rosenfeld, A., Kak, A.: 10. In: Digital picture processing. Volume 2. Academic Press (1982) Section 10.4.2.d (region merging).
2. Pavlidis, T.: 4–5. In: Structural Pattern Recognition. Volume 1 of Springer Series in Electrophysics. Springer-Verlag (1977) 90–123 (segmentation techniques).
3. Beucher, S., Lantuéjoul, C.: Use of watersheds in contour detection. In: procs. Int Workshop on Image Processing Real-Time Edge and Motion Detection/Estimation. (1979)
4. Vincent, L., Soille, P.: Watersheds in digital spaces: An efficient algorithm based on immersion simulations. PAMI **13**(6) (1991) 583–598
5. Meyer, F.: Un algorithme optimal de ligne de partage des eaux. In: Actes du 8ème Congrès AFCET, Lyon-Villeurbanne, France (1991) 847–859
6. Couprie, M., Bertrand, G.: Topological grayscale watershed transform. In: SPIE Vision Geometry V Proceedings. Volume 3168. (1997) 136–146
7. Bertrand, G.: On topological watersheds. JMIV **22**(2-3) (2005) 217–230
8. Couprie, M., Najman, L., Bertrand, G.: Quasi-linear algorithms for the topological watershed. JMIV **22**(2-3) (2005) 231–249
9. Najman, L., Couprie, M., Bertrand, G.: Watersheds, mosaics and the emergence paradigm. DAM **147**(2-3) (2005) 301–324
10. Jasiobedzki, P., Taylor, C., Brunt, J.: Automated analysis of retinal images. IVC **1**(3) (1993) 139–144
11. Cousty, J., Bertrand, G., Couprie, M., Najman, L.: Fusion graphs: merging properties and watershed. CVIU (2006) submitted. Also in technical report IGM2005-04, http://igm.univ-mlv.fr/LabInfo/rapportsInternes/2005/04.pdf.
12. Beineke, L.: On derived graphs and digraphs. In Sachs, H., Voss, H., Walther, H., eds.: Beiträge zur graphen theorie. Teubner (1968) 17–23
13. Cousty, J., Couprie, M., Najman, L., Bertrand, G.: Grayscale watersheds on perfect fusion graphs. In: procs. IWCIA. (2006)

Revisiting Digital Straight Segment Recognition

François de Vieilleville and Jacques-Olivier Lachaud

LaBRI, Univ. Bordeaux 1
351 cours de la Libération,
33405 Talence Cedex, France
{devieill, lachaud}@labri.fr

Abstract. This paper presents new results about digital straight segments, their recognition and related properties. They come from the study of the arithmetically based recognition algorithm proposed by I. Debled-Rennesson and J.-P. Reveillès in 1995 [1]. We indeed exhibit the relations describing the possible changes in the parameters of the digital straight segment under investigation. This description is achieved by considering new parameters on digital segments: instead of their arithmetic description, we examine the parameters related to their combinatoric description. As a result we have a better understanding of their evolution during recognition and analytical formulas to compute them. We also show how this evolution can be projected onto the Stern-Brocot tree. These new relations have interesting consequences on the geometry of digital curves. We show how they can for instance be used to bound the slope difference between consecutive maximal segments.

1 Introduction

The study of digital straight lines is a fundamental topic in discrete geometry and several approaches have been proposed. Following the taxonomy of [2], we can divide them into three groups. The first one characterizes digital lines through the study of the *pre-image*: this aims at determining in a dual space the possible real lines whose digitization corresponds to a given set of pixels [3]. That kind of approaches has recently been used to define and recognize straight lines drawn on irregular isothetic grids [4].

A second group is related to combinatorics and relies on the link between continued fractions and recursive characterization of digital lines. It can cope with lines with rational or irrational slopes, as their digitization can be seen as a word over a finite alphabet. The tools developed to characterize and study those objects [5, 6] have for instance been used to study the asymptotic behavior of some digital segments over digitizations of $\mathcal{C}^3$ convex curves [7, 8].

The third group gathers arithmetic approaches, which are based on a formulation very similar to the one of real lines (Diophantine inequalities, slope and vertical shift). They have led to simple, incremental and optimal algorithms to recognize digital segments [1, 9]. For this approach the best known recognition algorithm is the above mentioned algorithm of Debled and Reveillès [1], referenced as **DR95** algorithm in the recent book of Klette and Rosenfeld [10]. This

A. Kuba, L.G. Nyúl, and K. Palágyi (Eds.): DGCI 2006, LNCS 4245, pp. 355–366, 2006.

algorithm extracts progressively the most simple digital line parameters of a finite connected sequence of pixels, updating the parameters at each pixel.

In this algorithm, the parameters refer to an arithmetic representation of digital straight lines: slope as a fraction, integer shift to origin, position of some specific limit points (upper and lower leaning points). Their evolution during the progressive steps of the recognition is governed by algorithmic computations: for instance, the new slope is computed from the last added point and some former leaning point. Although sufficient for recognizing digital lines, these parameters lack of descriptive content to fully understand what is digital straightness. For instance, they cannot answer a question like if two straight lines share a common part, how are related their slopes. Along the same lines, although it is known since Debled's thesis [11] that slope evolutions during recognition correspond to displacements in the Stern-Brocot tree, these parameters are nevertheless incomplete to fully describe it.

We propose here to use the combinatoric approach to give better insights about the **DR95** algorithm. A digital line is then characterized by the continued fraction of its slope and the number of patterns it contains. The evolution of these new parameters is then precisely stated with analytic formulas. We also give another interpretation of their evolution, as definite displacements on the Stern-Brocot tree. Afterwards we focus on a particular class of digital segments subset of digital curves, which are called *maximal segments* and which have interesting properties [12, 13, 14]. Informally, they form the inextensible digital straight segments on the curve. The preceding properties allow us to give an analytic writing of the minimal and maximal slope variation between two consecutive maximal segments. These bounds are fully described with our new parameters. Surprisingly, they show that consecutive maximal segments may not vary too much nor too little since both bounds are of the same order wrt parameters. On a long term, these quantitative relations will be crucial for designing digital curvature estimators based on slope variations.

This paper is organized as follows. First, basic arithmetic and combinatoric definitions and properties of digital lines are recalled (Section 2). Then we give a comprehensive explanation of the **DR95** algorithm, describing each possible evolution in terms of the new parameters (Section 3). Afterwards the connection between the **DR95** and the Stern-Brocot tree is explicited (Section 4). Eventually those properties have consequences on the geometry of maximal segments, namely bounds on their slope variations (Section 5). We conclude the paper by some perspectives to this work (Section 6).

2 Digital Straight Segments: Arithmetic and Combinatoric Approach

Given a compact set with rectifiable boundary we consider its Gauss digitization. The digital border of this digitization is chosen as the inter-pixel 4-connected path laying between its inner and outer digitization. This digital curve is referred as C. We consider that the points on the boundary are indexed increasingly, for instance with a counterclockwise order. Moreover, given two points on the boundary (say A and B), $C_{A,B}$ is the digital path from A to B. For convenience reasons we identify the index of a point on the boundary to the point itself. For example $A < B$ means that the point A is before the point B on the curve.

Let us recall the arithmetic definition of digital straight lines and explain the notations that will be used in the following of the paper. Following definitions hold in the first octant.

Definition 1. *The set of points (x, y) of the digital plane verifying $\mu \leq ax - by < \mu + |a| + |b|$, with a, b and μ integer numbers, is called the* standard line *with slope a/b and shift μ [15] (e.g. see Fig. 1).*

The *standard lines* are the 4-connected discrete lines. The quantity $r_{(a,b)}(P) = ax - by$ is the *remainder* of the point $P = (x, y)$ in the digital line of characteristics (a, b, μ). The points whose remainder is μ (resp. $\mu + |a| + |b| - 1$) are called *upper* (resp. *lower*) *leaning points.*

Definition 2. *A set of successive points $C_{i,j}$ of the digital curve C is a* digital straight segment (DSS) *iff there exists a standard line (a, b, μ) containing them. The predicate "$C_{i,j}$ is a DSS" is denoted by $S(i, j)$. When $S(i, j)$ the* characteristics *associated with the DSS $C_{i,j}$ (extracted with the* **DR95** *algorithm) [1] are the characteristics (a, b, μ) which minimize $a + b$.*

The original **DR95** algorithm recognizes naive digital straight line but it is easily adapted to standard lines. It extracts the characteristics (a, b, μ), with minimal $a+b$. The evolution of the characteristics is based on a simple test: each time we try to add a new point 4-connected to the current digital straight segment, we compute its *remainder* with respect to the DSS parameters. According to this value the point can be added or not. If it is greater than or equal to $\mu+a+b+1$ or less than or equal to $\mu - 2$ the point is said to be *exterior* to the digital straight segment and cannot be added. Otherwise the point can be added to the segment to form a longer DSS and falls into two categories:

- *interior* points, with a remainder between μ and $\mu+a+b-1$ both included;
- *weakly exterior* points, with a remainder of $\mu - 1$ for *upper weakly exterior* points and $\mu + a + b$ for *lower weakly exterior* points. Only in this case are the characteristics updated.

Even if the arithmetic approach is a powerful tool for digital straight segment recognition, other approaches may reveal useful to get analytic properties. We here recall one of those approaches which is connected to continued fractions.

Definition 3. *Given a standard line of characteristics (a, b, μ), we call* pattern *of characteristics (a, b) the word formed by the Freeman codes between any two consecutive upper leaning points. The Freeman codes defined between any two consecutive lower leaning points is the previous word read from back to front and is called the* reversed pattern *of characteristics (a, b).*

Since a DSS has at least either two upper or two lower leaning points, a DSS (a, b, μ) contains at least one pattern *or* one reversed pattern of characteristics

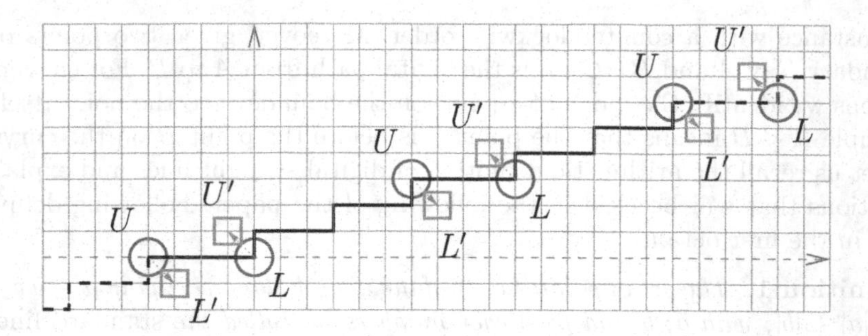

Fig. 1. Positions of weakly exterior points on a digital straight line of characteristics (3, 10, 12). Weakly exterior points are boxed and leaning points are circled.

(a, b). It is important to note that a DSS (a, b, μ) contains δ pattern (a, b) (resp. δ' reversed-pattern (a, b)) iff it has $\delta + 1$ upper leaning points (resp. $\delta' + 1$ lower leaning points). Moreover for any DSS(a, b, μ), the number of pattern (a, b) and reversed-pattern (a, b) differ from one.

There exists recursive transformations for computing the *pattern* of a standard line from the *simple continued fraction* of its slope ([5], [10] Chap. 9 or [6] Chap. 4). We chose to focus on Berstel's approach, which better suits our purpose. A *continued fraction* z will be conveniently denoted by $[0, u_1 \ldots, u_n, \ldots]$. The u_i are called *elements* or *partial coefficients* and the continued fraction formed with the $k+1$ first *partial coefficients* of z is said to be a *k-th convergent* of z and is denoted z_k. The *depth* of a k-th convergent equals k. We conveniently denote by p_k the numerator and by q_k the denominator of a k-th convergent.

We recall a few more relations regarding the way convergents can be formed:

$$\forall k \geq 1 \quad p_k q_{k-1} - p_{k-1} q_k = (-1)^{k+1}, \tag{1}$$

$$p_0 = 0 \quad p_{-1} = 1 \quad \forall k \geq 1 \quad p_k = u_k p_{k-1} + p_{k-2}, \tag{2}$$

$$q_0 = 1 \quad q_{-1} = 0 \quad \forall k \geq 1 \quad q_k = u_k q_{k-1} + q_{k-2}. \tag{3}$$

Continued fractions can be finite or infinite, we focus on the case of rational slopes of lines in the first octant, that is finite continued fractions between 0 and 1. For each i, u_i is assumed to be a strictly positive integer. In order to have a unique writing we consider that the last *partial coefficient* is greater or equal to two; except for slope $1 = [0, 1]$.

Let us now explain how to compute the *pattern* associated with a rational slope z in the first octant (i.e. $z = \frac{a}{b}$ with $0 \leq a < b$). Horizontal steps are denoted by 0 and vertical steps are denoted by 1. Let us define E a mapping from the set of positive rational number smaller than one onto the Freeman-code's words, more precisely $E(z_0) = 0$, $E(z_1) = 0^{u_1}1$ and others values are expressed recursively:

$$E(z_{2i+1}) = E(z_{2i})^{u_{2i+1}} E(z_{2i-1}), \tag{4}$$

$$E(z_{2i}) = E(z_{2i-2}) E(z_{2i-1})^{u_{2i}}. \tag{5}$$

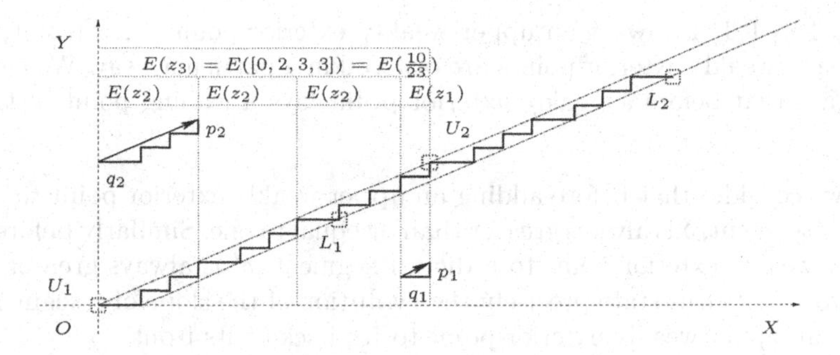

Fig. 2. A digital straight segment of characteristics (10, 23, 0) with an odd slope, taken between origin and its second lower leaning point

It has been shown that this mapping constructs the pattern (a, b) for any rational slope $z = \frac{a}{b}$. Fig. 2 exemplifies the construction.

There exists other equivalent relations for computing numerators and denominators (see [10] Chap. 9 and [6] Chap. 4) and the *splitting formula* can be used to obtain patterns. However the splitting formula uses two k-th convergent with the same depth, whereas we here use two k-th convergent of consecutive depth. The *parity* of a slope is defined as the parity of the depth of its development in continued fractions.

3 Combinatoric View of DR95 Algorithm

Changes in the slope with the **DR95** algorithm occur when weakly exterior points are added to the segment. We propose here to explain the different classes of parameters that rule the evolution process, that is, the characteristics of the straight segment (a, b, μ), the numbers of patterns and reversed-pattern (a, b) that constitute it, the depth of the rational fraction $\frac{a}{b}$, the type of weakly exterior point that is added (upper or lower) and if it is added to the front or to the back.

If a digital straight segment with characteristics (a, b, μ) does not contain any *pattern* (a, b) then it only contains a *reversed-pattern* (a, b), that is, two lower leaning points and one upper leaning point.

Lemma 1. *If a digital straight segment does not contain any pattern, then there is necessarily one upper leaning point laying on the digital path before an upper weakly exterior point. Similarly, if a digital straight segment does not contain any reversed-pattern, then there is necessarily one lower leaning point laying on the digital path before a lower weakly exterior point.*

Proof. Due to the values of the remainder, one can see that weakly exterior points and leaning points are connected (Fig. 1). Consider a digital straight line of characteristics (a, b, μ), and let U (resp. L) be an upper (resp. lower) leaning point of that line. The point $U + (1, -1)$ has a remainder of $\mu + a + b$ which means it is a lower weakly exterior point. A similar reasoning shows us that $L + (-1, 1)$ is always an upper weakly exterior point. As a result, leaning points and weakly exterior points are ordered in a particular way. We can therefore state that before a weakly exterior point lays a leaning point of the same type. □

Thus we consider that before adding an upper weakly exterior point to a digital straight segment, δ is always greater than or equal to one. Similarly before adding a lower weakly exterior point to a digital segment, δ' is always greater than or equal to one. Let us state precisely the evolution of the slope of a segment when adding an upper weakly exterior point to its back or its front.

Proposition 1. *The evolution of the slope of a DSS recognized with the* **DR95** *algorithm depends of the parity of its depth, the type of weakly exterior point added and the side where it is added. This process can be summed up as follows:*

– *slope with even depth* $[0, u_1, \ldots, u_{2i}]$, δ *pattern(s) and* δ' *reversed-pattern(s):*

	Back side	*Front side*
Upper weakly exterior	$[0, u_1, \ldots, u_{2i} - 1, 1, \delta]$	$[0, u_1, \ldots, u_{2i}, \delta]$
Lower weakly exterior	$[0, u_1, \ldots, u_{2i}, \delta']$	$[0, u_1, \ldots, u_{2i} - 1, 1, \delta']$

– *slope with odd depth* $[0, u_1, \ldots, u_{2i+1}]$, δ *pattern(s) and* δ' *reversed-pattern(s):*

	Back side	*Front side*
Upper weakly exterior	$[0, u_1, \ldots, u_{2i+1}, \delta]$	$[0, u_1, \ldots, u_{2i+1} - 1, 1, \delta]$
Lower weakly exterior	$[0, u_1, \ldots, u_{2i+1} - 1, 1, \delta']$	$[0, u_1, \ldots, u_{2i+1}, \delta']$

Proof. We give the proof in the case of an even slope when an upper weakly exterior point is added to the right side. Other cases are deduced with a similar reasoning or considering the segment upside-down.

Consider we have $\delta \geq 1$. Let U_L and U_R be the leftmost and rightmost upper leaning point of the DSS. We choose U_L as the origin. Let $\frac{p_{2i}}{q_{2i}} = [0; u_1, \ldots, u_{2i}]$.

The added point U' has a remainder equal to -1. Eq. (1) indicates that q_{2i-1} and p_{2i-1} are the smallest positive Bezout coefficient verifying $p_{2i}x - q_{2i}y = -1$. As a result: $\mathbf{U_R U'} = (q_{2i-1}, p_{2i-1})$.

From the recognition algorithm **DR95** the slope of the segment $C_{U_L,U'}$ is given by the vector $\mathbf{U_L U'}$. Since $\mathbf{U_L U'} = \mathbf{U_L U_R} + \mathbf{U_R U'}$, $\mathbf{U_L U'}$ equals $\delta(q_{2i}, p_{2i}) + (q_{2i-1}, p_{2i-1})$. From Eq. (2) and (3) this slope equal $\frac{p_{2i+1}}{q_{2i+1}} = [0, u_1, \ldots, u_{2i}, \delta]$. If δ equals one then $[0, u_1, \ldots, u_{2i}, 1] = [0, u_1, \ldots, u_{2i} + 1]$. □

The slope depth of a DSS when adding a weakly exterior point remains the same or is increased by one or two.

4 Connection with the Stern-Brocot Tree

We now show that the evolution of the DSS parameters during the recognition process, analytically given in Proposition 1, can be traced on the Stern-Brocot tree and has then a more intuitive interpretation. The Stern-Brocot tree represents all positive rational fractions. It was already observed that the **DR95** recognition algorithm has some connection with it [11]. More precisely the successive values of the slope taken by a segment are deeper and deeper nodes of this tree. Note that this tree has other connections with discrete geometry, like determining the minimal characteristics of the intersection of two digital straight lines [16,17].

The Stern-Brocot tree is a binary tree constructed by starting with the fractions $\frac{0}{1}$ and $\frac{1}{0}$ and iteratively inserting $\frac{m+m'}{n+n'}$ between each two adjacent fractions $\frac{m}{n}$ and $\frac{m'}{n'}$ (Fig. 3). Any node with a value between 0 (excluded) and 1 (included) is obtained by finite successive moves from the $\frac{1}{1}$ node. Those moves can be of two types: L is a move toward the left child, R is a move toward the right child. Those moves determine the type of node when they end paths: even (resp. odd) nodes end with a R (resp. L) move. It is known that those nodes have a development in continued fraction, and is such that:

- even nodes:
  $[0, u_1, \ldots, u_{2k}, u_{2k+1}, \ldots, u_{2i-1}, u_{2i}] \equiv R^0 L^{u_1} \ldots R^{u_{2k}} L^{u_{2k+1}} \ldots L^{u_{2i-1}} R^{u_{2i}-1}$
- odd nodes:
  $[0, u_1, \ldots, u_{2k}, u_{2k+1}, \ldots, u_{2i}, u_{2i+1}] \equiv R^0 L^{u_1} \ldots R^{u_{2k}} L^{u_{2k+1}} \ldots R^{u_{2i}} L^{u_{2i+1}-1}$

Of course odd nodes have an odd depth in their development in continued fractions, similarly even nodes have an even depth. When descending the tree the depth of a child changes if the move used to reach it differs from the last move used to reach its father. Consider the node $\frac{1}{2}$ whose depth equals one, the last move used to reach it is a L move, its left child $\frac{1}{3}$ has the same depth. The right child of $\frac{1}{2}$ is obtained by the successive moves $R^0 L^1 R^1$ and has a depth that equals two. We can classify the nodes of the Stern-Brocot tree according to the depth of their development in continued fraction (Fig. 3).

Nodes of this tree may also be viewed as slopes of a digital straight segment being recognized. As we consider rational slopes between zero and one, we only consider nodes whose value is between zero and one. All those nodes are derived (except for the zero node) from the $\frac{1}{1}$ node with a L move first. This implies that u_0 equals zero. It is possible to trace the slope evolution of a digital straight segment during recognition on the Stern-Brocot tree, as exemplified in Fig. 3.

Since each node has a particular development in continued fraction, the results shown in Proposition 1 can be reinterpreted in terms of descending moves on the Stern-Brocot tree. Thus the slope evolution of a DSS during recognition can be fully described with L and R moves as shown on Fig. 4. We then see that the number of successive moves of the same type directly depends on the number of patterns or reversed-patterns.

From Fig. 4 we can see that the left child nodes are always reached when we add to the right a lower weakly exterior point or to the left an upper weakly exterior point whatever the parity of the node depth. Same reasoning applies for right child nodes. We can now see how the slope evolution is translated as moves on the Stern-Brocot tree when adding a point. Fig. 5 pictures the possible slope evolutions from the $\frac{1}{2}$ node.

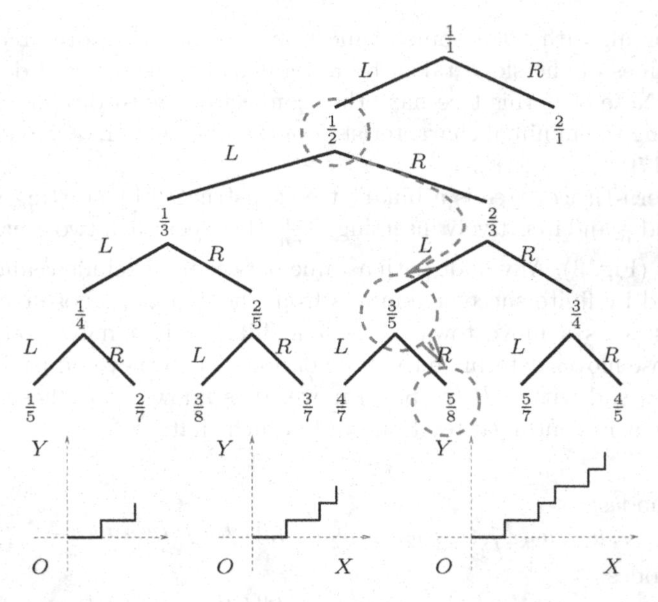

Fig. 3. Evolution of the slope of a digital straight segment. Successive values are $\frac{1}{2}$, $\frac{3}{5}$ and $\frac{5}{8}$. Modifications are triggered by the addition of an upper weakly exterior point. The successive slope depths are one, three and four. Nodes with depth one are $\frac{1}{1}, \frac{1}{2}, \frac{1}{3}, \frac{1}{4}, \frac{1}{5}$, nodes with depth two are $\frac{2}{3}, \frac{3}{4}, \frac{4}{5}, \frac{2}{5}, \frac{3}{7}, \frac{2}{7}$ nodes with depth three are $\frac{3}{5}$, $\frac{4}{7}, \frac{3}{8}, \frac{5}{7}$ and node with depth four is $\frac{5}{8}$.

5 Application to Maximal Segments

We now apply previous properties to get a better understanding of the geometry of maximal segments. Maximal segments form a particular class of digital straight segments on the digital curve. Their study is related to many discrete geometry problems such as digital convexity [14], polygonalization [12] or tangent computation [13]. The main result of this section is Theorem 1 which bounds the slope difference between two consecutive maximal segments. Let us first explain how to characterize them.

Given a point C_i on a digital curve C, the first index j greater than i such that $S(i,j)$ and $\neg S(i,j+1)$ is called the *front* of i. The map associating any i to its front is denoted by F. Symmetrically, the first index i such that $S(i,j)$ and $\neg S(i-1,j)$ is called the *back* of j and the associated mapping is denoted by B.

Definition 4. *Any set of points $C_{i,j}$ is called a* maximal segment *iff any of the following equivalent characterizations holds: (1) $S(i,j)$ and $\neg S(i,j+1)$ and $\neg S(i-1,j)$, (2) $B(j)=i$ and $F(i)=j$.*

Maximal segments form the set of DSS on the digital curve that cannot be extended on any side. They can be ordered along the curve. Consecutive maximal segments overlap and often on more than two points. The digital path that belongs to two consecutive maximal segments is called a *common part*, its as-

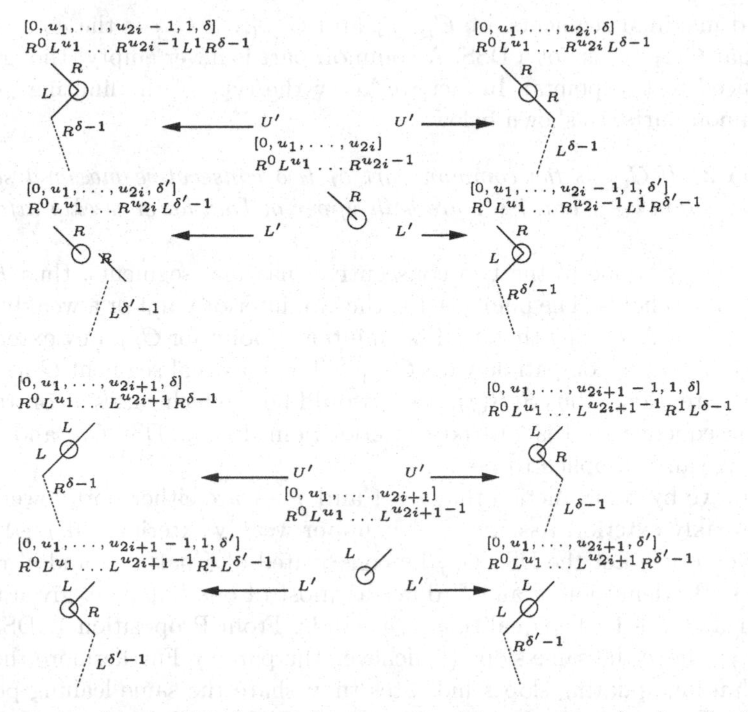

Fig. 4. Evolution from a digital straight segment with even slope (top) and odd slope (bottom) using the **DR95** algorithm, represented in terms of move on the Stern-Brocot tree. The point U' (resp. L') is an upper (resp. lower) weakly exterior point added to the back (left column) or to the front (right column) of the DSS.

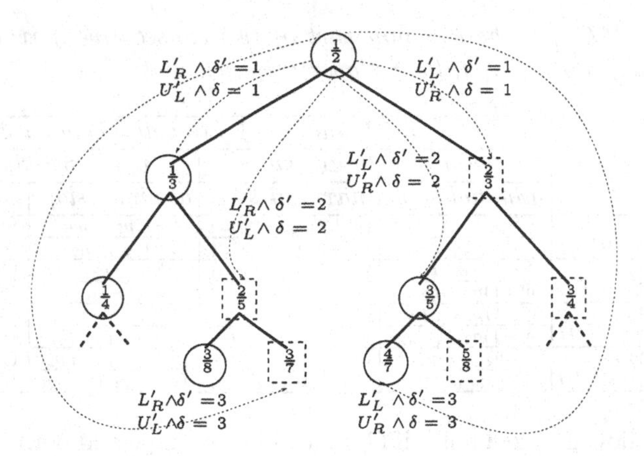

Fig. 5. Reachable slopes when a weakly exterior point is added to a digital straight segment of slope $\frac{1}{2}$. Nodes with an odd depth are circled, while those with even depth are dash-boxed. L'_L and U'_L (resp. L'_R and U'_R) stand for lower and upper weakly exterior point added to the back (resp. front). δ and δ' count the number of patterns and reversed-patterns respectively.

sociated maximal segments are $C_{B(j),j}$ and $C_{i,F(i)}$ if $C_{i,j}$ is the common part. Note that $C_{i-1,j+1}$ is not a DSS. A common part is never empty (though it may be reduced to two points). In fact we know the type of the limiting points for all common parts, as shown below:

Lemma 2. *If $C_{i,j}$ is the common part of two consecutive maximal segments, then the points $i-1$ and $j+1$ are both upper or both lower weakly exterior.*

Proof. $C_{i,F(i)}$ is one of the two consecutive maximal segments, thus $F(i) > j$ and $S(i, j+1)$ holds. The point $j+1$ is thus an interior point or a weakly exterior point for $C_{i,j}$. Assuming that $j+1$ is an interior point for $C_{i,j}$, any extensions to the back of $C_{i,j}$ is compatible with $C_{i,j+1}$. The maximal segment $C_{B(j),j}$ is one of these extensions, thus $S(B(j), j+1)$ would hold which raises a contradiction. As a consequence $j+1$ is a weakly exterior point for the DSS $C_{i,j}$ and a similar reasoning can be applied to $i-1$.

We prove by contradiction that $i-1$ and $j+1$ are either both lower or both upper weakly exterior. Assume $i-1$ is upper weakly exterior and $j+1$ is lower weakly exterior. Let the DSS $C_{i,j}$ be constituted of δ patterns and δ' reversed-patterns. By definition, δ and δ' differ at most of one but here, given the type of $i-1$ and $j+1$, the equality $\delta = \delta'$ holds. From Proposition 1, DSS $C_{i-1,j}$ and $C_{i,j+1}$ have the same slope (whichever the parity). Furthermore the **DR95** algorithm for updating slopes indicates they share the same leaning points L_L and U_R. These two assertions, combined together, entail $C_{i-1,j+1}$ is a DSS too, which contradicts the hypothesis that it is a common part. □

We give now analytic bounds on slopes of two consecutive maximal segments.

Theorem 1. *If $C_{i,j}$ is the common part of two consecutive maximal segments (namely $C_{B(j),j}$ and $C_{i,F(i)}$), their slopes are such that:*

		$i-1$ and $j+1$ are both (Lemma 2)			
		lower weakly exterior		*upper weakly exterior*	
		minimal slope	*maximal slope*	*minimal slope*	*maximal slope*
$C_{i,j}$ has an even slope	$C_{i,F(i)}$	$\frac{\delta' p_{2i}-p_{2i-1}}{\delta' q_{2i}-q_{2i-1}}$	$\frac{(\delta'+2)p_{2i}-p_{2i-1}}{(\delta'+2)q_{2i}-q_{2i-1}}$	$\frac{(\delta+1)p_{2i}+p_{2i-1}}{(\delta+1)q_{2i}+q_{2i-1}}$	$\frac{(\delta-1)p_{2i}+p_{2i-1}}{(\delta-1)q_{2i}+q_{2i-1}}$
	$C_{B(j),j}$	$\frac{(\delta'+1)p_{2i}+p_{2i-1}}{(\delta'+1)q_{2i}+q_{2i-1}}$	$\frac{(\delta'-1)p_{2i}+p_{2i-1}}{(\delta'-1)q_{2i}+q_{2i-1}}$	$\frac{\delta p_{2i}-p_{2i-1}}{\delta q_{2i}-q_{2i-1}}$	$\frac{(\delta+2)p_{2i}-p_{2i-1}}{(\delta+2)q_{2i}-q_{2i-1}}$
$C_{i,j}$ has an odd slope	$C_{i,F(i)}$	$\frac{(\delta'+1)p_{2i+1}+p_{2i}}{(\delta'+1)q_{2i+1}+q_{2i}}$	$\frac{(\delta'-1)p_{2i+1}+p_{2i}}{(\delta'-1)q_{2i+1}+q_{2i}}$	$\frac{\delta p_{2i+1}-p_{2i}}{\delta q_{2i+1}-q_{2i}}$	$\frac{(\delta+2)p_{2i+1}-p_{2i}}{(\delta+2)q_{2i+1}-q_{2i}}$
	$C_{B(j),j}$	$\frac{\delta' p_{2i+1}-p_{2i}}{\delta' q_{2i+1}-q_{2i}}$	$\frac{(\delta'+2)p_{2i+1}-p_{2i}}{(\delta'+2)q_{2i+1}-q_{2i}}$	$\frac{(\delta+1)p_{2i+1}+p_{2i}}{(\delta+1)q_{2i+1}+q_{2i}}$	$\frac{(\delta-1)p_{2i+1}+p_{2i}}{(\delta-1)q_{2i+1}+q_{2i}}$

Proof. The following proof holds if $C_{i,j}$ has an even slope and both $i-1$ and $j+1$ are lower weakly exterior points. Other cases are deduced from Proposition 1.

We bound the slopes obtained by extending $C_{i,j}$ to the front then extending $C_{i,j}$ to the back. Since $j+1$ is a lower weakly exterior point, $C_{i,j+1}$ has slope $z^R_{2i+2} = [0, u_1, \ldots, u_{2i} - 1, 1, \delta']$ (Proposition 1). Assuming that $C_{i,F(i)}$ has a slope that equals $[0, u_1, \ldots, u_{2i} - 1, 1, \delta' + \epsilon, u_{2i+3}, \ldots, u_p]$ with ϵ being -1 or zero (from Proposition 1). Simple calculation brings:

$$\cfrac{1}{u_{2i+3} + \cfrac{1}{\ldots + \cfrac{1}{u_p}}} = \epsilon' \quad \text{with } \epsilon' \in]0,1]$$

As a result, the slope of $C_{i,F(i)}$ equals $z^R(\epsilon_R) = [0, u_1, \ldots, u_{2i} - 1, 1, \delta' + \epsilon_R]$ with $\epsilon_R \in]-1, 1]$. Eq. (2) and Eq. (3) still hold when partial coefficient are real values, we thus get

$$z^R(\epsilon_R) = \frac{(\delta'+\epsilon_R+1)p_{2i}-p_{2i-1}}{(\delta'+\epsilon_R+1)q_{2i}-q_{2i-1}}$$

We bound this slope for extremal values of ϵ_R, giving

$$\frac{\delta' p_{2i}-p_{2i-1}}{\delta' q_{2i}-q_{2i-1}} \leq z^R(\epsilon_R) \leq \frac{(\delta'+2)p_{2i}-p_{2i-1}}{(\delta'+2)q_{2i}-q_{2i-1}}$$

Same reasoning applied to the back of the common part brings: $z^L_{2i+2} = [0, u_1, \ldots, u_{2i}, \delta']$ and $z^L(\epsilon_L) = [0, u_1, \ldots, u_{2i}, \delta' + \epsilon_L]$ with $\epsilon_L \in]-1, 1]$. Bounds are:

$$\frac{(\delta'+1)p_{2i}+p_{2i-1}}{(\delta'+1)q_{2i}+q_{2i-1}} \leq z^L(\epsilon_L) \leq \frac{(\delta'-1)p_{2i}+p_{2i-1}}{(\delta'-1)q_{2i}+q_{2i-1}}. \qquad \square$$

Furthermore, Theorem 1 give bounds on the slope difference Δz of two consecutive maximal segments as a function of the parameters of their common part. For instance the case of an even slope with lower weakly exterior points give the tight bound

$$\frac{2\delta'+3}{(\delta'^2+3\delta'+2)q_{2i}^2 + q_{2i}q_{2i-1} - q_{2i-1}^2} < |\Delta z| < \frac{2\delta'-1}{(\delta'^2-\delta')q_{2i}^2 + q_{2i}q_{2i-1} - q_{2i-1}^2}.$$

We give below a coarser bound for Δz, but which is expressed only in terms of the slope denominator and the number of reversed-patterns:

$$\frac{2\delta'+3}{(\delta'^2+3\delta'+3)q_{2i}^2} < |\Delta z| < \frac{2\delta'-1}{(\delta'^2-\delta')q_{2i}^2 + \frac{1}{2}q_{2i} + \frac{1}{2}}. \tag{6}$$

In the other cases, similar formula are obtained. These formulas induce that the average slope difference between consecutive maximal segments could be determined, provided the average behaviour of δ, δ' and q_n is known.

6 Conclusion

We have revisited a classical arithmetically-based DSS recognition algorithm with new parameters related to a combinatoric representation of DSS. New analytic relations have been established and the relation with the Stern-Brocot tree has been made explicit. At last, we have shown new geometric relations on maximal segments. The new parameters introduced in this paper seem to be good candidates to describe DSS and obtain new properties. It would be interesting to investigate the average asymptotic behavior of these parameters, that is δ, δ' and q_n, as functional of the grid step. This would lead us to estimate the asymptotic angle difference between consecutive maximal segments, a quantity related to curvature, and therefore to address the problem of finding a multigrid convergent curvature estimator. This study is thus a first step in this direction.

References

1. Debled-Renesson, I., Reveillès, J.P.: A linear algorithm for segmentation of discrete curves. Int. J. Pattern Recognit. Artif. Intell. **9** (1995) 635–662
2. Klette, R., Rosenfeld, A.: Digital straightness: a review. Discrete Appl. Math. **139**(1-3) (2004) 197–230
3. Dorst, L., Smeulders, A.W.M.: Discrete representation of straight lines. IEEE Trans. Pattern Anal. Mach. Intell. **6** (1984) 450–463
4. Coeurjolly, D.: Supercover model and digital straight line recognition on irregular isothetic grids. In Andrès, E., Damiand, G., Lienhardt, P., eds.: Proc. DGCI'2005. Volume 3429 of LNCS., Springer (2005) 311–322
5. Berstel, J., De Luca, A.: Sturmian words, lyndon words and trees. Theoret. Comput. Sci. **178**(1-2) (1997) 171–203
6. Voss, K.: Discrete Images, Objects, and Functions in $\mathbb{Z}^n$. Springer-Verlag (1993)
7. de Vieilleville, F., Lachaud, J.O., Feschet, F.: Maximal digital straight segments and convergence of discrete geometric estimators. In Kalviainen, H., Parkkinen, J., Kaarna, A., eds.: 14th Scandinavian Conference on Image Analysis. Number 3540 in LNCS, Springer-Verlag (2005) 988–997
8. F. de Vieilleville, J.O.L., Feschet, F.: Maximal digital straight segments and convergence of discrete geometric estimators. Research Report 1350-05, LaBRI, University Bordeaux 1, Talence, France (2005)
9. Feschet, F., Tougne, L.: Optimal time computation of the tangent of a discrete curve: application to the curvature. In: Proc. DGCI'1999. Number 1568 in LNCS, Springer Verlag (1999) 31–40
10. Klette, R., Rosenfeld, A.: Digital Geometry - Geometric Methods for Digital Picture Analysis. Morgan Kaufmann, San Francisco (2004)
11. Debled-Rennesson, I.: Etude et reconnaissance des droites et plans discrets. PhD thesis, Université Louis Pasteur, Strasbourg, France (1995)
12. Feschet., F., Tougne, L.: On the min DSS problem of closed discrete curves. In Lungo, A.D., Gesù, V.D., Kuba, A., eds.: IWCIA. Volume 12 of Electonic Notes in Discrete Math., Elsevier (2003)
13. Lachaud, J.O., Vialard, A., de Vieilleville, F.: Analysis and comparative evaluation of discrete tangent estimators. In Andrès, E., Damiand, G., Lienhardt, P., eds.: Proc. DGCI2005. LNCS 3429, Springer-Verlag (2005) 240–251
14. Reiter-Doerksen, H., Debled-Rennesson, I.: Convex and concave parts of digital curves. In: Dagstuhl Seminar "Geometric Properties from Incomplete Data". (2004)
15. Reveillès, J.P.: Géométrie discrète, calcul en nombres entiers et algorithmique. Thèse d'etat, Université Louis Pasteur, Strasbourg (1991)
16. Sivignon, I., Dupont, F., Chassery, J.M.: New results about digital intersections. In: Proc. DGCI'03. Volume 2886 of LNCS., Springer (2003) 102–113
17. Sivignon, I.: De la caractérisation des primitives à la reconstruction polyédrique de surfaces en géométrie discrète. PhD thesis, Institut national polytechnique de Grenoble, France (2004)

On Discrete Moments of Unbounded Order

Reinhard Klette[1] and Joviša Žunić[2,*]

[1] Computer Science, The University of Auckland
Auckland, New Zealand
[2] Computer Science, Exeter University
Exeter EX4 4QF, UK

Abstract. Moment-based procedures are commonly used in computer vision, image analysis, or pattern recognition. Basic shape features such as size, position, orientation, or elongation are estimated by moments of order ≤ 2. Shape invariants are defined by higher order moments. In contrast to a theory of moments in continuous mathematics, shape moments in imaging have to be estimated from digitized data. Infinitely many different shapes in Euclidean space are represented by an identical digital shape. There is an inherent loss of information, impacting moment estimation.

This paper discusses accuracy limitations in moment reconstruction in dependency of order of reconstructed moments and applied resolution of digital pictures. We consider moments of arbitrary order, which is not assumed to be bounded by a constant.

Keywords: moments, discrete moments, accuracy of estimation, multigrid convergence, digital shapes.

1 Introduction

Moments are widely used in computer vision, image analysis, or pattern recognition (since Hu [1]). A variety of types of moments and moment-based methods has been developed and studied, for example, for object recognition [2], reconstruction of geometric properties of regions [3], or determination of invariants [4]. The (p,q)-*moment* $m_{p,q}(S)$ of a planar set S is defined by the following:

$$m_{p,q}(S) \quad = \quad \iint_S x^p y^q \, dx \, dy$$

It has the order $p+q$.

Basic shape features (e.g., size, position, orientation, elongation) are computed from moments of order less or equal to two. Higher order moments are needed for computing, for example, the orientation of 3D rotationally symmetric shapes (see [5]) or moment invariants (see [1]). In imaging applications we have to

* The author is also with the Mathematical institute of Serbian Academy of Sciences and Arts, Belgrade.

A. Kuba, L.G. Nyúl, and K. Palágyi (Eds.): DGCI 2006, LNCS 4245, pp. 367–378, 2006.

deal with digitized shapes (objects); consequently, exact moment computation is impossible. The accuracy of moment estimation is limited by many factors, dominated by shape complexity, applied resolution of digital pictures, and the order of reconstructed moments.

Obviously, higher picture resolution enables a higher precision in moment reconstruction. Also, if picture resolution is fixed, then accuracy would decrease if the moment's order increases. Thus, if high-order moments are needed for a particular application, reconstruction accuracy can be improved by an increase in applied picture resolution. This is formally studied as multigrid convergence in digital geometry (see [6]).

Situations, where the order of moments is bounded while picture resolution is allowed to increase (to infinity), have been discussed in [7]. The case of unboundedly increases of orders of moments remained an open problem in that publication.

This paper also covers the case where the order of moments is allowed to tend to infinity. Furthermore, for this situation we consider the special case where the order of computed moments is at most logarithmic in applied picture resolution. We prove an upper bound for the resulting error in estimation which improves the best known upper bound to date (that follows from general tools provided in [8]).

We give definitions and notations as used in this paper. Center points of grid squares are assumed to have integer coordinates (i.e., to be grid points in $\mathbb{Z}^2$). In the diversity of different models for digitizing shapes in Euclidean spaces, we decide for the set of grid points contained in the given shape (analogous to Gauss digitization in [6]). That means, for a set $S \subset \mathbb{R}^2$, its digitization $G(S)$ is defined to be the set of all grid points which are contained in S.

Let $h > 0$ be the picture resolution (i.e., the number of grid points per unit). Instead of considering a digitization of S in a picture of resolution h, we prefer here (as standard in number theory) to use a digitization of the dilated set $h{\cdot}S = \{(h{\cdot}x, h{\cdot}y) \mid (x, y) \in S\}$ in the grid of resolution $h = 1$. We consider $G(h{\cdot}S)$ to be (under number-theoretical aspects) the *shape S digitized in a binary picture of resolution h*. Gauss digitization is defined analogously in 3D. If $S \subset \mathbb{R}^3$, the Gauss digitization $G(S)$ is the set of all 3D grid points contained in S.

The exact value of $m_{p,q}(S)$ remains unknown in digital imaging (because the exact Euclidean shape of S remains unknown). The following estimation is used:

$$m_{p,q}(S) = \frac{1}{h^{p+q+2}} \cdot \iint\limits_{h\cdot S} x^p y^q \, dx \, dy \approx \frac{1}{h^{p+q+2}} \cdot \sum_{(i,j)\in G(h\cdot S)} i^p \cdot j^q \qquad (1)$$

For a given *digital planar shape* A (i.e., a finite subset of $\mathbb{Z}^2$) and non-negative integers p and q, define the *discrete moment* $\mu_{p,q}(A)$ as follows:

$$\mu_{p,q}(A) = \sum_{(i,j)\in A\cap\mathbb{Z}^2} i^p \cdot j^q$$

3D discrete moments are defined analogously. For a finite set $B \subset \mathbb{Z}^3$ and non-negative integers p, q and t, we have

$$\mu_{p,q,t}(B) = \sum_{(i,j,k)\in B\cap\mathbb{Z}^3} i^p \cdot j^q \cdot k^t$$

Let $\mathcal{C}(S)$ denote the content of set S, which is the area $\mathcal{A}(S)$ for 2D, or the volume $\mathcal{V}(S)$ for 3D. We have $\mu_{0,0}(A) = \mathcal{A}(S)$ and $\mu_{0,0,0}(B) = \mathcal{V}(S)$, and both values are simply defined by cardinalities $\#A$ and $\#B$, respectively. The orders of $\mu_{p,q}(A)$ or $\mu_{p,q,t}(B)$ are $p+q$ and $p+q+t$, respectively. Throughout the paper we assume that all pixels (i.e., grid points) have nonnegative coordinates (i.e., the origin of the assumed coordinate system is at the lower left corner of a considered picture).

Under these assumptions, for a real shape S, $\mu_{p,q}(G(S))$ equals the number of integer points inside of the $3D$-body $B_{p,q}(G)$ defined as

$$B_{p,q}(S) = \{(x,y,z) : (x,y) \in S \wedge 0 < z \leq x^p \cdot y^q\} \tag{2}$$

In other words,

$$\mu_{p,q}(G(S)) = \#\left(B_{p,q}(S) \cap \mathbb{Z}^3\right) \tag{3}$$

This paper is about an analysis of the maximum error in the approximation $m_{p,q}(S) \approx h^{-(p+q+2)} \cdot \mu_{p,q}(G(h \cdot S))$, when real moments are estimated by corresponding discrete moments. Obviously, this problem is equivalent [see Equation (2)] to the study of the order of magnitude of

$$|m_{p,q}(h \cdot S) - \mu_{p,q}(G(h \cdot S))| \tag{4}$$

This paper deals with planar convex shapes, but due to the given moment definition the result can easily be extended to sets which are unions, intersections or set differences of a finite number of convex sets. Also, since the estimate of (4) becomes trivial if there are any straight sections on the frontier of S, we focus on shapes that have a strictly positive curvature at all points of their frontier. Precise (formal) conditions are given below.

2 Related Results

The number of grid points, contained in convex bodies, is intensively studied in number theory. Regarding (4), a direct application of Davenport's result in [8] (to our case) says that $|m_{p,q}(h \cdot S) - \mu_{p,q}(h \cdot S)|$ is upper bounded by the total sum of projections of $B_{p,q}(h \cdot S)$ onto xy-, xz-, and yz-plane, onton x-, y-, and z-axis, and finally increased by 1. In other words, we have

$$\begin{aligned}|m_{p,q}(h \cdot S) - \mu_{p,q}(h \cdot S)| &= \left|m_{p,q}(h \cdot S) - \#\left(B_{p,q}(h \cdot S) \cap \mathbb{Z}^3\right)\right| \\ \leq \left(\frac{x_{max}^{p+1} \cdot y_{max}^q}{p+1} + \frac{x_{max}^p \cdot y_{max}^{q+1}}{q+1}\right) &\cdot h^{p+q+1} + h^2 \cdot x_{max} \cdot y_{max} \\ +x_{max}^p \cdot y_{max}^q \cdot h^{p+q} &+ (x_{max} + y_{max}) \cdot h + 1 \end{aligned} \tag{5}$$

A better estimate than (5) is derived in [7] for bounded orders $p+q$. This paper shows that exploiting Huxley's result in [9] allows to obtain an estimate for (4) which improves estimate (5) even for orders of unbounded values of p and q.

We assume that frontiers γ of convex shapes S are composed of finitely many *smooth arcs* γ_i, either given by an equation $y = \phi(x)$, or by $x = \theta(y)$, functions $\phi(x)$ and $\theta(y)$ have at least continuous derivatives up to the third order, also satisfying the following (for $\psi = \phi$ or $\psi = \theta$):

(i) The radius ρ of curvature and its derivative $\frac{d\rho}{d\psi}$ exist on each arc γ_i, and both are continuous functions of ψ on γ_i.
(ii) On each arc γ_i, the radius of curvature ρ has a maximum value and a non-zero minimum value.
(iii) On each arc γ_i, the radius of curvature has a bounded number of local maxima and minima.

The following theorem is of major importance for this paper.

Theorem 1. (Huxley 2003). *Suppose that γ consists of finitely many smooth arcs, each of which satisfies conditions (i), (ii), and (iii). Then there is a constant c, calculated from the arcs γ_i of γ (where c is independent of the chosen length unit), such that, if the minimum radius of curvature of each γ_j is at last c, then the number of grid points in S is upper bounded by*

$$\mathcal{A}(S) + \mathcal{O}\left(R^{\frac{131}{208}} \cdot (\log R)^{\frac{18627}{8320}}\right)$$

where R is the maximum radius of curvature of γ. The constant implied in the order of magnitude notation is also calculated from the arcs of γ, and it is independent of the chosen length unit.

A planar convex set S, satisfying the preconditions of Theorem 1, is said to have a *sufficiently smooth frontier.* A direct consequence of Theorem 1 is the following:

Corollary 1. *Let S be a planar convex set with a sufficiently smooth frontier. Then it follows that*

$$\#G(h \cdot S) = h^2 \cdot \mathcal{A}(S) + \mathcal{O}\left(h^{\frac{131}{208}+\varepsilon}\right) \tag{6}$$

for any $\varepsilon > 0$.

This is a very strong result. It even improves the previously best known upper bound for the circle problem (i.e., if S is assumed to be a circle).

The following studies are divided into two different cases. The case where either p or q is zero, is studied in the next section. The case where both p and q are strictly positive, is studied in Section 4.

3 Error Estimate if Either $p = 0$ or $q = 0$

Obviously (due to symmetry), estimates for $\mu_{p,0}(h \cdot S)$ and $\mu_{0,q}(h \cdot S)$ can be derived in identical ways. We consider $\mu_{p,0}(h \cdot S)$.

For a compact set S, let $x_{min} = \min\{x : (x,y) \in S\}$, $x_{max} = \max\{x : (x,y) \in S\}$, $y_{min} = \min\{y : (x,y) \in S\}$, and $y_{max} = \max\{y : (x,y) \in S\}$.

Without loss of generality we can assume that the studied convex set S is a subset of $[0,1] \times [0,1]$. Consequently, we have $\{x_{min}, x_{max}, y_{min}, y_{max}\} \subset [0,1]$ in what follows.

Definition 1. *For a planar set S, integer k, and real $h > 0$, let*

$$(h \cdot S)(k) = \{(x,y) \ : \ (x,y) \in (h \cdot S) \ \wedge \ x \geq k\}$$

Consequently, $G((h \cdot S)(k))$ is the set of grid points in the digitization of $h \cdot S$ lying in the closed half plane determined by $x \geq k$.

Definition 2. *For a planar set S, integer k, and real $h > 0$, let*

$$L(h \cdot S, k) = \{(k,j) \ : \ (k,j) \in G(h \cdot S)\}.$$

In other words, $L(h \cdot S, k)$ is the set of those grid points in the Gauss digitization of $h \cdot S$ that belong to the line $x = k$. We have the following lemma [7].

Lemma 1. *Let S be a planar convex set and k an integer. We have*

$$\#G((h \cdot S)(k)) \ = \ \mathcal{A}((h \cdot S)(k)) \ + \ \frac{1}{2} \cdot \#L(h \cdot S, k) \ + \ \mathcal{O}(h^{\frac{131}{208}+\varepsilon})$$

We use the following definitions of 3D-sets W_i and W_i':

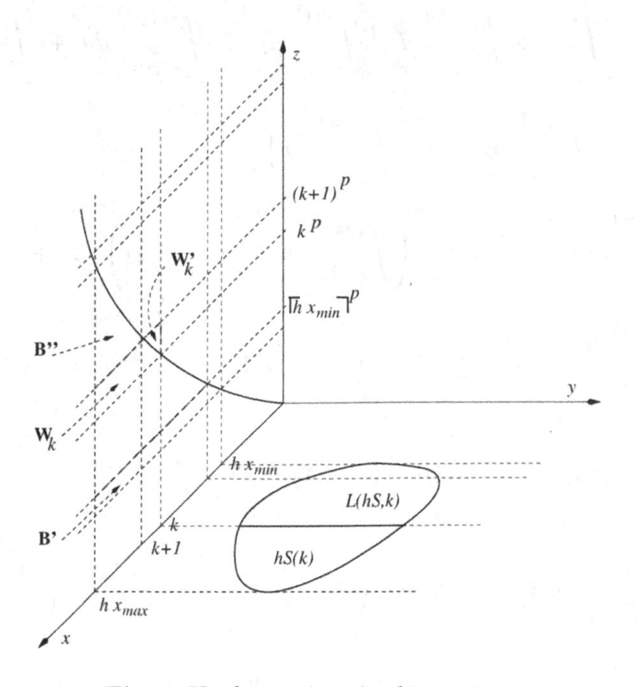

Fig. 1. Used notations in this section

Definition 3. *For planar convex set S and integer $i \in \{\lceil h \cdot x_{min} \rceil, \lceil h \cdot x_{min} \rceil + 1, \ldots, \lfloor h \cdot x_{max} \rfloor - 1\}$, we define 3D sets (see Figure 1)*

$$W_i = \{(x,y,z) \;:\; (x,y) \in h \cdot S \;\wedge\; x \geq i \;\wedge\; i^p < z \leq (i+1)^p\}$$

and

$$W_i' = \{(x,y,z) \;:\; (x,y) \in h \cdot S \;\wedge\; x \geq i \;\wedge\; x^p < z \leq (i+1)^p\}$$

Now we calculate $\mu_{p,0}(h \cdot S)$. As a reminder, $\mathcal{V}(B)$ is the volume of a 3D set B, and $\mathcal{A}(S)$ is the area of a 2D set S.

Lemma 2. *Let S be a convex set. Then*

$$\sum_{i=\lceil h \cdot x_{min} \rceil}^{\lfloor h \cdot x_{max} \rfloor - 1} \mathcal{V}(W_i')$$

$$= \sum_{i=\lceil h \cdot x_{min} \rceil}^{\lfloor h \cdot x_{max} \rfloor - 1} \#L(h \cdot S, i) \cdot \left((i+1)^p - i^p - \frac{p}{2} \cdot i^{p-1}\right) + \mathcal{O}\left(\frac{h^p}{p+1} \cdot \binom{p+1}{\lceil \frac{p+1}{2} \rceil}\right)$$

Proof. The frontier of $h \cdot S$ can be divided into two arcs of the form $y = y_1(x)$ and $y = y_2(x)$, such that $y_1(x) \leq y_2(x)$. Then we have that

$$\sum_{i=\lceil h \cdot x_{min} \rceil}^{\lfloor h \cdot x_{max} \rfloor - 1} \mathcal{V}(W_i') = \sum_{i=\lceil h \cdot x_{min} \rceil}^{\lfloor h \cdot x_{max} \rfloor - 1} \int_i^{i+1} dx \int_{x^p}^{(i+1)^p} dz \int_{y_1(x)}^{y_2(x)} dy$$

$$= \sum_{i=\lceil h \cdot x_{min} \rceil}^{\lfloor h \cdot x_{max} \rfloor - 1} \int_i^{i+1} dx \int_{x^p}^{(i+1)^p} dz \left(\int_{y_1(x)}^{y_1(i)} dy + \int_{y_1(i)}^{\lceil y_1(i) \rceil} dy + \int_{\lceil y_1(i) \rceil}^{\lfloor y_2(i) \rfloor} dy \right.$$

$$\left. + \int_{\lfloor y_2(i) \rfloor}^{y_2(i)} dy + \int_{y_2(i)}^{y_2(x)} dy \right)$$

$$= \sum_{i=\lceil h \cdot x_{min} \rceil}^{\lfloor h \cdot x_{max} \rfloor - 1} \int_i^{i+1} dx \int_{x^p}^{(i+1)^p} \left(\int_{\lceil y_1(i) \rceil}^{\lfloor y_2(i) \rfloor} dy + \mathcal{O}(1) \right) dz + \mathcal{O}(h^p)$$

$$= \sum_{i=\lceil h \cdot x_{min} \rceil}^{\lfloor h \cdot x_{max} \rfloor - 1} \int_i^{i+1} (\lfloor y_2(i) \rfloor - \lceil y_1(i) \rceil) \cdot ((i+1)^p - x^p) dx + \mathcal{O}(h^p)$$

$$= \sum_{i=\lceil h \cdot x_{min} \rceil}^{\lfloor h \cdot x_{max} \rfloor - 1} (\lfloor y_2(i) \rfloor - \lceil y_1(i) \rceil) \cdot \left((i+1)^p - i^p - \frac{p}{2} \cdot i^{p-1}\right) +$$

$$+ \sum_{i=\lceil h \cdot x_{min} \rceil}^{\lfloor h \cdot x_{max} \rfloor - 1} (\lfloor y_2(i) \rfloor - \lceil y_1(i) \rceil) \left(i^p + \frac{p}{2} \cdot i^{p-1} - \frac{(i+1)^{p+1} - i^{p+1}}{p+1} \right) + \mathcal{O}(h^p)$$

$$= \sum_{i=\lceil h \cdot x_{min} \rceil}^{\lfloor h \cdot x_{max} \rfloor - 1} \#L(h \cdot S, i) \cdot \left((i+1)^p - i^p - \frac{p}{2} \cdot i^{p-1}\right) + \mathcal{O}\left(\frac{h^p}{p+1} \cdot \binom{p+1}{\lceil \frac{p+1}{2} \rceil}\right)$$

The following estimate was used:

$$\frac{(i+1)^{p+1} - i^{p+1}}{p+1} - i^p - \frac{p}{2} \cdot i^{p-1}$$
$$= \frac{1}{p+1} \cdot \left(\binom{p+1}{3} \cdot i^{p-2} + \binom{p+1}{4} \cdot i^{p-3} + \ldots + \binom{p+1}{p+1} \cdot i^0 \right)$$
$$\leq \frac{p-1}{p+1} \cdot \binom{p+1}{\lceil \frac{p+1}{2} \rceil} \cdot i^{p-2}$$

□

Finally, Lemma 3 evaluates the discrete moments $\mu_{p,0}(h \cdot S)$ and $\mu_{0,q}(h \cdot S)$.

Lemma 3. *The following asymptotic expressions are satisfied:*

$$\mu_{p,0}(h \cdot S) = \sum_{(i,j) \in G(h \cdot S)} i^p = \iint_{h \cdot S} x^p dx dy + \mathcal{O}\left(h^p \cdot \left(\binom{p}{\lceil \frac{p}{2} \rceil} + h^{\frac{131}{208}+\varepsilon} \right) \right)$$

$$\mu_{0,q}(h \cdot S) = \sum_{(i,j) \in G(h \cdot S)} j^q = \iint_{h \cdot S} y^q dx dy + \mathcal{O}\left(h^q \cdot \left(\binom{q}{\lceil \frac{q}{2} \rceil} + h^{\frac{131}{208}+\varepsilon} \right) \right)$$

Proof. According to (3), $\mu_{p,0}(G(h \cdot S))$ equals the number of grid points belonging to the 3D set B given by

$$B = \{(x,y,z) : (x,y) \in h \cdot S \wedge 0 < z \leq x^p\} = B' \cup B''$$

where B' and B'' are defined as follows:

$$B' = \{(x,y,z) : (x,y) \in h \cdot S \wedge 0 < z \leq \lceil h \cdot x_{min} \rceil^p\}$$
$$B'' = \{(x,y,z) : (x,y) \in h \cdot S \wedge \lceil h \cdot x_{min} \rceil^p < z \leq x^p\}$$

First, consider the number of grid points which belong to B'. It follows that

$$\#G(B') = \lceil h \cdot x_{min} \rceil^p \cdot \left(\mathcal{A}(h \cdot S) + \mathcal{O}\left(h^{\frac{131}{208}+\varepsilon} \right) \right) = \mathcal{V}(B') + \mathcal{O}\left(h^{p+\frac{131}{208}+\varepsilon} \right)$$

Now we calculate the number of grid points which belong to B''. By Definition 3 and also using the (obvious) estimate

$$\mathcal{V}(\{ (x,y,z) : (x,y) \in h \cdot S \wedge x \geq \lfloor h \cdot x_{max} \rfloor \wedge z \leq x^p \}) = \mathcal{O}(h^p)$$

we derive

$$\mathcal{V}(B'') = \sum_{i=\lceil h \cdot x_{min} \rceil}^{\lfloor h \cdot x_{max} \rfloor - 1} (\mathcal{V}(W_i) - \mathcal{V}(W_i')) + \mathcal{O}(h^p)$$

$$= \sum_{i=\lceil h \cdot x_{min} \rceil}^{\lfloor h \cdot x_{max} \rfloor - 1} \mathcal{V}(W_i) - \sum_{i=\lceil h \cdot x_{min} \rceil}^{\lfloor h \cdot x_{max} \rfloor - 1} \mathcal{V}(W_i') + \mathcal{O}(h^p)$$

$$= \sum_{i=\lceil h\cdot x_{min}\rceil}^{\lfloor h\cdot x_{max}\rfloor -1} ((i+1)^p - i^p)\cdot \mathcal{A}((h\cdot S)(i)) - \sum_{i=\lceil h\cdot x_{min}\rceil}^{\lfloor h\cdot x_{max}\rfloor -1} \mathcal{V}(W_i') + \mathcal{O}(h^p)$$

(by using Lemata 1 and 2, it follows)

$$= \sum_{i=\lceil h\cdot x_{min}\rceil}^{\lfloor h\cdot x_{max}\rfloor -1} ((i+1)^p - i^p)\cdot \left(\#G((h\cdot S)(i)) - \frac{1}{2}\cdot \#L(h\cdot S,i) + \mathcal{O}(h^{\frac{131}{208}+\varepsilon})\right)$$

$$- \sum_{i=\lceil h\cdot x_{min}\rceil}^{\lfloor h\cdot x_{max}\rfloor -1} \#L(h\cdot S,i)\left((i+1)^p - i^p - \frac{p}{2}\cdot i^{p-1}\right) + \mathcal{O}\left(\frac{h^p}{p+1}\cdot \binom{p+1}{\lceil \frac{p+1}{2}\rceil}\right)$$

$$= \sum_{i=\lceil h\cdot x_{min}\rceil}^{\lfloor h\cdot x_{max}\rfloor -1} ((i+1)^p - i^p)\cdot \left(\#G((h\cdot S)(i)) - \#L(h\cdot S,i) + \mathcal{O}\left(h^{\frac{131}{208}+\varepsilon}\right)\right)$$

$$- \sum_{i=\lceil h\cdot x_{min}\rceil}^{\lfloor h\cdot x_{max}\rfloor -1} \frac{\#L(h\cdot S,i)}{2}\cdot ((i+1)^p - i^p - p\cdot i^{p-1}) + \mathcal{O}\left(\frac{h^p}{p+1}\cdot \binom{p+1}{\lceil \frac{p+1}{2}\rceil}\right)$$

$$= \sum_{i=\lceil h\cdot x_{min}\rceil}^{\lfloor h\cdot x_{max}\rfloor -1} ((i+1)^p - i^p)\cdot (\#G((h\cdot S)(i)) - \#L(h\cdot S,i))$$

$$+\mathcal{O}\left(h^{\frac{131}{208}+\varepsilon}\cdot ((\lfloor h\cdot x_{max}\rfloor)^p \;-\; (\lceil h\cdot x_{min}\rceil)^p)\right)$$

$$- \sum_{i=\lceil h\cdot x_{min}\rceil}^{\lfloor h\cdot x_{max}\rfloor -1} \frac{1}{2}\cdot \#L(h\cdot S,i)\cdot \left(\binom{p}{2}\cdot i^{p-2} + \binom{p}{3}\cdot i^{p-3} + \ldots + \binom{p}{p}\cdot i^0\right)$$

$$+\mathcal{O}\left(\frac{1}{p+1}\cdot \binom{p+1}{\lceil \frac{p+1}{2}\rceil}\cdot h^p\right) = \#G(B'') + \mathcal{O}\left(h^p\cdot \left(\binom{p}{\lceil \frac{p}{2}\rceil} + h^{\frac{131}{208}+\varepsilon}\right)\right)$$

The following inequalities are used:

a) $$\sum_{i=\lceil h\cdot x_{min}\rceil}^{\lfloor h\cdot x_{max}\rfloor -1} \left(\binom{p}{2}\cdot i^{p-2} + \binom{p}{3}\cdot i^{p-3} + \ldots + \binom{p}{p}\cdot i^0\right) \leq$$

$$\leq \sum_{i=\lceil h\cdot x_{min}\rceil}^{\lfloor h\cdot x_{max}\rfloor -1} (p-1)\cdot \binom{p}{\lceil \frac{p}{2}\rceil}\cdot i^{p-2} = \mathcal{O}\left(\binom{p}{\lceil \frac{p}{2}\rceil}\cdot h^{p-1}\right)$$

b) for a large p: $$\binom{p}{\lceil \frac{p}{2}\rceil}\cdot h^{p-1} \leq \frac{1}{p+1}\cdot \binom{p+1}{\lceil \frac{p+1}{2}\rceil}\cdot h^{p-1}$$

Note that, if an integer i with $h\cdot x_{min} \leq i \leq h\cdot x_{max}$ is fixed, then

$$((i+1)^p - i^p)\cdot (\#G((h\cdot S)(i)) - \#L(h\cdot S,i))$$

equals the number of grid points contained in W_i, and, consequently,

$$\sum_{i=\lceil h\cdot x_{min}\rceil}^{\lfloor h\cdot x_{max}\rfloor -1} ((i+1)^p - i^p)\cdot(\#G((h\cdot S)(i)) - \#L(h\cdot S, i)))$$

equals the number of grid points contained in B''.

Finally, the sum of $\#G(B')$ and $\#G(B'')$ is the number of grid points in B. Together with the already derived expression for $\#G(B')$, we have

$$\mu_{p,0}(G(h\cdot S)) = \#G(B) = \#G(B') + \#G(B'') \;=\; \mathcal{V}(B') + \mathcal{O}\left(h^p \cdot h^{\frac{131}{208}+\varepsilon}\right)$$

$$+\mathcal{V}(B'') + \mathcal{O}\left(h^p\cdot\left(\binom{p}{\lceil\frac{p}{2}\rceil} + h^{\frac{131}{208}+\varepsilon}\right)\right) \;=\; \mathcal{V}(B) \;+$$

$$\mathcal{O}\left(h^p\cdot\left(\binom{p}{\lceil\frac{p}{2}\rceil} + h^{\frac{131}{208}+\varepsilon}\right)\right) = m_{p,0}(h\cdot S) + \mathcal{O}\left(h^p\cdot\left(\binom{p}{\lceil\frac{p}{2}\rceil} + h^{\frac{131}{208}+\varepsilon}\right)\right) \;\Box$$

4 Error Estimate if $p > 0$ and $q > 0$

It remains to estimate $\mu_{p,q}(h\cdot S)$, if $p > 0$ and $q > 0$. (The next definition and lemma are analogous to Definition 1 and Lemma 1.)

Definition 4. *For a convex set S, integers k, p, q, and a real $r > 0$, let*

$$(h\cdot S)(k,p,q) = \{(x,y)\;:\;(x,y)\in(h\cdot S)\;\wedge\;x^p\cdot y^q \geq k\}$$

$G((h\cdot S)(k,p,q))$ is the set of grid points in the digitization of $h\cdot S$ lying in the closed part of the plane determined by $x^p\cdot y^q \geq k$. Since both S and $(h\cdot S)(k,p,q)$ satisfy the preconditions of Theorem 1, we have the following lemma:

Lemma 4. *For a convex set S with a sufficiently smooth frontier, and integers r, p, q, we have*

$$\#G((h\cdot S)(p,q,k)) \;=\; \mathcal{A}((h\cdot S)(k)) \;+\; \mathcal{O}\left(h^{\frac{131}{208}+\varepsilon}\right) \tag{7}$$

Lemma 5. *Let S be a convex set with a sufficiently smooth frontier, and p, $q > 0$. Then we have the following:*

$$\mu_{p,q}(h\cdot S) = \iint_{h\cdot S} x^p\cdot y^q dxdy + \mathcal{O}\left(h^{p+q}\cdot h^{\frac{131}{208}+\varepsilon}\right) \tag{8}$$

Proof. Note that $\mu_{p,q}(h\cdot S)$ is equal to the number of grid points belonging to the 3D set E given by

$$E = \{(x,y,z)\;:\;(x,y)\in h\cdot S\;\wedge\;0 < z \leq x^p\cdot y^q\} \;=\; E'\,\cup\,E''$$

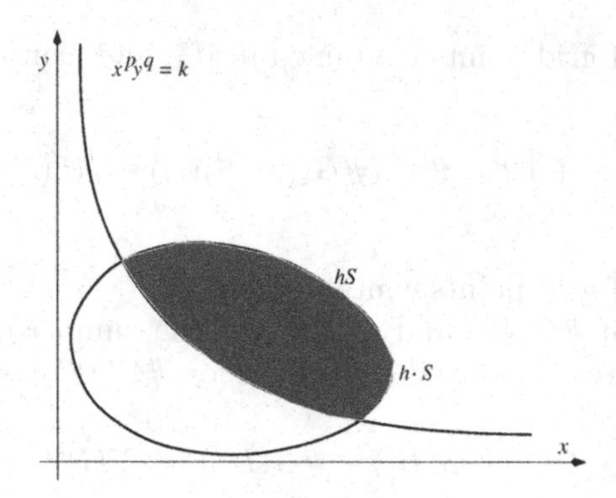

Fig. 2. The shaded area is $(h \cdot S)(k, p, q)$

where E' and E'' are defined as follows:

$$E' = \{(x, y, z) \; : \; (x, y) \in h \cdot S \; \wedge \; 0 < z < h^{p+q} \cdot z_{min}\}$$
$$E'' = \{(x, y, z) \; : \; (x, y) \in h \cdot S \; \wedge \; h^{p+q} \cdot z_{min} \leq z \leq x^p \cdot y^q\}$$

where $z_{min} = \min\{z \, : \, z = x^p \cdot y^q \; \wedge \; (x, y) \in S\}$ and $z_{max} = \max\{z \, : \, z = x^p \cdot y^q \; \wedge \; (x, y) \in S\}$.

Furthermore, from (9) we have

$$\#G(E') = (\lceil h^{p+q} \cdot z_{min} \rceil - 1) \cdot \left(\mathcal{A}(h \cdot S) + \mathcal{O}\left(h^{\frac{131}{208}+\varepsilon}\right)\right) = \mathcal{V}(E')$$
$$- \; h^{p+q} \cdot z_{min} \cdot \mathcal{A}(h \cdot S) + (\lceil h^{p+q} \cdot z_{min} \rceil - 1) \cdot \left(\mathcal{A}(h \cdot S) + \mathcal{O}\left(h^{\frac{131}{208}+\varepsilon}\right)\right)$$
$$= \; \mathcal{V}(E') + \mathcal{A}(h \cdot S) \cdot (\lceil h^{p+q} \cdot z_{min} \rceil - h^{p+q} \cdot z_{min}) + \mathcal{O}\left(h^{p+q} \cdot h^{\frac{131}{208}+\varepsilon}\right)$$

(Note that $\mathcal{A}(h \cdot S) = \mathcal{O}(h^2)$ and $p + q \geq 2$ have been used in this derivation.)

Now, let us calculate the number of grid points belonging to the set E''. What follows is a definition of 3D-sets ω_i and ω_i', for $i \in \{\lceil h^{p+q} \cdot x_{min} \rceil, \lceil h^{p+q} \cdot x_{min} \rceil + 1, \ldots, \lfloor h^{p+q} \cdot x_{max} \rfloor\}$:

$$\omega_i = \{(x, y, z) | (x, y) \in h \cdot S \; \wedge \; x^p \cdot y^q \geq i \; \wedge \; i < z < \min\{x^p \cdot y^q, i + 1\}\}$$
$$\omega_i' = \{(x, y, z) | (x, y) \in h \cdot S \; \wedge \; i < x^p \cdot y^q \leq i + 1 \; \wedge \; x^p \cdot y^q < z < i + 1\}$$

Now, we can estimate the volume of E''. By using $\mathcal{O}(h^2)$ as a trivial upper bound for the volume of

$$\{(x, y, z) \; : \; (x, y) \in h \cdot S \; \wedge \; x^p \cdot y^q \leq \lceil h^{p+q} \cdot z_{min} \rceil \; \wedge \; x^p \cdot y^q \leq z \leq \lceil h^{p+q} \cdot z_{min} \rceil\}$$

it follows that

$$\begin{aligned}
&\mathcal{V}(E'') \\
&= \sum_{i=\lceil h^{p+q}\cdot z_{min}\rceil}^{\lfloor h^{p+q}\cdot z_{max}\rfloor} \mathcal{V}(\omega_i) \;+\; \left(\lceil h^{p+q}\cdot z_{min}\rceil - h^{p+q}\cdot z_{min}\right)\cdot \mathcal{A}(h\cdot S) + \mathcal{O}(h^2) \\
&= \sum_{i=\lceil h^{p+q}\cdot z_{min}\rceil}^{\lfloor h^{p+q}\cdot z_{max}\rfloor} \left(\mathcal{A}((h\cdot S)(i,p,q)) - \mathcal{V}(\omega_i')\right) \\
&\qquad + \left(\lceil h^{p+q}\cdot z_{min}\rceil - h^{p+q}\cdot z_{min}\right)\cdot \mathcal{A}(h\cdot S) \;+\; \mathcal{O}(h^2) \\
&= \sum_{i=\lceil h^{p+q}\cdot z_{min}\rceil}^{\lfloor h^{p+q}\cdot z_{max}\rfloor} \mathcal{A}((h\cdot S)(i,p,q)) \;-\; \sum_{i=\lceil h^{p+q}\cdot z_{min}\rceil}^{\lfloor h^{p+q}\cdot z_{max}\rfloor} \mathcal{V}(\omega_i') \\
&\qquad + \left(\lceil h^{p+q}\cdot z_{min}\rceil - h^{p+q}\cdot z_{min}\right)\cdot \mathcal{A}(h\cdot S) \;+\; \mathcal{O}(h^2) \\
&\left(\text{note that } \sum_{i=\lceil h^{p+q}\cdot z_{min}\rceil}^{\lfloor h^{p+q}\cdot z_{max}\rfloor} \mathcal{V}(\omega_i') \le h^2\cdot \mathcal{A}(S) \text{ because the projections of } \omega_i' \text{ onto the xy-plane belong to } h\cdot S\right) \\
&= \sum_{i=\lceil h^{p+q}\cdot z_{min}\rceil}^{\lfloor h^{p+q}\cdot z_{max}\rfloor} \left(\#G((h\cdot S)(i,p,q)) \;+\; \mathcal{O}\left(h^{\frac{131}{208}+\varepsilon}\right)\right) \\
&\qquad + \left(\lceil h^{p+q}\cdot z_{min}\rceil - h^{p+q}\cdot z_{min}\right)\cdot \mathcal{A}(h\cdot S) + \mathcal{O}(h^2) \\
&= \#G(E'') + \left(\lceil h^{p+q}\cdot z_{min}\rceil - h^{p+q}\cdot z_{min}\right)\cdot \mathcal{A}(h\cdot S) + \mathcal{O}\left(h^{p+q+\frac{131}{208}+\varepsilon}\right).
\end{aligned}$$

Thus,

$$\#G(E'') = \mathcal{V}(E'') - \left(\lceil h^{p+q}\cdot z_{min}\rceil - h^{p+q}\cdot z_{min}\right)\cdot \mathcal{A}(h\cdot S) + \mathcal{O}\left(h^{p+q+\frac{131}{208}+\varepsilon}\right).$$

The proof of the lemma is finished by summing up $\#G(E')$ and $\#G(E'')$:

$$\begin{aligned}
&\mu_{p,q}(h\cdot S) \;=\; \#G(E') \;+\; \#G(E'') \\
&= \mathcal{V}(E') + \mathcal{A}(h\cdot S)\cdot\left(\lceil h^{p+q}\cdot z_{min}\rceil - h^{p+q}\cdot z_{min}\right) + \mathcal{O}\left(h^{p+q+\frac{131}{208}+\varepsilon}\right) \\
&\quad + \mathcal{V}(E'') - \left(\lceil h^{p+q}\cdot z_{min}\rceil - h^{p+q}\cdot z_{min}\right)\cdot \mathcal{A}(h\cdot S) + \mathcal{O}\left(h^{p+q+\frac{131}{208}+\varepsilon}\right) \\
&= \mathcal{V}(E) \;+\; \mathcal{O}\left(h^{p+q+\frac{131}{208}+\varepsilon}\right) \;=\; m_{p,q}(h\cdot S) + \mathcal{O}\left(h^{p+q+\frac{131}{208}+\varepsilon}\right). \qquad \square
\end{aligned}$$

Our theorem summarizes the accuracy in estimating real moments of an arbitrary order based on digitized sets.

Theorem 2. *Let S be a convex set that satisfies the preconditions of Theorem 1. Then we have the following:*

$$\left| m_{p,q}(S) - \frac{\mu_{p,q}(h \cdot S)}{h^{p+q+2}} \right| = \begin{cases} \mathcal{O}\left(h^{-\frac{285}{208}+\varepsilon} + \frac{1}{h^2} \cdot \binom{p}{\lceil \frac{p}{2} \rceil}\right) & \text{for } p = 0 \text{ or } q = 0 \\ \mathcal{O}\left(h^{-\frac{285}{208}+\varepsilon}\right) & \text{for } p > 0 \text{ and } q > 0. \end{cases}$$

Stirling's formula gives $\binom{p}{\lceil \frac{p}{2} \rceil} = \mathcal{O}(2^n)$ and implies the following:

Corollary 2. *Let S be a convex set with sufficiently smooth frontier, and let $p + q = o(\log h)$. Then we have the following:*

$$\left| m_{p,q}(S) - \frac{\mu_{p,q}(h \cdot S)}{h^{p+q+2}} \right| = \mathcal{O}\left(h^{-\frac{285}{208}+\varepsilon}\right) \qquad \text{for any} \quad \varepsilon > 0.$$

Corollary 2 shows that the error in approximating $m_{p,q}(S) \approx \frac{\mu_{p,q}(h \cdot S)}{h^{p+q+2}}$ can be reduced to any fraction of the pixel size (what is $1/h$) if a moment's order $p+q$ is not to large compared to the applied picture resolution. The assumed relation $p+q = o(\log h)$ is reasonable for practical applications. In such a case, Corollary 2 gives a better estimate than the estimate $\frac{1}{c^{o(\log h)} \cdot o(h \cdot \log h)}$ ($c > 0$ is computable from x_{max} and y_{max}) that follows from (5) (i.e., from Davenport's result).

References

1. Hu, M.: Visual pattern recognition by moment invariants. IRE Trans. Inf. Theory. **8** (1962) 179-187
2. Dudani, S. A., Breeding, K. J., McGhee, R. B.: Aircraft identification by moment invariants. IEEE Trans. Comp. **26** (1977) 39-46
3. Jain, R., Kasturi, R., Schunck B. G.: Machine Vision. McGraw-Hill, New York (1995)
4. Pei, S.-C., Liou, L.-G.: Using moments to acquire the motion parameters of a deformable object without correspondence. Image Vision Computing. **12** (1994) 475-485
5. Tsai, W. H., Chou, S. L.: Detection of generalized principal axes in rotationally symmetric shapes. Pattern Recognition. **24** (1991) 95-104
6. Klette, R., Rosenfeld, A.: Digital Geometry. Morgan Kaufmann, San Francisco (2004)
7. Klette, R., Žunić, J.: Multigrid convergence of calculated features in image analysis. J. Mathematical Imaging Vision. **13** (2000) 173-191
8. Davenport, H.: On a principle of Lipschitz. J. London Math. Soc. **26** (1951) 179-183
9. Huxley, M. N.: Exponential sums and lattice points III. Proc. London Math. Soc. **87** (2003) 591-609
10. Jiang, X. Y., Bunke, H.: Simple and fast computation of moments. Pattern Recognition. **24** (1991) 801-806
11. Leu, J.-G.: Computing a shape's moments from its frontier. Pattern Recognition. **24** (1991) 949-957
12. Singer, M. H.: A general approach to moment calculation for polygons and line segments. Pattern Recognition. **26** (1993) 1019-1028

Feature Based Defuzzification in $\mathbb{Z}^2$ and $\mathbb{Z}^3$ Using a Scale Space Approach

Joakim Lindblad[1], Nataša Sladoje[2,*], and Tibor Lukić[2]

[1] Centre for Image Analysis, Swedish University of Agricultural Sciences, Uppsala, Sweden
joakim@cb.uu.se
[2] Faculty of Engineering, University of Novi Sad, Novi Sad, Serbia
{sladoje, tibor}@uns.ns.ac.yu

Abstract. A defuzzification method based on feature distance minimization is further improved by incorporating into the distance function feature values measured on object representations at different scales. It is noticed that such an approach can improve defuzzification results by better preserving the properties of a fuzzy set; area preservation at scales in-between local (pixel-size) and global (the whole object) provides that characteristics of the fuzzy object are more appropriately exhibited in the defuzzification. For the purpose of comparing sets of different resolution, we propose a feature vector representation of a (fuzzy and crisp) set, utilizing a resolution pyramid. The distance measure is accordingly adjusted. The defuzzification method is extended to the 3D case. Illustrative examples are given.

1 Introduction

The advantages of representing objects in images as fuzzy spatial sets are numerous and have lead to increased interest for fuzzy approaches in image analysis [1]. Preservation of fuzziness by utilizing fuzzy segmented images implies preservation of important information about objects. However, a crisp representation of objects in the images may still be needed. Reasons for that are, e.g., to facilitate easier visualization and interpretation. Even though it contains less information, a crisp representation is often easier to interpret and understand, especially if the spatial dimensionality of the image is higher than two. Moreover, analogues for many tools available for the analysis of binary images are still not developed for fuzzy images. This may force us to perform at least some steps in the analysis process by using a crisp representation of the objects.

In our previous work related to defuzzification, i.e., the process of generating a crisp representation of a fuzzy digital object, we introduced a distance

* The author is financially supported by the Ministry of Science of the Republic of Serbia through the Project ON144029 of the Mathematical Institute of the Serbian Academy of Science and Arts.

A. Kuba, L.G. Nyúl, and K. Palágyi (Eds.): DGCI 2006, LNCS 4245, pp. 379–390, 2006.

measure between fuzzy sets incorporating a number of quantitative global and local features of the sets [2]. We have suggested to perform defuzzification by choosing the crisp representation that is closest to the given fuzzy set in terms of the proposed distance measure. The resulting crisp set can be generated at higher spatial resolution, compared to the spatial resolution of the fuzzy object, as shown in [3]. In that way, a (crisp) segmentation technique that provides crisp objects represented at a higher spatial resolution than the given image resolution has been proposed.

In this paper, we consider matching additional features in the defuzzification. We have noticed that the lack of requirement for feature preservation at meso-scale, i.e., scales in-between the local (pixel-size) and the global (the whole object) scale, may lead to rather inappropriate defuzzification solutions in spite of successful matching of both local and global features in the optimization algorithm. Therefore, we introduce meso-scale area components in the feature representation of a set to be a subject of optimization, in addition to already existing membership values of pixels, seen as local area components, and the global area of a set. Global features, perimeter and centroid, are considered in defuzzification, as it is suggested in [2]. Global features are important for successful defuzzification at increased spatial resolution, [3]. To facilitate the use of meso-scale area components (their calculation and updating during the optimization process), we utilize representations of the sets at a range of spatial resolution, i.e., we generate a resolution pyramid representations. We implement the defuzzification method in both the 2D and the 3D case.

The paper is organized as follows: Section 2 gives an overview of existing results related to the proposed method and lists the main definitions used in the paper. In Section 3, the main contribution of the paper, defuzzification by minimizing feature distance using a scale space approach, is presented. Examples of defuzzification of both synthetic and real 2D and 3D images are given in Section 4, and the positive effect of the suggested scale space approach to defuzzification is illustrated. Concluding remarks are given in Section 5.

2 Background

We give a list of definitions and notions used in the paper and present existing results related to defuzzification. In this paper, we extend our own work on defuzzification based on feature distance minimization, proposed originally in [2]; therefore, most of related work is referenced to our own results. Moreover, Section 2.3 particularly recalls the necessary framework derived in [2] and [3].

2.1 Definitions

A *fuzzy set* S on a reference set X is a set of ordered pairs $S = \{(x, \mu_S(x)) \mid x \in X\}$, where $\mu_S : X \to [0,1]$ is the *membership function* of S in X. We denote by $\mathcal{F}(X)$ the set of fuzzy sets on a reference set X and by $\mathcal{P}(X)$ the set of crisp subsets of a set (the power set).

Being interested in applications in digital image analysis, we consider digital fuzzy sets, where $X \subset \mathbb{Z}^n$. In addition, when using digital approaches (computers) to represent, store, and analyse images, the (finite) number, $\ell + 1$, of grey-levels available is a natural limitation to the number of membership values that can be assigned to a digital point.

An *α-cut* of a fuzzy set S, for $\alpha \in (0, 1]$, is the set $S_\alpha = \{x \in X \mid \mu_S(x) \geq \alpha\}$. The *support* of a fuzzy set S is the set $\text{Supp}(S) = \{x \in X \mid \mu_S(x) > 0\}$. The *core* of a fuzzy set S is the set $\text{Core}(S) = \{x \in X \mid \mu_S(x) = 1\}$. The *fuzzification principle*, based on the following equation:

$$f(S) = \int_0^1 \hat{f}(S_\alpha)\, d\alpha, \tag{1}$$

can be used to generalize properties $\hat{f}$ defined for crisp sets (here, α-cuts) to fuzzy sets. In order to generalize a function $\hat{f}$, defined on discrete crisp sets, the equation

$$f(S) = \frac{1}{\ell} \sum_{\alpha=1}^{\ell} \hat{f}(S_\alpha)\,, \tag{2}$$

can be used. In this paper, we use Equation (2) to define *perimeter* $P(S)$, and *surface area* $Surf(S)$, of a (2D and 3D) fuzzy set S, respectively (more detailed definition and properties are given in [4]). Equation (2) is also used to define *moments of zero and first order* of a discrete spatial fuzzy set S, denoted by $m_{p,q}(S)$, where for integers p, q it holds $p + q \leq 1$ (in 2D case), and $m_{p,q,r}(S)$, where for integers p, q, r it holds $p + q + r \leq 1$ (in 3D case) (more detailed definitions and properties are described in [5]).

The zero-order moment of a set S is equal to its area $A(S)$ (if $S \in \mathcal{F}(Z^2)$), or its volume $V(S)$ (if $S \in \mathcal{F}(Z^3)$). The centroid of a set S is also defined by the moments of a set S, [6].

2.2 Related Work

Defuzzification is the process of replacing a fuzzy set with an appropriately chosen crisp set. In image analysis, fuzzification of an image is a consequence of the combination of properties of the continuous original, discretization effects, and imaging conditions. Defuzzification should be performed by utilizing the fuzzy representation as a source of valuable information about geometric properties of the object of interest (fuzzy, or crisp), and by defining and following some criteria related to properties which characterize a satisfactory defuzzification result.

In our previous work [2], we present a defuzzification method based on minimizing the feature distance between a fuzzy set and its defuzzification. Feature distance is defined so that the distance between two sets is expressed in terms of the distance between their feature-based vector representations in some defined feature space. A selection of local and global numerical features can be included in the distance measure and considered in defuzzification. By using some optimization procedure, the crisp object fulfilling the minimization criterion is generated.

The results presented in [5,4] show that the precision of estimates for perimeter, area, and higher order moments of a continuous shape, is significantly higher if a fuzzy discrete shape representation, where the membership of a pixel is proportional to the part of its area covered by the observed object, is used instead of a crisp discrete representation. By including these estimates in the feature based representation of the fuzzy set, we generate a crisp digital object which is a good crisp representation of a fuzzy set and, at the same time, highly resembles the original continuous object. Furthermore, it is shown in [5,4], either theoretically or through statistical studies, that a fuzzy approach can provide an alternative to increasing the spatial resolution of the image. This observation motivated the study presented in [3], where we suggested to generate a crisp shape representation of a given fuzzy object at an r times increased spatial resolution. The features, estimated with a high precision from the fuzzy representation, are preserved in the defuzzification by minimizing the corresponding feature distance between the sets. We let each pixel in the fuzzy low resolution representation correspond to a block of $r \times r$ pixels in the crisp high resolution representation. Local (point-size) features of a fuzzy set are compared to the features derived from a block in the defuzzified set. This correspondence is used as a basis for the high-resolution object reconstruction.

2.3 Defuzzification by Feature Distance Minimization

The feature distance measure and the defuzzification method based on its minimization, proposed in [2], are further explored in this paper. The following notation and definitions, introduced in [2], are used in the sequel:

Defuzzification. An optimal defuzzification $\mathcal{D}(A)$ of a fuzzy set $A \in \mathcal{F}(X)$, with respect to the distance d, is

$$\mathcal{D}(A) \in \{C \in \mathcal{P}(X) \mid d(A,C) = \min_{B \in \mathcal{P}(X)} [d(A,B)]\} . \tag{3}$$

Distance Measure. Given an injective function Φ from $\mathcal{F}(X)$ to a metric space H, we define a metric on $\mathcal{F}(X)$ by requiring that Φ is an isometry. That is, the *feature distance* between fuzzy sets A and B is

$$d^{\Phi}(A,B) = d(\Phi(A), \Phi(B)) . \tag{4}$$

The vector $\Phi(A) \in H$ is understood as a feature representation of the set A.

We use $H \subset \mathbb{R}^n$, with the Minkowski distance of order $p = 1$ as our main choice of distance functions used in H, providing a corresponding feature distance d^{Φ} in $\mathcal{F}(X)$.

By suitably designing the mapping Φ, the distance measure can be tuned to provide defuzzifications where both shape characteristics and membership values are taken into account. This enables defuzzification that fits the individual problem well, and provides a powerful family of defuzzification methods. In this paper, we further explore the appropriate choices of features incorporated in Φ, in order to improve the preservation of relevant characteristics of the fuzzy set.

Optimization. In general, Equation (3) cannot be solved analytically. In addition, the search space $\mathcal{P}(X)$ is too big to be exhaustively traversed. As a consequence, we are forced to rely on heuristic search methods. In [2], two methods, floating search and simulated annealing, are used to find an approximate solution for Equation (3). Simulated annealing, starting from the α-cut closest to the given fuzzy set in terms of the considered distance, showed the most satisfactory behaviour, and is in this paper used as the optimization method of choice. Since the optimization task is a well separated problem, many other search methods can be used to approximatively solve Equation (3).

3 Feature Based Defuzzification Using a Scale Space Approach

In this section, we present our main contribution, a scale space approach to high-resolution defuzzification based on feature distance minimization using meso-scale features. We give a motivation for using such an approach, and we describe the suggested method, both in the 2D and the 3D case.

3.1 Motivation

Characteristics of a given fuzzy set often have different importance at different scales. When the feature distance is optimized so that point-wise and global features are preserved as well as possible in the defuzzification, it is still possible that meso-scale features are not matched sufficiently well, even though the achieved distance is satisfactory low. The resulting crisp set may, e.g., have area (number of points) perfectly well matched with the area of the fuzzy set, while the pixels are not distributed in an intuitive way over the crisp set. A synthetic example illustrating this is presented in Figure 1. An object, Figure 1(a), is composed of four discrete fuzzy disks, where all non-zero membership values of points are equal to 0.5. Such a (homogeneous) distribution of local features does not provide any information about preferable local distribution of pixels in a defuzzification. A globally optimal solution, Figure 1(b), appears rather non-intuitive, in spite of its well matched features; it is more appealing that defuzzification of individual parts of an image resembles (as much as possible) the individual parts of an image defuzzified as a whole. We conclude that a main problem is that the method may "transport" area from one part of the image to another part.

This observation leads us to investigate the incorporation of features calculated over a range of scales into the distance measure and the defuzzification. Use of a scale space approach provides an appropriate treatment of details in images, where the details are usually relevant only in some range of scales.

An alternative approach would be to include additional higher order moments, to provide a more complete description from a global level. We consider the scale space approach to be more appealing in terms of generality, and also probably in terms of robustness, since higher order moments are in general sensitive to noise.

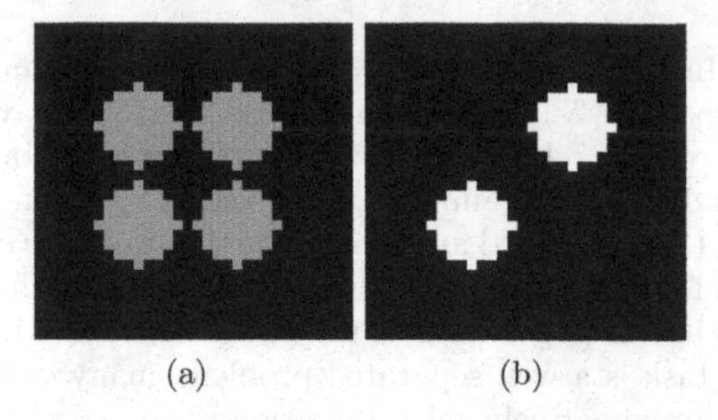

Fig. 1. (a) Synthetic image. (b) Defuzzification of (a) based on (local and global) feature distance minimization, as proposed in [2].

A good way to introduce scale dependent defuzzification is to use a resolution pyramid. Representations at different levels in the resolution pyramid correspond to objects made by mapping blocks of pixels (voxels) of an image at a given resolution into one pixel (voxel) in the image at some other (lower) resolution. By comparing the features of interest at corresponding resolution levels, blocks in the fuzzy and the defuzzified image are compared and the features are thereby considered, not only at the local and the global scale, but also at in between meso-scales. In this paper, our choice of feature to observe at meso-scale is area (volume), corresponding to membership values of pixels (voxels) in the representations in the resolution pyramid.

3.2 Scale Space Defuzzification of 2D Fuzzy Sets

For a given fuzzy set $F \in \mathcal{F}(X)$ of size $2^m \times 2^m$ pixels, we generate $m+1$ partitions of the set into square blocks of $2^{m-i} \times 2^{m-i}$ neighbouring pixels, for $i = 0, \ldots, m$. Each partition i consists of 2^{2i} blocks. (If the original image is not of size $2^m \times 2^m$ pixels, we pad it with zeros.) We use a feature representation $\Phi(F)$ consisting of the areas of all the blocks of all the partitions. Obviously, the membership values of all the pixels are included in such a representation, being local areas of one-pixel-size blocks ($i = m$), while the global area of the set is included as the area of the single block of the size $2^m \times 2^m$ (for $i = 0$). In addition, the perimeter of the set F, as well as the coordinates of its centroid, are included in the feature representation.

Weighting of Features. In order to provide that the effect of the total contribution of all measures of one (type of) feature, observed at one particular scale, is approximately the same size as the effect of one global feature, features of multiplicity h are scaled with $\frac{1}{\sqrt[p]{h}}$, where p is the exponent of the Minkowski distance in Equation (4) (we use $p = 1$ throughout this paper).

To compare features calculated at different scales, measures also have to be rescaled with respect to the spatial resolution of the image and the dimensionality of the particular feature. It is taken into account that $P(S) = \mathcal{O}(r_S)$, $A(S) = \mathcal{O}(r_S^2)$, $m_{1,0}(S) = \mathcal{O}(r_S^3)$, $m_{0,1}(S) = \mathcal{O}(r_S^3)$ for a set S inscribed into a grid with spatial resolution r_S. To get resolution invariant global features, we use

$$\tilde{P}(S) = \frac{P(S)}{P(X)}, \quad \tilde{A}(S) = \frac{A(S)}{A(X)}, \quad \tilde{C}_x(S) = \frac{C_x(S)}{C_x(X)}, \quad \tilde{C}_y(S) = \frac{C_y(S)}{C_y(X)},$$

for a fuzzy set $S \in \mathcal{F}(X)$. In this way, it is provided that $\tilde{P}(S) = \mathcal{O}(1)$, $\tilde{A}(S) = \mathcal{O}(1)$, $\tilde{C}_x(S) = \mathcal{O}(1)$, $\tilde{C}_y(S) = \mathcal{O}(1)$ for any grid resolution.

Feature Vector Representation. For a given fuzzy (or crisp) set S of size $2^n \times 2^n$ pixels, let B_j^i represent the jth block of $2^{n-i} \times 2^{n-i}$ pixels, where $j = 1, \dots, 2^{2i}$, $i = 0, \dots, n$. Block B_1^0 is equal to the set S and, correspondingly, $\tilde{A}(S) = \tilde{A}(B_1^0)$. The feature representation $\Phi_m(S)$ of S, for $m \leq n$, is then

$$\begin{aligned} \Phi_m(S) = \Bigg(& \frac{1}{\sqrt[p]{2^{2m}}}\tilde{A}(B_1^m), \dots, \frac{1}{\sqrt[p]{2^{2m}}}\tilde{A}(B_{2^{2m}}^m), \\ & \frac{1}{\sqrt[p]{2^{2(m-1)}}}\tilde{A}(B_1^{m-1}), \dots, \frac{1}{\sqrt[p]{2^{2(m-1)}}}\tilde{A}(B_{2^{2(m-1)}}^{m-1}), \\ & \dots \\ & \frac{1}{\sqrt[p]{2^0}}\tilde{A}(B_1^0), \\ & \tilde{P}(S), \tilde{C}_x(S), \tilde{C}_y(S) \Bigg) . \end{aligned} \tag{5}$$

Resolution Pyramids. We use two resolution pyramids for storing the areas of the blocks B_j^i of the fuzzy original set, and of the crisp defuzzification. Pyramids are built by grouping 2×2 neighbouring (*children*) pixels in the image at the current resolution level, and create one (*parent*) pixel at the next, lower, resolution level, where the value of the parent pixel is assigned to be the sum of the values of the children pixels. The process is repeated at every newly created resolution level, until the lowest possible resolution. The value assigned to the single element at the lowest level in the pyramid is the area of the starting image. To obtain rescaled areas of blocks, used in the feature representation $\Phi(S)$, each value (area of a block) is divided by the number of elements in the block.

Defuzzification. For a given fuzzy set F, containing $2^m \times 2^m$ pixels, a resolution pyramid representation with $m+1$ resolution levels is build. The α-cut of F at minimal distance d^Φ to F is used as the starting configuration for defuzzification. In order to obtain the initial configuration K at 2^r times increased resolution, each pixel in the α-cut is subdivided into 2^{2r} sub-pixels. A resolution pyramid for the crisp set K, with $m+r+1$ resolution levels, is created and defuzzification is performed by minimizing the feature distance

$$d^\Phi(F, K) = d(\Phi_m(F), \Phi_m(K)) , \tag{6}$$

where d is the Minkowski distance with $p = 1$.

During the search process, when changing one pixel in the crisp set K, all levels of the pyramid representations of K are locally updated.

Simulated Annealing. The simulated annealing search ([2]) was refined by applying a re-annealing scheme where the search was restarted 20 times. An automatic tuning of the temperature was achieved by restarting each annealing at twice the temperature where the current best solution of the previous annealing was accepted; did the previous annealing not provide any improvement, the new temperature was set to be twice the staring temperature of the previous one. Within one annealing, 5 000 random perturbations where tried at each temperature level, before the temperature was lowered so that $T_{new} = 0.995\,T_{old}$. The temperature was successively reduced until 50 000 successive perturbations did not manage to provide any step that gave a reduction in distance, after which a new re-annealing was restarted from the currently best found solution.

3.3 Scale Space Defuzzification of 3D Fuzzy Sets

The defuzzification method, suggested for 2D discrete spatial fuzzy sets is straightforwardly generalized to the 3D case. The features selected to be included in the feature distance are local, meso-scale, and global volumes, obtained by iterative grouping of blocks of $2 \times 2 \times 2$ voxels, and surface area and centroid, as additional global features.

Once when the feature representation is generated, the defuzzification process is exactly the same as in the 2D case.

4 Examples

We show three examples of defuzzification using the suggested method. We test its behaviour on the synthetic image, Figure 1(a) and we show two examples of defuzzification of parts of real 3D (medical) images.

4.1 Four Disks

The example presented in Figure 1 is repeated in Figure 2. The result of defuzzification of the object in Figure 2(a), using the proposed scale space approach, is shown in Figure 2(c). Even though the global features are perfectly matched in the solution presented in Figure 2(b) (obtained without meso-scale features), we consider the solution in Figure 2(c) to better preserve the properties of the original set. The contributions of the different features to the overall distance are given in Table 1.

4.2 Bone

An example of defuzzification of a 3D object is presented in Figure 3. The data volume here is a CT image of a bone implant (inserted in a leg of a rabbit). We applied the method to a part of the image ($51 \times 44 \times 59$ voxels) (Figure 3(a) shows a slice through the volume) containing a connected piece of bone area (dark grey), surrounded by a non-bone area (light grey). Figure 3(b) shows a slice through a 3D fuzzy set representing the bone region.

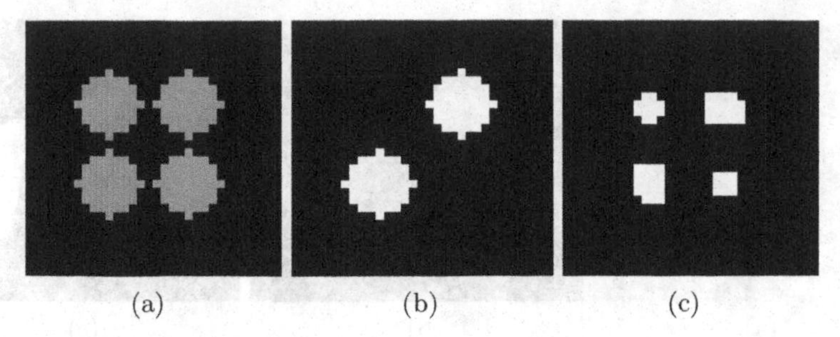

Fig. 2. **(a)** Four discrete disks of radius 4 and membership 0.5. **(b)** Optimal defuzzification using feature distance without meso-scale area components. **(c)** Defuzzification using feature distance including meso-scale area components.

Table 1. The contribution of the different features to the feature distance, and the total distance, without (Dist 1), and with (Dist 2), the meso-scale area features

Figure	Perimeter	Area	Centroid	Membership	Meso-scale	Dist 1	Dist 2
2(b)	0.0000	0.0000	0.0000	0.0957	0.3828	0.0957	0.4785
2(c)	0.0015	0.0381	0.0000	0.0957	0.1758	0.1353	0.3111

All features are matched well in this example; there are no large regions of high fuzziness, and the global features do not provide any reason for "transportation" of volume as in the example in Section 4.1. Defuzzifications with or without meso-scale features are therefore practically identical.

4.3 Vessels

Fuzzy representations of image objects are especially useful when the spatial resolution is too low to provide a good crisp representation. One such situation can be seen in Figure 4(b), which displays a maximum intensity projection of a part of a rotational b-plane x-ray scan of the arteries of the right half of a human head (provided by Philips Research, Hamburg, Germany), shown in Figure 4(a). A contrast agent is injected into the blood and an aneurism is shown to be present. The intensity values of the image voxels correspond fairly well with partial volume coverage, and are therefore used directly as fuzzy membership values.

This example image violates the sampling theorem; the vessels imaged are not resolved since they are smaller than one voxel thick. This fact causes a number of problems related to information extraction. Using a priori knowledge about the image, it is still possible to obtain a reasonable defuzzification. One such a priori piece of information is the knowledge that the vessel tree is simply connected. Starting from one simply connected component, and preserving topology ([7]) throughout the search, it is provided that the defuzzification is also simply connected.

Centroid position is not an intuitive feature to use for defuzzification of a vessel tree. It may interfere in undesirable ways with the topology preservation during the search procedure, so we exclude centroid from the feature representation in this example.

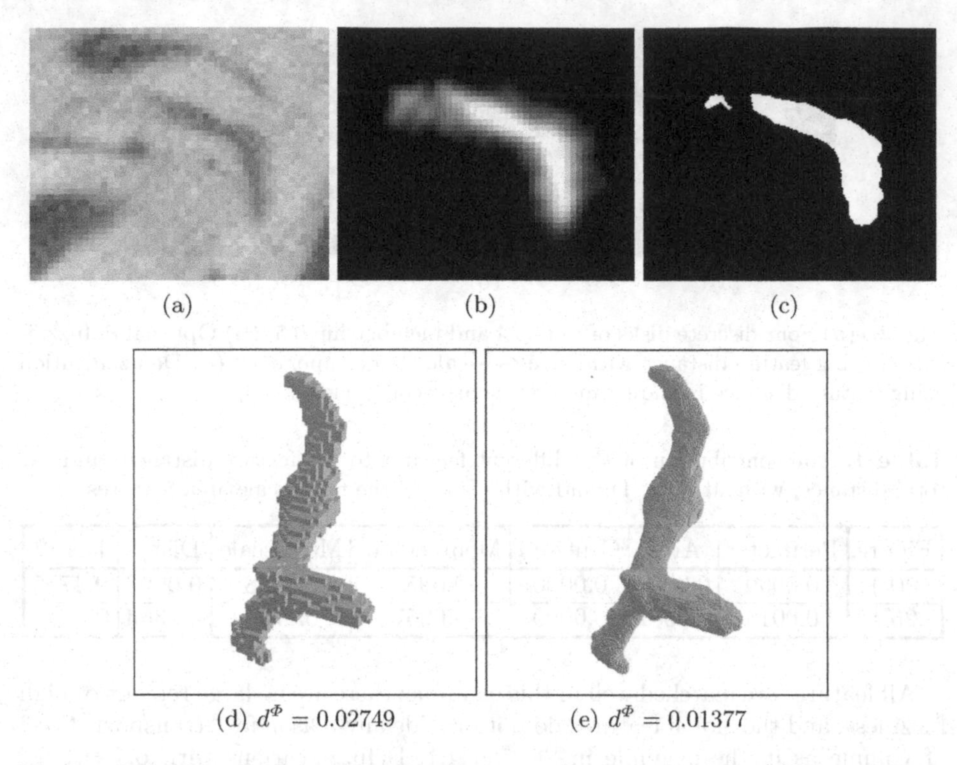

Fig. 3. Defuzzification of a part of a 3D image of a bone implant. (a) Slice through the image volume. The dark grey area is bone, the light parts are non-bone areas. (b) Slice through a fuzzy segmentation of the bone region in the image volume. (c) Slice through a defuzzification, using meso-scale volume features, of the fuzzy segmented image volume. (d) 3D rendering of the α-cut at smallest feature distance to the fuzzy object. (e) 3D rendering of a high resolution defuzzification of the fuzzy segmented object. A four times scaled up version of the best α-cut (d) was used as starting set for the simulated annealing search.

It is clear that high resolution reconstruction is really needed here; any crisp representative at the same resolution as the original image would be a rather bad representation; to preserve the volume of the fuzzy image, many parts of the vessel would not be included in the crisp set.

Performing defuzzification at two times the original resolution, we get the result presented in Figure 4(c). The result is not visually appealing, due to severe under-estimation of the surface area of a crisp thin (less than one voxel thick) structure by the surface area of the fuzzy set. This problem is not present for a crisp object whose fuzzy representation is obtained at sufficiently high resolution and contains points with memberships equal to one in the interior of the object. In the case presented in Figure 4, however, the defuzzification using the inaccurate surface area estimate fails to preserve the vessel structure.

It would be of high interest to have a better surface area estimate for the defuzzification. In the absence of such, we attempt defuzzification without

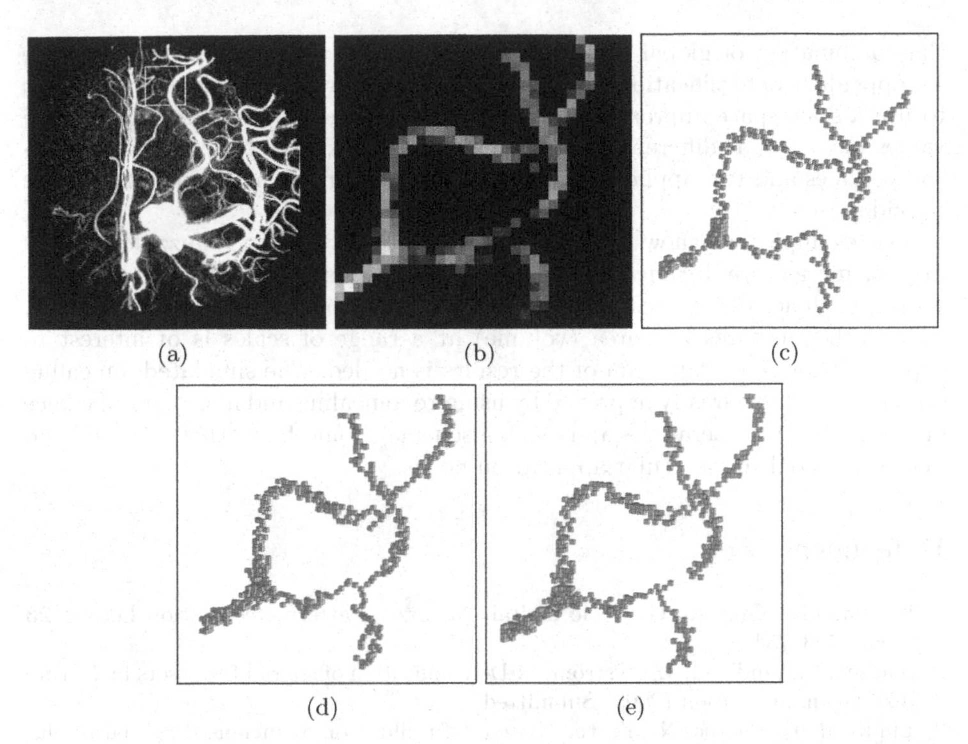

Fig. 4. Defuzzification of a selected part of an angiography 3D image, showing an X-ray scan of the arteries of the right half of a human head. (a) Maximum intensity projection through the image volume; the white square in the upper part of the image indicates the location of the selected part of the volume that is defuzzified in this example. (b) Maximum intensity projection through the selected part the volume. (c) 3D rendering of a defuzzification at twice the resolution using volumes of all scales and surface area. (d) 3D rendering of a defuzzification at twice the resolution using only volumes of all scale. (e) 3D rendering of a defuzzification at twice the resolution using only global and local volumes.

surface area feature. Using only volume based information (at a range of scales) the high resolution reconstruction is fairly unconstrained, which leads to the rather jagged result of Figure 4(d). Dropping the meso-scale feature from the feature representation, we get the result presented in Figure 4(e).

We note that, although not visible in Figure 4, the topology is in deed preserved; all the resulting objects are simply connected. However, the vessels are not always connected in a correct way, so some additional information on how vessels branch and bend may be required in this case.

5 Summary

We have presented an improvement to previously presented work on defuzzification by feature distance minimization. On a synthetic example we show that the combination of global and local features is not always enough to provide an appealing defuzzification. To overcome the described problem, we suggest to use a scale space approach, incorporating into the distance function feature values measured at different scales. The method is extended to work on 3D image volumes and two applications of the method on medical image volumes are provided.

The examples we show indicate possible future work on defuzzification by minimizing feature distance. Better feature estimates, and optionally more appropriate choice of features for thin elongated (tree-like) structures are needed. Using other features but area (volume) at a range of scales is of interest to explore. Thorough evaluation of the results is needed. The simulated annealing search algorithm, already improved by using re-annealing and more careful choice of parameters (temperatures and cooling scheme), could be further explored and better adjusted to particular application needs.

References

1. Udupa, J.K., Grevera, G.J.: Go digital, go fuzzy. Pattern Recognition Letters **23** (2002) 743–754
2. Sladoje, N., Lindblad, J., Nyström, I.: Defuzzification of spatial fuzzy sets by feature distance minimization (2005) Submitted.
3. Lindblad, J., Sladoje, N.: Feature based defuzzification at increased spatial resolution. In Reulke, R., Eckardt, U., Flach, B., Knauer, U., Polthier, K., (eds.): Proc. of 11th International Workshop on Combinatorial Image Analysis (IWCIA 2006). Volume 4040 of LNCS., Berlin, Germany, Springer-Verlag (2006) 131–143
4. Sladoje, N., Nyström, I., Saha, P.: Measurements of digitized objects with fuzzy borders in 2D and 3D. Image and Vision Computing **23** (2005) 123–132
5. Sladoje, N., Lindblad, J.: Estimation of moments of digitized objects with fuzzy borders. In Roli, F., Vitulano, S., (eds.): Proc. of 13th International Conference on Image Analysis and Processing (ICIAP 2005). Volume 3617 of LNCS., Cagliari, Italy, Springer-Verlag (2005) 188–195
6. Zadeh, L.: Fuzzy sets. Information and Control **8** (1965) 338–353
7. Borgefors, G., Nyström, I., Sanniti di Baja, G.: Connected components in 3D neighbourhoods. In: Proc. of 10th Scandinavian Conference on Image Analysis (SCIA 1997), Pattern Recognition Society of Finland (1997) 567–572

Improving Difference Operators by Local Feature Detection

Kristof Teelen and Peter Veelaert

University College Ghent - Ghent University Association,
Schoonmeersstraat 52, B9000 Ghent, Belgium
{Kristof.Teelen, Peter.Veelaert}@hogent.be

Abstract. Differential operators are required to compute several characteristics for continuous surfaces, as e.g. tangents, curvature, flatness, shape descriptors. We propose to replace differential operators by the combined action of sets of feature detectors and locally adapted difference operators. A set of simple local feature detectors is used to find the fitting function which locally yields the best approximation for the digitized image surface. For each class of fitting functions, we determine which difference operator locally yields the best result in comparison to the differential operator. Both the set of feature detectors and the difference operator for a function class have a rigid mathematical structure, which can be described by Groebner bases. In this paper we describe how to obtain discrete approximates for the Laplacian differential operator and how these difference operators improve the performance of the Laplacian of Gaussian edge detector.

1 Introduction

A delicate and often reoccurring problem in digital image processing is the application of operators from differential geometry to digital representations of curves and surfaces. For continuous surfaces well defined differential operators can be used to compute standard functions as e.g. curvatures, tangent planes, normals, shape operators. These differentials cannot be applied directly to digitized surfaces or digitized curves. Consider as a simple example how to compute the tangent in the point x_0 for the digitized curve f shown in Figure 1. For a continuous curve we would simply calculate the first derivative dy/dx. But for the digitized function, the solution is less obvious.

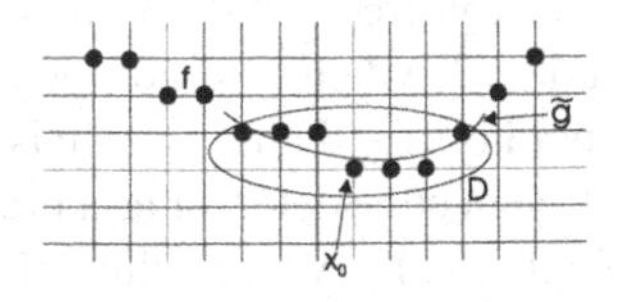

Fig. 1. Finding a tangent to a digitized curve

One could use a continuous fitting function $\tilde{g}$ to approximate the digitized function f in some neighborhood D of x_0, and then apply the differential operator to that continuous curve. We propose to digitize the differential operator, i.e. to replace it by a

A. Kuba, L.G. Nyúl, and K. Palágyi (Eds.): DGCI 2006, LNCS 4245, pp. 391–402, 2006.

difference operator that can be applied without problem to digitized as well as continuous functions. This approach also involves the choice of a class of fitting functions, since the selection of the difference operator involves the implicit assumption that there are some continuous functions for which the difference operator yields the correct result. Haralick [1], Langridge [2], Fleck [3], Karabassis and Spetsakis [4] all discuss possible methods to find the best possible fitting functions and difference operators to compute derivatives for a given application.

Digitizing differential operators requires two important choices. The first choice considers which class of fitting functions to use. This choice is not unique, e.g. if we choose linear fitting functions, then all second and higher order differences will vanish. Second, we must choose an appropriate difference operator for each particular class of fitting functions and also this choice is not unique. If we choose a linear fitting function $\tilde{g}(x) = ax + b$, and if we let σ denote the shift operator defined by $\sigma^j \tilde{g}(x) = \tilde{g}(x + j)$, then we can replace the differential d/dx by one of the three difference operators $\Delta = (\sigma^1 - 1)$, $(\sigma^2 - 1)/2$, or $(\sigma^3 - 1)/3$. These operators will all yield the same result when applied to $\tilde{g}(x)$: $d\tilde{g}/dx = \Delta\tilde{g} = a$. Hence there is no straightforward choice for the fitting functions and the difference operator.

In this paper, we show how to choose the appropriate difference operator for each class of fitting functions. We actually avoid the problem of computing the fitting functions by verifying whether the digitized function has the right features to be categorized in one of the function classes. Moreover, feature detection arises in a natural way if we want to compute differentials for digitized functions. The computation of the difference operator and the feature detectors for the function classes both fit in a rigid mathematical framework [5].

We propose a method to choose the most appropriate difference operator with a decision tree, considering the local features of the image. The framework of the method is illustrated by the computation of a digitized difference operator for the Laplacian. Lachaud *et al* [6] discuss how to estimate the tangent of a digital curve. Lindeberg [7] discusses how to define discrete derivative approximations for the computations of multi-scale low-level feature extraction, and their use in edge detection. Gunn [8] and Demigny *et al* [9] consider discrete versions of edge detection algorithms. Lowe [10] uses an approximation for the Laplacian of Gaussian to detect stable keypoints. In this paper we show how the Laplacian of Gaussian edge detector can benefit from locally applying the appropriate version of the Laplacian difference operator.

In section 2, we show how to digitize the differential operator for different classes of fitting functions. Next, a decision tree for the practical computation of difference operators is introduced in Section 3, where the computation of the feature detection templates and the difference operators for the Laplacian is given as an example. Section 4 presents the Laplacian of Gaussian edge detector as a practical application to illustrate the improvements by our method. Finally, we conclude this paper in Section 5.

2 Digitizing Differential Operators

When choosing a class of fitting functions for a digitized function, we choose in fact a class of features. We use features to examine how the digitized function behaves locally, that is to verify how smooth the function is. Next, we choose one particular difference operator, depending on the detected features. In the remainder of this section we show how to make the appropriate choice of fitting function and its corresponding difference operator. We start by introducing the notations and the conventions used in this paper.

Fitting Functions and Difference Operators. We will use a continuous real function $\tilde{g} : \mathbb{R}^m \to \mathbb{R}$ to approximate a digitized function $f : \mathbb{Z}^m \to \mathbb{Z}$. To approximate the value of a differential at a point x_0, it is sufficient to approximate f in a finite subset $D \subset \mathbb{Z}^m$ containing x_0. We write $|f - \tilde{g}| < \epsilon$ as a shorthand for $|f(x) - \tilde{g}(x)| < \epsilon$ for all $x \in D$.

The shift operator σ^j is defined by $\sigma^j f(x) = f(x+j)$, for $x, j \in \mathbb{Z}^m$. The functional composition of shift operators can be expressed as a multiplication of polynomials, i.e. $\sigma^j \sigma^k f = \sigma^{j+k} f$. A difference operator P can be represented as a polynomial in σ, that is $P = \sum_{j=0}^{l} p_j \sigma^j$. If we write that $P\tilde{g} = 0$, $\sum p_j \sigma^j \tilde{g}(x) = 0$ for all $x, j \in \mathbb{Z}^m$. If we write that $|Pf - P\tilde{g}| < \epsilon$, this means that $|Pf(x) - P\tilde{g}(x)| < \epsilon$ for all x for which $Pf(x)$ is well defined, that is $(x + j) \in D$ for every non-vanishing coefficient p_j of the difference operator P.

We will use the ideal I generated by a set of difference operators P_i. We write $I =< P_1, P_2, \ldots >$, and we say that the polynomials P_i form a basis for the ideal I. I consists of all operators $P = \sum_i S_i P_i$, where the S_i are arbitrary polynomials in σ.

Difference operators can also be represented by their templates. A two-dimensional difference operator $P = \sum p_j \sigma^j = \sum_{j_x, j_y} p_{j_x j_y} \sigma_x^{j_x} \sigma_y^{j_y}$, $j \in Z^2$ is represented by a two-dimensional template:

p_{00}	p_{10}	p_{20}	$\cdots$
p_{01}	p_{11}	p_{21}	
p_{02}	$\cdots$		

(1)

We use the convention that the box at the upper left corner corresponds to p_{00}. Boxes with vanishing coefficients are either not drawn, or drawn as empty boxes.

Approximation of Differentials. Let L be the differential operator that we want to approximate by a difference operator, which can be applied to some digitized function f. We find such an approximation by selecting a class G of fitting functions $\tilde{g}$, and a difference operator Q that works well for this class. We must choose in some way a class G of continuous functions and a difference operator $Q = \sum q_j \sigma^j$, such that $Q\tilde{g} = L\tilde{g}$ for every $\tilde{g} \in G$.

If $\tilde{g}$ is an approximation for f such that $|f - \tilde{g}| < \epsilon$, then we have $|Qf - L\tilde{g}| < \epsilon \sum |q_j|$ [5]. Hence, the difference operator Q will be a good approximation for the differential operator L provided G contains at least one function $\tilde{g}$ that is a good approximation for f.

Additional Constraints on Fitting Functions. It remains uncertain whether the class G of fitting functions contains a good approximation $\tilde{g}$ for the digitized function f. The uncertainty is removed by imposing additional constraints on G. We demand that fitting functions $\tilde{g}$ satisfy a (possibly infinite) set of difference equations $P_i\tilde{g} = 0$, for

$i = 1, 2, \ldots$. The operators P_i will be used to eliminate the fitting function $\tilde{g}$ from the inequality $|f - \tilde{g}| < \epsilon$. However, the set of operators that can be used to eliminate $\tilde{g}$ is much larger. For any operator P in the ideal $I = < P_1, P_2, \ldots >$, $P\tilde{g} = 0$. So also $P_1 P_2 \tilde{g} = 0$, $(P_1 + P_2)\tilde{g} = 0$, and $(\sum S_i P_i)\tilde{g} = 0$ for arbitrary difference operators $S_i \in \mathbb{R}[\sigma]$.

We will assume from now on that we have chosen a class of fitting functions G, by specifying a set of difference operators P_i, i.e. $G = \{\tilde{g} | P_i \tilde{g} = 0, \text{ for } i = 1, 2, \ldots\}$. We can then prove that there is a fitting function $\tilde{g} \in G$ such that $|f - \tilde{g}| < \epsilon$ if and only if

$$|Pf| < \epsilon \sum |p_j| \tag{2}$$

for every operator P in the ideal I generated by all P_i, i.e. $I = < P_1, P_2, \ldots >$ [5].

The ideal I of difference operators is generated by a possibly infinite set of operators P_i. Hilbert's Basis Theorem for polynomial ideals states that any ideal of polynomials in the ring $\mathbb{R}[\sigma_x, \sigma_y]$ can always be generated by a finite basis of polynomials. Even if a system has infinitely many difference equations, these can all be obtained by multiplying, adding and translating a finite set of basis equations. We can compute a Groebner basis for any ideal if we impose an ordering on the shift operators σ_x, σ_y [12]. For each different class of fitting functions, a Groebner basis completely characterizes all different templates that will recognize a function of that class. The Groebner basis is used in efficient algorithms for determining whether a given polynomial belongs to an ideal. The functional decomposition and combination of feature detectors can also be investigated using these bases [5].

How to Compute Feature Detectors for a Function Class? The introduction of the difference equations $P_i \tilde{g} = 0$ has an important consequence: by eliminating the explicit occurrence of $\tilde{g}$, the use of fitting functions will be replaced by the use of feature detectors P_i. Instead of actually fitting a good continuous approximation $\tilde{g}$ to f, we must verify whether Eq. 2 holds for every difference operator in I, i.e. whether f has the right features. So we do not bother which fitting function would actually yield the closest fit.

Features can be detected without error by verifying only a finite number of inequalities, when the solution space of the difference equations is a finite linear vector space. Assume that the solution set of the partial difference equations $P_1 g = 0, \ldots, P_n g = 0$ can be written as a linear vector space with $g_1, \ldots, g_l$ as a basis:

$$\alpha_1 g_1 + \cdots + \alpha_l g_l. \tag{3}$$

Let K_D be the set of all difference operators P_i of the form

$$\begin{vmatrix} g_1(x_1) & \ldots g_l(x_1) & \sigma^{x_1} \\ \ldots & & \\ g_1(x_{l+1}) & \ldots g_l(x_{l+1}) & \sigma^{x_{l+1}} \end{vmatrix} \tag{4}$$

with the points $x_j \in D$. The operators of K_D are written as determinantal expressions of the coefficients $g_j(x_j)$ and the shift operators σ^{x_j}. Let I_D denote the set of all the difference operators in I for which $P_i f(x)$ is well defined for at least one x in D, i.e. $(x + j) \in D$ for every non-vanishing coefficient p_j of the difference operator P_i.

Thus the neighborhood D must be large enough such that I_D is equal to I, i.e. I_D must contain a basis for I. Then the polynomials of K_D form a finite basis for the ideal generated by I_D. Furthermore, if the function f satisfies the inequality $|P_i f| < \epsilon \sum |p_j|$ for every polynomial P_i in K_D, then f will satisfy this inequality for all polynomials of I_D [11]. In general, a small sample of K_D suffices to calculate a Groebner basis for the ideal I.

How to Choose a Difference Operator for a Function Class? For any difference operator Q chosen such that $Q\tilde{g} = L\tilde{g}$, and f satisfying $|Pf| < \epsilon \sum |p_j|$ for all $P \in I =< P_1, \ldots, P_n >$, there exists a function $\tilde{g}$ satisfying $P\tilde{g} = 0$ such that

$$|Qf - L\tilde{g}| < \epsilon \sum |q_j|. \tag{5}$$

This result states exactly what we propose: the combined use of feature detectors P and a difference operator Q, linked to each other by the fitting functions $\tilde{g}$. It is sufficient to verify whether f has the right features so that we can apply this particular difference operator Q. However, as mentioned before, the choice of Q is not unique.

How to Choose the Best Difference Operator Q? Once the class of fitting functions has been chosen by verifying Eq. 2, we must choose a difference operator Q, which satisfies $Q\tilde{g} = L\tilde{g}$. Since $(Q + P)\tilde{g} = Q\tilde{g}$, there seem to be many possible ways to choose Q. There are however no other possibilities than those provided by the ideal I. Every operator R satisfying $R\tilde{g} = Q\tilde{g}$ can be written as $R = Q + P$.

Among all operators R, we can in fact look for the best candidate. For any operator $P \in I$ we have

$$|(Q + P)f - L\tilde{g}| < \epsilon \sum |p_j + q_j|. \tag{6}$$

The difference operator P for which the right side of the above inequality becomes minimal, gives the lowest error in $P + Q$ when used to approximate the differential L.

As $P_i\tilde{g} = 0$ for every $P_i \in I$, we can look for an operator of the form

$$O = Q + \sum (S_i P_i) \tag{7}$$

where $S(\sigma_x)$ is an arbitrary difference operator. Then the following inequality

$$|Of - L\tilde{g}| < \epsilon \sum |o_j|. \tag{8}$$

must be satisfied and the best difference operator is the one for which $\epsilon \sum |o_j|$ is minimized. This gives a systematic method for computing the difference operator to use for a particular fitting function class.

3 A Decision Tree for the Laplacian Operator

In this section we apply the above theory in the design of difference operators for the computation of the Laplacian $\partial^2/\partial x^2 + \partial^2/\partial y^2$. We want to adapt the widely used classical Laplacian kernel to yield better results once the local image characteristics are known. To determine the correct fitting function class in each point on the image surface, we use a decision tree as shown in Figure 2. Each node of the decision tree shows a Groebner basis for the feature detectors for a particular function class. Once the correct fitting function is chosen, the appropriate difference operator will be selected in the underlying leaf of the tree.

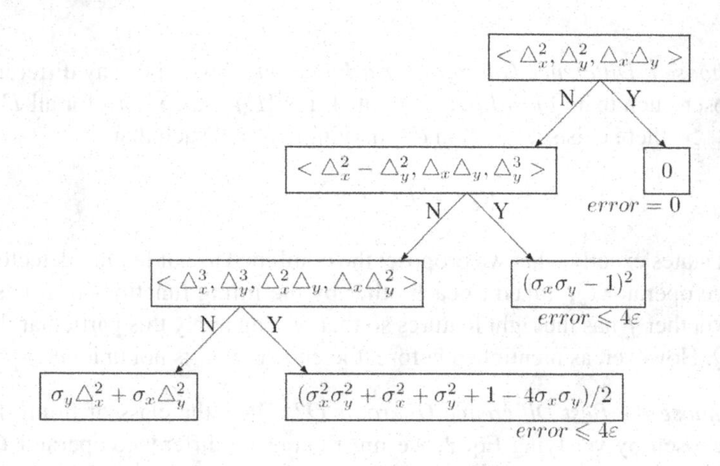

Fig. 2. A decision tree for the Laplacian

3.1 Template Bases for Feature Detectors

The first goal is to determine which fitting function is locally the best approximate for the digitized image surface. The decision tree uses three fitting function classes:

$$\begin{array}{l} \alpha_1 x + \alpha_2 y + \alpha_3 \\ \alpha_1 (x+y)^2 + \alpha_2 x + \alpha_3 y + \alpha_4 \\ \alpha_1 x^2 + \alpha_2 y^2 + \alpha_3 xy + \alpha_4 x + \alpha_5 y + \alpha_6 \end{array} \tag{9}$$

A set of feature detection templates is computed for each fitting functions as the polynomials of K_D in Eq. 4. This is illustrated for the function class used in the second node of the decision tree: $\alpha_1(x^2+y^2)+\alpha_2 x+\alpha_3 y+\alpha_4$. In this case, the polynomials of the set K_D have the form

$$\begin{vmatrix} 1 & x_1 & y_1 & x_1^2 + y_1^2 & \sigma_x^{x_1}\sigma_y^{y_1} \\ \dots \\ 1 & x_5 & y_5 & x_5^2 + y_5^2 & \sigma_x^{x_5}\sigma_y^{y_5} \end{vmatrix} \tag{10}$$

For example, for $\{(x_1,y_1),\dots,(x_5,y_5)\} = \{(0,0),(1,1),(2,0),(3,1),(4,0)\}$, the above determinant is equal to $1 - 2\sigma_x\sigma_y + 2\sigma_x^3\sigma_y - \sigma_x^4$ which corresponds to the template

1				-1
	-2		2	

(11)

Thus a basis for the ideal I_D for each function class can be generated systematically by computing all polynomials of K_D. Based on a limited set of polynomials we can compute $< (\sigma_x - 1)^2 - (\sigma_y - 1)^2, (\sigma_x - 1)(\sigma_y - 1), (\sigma_y - 1)^3 >$, i.e. $< \Delta_x^2 - \Delta_y^2, \Delta_x\Delta_y, \Delta_y^3 >$ as a Groebner basis for this function class. The corresponding templates are

$$\begin{array}{|c|c|c|} \hline & -2 & 1 \\ \hline 2 & & \\ \hline -1 & & \\ \hline \end{array} \;,\; \begin{array}{|c|c|} \hline 1 & -1 \\ \hline -1 & 1 \\ \hline \end{array} \;,\; \begin{array}{|c|} \hline -1 \\ \hline 3 \\ \hline -3 \\ \hline 1 \\ \hline \end{array} \;. \tag{12}$$

For each of the fitting function classes, the Groebner basis completely characterizes the templates that will recognize the function. Table 1 lists a selection of functions and their corresponding Groebner bases. Figure 3 shows the decision tree of Figure 2 with templates for the Groebner basis for the ideal of each function class. The nodes of the decision tree are practically implemented as a finite set of inequalities to be verified for the function of the required form. Or, equivalently, templates are generated that can detect features for each of the fitting functions. Now that we have an implementation for the nodes, the leaves of the tree remain to be filled in with the appropriate difference operators.

Table 1. Functions and corresponding Groebner bases with lexicographic ordering $\Delta_x > \Delta_y$

α_1	$< \Delta_x, \Delta_y >$
$\alpha_1 x + \alpha_2$	$< \Delta_x^2, \Delta_y >$
$\alpha_1(x+y) + \alpha_2$	$< \Delta_x - \Delta_y, \Delta_y^2 >$
$\alpha_1 x + \alpha_2 y + \alpha_3$	$< \Delta_x^2, \Delta_x\Delta_y, \Delta_y^2 >$
$\alpha_1 xy + \alpha_2 x + \alpha_3 y + \alpha_4$	$< \Delta_x^2, \Delta_y^2 >$
$\alpha_1(x+y)^2 + \alpha_2(x+y) + \alpha_3$	$< \Delta_x - \Delta_y, \Delta_y^3 >$
$\alpha_1(x^2+y^2) + \alpha_2 x + \alpha_3 y + \alpha_4$	$< \Delta_x^2 - \Delta_y^2, \Delta_x\Delta_y, \Delta_y^3 >$
$\alpha_1(x+y)^2 + \alpha_2 x + \alpha_3 y + \alpha_4$	$< \Delta_x^2 - \Delta_y^2, \Delta_y(\Delta_x - \Delta_y), \Delta_y^3 >$
$\alpha_1 x^2 + \alpha_2 y^2 + \alpha_3 x + \alpha_4 y + \alpha_5$	$< \Delta_x^3, \Delta_x\Delta_y, \Delta_y^3 >$
$\alpha_1 x^2 + \alpha_2 y^2 + \alpha_3 xy + \alpha_4 x + \alpha_5 y + \alpha_6$	$< \Delta_x^3, \Delta_x^2\Delta_y, \Delta_x\Delta_y^2, \Delta_y^3 >$

3.2 Difference Operators

Now that the template bases for the polynomial ideals are known, the next step is to determine the best difference operator for each class of fitting functions. As an example we describe the computation of the difference operator for quadratic functions of the form $\alpha_1 x^2 + \alpha_2 y^2 + \alpha_3 xy + \alpha_4 x + \alpha_5 y + \alpha_6$. We must choose Q so that the requirement $L\tilde{g} = Q\tilde{g}$ is satisfied, that is

$$\left(\frac{\partial^2}{\partial x^2} + \frac{\partial^2}{\partial y^2}\right)\tilde{g} = (\Delta_x^2 + \Delta_y^2)\tilde{g} = 2\alpha_1 + 2\alpha_2. \tag{13}$$

For this function class, any differential operator of the form

$$O(\sigma_x, \sigma_y) = \Delta_x^2 + \Delta_y^2 + \Delta_x^3 S_1 + \Delta_y^3 S_2 + \Delta_x^2\Delta_y S_3 + \Delta_x\Delta_y^2 S_4 \tag{14}$$

in which S_i are arbitrary polynomials in the shift operators, yields the exact value for the Laplacian. For optimal results, one must choose the operator of the form (14) for which $\sum |o_j|$ is as small as possible. For this function class, the best choice is the symmetric operator

$$\frac{\sigma_x^2\sigma_y^2 + \sigma_x^2 + \sigma_y^2 + 1 - 4\sigma_x\sigma_y}{2}. \tag{15}$$

The difference operators for the other function classes are obtained in a similar way, and their templates are shown in Figure 3.

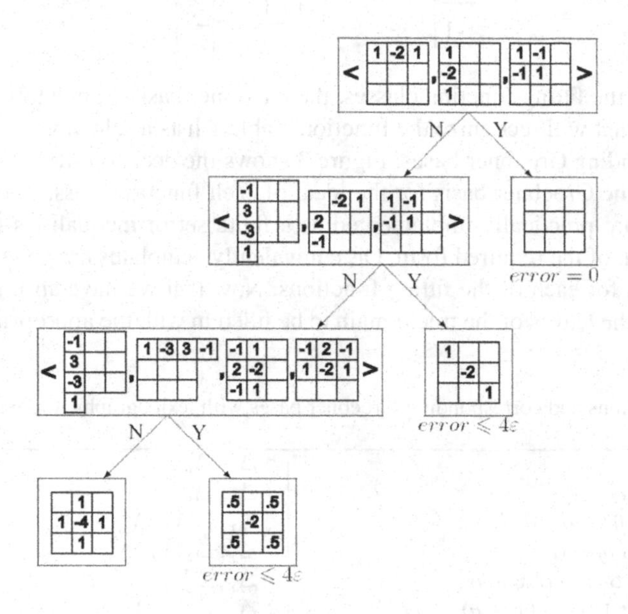

Fig. 3. A decision tree for the Laplacian

The obtained difference operators lead to some interesting conclusions. Without feature detection, we would always choose the classical discrete equivalent of the Laplacian as shown in the lower left leaf of the decision tree. When the digitized function is locally linear or close to linear, the best possible difference operator is the zero operator, yielding an error equal to zero. In fact, it is known without uncertainty that the Laplacian of a linear function vanishes. Since difference operators for linear functions do not have to compensate for quadratic terms, they perform better than the difference operators derived for quadratic functions. For both quadratic functions, we found a difference operator which yields a maximal error of 4ϵ on the computed value of the Laplacian. For quadratic functions with circular symmetry (the second level in the decision tree), it is sufficient to compute the second order difference in a diagonal direction. Finally, for quadratic functions the best difference operator has a template equal to the classical discrete Laplacian operator rotated over 45 degrees and divided by 2. If none of the feature detection tests succeed, we use the classical discrete Laplacian operator. In fact, the differences between the operators appear to be quite small, but the examples in section 4 show that considerable improvements are obtained by using the decision tree.

4 Laplacian of Gaussian Edge Detection

We illustrate the practical use of the decision tree for the computation of the Laplacian of Gaussian (LoG) for edge detection. The LoG is computed by first convolving the image with a Gaussian of a certain width and then passing a Laplacian filter kernel over the Gaussian smoothed image. The edges are computed as the zero-crossings of the Laplacian. We compare the results for the computation of the Laplacian by our method to the computation with the classical version of the kernel. Our method uses the decision tree designed above to select the most appropriate Laplacian operator in each image point. First, the local characteristics of the digitized image surface are determined by subsequently verifying a set of inequalities for each class of fitting functions. Once the correct fitting function is determined, the corresponding difference operator can be applied. For each image in the experiment, we constructed a scale space for different widths, i.e. different values for the standard deviation, of the Gaussian, so that we can compare both methods on increasing levels of smoothing. The results for the edge detection by both methods are shown in Figure 4.

Are the Fitting Function Classes Useful? For each function class, a set of inequalities is created to detect the local features of the digitized image. When the image surface

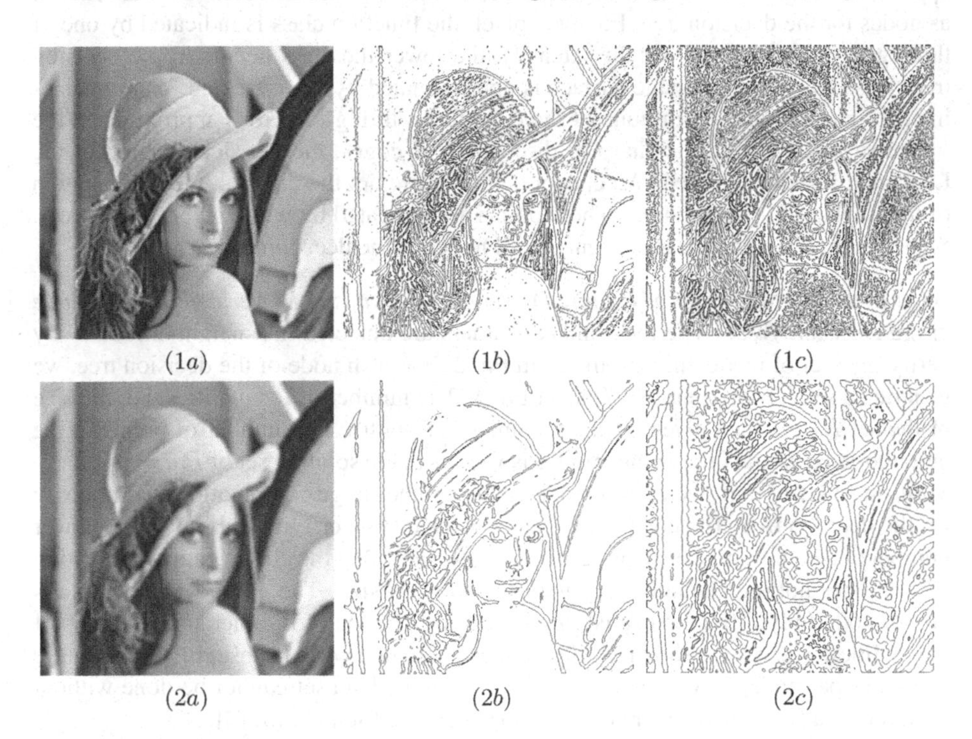

Fig. 4. LoG edge detection: In column a we see the Gaussian smoothed images on different scales, respectively for standard deviation $\sqrt{2}$ and $2\sqrt{2}$. Column b shows the result of edge detection after application of the decision tree to compute the Laplacian. Column c shows the result of edge detection after application of the classical Laplacian kernel.

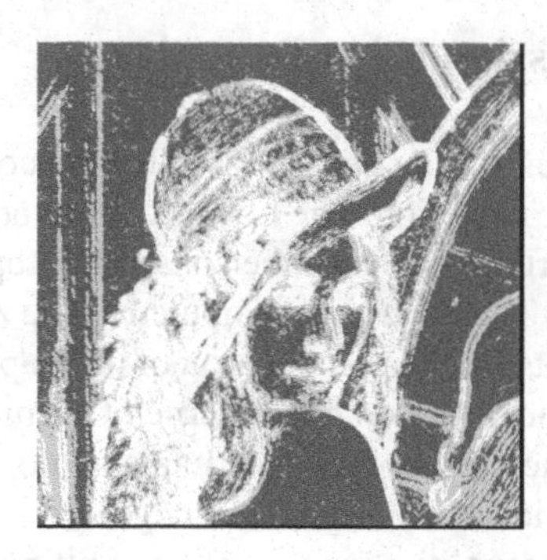

Fig. 5. The fitting function class in each image point is indicated: the higher (whiter) the point's intensity, the lower the node of the fitting function in the decision tree of Figure 3. When no correct approximation is found, points are colored white. The image was convolved first with a Gaussian of standard deviation $\sqrt{2}$.

locally fulfills the conditions posed by a limited subset of inequalities of the form Eq. 2, with $\epsilon = 1/2$ to account for digitization errors, we consider the fitting function a good approximation for the surface. Figure 5 shows that the three classes are well chosen as nodes for the decision tree. For each pixel, the function class is indicated by one of three gray values: the higher the intensity, the lower the class occurs in the decision tree. It is clear that the image surface is approximated by the expected function class in each image point. When none of the continuous fitting functions approximated the image surface well, e.g. at sharp or discontinuous edges, the classical version of the Laplacian kernel is applied. We cannot define or predict the error on the computation of the Laplacian in such points. These points are indicated by white pixels in Figure 5. Since all gray values appear in Figure 5, all nodes of the decision tree are useful.

What Is the Size of the Set of Feature Detection Templates to be Verified in Each Node of the Decision Tree? The local image features are detected in a neighborhood D by verifying a set of inequalities of the form Eq. 2. For each node of the decision tree, we generate a set K_D of operators P as in Eq. 4. The number of inequalities, i.e. the size of the set K_D, is then given by the binomial $\binom{n}{l}$, with n the number of points in the neighborhood D and l the number of basis vectors in the solution set of $P_i\tilde{g} = 0$ (Eq. 3). Verifying a set of that size at each node yields a rather large computational complexity for the decision tree. The complexity can be reduced by verifying subsets with a lower number of inequalities at each node. Simulations show that the classification in function classes does not significantly vary with the subset's size. The edge detection results do not noticeably differ, and satisfying results are obtained for rather small subsets of inequalities. The results shown in Figure 4 are obtained with a subset of seven feature templates per node. However, selecting only a limited subset cannot be done without introducing a small error. More on this subject can be found in [5, 13].

The computational complexity of the decision tree is further reduced by using feature templates in more than one neighborhood D. The computation of the templates is dependent on the size of D, but most of the polynomial's coefficients are zero, as can be seen in Eq. 11. So the majority of the results for the verification of inequalities can be reused in adjoining neighborhoods. To avoid systematic errors, introduced by using the same subset of templates over and over for all points, we randomly choose a new subset for every neighborhood to complement the reused results.

Comparison of Methods. The Laplacian computed with the proposed decision tree is compared to the result of applying the classical Laplacian kernel. The experiments are repeated in a scale space of Gaussians with increasing standard deviation. Figure 4 shows the results for Gaussians with standard deviation $\sqrt{2}$ and $2\sqrt{2}$. The results for the LoG edge detection are considerably better when the Laplacian is computed with the decision tree. The most significant edges are detected and the edges of important details are preserved, as the images in column (b) show. If we compare this to the result in column (c), we notice an abundance of edges in image regions considered homogeneous. Even on a higher scale, i.e. for even smoother images, the computation of the Laplacian with the classical kernel does not yield better results. First, the error on the localization of edges increases for higher levels of smoothing. Second, the edges of finer (and even coarser) details disappear on higher scales while edges are still detected in (noisy) homogeneous regions. Both problems are avoided in our method. Edges of details are already distinguished on finer scales. If linear fitting functions can locally approximate the image surface in homogeneous regions, the zero operator is used for the Laplacian so that zero crossings do not occur in these regions. Note that some drawbacks of LoG edge detection are apparent in the results of both methods, edges tend to form closed loops, and sharp corners are smoothed too much.

5 Conclusion

We present a mathematical framework from which both feature detection and difference operators arise in a natural way. By detecting local image features, we avoid the necessity of actually approximating the digitized image surface by fitting functions. For each function class, we define the appropriate difference operator which yields a minimal computational error when approximating the value we would have obtained by the differential operator.

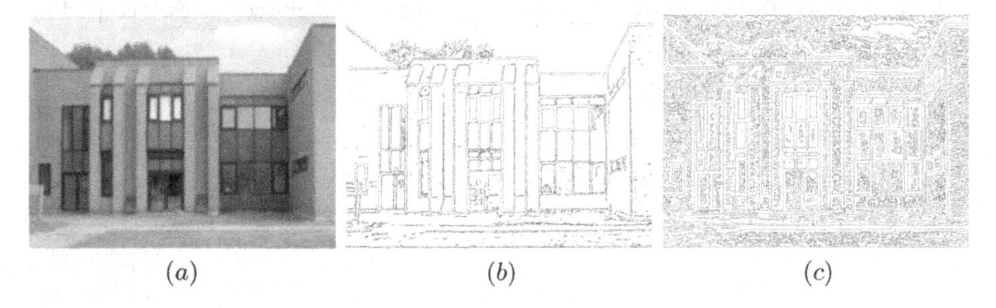

(a) (b) (c)

Fig. 6. LoG edge detection: Image (a) shows an Gaussian smoothed image with standard deviation $2\sqrt{2}$. Image (b) shows the result of edge detection after application of the decision tree to compute the Laplacian. Image (c) shows the result of edge detection after application of the classical Laplacian kernel

We conclude that the quality of the edge pixels detected by the LoG improves when the Laplacian difference operator is adapted to the knowledge about the local image features. Clearly, a practical application like the computation of straight lines and corners of a building as in Figure $6(a)$, is considerably easier when given the LoG edge pixels of Figure $6(b)$ (decision tree) as opposed to the information obtained by the default kernel, shown in Figure $6(c)$.

References

1. Haralick, R.: Digital step edges from zero crossing of second directional derivatives. IEEE Trans. Pattern Anal. Machine Intell. **6** (1984) 58–68
2. Langridge, D.: Detection of discontinuities in the first derivatives of surfaces. Comput. Vision Graphics Image Process. **27** (1984) 291–308
3. Fleck, M.: Multiple widths yield reliable finite differences. IEEE Trans. Pattern Anal. Machine Intell. **14** (1992) 412–429
4. Karabassis, E. and Spetsakis, M.E.: An analysis of image interpolation, differentiation, and reduction using local polynomial fits. CVGIP: Graphical Models and Image Processing **57** (1995) 183–196
5. Veelaert, P.: Local feature detection for digital surfaces. In: Proceedings of the SPIE Conference on Vision geometry V, SPIE Vol. 2826 (1996) 34–45
6. Lachaud, J.O., Vialard, A. and de Vieilleville, F.: Analysis and Comparative Evaluation of Discrete Tangent Estimators. In: Proc. DGCI'05, LNCS, Vol. 3429 (2005) 240–251
7. Lindeberg, T.: Discrete Derivative Approximations with Scale-Space Properties: A Basis for Low-Level Feature Extraction. J. of Mathematical Imaging and Vision **3** (1993) 349–376
8. Gunn, S.: On the discrete representation of the Laplacian of Gaussian. Pattern Recognition **32** (1999) 1463–1472
9. Demigny, D. and Kamlé, T.: A Discrete Expression of Canny's Criteria for Step Edge Detector Performances Evaluation. IEEE Trans. Patt. Anal. Mach. Intell. **19** (1997) 1199–1211
10. Lowe, D.G.: Distinctive image features from scale-invariant keypoints. International Journal of Computer Vision **60**(2) (2004) 91–110
11. Stoer, J. and Witzgall C.: Convexity and Optimization in Finite Dimensions I. Springer, Berlin (1970)
12. Cox, D., Little, J. and O'Shea, D.: Ideals, Varieties and Algorithms: an Introduction to Computational Algebraic Geometry and Commutative Algebra. Springer, New York (1992)
13. Veelaert, P. and Teelen K.: Fast polynomial segmentation of digitized curves, Accepted for DGCI 2006.

An Optimal Algorithm for Detecting Pseudo-squares*

Srečko Brlek and Xavier Provençal

Laboratoire de Combinatoire et d'Informatique Mathématique,
Université du Québec à Montréal,
CP 8888, Succ. Centre-ville, Montréal (QC) Canada H3C3P8
{brlek, provenca}@lacim.uqam.ca

Abstract. We consider the problem of determining if a given word, which encodes the boundary of a discrete figure, tiles the plane by translation. These words have been characterized by the Beauquier-Nivat condition, for which we provide a linear time algorithm in the case of pseudo-square polyominoes, improving the previous quadratic algorithm of Gambini and Vuillon.

1 Introduction

In discrete geometry many results are based on an arithmetic approach for characterizing and recognizing patterns having a certain shape. Here we take a combinatorics on words point of view that enable us with new tools for analyzing a shape in discrete planes.

The problem of deciding if a given polyomino tiles the plane by translation goes back to Wisjhoff and Van Leeuven [12] who coined the term *exact polyomino* for these, and also provided a polynomial $\mathcal{O}(n^4)$ algorithm for solving the problem. Polyominoes may be coded by words on a 4-letter alphabet $\Sigma = \{a, \overline{a}, b, \overline{b}\}$, also known as the Freeman chain codes [5,6] coding their boundaries (see [1] for further reading). For instance, the boundary $\mathbf{b}(P)$ of the polyomino in Figure 1 (a), in a counterclockwise manner, is coded by the word

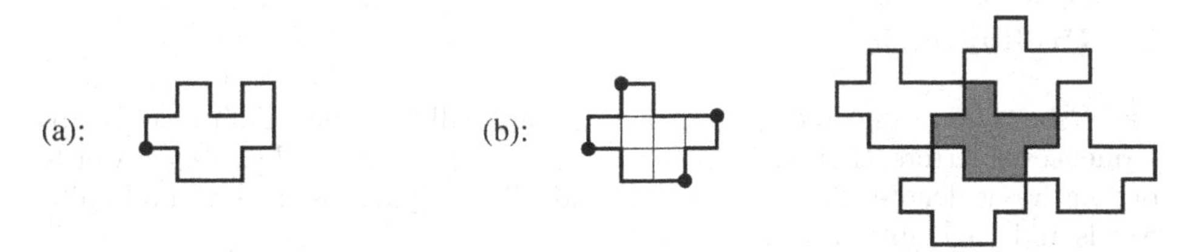

Fig. 1. (a) a polyomino; (b) an exact polyomino

$w = a\overline{b}aababb\overline{a}\overline{b}\overline{a}b\overline{a}\overline{b}\overline{a}\overline{b}$. Observe that we may consider the words as circular which avoids to fix an origin. The *perimeter* of a polyomino P is the length of

* With the support of NSERC (Canada).

A. Kuba, L.G. Nyúl, and K. Palágyi (Eds.): DGCI 2006, LNCS 4245, pp. 403–412, 2006.

its boundary word $\mathbf{b}(P)$ and is of even length $2n$. Beauquier and Nivat [2] gave a characterization stating that the boundary of such a polyomino P may be factorized (not necessarily in a unique way) as

$$\mathbf{b}(P) = A \cdot B \cdot C \cdot \widehat{A} \cdot \widehat{B} \cdot \widehat{C} \tag{1}$$

where at most one of the variables is possibly empty, and where $\widehat{(\ )}$ is defined by $\widehat{U} = \widetilde{\overline{U}}$, $\widetilde{(\ )}$ being the usual reversing operation and $\overline{(\ )}$ the transformation on $\Sigma = \{a, \overline{a}, b, \overline{b}\}$ sending each letter $\alpha \in \Sigma$ on its complement $\overline{\alpha}$. For instance, the exact polyomino in Figure 1 (b) is coded by the circular word

$$w = a\,a\,b\,a\,b\,\overline{a}\,\overline{a}\,b\,\overline{a}\,\overline{b}\,\overline{a}\,\overline{b}\,a\,\overline{b},$$

its semi-perimeter is 7, and its boundary may be factorized as

$$\mathbf{b}(P) = A \cdot B \cdot \widehat{A} \cdot \widehat{B} = a\overline{b}aa \cdot bab \cdot \overline{a}\overline{a}b\overline{a} \cdot \overline{b}\overline{a}\overline{b}.$$

Determining if a given word $w \in \Sigma^n$ is the boundary of a polyomino is computed in $\mathcal{O}(n)$. Therefore the problem reduces to find a factorization satisfying the Beauquier and Nivat condition. Recently, Gambini and Vuillon [7] improved the Wisjhoof-van Leeuven bound by designing an $\mathcal{O}(n^2)$ algorithm that checks the Beauquier-Nivat condition 1.

Our algorithms borrow from Lothaire [11] that the *Longest-Common-Factor*, the *Longest-Common-Prefix* and the *Longest-Common-Suffix* in two words may be computed in linear time. The approach is also inspired by the linear algorithm of Gusfield and Stoye [8] for detecting *tandem repeats* in a word, and by the linear algorithm used to detect repetitions with gaps, as shown in Lothaire [11]. More precisely, the computation of the *Longest-Common-Left-Extension* (LCLE) and *Longest-Common-Right-Extension* (LCRE) is achieved in constant time, provided a linear preprocessing is performed on u and v, by a clever utilization of suffix trees (see Gusfield [9]). Taking advantage of these algorithms we provide a linear algorithm, with respect to the length of words, for pseudo-square polyominoes.

2 Preliminaries

Let Σ be a finite alphabet whose elements are called *letters*. Finite words are sequences of letters, that is, functions $w : [0..n-1] \longrightarrow \Sigma$, and the set of words of length n is denoted Σ^n. The free monoid $\Sigma^* = \cup_{n=0}^{\infty} \Sigma^n$ is the set of all finite words and the empty word is denoted ϵ.

A morphism is a function $\sigma : \Sigma^* \longrightarrow \Sigma^*$ such that $\sigma(uv) = \sigma(u)\sigma(v)$. Clearly a morphism is defined by the image of the letters. A *factor* f of w is a word $f \in \Sigma^*$ satisfying

$$\exists x \in \Sigma^*, y \in \Sigma^*, w = xfy.$$

If $x = \epsilon$ (resp. $y = \epsilon$) then f is called *prefix* (resp. *suffix*). The set of all factors of w is denoted by $F(w)$, and those of length n is $F_n(w) = F(w) \cap \Sigma^n$. Finally

Pref(w) denotes the set of all prefixes of w. The length of a word w is $|w|$, and the number of occurrences of a factor $f \in \Sigma^*$ is $|w|_f$. A word is said to be *primitive* if it is not a power of another word. If $w = pu$, and $|w| = n, |p| = k$, then $p^{-1}w = w[k+1]..w[n-1] = u$ is the word obtained by erasing p. As a special case when $|p| = 1$ we have the shift operator σ defined by $\sigma(w) = w[1..(n-1)]$. Another useful operator is the circular permutation ρ defined by $\rho(w) = w[1..(n-1)] \cdot w[0]$.

Two words u and v are *conjugate* when there are words x, y such that $u = xy$ and $v = yx$. Equivalently, u and v are conjugate if and only if there exists an index k such that $u = \rho^k(v)$. Conjugaison is an equivalence relation written $u \equiv v$. The *mirror image* $\widetilde{u}$ of $u = u_1u_2\cdots u_n \in \Sigma^n$ is the word $\widetilde{u} = u_nu_{n-1}\cdots u_1$. A *palindrome* is a word p such that $p = \widetilde{p}$, and for a language $L \subseteq \Sigma^\infty$, we denote by Pal($L$) the set of its palindromic factors.

Paths on the square lattice $\mathbb{Z}\times\mathbb{Z}$ are encoded on the alphabet $\Sigma = \{a, \overline{a}, b, \overline{b}\}$ identified with the unit steps $\{\rightarrow, \leftarrow, \uparrow, \downarrow\}$. Parallel paths always define a translation and we say that two words are *homologue* when the corresponding paths define a translation. More precisely, two words u and v are said *homologue* when either

(i) $u = v$, or
(ii) $u = \widehat{v}$.

An exact polyomino P whose boundary is $\mathbf{b}(P) = A \cdot B \cdot C \cdot \widehat{A} \cdot \widehat{B} \cdot \widehat{C}$ is called a *pseudo-hexagon* if none of the variables is empty and a *pseudo-square* otherwise. In this factorization A (resp. B, C) and $\widehat{A}$ (resp. $\widehat{B}, \widehat{C}$) are homologue and define the respective translations. For instance, the translations defined by the homologue sides of the pseudo-square polyomino

$$\mathbf{b}(P) = A \cdot B \cdot \widehat{A} \cdot \widehat{B} = a\,\overline{b}\,a\,a \cdot b\,a\,b \cdot \overline{a}\,\overline{a}\,b\,\overline{a} \cdot \overline{b}\,\overline{a}\,\overline{b}$$

are shown in Figure 2 (a). In the case of a pseudo-hexagon, as in Figure 2(b),

Fig. 2. Translations defined by homologue sides of a polyomino tile

the translations are related by the relation $t_3 = t_1 + t_2$. Moreover, the relative positions of the starting and ending point of any path is completely determined by the sum of the unit vectors corresponding to each letter. By abuse of notation we write for a path $w : [0..n-1] \rightarrow \Sigma$

$$\overrightarrow{w} = \sum_{k=0}^{n-1} \overrightarrow{w_k}.$$

Note that $\overrightarrow{w} = 0$ if and only if w is a closed path, and that $\overrightarrow{u} = -\overrightarrow{\widehat{u}}$.

3 Searching the Homologue Factors

Since polyominoes are coded by circular words w, in order to find the homologue factors it is convenient to work with $w \cdot w$ since a pair of homologue factors might be split, depending on the starting point.

Therefore, finding the homologue factors amounts to look for the *longest common factor* of ww and $\widehat{ww}$ denoted $\mathrm{LCF}(ww, \widehat{ww})$.

For instance the longest common factors of the polyomino-tile P in Figure 1 (b) are

$$\mathrm{LCF}(ww, \widehat{ww}) = \{a\,\overline{b}\,a\,a, \overline{a}\,\overline{a}\,b\,\overline{a}\}$$

and they are necessarily homologue sides(!). Indeed, since we know the positions i and j of $a\,\overline{b}\,a\,a$ and $\overline{a}\,\overline{a}\,b\overline{a}$ in w, this is easy to check in linear time. Clearly the boundary of P may be written as

$$\mathbf{b}(P) = w = a\,\overline{b}\,a\,a \cdot u \cdot \overline{a}\,\overline{a}\,b\,\overline{a} \cdot v$$

and then one easily checks that $v = \widehat{u}$. Unfortunately the situation is not always that good. Indeed, let $w = a\,a\,b\,b\,b\,a\,a\,b\,\overline{a}\,\overline{a}\,b\,\overline{a}\,\overline{b}\,\overline{b}\,\overline{b}\,\overline{a}\,\overline{a}\,\overline{b}\,a\,\overline{b}$. Then the longest homologue factors of w are (see Figure 3)

$$\mathrm{LCF}(ww, \widehat{ww}) = \{a\,a\,b\,b\,b\,a, \overline{a}\,\overline{b}\,\overline{b}\,\overline{b}\,\overline{a}\,\overline{a}\},$$

but $w = a\,a\,b\,b\,b\,a \cdot a\,b\,\overline{a}\,\overline{a}\,b \cdot \overline{a}\,\overline{b}\,\overline{b}\,\overline{b}\,\overline{a}\,\overline{a} \cdot \overline{b}\,a\,\overline{b}$ does not satisfy the Beauquier-Nivat condition. A good factorization is $w \equiv b\,b \cdot b\,a\,a\,b \cdot \overline{a}\,\overline{a}\,b\,\overline{a} \cdot \overline{b}\,\overline{b} \cdot \overline{b}\,\overline{a}\,\overline{a}\,\overline{b} \cdot a\,\overline{b}\,a\,a$.

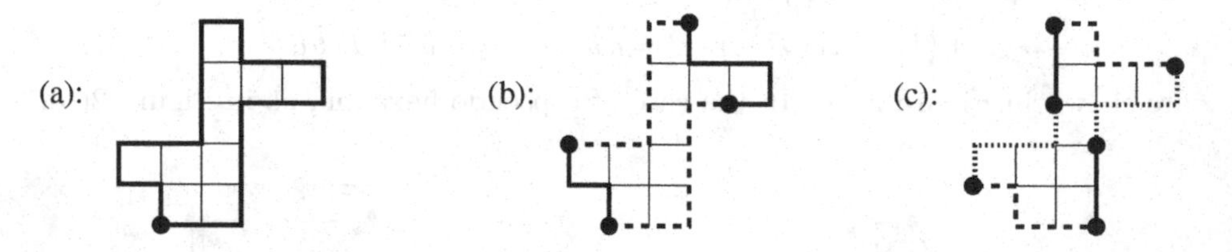

Fig. 3. (b) longest homologue factors; (c) a good factorization

This means that not all the homologue factors provide a factorization, and good candidates are those separated by factors of same length.

Definition 1. *Let $w \equiv \mathbf{b}(P)$ be the boundary word of a polyomino P. A factor A of w is* admissible *if*

(i) *$w \equiv Ax\widehat{A}y$, for some x, y such that $|x| = |y|$;*
(ii) *A is* saturated, *that is, $x_0 \neq \overline{x_{k-1}}$ and $y_0 \neq \overline{y_{k-1}}$ where $k = |x| = |y|$.*

Nevertheless, admissibility is ensured for words that code the boundary of polyominoes. Indeed, Gambini and Vuillon established the following property ([7], section 3.1) by using a geometric result of Daurat and Nivat [4].

Lemma 1. *Let $w \equiv ABC\widehat{A}\widehat{B}\widehat{C}$ be a Beauquier-Nivat factorization of the boundary* $\mathbf{b}(P)$ *of an exact polyomino P. Then A, B and C are admissible.*

Conversely, not all admissible factors lead to a Beauquier-Nivat factorization. For instance, in the polyomino $w \equiv a\,a\,a\,b\,a\,b\,\overline{a}\,b\,\overline{a}\,\overline{a}\,\overline{a}\,\overline{b}\,\overline{a}\,\overline{b}\,a\,\overline{b}$ shown below, the

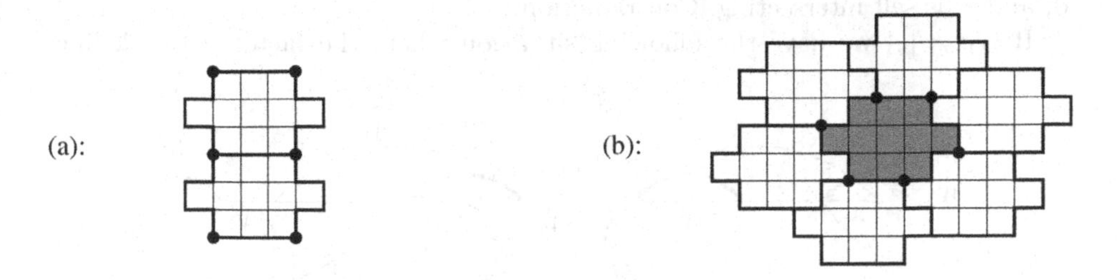

factor aaa is admissible but does not provide a correct factorization of w. Indeed, $A = aa$ is the admissible factor ($w = Ax\widehat{A}y$ with $x = abab\overline{a}b$, $y = \overline{a}\overline{b}\overline{a}\overline{b}a\overline{b}$) yielding a correct factorization with $B = aba$ and $C = b\overline{a}b$:

$$w \equiv a\,a \cdot a\,b\,a \cdot b\,\overline{a}\,b \cdot \overline{a}\,\overline{a} \cdot \overline{a}\,\overline{b}\,\overline{a} \cdot \overline{b}\,a\,\overline{b}.$$

The following proposition establishes a useful property.

Proposition 1. *Let $w = \mathbf{b}(P) \in \Sigma^{2n}$ be the contour of a polyomino P and let p be any fixed position in w. Let X be the set of all admissible factors overlapping the position p and $\widehat{X}$ be the set of their respective homologue factors. Then, there exist at least one position in w that is not covered by any element of $X \cup \widehat{X}$.*

Proof. By contradiction, assume that there is no such point. Let $A \in X$ be the factor that starts at the leftmost position and $B \in X$ be the one that ends at the righmost position as shown below. The homologue factors $A, \widehat{A}$ and $B, \widehat{B}$

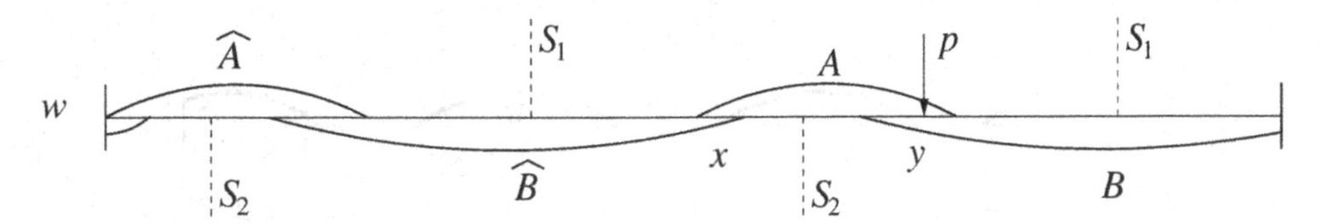

always define two symmetries denoted respectively by S_1 and S_2. Let x be the overlap between A and $\widehat{B}$, and y be the overlap between A and B. Without loss of generality we may consider that $|y| \geq |x|$. If $|x| = |y|$ the symmetry implies that $x = \widehat{y}$ and the factorization is

$$w \equiv x\,U\,\widehat{x}\,V\,x\,\widehat{U}\,\widehat{x}\,\widehat{V}\,. \tag{2}$$

We use a property proved in Brlek et all. ([3], DLT2005) that simplifies a result of Daurat and Nivat [4] on the number of salient and reentrant points of discrete sets: indeed, the number of right turns minus the number of left turns in a closed and non-intersecting path on a square lattice is 4. In equation 2, notice that all turns in a factor are cancelled by those of its homologue. Therefore we only have to consider the turns between consecutive factors. Reading, the word w from left to right, we see that each pair of consecutive factors is cancelled by its homologue: $x\,U$ is cancelled by $\widehat{U}\,\widehat{x}$, $U\,\widehat{x}$ by $x\,\widehat{U}$, $\widehat{x}\,V$ by $\widehat{V}x$ (the word w is circular), and $V\,x$ by $\widehat{x}\,\widehat{V}$. Hence the difference between right and left turns is 0, and w is self intersecting. Contradiction.

If $|x| \neq |y|$ we have the following situation where the factor y (thick line)

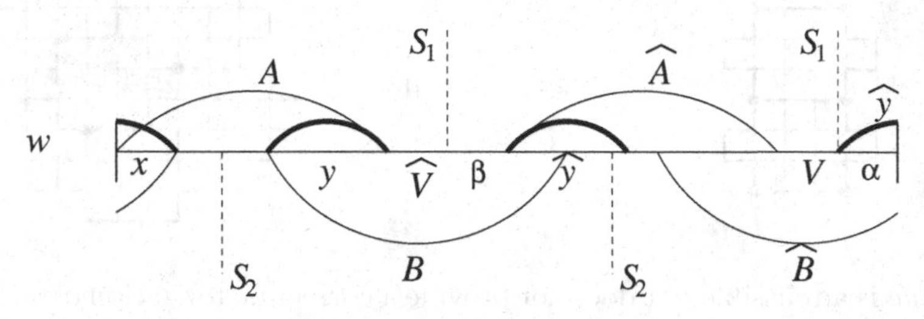

propagates as shown by using the symmetries S_1 and S_2. In this case $\widehat{y}$ does not overlap $\widehat{A}$ in $\widehat{B}$, so let V be the factor between $\widehat{A}$ and $\widehat{y}$. We have the following factorization

$$w \equiv A\widehat{V}\,\beta\,\widehat{A}\,V\,\alpha\,.$$

Passing to vectors, and using commutativity of addition, we have

$$\overrightarrow{w} = \overrightarrow{A} + \overrightarrow{\widehat{A}} + \overrightarrow{V} + \overrightarrow{\widehat{V}} + \overrightarrow{\beta} + \overrightarrow{\alpha} = \overrightarrow{\beta} + \overrightarrow{\alpha} = \overrightarrow{0}\,.$$

But $\widehat{y} = \alpha x$, so that β is followed by α in w. Therefore $\beta\alpha$ is a nonempty closed path on the boundary of P. Contradiction.

In the case where $\widehat{y}$ does overlap $\widehat{A}$ in $\widehat{B}$ we have the following situation where $\overrightarrow{\gamma} + \overrightarrow{\beta} = 0$ (by closure property $\overrightarrow{w} = 0 = \overrightarrow{A} + \overrightarrow{\gamma} + \overrightarrow{\widehat{A}} + \overrightarrow{\beta}$). Moreover,

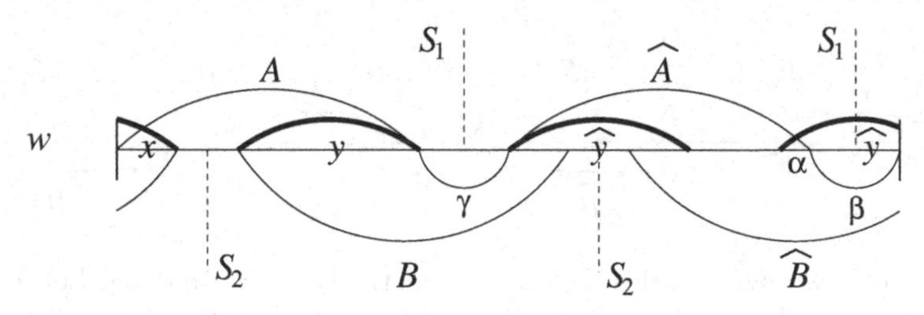

$\widehat{y} = \alpha\beta x$, so that $y\,\gamma\,\widehat{y}$ contains the nonempty factor $\widehat{\alpha}\,\gamma\,\alpha\,\beta$ corresponding to a closed path. Contradiction. ∎

Proposition 1 specializes for pseudo-squares as follows. Assume that a pseudo-square P has two factorizations

$$w = \mathbf{b}(P) \equiv AB\widehat{A}\widehat{B} \equiv XY\widehat{X}\widehat{Y}$$

where $A = sXt$. Then, by using the same argument as in the proof above, the boundary of P contains a loop yielding a contradiction.

Corollary 1. *If $w = \mathbf{b}(P) \equiv AB\widehat{A}\widehat{B} \equiv XY\widehat{X}\widehat{Y}$ are two distinct factorizations of the boundary of a pseudo-square P, then there exist α, β, γ such that $A = \alpha\beta$ and $X = \beta\gamma$.*

As an exemple we have the following pseudo-square

$$aba \cdot \overline{b}a\overline{b} \cdot \overline{a}\overline{b}\overline{a} \cdot b\overline{a}b \equiv bab \cdot a\overline{b}a \cdot \overline{b}\overline{a}\overline{b} \cdot \overline{a}b\overline{a},$$

showing two distinct factorizations. The problem of enumerating all the factorizations of a given pseudo-square will be addressed in a forthcoming paper.

3.1 A Linear Time Algorithm for Detecting Pseudo-squares

The main idea used to achieve linear time factorization, is to choose a position p in w and then list all the *admissible factors* A that overlap this fixed position. The following auxiliary functions are useful. The *Longest-Common-Right-Extension* (LCRE) and *Longest-Common-Left-Extension* (LCLE) of two words u and v at positions respectively m and n are partial functions

$$\mathrm{LCRE}, \mathrm{LCLE} : \Sigma^* \times \Sigma^* \times \mathbb{N} \times \mathbb{N} \longrightarrow \mathbb{N}$$

defined as follows. For $u, v \in \Sigma^*$, let m and n be such that $0 \leq m \leq |u|$ and $0 \leq n \leq |v|$, then

$$\mathrm{LCRE}(u, v, m, n) = \mathrm{LCP}(\rho^m(u), \rho^n(v))$$
$$\mathrm{LCLE}(u, v, m, n) = \mathrm{LCS}(\rho^{|u|-m}(u), \rho^{|v|-n}(v))$$

Remark 1. It is clear from the definition above that LCRE and LCLE may be computed in linear time. Their computation may also be performed directly by the following formulas. Since we use circular words w, denote $\underline{m} = m \mod |w|$. If $u[\underline{m}] = v[\underline{n}]$ then

(i) $\mathrm{LCRE}(u, v, \underline{m}, \underline{n}) = \max\{k \in \mathbb{N} \mid u[\underline{m}..(\underline{m} + \underline{k})] = v[\underline{n}..(\underline{n} + \underline{k})]\} + 1,$
(ii) $\mathrm{LCLE}(u, v, \underline{m}, \underline{n}) = \max\{k \in \mathbb{N} \mid u[(\underline{m} - \underline{k})..\underline{m}] = v[(\underline{n} - \underline{k})..\underline{n}]\} + 1,$

and, otherwise, $\mathrm{LCRE}(u, v, \underline{m}, \underline{n}) = \mathrm{LCLE}(u, v, \underline{m}, \underline{n}) = 0$.

For example, if $u = aabbbaabaababababa$, $v = babaabbbaabbababab b$, $i = 4$ and $j = 7$ then (note that the words all starts at position 0) we have

$$u = \underline{a\,a\,b\,b}\,\mathbf{b}\,\underline{a\,a\,b}\,a\,a\,b\,a\,b\,a\,b\,a\,b\,a,$$
$$v = b\,a\,b\,\underline{a\,a\,b\,b}\,\mathbf{b}\,\underline{a\,a\,b}\,b\,a\,b\,a\,b\,a\,b\,b,$$

and $\mathrm{LCRE}(u, v, 4, 7) = 4, \mathrm{LCLE}(u, v, 4, 7) = 5$. On the other hand LCRE(u,v,4,1) = LCLE(u,v,4,1)=0.

Later we will need to perform these computations $\mathcal{O}(n)$ times. Fortunately, the computation of LCLE and LCRE is achieved in constant time, provided a linear preprocessing is performed on u and v, by a clever utilization of suffix trees.

Indeed, for computing $\mathrm{LCRE}(u, v, i, j)$, one proceeds by building a common suffix tree for both u and v, and a second pass consists in labeling each node in such a way that the least common ancestor of two nodes is computed in constant time. Both building the suffix tree and its preprocessing is performed in $\mathcal{O}(n)$. For a more detailed description see Gusfield [9], section 9.1, or Gusfield and Stoye [8], page 531.

For computing $\mathrm{LCLE}(u, v, i, j)$, it suffices to process the mirror images and compute $\mathrm{LCRE}(\widetilde{u}, \widetilde{v}, |u| - i - 1, |v| - j - 1)$.

Lemma 2. *Let $w = \mathbf{b}(P)$ be the boundary of P. For each occurrence of A in w and each occurrence of A in $\widehat{w}$, whether A is admissible or not is decidable in constant time.*

Proof. Given an occurrence of A in $\widehat{w}$, one computes in constant time the corresponding position of $\widehat{A}$ in w. If $\widehat{A}$ overlaps A in w is decidable in constant time. If $\widehat{A}$ and A do not overlap then, $u \equiv Ax\widehat{A}y$ and A is an admissible factor, by definition, if and only if the three following conditions are verified : $|x| = |y|$, $x_0 \neq \overline{x_{k-1}}$ and $y_0 \neq \overline{y_{k-1}}$ where $k = |x| = |y|$. ■

Lemma 3. *Let $w = \mathbf{b}(P) \in \Sigma^{2n}$ be the boundary of P. For any position p in w, listing all the admissible factors overlapping p is computed in linear time.*

Proof. The following algorithm list all admissible factors containing the p-th letter w. Since the longest common right and left extension problem can be solved in constant time after linear time preprocessing.

Algorithm 1
Input : $w = \mathbf{b}(P) \in \Sigma^{2n}$
1 : **For** $i := 0$ *to* $2n - 1$ **do**
2 : $l := \mathrm{LCLE}(w, \widehat{w}, p, i) - 1$
3 : $r := \mathrm{LCRE}(w, \widehat{w}, p, i) - 1$
4 : $A := w[p - l, \ldots, p + r]$
5 : **If** $w \equiv Ax\widehat{A}y$ *with* $|x| = |y|$ **then**
6 : *Add A to the list of admissible factors.*
7 : **end if**
8 : **end for**

Using the modulo in managing the positions is superfluous because we may assume, without loss of generality (since w is a circular word) that $p = n$. Note that, by definition of LCRE and LCLE, the factor A in this algorithm is necessarily saturated. As shown in Lemma 2, the condition can be tested in constant time by direct computation of positions in w. Finally, the loop is performed exactly $2n$ times. ■

Remark 2. This lemma implies that the number of admissible factors in a word is linear. To determine a precise upper bound remains an open problem which is similar to the problem of determining a tight upper bound for the number of distinct squares in a word (see for instance Lothaire [11] or Ilie [10]).

Theorem 1. *Let* $w = \mathbf{b}(P) \in \Sigma^{2n}$ *be the boundary of* P*. Determining if* w *codes a* pseudo-square *is decidable in linear time.*

Proof. If w encodes an exact polyomino, any position belongs to some admissible factor of the Beauquier-Nivat factorization. Therefore, it suffices to apply Lemma 3 to an arbitrary position p. Then, Algorithm 1 provides the list of all admissible factors overlapping the position p, and it only remains to check, for each admissible factor, if $x = \widehat{y}$. Lemma 1 ensures that if $w \equiv AB\widehat{A}\widehat{B}$ then B is saturated, as shown below.

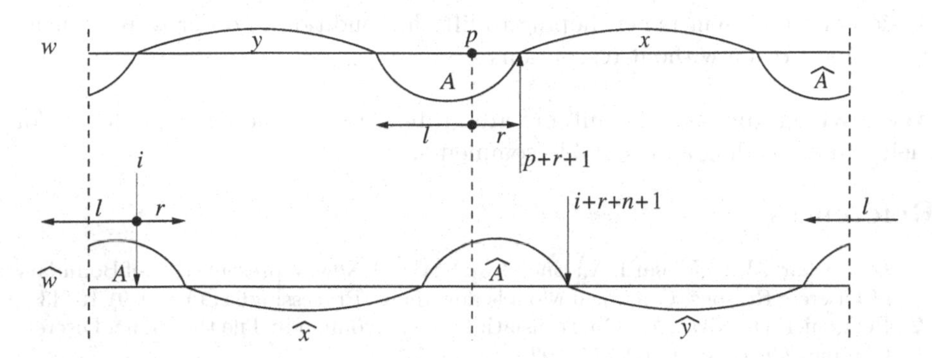

It suffices now to replace step 6 in Algorithm 1 by:

$6a$: **If** LCRE$(w, \widehat{w}, p + r + 1, \underline{i + r + n + 1}) = |x|$ **then**
$6b$: P is a *pseudo-square.*
$6c$: **End if**

Since LCRE is computed in constant time, the overall algorithm is linear. ∎

4 Concluding Remarks

Using combinatorics on words as a tool for analyzing shapes proved powerful as shown in a recent paper by Brlek et al. [3], where an elementary proof of a result of Daurat and Nivat [4] establishing a relation between convex and concave points in discrete figures. In addition, this approach is algorithmic and brings some new insight for addressing geometrical problems.

The results presented here can be extended to more general tilings. Indeed, since the Beauquier-Nivat factorization involves path properties, there is no need for a tile to by a polyomino. For instance, the tile T below might be also called pseudo-square and its Beauquier-Nivat factorization (starting from $\mathbf{S}$) is

$$\mathbf{b}(T) = a\,a\,a\,a\,b\,\bar{a}\,b\,a\,a\,\bar{b}\,\bar{a}\,\bar{b}\,a\,\bar{b}\,a \cdot b\,a\,b\,b\,a\,b \cdot \bar{a}\,b\,\bar{a}\,b\,a\,b\,\bar{a}\,\bar{a}\,\bar{b}\,a\,\bar{b}\,\bar{a}\,\bar{a}\,\bar{a}\,\bar{a} \cdot \bar{b}\,\bar{a}\,\bar{b}\,\bar{b}\,\bar{a}\,\bar{b}\,.$$

In another direction, it is quite natural to extend the method for pseudo-hexagons. The situation reveals more complicated, and we designed an optimal algorithm for detecting pseudo-hexagons for a restricted class of closed paths, namely those not having too large squares. This will be described in an extended

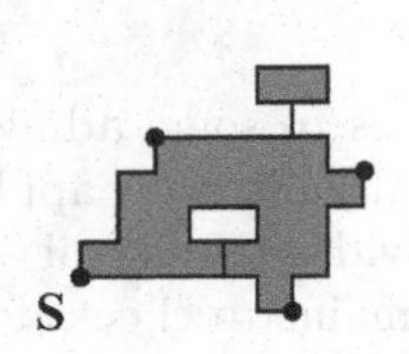

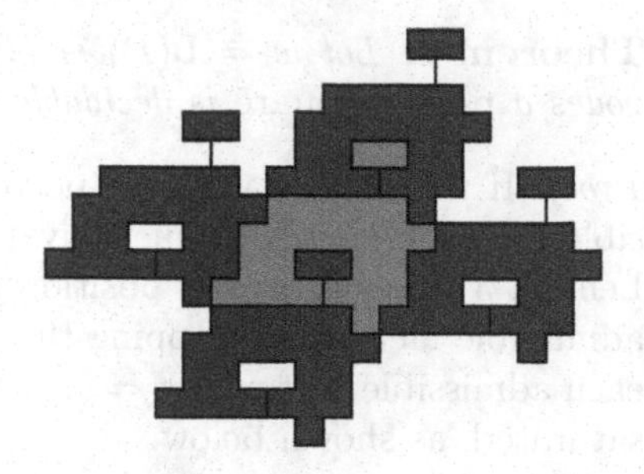

version of the present paper, hoping to lift the condition in order to provide an optimal algorithm without restrictions.

Acknowledgements. The authors are grateful to the anonymous referees for their careful reading and valuable comments.

References

1. Braquelaire, J.-P., Vialard, A.: Euclidean Paths: A New Representation of Boundary of Discrete Regions, Graphical Models and Image Processing**61(1)** (1999) 16–43
2. Beauquier, D., Nivat, M.: On Translating one Polyomino to Tile the Plane, Discrete Comput. Geom. **6** (1991) 575–592
3. Brlek, S., Labelle, G., Lacasse, A.: A note on a result of Daurat and Nivat, in C. de Felice and A. Restivo (Eds.) *Proc. DLT 2005, 9-th International Conference on Developments in Language Theory* (Palermo, Italia, July 4–8, 2005) Springer-Verlag LNCS 3572, 189–198
4. Daurat, A., Nivat, M.: Salient and Reentrant Points of Discrete Sets, In Del Lungo, A., Di Gesu, V., Kuba, A., (Eds.) *Proc. International Workshop on Combinatorial Image Analysis (IWCIA'03)* (Palermo, Italy, 14–16 May 2003), Electronic Notes in Discrete Mathematics **12**, Elsevier Science Publishers (2003)
5. Freeman, H.: On the Encoding of Arbitrary Geometric Configurations, IRE Trans. Electronic Computer **10** (1961) 260–268
6. Freeman, H.: Boundary encoding and processing, in B.S. Lipkin and A. Rosenfeld (Eds.) *Picture Processing and Psychopictorics*, (Academic Press, New York, 1970) 241–266
7. Gambini, I., Vuillon, L. : An algorithm for deciding if a polyomino tiles the plane by translations, LAMA research report (2003)
8. Gusfield, D., Stoye, J.: Linear time algorithms for finding and representing all the tandem repeats in a string, Journal of Computer and System Sciences **69** (2004) 525–546
9. Gusfield, D. : Algorithms on Strings, Trees and Sequences, Cambridge University Press, Cambridge (UK) 1997
10. Ilie, L.: A note on the number of distinct squares in a word, In S. Brlek and C. Reutenauer (Eds.) *Proc. 5th International Conference ion Words, Words2005* (Montreal, Canada, 13–17 September 2005) Publications du LaCIM **36** (2005) 289–294
11. Lothaire, M.: Applied Combinatorics on words, Cambridge University Press, Cambridge (UK) 2005
12. Wijshoff, H.A.G., Van Leeuven, J.: Arbitrary versus periodic storage schemes and tesselations of the plane using one type of polyomino, Inform. Control **62** (1984) 1–25

Optimization Schemes for the Reversible Discrete Volume Polyhedrization Using Marching Cubes Simplification

David Coeurjolly, Florent Dupont, Laurent Jospin, and Isabelle Sivignon

Laboratoire LIRIS/ UMR CNRS 5205 - Université Claude Bernard Lyon 1
Bâtiment Nautibus - 8, boulevard Niels Bohr
F-69622 Villeurbanne cedex, France
{david.coeurjolly, florent.dupont, isabelle.sivignon}@liris.cnrs.fr

Abstract. The aim of this article is to present a reversible and topologically correct construction of a polyhedron from a binary object. The proposed algorithm is based on a Marching Cubes (MC) surface, a digital plane segmentation of the binary object surface and an optimization step to simplify the MC surface using the segmentation information.

1 Introduction

3D discrete volumes are more and more used especially in the medical area since they result from MRI and CT scanners for example. As 2D images are composed of pixels, these 3D images are composed of voxels. This structure induces many difficulties in the exploitation and study of these objects: for each voxel a value is stored, thus the volume of data for an image is huge which is a problem to get a fluent interactive visualization ; the facet structure (voxels' faces) of the discrete object induces many problems to get a nice visualization that is necessary for medicines, as no rendering nor texture algorithm can be applied. The general idea to solve these problems is to transform discrete volumes into polyhedra with vertices in $\mathbb{R}^3$. An important property that must fulfill the Euclidean polyhedron is its reversibility up to a given digitization process (e.g. the result of the digitization must be the original discrete volume itself). In other words, no information are neither created nor lost during the transformation.

Many research activities have already been achieved to find solutions to compute this reversible transformation, using Euclidean geometry or discrete geometry [1,2,3,4,5,6]. To get a good visualization of discrete volumes, classical methods use the Marching Cubes algorithm [7,8], which considers local voxel configurations to replace them by small triangles. Even if these methods offer a good visualization, it does not provide a good data compression (huge number of facets) but we have a first reversible solution. Digital geometry solutions deal with a first step that segments the object boundary into pieces of digital plane [9,2,10,1,11,3]. The digital plane is a fundamental object for this problem because reversibility properties exist. The next step consists in associating a polygon to each piece of digital plane and finally to construct the Euclidean

A. Kuba, L.G. Nyúl, and K. Palágyi (Eds.): DGCI 2006, LNCS 4245, pp. 413–424, 2006.

polyhedron while sewing the polygons. The major problem of these methods is to ensure both the reversibility and the correct topology of the polyhedron.

In [4], we have proposed a polyhedrization algorithm with the following properties: it computes a reverse polyhedrization of the input digital object with the warranty that the obtained polyhedron is topologically correct. More precisely, the final polyhedron is a combinatorial 2-manifold. This algorithm is based on a simplification of the Marching-Cubes surface with digital plane segmentation information. In the following, we extend this algorithm using linear programming techniques to reduce the number of facets of the final object while preserving both the reversibility of the surface and its topology.

In section 2, we describe the preliminaries with a review of existing algorithms. In section 3, we detail the Marching-Cubes based simplification algorithm and its optimizations to obtain a polyhedron from a discrete object with a reduced number of facets.

2 Preliminaries

2.1 The Marching-Cubes Algorithm

Let us assume a discrete 3D image that maps a value $V(x, y, z) \in \mathbb{R}$ to each grid point $(x, y, z) \in \mathbb{Z}^3$. The image V can also be considered as a density function on a subset of $\mathbb{Z}^3$. The Marching-Cubes (MC) algorithm was first introduced by Lorensen and Cline [7] to extract a triangulated surface from V corresponding to an iso-density value. The first application of this work was the visualization of iso-density surfaces in medical imaging. We first consider cubic cells of coordinate (x, y, z) whose vertices are placed on the 8 input samples $(x+i, y+j, z+k)$ of the volume data, with $i,\ j,\ k \in \{0, 1\}$. The triangulated iso-surface given by the Marching-Cubes algorithm is locally computed according to the way the surface intersects each cell of V using a look-up table with 14 possible configurations (see Figure 1). The coordinates of the MC vertices along an edge of a cell is given by an interpolation process between the values of V and the chosen iso-level.

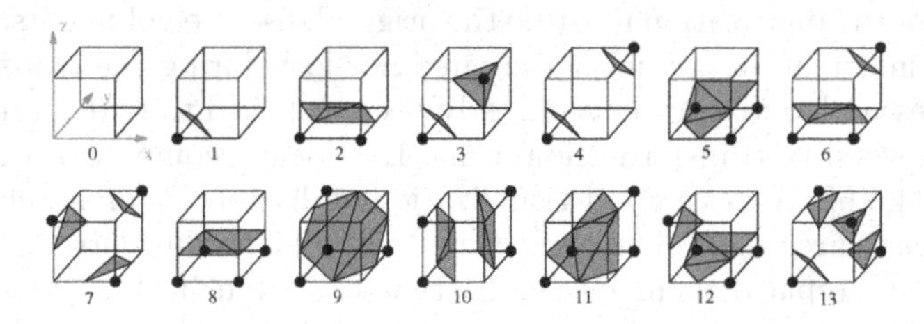

Fig. 1. The 14 different standard triangulations of the Marching-Cubes algorithm

Note that some of original Lorensen and Cline's configurations may lead to ambiguities in the reconstruction and thus construct surfaces with holes. To have properties on the topology of the reconstruction, we need a process that disambiguates the configurations according to the topology of the input discrete surface. The configurations presented in the Figure 1 correspond to a $(18,6)-$surface [8,12]. Hence, if the binary object is 6-connected, the triangulated surface is a combinatorial 2-manifold, *i.e.* closed, oriented and without self crossing [8,12]. In the following, we consider the Object Boundary Quantization (OBQ) scheme, also called *Gauss* digitization [13]: given an region $P \subset \mathbb{R}^3$ in $\mathbb{R}^3$, the OBQ digitization of P is the set of voxels $P \cap \mathbb{Z}^3$. If a binary object is considered, *i.e.* if $V(x,y,z) \in \{0,1\}$, for all x, y and z, from [4], we have the following lemma (see Figure 2):

Lemma 1 ([4]). *The Marching-Cubes surface of a digital object, obtained with a an iso-level in* $]0,1[$, *is a reversible polyhedrization of the binary object according to the Object Boundary Quantization model.*

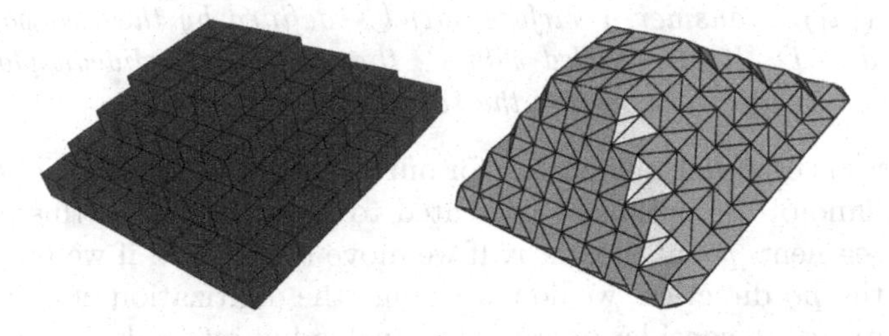

Fig. 2. A binary 3D object and the obtained Marching-Cubes surface

Given a discrete object, the *surfels* are cellular elements of the unit cube and are defined as the *square* shared by a voxel p in the binary object and a voxel q in its complementary, denoted $\{p,q\}$. Hence, according to the MC configurations, we have a one-to-one and onto mapping between MC vertices and surfels of the binary object. Indeed, vertices of the MC surface belongs to the $]pq[$ straight line segment (the vertex cannot be neither p nor q). Furthermore, it is easy to see that moving a MC vertex along its $]pq[$ intervals do not change the result of Lemma 1.

In the following, we propose a reversible polyhedrization based on a simplification of the Marching-Cubes surface.

2.2 Digital Plane Segmentation of a Discrete Surface

In order to simplify the Marching-Cubes surface, we compute a decomposition of the digital surface into coplanar set of surface elements. Consider a set of voxels $\mathcal{V}$, this set is a piece of digital plane with $x \geq z, y \geq z$ and $z > 0$ if and only if there exists a Euclidean plane containing $\mathcal{V}$ in its digitization. In other words, there exists (α,β,γ) in $[0,1]^2 \times [0,1[$ such that $\mathcal{V}$ is included in $P = \{(x,y,z) \in \mathbb{Z}^3 \mid 0 \leq \alpha x + \beta y + \gamma + z < 1\}$ [2,14,13]. Thus we can define the preimage of $\mathcal{V}$ as the set of (α,β,γ) parameters fulfilling this condition [10,3,14]. In the following, we call digital plane segments (DPS) coplanar sets of voxels.

This preimage is an efficient tool for the recognition process: given a set of voxels $\mathcal{V}$, decide if $\mathcal{V}$ is a DPS and if so, compute its parameters [10,3,15].

The definition given previously assumes that a direction is chosen before the recognition process. Generalizing the directional constraint, each direction of the set $D = \{(1,0,0),(-1,0,0),(0,1,0),(0,-1,0),(0,0,1),(0,0,-1)\}$ defines a preimage associated to a given set of voxels. In most cases, only one of those preimages is not empty for a given set of voxels. Considering a direction $\boldsymbol{d} \in D$ and according to the preimage definition, the preimage is the set of Euclidean planes crossing all the segments $[pq[$ where p is a voxel of $\mathcal{V}$ and q is the voxel of coordinates $p + \boldsymbol{d}$. Note that in practice, p is a voxel of the object while q belongs to the background.

The decomposition algorithm we use is the one presented in [4], which consists in labeling every surfel of the object's surface such that the following property is fulfilled:

Lemma 2 ([4]). *Consider a surface surfel s defined by the two voxels p and $q = p + d$, $d \in D$. If s is labeled with P, then all the Euclidean planes of the preimage of P in direction d cross the segment $[pq[$.*

This property is of major importance for our problem. Indeed, let (α, β, γ) be an Euclidean plane of the preimage associated to s. According to this property, it crosses the segment $[pq[$ at a point r. If we move v to r, *i.e.*, if we project v onto (α, β, γ) in the $\boldsymbol{pq}$ direction, we do not change the digitization of v. Note that it is straightforward to consider intervals $]pq[$ instead of intervals $[pq[$, we just have to handle strict inequalities in the digital plane definition without changing the algorithms.

3 Marching-Cubes Simplification and Optimization

Since there is a one-to-one and onto mapping between the MC vertices and the surfels of the input discrete object, we introduce a label on MC triangles as follows:

Definition 1 (Homogeneous and non-homogeneous triangle). *Let T be a triangle of the MC surface, T is homogeneous (H) if its three vertices are associated to surfels belonging to the same digital plane. Otherwise, T is called non-homogeneous (NH). If T is homogeneous, T is labeled with the digital plane segment label of its vertices.*

Furthermore, we can define the 2-NH triangle (resp. 3-NH triangle) if the number of distinct discrete plane segments associated to its vertices is exactly 2 (resp. 3).

In the following, we introduce a projection process of a MC vertex onto an Euclidean plane: let v be a MC vertex and p, q be the two voxels (p belongs to the object and q to the background) such that v is associated to the surfel $\{p, q\}$. Thus, only the projection of v onto an Euclidean plane P according to the $\boldsymbol{pq}$ direction is considered.

3.1 Homogeneous Triangles Case

Using [4], we have the following result on H-triangles:

Lemma 3 ([4]). *Let v be a vertex of an H-triangle, let P be an Euclidean plane from the preimage of the discrete plane associated to the triangle. The projection of v onto P does not change neither the reversibility nor its topological properties of the global surface.*

This lemma can easily be proven by definition and properties of the discrete plane segmentation process and using Lemma 2.

In [4], the authors design a simplification algorithm based on the previous lemma to remove the homogeneous triangles: let S be a connected set of H-triangle with the same label, they extract from the DPS preimage associated to S an Euclidean plane P. Then, if we project all vertices of S onto P, triangles in S become coplanar. Finally, a post-processing step converts all connected sets of H-triangles with the same label into a single facet. At each step of this algorithm, we ensure the reversibility property and the final surface is still a combinatorial 2-manifold. Note that no assumption is needed during the choice of the plane P.

As presented in Figure 8, for each connected set of H-triangle with the same label, we have obtained a facet. NH-triangles allow to sue together all the facets maintaining the topological property of the polyhedron.

In the next section, we present a linear programming framework to extract, from the preimage, an appropriate Euclidean plane P in order to remove NH-triangles.

3.2 Non-homogeneous Triangles Case

The basic idea to remove the NH-triangles consists in adding linear constraints in the DPS preimages. Then, the choice of the Euclidean plane P is made by a linear inequality system solver.

However, to have an efficient algorithm, we restrict the problem using the following two heuristics:

Local analysis: let us examine the 2D reconstruction presented in the Figure 3. If we consider the OBQ scheme, both polygons are correct regarding to the reversibility property. However, the visual aspect of the dashed polygon compared to the initial binary object is worse than the bold one. Hence, our reconstruction is restricted to a polyhedron defined in the cells defined by the MC surface. More precisely, when a modification of an NH triangle is performed, the result must belong to the MC cell associated to the triangle. This heuristic is a restriction on the possible reconstruction but it allows to design efficient algorithms since the surface properties (reversibility and topology) can be ensure using local analysis. Other arguments justifying this approach are based on the fact that the OBQ digitization scheme associated to MC surfaces is not a complete digitization model [16].

Linear programming problem in dimension 3: during the DPS recognition process, we have used linear programming algorithms in dimension 3 to compute the preimages[10,3,15]. In this optimization process, the dimension of the linear constraint system that conducts the NH triangle simplification must be bounded by 3. Even if this choice influences and reduces the scope of the algorithm, we limit the computational cost of the linear programming solver this way. Furthermore this process is still consistent with the DPS preimage parameter space.

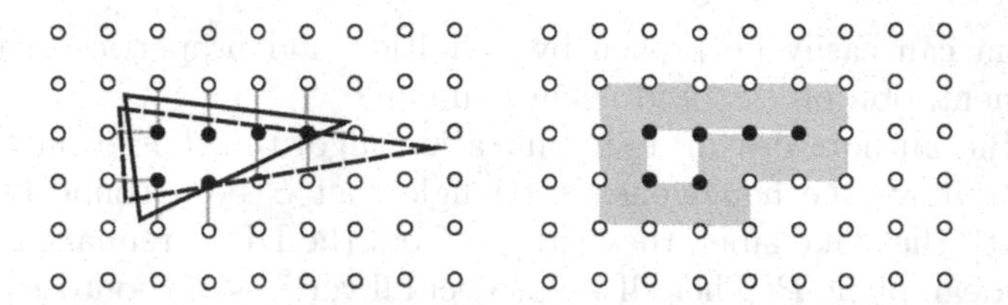

Fig. 3. *(left):* Two possible polygonalizations of a binary object (dark grey dots). The grey segment represent the $]pq[$ intervals in the OBQ scheme. *(right):* The light grey area define the allowed location of the polygon vertices we use (hence only the bold polygon in the left figure would be considered in our algorithm).

Using these heuristics, the process can be summarized as follows: when a NH triangle T is considered, two different cases may occur during the simplification process (see Figure 4):

- remove an edge from T : in this case, the edge is collapsed into a point. Furthermore, such a point belongs to a face of the MC cell containing T. Hence, a 2D processing is used to constraint the new point to be in the MC cell (see Figure 5).
- Remove a triangle : the triangle is collapsed into a single point and we have to ensure that the point belongs to the MC cell.

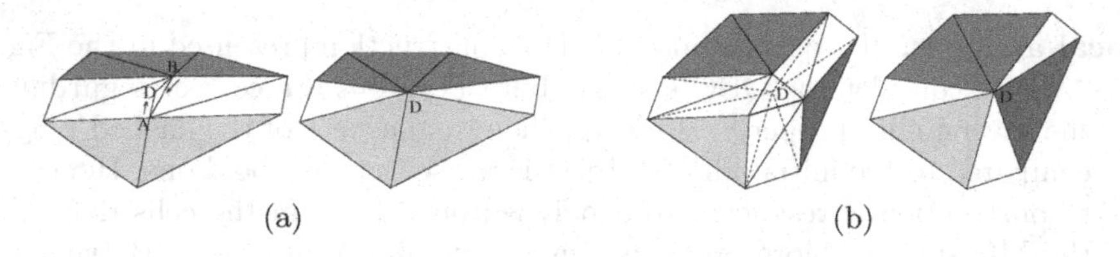

Fig. 4. Illustration of the removal of an edge (a) and a triangle (b) of the MC surface

Let T be a NH-triangle, to check if an edge of T can be removed, we consider the three MC cell faces on which T edges are defined (see Figure 5). From the three edges of T, at least one out of the three edges of T is such that its vertices do not belong to the same discrete plane segment. Let $\mathcal{P}_1$ and $\mathcal{P}_2$ be the two preimages associated to such edge e. The edge e can be removed if for all $P_1 \in \mathcal{P}_1$

and $P_2 \in \mathcal{P}_2$, the intersection of P_1 and P_2 belongs to the MC cell face associated to e. It is not possible to linearly express those conditions without changing the dimension of the linear programming problem. To solve that point, we consider two approaches to obtain sufficient conditions on the intersection of P_1 and P_2.

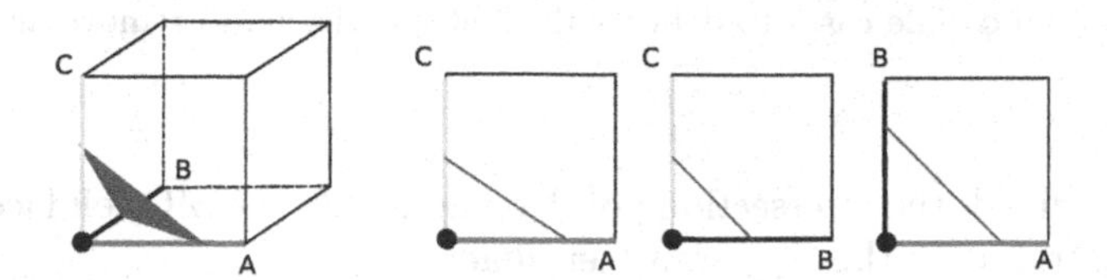

Fig. 5. Illustration of the 2-D decomposition of a MC cell into its faces in order to decide if an edge of a cell triangle can be removed

Global Simplification. First of all, we have a global simplification process to remove NH elements. In this step, we only consider the simple MC configurations, *i.e.* the configurations with a single surface patch (1, 2, 5, 8, 9, 11 in the Figure 1). In the other configurations, we have to check the intersection of the two surface patches and we cannot add linear constraints to ensure the topology during the global simplification. The analysis of these configurations is done during the greedy simplification.

To obtain sufficient conditions on the intersection of P_1 and P_2, we can list three cases (see Figure 6), depending of how many voxels belong to the object on the considered face of the MC cell. If only one voxel A belongs to the object on a face $ABCD$, then the plane P_1 associated to the surfel $\{A, B\}$ crosses the segment CD and P_2 associated to the surfel $\{A, D\}$ crosses the segment BC. Thus we ensure that the intersection of P_1 and P_2 is inside the square. If we consider the case where two voxels belongs to the object on a face, then there is no interesting linear constraints. If we consider the case where only one voxel C does not belong to the object on a face $ABCD$, then we will have the plane P_1 associated to surfel $\{D, C\}$ cross the segment AB and the plane P_2 associated to surfel $\{B, C\}$ cross the segment AD. As in the first case, those conditions ensure that the intersection of the two planes is inside the square (see Figure 6). Finally, these constraints lead to simple linear constraints in dimension 3 that reduce both the preimages $\mathcal{P}_1$ and $\mathcal{P}_2$ to preimages $\mathcal{P}'_1$ and $\mathcal{P}'_2$. Hence, if $\mathcal{P}'_1$ and $\mathcal{P}'_2$ are not empty, whatever $P_1 \in \mathcal{P}'_1$ and $P_2 \in \mathcal{P}'_2$, the intersection of P_1 and P_2 belongs to the face $ABCD$ of the MC cell, ensuring the reversibility of the modified surface. If one of the two preimages is empty, the edge is not removed.

Greedy Simplification. This step consists in fixing planes one by one, to have more flexible constraints on the preimage of the remaining planes, and to be able to handle more cases. So the scheme is to fix one Euclidean plane P_1 (arbitrarily chosen in its associated preimage $\mathcal{P}_1$). Then, if T is a NH triangle associated to the DPS represented by the Euclidean plane P_1 and another DPS with preimage $\mathcal{P}_2$, we insert linear constraints on $\mathcal{P}_2$ to control the intersection between P_1 and

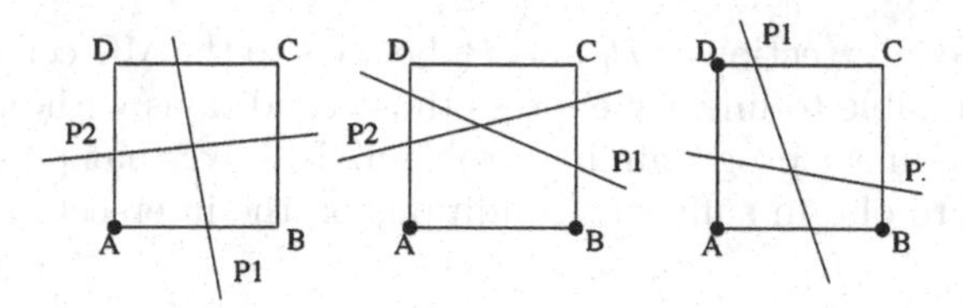

Fig. 6. The three possible cases to define sufficient conditions to remove an edge of a NH triangle

P_2. Since P_1 is fixed, the intersection point being inside the MC cell face is given by linear constraints in the $\mathcal{P}_2$ parameter space.

Indeed let us consider a plane P_1 and a mobile plane P_2 on a face $ABCD$ (see Figure 7), if I is the intersection of P_1 and P_2, to ensure that I is inside the square $ABCD$, we have the constraints:

$$\begin{cases} x_A < x_I < x_A + 1 \\ y_A < y_I < y_A + 1 \end{cases}$$

As x_A and y_A are constants and x_I, y_I only depend on the $\mathcal{P}_2$ parameters, these inequalities result in linear constraints.

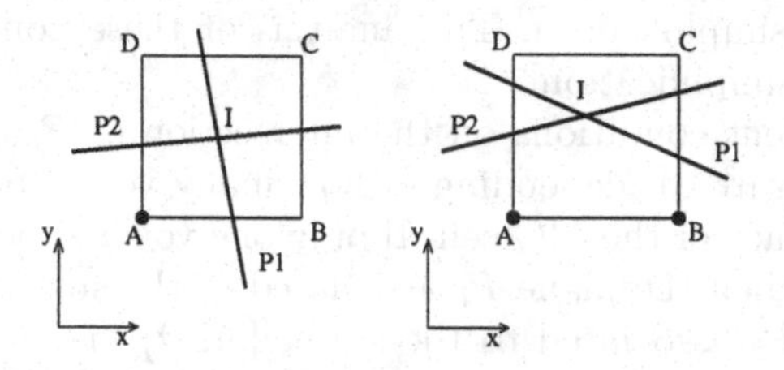

Fig. 7. Illustration of the greedy simplification approach

Finally, if we fix a plane P_1 for a DPS, we propagate this piece of information to each neighboring DPS preimages. This process is greedy since we do not backtrack on the choice of P_1. Once all neighboring DPS have been considered, the greedy step can choose another Euclidean plane in another preimage and the process starts over.

Concerning 3-NH triangles, we need to ensure that the intersection of the 3 planes is inside the MC cell associated to the triangle. To do so, we need 2 of the 3 associated planes to be fixed to get linear constraints from the inequalities:

$$\begin{cases} x_A < x_I < x_A + 1 \\ y_A < y_I < y_A + 1 \\ z_A < z_I < z_A + 1 \end{cases}$$

Furthermore, we need to constrain the planes such that their intersection two by two with the associated face of the MC cell is inside that face. This leads to the same constraints as in the 2-NH case.

3.3 Overall Algorithm

In this section, we sketch the overall simplification algorithm based on the two approaches presented above.

The first step is to convert the discrete object into a triangular polyhedron, this is done with the Marching Cubes algorithm previously presented. The next step is to segment the discrete object surface into DPS, each of these segments being associated with their preimage. Note that we only consider the NH triangles such that their associated MC configuration is in $(1, 2, 5, 8, 9, 11)$ (see Figure 1). Indeed, other configurations lead to two or three components of the MC surface and defining sufficient conditions to avoid self-crossings of the surface using constraints in dimension 3 would have led to too restrictive conditions. Hence these triangles are not optimized.

In the first place, we perform a global optimization. We have an unsorted list of all NH triangles processed one by one. If we have a 2-NH triangle, we can arbitrarily remove any of the two edges since it does not change anything on the final number of facets. The removal consists in adding constraints over the two planes. If a constraint makes one of the preimage empty, then the constraints are removed, and the removal is handled in the next optimization. In this step, we do not handle the 3-NH because we cannot write linear constraints for a triangle removal.

When all triangles have been processed, we start with the second step of the NH removal. We arbitrarily choose a 2-NH triangle and fix one of its planes with the barycenter of its preimage. When this is done, we add constraints over the second plane to remove that triangle, if possible. Then we move to an adjacent triangle and repeat the process. If it is a 3-NH, we skip it until two out of its planes are fixed. When a triangle cannot be removed and the list of adjacent triangles is empty, we choose a new 2-NH triangle and apply the same process on that one. At the end of the process, all planes have been fixed and we can displace all vertices on the intersection of their euclidean plane and their $]pq[$ segment. Finally, we group coplanar triangles into polygons.

This algorithm can be sketched as follows:

1. Computation of the MC surface
2. Decomposition of the discrete object surface into DPS
3. Optimization on NH triangles, i.e. find an Euclidean plane in each DP preimage:
   (a) Step 1: global optimization, processing of all 2-NH triangles
   (b) Step 2: greedy optimization, fixing planes one by one to try to remove remaining NH triangles
4. Vertices displacements and simplification of coplanar triangles.

Lemma 4. *The algorithm presented above constructs a reversible polyhedron which is a combinatorial 2-manifold.*

Proof. The proof is straightforward according to Property 2. To prove the topology, since the MC surface is a combinatorial 2-manifold [8,12] and we can locally prove that treatments on both H triangle and NH triangle do not change the topology: no holes are created, no self-crossings are introduced since we remain on the MC cell, and both the orientation and the combinatorial aspects of the surface are maintained. Hence, the final overall surface is still a combinatorial 2-manifold (see [4] for details on the H triangle treatment). Furthermore, since each new element (facets and vertices) belongs to the MC cells in which the surface is defined, the OBQ digitization of the final polyhedron exactly corresponds to the input set of voxels. Note that since the topology is preserved, the polyhedral surface is still oriented and the OBQ digitization scheme is still well defined. □

The computational cost of the algorithm is highly dependent on the size of the DPS preimages [15,17]. Indeed, a bound on these objects allow us to have computational costs for the DPS decomposition step and for the efficiency of the linear programming solver used to reduce the preimages.

4 Experiments and Results

In the experimentation, the digital plane segmentation has been performed using an implementation of the algorithm proposed in [3]. The output of this algorithm is a labeling of each surfel with a digital plane segment label, associated to a

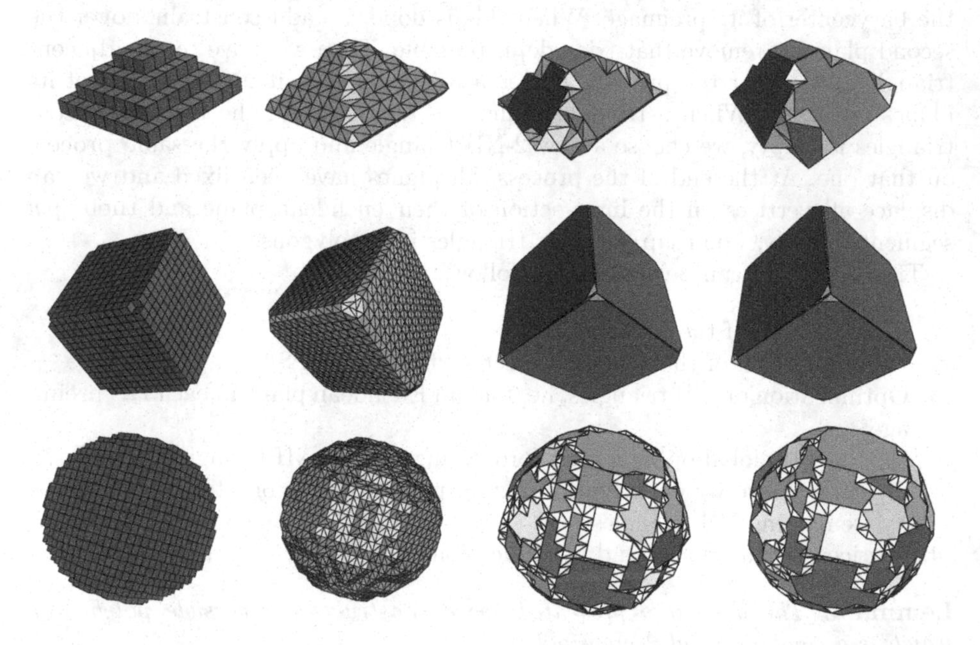

Fig. 8. Comparison between the normal simplification, and the simplification including NH triangles removal. From left to right : facetized voxel representation, MC surface, normal simplfication, and simplification using NH triangles removal.

Table 1. Some results of the presented work

	MC			Removal rate	
object	# MC triangles	# H triangles	# NH triangles	NH triangles	global
pyramid_4	512	342	170	62%	87%
rd_cube_7	2024	1720	304	89%	98%
sphere_10	3656	2200	1456	37%	75%

preimage. The modification of the preimages during the NH-triangle simplification have been performed using a linear programming library in dimension 3[1].

Figure 8 and Table 1 show some experiments. We can notice that in all presented cases, the global removal rate is always greater than 75% which also holds for most experimented objects. The NH triangle removal shows an improvement of at least 30% and up to 60% compared to the initial algorithm. On the rounded cube, the algorithm could remove almost all triangles keeping only one polygon for each face, one for some edges, and some triangles for corners where the algorithm was not really efficient. On the sphere we see one of the worst result of the algorithm which is still decent, the digital plane recognition showed pretty good result considering we were processing a sphere and the NH simplification could remove a good part of the remaining triangles.

5 Conclusion

In this article, we have presented an algorithm to construct a reversible polyhedron from a digital plane segmentation of a binary object. Once the digital plane segmentation is computed, the proposed algorithm is based on a simplification and an optimization of the Marching-Cubes surface. The next step for this work would be to perform exhaustive comparisons between this algorithm and classical simplification schemes of MC surfaces in the Modeling community according to the number of remaining facets. Note that compared to these algorithms, the reversible property ensured by our technique is an important advantage. Note that to extend this work to handle large volumes, we only have the bottleneck implied by the digital plane decomposition step: to have exact computations, a rational arithmetic must be used to recognize the DPS. When large digital plane segments are considered, the arithmetical size of internal rational numbers quickly increases. Hence, further preliminary analysis on the computational aspects of the DPS recognition are required.

References

1. Françon, J., Papier, L.: Polyhedrization of the boundary of a voxel object. In: 8th International Workshop in Discrete Geometry for Computer Imagery, Springer-Verlag, LNCS, 1568 (1999) 425–434

[1] http://www.cs.washington.edu/research/constraints/cassowary/

2. Debled-Rennesson, I.: Etude et reconnaissance des droites et plans discrets. PhD thesis, Université Louis Pasteur (1995)
3. Sivignon, I., Dupont, F., Chassery, J.M.: Decomposition of a three-dimensional discrete object surface into discrete plane pieces. Algorithmica **38**(1) (2004) 25–43
4. Coeurjolly, D., Guillaume, A., Sivignon, I.: Reversible discrete volume polyhedrization using marching cubes simplification. In: SPIE Vision Geometry XII. Volume 5300., San Jose, USA (2004) 1–11
5. Sivignon, I., Dupont, F., Chassery, J.M.: Reversible polygonalization of a 3D planar discrete curve: Application on discrete surfaces. In: International Conference on Discrete Geometry for Computer Imagery. (2005) 347–358
6. Dexet., M.: Design of a topology based geometrical discrete modeler and reconstruction methods in 2D and 3D. PhD thesis, Laboratoire SIC, Université de Poitiers (2006) (in French).
7. Lorensen, W.E., Cline, H.E.: Marching cubes: a high resolution 3D surface construction algorithm. In Stone, M.C., ed.: SIGGRAPH '87 Conference Proceedings (Anaheim, CA, July 27–31, 1987), Computer Graphics, Volume 21, Number 4 (1987) 163–170
8. Lachaud, J.O.: Topologically defined isosurfaces. In: 6th Discrete Geometry for Computer Imagery, LNCS 1176, Springer-Verlag (1996) 245–256
9. Borianne, P., Francon, J.: Reversible polyhedrization of discrete volumes. In: 4th Discrete Geometry for Computer Imagery, Grenoble, France (1994) 157–168
10. Vittone, J., Chassery, J.M.: Recognition of digital naive planes and polyhedization. In: Discrete Geometry for Computer Imagery. Number 1953 in LNCS, Springer (2000) 296–307
11. Klette, R., Sun, H.J.: Digital planar segment based polyhedrization for surface area estimation. In Arcelli, C., Cordella, L.P., Sanniti di Baja, G., eds.: International Workshop on Visual Form 4. Volume 2059 of Lect. Notes Comput. Sci., Springer-Verlag (2001) 356–366
12. Lachaud, J.O., Montanvert, A.: Continuous analogs of digital boundaries: A topological approach to iso-surfaces. Graphical Models and Image Processing **62** (2000) 129–164
13. Klette, R., Rosenfeld, A.: Digital Geometry: Geometric Methods for Digital Picture Analysis. Series in Computer Graphics and Geometric Modelin. Morgan Kaufmann (2004)
14. Brimkov, V., Coeurjolly, D., Klette, R.: Digital planarity - a review. Technical Report RR-2004-024, Laboratoire LIRIS, Université Claude Bernard Lyon 1 (2004)
15. Coeurjolly, D., Sivignon, I., Dupont, F., Feschet, F., Chassery, J.M.: On digital plane preimage structure. Discrete Applied Mathematics **151**(1–3) (2005) 78–92
16. Andrès, E.: Discrete linear objects in dimension n: the standard model. Graphical models **65**(1–3) (2003) 92–111
17. Coeurjolly, D., Brimkov, V.: Computational aspects of digital plane and hyperplane recognition. In: 11th International Workshop on Combinatorial Image Analysis. Volume 4040 of Lecture Notes in Computer Science., Springer-Verlag (2006) 291–306

Arithmetic Discrete Hyperspheres and Separatingness

Christophe Fiorio and Jean-Luc Toutant

LIRMM - CNRS UMR 5506 - Université de Montpellier II
161 rue Ada - 34392 Montpellier Cedex 5, France
{fiorio, toutant}@lirmm.fr

Abstract. In the framework of the arithmetic discrete geometry, a discrete object is provided with its own analytical definition corresponding to a discretization scheme. It can thus be considered as the equivalent, in a discrete space, of a Euclidean object. Linear objects, namely lines and hyperplanes, have been widely studied under this assumption and are now deeply understood. This is not the case for discrete circles and hyperspheres for which no satisfactory definition exists. In the present paper, we try to fill this gap. Our main results are a general definition of discrete hyperspheres and the characterization of the k-minimal ones thanks to an arithmetic definition based on a non-constant thickness function. To reach such topological properties, we link adjacency and separatingness with norms.

1 Introduction

Discrete lines, namely the analogue of Euclidean lines in the discrete space $\mathbb{Z}^2$, have been widely studied. At first, J. Bresenham [1], H. Freeman [2] and A. Rosenfeld [3] have followed an algorithmic approach and have defined them as digitizations of Euclidean lines. They have provided tools for drawing and recognition [1, 2, 3]. Later, J.-P. Reveillès has initiated the arithmetic discrete geometry [4] and has introduced the notion of arithmetic discrete line as solutions of a system of Diophantine inequalities, that is, as the subset of $\mathbb{Z}^2$ contained in a band. Such an approach enhances the knowledge of discrete lines. In addition to give new drawing [4] and recognition [5] algorithms, it directly links topological and geometrical properties of an arithmetic discrete line with its definition. For instance, its connectedness is entirely characterized by the width of the band, that is, its arithmetic thickness. In d-dimensional discrete spaces, the arithmetic discrete hyperplane [6] is a natural generalization of the arithmetic discrete line [4].

Similarly, first investigations into discrete circles have been algorithmic ones [7, 8, 9]. Discrete circles were only considered as digitizations of Euclidean circles. It is thus natural to ask whether or not J.-P. Reveillès' arithmetic approach is extendable to discrete circles and can supply them with an arithmetic definition. Such an extension has been proposed by É. Andres [10]. He

A. Kuba, L.G. Nyúl, and K. Palágyi (Eds.): DGCI 2006, LNCS 4245, pp. 425–436, 2006.

has defined the discrete analytical hypersphere as solutions of a system of Diophantine inequalities, or in other words, as the subsets of $\mathbb{Z}^d$ contained in a ring of width its arithmetic thickness. Concentric discrete analytical hyperspheres tile the discrete space, but one does not retrieve topological properties as strict k-connectedness or k-minimality. Recently, an arithmetic definition implying a non-constant thickness function was proposed in [11]. It provides discrete circles of integer parameters with 0-minimality or 1-minimality and gives an arithmetic characterization of the well-known Bresenham's circle [7].

In the present paper, we focus on discrete hyperspheres and on the arithmetic approach to generalize results of [11]. We deeper study the notion of thickness. In particular, we characterize the k-minimal arithmetic discrete hypersphere by relating the thickness function to a particular norm, the k-minimality one, and to the local behaviour of the hypersphere.

The paper is organized as follows. First, we begin with some recalls on discrete geometry useful to fully understand the matter. Second, works already done on the discrete analytical hypersphere [10] and the arithmetic discrete circle [11] are presented. A general definition is then proposed to unify the both approaches. Next, we focus on the topological properties of d-dimensional discrete sets to characterize the k-minimal arithmetic discrete hypersphere.

2 Basic Notions

The aim of this section is to introduce the basic notions of discrete geometry used throughout the present paper. Let d be an integer greater than 1 and let $\{\mathbf{e_1}, \ldots, \mathbf{e_d}\}$ denote the canonical basis of the Euclidean vector space $\mathbb{R}^d$. Let us call *discrete set* any subset of the *discrete space* $\mathbb{Z}^d$. The point $\mathbf{x} = \sum_{i=1}^{d} x_i \mathbf{e_i} \in \mathbb{R}^d$, with $x_i \in \mathbb{R}$ for each $i \in \{1, \ldots, d\}$, is represented by $(x_1, \ldots, x_d)$. A point $\mathbf{v} \in \mathbb{Z}^d$ is a *voxel* in a d-dimensional space.

Definition 1 (k-adjacency or k-neighborhood). *Let d be the dimension of the discrete space and $k \in \mathbb{N}$ such that $k < d$. Two voxels $\mathbf{v} = (v_1, \ldots, v_d)$ and $\mathbf{w} = (w_1, \ldots, w_d)$ are* k-neighbors *or* k-adjacent *if and only if:*

$$\|\mathbf{v}-\mathbf{w}\|_\infty = \max\{|v_1-w_1|, \ldots, |v_d-w_d|\} = 1 \text{ and } \|\mathbf{v}-\mathbf{w}\|_1 = \sum_{i=1}^{d} |v_i-w_i| \leq d-k \,.$$

Let $k \in \{0, \ldots, d-1\}$. A discrete set E is said to be *k-connected* if for each pair of voxels $(\mathbf{v}, \mathbf{w}) \in \mathrm{E}^2$, there exists a finite sequence of voxels $(\mathbf{s_1}, \ldots, \mathbf{s_p}) \in \mathrm{E}^p$ such that $\mathbf{v} = \mathbf{s_1}$, $\mathbf{w} = \mathbf{s_p}$ and the voxels $\mathbf{s_j}$ and $\mathbf{s_{j+1}}$ are k-neighbors, for each $j \in \{1, \ldots, p-1\}$.

Let E be a discrete set, $\mathbf{v} \in \mathrm{E}$ and $k \in \{0, \ldots, d-1\}$. The *k-connected component* of $\mathbf{v}$ in E is the maximal k-connected subset of E (with respect to set inclusion) containing $\mathbf{v}$.

Definition 2 (k-separating set). *A discrete set* E *is* k-separating *in a discrete set* F *if its complement in* F, $\overline{\mathrm{E}} = \mathrm{F} \setminus \mathrm{E}$, *has two distinct k-connected components.* E *is called a* separator *of* F.

Definition 3 (k-simple point, k-minimality). *Let d be the dimension of the space and $k \in \mathbb{N}$ such that $k < d$. Let also* F *and* E *be two discrete sets such that* E *is k-separating in F. A voxel $\mathbf{v} \in$ E is said to be k-*simple *if* $\mathrm{E} \setminus \{\mathbf{v}\}$ *remains k-separating in* F*. Moreover, a k-separating discrete set in* F *without k-simple points is said to be k-*minimal *in* F.

3 Arithmetic Discrete Hyperspheres

Initiated by J.-P. Reveillès with the arithmetic definition of the discrete line [4], the arithmetic discrete geometry has led to a wide literature, mainly about linear objects. Discrete lines [4], planes and hyperplanes [6] are now well characterized and only one parameter, the *arithmetic thickness* ω, controls their topological properties. As far as we know, no satisfactory generalization to non-linear objects exists. É. Andres has defined the *discrete analytical hypersphere* [10] so as to obtain concentric hyperspheres tiling the space.

Definition 4 (Discrete analytical hypersphere [10]). *Let d be the dimension of the space. Let $r \in \mathbb{R}_+^\star$, $\mathbf{o} = (o_1, \ldots, o_d) \in \mathbb{R}^d$ and $\omega \in \mathbb{R}_+^\star$. The* discrete analytical hypersphere $\mathbb{S}(\mathbf{o}, r, \omega)$ *of* center $\mathbf{o}$, radius r *and* arithmetic thickness ω*, is the subset of $\mathbb{Z}^d$ defined by:*

$$\mathbb{S}(\mathbf{o}, r, \omega) = \left\{ \mathbf{v} \in \mathbb{Z}^d \mid \left(r - \frac{\omega}{2}\right)^2 \leq \sum_{i=1}^{d} (v_i - o_i)^2 < \left(r + \frac{\omega}{2}\right)^2 \right\} . \quad (1)$$

However, since no topological characterization is possible with this definition, the notion of arithmetic discrete circle was proposed in [11]. Its analytical description is not based on the usual constant arithmetic thickness ω, but on a *thickness function* $\omega : \mathbb{R}^2 \longrightarrow \mathbb{R}_+^\star$. The importance of keeping apart the analytical expression of the considered curve and the approximation induced by the discrete space, that is, the thickness, was also highlighted. Finally, those considerations lead to partial topological results on discrete circles. The *naive* and *standard* circles with integer parameters were characterized and Bresenham's circle [7] was provided with an arithmetic definition. For that, the thickness function is regarded as a measurement, by the usual discrete norms $\|\cdot\|_1$ and $\|\cdot\|_\infty$, of the local behaviour of the circle.

Definition 5 (Arithmetic discrete circle [11]). *Let $\mathbf{o} = (o_1, o_2) \in \mathbb{R}^2$, $r \in \mathbb{R}_+$. Let $\omega : \mathbb{R}^2 \longrightarrow \mathbb{R}$ be a map. The* arithmetic discrete circle $\mathrm{C}(\mathbf{o}, r, \omega)$ *of* center $\mathbf{o}$, radius r *and* thickness function ω *is the following set:*

$$\mathrm{C}(\mathbf{o}, r, \omega) = \left\{ \mathbf{v} \in \mathbb{Z}^2 \mid -\frac{\omega(\mathbf{v})}{2} \leq (v_1 - o_1)^2 + (v_2 - o_2)^2 - r^2 < \frac{\omega(\mathbf{v})}{2} \right\} . \quad (2)$$

In the present paper, we improve this last definition and generalize it to discrete hyperspheres, namely the d-dimensional case.

Definition 6 (Arithmetic discrete hypersphere). *Let d be the dimension of the space, $r \in \mathbb{R}_+^*$ and $\mathbf{o} = (o_1, \ldots, o_d) \in \mathbb{R}^d$. Let $\omega_1 : \mathbb{R}^d \longrightarrow \mathbb{R}_-$ and $\omega_2 : \mathbb{R}^d \longrightarrow \mathbb{R}_+$ be maps. The* arithmetic discrete hypersphere $\mathrm{S}(\mathbf{o}, r, \omega_1, \omega_2)$ *with* center $\mathbf{o}$, radius r *and* thickness functions ω_1 *and* ω_2 *is:*

$$\mathrm{S}(\mathbf{o}, r, \omega_1, \omega_2) = \left\{ \mathbf{v} \in \mathbb{Z}^d \mid \omega_1(\mathbf{v}) \leq \sum_{i=1}^{d} (v_i - o_i)^2 - r^2 < \omega_2(\mathbf{v}) \right\} . \quad (3)$$

Contrary to Definition 5, Definition 6 includes two distinct thickness functions ω_1 and ω_2. That way, the global thickness can be distributed more or less inside or outside the hypersphere. This feature is interesting when one consider non-linear analytical expression and when thickness should not be the same inside or outside the curvature.

Such a definition includes all previous attempts to define arithmetically discrete hyperspheres. In particular, the set of discrete analytical hyperspheres (see Definition 4) introduced by É. Andres [10] is included in the set of arithmetic discrete hyperspheres. They are defined using two thickness functions ω_1 and ω_2 such that $\omega_1 \neq -\omega_2$.

Remark 1 (Discrete Analytical Hypersphere [10] and Arithmetic Discrete Hypersphere). The discrete analytical hypersphere $\mathbb{S}(\mathbf{o}, r, \omega)$ is the arithmetic discrete hypersphere $\mathrm{S}(\mathbf{o}, r, \omega_1, \omega_2)$ such that:

$$\begin{array}{rlrl} \omega_1 : \mathbb{R}^d & \longrightarrow \mathbb{R}_- & \qquad \omega_2 : \mathbb{R}^d & \longrightarrow \mathbb{R}_+ \\ \mathbf{x} & \longmapsto -r\omega + \dfrac{\omega^2}{4} & \qquad \mathbf{x} & \longmapsto r\omega + \dfrac{\omega^2}{4} . \end{array}$$

In the same way, the arithmetic discrete circle introduced in [11] is an arithmetic discrete hypersphere.

Remark 2 (Arithmetic Discrete circles [11] and Arithmetic Discrete Hyperspheres). An arithmetic discrete circle $\mathrm{C}(\mathbf{o}, r, \omega)$ is an arithmetic discrete hypersphere $\mathrm{S}(\mathbf{o}, r, \omega_1, \omega_2)$ such that:

$$\begin{array}{rlrl} \omega_1 : \mathbb{R}^d & \longrightarrow \mathbb{R}_- & \qquad \omega_2 : \mathbb{R}^d & \longrightarrow \mathbb{R}_+ \\ \mathbf{x} & \longmapsto -\dfrac{\omega(\mathbf{x})}{2} & \qquad \mathbf{x} & \longmapsto \dfrac{\omega(\mathbf{x})}{2} . \end{array}$$

Definition 6 allows to build a considerable amount of discrete objects and some do not look like what is expected from an hypersphere. Consequently, as in the case of the arithmetic discrete circle [11], suitable thickness functions are needed to characterize discrete hyperspheres with basic topological properties.

4 Topology of d-Dimensional Discrete Sets

Before defining suitable thickness functions, we have to determine which topological properties are meaningful for hyperspheres. For that, we focus on the more general case of d-dimensional discrete sets. The most studied one is the arithmetic discrete hyperplane [6] and the minimality (or the separatingness) is its best characterized topological property [6]. It seems anyway to be the most evident property for all d-dimensional discrete objects since it intuitively refers to the notion of objects without holes.

4.1 k-Adjacency Norm

The k-separatingness, and so the k-minimality, are reached when none of the voxels on one side of the separator is k-adjacent with a voxel on the other side. The definition of k-adjacency given in Section 2 is not easy to handle. Hence, we propose an equivalent and more formal expression. Indeed, two voxels are k-adjacent if they share at least k coordinates and among the others, the greatest difference is equal to 1.

Theorem 1. *Let d be the dimension of the space. Let $k \in \mathbb{N}$ such that $k < d$. Two voxels $\mathbf{v} = (v_1, \dots, v_d)$ and $\mathbf{w} = (w_1, \dots, w_d)$ are* k-neighbor *or* k-adjacent *if and only if:*

$$|v_{\sigma(d)} - w_{\sigma(d)}| + \sum_{i=1}^{k} |v_{\sigma(i)} - w_{\sigma(i)}| = 1 , \tag{4}$$

with σ the permutation of the set $\{1, \dots, d\}$ such that, for all $i \in \{1, \dots, d-1\}$, $|v_{\sigma(i)} - w_{\sigma(i)}| \leq |v_{\sigma(i+1)} - w_{\sigma(i+1)}|$.

Proof

$$\begin{aligned}
\mathbf{v} \text{ and } \mathbf{w} \text{ are } k\text{-adjacent.} &\Leftrightarrow \max_{1 \leq i \leq d} \{|v_i - w_i|\} = 1 \text{ and } \sum_{i=1}^{d} |v_i - w_i| \leq d - k. \\
&\Leftrightarrow |v_{\sigma(d)} - w_{\sigma(d)}| = 1 \text{ and } \sum_{i=1}^{k} |v_{\sigma(i)} - w_{\sigma(i)}| = 0. \\
&\Leftrightarrow |v_{\sigma(d)} - w_{\sigma(d)}| + \sum_{i=1}^{k} |v_{\sigma(i)} - w_{\sigma(i)}| = 1.
\end{aligned}$$

In fact, the k-adjacency can be characterized by a norm that we call *k-adjacency norm.*

Proposition 1. *Let d be the dimension of the space and $k \in \mathbb{N}$ such that $k < d$. The map $[\cdot]_k$ defined by:*

$$\begin{aligned}
[\cdot]_k : \mathbb{R}^d &\longrightarrow \mathbb{R} \\
\mathbf{x} &\longmapsto |x_{\sigma(d)}| + \sum_{i=1}^{k} |x_{\sigma(i)}| ,
\end{aligned} \tag{5}$$

with σ the permutation of the set $\{1, \dots, d\}$ such that, for all $i \in \{1, \dots, d-1\}$, $|x_{\sigma(i)}| \leq |x_{\sigma(i+1)}|$, is a norm.

Proof. Here, *positivity*, *scalability* and *triangle inequality* are evident properties.

Definition 7 (k-adjacency norm). *Let d be the dimension of the space and $k \in \mathbb{N}$ such that $k < d$. We call* k-adjacency norm, $[\cdot]_k$.

Notice that the k-adjacency norm is related to usual discrete norms $\|\cdot\|_1$ and $\|\cdot\|_\infty$. Let d be the dimension of the space and $\mathbf{v} \in \mathbb{R}^d$, then $[\mathbf{v}]_0 = \|\mathbf{v}\|_\infty$ and $[\mathbf{v}]_{(d-1)} = \|\mathbf{v}\|_1$.

4.2 (λ, k)-Adjacency, (λ, k)-Separating Set and (λ, k)-Hull of a Voxel Set

In Definition 6, we propose to distribute the thickness inside and outside the hypersphere. From a practical point of view, we can then consider the arithmetic discrete hypersphere as the union of two discrete hyperspheres, an outer one and an inner one. Since we are interesting in k-minimal discrete hyperspheres, inner and outer constituting hyperspheres can be thinner than k-minimal ones. So we need a notion more general than k-separatingness to define them. Since we characterize k-adjacency with a norm, we can extend it from the discrete space to the continuous one and define the (λ, k)-adjacency where distance between adjacent points can be a real number. It follows a generalization of the k-separating sets, the (λ, k)-separating sets. We thus control precisely the thickness of the set of voxels and can distribute the thickness as mentioned above.

Definition 8 ((λ, k)-adjacency). *Let d be the dimension of the space and $k \in \mathbb{N}$ such that $k < d$. Let $\mathbf{x} \in \mathbb{R}^d$ and $\mathbf{y} \in \mathbb{R}^d$. The d-dimensional points $\mathbf{x}$ and $\mathbf{y}$ are (λ, k)-adjacent if and only if $[\mathbf{x} - \mathbf{y}]_k \leq \lambda$.*

From (λ, k)-adjacency, the notion of (λ, k)-connected sets follows. It is thus natural to define (λ, k)-separating sets from (λ, k)-adjacency.

Definition 9 ((λ, k)-separating set). *A set E is (λ, k)-separating in a set F if its complements in F, $\overline{\mathrm{E}} = \mathrm{F} \backslash \mathrm{E}$, has two distinct (λ, k)-connected components.*

With Definition 9, we can distinguish two parts in a set of voxels $\mathrm{E} \subset \mathrm{U}$, its discrete (λ, k)-hull, which contains its voxels (λ, k)-adjacent with the exterior, and its (λ, k)-interior, (λ, k)-separated from the exterior.

Definition 10 ((λ, k)-hull and (λ, k)-interior). *Let d be the dimension of the space, $k \in \mathbb{N}$ such that $k < d$ and $\lambda \in \mathbb{R}$. Let also O be a set of voxels. The (λ, k)-hull, $\mathrm{H}_{(\lambda,\mathrm{k})}(\mathrm{O})$, of O based on the* normal thickness λ *relatively to the k-adjacency norm $[\cdot]_k$ is defined as follows:*

$$\mathrm{H}_{(\lambda,\mathrm{k})}(\mathrm{O}) = \left\{\mathbf{v} \in \mathrm{O} \mid \exists \mathbf{w} \in \mathbb{Z}^d \backslash \mathrm{O}, [\mathbf{v} - \mathbf{w}]_k \leq \lambda\right\} . \quad (6)$$

The (λ, k)-interior of O is then $\mathrm{I}_{(\lambda,\mathrm{k})}(\mathrm{O}) = \mathrm{O} \backslash \mathrm{H}_{(\lambda,\mathrm{k})}(\mathrm{O})$.

We focus on particular sets of voxels, the ones described by an analytical expression. So we now take care of the discrete hull of a discrete set defined in such a way. Let $f : \mathbb{R}^d \longrightarrow \mathbb{R}$ be a map. Then we define the discrete set $\mathrm{O}(f,+)$ such that:

$$\mathrm{O}(f,+) = \left\{\mathbf{v} \in \mathbb{Z}^d \mid f(\mathbf{v}) \geq 0\right\} .$$

Taking into account the map f, the (λ, k)-hull of the discrete set $\mathrm{O}(f,+)$ can be rewritten as follows:

$$\mathrm{H}_{(\lambda,\mathrm{k})}(\mathrm{O}(f,+)) = \left\{\mathbf{v} \in \mathbb{Z}^d \mid \exists \mathbf{x} \in \mathbb{R}^d, f(\mathbf{x}) = 0 \wedge f(\mathbf{v}) \geq 0 \wedge [\mathbf{v}-\mathbf{w}]_k < \lambda\right\} . \quad (7)$$

According to Equation (7), a (λ, k)-hull can be seen as a discrete object, based on a map f, (λ, k)-separating the discrete space. Unfortunately the set of voxels it contains is difficult to determine since the definition brings into play two different measurements, namely the norm $[\cdot]_k$ and the function f.

5 (λ, k)-Separating Discrete Hyperspheres

The notion of (λ, k)-hull allows to define discrete d-dimensional hyperspheres of center $\mathbf{o} = (o_1, \ldots, o_d) \in \mathbb{Z}^d$ and radius $r \in \mathbb{R}_+^*$ with the particular map s defined by:

$$\begin{aligned} s : \mathbb{R}^d &\longrightarrow \mathbb{R} \\ \mathbf{x} = (x_1, \ldots, x_d) &\longmapsto \textstyle\sum_{i=1}^d (x_i - o_i)^2 - r^2 . \end{aligned}$$

To go further, we consider the restriction of s, $s_E : E \longrightarrow \mathbb{R}$, to the subspace $E \subset \mathbb{R}^d$ such that:

$$E = \left\{\mathbf{x} = (x_1, \ldots, x_d) \in \mathbb{R}^d \mid \forall i \in \{1, \ldots, d\}, x_i - o_i \geq 0\right\} .$$

The (λ, k)-hull $\mathrm{H}_{(\lambda,\mathrm{k})}(\mathrm{O}(s_E,+))$ of the discrete set $\mathrm{O}(s_E,+)$ in E is then:

$$\begin{aligned} \mathrm{H}_{(\lambda,\mathrm{k})}(\mathrm{O}(s_E,+)) = \{\mathbf{v} \in (\mathbb{Z}^d \cap E) \mid \exists x \in E, s_E(\mathbf{x}) = 0 \wedge s_E(\mathbf{v}) \geq 0 \\ \wedge [\mathbf{v}-\mathbf{w}]_k < \lambda\} . \end{aligned}$$

Since $s_E(\mathbf{x}) = 0$ and $s_E(\mathbf{v}) \geq 0$, we have the following equalities:

$$\begin{aligned} s_E(\mathbf{v}) &= |s_E(\mathbf{v}) - s_E(\mathbf{x})| , \\ &= \left|\sum_{i=1}^d 2(v_i - o_i)(v_i - x_i) - (v_i - x_i)^2\right| . \end{aligned}$$

Since $\forall i \in \{1, \ldots, d\}, v_i - o_i \geq 0$, the absolute values are not required:

$$s_E(\mathbf{v}) = \sum_{i=1}^d 2(v_i - o_i)(v_i - x_i) - (v_i - x_i)^2 .$$

With the last condition on $\mathrm{H}_{(\lambda,\mathrm{k})}(\mathrm{O}(s_E,+))$, $[\mathbf{v}-\mathbf{x}]_k < \lambda$, we give an upper bound on $s_E(\mathbf{v})$ and:

$$\mathrm{H}_{(\lambda,\mathrm{k})}(\mathrm{O}(s_E,+)) = \left\{\mathbf{v} \in (\mathbb{Z}^d \cap E) \mid 0 \leq s_E(\mathbf{v}) < \sum_{i=d-k}^{d} 2\lambda(v_{\sigma(i)} - o_{\sigma(i)}) - \lambda^2\right\} ,$$

with σ the permutation of the set $\{1, \dots, d\}$ such that, for all $i \in \{1, \dots, d-1\}$, $|v_{\sigma(i)} - o_{\sigma(i)})| \leq |v_{\sigma(i+1)} - o_{\sigma(i+1)})|$. Finally, thanks to the symmetries of s, we extend this last result to $\mathrm{H}_{(\lambda,k)}(\mathrm{O}(s,+))$:

$$\mathrm{H}_{(\lambda,k)}(\mathrm{O}(s,+)) = \left\{ \mathbf{v} \in \mathbb{Z}^d \mid 0 \leq s(\mathbf{v}) < \sum_{i=d-k}^{d} \left|2\lambda|v_{\sigma(i)} - o_{\sigma(i)}| - \lambda^2\right| \right\} . \quad (8)$$

According to Definition 6 and intrinsic separating properties, the discrete set $\mathrm{H}_{(\lambda,k)}(\mathrm{O}(s,+))$ is a good candidate for being an arithmetic discrete hypersphere. Before introducing such a definition, we propose notations to express the upper bound of the inequality in expression (8). Indeed, one can see it as a norm, depending on k, applied on a vector, depending on the local behaviour of s.

Proposition 2. *Let d be the dimension of the space and $k \in \mathbb{N}$ such that $k < d$. The map $]\cdot[_k$ defined by:*

$$\begin{aligned}]\cdot[_k : \mathbb{R}^d &\longrightarrow \mathbb{R} \\ \mathbf{x} &\longmapsto \sum_{i=d-k}^{d} |x_{\sigma(i)}| \,, \end{aligned} \quad (9)$$

with σ the permutation of the set $\{1, \dots, d\}$ such that, for all $i \in \{1, \dots, d-1\}$, $|x_{\sigma(i)}| \leq |x_{\sigma(i+1)}|$, is a norm.

Proof. Here, *positivity*, *scalability* and *triangle inequality* are evident properties.

Definition 11 (k-minimality norm). *Let d be the dimension of the space and $k \in \mathbb{N}$ such that $k < d$. We call* k-minimality norm, *the norm* $]\cdot[_k$.

Similarly to the k-adjacency norm, the k-minimality norm is related to usual discrete norms $\|\cdot\|_\infty$ and $\|\cdot\|_1$. One has $]\mathbf{v}[_{(d-1)} = \|\mathbf{v}\|_1$ and $]\mathbf{v}[_0 = \|\mathbf{v}\|_\infty$.

The upper bound in Equation (8) can be considered as the k-minimality norm of a particular vector. To achieve our goal to obtain a thickness depending on the local behaviour of s, and more generally of a function, we propose to define the *Discrete Variation Map* of a function, according to a thickness parameter:

Definition 12 (Discrete variation map). *Let d be the dimension of the space. Let $\lambda \in \mathbb{R}$ and $f : \mathbb{R}^d \longrightarrow \mathbb{R}$ be a function. The* discrete variation map $\Delta_\lambda f$ *of* normal thickness λ *related to the function f is:*

$$\begin{aligned} &\Delta_\lambda f : \mathbb{Z}^d \longrightarrow \mathbb{R}^d \\ &\mathbf{v} \longmapsto \left(f\left(\mathbf{v} + \frac{\partial_1 f(\mathbf{v})}{|\partial_1 f(\mathbf{v})|}\lambda \mathbf{e_1}\right) - f(\mathbf{v}), \dots, f\left(\mathbf{v} + \frac{\partial_d f(\mathbf{v})}{|\partial_d f(\mathbf{v})|}\lambda \mathbf{e_d}\right) - f(\mathbf{v}) \right) . \end{aligned} \quad (10)$$

Now, we can define the (λ, k)-*separating discrete hypersphere* according to the Definition 6 of an arithmetic discrete hypersphere.

Definition 13 ((λ, k)-separating discrete hypersphere). *Let d be the dimension of the space and $k \in \mathbb{N}$ such that $k < d$. Let $\mathbf{o} = (o_1, \ldots, o_d)$ and $r \in \mathbb{R}_+^\star$. Let $\lambda_1 \in \mathbb{R}_-$ and $\lambda_2 \in \mathbb{R}_+$. The* (λ, k)-separating discrete hypersphere $\mathrm{S}(\mathbf{o}, r, \omega_{(\lambda_1,k)}, \omega_{(\lambda_2,k)})$ *with center* $\mathbf{o}$, *radius* r, normal thickness $\lambda = \lambda_2 - \lambda_1$ *and related to the k-minimality norm is defined by:*

$$\mathrm{S}(\mathbf{o}, r, \omega_{(\lambda_1,k)}, \omega_{(\lambda_2,k)}) = \left\{\mathbf{v} \in \mathbb{Z}^d \mid \omega_{(\lambda_1,k)}(\mathbf{v}) \leq s(\mathbf{v}) < \omega_{(\lambda_2,k)}(\mathbf{v})\right\} \ , \qquad (11)$$

with $\omega_{(\lambda_1,k)}(\mathbf{v}) = -]\Delta_{\lambda_1} s(\mathbf{v})[_k$ *and* $\omega_{(\lambda_2,k)}(\mathbf{v}) =]\Delta_{\lambda_2} s(\mathbf{v})[_k$.

Remark 3. In Definition 13, we arbitrarily decide to consider a large inequality outside the hypersphere and a strict one inside it. We also can do the opposite choice without changing the properties stated below.

Due to the underlying notion of (λ, k)-hull, the (λ, k)-separating discrete hypersphere has separating properties.

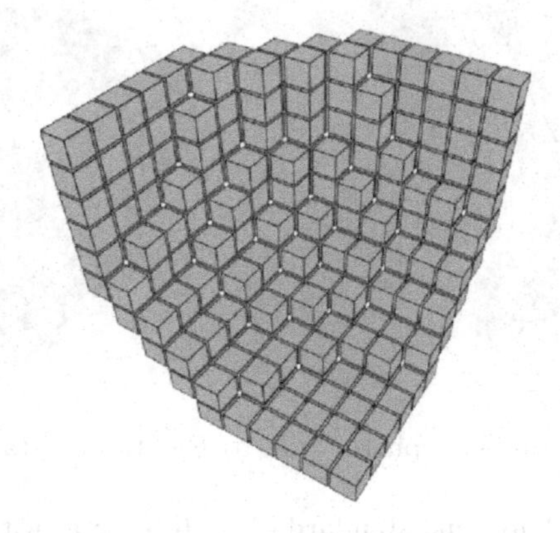

Fig. 1. Eighth of a 1-minimal discrete sphere with radius 9

Theorem 2 (k-separating discrete hypersphere). *The (λ, k)-separating discrete hypersphere* $\mathrm{S}(\mathbf{o}, r, \omega_{(\lambda_1,k)}, \omega_{(\lambda_2,k)})$ *with normal thickness $\lambda = \lambda_2 - \lambda_1$, such that $\lambda \in [1, +\infty[$, is k-separating in $\mathbb{Z}^d$.*

Proof (Sketch). The (λ, k)-hull of a discrete object (λ, k)-separates its interior and the remaining discrete space. So the union of the (λ_1, k)-hull of an object and of the (λ_2, k)-hull of its complement is $(\lambda_2 - \lambda_1, k)$-separating.

In particular, we characterize k-minimal arithmetic discrete hyperspheres. For instance, a part of a 1-minimal sphere is drawn in Figure 1. We notice that voxels in one of the 2-connected components of the exterior are 0-adjacent with voxels of the other 2-connected component as expected.

Theorem 3 (k-minimal discrete hypersphere). *The (λ, k)-separating discrete hypersphere* $\mathrm{S}(\mathbf{o}, r, 0, \omega_{(1,k)})$ *is k-minimal in $\mathbb{Z}^d$ for $k \in \{0, \ldots, d-1\}$. (λ, k)-separating discrete hyperspheres* $\mathrm{S}(\mathbf{o}, r, \omega_{(\varepsilon-1,k)}, \omega_{(\varepsilon,k)})$ *are also k-minimal for $\varepsilon \in [0, 1[$ and $k \in \{1, \ldots, d-1\}$.*

Proof (Sketch). In a discrete hypersphere $\mathrm{S}(\mathbf{o}, r, \omega_{(\varepsilon-1,k)}, \omega_{(\varepsilon,k)})$, a k-simple point is a voxel which is just $(k-1)$-connected with one of both k-connected components of the exterior. Due to the symmetry of the hypersphere, if such a configuration does not appear in the neighborhood of the voxels $\mathbf{v}$ such that $|v_1 - o_1| \simeq \cdots \simeq |v_d - o_d| \simeq (r/\sqrt{d})$, it then appears nowhere and the discrete hypersphere is k-minimal. Finally, we verify that $(1,k)$-separating discrete hyperspheres $\mathrm{S}(\mathbf{o}, r, 0, \omega_{(1,k)})$ with $k \in \{0, \ldots, d-1\}$ and $\mathrm{S}(\mathbf{o}, r, \omega_{(\varepsilon-1,k)}, \omega_{(\varepsilon,k)})$ with $k \in \{1, \ldots, d-1\}$ are k-minimal in $\mathbb{Z}^d$.

Now, we define naive and standard arithmetic discrete hyperspheres as already done for arithmetic discrete hyperplanes. In Figure 2(a), we notice, as expected, that some voxels in one side of the naive discrete hypersphere are 1-adjacent with voxels on the other side. On the contrary, the standard discrete sphere in Figure 2(b) does not have any hole.

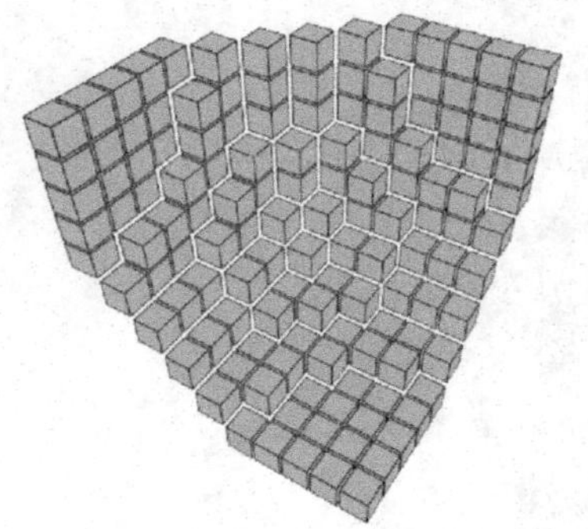

(a) Eighth of a naive discrete sphere

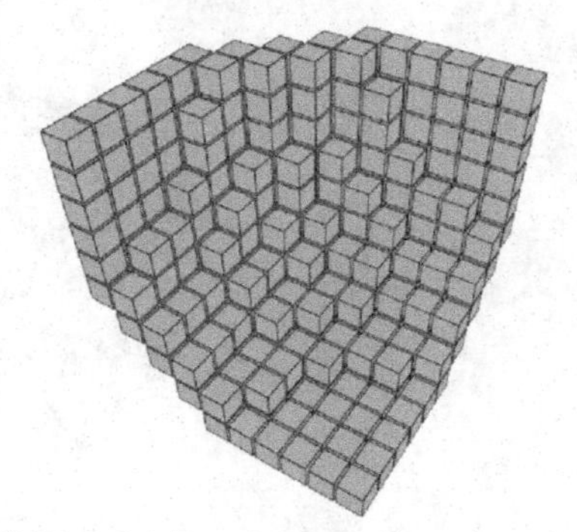

(b) Eighth of a standard discrete sphere

Fig. 2. Naive and standard discrete spheres with radius 9

Definition 14 (Naive and standard discrete hyperspheres). *Let d be the dimension of the space. A* naive *(respectively* standard*) discrete hypersphere is a $(d-1)$-minimal (respectively 0-minimal) one.*

In the d-dimensional discrete space, the most studied object is the discrete hyperplane [6]. It is a generalization of the arithmetic discrete line [4]. Provided with a particular arithmetic thickness, the arithmetic discrete hyperplane presents basic topological properties, that is, the k-minimality [6]. Such a thickness can be defined by combining the k-minimality norm and the discrete variation map associated to the hyperplane. In Figure 3, k-minimal discrete planes are drawn to be compared with k-minimal discrete spheres previously presented.

Proposition 3. *Let $\mu \in \mathbb{R}$, $\mathbf{n} \in \mathbb{R}^d$. Let also $p : \mathbb{R}^d \longrightarrow \mathbb{R}$ be a map such that: $\forall \mathbf{x} \in \mathbb{R}^d, p(\mathbf{x}) = \mathbf{n} \cdot \mathbf{x} + \mu$. The k-minimal arithmetic discrete hyperplane $\mathrm{P}(\mathbf{n}, \mu, \omega_k)$ of normal vector $\mathbf{n}$ and translation parameter μ is defined as:*

$$\mathrm{P}(\mathbf{n}, \mu, \omega_k) = \mathrm{P}(\mathbf{n}, \mu, 0, \omega_{(1,k)}) = \left\{\mathbf{v} \in \mathbb{Z}^d \mid 0 \leq p(\mathbf{v}) <]\Delta_1 p(\mathbf{v})[_k\right\} . \quad (12)$$

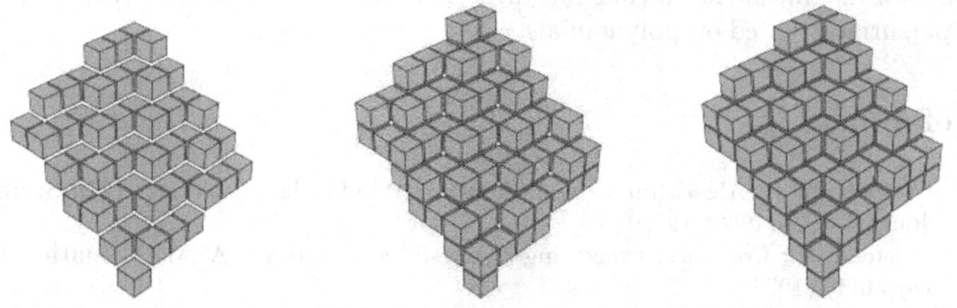

(a) 0-minimal discrete plane (b) 1-minimal discrete plane (c) 2-minimal discrete plane

Fig. 3. k-minimal discrete planes with normal vector $\mathbf{n} = (1, 2, 3)$

Consequently, our characterization is not specific to hypersphere but rather a generalization of what we already known about the discrete hyperplane.

6 Conclusion and Further Research

In the present paper, we investigated the discrete hypersphere in order to provide it with a definition related to thickness and verifying basic topological properties. The first result is a general arithmetic definition of the discrete hypersphere. In this definition, usual constant arithmetic thickness is replaced by two thickness functions in order to take into account the non-linearity of hyperspheres and to distribute global thickness more or less inside or outside the hypersphere. By relating the k-adjacency and the k-minimality to norms we control precisely the thickness for discrete sets with non linear analytical description. We thus based thickness functions on the measurement by the k-minimality norm of a vector related to the local behavior of the hypersphere and characterized the k-minimal arithmetic discrete hypersphere.

This characterization could be helpful to infer drawing and recognition algorithms. In the case of the discrete line, the arithmetic definition has improved the understanding of such algorithms. Why the same situation does not apply to the hypersphere or at least to the circle?

Moreover, beyond the particular case of the hypersphere, we go further in the understanding of the arithmetic discrete geometry. Focusing on simple objects, namely lines or circles, reduces the number of parameters in play and allows to better study each of them. Two points seem to be essential and common to all objects arithmetically studied: the use of norms and the study of the local discrete behaviour of the object. As far as we know, remarkable arithmetic discrete objects have always been characterized by measuring, with norms, their local variations. Lines or hyperplanes are linear objects and their normal vector is constant in magnitude and direction. For circles and hyperspheres, the normal vector is only constant in magnitude. The next natural step would be the arithmetic study of hypersurfaces for which the normal vector would have neither a constant magnitude nor a constant direction, in other words, the general case of hypersurfaces based on polynomials.

References

1. Bresenham, J.: Algorithm for computer control of a digital plotter. IBM Systems Journal **4**(1) (1965) 25–30
2. Freeman, H.: Computer processing of line-drawing images. ACM Computing Surveys **6**(1) (1974) 57–97
3. Rosenfeld, A.: Digital straight lines segments. In: IEEE Transactions on Computers. (1974) 1264–1369
4. Reveillès, J.P.: Géométrie discrète, calcul en nombres entiers et algorithmique. Thèse d'Etat, Université Louis Pasteur, Strasbourg (1991)
5. Debled-Rennesson, I., Reveillès, J.P.: A linear algorithm for segmentation of digital curves. IJPRAI **9**(4) (1995) 635–662
6. Andres, E., Acharya, R., Sibata, C.: Discrete analytical hyperplanes. CVGIP: Graphical Models and Image Processing **59**(5) (1997) 302–309
7. Bresenham, J.: A linear algorithm for incremental digital display of circular arcs. Communication of the ACM **20**(2) (1977) 100–106
8. McIlroy, M.: Best approximate circles on integer grids. ACM Transactions on Graphics **2**(4) (1983) 237–263
9. Pham, S.: Digital circles with non-lattice point centers. The Visual Computer **9**(1) (1992) 1–24
10. Andres, E., Jacob, M.A.: The discrete analytical hyperspheres. IEEE Transactions on Visualization and Computer Graphics **3**(1) (1997) 75–86
11. Fiorio, C., Jamet, D., Toutant, J.L.: Discrete circles: an arithmetical approach with non-constant thickness. In: Vision Geometry XIV, Electronic Imaging, SPIE, San Jose (CA), USA (2006)

The Eccentricity Transform (of a Digital Shape)*

Walter G. Kropatsch, Adrian Ion, Yll Haxhimusa, and Thomas Flanitzer

Pattern Recognition and Image Processing Group
Vienna University of Technology, Favoritenstr. 9/1832, A-1040 Vienna, Austria
{krw, ion, yll, flanitzt}@prip.tuwien.ac.at

Abstract. Eccentricity measures the shortest length of the paths from a given vertex v to reach any other vertex w of a connected graph. Computed for every vertex v it transforms the connectivity structure of the graph into a set of values. For a connected region of a digital image it is defined through its neighbourhood graph and the given metric. This transform assigns to each element of a region a value that depends on it's location inside the region and the region's shape. The definition and several properties are given. Presented experimental results verify its robustness against noise, and its increased stability compared to the distance transform. Future work will include using it for shape decomposition, representation, and matching.

1 Introduction

Recognition, manipulation and *representation* of visual objects can be simplified significantly by "abstraction". Abstraction extracts essential features and properties while it neglects unnecessary details. Shape is one such form of visual abstraction, which describes distinctive features of the object's appearance i.e. its projection on the surface of a 2D sensor (in our case the retina). If shape matching is done invariant with respect to certain deformation classes (e.g. articulated motion), shape based object recognition can be used for generic object recognition, a much desired ability of humans.

Different approaches that use shape for recognition exist [1,2,3,4], with many of them using the distance transform [5] derived skeletons [6,7] as a basis for shape description. Skeletons have proved themselves to be the basis of powerful shape descriptors [2] with the main advantages including their 'cue' for a natural decomposition of shapes into parts (e.g. usually the parts of the skeleton of a human decompose its shape into body, limbs, and head) and their invariance to certain types of movement including the very important articulated motion. On the other hand, one of their weak points come from their apparent locality and the fact that they are derived from the distance transform which is known to be unstable with respect to small perturbation of the shape (e.g. spurious branches can appear in the skeleton if a few pixels are added at the border of the region).

* Supported by the Austrian Science Fund under grants S9103-N04 and P18716-N13.

A. Kuba, L.G. Nyúl, and K. Palágyi (Eds.): DGCI 2006, LNCS 4245, pp. 437–448, 2006.

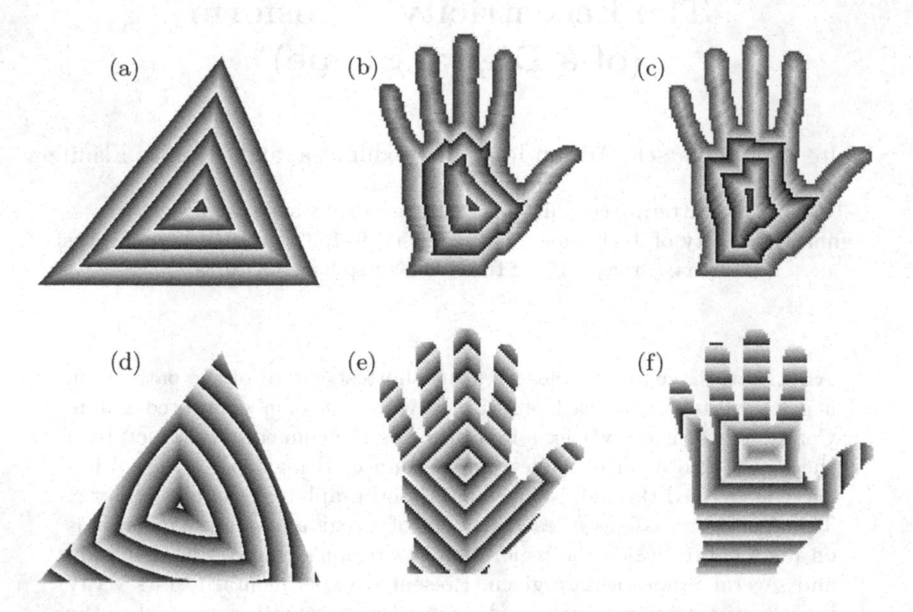

Fig. 1. Isoheight lines for distance (a-c) and eccentricity (d-f) transform of 2 images, using the euclidean (a,d), 4− (b,e), and 8− (c,f) neighbourhood (where continuous, lighter means higher value)

The distance transform associates to each point of the shape, the minimum distance from it to the border of the shape (see Fig. 1a: gray values are independent between the two images, and where continuously changing: lighter means higher value), which makes it very unstable with respect to Salt and Pepper noise and certain kind of segmentation errors. Approaches like removing regions below a certain size or pruning spurious branches of the obtained skeleton have been used to cope with these kinds of problems, but this has been shown not to be the optimal way and should be avoided mainly because the size of a region does not tell anything about its importance [8].

Instead of minimum distance, other measures have also been used, e.g. the mean time for a randomly moving particle to hit the border [9].

Inspired from graph theory, we present a new transform which associates to each point the longest distance (geodesic) from it to the points on the border of the shape (see Fig. 1d). We show that it is robust against the types of noise mentioned above, and comment about it's applicability to shape description and matching.

We recall the distance transform in Sec. 2, including a formulation for the distance transform of a graph (Sec. 2.2). The eccentricity transform is presented in Sec. 3, beginning with a recall of the graph theory based definition for eccentricity (Sec. 3.1). The properties of the transform are discussed in Sec. 4,

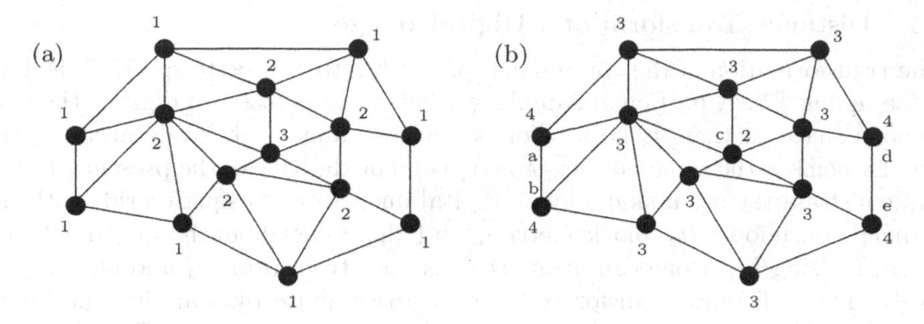

Fig. 2. Distance transform (a) and eccentricity transform (b) of a graph

followed by computation strategies in Sec. 5. Experimental results in Sec. 6 will complete the presentation, summed up in Sec. 7 with conclusions and outlook.

2 Distance Transform

The distance transform assigns to each point in the binary image a value of a distance to the closest point on its border (obstacles). Let $I = B \cup \overline{B}$ be a binary image and let a point $p \in B$. We adapt the definition of the general distance function [10] for the rest of the section. A neighbourhood N_i is a pair of (P_i, d_i), where P_i is a finite subset of $\mathbb{Z}^K$ and d_i is a function $d_i : P \to \mathbb{R}^+$, for $i = 0, ..., T-1$ and $T, K \in \mathbb{Z}^+$. We say that p_i is adjacent to p_{i+1} iff $p_{i+1} = p_i + r$ for some $r \in P_i$. Let α be a finite sequence of neighbourhoods $N_0 N_1 ... N_{T-1}$, where T is called the period of the sequence. The distance transform dt_α of I associates to every point $p \in B$ the minimal distance from p to $\overline{B}$, formally we write:

$$dt_\alpha(p) = \min\{\lambda(\pi_\alpha(p, q)) \mid q \in B \wedge q \in N_\alpha(\overline{q}), \overline{q} \in \overline{B}\}, \tag{1}$$

where N_α is an α-adjacency, $\pi_\alpha(p, q)$ is the set of all α-paths from point p to q, and $\lambda(\pi_\alpha(p, q))$ is the length of one of the paths $\pi_\alpha(p, q)$. The α-path is a sequence of points $(p_0, p_1, ..., p_n)$, such that end points are $p_0 = p$, $p_n = q$, and p_{i+1} is α-adjacent to p_i $(0 \leq i \leq n-1)$, then the length of this path is the sum of $d_i(p_{i+1} - p_i)$ for all $i = 1, ..., n$. If $d_i(r) = 1$ then the length of the path is n, the number of points. To define the chessboard distance (dt_8) or the square distance (dt_4) one takes the sequence of neighbourhood with $T = 1$ ($\alpha = N_0$) and defines the neighbourhood P_0 as in [10, page 239]. Note that there may be many shortest paths. If there is no other shorter α-path between the same end points, then this path is called α-geodesic [11]. The border point $q \in B$ is α-neighbour of a point not being in B. In the Euclidean space there is always a unique path between two points, which is the straight line between the points. This straight line does not exist in digital images and thus the distance transform computed is dependent on the way the neighbourhood is defined, i.e. how the Euclidean distance is approximated. In the section below we use the definitions above as the basis of defining the distance transform of digital images and graphs.

2.1 Distance Transform of a Digital Image

The transformation of the continuous space $\mathbb{R}^n$ into a discrete space $\mathbb{Z}^n$ is done by sampling $\mathbb{R}^n$. A particular sampling scheme can be used to digitise the continuous space. If there is no a priori knowledge about the local variation, the usual scheme is the square or hexagonal grid. For the sake of the presentation we will constrain the discussion only on digital images on 2D square grids with the 4−neighbourhood (city block metric), and the 8−neighbourhood (chessboard metric). Using Eq. 1 one can define the distance transform dt_4 and dt_8, respectively. These distance transforms are easy to compute by scanning the image twice [12,11], although they are not a good approximation of the Euclidean distance. In Fig. 1b,c) distance transforms dt_4 and dt_8 of a binary image are shown. Better approximations can be found by chamfer distances [13].

2.2 Distance Transform of a Graph

If the sampling grid is not regular, one could use graph representation for the sampling points. Let $G = (V, E, a, w)$ be the undirected weighted graph with vertices $v \in V$ representing sampling points, edge set $e = (v, w) \in E$ representing the connection between vertices; and $a : V \to \mathbb{Z}^+$ and $w : E \to \mathbb{Z}^+$ are attributes on vertices and edges respectively. Let the weights on edges represent the cost of going from one vertex to the other. In order to define the distance transform one should define the boundary vertices [14]. Any bounded region has a boundary that separates it from the background. The background can be considered as the complement of the region with respect to the embedding space. Border faces are faces of the dual graph that are surrounded by both vertices of $G \subset G'$ and G'. The boundary of a subgraph $G = (V, E) \subset G' = (V', E')$ collects all the vertices $C \subset V$ which bound border faces.

A path $\pi_G(v, w)$ is a sequence of vertices $(v_0, v_1, ..., v_n)$ in G such that the end vertices are $v = v_0$, $w = v_n$, all vertices are distinct and $\exists e = (v_i, v_{i+1}) \in E, i = 0, 1, ..., n-1$. The length $\lambda(\pi_G)$ of path λ_G is the sum of the edge weight in the sequence:

$$dt_G(v) = \min\{\lambda(\pi_G(v, w)), \; v \in G \setminus C \wedge w \in C\}. \tag{2}$$

Usually, the border vertices are set to 1. If vertices v and w are not connected, we say that the $\lambda(\pi_G(v, w))$ is infinite. If the graph G is connected then this distance is a graph metric [11]. A simple example of the distance transform on a graph is given in Fig. 2a, where the edge cost is set to 1. Note that square grid can be easily represented by graphs. In this case the weight on edges could be set to 1 (but not necessarily). Similar to the square grid, we can define the 4-, 8-neighbourhood of vertices.

3 Eccentricity Transform

The eccentricity transform assigns to each point in the binary image the shortest distance to the point farthest away from it. Analogously, to the notation presented in Sec. 2 we define the eccentricity transform $ecc_\alpha(p)$ of $I = B \cup \overline{B}$ such that it associates to every point $p \in B$ the longest of the distances to any other point $q \in B$, formally we write:

$$ecc_\alpha(p) = \max\{\lambda(\pi_\alpha(p,q)) \mid \forall q \in B\}, \quad (3)$$

where $\pi_\alpha(p,q)$ is the shortest α-path from point p to q, and $\lambda(\pi_\alpha(p,q))$ is the length of the path $\pi_\alpha(p,q)$. In the section below we use the Eq. 3 as the basis in defining the eccentricity transform of graphs and digital images.

3.1 Eccentricity Transform of a Graph

Let $G = (V, E, w)$ be an attributed undirected and connected graph with vertex set V, edge set E and with edge weights $w : E \to \mathbb{Z}^+$ as the cost of going from one vertex to the other. Let v be a vertex in V. The eccentricity $ecc_G(v)$ of v is the distance to a vertex farthest from v and it is defined as [15, Page 31]:

$$ecc_G(v) = \max\{\lambda(\pi_G(v,w)) \mid \forall w \in V\}, \quad (4)$$

where $\lambda(\pi_G(v,w))$ is the length of the shortest path between the two vertices v and w. One could say that eccentricity of a vertex is the longest shortest path to any other vertex in the graph. A simple example of the eccentricity transform is given in Fig. 2b), where we set the edge's cost to 1.

Some definitions concerning the eccentricity transform are of importance [15]:

- the *eccentric* vertices of v are all the vertices w at a distance $ecc_G(v)$;
- the *radius* $r(G)$ of G is the minimum eccentricity;
- the *diameter* $d(G)$ of G is the maximum eccentricity;
- v is a *central vertex* of G if $ecc_G(v) = r(G)$;
- the *center* $C(G)$ is the set of all central vertices;
- v is a *peripheral vertex* of G if $ecc_G(v) = d(G)$;
- the *periphery* $P(G)$ is the set of all peripheral vertices;

For the graph in Fig. 2b) the radius $r(G) = 2$, the diameter $d(G) = 4$, central vertex c, the center $C(G) = \{c\}$, peripheral vertices a, b, d and e, and the periphery $P(G) = \{a, b, d, e\}$. Sec. 4 presents a detailed discussion of the properties.

3.2 Eccentricity Transform of a Digital Image

Similarly to distance transform, a particular sampling scheme can be used to discretize an image. We constrain our discussion only on a 2D square grid, and two classical pixel adjacencies; the 4− and 8−neighbourhood. Let I be a binary image $I = B \cup \overline{B}$, and let a pixel p be in B. Now we can use Eq. 3 to define the eccentricity transform ecc_α on a square grid digital image for the connected set B. One can say that eccentricity of a pixel is the longest shortest path to any other pixel in the same connected region. Similarly to the distance transform also the eccentricity transform is affected by how well the Euclidean distance can be approximated. The same concerns made in Sec. 2 with respect to the Euclidean plane apply for eccentricity transform as well. Thus eccentricity transform is also dependent on the way one defines the pixel neighbourhood. In Fig. 1e,f eccentricity transforms ecc_4 and ecc_8 are shown on the same hand image as in Fig. 1b,c.

4 Properties of the Eccentricity Transform

We shortly discuss some of the properties of the eccentricity transform, some of them known from graph theory and extended to the discrete domain, some are interesting and useful in the context of describing the shape of a region.

Center: The vertices with the minimum value of the eccentricity transform are called the center of the graph. They lie in a block of the graph, i.e. the corresponding subset of vertices containing the center is connected and does not contain a cut vertex. We notice that the center of a discrete region is always a part of the region in contrast to the center of gravity which can be located outside the region in case of a concavity or a hole in the middle of the region. This may be useful in several applications, e.g. in tracking where the center of a tracked region may be used as the start for searching the region in the next frame of the sequence.

Robustness: The eccentricity transform is robust with respect to (salt and pepper) noise. This is due to the fact that a noise vertex on the path between two distant points '*just goes around*' the obstacle without prolongating the length by much. In the case of discrete metrics (like 4− or 8−connectivity) the likelihood of finding many paths with the same (shortest) length is very high. In such a case the eccentricity is affected only if all shortest paths between the vertex and its farthest vertex are interrupted by noise or a noisy pixel (vertex) p is added to B such that p is at maximum distance from the vertex.

4-connectivity: In the Euclidean space two points are connected by a unique straight line. In the discrete space with 4-connectivity this is only the case if the two points are along the two coordinate axes, in all other cases there are more than one shortest paths. In fact any permutation of the two primitive steps to connect the two end points is also a shortest path.

This fact increases the robustness of the eccentricity transform but has also two other consequences:

1. There is not a single midpoint between the two endpoints making the center of an elongated region a diagonal line. In fact the length of this line is as long as the smaller coordinate differences of the two end points.
2. Since the number of midpoints depends on the angle of the discrete line the resulting centers are no more rotationally invariant (which they are in the Euclidean case).

Maxima are all on the boundary if the graph has no inner pending vertex: (See Sec. 2.2 for the boundary of a graph) If G' is connected there are paths between any pairs of vertices $v, w \in V'$. Any non-border vertex has a degree greater than 1. If none of the neighbours of a vertex of degree greater than 1 belongs to a border face it cannot be extremal since any path leading to it can be continued.

Complementarity between distance transform and eccentricity transform: In the distance transform the smallest values are on the boundary and the highest values can be found where a circle with maximum radius

Algorithm 1 – Eccentricity Transform-naive implementation

Input: Attributed graph $G = (V, E)$.
1: **for all** $v \in V$ **do**
2: $ecc(v) \leftarrow 0$
3: **for all** $u \in V - \{v\}$ **do**
4: $ecc(v) \leftarrow \max\{ecc(v), \text{shortestPathLength}(v, u)\}$
5: **end for**
6: **end for**

Output: Eccentricity $ecc(v)$ for all vertices $v \in V$ of G.

touches the boundary in at least two opposite points. These local maxima form the skeleton/medial axis/symmetry transform. Local maxima of the eccentricity transform are on the boundary while the minimum defines the center. However there are local minima along the boundary and discontinuities inside the region which give rise to interesting partitionings of the region.

Invariance: The eccentricity transform computes the lengths of paths inside a given region. It is therefore invariant to any translation and invariant to rotation for the Euclidean metric. There is some dependency on the orientation for discrete metrics but not for all shapes. Furthermore in the case of thin regions, it is robust with respect to articulated motion, it may differ by the thickness of the shape at the articulation point which in many natural cases is thin in relation to the length (arms, legs, fingers).

5 Computation

Two algorithms for computing the eccentricity transform are given here. They are both defined for graphs, but the adaptation to digital images is straight forward. One has just to decide for a neighbourhood ($\alpha = \{4, 8\}$) and choose the pixels that make up the connected region for which the transform will be applied. Note that Floyds [16] algorithm, that produces the minimum path length from all vertices to all other vertices can also be used to obtain the eccentricity (for each vertex, one has just to take the maximum of the values obtained for it).

Naive Alg. 1 iterates through all the vertices of G and for each, it calculates the maximum of the length of the shortest paths to all other vertices in the graph. Lines 3 - 5 can be implemented by taking the maximum of the lengths calculated using Dijkstra's single-source shortest path algorithm [16]. The complexity of the naive implementation is between $O(|V|^3)$ and $O(|V||E|+|V|^2 \log |V|)$ depending on the implementation of the shortest path problem.

Alg. 2 uses the fact that the set of *eccentric vertices* is a subset of V and that calculating the shortest path for each of these vertices to all the other vertices in V and combining the results (i.e. taking the maximum) is enough to obtain the eccentricity transform for the whole graph. The eccentric vertices can be found by calculating the shortest path from the center of the graph to all the other vertices and looking at the local maximum. To find the center of a graph we find it's diameter, which is connecting the vertices with the highest eccentricity.

Algorithm 2 – Eccentricity Transform-optimised implementation

Input: Attributed graph $G = (V, E)$.

1: $\forall v \in V, ecc(v) \leftarrow 0$ /* initialise eccentricity cumulation table with 0 */
2: $vp \leftarrow$ random vertex of V
3: **repeat**
4: mark vp as visited
5: $\forall v \in V, ecc(v) = \max\{ecc(v), \text{shortestPathLength}(v, vp)\}$
6: $vp \leftarrow$ vertex with maximum ecc
7: **until** (ecc not changed) or (vp allready visited) /* iterate until the endpoints of a diameter found */
8: **repeat**
9: $m \leftarrow$ **one** (random) or **all** unvisited vertices with minimum ecc
10: **for all** $v_m \in m$ **do**
11: $\forall v \in V, ecc(v) = \max\{ecc(v), \text{shortestPathLength}(v, v_m)\}$
12: mark v_m as visited
13: **end for**
14: $M \leftarrow$ all unvisited vertices with local maximum ecc
/* M includes non monotonic maxima */
15: **for all** $v_M \in M$ **do**
16: $\forall v \in V, ecc(v) = \max\{ecc(v), \text{shortestPathLength}(v, v_M)\}$
17: mark v_M as visited
18: **end for**
19: **until** ecc not changed /* repeat until converged */

Output: Eccentricity $ecc(v)$ for all vertices $v \in V$ of G.

Lines 3 - 7 start with a random point and iterate to find the vertices with the highest eccentricity (diameter endpoints). The calculated shortest path lengths are added to the cumulation table ecc and the vertex is marked as visited. Lines 8 - 19 iterate finding the center vertices and the local maximum until the ecc cumulation table converges. On line 9 (approximate the center), two options have been tried, taking one or all the existing minima. On shapes without holes, both have produced the correct solution, while on shapes with holes neither of them did. In our experiments, the first loop (lines 3 - 7) converged after 3 cycles (random point, first diameter end, second diameter end). The second loop is bounded by the number of vertices on the border of the graph.

Alg. 2 is much faster than Alg. 1 but gives correct results only on simply connected shapes (no holes). On shapes with holes, complex forms of the center appear e.g. for a disc with a circular hole in the middle, the center consists of a circle for euclidean distance, and a set of disconnected points for 4 connectivity, all concentrated around the hole. In such cases, Alg. 2 produces results close to the correct one, but we cannot give any upper bound for the error.

6 Experiments

We have conducted experiments to test the properties of the eccentricity transform and find the important differences compared to the distance transform.

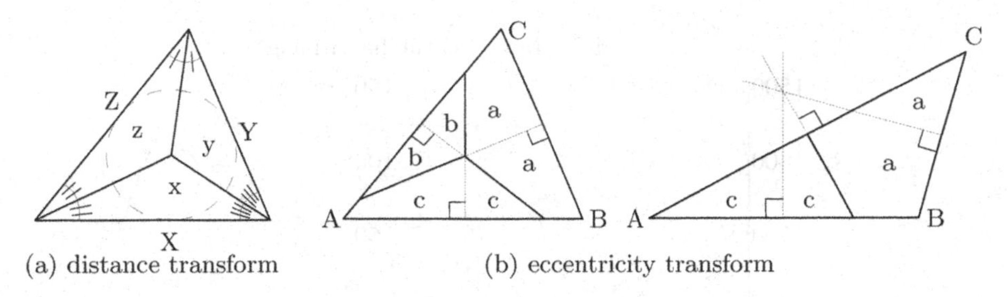

Fig. 3. Transforms on a triangle. Capital letters denote triangle elements (A,B,C points; X, Y, Z segments) and lower case letters denote their respective influence area.

6.1 Eccentricity Transform of a Triangle

First we have looked at a simple shape, the triangle, for which both transforms, when using the euclidean distance, can be solved analytically. In the case of the distance transform (see Fig. 1a and 3a), the 3 angle bisectors divide the triangle into 3 parts (x, y, z in Fig. 3a), with all the points inside the same part having the distance transform equal to the distance to the same side of the triangle. The point with the highest distance transform is the intersection of the 3 angle bisectors i.e the incenter. The isoheight lines are all polygonal lines (triangles).

In the case of the eccentricity transform, the 3 perpendicular bisectors divide the triangle into 2 or 3 parts (see Fig. 1d and 3b) depending on whether the triangle is an optuse one or not (i.e the circumcenter lies outside or within the triangle). All points inside the same part have the eccentricity transform equal with the distance to the same point. The isoheight lines are made out of arcs.

6.2 Properties Depending on Connectivity/Metric

Fig. 1 shows the isoheight lines of the eccentricity transform for the 4 and 8 connectivities. One can see that the place of the center (global minima) and the form of the isoheight lines changed. Depending on the shape, the positions of the diameter ends/global maxima also change.

6.3 Robustness Against Salt and Pepper Noise

To test the robustness against Salt and Pepper noise, we have calculated the eccentricity and distance transforms (using both 4− and 8− neighbourhood) for 89 randomly selected shapes from [2] (for some example shapes see the top row from Tab. 1). We applied 5% Salt and Pepper noise to the images and calculated the two transforms again.

To measure the robustness, for each image, each neighbourhood, and each transform, we have calculated the root mean square error (RMSE) between the values obtained for the original and noisy images (calculation was done using the values, of the pixels part of the shape, in both images i.e. noisy pixels are excluded). We have calculated for each image and each neighbourhood, the ratio

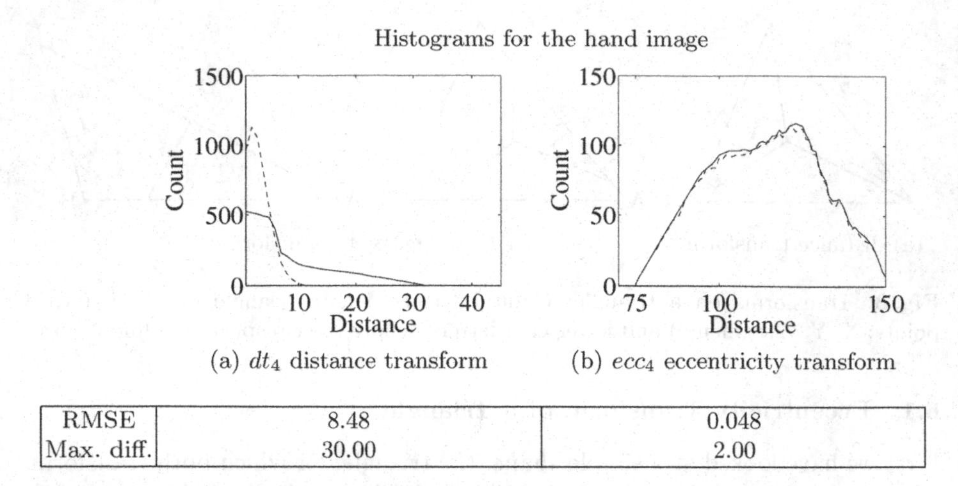

(a) dt_4 distance transform (b) ecc_4 eccentricity transform

	(a) dt_4	(b) ecc_4
RMSE	8.48	0.048
Max. diff.	30.00	2.00

Fig. 4. Distance and eccentricity transform histograms, RMSE and Max. Diff. (solid - original image, dotted - noisy image)

between the RMSE for the eccentricity and distance transforms. Then, for each neighbourhood, we have calculated the mean of these ratios and inverted the result ($1/x$) i.e. we obtain *dt* error divided by *ecc* error. The error of the distance transform is **8.07** times higher in the case of the 4−neighbourhood and **22.63** times higher for the 8−neighbourhood, then the one of the eccentricity transform.

Fig. 4 shows the histogram of the eccentricity and distance transforms for one of the images, the hand (original and noisy) using the 4 neighbourhood. Also shown is the RMSE between the values of the transforms for the original and noisy images, and the maximum difference value for each transform. One can see that the error and maximum deviation of the eccentricity transform is much smaller than that of the distance transform. Note that in the case of the noisy image, a valid transform value has been calculated for less pixels. This makes the histogram of the eccentricity transform of the noisy image lie below the histogram of the original one.

6.4 Minor Segmentation Errors

For this experiment, we have selected a few shapes and simulated segmentation errors and partial occlusion by removing some parts of the shapes i.e. simulated noise on the border of the shape. We have calculated the correlation between the local maxima of the eccentricity transforms of the original and the images with partial occlusion (for each image, original and partially occluded, we have created a matrix where the positions of the eccentricity transfrom regional maxima were marked with 1, and the rest with 0, and calculated the correlation between the 2 matrices - only maxima that where located inside the partially occluded shape

Table 1. Correlation results for local maxima in eccentricity transform of original (top row) and partially occluded shapes (middle and bottom rows)

	Original Shapes									
	Partially occluded set 1									
4–nb.	0.73	1.00	1.00	0.96	0.96	1.00	0.77	1.00	1.00	0.95
8–nb.	0.93	1.00	0.72	0.97	1.00	0.82	1.00	1.00	1.00	0.98
	Partially occluded set 2									
4–nb.	0.71	0.79	0.97	0.96	0.89	0.97	0.71	0.98	0.87	0.92
8–nb.	0.48	0.45	0.90	0.96	0.72	0.65	0.98	0.97	0.73	0.98

were taken into consideration). The correlation Tab. 1 shows these shapes and the obtained correlation values.

6.5 Articulated Motion

To simulate articulated motion, two elongated parts have been overlapped at one of their ends in a way in which they approximate a joint (the angle between the two parts is a parameter, see Fig. 5a for some examples).

For each angle (in our experiment we have used 90°, 105°, 120°, 135°, 150°, 165°, and 180°) we have applied the eccentricity transform and calculated the minimum, maximum, and average eccentricity. Fig. 5b shows the mean and standard deviation of the 3 values over the whole spectrum of joint angles tested. Note that the values are stable under these conditions.

(a)

(b)

	Min		Max		Average	
neighbourhood	mean	std	mean	std	mean	std
4	66.00	4.63	131.88	9.11	97.86	7.06
8	54.33	3.39	108.11	6.60	80.33	5.37

Fig. 5. Example of images used for testing the variation under articulated motion (a), and mean and standard deviation of eccentricity value for the simulated joint (b)

7 Conclusion and Outlook

We propose a new transform for a digital image called the eccentricity transform. This transform associates to every pixel the maximum length of the shortest paths connecting it with all the other vertices. The definition, several properties, and algorithms have been given. Presented experimental results verify its robustness against noise, and its increased stability compared to the distance transform, e.g. in the case of 5% Salt and Pepper noise, we obtained changes about 10 times higher of the distance transform compared to the eccentricity transform (a distance change of 30 in the case of the distance transform is 3 in the case of the eccentricity transform). Behaviour under minor segmentation errors and articulated motion has also been tested and shows promising results. Future work will include using it for shape decomposition, representation, and matching.

References

1. Belongie, S., Malik, J., Puzicha, J.: Shape matching and object recognition using shape contexts. IEEE Trans. Pattern Anal. Mach. Intell. **24**(4) (2002) 509–522
2. Siddiqi, K., Shokoufandeh, A., Dickinson, S., Zucker, S.W.: Shock graphs and shape matching. International Journal of Computer Vision **30** (1999) 1–24
3. Felzenszwalb, P.F.: Representation and detection of deformable shapes. IEEE Trans. Pattern Anal. Mach. Intell. **27**(2) (2005) 208–220
4. Mori, G.: Guiding model search using segmentation. In: ICCV. (2005) 1417–1423
5. Rosenfeld, A.: A note on 'geometric transforms' of digital sets. Pattern Recognition Letters **1**(4) (1983) 223–225
6. Ogniewicz, R.L., Kübler, O.: Hierarchic voronoi skeletons. Pattern Recognition **28**(3) (1995) 343–359
7. Borgefors, G., Nyström, I., Sanniti Di Baja, G.: Computing skeletons in three dimensions. Pattern Recognition **37**(7) (1999) 1225–1236
8. Haxhimusa, Y., Ion, A., Kropatsch, W.G., Illetschko, T.: Evaluating minimum spanning tree based segmentation algorithms. In: CAIP. (2005) 579–586
9. Gorelick, L., Galun, M., Sharon, E., Basri, R., Brandt, A.: Shape representation and classification using the poisson equation. In: CVPR (2). (2004) 61–67
10. Yamashita, M., Ibaraki, T.: Distances defined by neighborhood sequences. Pattern Recognition **19** (1986) 237–246
11. Klette, R., Rosenfeld, A.: Digital Geometry. Morgan Kaufmann (2004)
12. Rosenfeld, A., Pfaltz, J.L.: Sequential operations in digital picture processing. Journal of Association of Computer Machinery **13**(4) (1966) 471–494
13. Borgefors, G.: Distance transformation in digital images. Computer Vision, Graphics, and Image Processing **34** (1986) 344–371
14. Kropatsch, W.G., Haxhimusa, Y., Pizlo, Z.: Integral trees: Subtree depth and diameter. In: IWCIA 2004. Volume 3322 of LNCS., Springer (2004) 77–87
15. Buckley, F., Harary, F.: Distances in Graphs. Addison-Wesley Publishing Company (1990)
16. Thulasiraman, K., Swamy, M.N.S.: Graphs: Theory and Algorithms. Wiley-Interscience (1992)

Projected Area Based 3D Shape Similarity Evaluation

Tetsuo Miyake[1], Naoya Iwata[2], Satoshi Horihata[1], and Zhong Zhang[1]

[1] Dept. of Prod. Sys. Eng., Toyohashi University of Technology, 1-1, Hibarigaoka, Tenpaku, Toyohashi, 441-8580, Japan
[2] Toyohashi University of Technology. Now in Aishin AW Co., LTD, 10, Takane, Fujii-cho, Anjo City, Aichi 444-1192, Japan
{miyake, horihata, zhang}@is.pse.tut.ac.jp

Abstract. Because the appearance of 3D objects changes according to viewing directions, it is not easy to evaluate similarity between two objects in a few appearances. In this paper we propose similarity measure between two shapes of 3D objects. The feature of a shape is represented by a distribution of a projected area on a unit sphere, and the distribution is expanded in spherical harmonics. The degree of similarity between several kinds of shape is calculated and is compared with human sense. The results of computer simulation demonstrate the validity of our method.

1 Introduction

We recognize a 3D shape of an object and can evaluate similarity between two objects without much difficulty. For example, we say on occasion that this cup is similar to that. But if we are asked which body a cylinder or a pyramid is similar to a prism, what answer shall we return? Generally speaking, because the appearance of an object changes according to viewing directions, it is not easy to evaluate similarity between two objects in a few appearances. Even if two objects are exactly same, it is hard to draw the right conclusion at a glance.

Many methods are proposed in the published papers that have addressed shape similarity evaluation for shape searching [1]. Recently similarity search between a query object and a target in a database is actively studied in data retrieval of 3D objects. On the other hand, it is well known that expansion into a set of orthonormal basis is strong means in analyzing a given function, and spherical harmonics has performed the same role in some of these studies. Our proposed method also uses a set of spherical harmonics, but it is some different in defining a function that expresses the feature of a shape of an object.

In order to retrieve a target object with high reliability, a shape-based search method must be presented. There are various kinds of 3D structural model including CAD models, voxel data and polygonal meshes in the previous works that applied orthogonal expansion to similarity evaluation. Princeton Shape Retrieval and Analysis Group have got successful results using several structural models [2, 3, 4]. Saupe et al. developed a method that was applied to a polygonal mesh model [5]. Proriol tried shape recognition using one or more than

A. Kuba, L.G. Nyúl, and K. Palágyi (Eds.): DGCI 2006, LNCS 4245, pp. 449–459, 2006.

one silhouettes of simple bodies [6]. The body with a certain solid structure was needed in calculating expansion coefficients. Tanaka et al. represented polyhedral objects using extended Gaussian images [7].

The aim of our study is not to retrieve an object that is most similar to the given object, but to define a similarity measure that simulate a human sense of the similarity evaluation. If we can realize it, we can clarify what kind of feature of appearance a human perceives in evaluating shape similarity. It is not difficult in a sense to judge whether or not a query shape is equal to or similar to a target using detailed geometrical description. However, in practice, we usually have no structural information about an object's shape except its appearances. Therefore we define the feature function of an object's shape based on a projected area, a silhouette of an object, which must be one of the simplest feature of an object.

In this paper, we first define similarity measure for evaluating the 3D shape of an object. The measure is determined based on the spherical harmonics. Next we evaluate the similarity of several kinds of shape and finally we compare the measure with human sense. The results of computer simulation demonstrate the validity of our method.

2 Principle of Similarity Evaluation

It is necessary to solve the following problems so as to perform similarity evaluation.

- What feature of an object in appearance is observed.
- How viewing directions are determined.
- How the observed features are transformed to a numerical value.
- How the numerical values are analyzed.

We discuss each problem in the following sections.

2.1 Observation of the Object

The object is assumed to be placed at the center of a sphere. We observe the object from a point on the surface of a unit sphere, which position is expressed by (θ, ϕ). Figure 1 shows the object and the viewpoint on the surface. Since the appearance of an object changes according to the viewpoint, we can not determine its 3D shape based only on one appearance. Therefore we need to observe it from not a few number of viewpoints.

It is preferable to observe the object in the directions uniformly distributed in space, but it is known that such directions can not be realized. Though the vertices of regular polyhedra can be applied, the maximum number is at most 20. In this paper we determine the direction where the object is observed using a geodesic dome. The geodesic dome is one solution for generating almost uniformly distributed directions [8]. Initially the dome has a shape of a regular icosahedron. Each facet of the polyhedron, whose shape is a triangle, is iteratively divided as shown in Fig. 2. After the facet is divided two times, 162 vertices are generated on the spherical surface.

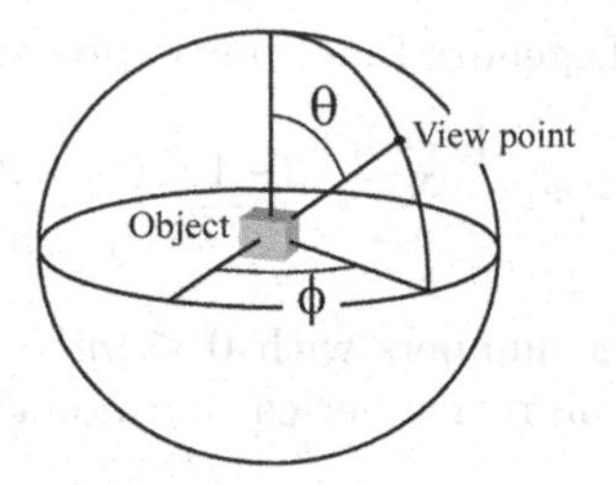

Fig. 1. The object is observed from (θ, ϕ)

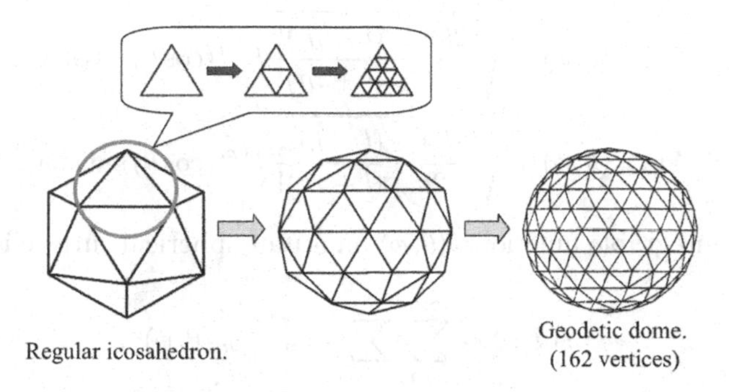

Fig. 2. Geodesic dome is generated based on a regular icosahedron

2.2 Feature of the Object

Several kinds of feature including colors, textures, a number of vertices and outline shapes are observed in the projected image of the object. Because of their change in appearance according to the viewing direction, it is not easy to compare two objects and judge whether they are similar or not.

For doing that, we propose to use a projected area of the image as a clue of the similarity measure. Suppose that we can only observe a silhouette of an unknown object. In spite of the particular condition, we conclude that the shape of the object must be a sphere if the image projected to any direction is a circle, for example. Therefore it is considered that the shape similarity can be evaluated using the distributed function of the projected area on a spherical surface.

2.3 Similarity Measure

Spherical harmonics are strong means for analyzing a function $f(\theta, \phi)$ defined on a unit sphere. It is to the function what a trigonometric function is to a single-valued function with respect to time.

Spherical harmonics denoted $Y_{l,m}$ is defined as

$$Y_{l,m}(\theta, \phi) = \sqrt{\frac{2l+1}{4\pi}\frac{(l-m)!}{(l+m)!}} P_l^{\,m}(\cos\theta)e^{im\phi} \ . \tag{1}$$

$P_l^m(x)$ is the associated Legendre functions expressed by

$$P_l^m(x) = (1 - x^2)^{m/2} \sum_{k=0}^{[(l-m)/2]} \frac{(-1)^k}{2^k\, k!} \frac{(2l - 2k - 1)!!}{(l - m - 2k)!} x^{l-m-2k} ,$$

where l and m are integer numbers with $0 \le |m| \le l$. In this paper, we define similarity measure based on real spherical harmonics, defined as

$$\begin{aligned} \tilde{Y}_{l,0}(\theta, \phi) &= \sqrt{\frac{2l+1}{4\pi}} P_l^0(\cos\theta) , \\ \tilde{Y}_{l,m}(\theta, \phi) &= \sqrt{\frac{2l+1}{2\pi} \frac{(l-m)!}{(l+m)!}} P_l^m(\cos\theta) \cos m\phi , \\ \tilde{Y}_{l,-m}(\theta, \phi) &= \sqrt{\frac{2l+1}{2\pi} \frac{(l-m)!}{(l+m)!}} P_l^m(\cos\theta) \sin m\phi . \end{aligned} \tag{2}$$

The square integrable function $f(\theta, \phi)$ on a unit spherical surface is expanded to

$$f(\theta, \phi) \sim \sum_{l=0}^{\infty} \sum_{m=-l}^{l} a_{l,m}\, \tilde{Y}_{l,m}(\theta, \phi) , \tag{3}$$

in terms of completeness of the spherical harmonics, where $a_{l,m}$ is a coefficient of the expansion expressed by

$$a_{l,m} = \int_0^{2\pi} d\phi \int_0^{\pi} \tilde{Y}_{l,m}(\theta, \phi)\, f(\theta, \phi) \sin\theta\, d\theta . \tag{4}$$

When calculating (4), the equation is discretized to

$$a_{l,m} = \sum_{i=1}^{N} \tilde{Y}_{l,m}(\theta, \phi)\, f(\theta, \phi)\, \Delta A , \tag{5}$$

where N is the number of view points and ΔA is an infinitesimal area at each view point (θ, ϕ). The value is equal to a surface area of a unit sphere divided by N, then $\Delta A = 4\pi/N$. Since these coefficients are unique to the function $f(\theta, \phi)$, the difference of the coefficients can be used for the similarity measure.

We should note, however, that distribution of $f(\theta, \phi)$ changes according to the pose of an object and as a matter of course the coefficients change. Therefore we consider all coefficients together that have a same number of degree l so as to get rid of the effect of the pose. Letting

$$\boldsymbol{v}^{(l)} = \left[a_{l,l}\, a_{l,l-1} \cdots a_{l,-l} \right] \tag{6}$$

define a coefficient vector $\boldsymbol{v}^{(l)}$, the norm of $\boldsymbol{v}^{(l)}$ is represented by

$$||\boldsymbol{v}^{(l)}|| = \sqrt{\sum_{k=-l}^{l} {a_{l,k}}^2} . \tag{7}$$

Though the vector changes according to the pose, its norm does not [7].

Finally we construct a feature vector $\boldsymbol{s}$ such that

$$\boldsymbol{s} = \left[c^{(0)}c^{(1)}\cdots c^{(l)}\cdots\right] , \tag{8}$$

where each component $c^{(l)}$ is expressed by

$$c^{(l)} = \frac{\| \boldsymbol{v}^{(l)} \|}{\sqrt{\sum_{k=0}^{\infty} \| \boldsymbol{v}^{(k)} \|^2}} . \tag{9}$$

Because each value of $||\boldsymbol{v}^{(l)}||$ depends on the size of the image, normalization is then necessary. By normalization, all geometrically similar shapes can have the same feature vector $\boldsymbol{s}$ in (8). It is thought that $\boldsymbol{s}$ represents the feature of the shape.

The degree of similarity, denoted $S_{A,B}$, between the object A and B is calculated by

$$S_{A,B} = \boldsymbol{s}_A \cdot \boldsymbol{s}_B = \sum_{k=0}^{\infty} c_A^{(k)} c_B^{(k)} , \tag{10}$$

where $c_X^{(k)}$ represents the k-th component of the feature vector $\boldsymbol{s}_X$. Since (10) means a natural inner product, it is obvious that $0.0 < S_{A,B} \leq 1.0$. The value of $c^{(0)}$ is based on the normalized value of $a_{0,0}$, which is the integral of $f(\theta, \phi)$ over the surface. Therefore $0.0 < c_A^{(0)}$ and $0.0 < c_B^{(0)}$, and it always holds that $0.0 < S_{A,B}$. If and only if the two objects are exactly the same from the view point of its silhouette, except the mirror symmetry, $S_{A,B}$ is equal to 1.

3 Numerical Simulation of Similarity Evaluation

Numerical simulation of similarity evaluation is conducted using virtual objects defined in a computer. Firstly it is demonstrated that the proposed similarity measure is pose invariant. Secondary the degree of similarity is evaluated between two objects with various kinds of shape.

3.1 Generation of Distribution Function

We use polyhedra and quadric surfaces as objects in the similarity evaluation. These shape models are convenient for mathematical treatment in the computer and can represent complicated shapes and curved surfaces. The polyhedral object is defined by boundary representation.

The object is placed at the center of the geodesic dome such that the center of gravity of it coincides with the center. If the center of gravity is outside of the object such as a torus, no problem occurs in the evaluation. One of the vertices of the dome is chosen as a viewpoint and the projected image of the object is generated on a virtual image plane. The projection is assumed to be parallel to the line from the center to the viewpoint and the image plane is assumed to be perpendicular to the line. Then the projected area of the image is calculated. By calculating the areas at all viewpoints in the half region of the dome, the distributed function of the projected area on a unit sphere is generated.

3.2 Pose Invariance

The similarity measure must satisfy pose invariance. It means that if the object A is exactly the same as B but its pose is different from B as shown in Fig. 3, the degree of similarity $S_{A,B}$ must become 1.0.

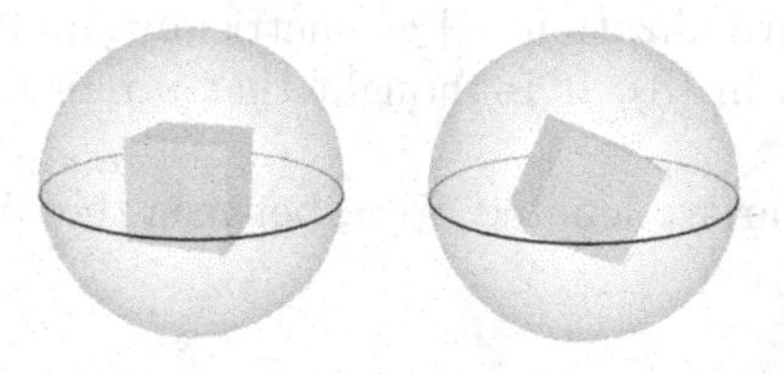

Fig. 3. The same two objects in the different pose

We evaluated the similarity between the same two objects in the different pose using three polyhedra, one convex and two concave. These bodies are shown in Fig. 4. A cube that belongs among a convex polyhedron has no occlusion by itself, but a concave polyhedron has. The projected area of the thin body changes much according to the viewing direction.

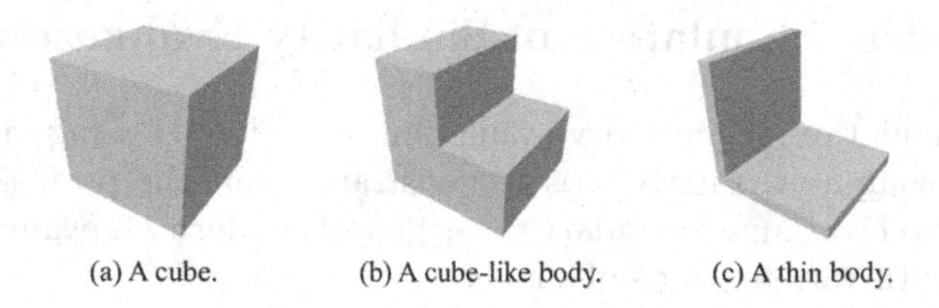

(a) A cube.
(b) A cube-like body.
(c) A thin body.

Fig. 4. Three polyhedra; (a) is convex and (b) and (c) are concave

A body to be compared was turned on the vertical and the horizontal axis, and was compared to the original one. The results of similarity evaluation for the thin body are shown in Fig. 5. Here the maximum degree of expansion in spherical harmonics is limited to 10. It can be seen that the degree of similarity $S_{A,A}$ is always 1.0 irrespective of its pose. Because almost the same results were obtained using the other bodies, it is concluded that the proposed similarity evaluation satisfies pose invariance.

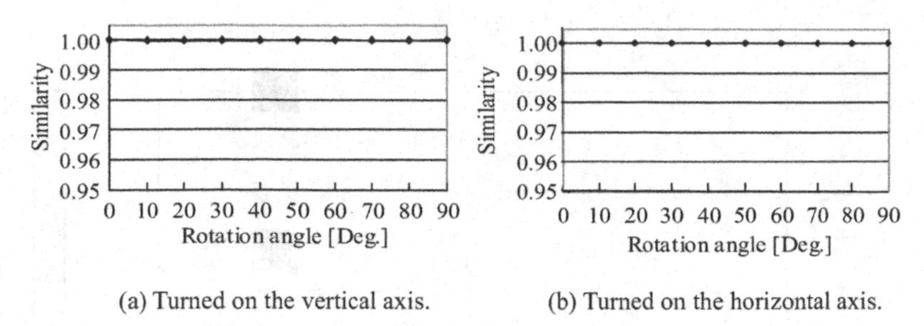

Fig. 5. The results of similarity evaluation according to the pose change

3.3 Similarity Evaluation

We evaluate the degree of similarity between two objects among twelve bodies shown in Fig. 6. These bodies are divided into three types of shape according to its appearance; a convex shape, a concave shape and a concave and thin shape. The shape of all bodies is defined by the boundary between inside and outside.

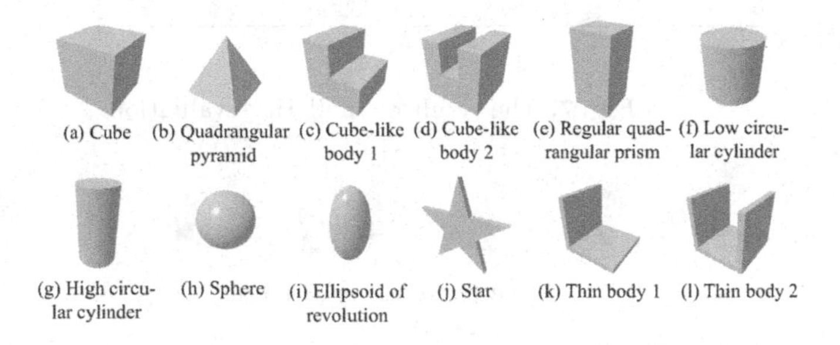

Fig. 6. Twelve bodies used in the similarity evaluation

Figure 7 shows the result of similarity evaluation. Each gray patch represents the degree of similarity between the two objects in the row and column position. Lighter gray means that the degree is higher.

It can be seen from the light and dark gray distribution that the bodies are divided into thick and thin ones. The degree of similarity of almost all bodies against a star is lowest. Since the star has almost a 2D shape compared with the other bodies, it is considered that the distribution pattern of the projected area of the star is much different.

Figure 8 shows the bodies similar to a cube and similar to a star in order of the degree of similarity from highest to lowest. The degree of similarity of a thick body to a thin body is low and the reverse is also true. It can be said that the similarity measure represents the feature of the shape well. It is a little strange from the point of view of human sense that the ellipsoid of revolution has lower degree than the thin body 2 in Fig. 8(a). On the other hand, the order shown in fig. 8(b) is fully understandable.

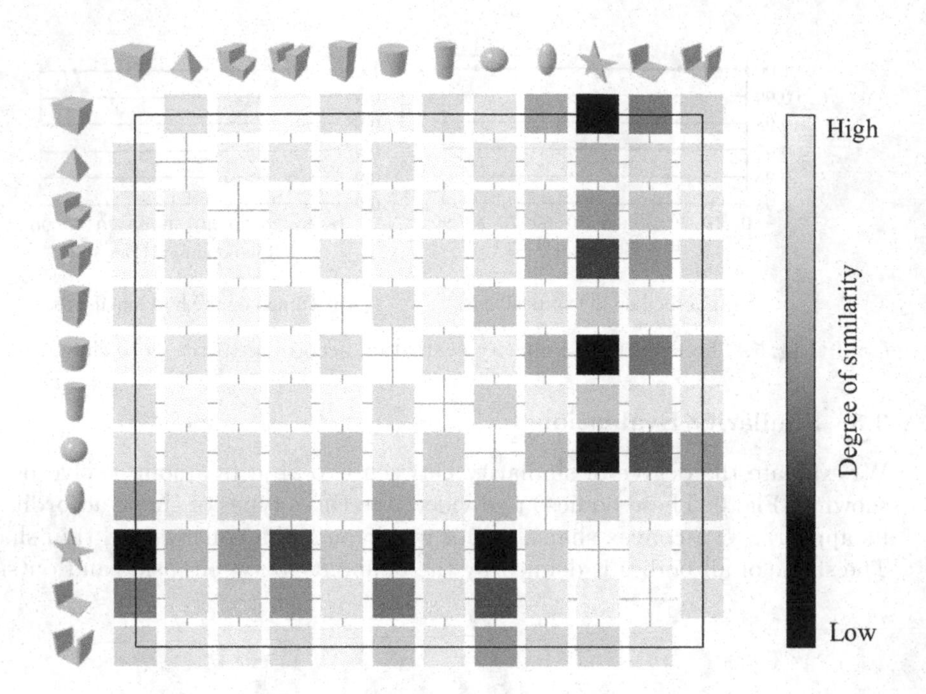

Fig. 7. The result of similarity evaluation

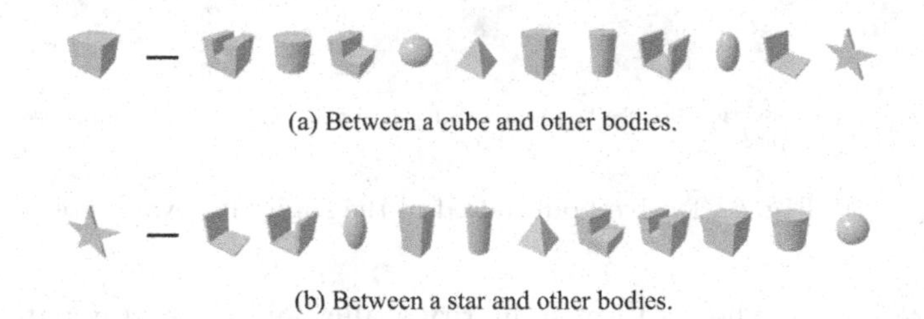

(a) Between a cube and other bodies.
(b) Between a star and other bodies.

Fig. 8. The order according to the degree of similarity

It may seem curious that though a star indicates the lowest degree of similarity to a cube, a sphere to a star. But because the feature vector defined by (8) is a 11-dimensional vector, the result is possible. Calculating an inner product of the two 11-dimensional feature vectors for the similarity evaluation means that the degree of similarity is measured by the angle between two vectors, which has one dimension.

Figure 9 shows the shape similarity in two dimensions. The vertical axis indicates the degree of similarity to a sphere, and the horizontal axis to a circle. The figure shows that an ellipsoid of revolution is similar both to a sphere and to a circle, which expresses the characteristic of its shape properly.

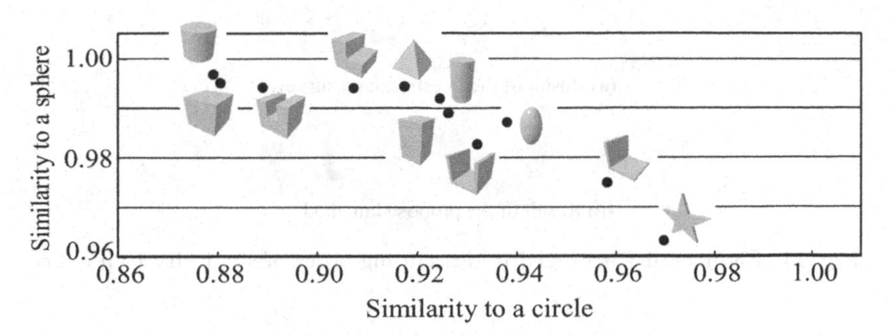

Fig. 9. 2D representation of the degree of similarity

4 Comparison with Human Sense

Human sense in shape similarity evaluation was investigated through a questionnaire survey. The survey was conducted on the internet web site and 127 visitors answered the question whether a cube is similar to a pyramid, a cylinder and so on. Figure 10(a) shows the percentage of the number of who answered "yes", and Fig. 10(b) shows the results of our method. The eight bodies arranged in descending order of the percentage and the similarity measure are shown in Figure 11(a) and (b), respectively.

The proposed similarity measure agrees well with human sense except for a regular quadrangular prism. Since a cube and a regular prism have only straight line edges and they resemble each other from the point of view of topological geometry, it is natural for a human to feel that a prism is similar to a cube, even if the height and width ratio is different. Information about whether the body is convex or concave, or whether it has plain surfaces or curved ones gives us intuitive impression, and affects our sense of similarity evaluation.

We need to consider topology as well as geometry, but topology itself is hard to be perceived. A human stores much knowledge about appearances of an object, and it is thought to be the reason why a human can describe its topological features. Object recognition may have to be done first before similarity evaluation

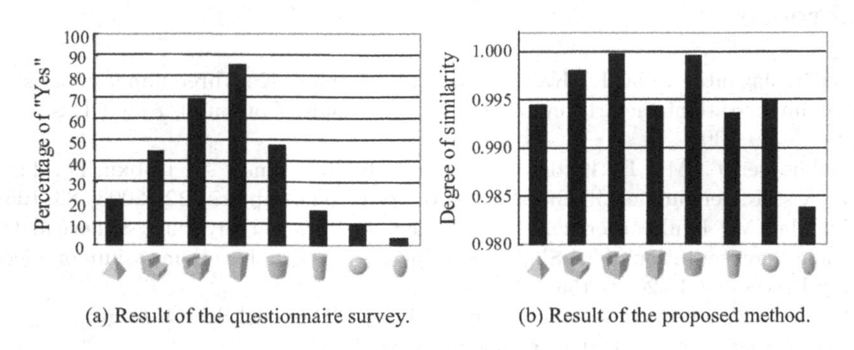

(a) Result of the questionnaire survey.

(b) Result of the proposed method.

Fig. 10. A cube is compared to the other shapes

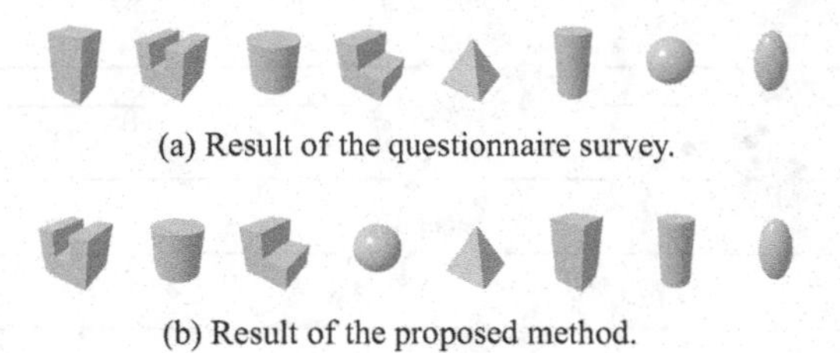

(a) Result of the questionnaire survey.

(b) Result of the proposed method.

Fig. 11. Eight bodies arranged in descending order of similarity to a cube

by the computer. Therefore we need to collect as much visible information as possible in order to simulate a human brain. The number of vertices, the length of inner edges or an outline and various kinds of 2D shape descriptors are valid features which can be defined in the same spherical function as the projected area.

5 Conclusion

We proposed 3D shape similarity evaluation using the projected area in multiple directions. A spherical function defined by the distribution of the projected area on a unit sphere is expanded in spherical harmonics, and the similarity measure is determined by coefficients of the expansion. The numerical simulation has demonstrated that the similarity measure well expresses the features of the shape of an object. After that the proposed measure has been compared with human sense through the questionnaire survey.

Finally the proposed measure almost agrees with human sense but some difference became clear. Though only the projected area was used as the feature of the object, the simulation has shown satisfactory results. It is considered that more information can reduce the present gap between the simulation and the human sense. The future study will include the distribution of the length of edges, which supply some topological information.

References

1. Iyer N., Jayanti S., Low K., Kalyanaraman Y., Ramani K.: Three-dimensional shape searching: state-of-the-art review and future trends. Computer Aided Design, **37** (2005) 509–530
2. Funkhouser T., Min P., Kazhdan M., Chen J., Halderman A., Dobkin D., Jacobs D.: A search engine for 3D models. ACM Trans. on Graphics, **22** (2003) 83–105
3. Kazhdan M., Funkhouser T., Rusinkiewicz S.; Rotation Invariant Spherical Harmonic Representation of 3D Shape Descriptors. Eurographics Symposium on Geometry Processing **1** (2003) 156–165
4. Princeton Shape Retrieval and Analysis Group: 3D Model Search Engine `http://shape.cs.princeton.edu/search.html`

5. Saupe D. and Varanić D. V.: 3D model retrieval with spherical harmonics and moments. Proc. of DAGM2001 **23** (2001) 392–397
6. Proriol J.: Definition of new 3D invariants. Intl. J. of Neural Systems **7** (1996) 709–714
7. Tanaka K, Sano M, Mukawa N, Kaneko H.: 3D object representation using spherical harmonic functions. Proc. of the IEEE/RSJ (1993) 1873–1880
8. Horn B. k. P.: Robot Vision. The MIT Press (1986)

Continuous Level of Detail on Graphics Hardware

Francisco Ramos, Miguel Chover, Oscar Ripolles, and Carlos Granell

Universitat Jaume I, Depto. Lenguajes y Sistemas Informaticos
12071 Castellon, Spain
{Francisco.Ramos, chover, oripolle, canut}@uji.es

Abstract. Recent advances in graphics hardware provide new possibilities to successfully integrate and improve multiresolution models. In this paper, we present a new continuous multiresolution model that maintains its geometry, based on triangle strips, in high-performance memory in the GPU. This model manages the level of detail by performing fast strip updating operations. We show how this approach takes advantage of the new capabilities of GPUs in an efficient manner.

1 Introduction

One of the main problems of interactive graphic applications, such as computer games or virtual reality, is the geometric complexity of the scenes they represent. In order to solve this problem, different techniques for modeling by level of detail have been developed that attempt to adapt the number of polygons of the objects to their importance within the scene.

The application of these techniques is common in standards such as X3D, graphic libraries such as OpenInventor, OSG, and even in game engines such as Torque, CryEngine, and so forth, where models with continuous levels of detail, based mainly on Progressive Meshes [1], are introduced.

The tendency in recent years has been to improve the features of continuous models by using the possibilities offered by the graphics hardware to the maximum, with the intention of competing with the discrete models that, although more limited, are perfectly adapted to current graphics hardware. Specifically, they have worked on the representation of multiresolution models which use triangle strips to accelerate visualization by means of vertex arrays located in the GPU. The fundamental problem of these techniques is the fact that a continuous model needs to make changes in the list of indexes of the primitives it draws and carrying out this kind of operations causes graphics hardware to lower its performance.

1.1 Related Work

In recent years, multiresolution models have progressed substantially. At the beginning, discrete models were employed in graphics applications, due mainly

A. Kuba, L.G. Nyúl, and K. Palágyi (Eds.): DGCI 2006, LNCS 4245, pp. 460–469, 2006.

to the low degree of complexity involved in implementing them, which is the reason why nowadays they are still used in applications without high graphics requirements. Nevertheless, the increase in realism in graphics applications make it necessary to use multiresolution models which are more exact in their approximations, which do not call for high storage costs and which are faster in visualization. This has given way to continuous models, where two consecutive levels of detail only differ by a few polygons and where, additionally, the duplication of information is avoided to a considerable extent, thus improving on the spatial cost offered by the discrete models.

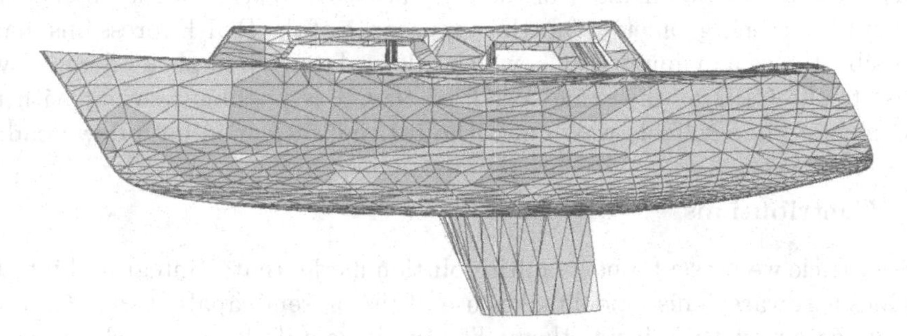

Fig. 1. Boat model in triangle strips

The best known continuous multiresolution model is Progressive Meshes [1], included in Microsoft Corporation's graphic library DirectX. This model offers excellent results in visualization in real time, although it is based on triangle primitives.

Advances have been made in the use of new graphics primitives which minimize the data transfer between the CPU and the GPU, apart from trying to make use of the connectivity information given by a polygonal mesh. For this purpose, graphics primitives with implicit connectivity, such as triangle strips (see Figure 1) and triangle fans, have been developed. Many continuous models based on this type of primitives have been recently developed [2,3,4,5,6,7].

In these last few years, graphics hardware performance has evolved outstandingly, giving rise to new techniques which allow the continuous models to accelerate even more. The use of stripification algorithms, which try to take the maximum advantage of the GPU cache, and the new extensions of graphics libraries that allow visualization of a whole mesh with only a few instructions are examples of these new techniques.

Nowadays GPUs offer new capabilities that, when exploited to the maximum, can offer very good results in several aspects. One of them involves storing information directly in the high speed memory located in the GPU. This characteristic allows information to be managed in the GPU while avoiding data transfer between the CPU and the GPU, and taking the maximum advantage of the proximity of the memory and the graphics processor. There are a number of related works which make use of the new capabilities of the current GPUs, such as [8], which implements a discrete model manager that puts geomorphing into practice by using vertex shaders; another work is [9], which creates different shaders depending on the level of detail.

1.2 Motivation

In general, the main problem with continuous models lies in the high cost of extracting the level of detail, which usually takes about 20% of the total visualization cost. Apart from extraction, the use of AGP buses poses the problem of their being much better optimized to upload data than to download it, thus favoring the use of the memory of the graphics card to store static objects that do not change their geometry. But the appearance of the PCI-Express bus makes it possible to use a symmetric bus, which allows data to be uploaded and downloaded to the GPU at the same speed, so that it is possible to work with the GPU memory in a reliable way and without penalizations in data download.

1.3 Contributions

In this article we present a new multiresolution model that is integrated into the graphics hardware. This model makes use of the present capabilities of GPUs to store its data structures inside them. The fundamental idea on which the model is based is the creation of efficient data structures that can be integrated into the GPU and which, at the same time, offer an optimum performance with respect to both visualization and spatial cost. The model works directly with the GPU memory, appreciable improvements being obtained, as can be seen in the results section.

Hence, what this model offers is complete integration into the graphics hardware, a low cost of extraction of the level of detail, by exploiting the coherence between levels of detail, and a low spatial cost.

The implemented model features different characteristics:

- Wholly based on triangle strips.
- Simplification based on progressive edge collapses.
- Static stripification. Triangle strips are only generated once, at the highest level of detail, using a method that takes advantage of the GPU cache.
- Geometric information of the model is maintained and stored in the GPU.
- Level of detail management is performed by a data structure, LOD-Manager, which allows fast updating of strips and removal of degenerated triangles.

2 Fundamentals

2.1 Multiresolution Models

To construct a continuous multiresolution model based on primitives with implicit connectivity, such as triangle strips, certain requirements must be fulfilled. On the one hand, a mesh made up of this kind of primitives must be available

Fig. 2. Three levels of detail from the AL model (LOD=1,0.5 and 0, respectively)

and, on the other hand, the simplification method that should be employed in order to generate the different levels of detail must be selected, an example is shown in Figure 2.

There are several mesh simplification methods [10, 11], but one of the most important in progressive mesh simplification is [1]. This method is based on iterative edge contractions, and it is the one employed in well-known multiresolution models such as [2, 3, 4, 5, 6, 7].

Many works can be found in the literature where the problem of converting a polygonal mesh made up of triangles into triangle strips is solved [12, 13]. This process is commonly called stripification, and it can be carried out in a dynamic or static way. Dynamic stripification involves generating the triangle strips in real time, that is, for each level of detail new strips are generated. On the other hand, static stripification entails first creating triangle strips and then working with versions of the original strips. There are several models that use dynamic stripification [3,4], especially variable resolution models. Other models such as [2, 5, 6, 7], however, use static stripification techniques.

The main problem of static stripification models can be observed in Figure 3. As the model reaches lower levels of detail, it presents vertex repetitions that do not add any information to the final scene but nevertheless involve higher data traffic between the CPU and the GPU. Models like [2, 7] solve this problem by applying filters to eliminate degenerated triangles. The first employs filters in visualization, thus avoiding sending those vertices at the moment of rendering, and the second runs a preprocess that detects them early on, and then stores that information and eliminates them from the strips before visualizing them.

Given the architecture of present-day GPUs, it is better to employ static stripification techniques since we thereby avoid strip creation and destruction in the GPU, which would imply an additional cost that would make the model much less competitive. Furthermore, there is an additional cost stemming from the calculation of the new triangle strips at each level of detail, which also penalizes the

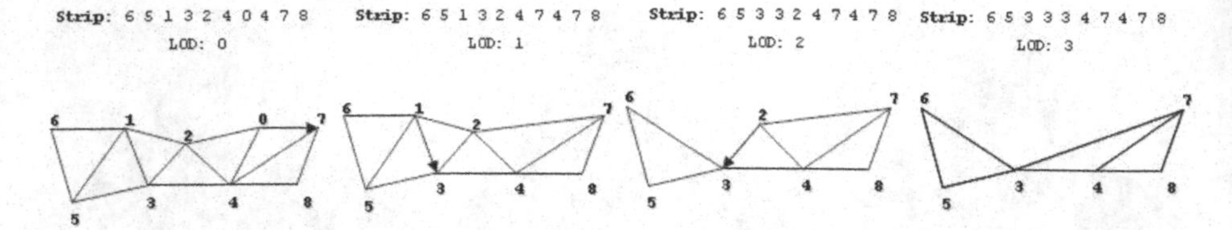

Fig. 3. Multirresolution triangle strips

use of these techniques. Moreover, it is preferable to eliminate degenerated triangles before visualization, which allows a considerable degree of acceleration to be accomplished by resizing strips, apart from also enabling a better implementation of the model in the GPU by avoiding the need to create a specific code for the filters. Nowadays, a variety of acceleration techniques have appeared, which, if integrated into a multiresolution model, would also become essential to improving its performance. Basically, we can observe stripification techniques oriented toward exploiting vertex caches [12] and hardware acceleration techniques based on graphics library extensions [14].

2.2 High-Performance Memory in GPUs

A vertex buffer object is a feature that enables us to store data in high-performance memory in the GPU. The basic idea is to provide some buffers, which will be available through identifiers. There are different ways to interact with buffers:

- Bind a buffer: this activates the buffer in order to be used by the application.
- Put and get data: this allows us to copy data between a client's area and a buffer object in the GPU.
- Map a buffer: you can get a pointer to a buffer object in the client's area, but this can lead to the driver's waiting for the GPU to finish its operations.

There are two kinds of vertex buffer objects: array buffers and element array buffers. On the one hand, array buffers contain vertex attributes, such as vertex coordinates, texture coordinates data, per-vertex color data and normals. On the other hand, element array buffers contain only indexes to elements in array buffers. The ability to switch between various element buffers while keeping the same vertex array allows us to implement level of detail schemes by changing the elements buffer while working on the same array of vertices.

In order to implement the model on graphics hardware, we used different functions which interact with buffer objects. Among them, we can highlight:

- glBindBufferARB: this function sets up internal parameters so that the next operations work on this current buffer object.
- glBufferDataARB: this function is an abstraction layer between the memory and the application. Basically, this function copies data from the client memory to the buffer object bound.

– glBufferSubDataARB and glGetBufferSubDataARB: its purpose consists in replacing or obtaining, respectively, data from an existing buffer.

3 Implementation Details

3.1 General Framework

A brief outline of the model is shown in Figure 4. At the beginning, information about vertices and strips, at the highest level of detail, is uploaded into the GPU. Later, by means of the LOD-Manager data structure, strips are updated in accordance with the current level of detail.

In our approach we first perform two essential tasks: generation of triangle strips at the highest level of detail and calculation of vertex-collapse simplification.

At runtime, we upload information about vertices and strips into the GPU. Then, depending on application demands, we perform vertex-split or edge-collapse operations directly on the strips. This task is executed by the LOD-Manager. More specifically, when a level of detail transition is required, it downloads the strips affected by these changes from the GPU. Later, it modifies and uploads the updated strips to the graphics system. Lastly, strip information in the GPU is then used for display.

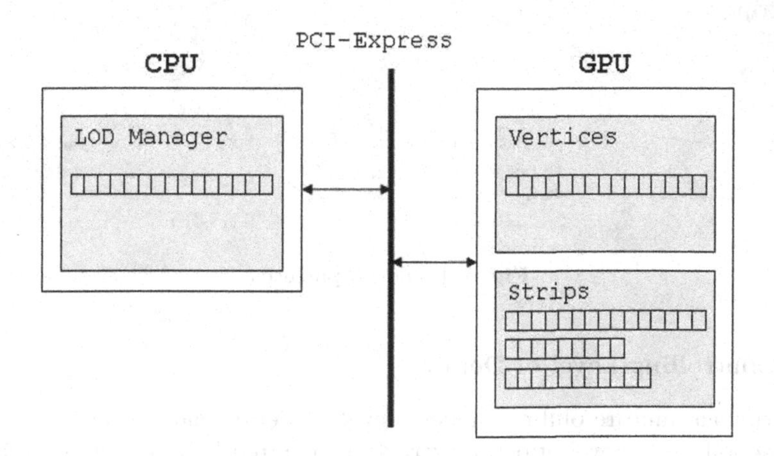

Fig. 4. Model architecture

3.2 LOD-Manager Data Structures

The main function of LOD-Manager consists in serving the level of detail demands required by applications. It is able to quickly change the geometric information located in the GPU by applying a series of pre-calculated records. These records store mainly two kinds of information: simplifications and filters.

Simplification information contains data about which strips change for each level of detail, and where the vertices to be split or collapsed are located. It allows us to quickly locate information to be modified when we move from one level of detail to another. However, as the model moves to coarse LODs, an accumulation of identical vertices is produced. Sending these vertex repetitions to the graphics hardware does not contribute at all to the final scene because it is equivalent to send degenerated triangles, as is shown in Figure 3. We have proved that most vertex repetitions can be removed, following patterns like aa(a)+ or ab(ab)+. Patterns aa(a)+ are replaced by aa, and ab(ab)+ by ab. Figure 5 shows an example for each kind of pattern, and it can be observed that the final geometry of strips does not change after removing these patterns.

3.3 GPU Data Structures

Two essential data structures for the performance of the model are stored in the GPU: vertices and strips, which constitute the polygonal mesh. On the one hand, vertices are stored in a vertex array buffer. On the other hand, we might allocate each strip in an element buffer. However, we have observed that creating as many buffers as strips leads to noticeable decreases in performance due to bind operations. A solution to this problem, with optimum results, consists in creating a single element buffer, where every strip to be rendered is located. In this way, we avoid the need for continuous bind operations to assign an element buffer for each strip.

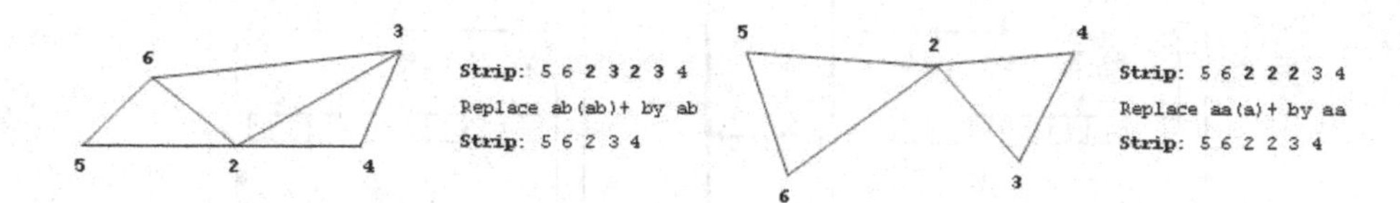

Fig. 5. Removed patterns

3.4 Controlling Level of Detail

In continuous multiresolution models, level of detail management entails two fundamental tasks: level of detail extraction required by applications and visualization of resulting geometry.

Level of Detail Extraction. At a high level, the pseudo algorithm for moving from LOD n to LOD n+1 would consist in downloading, from the GPU, the chunks of memory corresponding to the strips affected by the change in the level of detail. After that, we replace vertex n by the vertex it collapses to, in every strip where it appears. Later, derived vertex repetitions must be removed. Finally, the strip is uploaded to the GPU for visualization. Figure 6 shows the algorithm.

```
for LOD = currentLOD to demandedLOD
  for Strip = StripsAffected(LOD).Begin() to StripsAffected(LOD).End()
    auxStrip=DownloadFromGPU(Strip);
    CollapseOrSplit(auxStrip,LOD);
    UploadToGPU(auxStrip);
  end for
end for
```

Fig. 6. Level of detail extraction from a LOD to a coarse one

Visualization. Figure 7 shows the visualization algorithm. This algorithm takes advantage of the capabilities of the latest GPUscapabilities. It stores and manages strips to be visualized directly from the graphics hardware memory.

```
for IndexStrip = 0 to NumberOfStrips - 1
  glDrawRangeElements (
    GL_TRIANGLE_STRIP,
    currentLOD,
    NumberOfVertices - 1,
    StripBufferManager(IndexStrip).size(),
    GL_UNSIGNED_INT,
    (const void*)(StripBufferManager(IndexStrip).Offset()*sizeof(EnteroUn)),
end for
```

Fig. 7. Visualization algorithm

4 Results

Figure 8 shows a comparison of spatial costs. On average, the model presented in this paper fits in 1.5 times the original mesh in triangles and 2.3 times in triangle strips.

Two well-known utilities to generate strips were tested in this multiresolution model: Stripe Utility [13] and NVTriStrip Library [12]. Triangle strips for different objects were generated using both utilities. The model generated from the

	Cow	Al	Bunny	Panther	Dragon	Phone	Buddha
Vertices	2904	3618	34834	38911	54294	83044	543699
Faces	5804	7124	69451	69397	108588	165963	1085634
Size Tris kb	113.4	140.0	1358.2	1421.2	2120.9	3242.5	21217.6
Size Strips kb	73.5	91.4	867.6	971.1	1387.6	1999.5	14107.9
Model Cost Mb	0.16	0.20	2.07	2.00	3.59	4.71	33.53
Ratio Triangles	1.5	1.5	1.6	1.4	1.7	1.5	1.6
Ratio Strips	2.3	2.2	2.4	2.1	2.7	2.4	2.4

Fig. 8. Spatial cost comparison

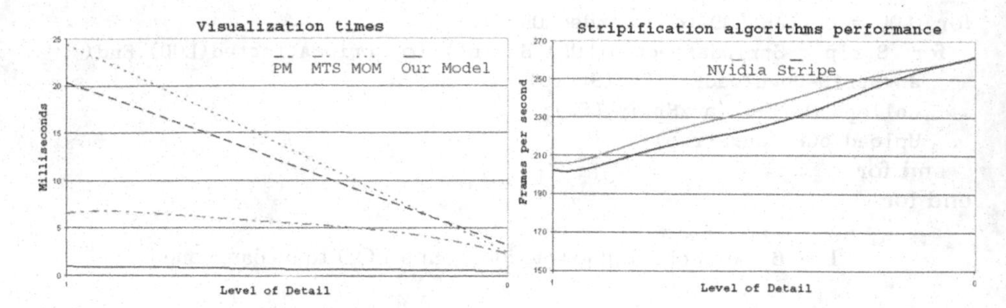

Fig. 9. Results obtained from the bunny object. On the left, multiresolution models comparison of PM [1], MTS [5], MOM [6] and our model. On the right, stripification techniques perfomance in our approach.

NVTriStrip Library shows better frame-per-second rates than the Stripe object when the level of detail is higher; this behavior is shown in Figure 9(right).

Results of visualization are shown in Figure 9(left), where our approach is compared to other models. It can be seen that our model offers the best visualization times due to its being integrated into the hardware.

5 Conclusions

We have presented a uniform resolution model that noticeably improves existing models in terms of both storage and visualization cost. This model features: total graphics hardware integration with implementation in high-performance memory, optimized hardware primitives, vertex cache exploitation and low spatial cost.

References

1. Hoppe, H.: Progressive Meshes. Computer Graphics (SIGGRAPH) **30** (1996) 99-108
2. El-Sana, J., Azanli, E., Varshney, A.: Skip strips: maintaining triangle strips for view-dependent rendering. In: Proceedings of Visualization 99 (1999) 131-137
3. Shafae, M., Pajarola, R.: DStrips. Dynamic Triangle Strips for Real-Time Mesh Simplification and Rendering. Proceedings Pacific Graphics Conference (2003)
4. Stewart, A.J.: Tunneling for Triangle Strips in Continuous Level-of-Detail Meshes. Graphics Interface (2001) 91-100
5. Belmonte, O., Remolar, I., Ribelles, J., Chover, M., Fernndez, M.: Efficient Use Connectivity Information between Triangles in a Mesh for Real-Time Rendering, Future Generation Computer Systems, Special issue on Computer Graphics and Geometric Modeling (2003)
6. Ribelles, J., Lpez, A., Remolar, I., Belmonte, O., Chover, M.: Multiresolution Modeling of Polygonal Surface Meshes Using Triangle Fans. Proc. of 9th DGCI 2000, (2000) 431-442

7. Ramos, J.F., Chover, M.: LodStrips. Level of Detail Strips, Lecture notes in Computer Science, Proc. of Computational Science ICCS 2004, Springer, ISBN/ISSN 3-540-22129-8, Krakow (Poland), vol. 3039, (2004) 107-114
8. Olano, M., Kuehne, B., Simmons, M.: Automatic Shader Level of Detail. Proceedings of Graphics Hardware 2003, Eurographics/ACM SIGGRAPH, July 2003
9. Gain, J., Southern, R.: Creation and Control of Real-time Continuous Level of Detail on Programmable Graphics Hardware. Computer Graphics Forum, March 2003
10. Garland, M., Heckbert, P.: Surface simplification using quadric error metrics. In: Proceedings of SIGGRAPH '97 (Los Angeles, CA), Computer Graphics Proceedings, Annual Conference Series, ACM SIGGRAPH, ACM Press, (1997) 209–216
11. Luebke, P.:A Developer's Survey of Polygonal Simplification Algorithms, IEEE CGA, June, 2001
12. NvTriStrip Library, NVIDIA Corporation (2002). Available in Internet at following URL http://developer.nvidia.com/object/ nvtristrip_library.html.
13. Evans, F., Skiena, S., Varshney, A.: Optimising Triangle Strips for Fast Rendering, IEEE Visualization '96, (1996) 319-326 http://www.cs.sunysb.edu/ stripe
14. ARB_vertex_buffer_object Specification. http://oss.sgi.com/projects/ogl-sample/registry/ARB/vertex_buffer_object.txt

Topological and Geometrical Reconstruction of Complex Objects on Irregular Isothetic Grids

Antoine Vacavant[1], David Coeurjolly[2], and Laure Tougne[1]

[1] LIRIS - UMR 5205, Université Lumière Lyon 2
5, avenue Pierre Mendès-France
69676 Bron cedex, France
[2] LIRIS - UMR 5205, Université Claude Bernard Lyon 1
43, boulevard du 11 novembre 1918
69622 Villeurbanne cedex, France
{antoine.vacavant, david.coeurjolly, laure.tougne}@liris.cnrs.fr

Abstract. In this paper, we address the problem of vectorization of binary images on irregular isothetic grids. The representation of graphical elements by lines is common in document analysis, where images are digitized on (sometimes very-large scale) regular grids. Regardless of final application, we propose to first describe the topology of an irregular two-dimensional object with its associated Reeb graph, and we recode it with simple irregular discrete arcs. The second phase of our algorithm consists of a polygonal reconstruction of this object, with discrete lines through the elementary arcs computed in the previous stage. We also illustrate the robustness of our method, and discuss applications and improvements.

1 Introduction

The character and symbol representation, description and classification are necessary tasks in many current applications, and concern both research and industrial challenges. Those tasks are applied on images generally designed within a regular grid, *i.e.* all the pixels have the same size, and their position can be easily indexed. However, it is now common to successively divide an image into subimages, as in *quadtree* decomposition [1, 2], to represent a part of an image in a more compact and adapted structure. These techniques describe interesting parts of an image, from different points of view, through a set of irregular pixels. In this paper, we introduce the concept of shape representation within an *irregular isothetic grid* ($\mathbb{I}$-grid for short) [3]. The pixels are defined by variable sizes and positions, and may be determined by subdivision rules. We propose to represent the topology of the elements contained in the irregular two-dimensional (2-D) image by constructing their associated Reeb graph [4], then we represent them by a simple polygonal structure that respects the extended supercover digitization model defined in [3]. This structure also preserves the topology that we reveal in the previous stage. We clearly address the problem of *vectorization* (or

A. Kuba, L.G. Nyúl, and K. Palágyi (Eds.): DGCI 2006, LNCS 4245, pp. 470–481, 2006.

raster-to-vector) on irregular isothetic grids, and not only in the scope of document analysis. In our framework, we are interested in binary images containing irregular objects, *i.e.* $k-$objects in respect to the definition given in [3], where k represents the considered relation of adjacency (see Section 2 for further details). Those complex objects may contain holes, and could represent characters, symbols, lines, etc. An application of such binary image processing is clearly document and line drawings analysis, but we can also consider a discrete subdivision of a part of $\mathbb{R}^2$ representing the solutions of a given function $f : \mathbb{R}^2 \to \mathbb{R}$. The algorithms designed in interval arithmetic are interesting approaches to address those problems [5, 6, 7].

The techniques of vectorization developed until now on the discrete regular domain can be divided into several classes, up to the final application of the method [8, 9, 10, 11]. We will only focus on a few kinds of raster-to-vector methodologies, largely developped for document analysis applications. To our knowledge, there exists no generic extension of those approaches on irregular isothetic grids. The *run length encoding* (RLE) *based methods* first build a decomposition into elongated cells along an axis of the image where we can build a line adjacency graph (LAG) [12,13]. Those methods aim to describe the topology of the encountered objects in the image, but the geometrical structure deduced from it has to be improved by many post-treatment processings. The *skeletonization* and *thinning methods* are surely the most widely employed methods in vectorization. We can notice that tools designed in mathematical morphology [14] are a frequent choice to prepare the images before processing the skeletonization. A survey of vectorization methods based on skeleton can be found in [15], and another one about such techniques not using it in [16]. The aim is to compute a medial axis of the object that minimally represents its shape [17]. However, those techniques modify the original geometry of the object to obtain a minimal representation of it. Besides, they need filtering or smoothing pre-treatment processings to reduce the noise that could pertubate the final medial axis. The $k-$object can contain holes, and so may be composed by *thick arcs.* In the work of Debled et al. [18, 19, 20], the definitions of *discrete lines* and *blurred segments* join the concept of thick regular arcs. But, beyond this geometrical representation of arcs, the global structure is not aborded, and thus there are no description of the topology of the recognized objects.

In this article, we first introduce the concepts of $k-$arcs and $k-$objects by recalling some definitions, then we present the extended supercover model on an $\mathbb{I}-$grid. We also recall the invertible reconstruction of $k-$arcs described in [21]. In the third part, we give details about the two main phases of our system: the description of the topology of a complex object based on the Reeb graph [4], and its polygonal reconstruction. Then, we present some experiments and revealing results to illustrate the two phases of our algorithm. We also proove the robustness of the polygonal reconstruction by a test on a large image of technical drawing. We finally discuss the applications of our contribution, and the improvement on its global performance.

2 Preliminaries

We first define an irregular isothetic grid, denoted $\mathbb{I}$, as a tiling of the plane with isothetic rectangles. We shortly recall that each rectangle P (also called *cell*) of $\mathbb{I}$ is defined by its center $(x_P, y_P) \in \mathbb{R}^2$ and a size $(l_P^x, l_P^y) \in \mathbb{R}^2$. The position and the size of P may be controlled by different level of constraints; *e.g.* in the case of quadtree decomposition [1, 2], for a cell of level k, $(x_P, y_P) = (\frac{m}{2^k}, \frac{n}{2^k})$ and $l_P^x = l_P^y = \frac{1}{2^{k-1}}$ for some $m, n \in \mathbb{Z}$ [3, 21].

In our framework, adjacency relation is an important feature that we depict through the following definitions.

Definition 1 (ve−adjacency and e−adjacency). *Let P and Q be two cells. P and Q are ve−adjacent (vertex and edge adjacent) if :*

$$\begin{cases} |x_P - x_Q| = \frac{l_P^x + l_Q^x}{2} \text{ and } |y_P - y_Q| \leq \frac{l_P^y + l_Q^y}{2} \\ \text{or} \\ |y_P - y_Q| = \frac{l_P^y + l_Q^y}{2} \text{ and } |x_P - x_Q| \leq \frac{l_P^x + l_Q^x}{2} \end{cases}$$

P and Q are e−adjacent (edge adjacent) if we consider an exclusive "or" and strict inequalities in the above ve−adjacency definition. k may be interpreted as e or ve in the following definitions.

Definition 2 (k−arc). *Let $\mathcal{E}$ be a set of cells, $\mathcal{E}$ is a k−arc if and only if for each element of $\mathcal{E} = \{P_i, i \in \{1, ..., n\}\}$), P_i has exactly two k−adjacent cells, except P_1 and P_n which are called extremities of the k−arc.*

Definition 3 (k−object). *Let $\mathcal{E}$ be a set of cells, $\mathcal{E}$ is a k−object if and only if for each couple of cells (P, Q) belonging to $\mathcal{E} \times \mathcal{E}$, there exists a k−path between P and Q in $\mathcal{E}$.*

We now consider the extension of the supercover model from [22] on irregular isothetic grids [3] to digitize Euclidean objects on $\mathbb{I}$.

Definition 4 (Supercover on irregular isothetic grids). *Let F be an Euclidean object in $\mathbb{R}^2$. The supercover $\mathbb{S}(F)$ is defined on an irregular isothetic grid $\mathbb{I}$ by:*

$$\begin{aligned} \mathbb{S}(F) &= \{P \in \mathbb{I} \mid \mathbb{B}^\infty(P) \cap F \neq \emptyset\} \\ &= \{P \in \mathbb{I} \mid \exists (x, y) \in F, |x_P - x| \leq \frac{l_P^x}{2} \text{ and } |y_P - y| \leq \frac{l_P^y}{2}\} \end{aligned}$$

where $\mathbb{B}^\infty(P)$ is the rectangle centered in (x_P, y_P) of size (l_P^x, l_P^y) (if $l_P^x = l_P^y$, $\mathbb{B}^\infty(P)$ is the ball centered in (x_P, y_P) of size l_P^x for the L_∞ norm).

This model has several interesting properties, *e.g.* for F, G two Euclidean objects in $\mathbb{R}^2$, we have $\mathbb{S}(F \cup G) = \mathbb{S}(F) \cup \mathbb{S}(G)$ or $\mathbb{S}(F \cap G) \subseteq \mathbb{S}(F) \cap \mathbb{S}(G)$ (see proposition 2 in [3] for more details).

We now present the k−arc reconstruction algorithm we use in our complex object geometrical representation phase (Section 3.2). Moreover, this approach respects the supercover model we have just presented. The algorithm proposed in [21] to decompose a curve into segments is first based on the following definition of an irregular digital line.

Definition 5 (Irregular isothetic digital straight line). *Let S be a set of cells in $\mathbb{I}$, S is called a piece of irregular digital straight line (IDSL for short) iff there exists an Euclidean straight line l such that:*

$$S \subseteq \mathbb{S}(l)$$

In other words, S is a piece of IDSL iff there exists l such that for all $P \in S$, $\mathbb{B}^\infty(P) \cap l \neq \emptyset$.

The algorithm inspired from [23] principally uses the construction and update procedures of a *visibility cone*, and can be sketched as follows. We first fix the extremity p_0 of the first segment such that $p_0 \in P_0$. We note e_0 the Euclidean segment shared by P_0 and P_1, and we consider the first cone $C_0(p_0, s, t)$ such that s and t coincide with the extremities of e_0 and $\{p_0, s, t\}$ is sorted counterclockwise. Then, for each cell P_i, we consider the shared segment e_i between P_{i-1} and P_i, and the current cone $C_j(p_j, s, t)$ is updated. When the update procedure fails, a new cone $C_{j+1}(p_{j+1}, s, t)$ is set up, and we add the point p_{j+1} to the reconstruction: to compute the new cone, authors of [21] consider the bisector of the cone and define p_{j+1} as the midpoint of the intersection between the bisector and the pixel P_{i-1}. The Figure 1 illustrates the progressive construction of cones in a $k-$arc, and the resulting segmentation into lines.

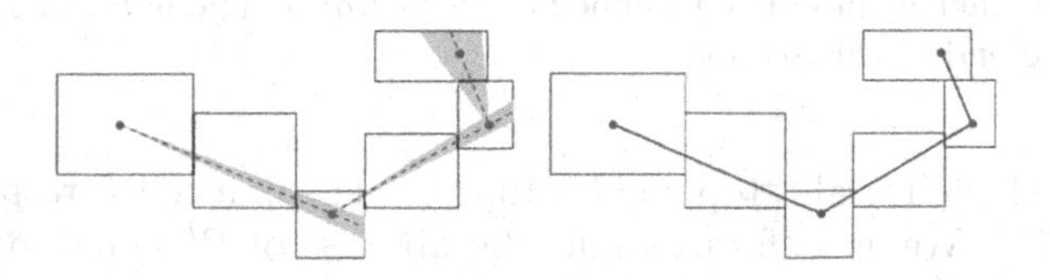

Fig. 1. An example of the progressive construction of cones in a $k-$arc *(left)*, and the reconstruction into segments we obtain *(right)*

3 Complex Objects Definition and Representation on Irregular Isothetic Grids

In this section, we present the two main phases of our system for object representation on irregular isothetic grids.

3.1 Representation of a Complex Topology

To represent the shape of a $k-$object $\mathcal{E}$, we have chosen an incremental directional approach to build its associated Reeb graph G, as in continuous space (see Figure 2). It is an interesting structure introduced by G. Reeb [4] based on the Morse theory [24, 25]. This graph is also used in many applications for surface and curve description [26, 27, 28]. The Reeb graph G is associated to a *height function* f defined on $\mathcal{E}$, and nodes of G represent the critical points of f. Moreover, to have a minimal representation of the topological information of $\mathcal{E}$, each edge of the Reeb graph corresponds to a $k-$arc. Those $k-$arcs will be segmented in the stage of polygonal description of $\mathcal{E}$ (Section 3.2).

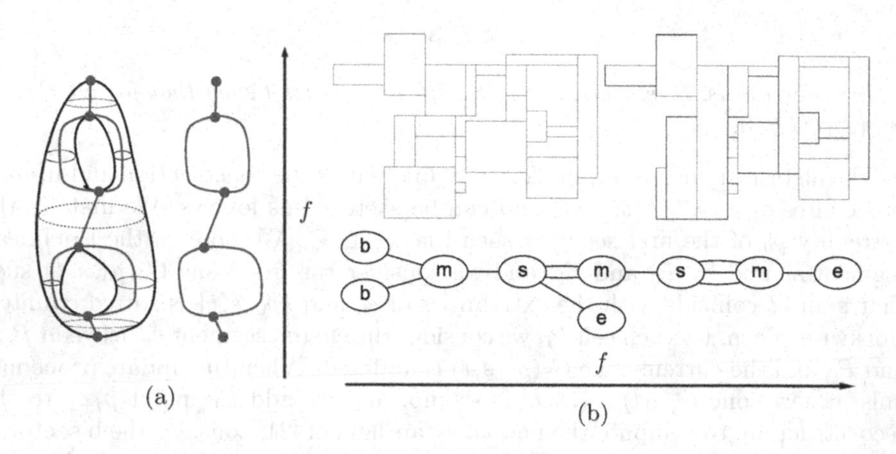

Fig. 2. *(a)*: an example of the Reeb graph G of a continuous object $\mathcal{E}$. The nodes of G represent the critical points of f (maxima, minima, inflection points), and an edge is a connected component of $\mathcal{E}$ between two critical points. *(b)*: an example of an irregular object $\mathcal{E}$ *(left)*, the final recoded structure with $k-$arcs *(right)* and the Reeb graph associated to the height function f defined on $\mathcal{E}$ *(bottom)*. The notations b, e, m and s are given at the end of this section.

We denote the left, right, top and bottom borders of a cell P respectively P^L, P^R, P^T and P^B. We have, for example, the abscissa of P^L equal to $x_P - (l_P^X/2)$ (that we denote $P^L = x_P - (l_P^X/2)$). We also abusively say that a $k-$arc A and a cell P are $k-$adjacent if there exists a cell Q in A such that P and Q are $k-$adjacent. Let $\mathcal{E} = \{P_i\}_{i=1,\ldots,n}$ be a given 2-D set of cells. We first choose a direction to treat the cells of $\mathcal{E}$. Without loss of generality, we can suppose that we choose the left-to-right orientation above X axis, *i.e.* the height function f is defined along X axis. At time $t = 0$, we merge together all the $k-$adjacent cells P of $\mathcal{E}$ with the smallest left border $x_{t=0} = x_0$, *e.g.* $P^L = x_0 = 0$. This merging task is processed by the update procedure described below. Those m collections of cells define the *begin cells* of the initial recognized $k-$arcs $A_1, A_2, ..., A_m$.

Update Procedure. Let A be a $k-$arc, and P_1 and P_2 two adjacent cells of $\mathcal{E}$ such that $P_1 \in A$, $P_1^L < P_2^L$, and P_2 should be added to A. If $P_2^L = P_1^R$, we just add P_2 to A, else the procedure *updates* the $k-$arc A with P_2, and may recode A. For that, we first build the *greatest common rectangle* F_2 of P_1 and P_2.

Definition 6 (Greatest common rectangle). *Let P_1 and P_2 be two adjacent rectangles. F_2 is the greatest common rectangle (or GCR) of P_1 and P_2 iff*

i) $F_2 \subseteq P_1 \cup P_2$,
ii) $F_2 \cap P_1 \neq \emptyset$,
iii) $F_2 \cap P_2 \neq \emptyset$,
iv) there is no rectangle greater than F_2 by inclusion respecting i), ii) and iii).

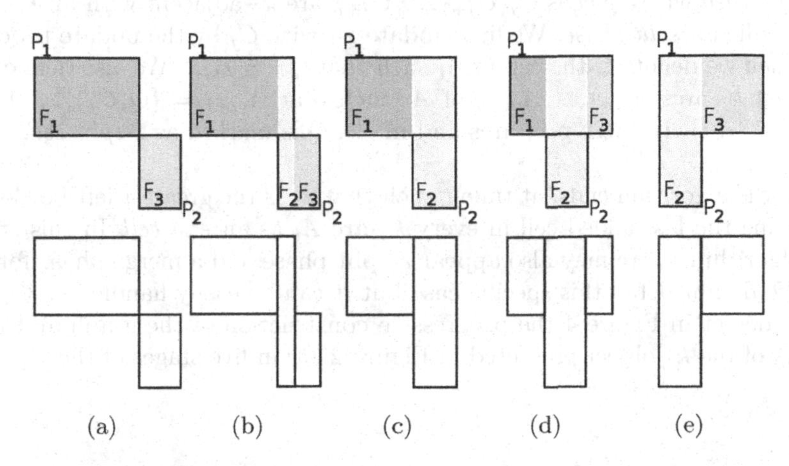

Fig. 3. Description of rectangles F_1, F_2 and F_3 in the update procedure *(top)*, and the associated cells as result *(bottom)*. When $P_1^R < P_2^R$ (*a* and *b*), $P_1 - F_2 = F_1$ and $P_2 - F_2 = F_3$, else $P_1 - F_2 = \{F_1, F_3\}$ (*d* and *e*). If $P_1^R = P_2^L$, $F_2 = \emptyset$, when $P_1^R = P_2^R$, $F_3 = \emptyset$ and finally $F_1 = \emptyset$ in the case $P_1^L = P_2^L$.

Then, we consider the rectangles $P_1 - F_2$ and $P_2 - F_2$. If $P_1^R < P_2^R$, we denote $P_1 - F_2 = F_1$ and $P_2 - F_2 = F_3$, else we prefer $P_1 - F_2 = \{F_1, F_3\}$. We can notice that those rectangles may be empty, *e.g.* $F_3 = \emptyset$ if $P_1^R = P_2^R$, since in that case $F_3^L = F_3^R$. Figure 3 presents the five general configurations of update procedure (there are also five other configurations, obtained by symetry when $P_2^T > P_1^T$), and the $k-$arc recoding that we have to consider. Besides, we propose to reduce the number of cells in A by joining the two rectangles F_1 and F_3 if $F_1^T = F_3^T$, $F_1^B = F_3^B$ and $F_2 = \emptyset$. This junction is processed by replacing F_1 and F_3 by the rectangle $F_1 \cup F_3$. Finally, the procedure ends by removing P_1 from A, and by adding the cells corresponding to the rectangles F_1 and F_2 to A. F_3 is also pushed in $\mathcal{E}$, and will be treated later; more exactely at time t such that $x_t = F_3^L$.

At time $t+1$, our algorithm consists first in merging the adjacent cells with the same left border x_{t+1} in k cells $C_1, C_2, ..., C_k$ (see update procedure for details). Those candidate cells may be added to one or more $k-$arcs among A_i, $i \in \{1, ..., m\}$ if they are adjacent to A_i. It is clear that only a cell Q built at time t and having its right border Q^R equal to x_{t+1} may be adjacent with a cell C_j, $j \in \{1, ..., k\}$. A cell C_j can be treated by several manners:

- C_j is not adjacent with any $k-$arc A_i. We initialize a new $k-$arc A_{m+1} with the cell C_j. C_j represents the *begin cell* of A_{m+1}.
- If C_j is adjacent with one $k-$arc A_i, then we just update A_i with C_j.
- When C_j is $k-$adjacent with p $k-$arcs $A_i, A_{i+1}, ..., Ai+p$, it is a *merge phase*. First, we update each $k-$arc with C_j. The cell C_j is marked as a *merge cell* and indicates that each $k-$arc $A_i, ..., A_{i+p}$ has a $k-$arc $A_{m+1} = \{C_j\}$ linked as a *next arc*.
- The case where p cells $C_j, C_{j+1}, ..., C_{j+p}$ are $k-$adjacent with an $k-$arc A_i is called a *split phase*. We first update A_i with C_j by the update procedure. Then we denote Q the cell in A_i such that $Q^R = x_{t+1}$. We also define p new next $k-$arcs $A_{m+1}, ..., A_{m+p}$ of A_i such that $A_{m+1} = \{Q, C_j\}, ..., A_{m+p} = \{Q, C_{j+p}\}$. In those p $k-$arcs and in A_i, Q is marked as a *split cell.*

When the algorithm ends, at time t such that x_t is the greatest left border in $\mathcal{E}$, we define the last added cell in every $k-$arc A_i as an *end cell.* In this stage of our algorithm, there may also appear a split phase and a merge phase for a cell C_j. We do not detail this specific case but it can be easily handled.

We depict in Figure 4 the progressive construction of the graph and the recoding of the $k-$object presented in Figure 2 *(b)* in five stages of the algorithm.

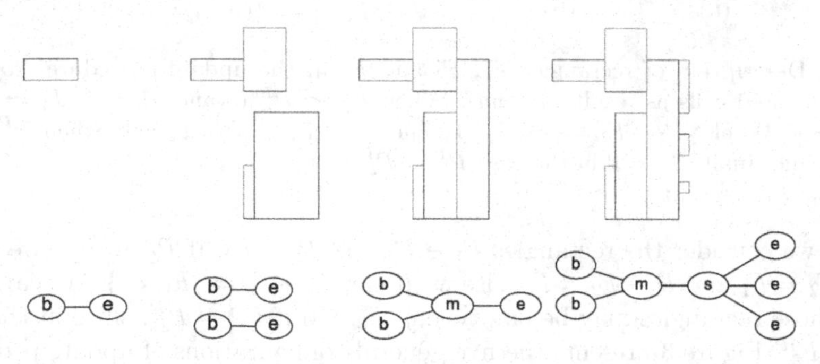

Fig. 4. The recognized $k-$arcs and the associated Reeb graph for some iterations of our algorithm on the object presented in Figure 2 *(b)*. First, we initialize a $k-$arc with the cell with the smallest left border. Then, we progressively update and recode $k-$arcs. The third and fourth images present merge and split phases. We can notice that in one hand the recoding stage is not detailed in this figure, and in the other hand the edges $m-s$ represent a $k-$arc with one cell in this example.

Our algorithm finally builds a complete topological representation of $\mathcal{E}$ with the Reeb graph G by recognizing and linking *begin* (b), *merge* (m), *split* (s) and *end* (e) cells in it. There are nine possible configurations of edges in G: $b-s$, $b-m$, $b-e$, $s-s$, $s-m$, $s-e$, $m-s$, $m-m$ and $m-e$. The number of critical points in f can be linked to the Euler number χ of $\mathcal{E}$ [4]. We consider the following equation, where G is denoted as the couple of sets of vertices and edges (V, E):

$$\chi = \sum_{n \in V, (n=b) \vee (n=e)} (deg(n)) - \sum_{n \in V, (n=s) \vee (n=m)} (deg(n) - 2)$$

where $deg(n)$ is the *degree* of the node n in G, so $deg(n) = 1$ if n is a *begin* or *end* node. The Euler number permits to describe the topology of an object by an unique value. For example, for a torus, $\chi = 0$, for a disc, $\chi = 2$, and the object described in Figure 2 *(b)* has a Euler number $\chi = -4$; we can also say that this shape is homeomorpheous to a torus with 3 holes where $\chi = 2 - 2 \times \#(holes) = -4$. Beside the topological invariants obtained by critical points, the structure of the graph clearly depends on the direction we choose for the height function f. A part of the nodes and the edges may change, but the information on the topology of $\mathcal{E}$, *i.e.* internal nodes of G, is not modified. The Euler number is an example of the use of the Reeb graph for shape description. Let us consider now $\mathcal{E}'$ as the object drawn in the fourth image of Figure 4. The three cells added during the last iteration could be noise modifying the contour of $\mathcal{E}'$. The Reeb graph is modified by a split phase, three nodes are created, whereas these cells are maybe noise. Actually, the problem of the perturbation of the contour of $\mathcal{E}'$ could be certainly reduced if the object was first filtered or smoothed. This kind of pre-treatment processings is often adopted, whatever the approach we may choose for shape representation, *e.g.* skeletonization. Finally, with the update procedure, we recode the cells in $\mathcal{E}$ so that a $k-$arc is always represented between two nodes of G. This geometrical rearrangement clearly depends on the direction of f, but does not change neither the topology nor the contour of the recognized $k-$arcs. The topological structure so described is simple, and prepares the next phase of our complex objects reconstruction system.

3.2 Polygonal Reconstruction of Thick Objects

Since the reconstruction into polylines always affects the first point p_0 as the center of the first treated cell, we propose to start the reconstruction of every $k-$arcs computed in the previous stage by the *merge* and *split* nodes detected in the Reeb graph G. This insures that each of those particular nodes of G will be represented by an unique point in the final polygonalization. The segments are recognized from intersections between several parts of the object $\mathcal{E}$ to its extremities, *i.e.* we consider the edges $m - e$, $s - e$, $m - b$ and $s - b$ of G. Moreover, since the recognition algorithm is greedy, the possible error induced by the visibility cone approach is propagated to the extremities of $\mathcal{E}$, instead of those intersections that represent the shape of the object. For the $m - s$, $m - m$, $s - s$ and $s - m$ configurations of edges in G we propose to process a bidirectional reconstruction that begins from each node of the edge, and ends in its center. Thus, the error may be concentrated in the midpoint of those edges. This approach confirms that the nodes m and s of G represent the places of an object where the description of its geometry must be precise. Finally, we choose to treat edges $b - e$ by the same bidirectional reconstruction, that seems to be the more efficient way to insure a robust reconstruction. We do not deal about the problem of linking the two reconstructions on the $k-$arc (reconstruction with *patch*), because an efficient and general joint technique between two discrete lines implies that our algorithm would not be linear anymore [29]. Hence we just add a segment between the two polylines. This phase of our system can not be handled without patch, since we use the internal points of the shape of $\mathcal{E}$ to guide the geometrical reconstruction.

In Figure 5 *b*, we illustrate the behaviour of our algorithm in the case of the object $\mathcal{E}$ presented in the previous section. We also show the interest of our approach for a symetrical complex object.

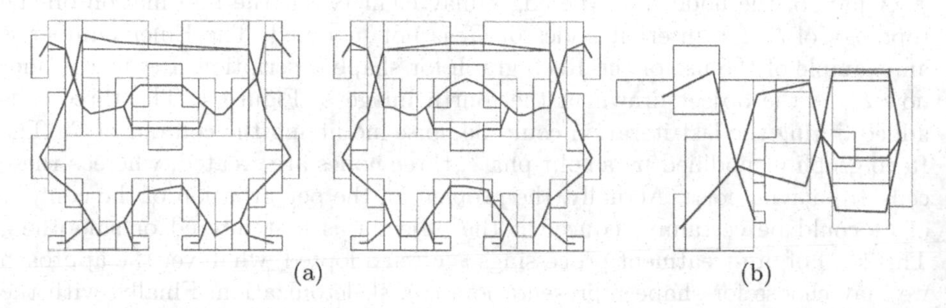

Fig. 5. If we consider the original orientation of the $k-$arcs, the shape of the $k-$object presented in the next section *(a left)* is not well defined since the symetry is not preserved. So, we propose to start the reconstruction by the nodes s and m *(a right)*. This structure respects the supercover model, and the symetrical shape of this object. We also show the result of our algorithm on the $k-$object presented in the previous section *(b)*.

Contrary to conventional vectorization methods, we propose a technique that respects the supercover model on an $\mathbb{I}-$grid. We do not address the quality of the global polygonal structure deduced from this second phase of our system. To introduce the concept of quality in the framework of document analysis, we may refer to [30].

4 Experimentation and Results

In Figure 6, we present the polygonal structure obtained on an image first rearranged by a quadtree-based approach. The reconstruction of $k-$arcs stands inside the object, and the *split* and *merge* nodes are represented with one point in the reconstruction. The polygonal representation also permits to measure geometrical features (*e.g.* length) of a complex function $f : \mathbb{R}^2 \to \mathbb{R}$ (Figure 7). f is first discretized by an interval computing algorithm through a set of cells $\mathcal{E}$, then we use our system to minimally describe the curves of $\mathcal{E}$. Finally, to show the robustness of our system, we present in Figure 8 the polygonal and topological reconstructions of a large image of technical drawing.

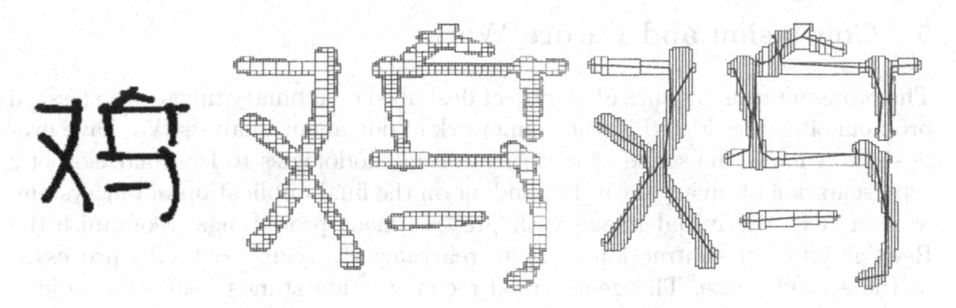

Fig. 6. An image of a chinese character *(left)*, compressed by a quadtree-based approach *(center)*. We show the final k−arcs recoding and the polygonalization *(right)*.

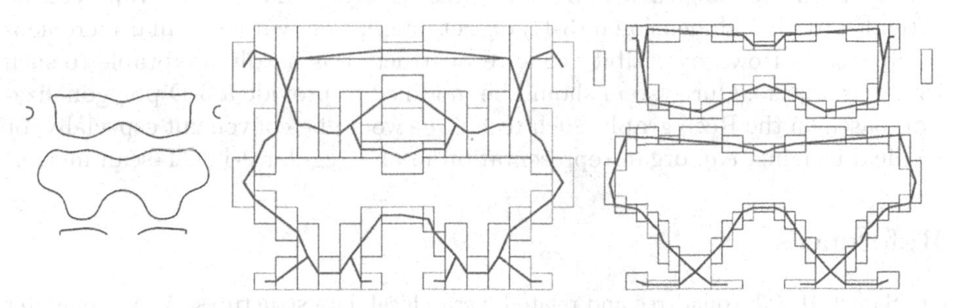

Fig. 7. The function $x^2 + y^2 + \cos(2\pi x) + \sin(2\pi y) + \sin(2\pi x^2)\cos(2\pi y^2) = 1$ on $[-1.1; 1.1] \times [-1.1; 1.1]$ *(left)* discretized by an algorithm described in [7] with two different resolutions, then recoded and polygonalized *(center* and *right)*

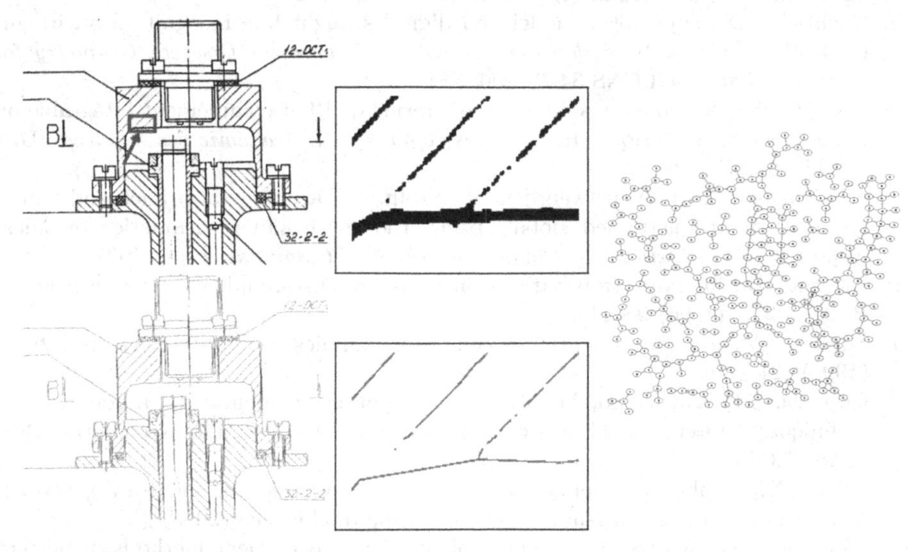

Fig. 8. An image of technical drawing of size 1765 x 1437 pixels we submit to our system, and a zoomed part of it, indicated by the arrow *(up)*. The polygonalization we obtain and the associated zoom are presented *(bottom)*. The complete Reeb graph (about 300 nodes) is also illustrated in a circular format *(right)*.

5 Conclusion and Future Work

The representation by lines of an object described on a binary image is a classical problem often considered in the framework of document analysis. We have proposed to enlarge the scope of vectorization methodologies to irregular isothetic representation of binary data. Depending on the final application of our system, we can treat the initial image with pre-treatment processings, reorganize the Reeb graph (edge contraction, etc.), or rearrange the segments finally processed in the second phase. The geometrical reconstruction stands inside the object, *i.e.* it respects the irregular digitization supercover model. Moreover, this reconstruction preserves the topology described by the Reeb graph. Thus, our system is robust, and topologically and geometrically correct. The Reeb graph can be extended to three-dimensional (3-D) object description, with a similar incremental approach. However, visibity cone reconstruction is hardly adaptable to such irregular objects. Our system should be modified to provide a 3-D polygonalization based on the Reeb graph. Such technique would be convenient especially for medical imaging, *e.g.* organ representation in an irregular 3-D CT-scan image.

References

1. Samet, H.: The quadtree and related hierarchical data structures. ACM Computer Survey **16**(2) (1984) 187–260
2. Samet, H.: Hierarchical spatial data structures. In *Design and Implementation of Large Spatial Databases, First Symposium SSD'89* **409**, published by Springer, Santa Barbara, California, July 17–18 (1989) 193–212
3. Coeurjolly, D.: Supercover model and digital straight line recognition on irregular Isothetic Grids. In *12th International Conference on Discrete Geometry for Computer Imagery*, LCNS 3429 (2005) 311–322
4. Reeb, G.: Sur les points singuliers d'une forme de Pfaff complément intégrable ou d'une fonction numérique. In *Comptes Rendus de L'Académie ses Séances*, Paris 222 (1946) 847–849
5. de Figueiredo, L.H., Van Iwaarden, R., Stolfi, J.: Fast interval branch-and-bound methods for unconstrained global optimization with affine arithmetic. Technical Report IC-9708, *Institute of Computing, Univ. of Campinas*, June (1997)
6. Kearfott, B.: Interval computations: introduction, uses, and resources. Euromath Bulletin **2**(1) (1996) 95–112
7. Snyder, J.M.: Interval analysis for computer graphics. Computer Graphics **26**(2) (1992) 121–130
8. Cordella, L.P. and Vento, M.: Symbol recognition in documents: a collection of techniques ? International Journal on Document Analysis and Recognition **3**(2) (2000) 73–88
9. Hilaire, X., Tombre, K.: Robust and accurate vectorization of line drawings. IEEE Transactions on Pattern Analysis and Machine Intelligence (2005)
10. Mertzios, B.G., Karras, D.A.: On applying fast and efficient methods in pattern recognition. In *Signal Processing for Multimedia*, published by IOS Press, J.S. Byrnes Ed. (1999)
11. Wenyin, L., Dori, D.: From raster to vectors: extracting visual information from line drawings. Pattern Analysis and Application **2**(1) (1999) 10–21

12. Burge, M., Kropatsch, W.G.: A minimal line property preserving representation of lines images. Computing **62** (1999) 355–368
13. Elgammal, A., Ismail, M.A.: Graph-based segmentation and feature-extraction framework for arabic text recognition. In *6th International Conference on Document Analysis and Recognition (ICDAR 01)*, Seattle, Washington, USA, September 10–13 (2001)
14. Soille, P.: Morphological image analysis, 2nd ed. Published by Springer, Berlin, Germany (2003)
15. Lam, L., Lee, S.W., Suen C.Y.: Thinning methodologies - a comprehensive survey. IEEE Transactions on Pattern Analysis and Machine Intelligence **14**(9) (1992) 869–885
16. Wenyin, L., Dori, D.: A Survey of non-thinning based vectorization methods. In *A. Amin, D. Dori, P. Pudil, and H. Freeman, editors, Advances in Pattern Recognition (Proceedings of Joint IAPR Workshops SSPR'98 and SPR'98)*, LNCS 1451, Sydney, Australia (1998) 230–241
17. Klette, R., Rosenfeld, A.: Digital geometry. Published by Elsevier, San Fransisco, USA (2004)
18. Debled, I., Reveillès, J.P.: A linear algorithm for segmentation of digital curves. In *Third International Workshop on Parallel Image Analysis*, June (1994)
19. Debled, I., Tabbone, S., Wendling, L.: Fast polygonal approximation of digital curves. In *International Conference on Pattern Recognition*, Volume 1, Cambridge, United Kingdom, August (2004) 465–468
20. Debled, I., Feschet, F and Rouyer-Degli, J.: Optimal blurred segments decomposition in linear time. In *12th International Conference on Discrete Geometry for Computer Imagery*, LCNS 3429 (2005) 311–322
21. Coeurjolly, D., Zerarga, L.: Supercover model, digital straight line recognition and curve reconstruction on the irregular isothetic grids. Computer and Graphics **30**(1) (2006) 46–53
22. Cohen-Or, D., Kaufman, A.: Fundamentals of surface voxelization. Graphical models and image processing : GIMP **57**(6) (1995) 453–461
23. Sivignon, I., Breton, R., Dupont, F., Andres, E.: Discrete analytical curve reconstruction without patches. Image and Vision Computing **23**(2) (2005) 191–202
24. Gramain, A.: Topologie des surfaces. *Presses Universitaires Françaises*, 1971
25. Hart, J.C.: Computational topology for shape modeling. In *Shape Modeling International SMI'99*, published by IEEE Computer Society, Aizu, Japan, March 1–4 (1999) 36–43
26. Hétroy, F.: Méthodes de partitionnement de surfaces. PhD thesis, *Institut National Polytechnique de Grenoble*, Grenoble, France, September (2003)
27. Tung T.: Indexation 3D de bases de données d'objets 3D par graphes de Reeb améliorés. PhD thesis, *Telecom Paris, ENST/TIC*, Paris, France, June (2005)
28. Xiao, Y., Siebert, P., Werghi, N.: A discrete Reeb graph approach for the segmentation of human body scans. In *4th International Conference on 3D Digital Imaging and Modeling*, Banff, Canada, October 6–10 (2003) 387–385
29. Breton, R: Reconstruction inversible d'objets discrets 2D. PhD thesis, *Université de Poitiers*, Poitiers, France, December (2003)
30. Wenyin, L., Dori, D.: A protocol for performance evaluation of line detection algorithms. Machine Vision and Applications **9**(5/6) (1997) 240–250

Fast Polynomial Segmentation of Digitized Curves

Peter Veelaert and Kristof Teelen

University College Ghent, Member of Association University Ghent, Belgium
{peter.veelaert, kristof.teelen}@hogent.be

Abstract. We propose a linear-time algorithm for curve segmentation which is based on constructive polynomial fitting. This work extends previous work on constructive fitting by taking the topological properties of a digitized curve into account. The algorithm uses uniform (or L_∞) fitting and it works for segments of arbitrary thickness. We illustrate the algorithm with the segmentation of contours into straight and parabolic segments.

1 Introduction

In this work we consider the segmentation of digitized curves into linear or parabolic segments. Curve segmentation is used for curve coding and representation, length and tangent estimation [1, 2], and shape matching. In discrete geometry one popular approach to curve segmentation is to segment a curve into digital straight segments (DSS). A rather complicated linear time-algorithm for DSS segmentation was found by Smeulders and Dorst [3]. Debled-Renesson and Réveillès gave a simple, easy to implement algorithm [4]. More recently, Buzer developed an algorithm for the recognition of straight lines of arbitrary thickness, based on convex hulls. Buzer's incremental algorithm can also be used for curve segmentation [5].

Since curve segmentation has many diverse applications in image processing it is worthwhile to look at some other approaches and their points of attention. Leclerc and Zucker [6] discuss how to segment a one-dimensional image through curve fitting, and they discuss how to avoid the misclassification of points near a discontinuity. Likewise, Dunham gives an algorithm that finds a piecewise linear approximation of a curve within uniform error and with a minimal number of vertices [7]. Some methods borrow techniques from Robust Analysis. Rosin and West propose a method that recursively subdivides a curve to form a binary tree, which is then traversed to select the best representation of the curve [8]. They use a least median of squares method to fit ellipses to the data, which is more robust than ordinary least squares. Boyer *et al* use a robust sequential estimator to parametrize and organize range images [9]. Also variable order fitting has shown its value in curve and image segmentation. An example is Besl and Jain's image segmentation algorithm which is based on variable-order surface fitting [10]. As for curve segmentation an important application of variable order fitting is feature point detection for shape matching. When a large circular arc is segmented into digital straight segments, the endpoints are unsuited as feature points, since their location is more or less arbitrary. The examples in this paper show that quadratic segmentation is more meaningful, and indicate that tangent and length estimation could also profit from variable order fitting. Also the thickness of the segments should be adjustable so that

A. Kuba, L.G. Nyúl, and K. Palágyi (Eds.): DGCI 2006, LNCS 4245, pp. 482–493, 2006.

the feature points necessary for matching are all present. One of the reasons variable order and variable thickness segmentation is not generally applied is its computational burden [11,10,12]. To summarize, the desirable properties for a segmentation algorithm are:

- low or even linear-time complexity;
- arbitrary segment thickness;
- more robust than least squares fitting;
- suited for variable-order fitting;
- take into account and/or benefit from the topological properties of a digitized curve;
- simple, and with predictable control and data flow for implementations on dedicated hardware.

In this paper we propose a simple, linear-time algorithm for segmentation with polynomials of arbitrary order, which we illustrate for linear and parabolic segments. Linear-time complexity, simplicity and generality do not come for free, however. The method is based on the estimation of the fitting error, not the exact computation, and therefore also the segmentation will not always be exact. The amount of errors can be reduced, however, by increasing the number of computations, and for most applications the errors are completely acceptable. Section 2 describes the mathematical basis of the segmentation algorithm. Section 3 examines the performance of the fitting cost estimators that are used in the algorithm en Section 4 shows segmentation results.

2 Constructive Fitting

In this paper we focus on one particular aspect of segmentation. Given an initial seed region, how can we find a region of maximal size where a curve or surface can be approximated sufficiently well by a polynomial of given order? And how can we find this region with minimal computational effort? In fact, although region growing and curve segmentation based on polynomial fitting is often a computationally expensive process [10, 11, 12], we propose a linear-time algorithm that solves some simple but basic segmentation problems, such as the segmentation of a curve in piecewise linear segments. This efficiency is based on a simple principle. The purpose of region growing and segmentation is to partition a curve or image into segments that satisfy a certain criterion, such as: being straight or smooth. To find such segments, a typical region growing algorithm starts from a small seed segment, and then repeatedly tries to add new points to this segment, while verifying whether the segmentation criterion is still satisfied for the enlarged segment. If not, a new segment is started, or another point is chosen. In general, the verification of the segmentation criterion requires increasingly more computation time when the segment gets larger. It is the purpose of this paper to show that this computation time can be reduced considerably by comparing the extension point with a small, but well-chosen set of reference points within the segment. Since the number of reference points is constant during the entire region growing process, the resulting algorithms have linear time complexity. This process has also been called constructive fitting.

We briefly review the basics of the particular method of fitting geometric primitives to data sets, on which the region growing processes proposed in this paper are based.

This method is called constructive fitting because it allows us to construct global fits to an entire data set from so-called elemental fits to small parts of the data [13, 14, 15]. We consider constructive fitting problems for one variable. The generalization to other dimensions is straightforward. Let $C = \{(x_0, y_0), \ldots, \} \subset \mathbb{Z}^2$ be the points of a finite curve. Let G be a vector space of fitting functions, for instance, the vector space of polynomial functions of the form

$$g(x) = \alpha_0 + \alpha_1 x + \cdots + \alpha_l x^l.$$

To simplify the exposition we impose the mild constraint that the curve C contains at least $n+1$ points with distinct x-coordinates, where n is the dimension of the space of fitting functions. The uniform (or L_∞) fitting cost of fitting $g(x)$ to the curve C is defined as

$$r_g(C) = \max_{(x_i, y_i) \in C} |g(x_i) - y_i|.$$

The *best fit* is the function $g(x)$ in G for which $r_g(C)$ is minimal. We denote this minimal cost as $r(C)$, and we call it the *fitting cost*. To be precise,

$$r(C) = \min_{g \in G} r_g(C).$$

We call $\tau = 2r(C)$ also the thickness of C, because the points of C are contained in a strip of thickness τ parallel to the graph of $g(x)$. This is also related to the notion of α-thickness in [5].

The best fit and its fitting cost can be computed from elemental fits to the so-called *elemental subsets* of C. These are subsets of the curve C that contain precisely $n+1$ points, where n denotes the dimension of the vector space of the fitting functions G. The fitting cost over an elemental subset itself can be computed in a straightforward manner. To be precise, let $D = \{(x_{i_1}, y_{i_1}), \ldots, (x_{i_{n+1}}, y_{i_{n+1}})\}$ be an elemental subset of R. Let E_j denote the cofactor of the element at the intersection of the last column and the jth row of the following matrix:

$$(A_D|B_D) = \left(\begin{array}{cccc|c} 1 & x_{i_1} & \ldots & x_{i_1}^l & y_{i_1} \\ \ldots & & & & \\ 1 & x_{i_{n+1}} & \ldots & x_{i_{n+1}}^l & y_{i_{n+1}} \end{array}\right)$$

Then one can show that the fitting cost for the elemental subset D is equal to

$$\begin{aligned} r(D) &= \det(A_D|B_D)/(|E_1| + \cdots + |E_{n+1}|) \\ &= |E_1 y_{i_1} + \cdots + E_{n+1} y_{i_{n+1}}|/(|E_1| + \cdots + |E_{n+1}|), \end{aligned} \tag{1}$$

provided the denominator at the right side is non-vanishing [14, 16]. Furthermore, one can prove that the fitting cost over C is the maximal value of the elemental fitting costs (1) over all elemental subsets of the curve C [14]:

Proposition 1. *Let $\mathcal{M}$ be the collection of all elemental subsets D of C for which $|E_1| + \cdots + |E_{n+1}| > 0$. If $\mathcal{M}$ is non-empty, then we have $r(C) = \max_{D \in \mathcal{M}} r(D)$.*

Although Prop. 1 yields a simple method to compute $r(C)$, the computational cost of evaluating $r(D)$ for all elemental subsets can be high. We can obtain a reliable estimate of the fitting cost, with far less computations if, instead of calculating $\max_{D \in \mathcal{M}} r(D)$, we compute $\max_{D \in \widetilde{\mathcal{M}}} r(D)$ where $\widetilde{\mathcal{M}}$ is some small subcollection of $\mathcal{M}$.

One can prove that the estimate is reliable provided $\widetilde{\mathcal{M}}$ forms a so-called *rigid subcollection* of elemental subsets [14]. The rigidity of $\widetilde{\mathcal{M}}$ depends on its elemental subsets, and can be verified as follows. Assume that the points of the curve C have been ordered in a fixed but arbitrary manner, i.e., $C = \{(x_1, y_1), \ldots, (x_k, y_k)\}$. Let $D = \{(x_{i_1}, y_{i_1}), (x_{i_2}, y_{i_2}), \ldots, (x_{i_{n+1}}, y_{i_{n+1}})\}$ be an elemental subset of C, with $1 \le i_j \le k$, and let E_j, $1 \le j \le n+1$, be the cofactors corresponding to the elemental subcountour D. In the vector space $\mathbf{R}^k$ let

$$v_D = \frac{(\ldots, E_1, \ldots, E_2, \ldots, E_{n+1}, \ldots)}{|E_1| + |E_2| + \cdots + |E_{n+1}|} = (d_1, \ldots, d_k),$$

denote a vector associated with D, where the jth cofactor E_j occurs as the i_jth element. The remaining $k - (n+1)$ entries are zero. The vectors v_D, where $D \in \mathcal{M}$, span a linear subspace L of $\mathbf{R}^k$. If we only consider vectors v_D that correspond to elemental subsets in the subcollection $\widetilde{\mathcal{M}}$, then these vectors span a linear subspace $\widetilde{L} \subset L$, where $\widetilde{L}$ can be smaller than L. Whether the elemental subsets $\widetilde{\mathcal{M}}$ can be used to accurately estimate the fitting cost depends on the question whether these two linear subspaces are equal or not, a notion formalized as follows.

Definition 1. We say that $\widetilde{\mathcal{M}}$ forms a *rigid subcollection of elemental subsets* if the linear space $\widetilde{L}$ spanned by the vectors v_D, where $D \in \widetilde{\mathcal{M}}$, is the same as the linear space L spanned by the vectors v_D, where $D \in \mathcal{M}$.

Apart from obeying the condition $r(C) \le \tau$, a digitized curve is also an ordered set of points with topological properties, the most common being that the curve must be either 4- or 8-connected. We incorporate this property by requiring that two subsequent points (x_i, y_i), (x_{i+1}, y_{i+1}) on the curve satisfy $|y_i - y_{i+1}| \le \epsilon$. To obtain 4- or 8-connectivity we choose $\epsilon = 1$. With the above concept of rigid subcollections we have the following proposition which extends previous results [13, 14].

Proposition 2. *Let $\widetilde{\mathcal{M}}$ be a rigid subcollection of elemental subsets chosen from the collection of all elemental subsets $\mathcal{M}$ for a curve length k. Let τ denote the maximum thickness of a curve segment, and ϵ the maximum absolute difference between y-coordinates of subsequent points on the curve. For the vectors $(z_1, \ldots, z_k) \in \mathbf{R}^k$, let S be the k-dimensional polytope defined by the inequalities*

$$|z_i - z_{i+1}| \le \epsilon, \text{ for } i = 1, \ldots, k-1, \tag{2}$$

and

$$\frac{|d_1 z_1 + \cdots + d_k z_k|}{|d_1| + \cdots + |d_k|} \le \frac{\tau}{2}, \quad \text{for all } D \in \widetilde{\mathcal{M}}. \tag{3}$$

Then for each elemental subset $P \in \mathcal{M}$, with associated vector $(p_1, \ldots, p_k)$, the values

$$\gamma_P(\widetilde{\mathcal{M}}, \tau, \epsilon) = \max_{(z_1, \ldots, z_k) \in S} \frac{2}{\tau} \frac{|p_1 z_1 + \cdots + p_k z_k|}{|p_1| + \cdots + |p_k|}, \tag{4}$$

and

$$\gamma(\widetilde{\mathcal{M}}, \tau, \epsilon) = \max_P \gamma_P(\widetilde{\mathcal{M}}, \tau, \epsilon) \tag{5}$$

are well defined.

Proof. It is sufficient to show that $\gamma_P()$ is well-defined for all elemental subsets P. By the rigidity assumption there is a finite set of weight vectors $v_A = (\ldots, a_1, \ldots, a_l, \ldots)$, $v_B = (\ldots, b_1, \ldots, b_l, \ldots)$, ..., $v_D = (\ldots, d_1, \ldots, d_l, \ldots)$ with $A, B, \ldots, D \in \widetilde{\mathcal{M}}$, such that $v_P = (\ldots, p_1, \ldots, p_l, \ldots)$ can be written as a linear combination $v_P = \alpha v_A + \beta v_B + \cdots + \omega v_D$. It follows that

$$\begin{aligned} |p_1 z_1 + \cdots p_l z_l| \leq\ & |\alpha||a_1 z_1 + \cdots + a_l z_l| + \\ & |\beta||b_1 z_1 + \cdots + b_l z_l| + \cdots + \\ & |\omega||d_1 z_1 + \cdots + d_l z_l|. \end{aligned}$$

As a consequence of the constraints (3), every term at the right side of this inequality has an upper bound. Hence $|p_1 z_1 + \cdots p_l z_l|/(|p_1| + \cdots |p_l|)$ will also have an upper bound, and it follows that $\gamma_P()$ is well-defined, and therefore $\gamma()$ is well defined.

The primary use of Prop. 2 is to determine the relative estimation error. Suppose we find the estimated cost $\widetilde{r}(C)$ of a curve from the elemental subsets in $\widetilde{\mathcal{M}}$, and that we have $\widetilde{r}(C) \leq \tau/2$. Then from Prop. 2 it follows that the outcome of any other elemental subset $D \in \mathcal{M} \setminus \widetilde{\mathcal{M}}$ satisfies $r(D) \leq \gamma(\widetilde{\mathcal{M}}, \tau, \epsilon)\tau/2$. Thus, although the thickness of the curve is not known exactly, it must be less than $\tau\gamma(\widetilde{\mathcal{M}}, \tau, \epsilon)$.

Prop. 2 yields an upper bound $\gamma()$ for the estimation error, which is found by solving linear programming problems for different cost functions (4), but over the same polytope S, defined by the inequalities (2) and (3). Each vector in S corresponds to a possible configuration of points satisfying the connectivity constraints (3) and with estimated fitting cost less than $\tau/2$. Although the fitting costs $r(C)$ and $\widetilde{r}(C)$ depend on the real point positions of the curve C, the upper bound $\gamma()$ only depends on the x-coordinates of the elemental subsets in $\widetilde{\mathcal{M}}$, the thickness τ and the connectivity parameter ϵ, since these parameters completely determine the polytope S and the linear cost functions (4). Thus the upper bound $\gamma(\widetilde{\mathcal{M}}, \tau, \epsilon)$ can be computed once and for all, since it does not depend on the curve itself. The rigidity of $\widetilde{\mathcal{M}}$ is easy to verify [13].

As a special case, $\gamma(\widetilde{\mathcal{M}}, \tau, \infty)$ denotes the value when no topological constraint is imposed. To compute $\gamma(\widetilde{\mathcal{M}}, \tau, \infty)$ we can apply Prop. 2 with ϵ a large positive number. The common case where we fit digital straight segments to connected curves corresponds to $\gamma(\widetilde{\mathcal{M}}, 1, 1)$. By observing the expressions (4), (2), and (3) in Prop. 2 one can see that $\gamma(\widetilde{\mathcal{M}}, \tau, \epsilon)$ only depends on the ratio $\mu = \tau/\epsilon$. In Section 3 we will see that when ϵ is small compared to the thickness τ, then the connectivity of the curve cannot be neglected when estimating fitting costs. Previous results in [13, 14] did not take into account connectivity and only considered the case $\gamma(\widetilde{\mathcal{M}}, \tau, \infty)$.

According to Prop. 2, given $\widetilde{\mathcal{M}}$, we can compute $\gamma(\widetilde{\mathcal{M}}, \tau, \epsilon)$ which gives us a precise measure for the maximal error of the estimated fitting cost $\widetilde{r}(R)$. Moreover, $\gamma(\widetilde{\mathcal{M}})$ depends only on the x-coordinates of the points on the subsets D in $\widetilde{\mathcal{M}}$ and not on the curve itself. In principle, Prop. 2 enables us to evaluate the maximal error for a particular estimator $\widetilde{\mathcal{M}}$, by performing an exhaustive search for the estimator that has the lowest maximal error. This has been explored in previous work, but without topological constraints [13]. With topological constraints the performance depends strongly on the ratio between the allowed thickness of a curve and the connectivity.

3 Performance of Fitting Cost Estimators

When $\widetilde{\mathcal{M}}$ satisfies the conditions of Definition 1, we refer to it as a fitting cost estimator $\widetilde{\mathcal{M}}$. Using Prop. 2 we can search, in a systematic way, for the estimator $\widetilde{\mathcal{M}}$ that yields the lowest possible estimation error.

3.1 Linear Fitting Cost Estimators

We will illustrate our search with a simple fitting problem in one variable. Fig. 1 shows a curve C. Let G be a two-dimensional vector space of fitting functions of the form $g(x) = \alpha_0 + \alpha_1 x$. Then the best fit to C is the function $g(x)$ for which $\max_{(x_i,y_i)\in C}(|g(x_i) - y_i|)$ is minimal. When C is a digital arc the fitting cost $r(C)$ is a measure for the straightness of the digital arc [14]. First we compute the fitting cost exactly. According to Prop. 1, the fitting cost $r(C)$ is equal to $\max_D r(D)$, where the maximum is taken over all 3-point subsets of C. The fitting cost $r(D)$ of a subset is equal to $|E_1 y_i + E_2 y_j + E_3 y_k|/(|E_1| + |E_2| + |E_3|)$, where the E_j are the cofactors of the last column of the matrix

$$\begin{pmatrix} 1 & x_{i_1} & y_i \\ 1 & x_{i_2} & y_j \\ 1 & x_{i_3} & y_k \end{pmatrix}$$

Thus, for the function shown in Fig. 1, we find $r(C) = 8/9 \approx 0.89$. Note that the exact computation of $r(C)$ involves the evaluation of 120 elemental fitting costs $r(D)$ (a 10-point curve has 120 distinct 3-point subsets).

Estimating $r(C)$ instead of computing it exactly, considerably reduces the number of elemental fitting costs that must be evaluated. The challenge is how to choose these elemental subsets so that the estimate is as good as possible. Table 1 shows the theoretical performance of different estimators when they are applied to a curve segment consisting of 10 subsequent points $(x_1, y_1), (x_1 + 1, y_2), \ldots, (x_1 + 9, y_{10})$. The first column describes the estimator, where the index i varies in each as $i = 2, \ldots, 9$. For example, (p_1, p_i, p_{i+1}) denotes the estimator $\widetilde{\mathcal{M}} = \{(p_1, p_2, p_3), (p_1, p_3, p_4), \ldots, (p_1, p_9, p_{10}))\}$. The estimator called *iterative subdivision* is constructed by iteratively subdividing the curve into shorter curves as follows: $\widetilde{\mathcal{M}} = \{(p_1, p_5, p_{10}), (p_1, p_3, p_5), (p_5, p_7, p_{10}), (p_1, p_2, p_3), \ldots, (p_7, p_9, p_{10}))\}$. For $\widetilde{\mathcal{M}}_7$, the index varies from $i = 3$ to $i = 9$, and also $(1, 2, 3)$ is included.

The first two estimators are general in the sense that they assume all points are known from the start. The estimators $\widetilde{\mathcal{M}}_3, \ldots, \widetilde{\mathcal{M}}_7$ can be used in incremental algorithms, where only one point at a time is added to a segment, while each time the fitting cost is evaluated. For curve segmentation the incremental algorithms are the most interesting. The estimator $\widetilde{\mathcal{M}}_4 \cup \widetilde{\mathcal{M}}_5$ use the conditions of both $\widetilde{\mathcal{M}}_4$ and $\widetilde{\mathcal{M}}_5$ to estimate the fitting cost.

Columns 2 to 5 show the worst case behavior $\gamma(\widetilde{\mathcal{M}}, \tau, \epsilon)$ for each estimator as computed by Prop. 2. Note that these values for $\gamma()$ do not depend on any particular curve, but that they are valid for any curve of length 10, with thickness not larger than τ and satisfying the connectivity constraint determined by ϵ. Column 6 on the other hand shows the estimated cost for one particular curve which is shown in Fig. 1. In all cases the real fitting cost of 0.89 is always less than product of the estimated cost and maximum relative error $\gamma()$. Columns 7 and 8 are discussed in Section 4.

Table 1. Performance of different estimators for varying values of the thickness and connectivity constraint, for the curve length $k = 10$

Estimators	$\gamma(\widetilde{\mathcal{M}}, 1, \infty)$	$\gamma(\widetilde{\mathcal{M}}, 1, 2)$	$\gamma(\widetilde{\mathcal{M}}, 1, 1)$	$\gamma(\widetilde{\mathcal{M}}, 2, 1)$	estim. cost	max cost	aver. cost
General estimators:							
$\widetilde{\mathcal{M}}_1$: (p_1, p_i, p_{10})	2.	2.	2	1.5	0.89		
$\widetilde{\mathcal{M}}_2$: iterative subdivision	3.1	2.94	2.65	1.71	0.72		
Incremental algorithms:							
$\widetilde{\mathcal{M}}_3$: (p_{i-1}, p_i, p_{i+1})	20.	7.11	4	2	0.5		
$\widetilde{\mathcal{M}}_4$: (p_1, p_i, p_{i+1})	4.13	3.92	3.47	2	0.5	36.	11.
$\widetilde{\mathcal{M}}_5$: $(p_1, p_{\lceil i/2 \rceil + 1}, p_{i+1})$	21.58	5.4	3.09	1.84	0.72	2.	1.03
$\widetilde{\mathcal{M}}_6$: $(p_1, p_{\lceil \frac{i+4}{3} \rceil}, p_{i+1})$	19.83	6.56	3.56	2.	0.66	3.96	1.41
$\widetilde{\mathcal{M}}_7$: (p_1, p_{i-1}, p_{i+1})	16.74	3.27	2.37	1.83	0.89	2.6	0.97
Combined incremental:							
$\widetilde{\mathcal{M}}_4 \cup \widetilde{\mathcal{M}}_5$	3.	2.94	2.76	1.84	0.72	2.	1.02
$\widetilde{\mathcal{M}}_4 \cup \widetilde{\mathcal{M}}_5 \cup \widetilde{\mathcal{M}}_6$	2.94	2.94	2.68	1.84	0.72	2.	1.02
$\widetilde{\mathcal{M}}_4 \cup \widetilde{\mathcal{M}}_5 \cup \widetilde{\mathcal{M}}_6 \cup \widetilde{\mathcal{M}}_7$	2.17	2.14	1.95	1.71	0.89	2.08	0.84

Table 1 clearly shows that the performance $\gamma(\widetilde{\mathcal{M}}, \tau, \epsilon)$ of an estimator depends strongly on the ratio between the thickness τ and the connectivity constraint ϵ. In fact, although according to the first column $\widetilde{\mathcal{M}}_4$ performs much better ($\gamma = 4.13$) than $\widetilde{\mathcal{M}}_5$ ($\gamma = 21.58$) for arbitrary points sets, columns 4 and 5 show that when we impose the topological constraint $\epsilon = 1$, $\widetilde{\mathcal{M}}_5$ ($\gamma = 1.84$) is better than $\widetilde{\mathcal{M}}_4$ ($\gamma = 2$). The reason can be understood as follows. The estimator $\widetilde{\mathcal{M}}_4$ compares the y-coordinate of the new point p_{i+1} with the y-coordinates of the points p_1 and p_i. However, if we impose the topological condition that the difference between the y-coordinates of subsequent points is never larger than one, then for $\widetilde{\mathcal{M}}_4$ the topological condition (2) coincides almost completely with the condition (3). For $\widetilde{\mathcal{M}}_5$, however, the topological constraint makes sense, and $\gamma(\widetilde{\mathcal{M}}_5, \tau, \epsilon)$ drops when ϵ gets smaller. Also note that for all estimators in Table 1 we have $\gamma(\widetilde{\mathcal{M}}, \tau = 1, \epsilon = 1) \geq \gamma(\widetilde{\mathcal{M}}, \tau = 2, \epsilon = 1)$. That is, for a 4- or 8-connected curve, the relative error on the estimated thickness is smaller for $\tau = 2$ than for $\tau = 1$. The reason is that for a curve with thickness $\tau = 2$, connectivity is a more important additional constraint than for a curve with thickness $\tau = 1$.

The best estimator is $\widetilde{\mathcal{M}_1}$, which performs well for curves and for disconnected point sets. However, $\widetilde{\mathcal{M}_1}$ is not suited for curve segmentation. The best estimator for curve segmentation is $\widetilde{\mathcal{M}_7}$, if we assume that the curves are either 4- or 8-connected. It makes sense to combine several estimators. The overall performance of estimator $\widetilde{\mathcal{M}_4} \cup \widetilde{\mathcal{M}_5} \cup \widetilde{\mathcal{M}_6} \cup \widetilde{\mathcal{M}_7}$ is better than the performance of its parts. The drawback is that the combined estimator evaluates four times more conditions than for example $\widetilde{\mathcal{M}_5}$. Nonetheless, the time complexity of this combined estimator is still $O(N)$, where N denotes the number of points on the segment, as we use 4 elemental subsets to examine the fitting cost for each new point.

Table 1 only shows the results for curves of length 10. For lengths other than 10 we get similar results. An estimator that performs well for length 10, such as $\widetilde{\mathcal{M}_4} \cup \widetilde{\mathcal{M}_5} \cup \widetilde{\mathcal{M}_6} \cup \widetilde{\mathcal{M}_7}$, also performs well for curves of arbitrary length.

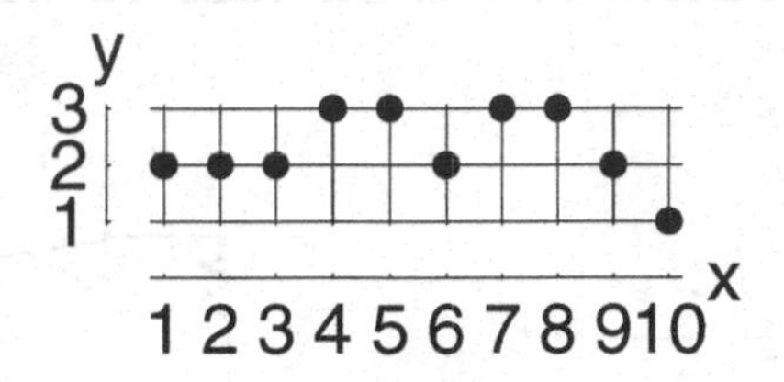

Fig. 1. Simple curve used to examine the performance of the estimators in Table 1

3.2 Quadratic Fitting Cost Estimators

Table 2 shows the results for fitting cost estimators that estimate the cost of fitting a quadratic polynomial to a curve of length 10 with points that have consecutive x-coordinates. As for line fitting, the ratio of the thickness τ and the topological constraint ϵ plays an important role. The first column describes the kind of estimator, where the index i varies in each as $i = 3, \ldots, 9$. Note that the topological constraints play an even larger role for quadratic fitting than for straight fitting. The requirement that a parabolic segment is connected already imposes strong restrictions.

4 Curve Segmentation

To illustrate the fitting cost estimators we discuss the decomposition of a digitized curve into digital straight or parabolic segments. For Table 1 and 2 we made the assumption that the curve consists of subsequent points with x-coordinates that increase regularly. In the general case, a curve can be quite different and subsequent points may have the same x-coordinate, e.g., $(x_1, y_1), (x_1 + 1, y_2), (x_1 + 1, y_3), (x_1 + 2, y_4), \ldots$. Without restriction, Prop. 2 can also be used to find good estimators for this case, by using an appropriate form for the curve C and its elemental subsets. The number of possibilities is unlimited, however, and for each possible configuration of x-coordinates we will find other values for $\gamma(\widetilde{\mathcal{M}}, \tau, \epsilon)$. The examples show that the best estimators of Table 1 and 2 also perform well for more general curve segments.

Table 2. Performance of different quadratic estimators for varying values of the thickness τ and connectivity constraint ϵ

Estimators	$\gamma(\widetilde{\mathcal{M}},1,\infty)$	$\gamma(\widetilde{\mathcal{M}},1,2)$	$\gamma(\widetilde{\mathcal{M}},1,1)$	$\gamma(\widetilde{\mathcal{M}},2,1)$
Region growers:				
$\widetilde{\mathcal{M}}_1$: $(p_1, p_{\lfloor\frac{i+1}{2}\rfloor}, p_i, p_{i+1})$	3.72	3.26	2.15	1.26
$\widetilde{\mathcal{M}}_2$: $(p_1, p_{\lfloor\frac{i+3}{3}\rfloor}, p_{\lfloor\frac{2i+3}{3}\rfloor}, p_{i+1})$	140.	4.34	2.31	1.26
$\widetilde{\mathcal{M}}_3$: $(p_1, p_{\lfloor\frac{i+1}{2}\rfloor}, p_{\lfloor\frac{3i+3}{4}\rfloor}, p_{i+1})$	154.	3.8	2.23	1.26
$\widetilde{\mathcal{M}}_4$: $(p_1, p_{\lfloor\frac{i+5}{4}\rfloor}, p_{\lfloor\frac{i+3}{2}\rfloor}, p_{i+1})$	124.	4.58	2.51	1.26
$\widetilde{\mathcal{M}}_5$: $(p_{\lfloor\frac{i+5}{5}\rfloor}, p_{\lfloor\frac{i+2}{2}\rfloor}, p_{\lfloor\frac{4i+5}{5}\rfloor}, p_{i+1})$	54.	3.21	1.93	1.25
Combined growers:				
$\widetilde{\mathcal{M}}_1 \cup \widetilde{\mathcal{M}}_5$	3.01	2.72	1.91	1.25
$\widetilde{\mathcal{M}}_1 \cup \widetilde{\mathcal{M}}_2 \cup \widetilde{\mathcal{M}}_3 \cup \widetilde{\mathcal{M}}_4 \cup \widetilde{\mathcal{M}}_5$	2.37	2.32	1.85	1.24

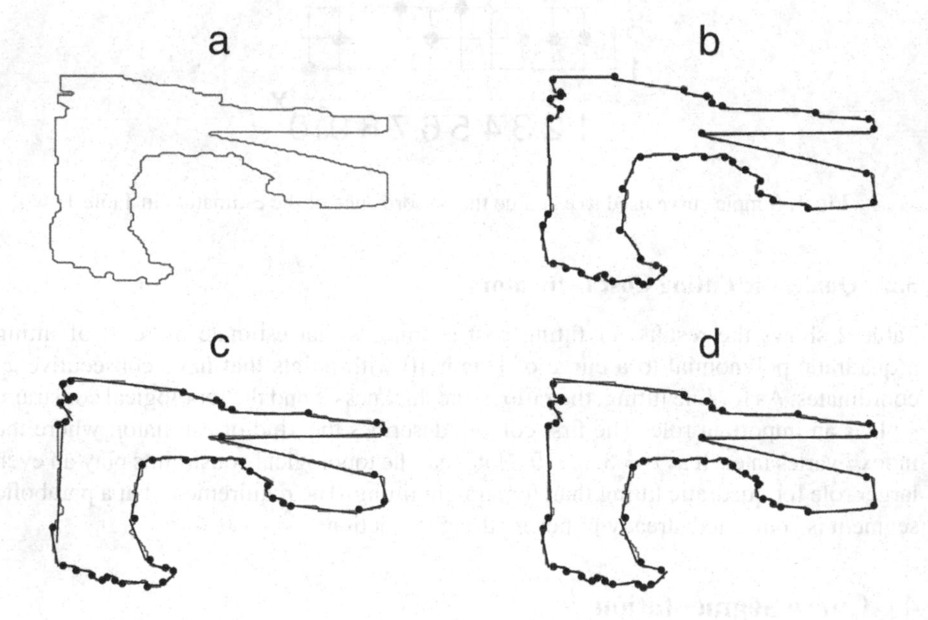

Fig. 2. (a) Original contour, (b) perfect segmentation, (c) segmentation by region grower $\widetilde{\mathcal{M}}_5$, (d) segmentation by the combined region grower $\widetilde{\mathcal{M}}_4 \cup \widetilde{\mathcal{M}}_5 \cup \widetilde{\mathcal{M}}_6 \cup \widetilde{\mathcal{M}}_7$

To find digital straight curve segments of S of maximal length, we use a simple linear-time algorithm. Each time a new point is added, we perform elemental fits to verify whether the enlarged segment is still straight. Since we estimate the fitting cost (instead of computing it exactly), the algorithm that we propose is not error-free. However, since an upper bound for the fitting cost is known instead of obtaining segments of thickness $\leq \tau$, in the worst case we have segments of thickness $\leq \gamma()\tau$. In fact, average case behavior will be much better than worst case behavior.

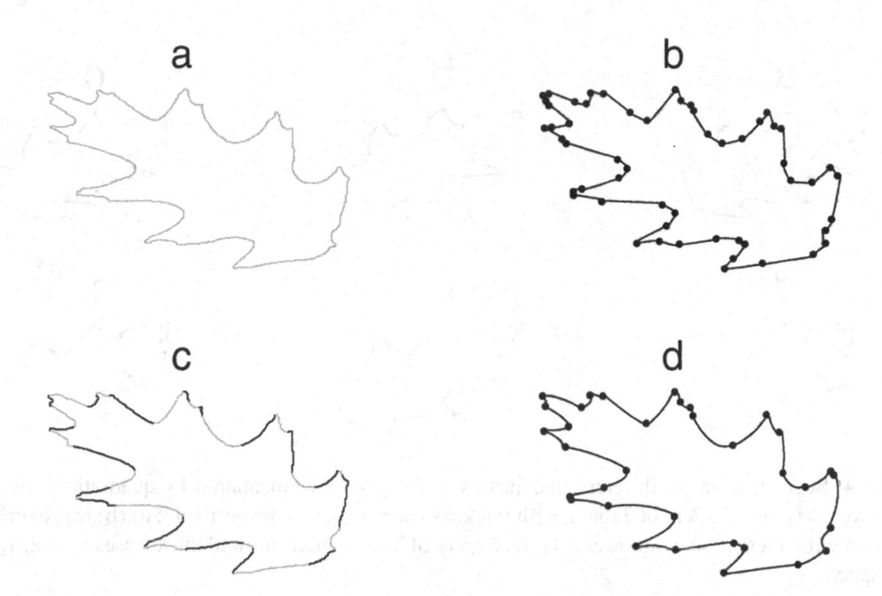

Fig. 3. Segmentation of Kelloggii plant leaf: (a) Original contour, (b) piecewise linear segmentation with segment grower $\widetilde{\mathcal{M}}_4 \cup \ldots \cup \widetilde{\mathcal{M}}_7$ of Table 1 with allowed thickness $\tau = 1$, (c) segmentation by quadratic segment grower $\widetilde{\mathcal{M}}_1 \cup \ldots \cup \widetilde{\mathcal{M}}_5$ also with allowed thickness $\tau = 1$, of Table 2, (d) the same segments as in (c), but now also the parabolas are shown and the endpoints of the segments

Fig. 2 shows the results for a more complex 8-connected curve. Fig. 2(a) shows the original contour, (b) the result of perfect segmentation (in which all elemental subsets are used to compute the fitting cost), (c) the result of segmentation by region grower $\widetilde{\mathcal{M}}_5$, and (d) the result of segmentation by the combined region grower $\widetilde{\mathcal{M}}_4 \cup \widetilde{\mathcal{M}}_5 \cup \widetilde{\mathcal{M}}_6 \cup \widetilde{\mathcal{M}}_7$, with $\tau = 1$ for all cases. For most applications, when extracting feature points from contours, the results in (d) are satisfactory and come with large computational gain.

Column 7 in Table 1 shows the maximal thickness of the segments when the curve in Fig. 2(a) is segmented using the corresponding estimator for $\tau = 1$. For perfect estimates the values in column 7 should not exceed 1. For curves where all x-coordinates are distinct and do not differ by more than one, the values should be less than $\gamma(\widetilde{\mathcal{M}}, 1, 1)$. This is not always the case since the topology of the curve can be different, as mentioned previously, and because the length of the curve is most of the time not equal to ten. Column 8 gives the average thickness of the segments, which shows that the average estimates are better than the worst case estimates. Again, $\widetilde{\mathcal{M}}_4 \cup \widetilde{\mathcal{M}}_5 \cup \widetilde{\mathcal{M}}_6 \cup \widetilde{\mathcal{M}}_7$ is clearly the best estimator. The average perfect fitting cost of the segments that it accepted is 0.84.

Fig. 3 shows the result of quadratic segment growers for plant leafs. One possible goal of splitting the 8-connected contour of a plant into segments is to locate the feature points that can be used for matching, i.e. the sharp corners of the leaf. Fig. 3 (a) shows

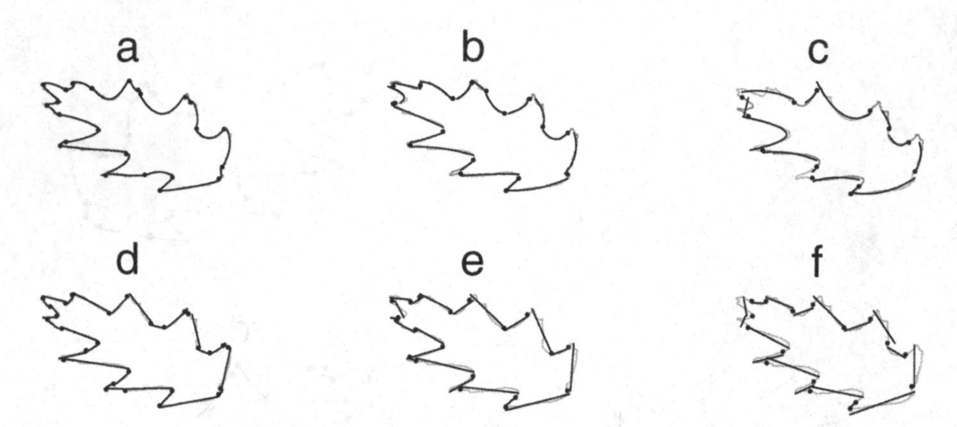

Fig. 4. Segmentation for different thicknesses τ. (a),(b),(c) segmentation by quadratic segment grower $\widetilde{\mathcal{M}}_1 \cup \ldots \cup \widetilde{\mathcal{M}}_5$ of Table 2 with thicknesses $\tau = 2, 4, 8$, respectively; a),(b),(c) segmentation with linear segment grower $\widetilde{\mathcal{M}}_4 \cup \ldots \cup \widetilde{\mathcal{M}}_7$ of Table 1 with allowed thicknesses $\tau = 2, 4, 8$, respectively.

the original contour of a plant leaf, (b) the result of linear segment growing and (c) and (d) the result of quadratic segment growing, with allowed thickness $\tau = 1$. Clearly, for this kind of curves it makes sense to use quadratic segments. The number of segments in (b) is almost twice the number of segments in (d) and there are more endpoints in (b) that cannot be used as feature points. Fig. 4 shows the results of fitting with different values for the thickness τ. For this kind of curve quadratic fitting remains more accurate for all thicknesses.

5 Concluding Remarks

We have applied constructive fitting techniques to develop segment growing algorithms for curve segmentation based on the following idea. When extending a curve C_k with a new point p_k it is not necessary to compare the point with the entire curve C_k. Instead it suffices to compare the point with one or more elemental subsets. Thus the time complexity of the contour growing process is greatly reduced, and the algorithm is simple enough to be implementable on dedicated hardware. The major challenge is how to choose the subsets so that the estimation of the fitting is as reliable as possible. Theoretical results allow us to select the elemental subsets in an optimal fashion. To be specific, there is a performance parameter $\gamma()$ that allows us to quantify the quality of the subsets. In this work we have shown that the choice of the subsets depends on the allowed thickness of the contour and the connectivity constraints. The emphasis in this paper was on how to develop computationally efficient segmentation algorithms through variable order fitting algorithms. However, also the robust estimation of curve or surface parameters may benefit from fast variable order fitting [1, 17].

References

1. Lachaud, J.-O., Vialard, A., de Vieilleville, F.:Analysis and Comparative Evaluation of Discrete Tangent Estimators. Lecture Notes in Computer Science **3429** (2005) 240–251
2. Coeurjolly, D., Klette, R.: Comparative evaluation of length estimators of digital curves. IEEE Trans. Pattern Anal. Machine Intell. **26** (2004) 252–258
3. Smeulders, A., Dorst, L.: Decomposition of discrete curves into piecewise straight segments in linear time. In: Vision Geometry (R. A. Melter, A. Rosenfeld, and P. Bhattacharya, eds.), Contemporary Mathematics Series, American Mathematical Society **119** (1991) 169–195
4. Debled-Renesson, I., Reveilles, J.-P.: A linear algorithm for segmentation of discrete curves. International Journal of Pattern Recognition and Artificial Intelligence **9** (1995) 635–662
5. Buzer, L.: An elementary algorithm for digital line recognition in the general case. Lecture Notes in Computer Science **3429** (2005) 299–310
6. Leclerc, Y., Zucker, S., The local structure of image discontinuities in one dimension. IEEE Trans. Pattern Anal. Machine Intell. **9** (1987) 341–355
7. Dunham, J.: Optimum uniform piecewise linear approximation of planar curves. IEEE Trans. Pattern Anal. Machine Intell. **8** (1986) 67–75
8. Rosin, P., West, G.: Nonparametric segmentation of curves into various representations. IEEE Trans. Pattern Anal. Machine Intell. **17** (1995) 1140–1153
9. Boyer, K., Mirza, M., Ganguly, G.: The robust sequential estimator: A general approach and its application to surface organization in range data. IEEE Trans. Pattern Anal. Machine Intell. **16** (1994) 987–1001
10. Besl, P., Jain, R.: Segmentation through variable-order surface fitting. IEEE Trans. Pattern Anal. Machine Intell. **10** (1988) 167–192
11. Cooper, D., Yalabik, N.: On the computational cost of approximating and recognizing noise-perturbed straight lines and quadratic arcs in the plane. IEEE Trans. Comput. **25** (1976) 1020–1032
12. Taubin, G.: Estimation of planar curves, surfaces and nonplanar space curves defined by implicit equations with applications to edge and range image segmentation. IEEE Trans. Pattern Anal. Machine Intell. **13** (1991) 1115–1138
13. Veelaert, P.: Linear-time algorithms for region growing with applications to image and curve segmentation. In: Proceedings of the SPIE conference on Vision Geometry VI **3168** (1997) 76-87
14. Veelaert, P.: Constructive fitting and extraction of geometric primitives. CVGIP: Graphical Models and Image Processing **59** (1997) 233–251
15. Stromberg, A.: Computing the exact least median of squares estimate and stability diagnostics in multiple linear regression. SIAM J. Sci. Comput. **14** (1993) 1289–1299
16. Veelaert, P.: On the flatness of digital hyperplanes. J. Math. Imaging and Vision **3** (1993) 205–221
17. Teelen, K., Veelaert, P.: Improving difference operators by local feature detection. To appear in proceedings DGCI 2006.

Fuzzy Segmentation of Color Video Shots

Bruno M. Carvalho, Lucas M. Oliveira, and Gilbran S. Andrade

Departamento de Informática e Matemática Aplicada,
Universidade Federal do Rio Grande do Norte,
59072-970, Natal, RN, Brazil
bruno_m_carvalho@yahoo.com, lmoliveira@gmail.com, gilbransilva@yahoo.com.br

Abstract. Fuzzy segmentation is a region growing technique that assigns a grade of membership to an object to each element in an image. In this paper we present a method for segmenting video shots by using a fast implementation of the fuzzy segmentation technique. The video shot is treated as a three-dimensional volume with different z slices being occupied by different frames of the video shot. The volume is interactively segmented based on selected seed elements, that will determine the affinity functions based on their intensity and color properties. Experiments with a synthetic video under different noise conditions are performed, as well as examples of two real video shot segmentations are presented, showing the applicability of our method.

1 Introduction

Digital image segmentation is the process of assigning labels to different objects in a digital image, where the level of detail indicated by the labeling is related to the application at hand. To perform object identification in digital or continuous, moving or still images, humans make use of different visual cues and high-level reasoning and knowledge. The difficulty of incorporating such type of reasoning into a computer program makes the task of segmenting out an object from its background a hard one. This task is even more challenging for a computer program when, instead of intensity values, what distinguishes the object from the background is some textural property, or when the image is corrupted by inhomogeneous illumination and/or noise.

Video digital segmentation consists of segmenting objects on sequences (shots) of images (frames). Conceptually, digital video segmentation is different from individually segmenting a sequence of digital images in the sense that there needs to be a consistency between the segmentations of a frame and of the frames that come before and after it on the segmented sequence. For example, the level of detail of the labels has to be consistent through out the segmented sequence.

Region growing algorithms segment an image by appending pixels to regions defined by seed pixels until all pixels in the image have been assigned to a set defined by one of the seed pixels, where the decision of to which object a pixel will be assigned is based on some predefined criteria. The selection of the seed pixels can be performed automatically, based on the nature of the problem. However, when no *a priori* information is available, these algorithms are semi-automatic, requiring the selection of the seed pixels by an user.

A. Kuba, L.G. Nyúl, and K. Palágyi (Eds.): DGCI 2006, LNCS 4245, pp. 494–505, 2006.

Fuzzy segmentation [1,2] is a region growing method that has been successfully used for segmenting images corrupted by inhomogeneous illumination and/or noise [2]. In this paper we propose a method for segmenting color video shots based on the fuzzy segmentation method of [2,3], that calculates simultaneously the fuzzy segmentation of multiple objects. The proposed method considers the frames of the video sequence as z slices of a 3D volume and uses the fast implementation of the algorithm presented in [3]. The speed of the segmentation allows an user to add and/or delete seed pixels, rerun the algorithm a few times, and still get the final segmentation in a reasonable amount of time.

The reason why we consider the video sequence as a 3D volume is that we want to stylize objects in pre-acquired video shots using Non-Photorealistic Rendering (NPR) techniques. This work was performed as a part of a project for providing computational tools to non-experienced users for generating animations using NPR techniques. NPR techniques aim to reproduce artistic techniques renderings, trying to express feelings and moods on the rendered scenes. Another way of defining NPR is that it is the processing of images or videos into artwork, generating images or videos that can have the visual appeal of pieces of art, expressing the visual and emotional characteristics of artistic styles such as impressionism and watercolor painting.

NPR techniques can be applied to still images, to 3D models or to video sequences, a task also called video stylization. If the input for the NPR video is a normal video, not maintaining temporal coherence of elements of the stylization, such as brush strokes, incurs in severe flickering on the output video [4]. This flickering comes not only from changed objects being rendered with elements that follow the object movement but also from static areas being rendered differently each time. Because of this problem, some authors have used optical flow techniques for enforcing temporal coherence [4,5]. However, the local characteristic of the optical flow techniques and their sensitivity to noisy images somehow limit their applicability. To overcome those problems, segmentation algorithms have been applied to video shot segmentation to produce end-to-end segmentations that are later used to enforce temporal coherence [6,7]. Here we propose the use of a fast implementation of fuzzy segmentation for segmenting color video shots as 3D volumes interactively.

2 Fuzzy Segmentation

Fuzzy segmentation is segmentation technique that computes the fuzzy connectedness, a concept introduced in [8], for every pixel in an image. The original fuzzy segmentation algorithm, introduced in [1], was generalized in [2], where the technique deals with an arbitrary finite set V, composed of *spels* (short for *spatial elements*). These spels can represent many different things, such as pixels of an image (as in [1,9]), voxels placed on a simple cubic grid (as in [3]) or on a face-centered cubic grid (as in [10]), dots in the plane (as in [11]) or feature vectors (as in [12]). The theory and the algorithms discussed in [2,3] are independent of the specifics of the application area, and so, can be applied to data clustering [13] in general.

The objective of the fuzzy segmentation algorithms of [2,3] is to produce a partition of the set V into a specified number of objects, but in a fuzzy way; i.e., in addition to assigning an object label for every spel, the algorithm also assigns a grade of membership for that object. A grade of membership is a number between 0 and 1, where 0 indicates that the spel definitely does not belong to the object, and 1 indicates that it definitely does). The formalization of such fuzzy partitioning is achieved by the concept of an M-semisegmentation (where M is the number of objects), defined below.

An *M-semisegmentation* of V is a function σ that maps each $c \in V$ into an $(M+1)$-dimensional vector $\sigma^c = (\sigma_0^c, \sigma_1^c, \cdots, \sigma_M^c)$, such that

1. $\sigma_0^c \in [0,1]$ (i.e., σ_0 is nonnegative but not greater than 1),
2. for each m ($1 \le m \le M$), the value of σ_m^c is either 0 or σ_0^c, and
3. for at least one m ($1 \le m \le M$), $\sigma_m^c = \sigma_0^c$.

In the definitions above, σ_m^c represents the grade of membership of the spel c in the mth object, and σ_0^c is always $\max_{1 \le m \le M} \sigma_m^c$. It is easy to see that this definition of M-semisegmentation allows the case where a spel belongs to more than one object, as long as it has the same grade of membership in all of them. An *M-segmentation* of V is an *M-semisegmentation* where $\sigma_0^c > 0$ for all spels $c \in V$.

In this context, a *chain* is defined as a sequence $\langle c^{(0)}, \cdots, c^{(K)} \rangle$ of distinct spels, and its *links* are the ordered pairs $(c^{(k-1)}, c^{(k)})$ of consecutive spels in the sequence. The strength of a link (c,d), or *ψ-strength of a link*, is also a fuzzy concept, with a real value between 0 and 1 being assigned to it by an appropriate *fuzzy spel affinity* function $\psi : V^2 \to [0,1]$. The *ψ-strength of a chain* is defined as the ψ-strength of its weakest link if the chain has two or more spels on it, and 1 if the chain has only one spel in it. A set $U (\subseteq V)$ is said to be *ψ-connected* if, for every pair of spels in U, there is a chain in U of positive ψ-strength from the first spel of the pair to the second. If one wants to segment multiple objects, it is reasonable to define different fuzzy spel affinities for each one of them.

A M-semisegmentation in our theory is determined by an *M-fuzzy graph*, that is a pair (V, Ψ), where V is a nonempty finite set and $\Psi = (\psi_1, \cdots, \psi_M)$ with ψ_m (for $1 \le m \le M$) being a fuzzy spel affinity. A *seeded* M-fuzzy graph is a triple $(V, \Psi, \mathcal{V})$ such that (V, Ψ) is an M-fuzzy graph and $\mathcal{V} = (V_1, \cdots, V_M)$, where $V_m \subseteq V$ for $1 \le m \le M$. We say that a seeded M-fuzzy graph $(V, (\psi_1, \cdots, \psi_M), (V_1, \cdots, V_M))$ is *connectable* if

1. the set V is ϕ_Ψ-connected, where $\phi_\Psi(c,d) = \min_{1 \le m \le M} \psi_m(c,d)$ for all $c, d \in V$, and
2. $V_m \ne \emptyset$, for at least one m, $1 \le m \le M$.

For an M-semisegmentation σ of V and for $1 \le m \le M$, we define the chain $\langle c^{(0)}, \cdots, c^{(K)} \rangle$ to be a *σm-chain* if $\sigma_m^{c^{(k)}} > 0$, for $0 \le k \le K$. Furthermore, for $W \subseteq V$ and $c \in V$, we use $\mu_{\sigma,m,W}(c)$ to denote the maximal ψ_m-strength of a σm-chain from a spel in W to c. (This is 0 if there is no such chain.)

Theorem 1. *If $(V, \Psi, \mathcal{V})$ is a seeded M-fuzzy graph (where $\Psi = (\psi_1, \cdots, \psi_M)$ and $\mathcal{V} = (V_1, \cdots, V_M)$), then*

(i) there exists an M-semisegmentation σ of V with the following property: for every $c \in V$, if for $1 \le n \le M$

$$s_n^c = \begin{cases} 1, & \text{if } c \in V_n, \\ \max_{d \in V}(\min(\mu_{\sigma,n,V_n}(d), \psi_n(d,c))), & \text{otherwise,} \end{cases} \tag{1}$$

then for $1 \le m \le M$

$$\sigma_m^c = \begin{cases} s_m^c, \text{ if } s_m^c \geq s_n^c, \text{ for } 1 \leq n \leq M, \\ 0, \quad \text{otherwise}; \end{cases} \tag{2}$$

(ii) this M-semisegmentation is unique; and
(iii) it is an M-segmentation, provided that $(V, \Psi, \mathcal{V})$ is connectable.

What Theorem 1 says in general terms, is that the mth object of an M-semisegmentation can "claim" a spel c as part of it if, and only if, s_m^c is maximal. We can see from the definition of an M-semisegmentation in shown previously that σ_m^c has a positive value only for such objects. The calculation of σ for a spel c (determined by a local condition) is performed by computing the values of the s_n^c using (1) and satisfying the conditions of (2) at c. Theorem 1 says that there is a unique M-semisegmentation which satisfies this reasonable property simultaneously everywhere, and that this M-semisegmentation is in fact an M-segmentation provided that the seeded M-fuzzy graph is connectable.

The original algorithm for computing the fuzzy segmentation according to the specifications above, called MOFS (multi object fuzzy segmentation) can be found in [2,3], as well as the proofs for Theorem 1.(ii) and 1.(iii). The proof of Theorem 1.(i) can be found on [3].

The affinities ψ_m can be specified in several ways. In [2], as well as in here, the affinities and the V_m, $1 \leq m \leq M$ are specified by using the information on the neighborhood of seed spels clicked by an user. When the user clicks on a spel and associates it with an object m, he/she is saying that it is certain that the spel belongs to the object m. Then, we collect information on a 3×3 neighborhood, defining g_m to be the mean and h_m to be the standard deviation of the average brightness for all edge-adjacent pairs of spels in V_m and a_m to be the mean and b_m to be the standard deviation of the absolute differences of brightness for all edge-adjacent pairs of spels in V_m. We then define $\psi_m(c,d)$ to be 0 if c and d are not edge-adjacent and to be $\left[\rho_{g_m,h_m}(g) + \rho_{a_m,b_m}(a)\right]/2$ if they are, where g,h,a,b are as defined above and the function $\rho_{r,s}(x)$ is the probability density function of the Gaussian distribution with mean r and standard deviation s multiplied by a constant so that the peak value becomes 1.

After the initialization steps, the central part of the MOFS algorithm updates the best guesses for the final values of the σ_m^c for all $c \in V$. A current value is replaced by a larger one if it is found that there is a σm-chain from a seed spel in V_m to c of ψ_m-strength greater than the old value (the previously maximal ψ_m-strength of the known σm-chains of this kind) and it is replaced by 0 if it is found that (for an $n \neq m$) there is a σn-chain from a seed spel in V_n to c of ψ_n-strength greater than the old value of σ_m^c.

The total computational complexity of the algorithm is $O(N(\log N + ML))$, where N denotes the number of elements of V, M is the number of objects, and L denotes the number of neighbor spels, usually a small number in the application of image segmentation (4 or 8 in 2D square grid images and 6, 18 or 26 in 3D cubic grid volumes).

2.1 Fast Fuzzy Segmentation

Even though we were able to segment a 3D image with more than 7,000,000 spels in approximately 4 minutes, as shown on [3], this response time may not be sufficiently fast for some applications. A fast implementation of the original MOFS algorithm for computing the simultaneous fuzzy segmentation of multiple objects, introduced in [2], was presented in [3]. (The segmentation of the same 3D image mentioned above using the fast implementation was achieved in 35 seconds.) This fast implementation is briefly described below.

The original MOFS algorithm belongs to the class of greedy algorithms [14], and was implemented using a binary heap to keep a partial ordering according to the spels' current σ^0 values. The values of the chains stored in the heap are updated as new chains with values greater than the ones currently stored are found. Since a heap is used to keep the partial ordering needed by the algorithm, the operations of spel insertion, deletion and σ^0 update take $O(\log N)$, where N is the number of spels in V. (A heap has to be used because the values of σ^0 can assume any real value between 0 and 1.)

However, suppose that the set of nonzero fuzzy spel affinities for a particular class of problems is always a subset of a fixed set A. Let K be the cardinality of the set $A \cup \{1\}$, and let $1 = a_1 > a_2 > \cdots > a_K > 0$ be the elements of A. For example, in many applications the quality of the fuzzy segmentation is not significantly affected if we round each fuzzy spel affinity to three decimal places. If we use such rounded spel affinities, then we can take $A = \{0.001, 0.002, \cdots, 0.999, 1.000\}$, so that $K = 1000$ and $a_k = 1.001 - k/1000$.

By restricting the affinity values to a fixed set of values, as above, we can use an $M \times K$ array $U[m][k]$ of sets of nodes that represent spels, where M is (as before) the number of objects. This array stores the spels according to their σ^0 values, thus, maintaining the partial ordering of the σ^0 values. Now, the cost of the operations of spel insertion and removal and the update of the σ^0 values becomes proportional to a constant, i.e., $O(1)$. (Similar ideas were used in [15] to speed up the algorithm of [1].) This implementation (shown below in pseudo-code first published in [3]) is most effective if all of its data structures (with space complexity $O(M(K+V))$) can be held in the main memory.

Fast implementation of the MOFS algorithm

1. **for** $c \in V$ **do**
2.   **for** $m \leftarrow 0$ **to** M **do**
3. $\sigma_m^c \leftarrow 0$
4. **for** $m \leftarrow 1$ **to** M **do**
5.   **for** $c \in V_m$ **do**
6. $\sigma_0^c \leftarrow \sigma_m^c \leftarrow 1$
7. $U[m][1] \leftarrow V_m$
8.   **for** $k \leftarrow 2$ **to** K **do**
9. $U[m][k] \leftarrow \emptyset$
10. **for** $k \leftarrow 1$ **to** K **do**
11.   **for** $m \leftarrow 1$ **to** M **do**
12.   **while** $U[m][k] \neq \emptyset$ **do**
13.   remove a spel d from the set $U[m][k]$
14. $C \leftarrow \{c \in V \mid \sigma_m^c < \min(a_k, \psi_m(d,c))$ **and** $\sigma_0^c \leq \min(a_k, \psi_m(d,c))\}$
15.   **while** $C \neq \emptyset$ **do**
16.   remove a spel c from C
17. $t \leftarrow \min(a_k, \psi_m(d,c))$
18.   **if** $\sigma_0^c < t$ **then do**

19. remove c from each set in U that contains it
20. **for** $n \leftarrow 1$ **to** M **do**
21. $\sigma_n^c \leftarrow 0$
22. $\sigma_0^c \leftarrow \sigma_m^c \leftarrow t$
23. insert c into the set $U[m][l]$ where l is the integer such that $a_l = t$

This fast version of the MOFS algorithm was chosen to be used in the color video segmentation because the small running times achieved allows an user to include or delete seed spels and rerun the algorithm a few times in a reasonable amount of time.

3 Color Video Segmentation

The fuzzy affinity function is now defined as an average of six components, two for each YUV channel, the color model chosen to code the input images. The two components of each channel used in the fuzzy affinity function are defined in the same way as before, i.e., using the mean and the standard deviation of the average value for all edge-adjacent pairs of spels in V_m and the mean and the standard deviation of the absolute differences of values for all edge-adjacent pairs of spels in V_m. Once more, $\psi_m(c,d)$ was defined to be 0 if c and d are not edge-adjacent and to be

$$\left[\rho_{gY_m,hY_m}(g) + \rho_{aY_m,bY_m}(a) + \rho_{gU_m,hU_m}(g) + \rho_{aU_m,bU_m}(a) + \rho_{gV_m,hV_m}(g) + \rho_{aV_m,bV_m}(a)\right]/6 \tag{3}$$

if they were, where g is the mean and a is the absolute difference of the Y, U or V channel values of c and d and the function $\rho_{r,s}(x)$ is the probability density function of the Gaussian distribution with mean r and standard deviation s multiplied by a constant so that the peak value becomes 1.

The affinity function defined above can be used for segmenting both 2D or 3D images. The volume composed by the video sequence frames is treated as a volume on a cubic grid with face adjacency, and the user can select seed spels on any slice (frame) of the volume. By doing this, the user can successfully segment objects that do not appear on the first frame. Moreover, the user can identify disconnected objects in the time direction as being the same one, as in the case of an object that is visible, occluded for some frames and visible again in the video sequence.

The idea of using the fast implementation of the fuzzy segmentation algorithm described here is that besides the ability of the fuzzy segmentation algorithm for segmenting images corrupted by noise and/or varying illumination, the small total segmentation time of the algorithm (not including the user interaction) allows the user to inspect the results and add or remove seed spels to achieve a better segmentation. This adjustment process of the segmentation can be repeated a few times and still be finished in a short period of time (in the order of seconds or a few minutes). The interactivity and the speed of the method makes it possible for the user to include high level knowledge in the segmentation process.

4 Experiments

First, we used the fast MOFS algorithm to segment a synthetic video shot (obtained from [16]) with and without added noise. The noisy sequence was corrupted by a

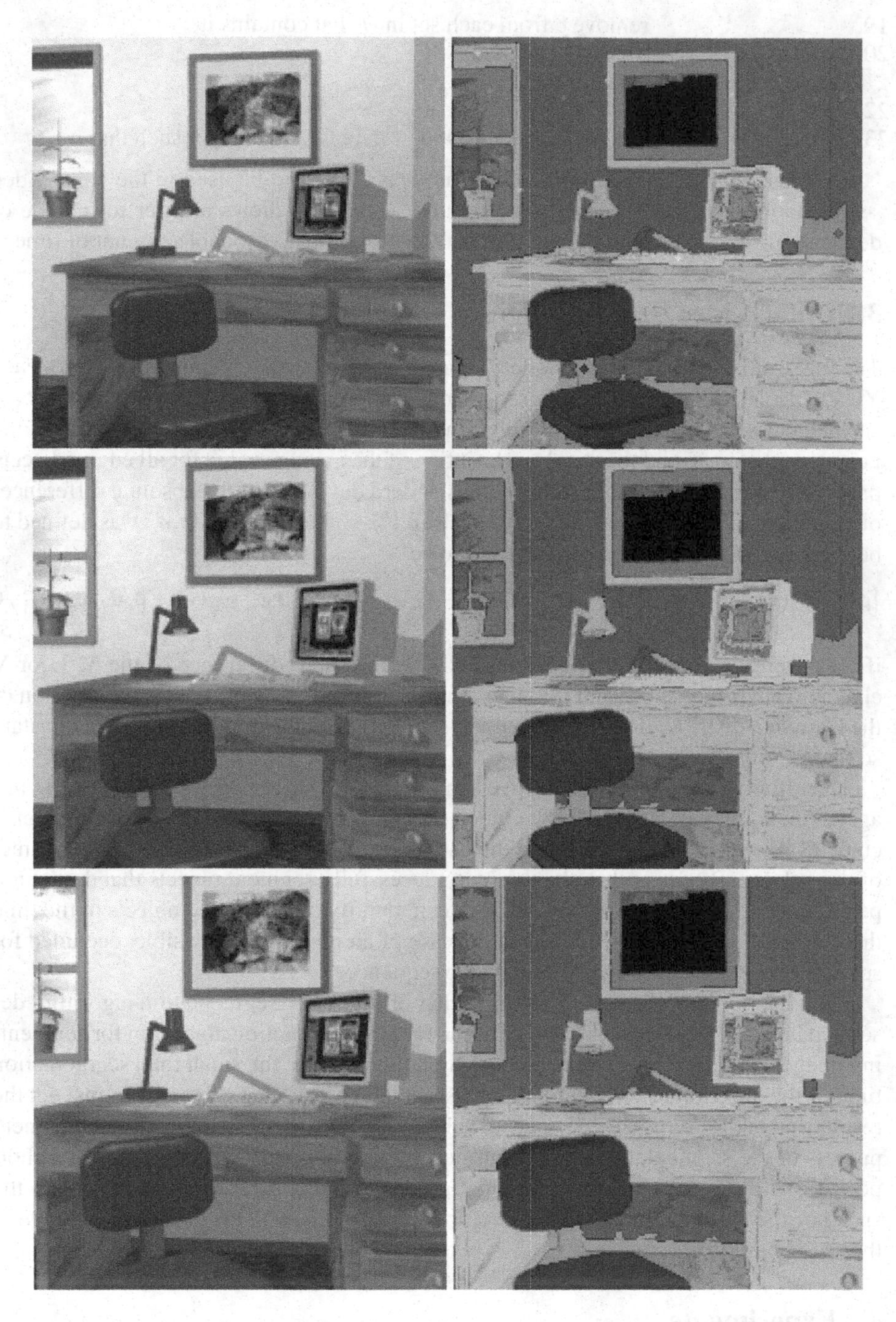

Fig. 1. Original frame images (left) and the corresponding fuzzy segmentation connectedness maps (right) for the 1st, 10th and 20th frames of the noiseless video sequence

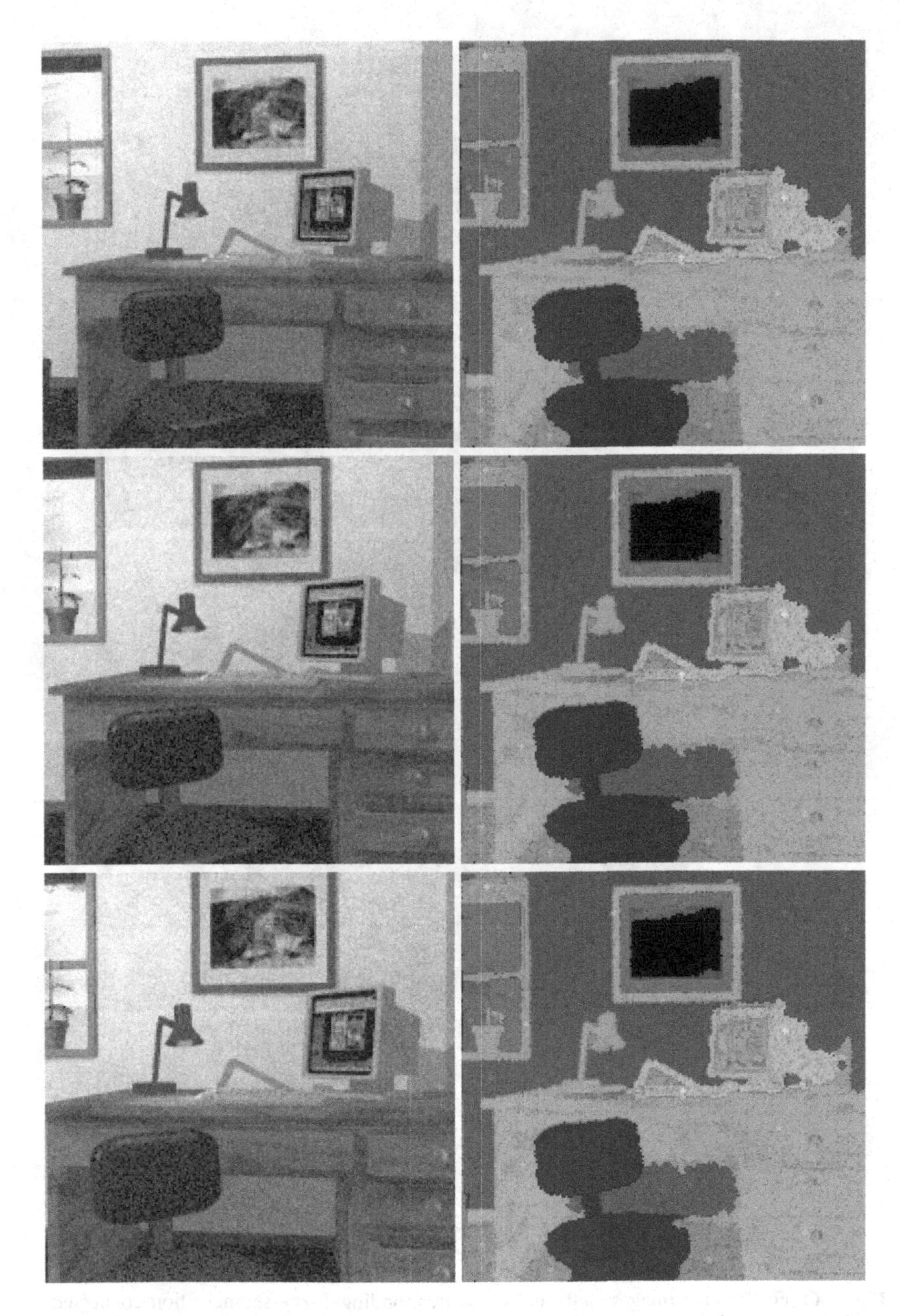

Fig. 2. Original frame images (left) and the corresponding fuzzy segmentation connectedness maps (right) for the 1st, 10th and 20th frames of the video sequence corrupted by noise with $\sigma = 10$ units in all three channels

Fig. 3. Original frame images (left) and the corresponding fuzzy segmentation connectedness maps (right). The rows correspond to the 1st, 6th, 11th, 16th and 21st frames (slices) of the video sequence (volume).

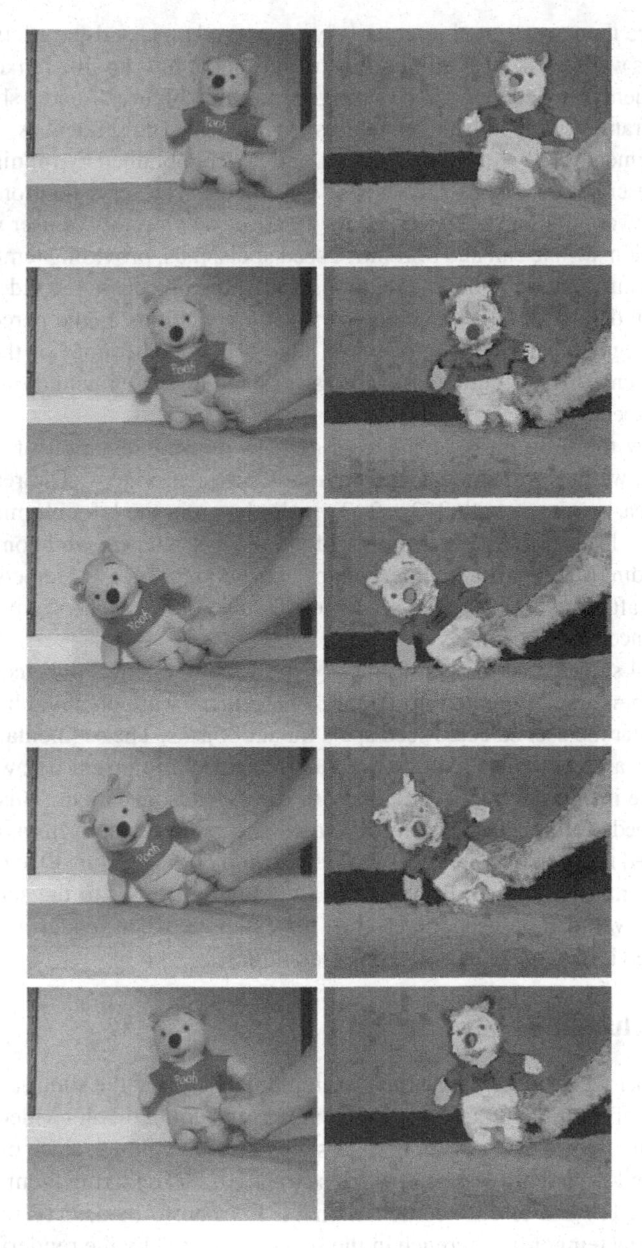

Fig. 4. Original frame images (left) and the corresponding fuzzy segmentation connectedness maps (right). The rows correspond to the 1st, 16th, 31st, 46th and 61st frames (slices) of the video sequence (volume).

Gaussian noise with mean 0 and standard deviation of 10 units (in the range $[0, 255]$) for all three channels, and pseudo-random values of this distribution were added to all three channels of the video sequence. The 20 frame long video sequences of 200×200

pixels were then segmented interactively on a single try, and the accuracy of the segmentations were measured by counting the number of misclassified pixels for all frames of the sequenced when compared to the ground truth. Figures 1 and 2 show the 1st, 10th and 20th frames of the noiseless and noisy synthetic video sequences, respectively.

The segmentations shown in Figures 1 and 2 were obtained by running the algorithm after the user selected seed spels for seven objects. The segmentation times for these sequences were the same, 2 seconds. As mentioned before, if the user wants to refine a segmentation, he/she can add or remove seed spels, thus, providing temporal and spatial high-level information about the relationships of the objects in the video sequence.

We then determined the accuracy of the segmentations as the percentage of pixels correctly segmented when compared to hand segmented frames of the noiseless synthetic sequence. The accuracy of the noiseless and noisy segmentations are 94.3% and 92.0%, respectively.

To show the usefulness of the algorithm to the segmentation of real color video sequences, we performed the segmentation of two real videos. The real images of the 21 frame car sequence with 352×240 pixels shown in the left column of Figure 3 are the 1st, 6th, 11th, 16th and 21st frames of the video sequence, while on the right are the corresponding fuzzy connectedness maps. These maps were obtained by running the algorithm after the user selected seed spels for five objects. The segmentation time for this sequence was 3 seconds.

Figure 4 shows 5 360×288 frames of the 71 frame long Pooh sequence (left) and the corresponding segmentations (right). These maps were obtained by running the algorithm after the user selected seed spels for six objects. The segmentation time for this sequence was 23 seconds. Notice the faulty segmentation right below the arm on the right of the image on the 3rd, 4th and 5th rows. After visualizing this result, the user can add seeds for a new object for that area and rerun the algorithm. (The interaction time needed for selecting the seeds in these examples range from 30 to 60 seconds.) The high level information that this new object actually belongs to the same object as the rest of the wall can be used afterwards, for example, when rendering the background wall of the video sequence using NPR techniques.

5 Conclusion

In this paper we propose to use the fast implementation of the simultaneous fuzzy segmentation of multiple objects for segmenting pre-acquired color video sequences. We consider the frames of the sequences as slices of a 3D volume, since our objective is to segment end-to-end pre-acquired video sequences, for rendering them later using NPR techniques. The end-to-end segmentation has to be performed, so its result can be used for enforcing temporal coherence in the scene generated by the rendering technique.

The results show a good accuracy rate of the segmented pixels for the synthetic sequences, with and without noise, and high quality segmentations of real video sequences, all with very small running times. This allows the user to rerun the algorithm after adding/removing seed spels, if the result of the segmentation was not satisfactory.

Future work will concentrate on incorporating motion cues (such as the algorithms presented in [17, 18]) in the fuzzy affinity functions for more accurate segmentation of different objects that have similar colors.

References

1. Udupa, J., Samarasekera, S.: Fuzzy connectedness and object definition: Theory, algorithms and applications in image segmentation. Graph. Models Image Proc. **58** (1996) 246–261
2. Herman, G., Carvalho, B.: Multiseeded segmentation using fuzzy conectedness. IEEE Trans. on Pattern Anal. Mach. Intell. **23** (2001) 460–474
3. Carvalho, B., Herman, G., Kong, T.: Simultaneous fuzzy segmentation of multiple objects. Disc. Appl. Math. **151** (2005) 55–77
4. Hertzmann, A., Perlin, K.: Painterly rendering for video and interaction. In: Proc. ACM Non-Photorealistic Animation and Rendering (NPAR). (2000) 7–12
5. Litwinowicz, P.: Processing images and video for an impressionist effect. In: Proc. ACM SIGGRAPH. (1997) 407–414
6. Collomosse, J., Rowntree, D., Hall, P.: Stroke surfaces: Temporally coherent artistic animations from video. IEEE Trans. Visualiz. and Comp. Graph. **11** (2005) 540–549
7. Wang, J., Xu, Y., Shum, H.Y., Cohen, M.: Video tooning. ACM Trans. on Graph. **23** (2004) 574–583
8. Rosenfeld, A.: Fuzzy digital topology. Inform. and Control **40** (1979) 76–87
9. Carvalho, B., Gau, C., Herman, G., Kong, T.: Algorithms for fuzzy segmentation. Pattern Anal. Appl. **2** (1999) 73–81
10. Carvalho, B., Garduño, E., Herman, G.: Multiseeded fuzzy segmentation on the face centered cubic grid. In Singh, S., Murshed, N., Kropatsch, W., eds.: Advances in Pattern Recognition: Second International Conference - ICAPR 2001 (LNCS 2013), Rio de Janeiro, Brazil, Springer-Verlag (2001) 339–348
11. Zahn, C.: Graph-theoretic methods for detecting and describing Gestalt clusters. IEEE Trans. Comp. **1** (1971) 68–86
12. Duda, R., Hart, P.: Pattern Classification and Scene Analysis. John Wiley & Sons, New York (1973)
13. Jain, A., Murty, M., Flynn, P.: Data clustering: a review. ACM Comput. Surveys **31** (1999) 264–323
14. Cormen, T., Leiserson, C., Rivest, R.: Introduction to Algorithms. MIT Press, Cambridge, MA (1990)
15. Nyul, L., Falcão, A., Udupa, J.: Fuzzy-connected 3D image segmentation at interactive speeds. Graph. Models **64** (2002) 259–281
16. Computer Vision Homepage: University of Otago, available at http://www.cs.otago.ac.nz/research/vision/Research/OpticalFlow/opticalflow.html (2006)
17. Galun, M., Apartsin, A., Basri, R.: Multiscale segmentation by combining motion and intensity cues. In: Proc. IEEE Comp. Soc. Conf. on Comp. Vision and Patt. Recog. (2005) 256–263
18. Khan, S., Shah, M.: Object based segmentation of video using color, motion and spatial information. In: IEEE Comp. Soc. Conf. on Comp. Vision and Patt. Recog. Volume 2. (2001) 746–751

Application of Surface Topological Segmentation to Seismic Imaging

Timothée Faucon[1,2], Etienne Decencière[1], and Cédric Magneron[2]

[1] Centre de Morphologie Mathémathique
Ecole des Mines de Paris,
35 rue Saint Honoré,
77305 Fontainebleau, France
{timothee.faucon, etienne.decenciere}@cmm.ensmp.fr
http://cmm.ensmp.fr

[2] Earth Resource Management Services (ERM.S),
16, rue du château,
77300 Fontainebleau, France
{timothee.faucon, cedric.magneron}@erms.fr
http://www.erms.fr

Abstract. An original and efficient method to segment and label horizontal structures in 3D seismic images is presented. It is based on a morphological hierarchical segmentation. The initial extracted surfaces are post-processed using the topological segmentation method proposed by Malandain et al [1]. A last post-processing step allows to separate remaining multi-layered surfaces.

1 Introduction

Seismic imaging has become an essential technique in seismic exploration and exploitation. Thanks to it, industrialists have a better understanding of the subsoil structure and save time in seismic data processing. It also brings them a large panel of tools to improve their exploration and exploitation process. However, information is hidden in 3D seismic images and one of the prerequisites to benefit from the interesting information contained in the 3D data is to carry out several processings to adequately modify original data. One of these consists in extracting quasi horizontal structures also called horizons. They correspond to the sediment layers and are useful for geophysicists to locate oil and gas reservoirs. Horizon extraction might seem relatively easy. In fact, it is difficult due to the presence of geological faults, noise and acquisition artifacts. Moreover, it must be accurate since many applications depend on it. One of the other constraints in horizon extraction is that it is really time consuming and often requires a geoscientist expertise to be done.

The extraction method described here is an original way of picking and labelling horizons from seismic data. After briefly presenting seismic imaging in section 2, we proceed with the description of the picking method: in section 3 we

A. Kuba, L.G. Nyúl, and K. Palágyi (Eds.): DGCI 2006, LNCS 4245, pp. 506–517, 2006.

describe briefly the horizon extraction part, which was presented in [2]. In section 4 we present the application of the surface segmentation method proposed by Malandain et al. [1] to our data. The result is not completly satisfactory in our framework due to surfaces composed of multiple layers. In section 5 a new method to solve this problem is introduced. Then before concluding on this work, we give some computation times to evaluate the efficiency of the processing chain.

2 Seismic Imaging

The first seismic data were acquired on tapes in a rudimentary way. Then, in the 60s, transistors and digital technologies appeared and allowed to sample the subsoil in 2D. The oil industry started to be interested in those methods of subsoil imaging to find and monitor oil fields. In the early 70s the first 3D seismic survey was achieved on a field near Houston. 3D seismic acquisition has been since then considered as a reliable and cost-effective method of optimizing oil field exploration, development and management.

2.1 Acquisition Mode

Like echography in the medical field, seismic images are acquired through an indirect technique using acoustic waves. The acquisition process can be implemented offshore as well as onshore. In both cases it requires a sonar source (air-gun for marine acquisition and vibrators for ground one) which sends waves across the subsoil and receptors that register the signal reflected by the interfaces of sediment layers. This limit between two layers with different acoustic properties is called a seismic horizon. A large number of preprocessings are necessary to make the acquired data ready for analyses and interpretations by geoscientists. Those preprocessings provide the interpreters with many images. Each image is the representation of a seismic attribute which is defined by everything that can be computed from the original data or from other attributes [3]. Up to now, we have only worked on images of wave amplitudes. As shown in Fig. 1 amplitude images are a bright and dark layers stacking. These layers correspond to the sediment deposits. We call inline the direction of the acquisition system (given by the alignment of source and receptors) and cross-line the perpendicular direction to the inline one. These two directions correspond to the horizontal planes. The third dimension is called time (which can be converted into depth) and corresponds to the vertical direction perpendicular to the two others.

2.2 Why Segmenting 3D Seismic Images?

Segmenting 3D seismic images, i.e. extracting coherent seismic events (horizons), addresses several issues. A first family of these issues may be related to seismic processing. A set of coherent horizons is sometimes used as a guide within particular seismic processing steps. It allows to introduce geological information into classical seismic oriented processing. A second family is related to the time to

Fig. 1. Two views of the same 3D seismic cube used to illustrate this paper. Dimensions: $201 \times 201 \times 150$.

depth conversion process. Seismic images in time are converted in depth. Major identified horizons are useful to achieve this conversion. They result from a structural interpretation. A third family can be identified when dealing with reservoir modelling. Horizons at a small vertical scale result from a stratigraphic interpretation. Other operational applications may be found as segmenting 3D seismic images is a really new approach. Some techniques already exist to segment seismic images. Among those, we can find the ones which take into account the image structure [4] or use wavelets [5] to realize the segmentation. Some other techniques are based on the division of space into a set of meshes and on a pairing of similar meshes [6]. We also find seismic images segmentation using fuzzy techniques [7]. A little number of papers deal with seismic data processed by mathematical morphology techniques ([8], [9]). Thus segmenting 3D seismic images with mathematical morphology appears to be an innovative method.

3 Seismic Horizon Extraction Using Morphological Segmentation

Segmenting an image can be achieved in several ways. The watershed invented in 1979 by S. Beucher and Ch. Lantuéjoul [10] is a morphological solution and is based on an imaginary flooding of a topography usually given by the gradient of the original image. The flooding starts from sources determined by the user or another algorithm and produces a fine partition of the gradient image.

Because of the seismic image structure and of the horizon picking extraction application, we do not use the gradient but the original amplitude image as an entry to the watershed algorithm. The horizons we are looking for are the image brighter surfaces. With the watershed, we directly obtain the principal horizons that correspond to the maximal amplitudes.

3.1 Hierarchical Segmentation

The fine segmentation of a seismic image gives however too many regions to be easily interpreted. To restrict the number of areas, we could have made a selection of the flooding sources but nothing can ensure that the brighter surfaces are situated between two selected minima. Another solution to reduce the final number of horizons, while preserving the more representative geological structures, is to select the best ones according to a certain criterion among the whole set of horizons. To reach this goal, we used the hierarchical segmentation introduced by F. Meyer in 1994 [11]. This algorithm creates a graph associated to the image. Each graph node represents one of the image regions i.e. the influence zone of a minimum with respect to the topography. During the flooding process, each time two basins merge, the corresponding nodes are linked by an edge valuated with the smallest value of a criterion computed from the basins characteristics. This criterion can be the depth, the area or the volume of the basin. We finally obtain a minimum spanning tree. By cutting the $k-1$ bigger edges, we obtain from this tree k regions, which are typically separated by the brightest horizons.

3.2 Segmentation with Cylinders

Some horizons that seem to be among the brightest ones do no appear in the results even if we ask for a large amount of horizons. This is due to some voxels of these horizons which have a lower gray level than the rest of the structure. These voxels lead to a "leak" during the flooding step and as the two bordering basins merge earlier than they should, the final surface appears lower in the hierarchy. To restrict this problem to the leak location and pick the rest of the horizon we used a method described in [2], [12]. This method consists in dividing the original image into many subimages called cylinders and in computing a hierarchical segmentation inside each of them. The leak problem is thus restricted to a small number of cylinders and the major part of each brightest horizon appears in the resulting image. To preserve the picked structures continuity, we make the cylinders overlap.

Each time a pixel within a given cylinder is considered as belonging to a horizon (i.e. belongs to the watershed line), an accumulator image is incremented at the corresponding position. Once all cylinders have been segmented, the accumulator contains a grey level image. After thresholding, the resulting binary image corresponds to the extracted horizons. Figure 2 shows the result of the application of the horizon extraction method to our test cube, after labelling (26-connexity). It contains only three connected components.

In practice, most structures appearing in the accumulator are thin, but, for several reasons, this is not always the case. Therefore, a simple threshold is not enough to binarise the accumulator if we want to obtain thin surfaces.

3.3 Thinning

To implement our labelling algorithm and avoid the errors due to the thickness of surfaces, we need thin surfaces. That means that surfaces are expected to

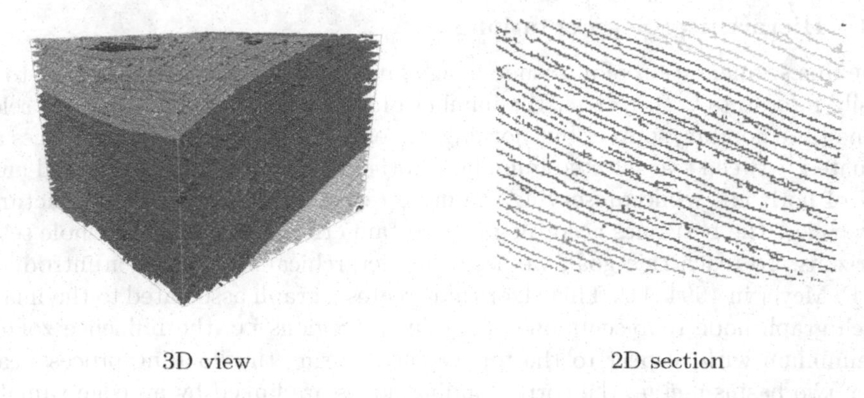

3D view 2D section

Fig. 2. Labelled horizons extracted from the test cube, without post-processing

have a thickness of one voxel. We have used an *a priori* knowledge about the application for which we are trying to segment the image and we work with a binarisation of the accumulation image obtained from the morphological segmentation.

As we are trying to extract the brightest voxels, which characterise the structures we are looking for (Fig. 3a), we identify the connected voxels in a vertical neighborhood ($1 \times 1 \times 3$) see Fig. 3b. Then we eliminate those which have the lowest amplitude (Fig. 3c). In the particular case of slopes higher than forty five degrees, this technique can introduce horizontal holes into the surface but since we are working in 3D the surface connectivity is preserved. Given the nature of the images, such high slopes are very uncommon, however, further developments of this thinning process are envisaged and will lead to the correction of these drawbacks.

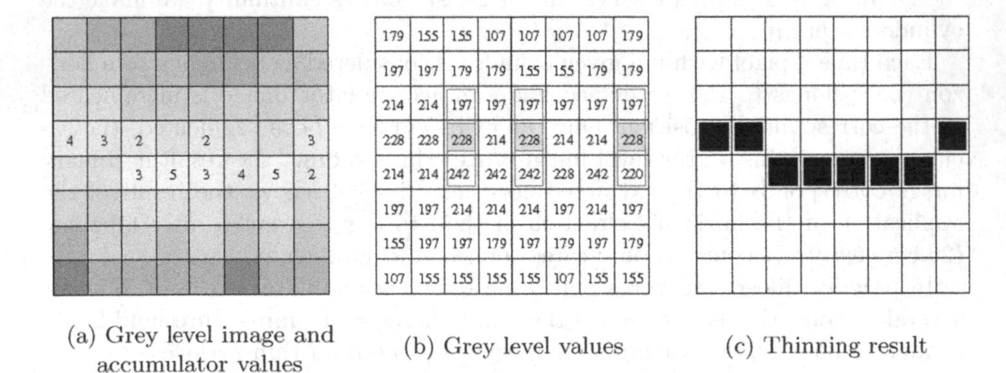

(a) Grey level image and accumulator values (b) Grey level values (c) Thinning result

Fig. 3. Thinning process

4 Topological Segmentation

Superimposed horizons which are connected (i.e. surfaces in which some voxels have the same geographical coordinates) are a real problem for the interpretation and the other applications depending on geological structure picking. To address this problem, we took as a starting point the approach of Malandain et al. [1]. This method creates a voxel classification using the topology of object and background voxel neighborhood.

4.1 Voxel Classification

Malandain et al. have implemented a voxel classification in order to segment surfaces in 3D images. Thus they have established two relevant numbers C^* and $\bar{C}$ that respectively describe the object connectivity and the background connectivity in a given neighborhood. Considering an object X, its background $\bar{X}$ and a neighborhood V of an object point x, the two numbers are:

- C^*: number of 26-connected components of $X \cap V^*$ 26-adjacent to x, and
- $\bar{C}$: number of 6-connected components of $\bar{X} \cap V$ 6-adjacent to x,

where V^* is $V \backslash \{x\}$.

Connectivity and Neighborhoods. According to [13], object and background adjacency have to be different. Usually, adjacency used for the object is the 26- or 18- adjacency and the one used for the background is the 6-adjacency. Like Malandain et al., we have followed this convention.

According to the definition of C^*, the connected components of the object have to be 26-adjacent to the central point x. Thus, using a 26-connected neighborhood to compute C^* allows to avoid the calculation of the 26-adjacency. We differ from Malandain et al. in the neighborhood choice for the computation of $\bar{C}$. Indeed, we have chosen to compute it rather in a 26-neighborhood than in a 18-neighborhood. This choice is motivated by an algorithmic reason. Computing C^* and $\bar{C}$ in the same neighborhood simplifies and accelerates the computation.

The resulting values of C^* and $\bar{C}$ according to their neighborhood are summarized in Table 1.

Table 1. Class of point according to the value of C^* and $\bar{C}$

Type A	interior point	$\bar{C} = 0$	
Type B	isolated point		$C^* = 0$
Type C	**border point**	$\bar{C} = 1$,	$C^* = 1$
Type D	curve point	$\bar{C} = 1$,	$C^* = 2$
Type E	curve junction	$\bar{C} = 1$,	$C^* > 2$
Type F	**surface point**	$\bar{C} = 2$,	$C^* = 1$
Type G	surface-curve junction	$\bar{C} = 2$,	$C^* \geq 2$
Type H	**surface junction**	$\bar{C} > 2$,	$C^* = 1$
Type I	surface-curve junction	$\bar{C} > 2$,	$C^* \geq 2$

4.2 Misclassification

Problem of Misclassification. After this topological classification, we select the voxels we are interested in. These voxels are C Type (border points) and F Type (surface points) ones. As we eliminate the other points, specially the junction points, we should have well disconnected and easy to label surfaces with a simple labelling algorithm. In fact, that does not occur like that, as shown in Fig. 4. Some horizons are correctly separated from the others but some surface junctions are not correctly detected. This phenomenon comes from some troubles in the detection of junction points. Some junctions, because of their thickness, are wrongly labelled. The two examples shown in Fig. 5 present two cases where the junction points are labelled as surface points (the two grids represent a slice ot the 3D object). In case a, $\bar{C} = 0$ for the center point. Thus, this point is considered as an interior point. In case b, $\bar{C} = 0$ and $C^* = 1$ thus the four center points are misclassified as surface points.

Processing of Misclassified Points. To tackle this problem, we refer again to [1] where a characterisation of simple surfaces is explained. Considering a surface point x, B_x and C_x are the two connected components of $\bar{X} \cap N_{26}^*$ 6-adjacent to x included in X. Two surface points x and y are said to be in relation if there is a 26-path $(x_0, x_1, ..., x_i, ..., x_n)$ included in X with $x_0 = x$ and $x_n = y$ such that for $i \in [0, ...n-1]$:

- $B_{x_i} \cap B_{x_{i+1}} \neq 0$ and $C_{x_i} \cap C_{x_{i+1}} \neq 0$, or
- $B_{x_i} \cap C_{x_{i+1}} \neq 0$ and $C_{x_i} \cap B_{x_{i+1}} \neq 0$.

Image corrected from misclassified points

Fig. 4. Topological segmentation

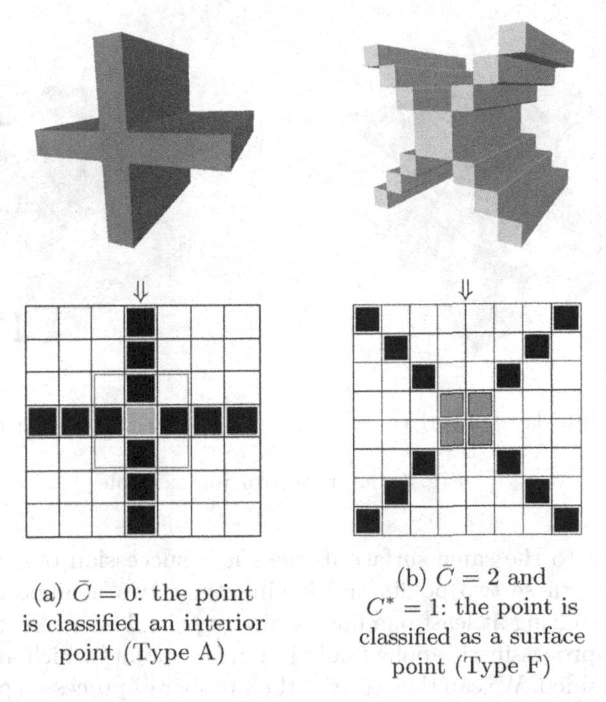

(a) $\bar{C} = 0$: the point is classified an interior point (Type A)

(b) $\bar{C} = 2$ and $C^* = 1$: the point is classified as a surface point (Type F)

Fig. 5. Misclassification of junction points. Each image represents a slice of two different binary images constituted of a succession of these images.

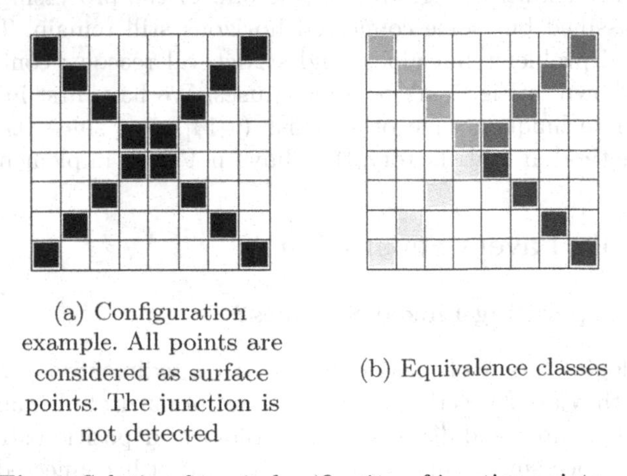

(a) Configuration example. All points are considered as surface points. The junction is not detected

(b) Equivalence classes

Fig. 6. Solution for misclassification of junction points

This relation is an equivalence relation (it is reflexive, symmetric and transitive) and its equivalence classes compose therefore a partition of the surface points of X. Each region of the partition is a simple surface. In short, two surface

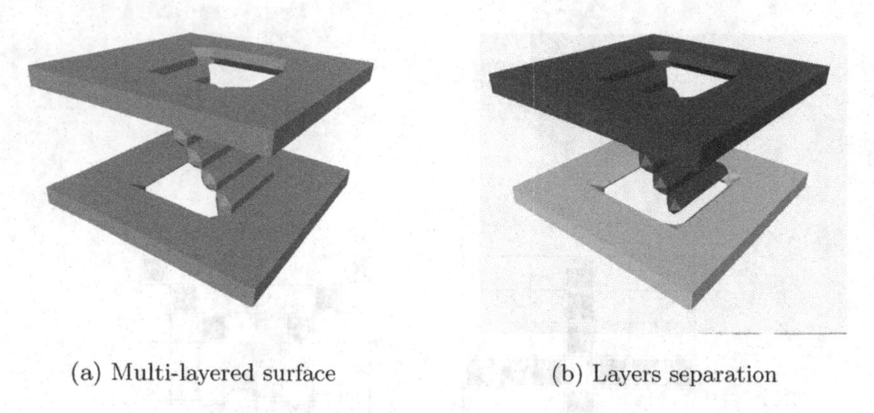

(a) Multi-layered surface (b) Layers separation

Fig. 7. Layers separation example

points belong to the same surface if there is a succession of connected surface points linking these two points and having their two neighborhood connected components sharing at least one point pair by pair. This is illustrated in Fig. 6.

This post-processing is applied only to surface points which are the only ones to be misclassified. We can thus restrict the number of processed points. However, during the processing, border points are considered and labelled too in order to be integrated to the surface but they are not processed as the surface points. The results are shown in Fig. 4b). At the end of the processing, all points are correctly classified but some connected horizons still remain. The reasons are no more local problems but global and structural reasons: configurations such as the one shown in Fig. 7 are possible causes. We have first implemented this classification technique. Some other exist ([14]) but since the results of the presented method are satisfactory, they have not been implemented.

5 Multiple Layers Separation

5.1 Why Separating Linked Surfaces?

From a geological point of view, differenciating sediments layers is interesting especially if they are correctly separated even if a natural phenomenon (erosion, faults, etc...) occured and disturbed the geological deposition structure. At this step of our processing chain, some surfaces are still connected whereas they should not. We say that a surface is made of several layers if two points of this surface have the same geographical coordinates (inline and crossline). The way of separating the different structure layers is described below.

In the example of Fig. 7a all the voxels are considered as border or surface voxels and they belong to the same connected component. They will thus receive the same label. This kind of configuration can be due to a fuzzy part of the image which implies a bad segmentation area. The resulting segmentation leads to this link between two surfaces which are not connected anywhere else. To solve this problem, we realize a last processing.

5.2 Structure Separation

To separate the linked surfaces, we have to give to each component of the multiple-layer surface a different label. Our new labelling process consists in maintaining a map of the current processed horizon (made of multiple layers) and to make a selection among the voxels candidates for stacking into the queue. Those for which there is already a voxel with the same geographical coordinates in the queue are not stacked in the same queue because they belong to another surface. Once the current queue is empty, we process the other ones. A point can be queued several times but it will never be labelled twice because a tag checking is done as it is pushed back from queue.

This leads to the results shown in Fig. 7b where the two different surfaces are separated.

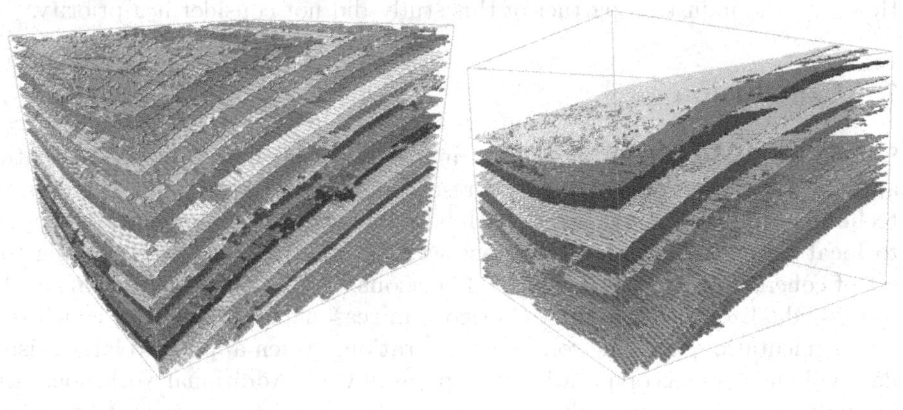

(a) Horizons correctly labelled (The multiple layer structure on the bottom right of the image is due to the visualisation software colormap capabilities (only 256 levels).)

(b) With area filtering (elimination of small surfaces)

Fig. 8. Final labellisation

According to specialists, final result is very interesting. The surfaces extracted from the original data follow the maxima values of the geological structures and the surfaces are correctly separated. These results are shown in Fig. 8.

6 Computation Times

The algorithms have been tested on a computer equipped with an Intel Pentium IV 3 GHz and 1 Gb of RAM. The test image is an 8 bits grey level image with

8 million points (501 × 400 × 400). With these elements, the computation times are the following:

- The segmentation with cylinders takes about 3.5 hours with 501 × 10 × 10 cylinders (i.e. 44100 cylinders) and with a step of one pixel in both directions. We have asked for forty regions in each cylinder (this parameter does not change the final computation time). Since each cylinder can be processed independently, this algorithm is strongly parallelizable.
- Topological surface labelling takes about forty minutes. This duration is not really representative because it depends directly on the number of required regions at the segmentation step i.e. the number of points contained in the object.
- Multi layer structure elimination. This section depends completely on the number of object points and the complexity of the structures. In our example its computation takes about 20 minutes.

Most of these times could be considerably improved through code optimization. However, the industrial partner of this study did not consider it a priority.

7 Conclusion

Segmenting and labelling 3D seismic images is a challenge. Both the 3D nature and the very large data volume represent major constraints for segmentation techniques to overcome. Our approach based on watershed segmentation applied to local volumes allows to derive fast segmentation results corresponding to a set of coherent seismic surfaces called horizons. Our labelling algorithms enable to individualize efficiently theses horizons in reasonable delays. As a conclusion, our segmentation/labelling solution is operational when applied to large seismic data volumes from a computation time point of view. Additional work, consisting in analysing the quality of the provided results, is planned in order to definitely validate the approach. Afterwards, several operational applications like geology guided seismic processing, structural interpretation, stratigraphic interpretation, etc. should benefit from it.

References

1. Malandain, G., Bertrand, G., Ayache, N.: Topological segmentation of discrete surfaces. International Journal of Computer Vision **10**(2) (1993) 183–197
2. Faucon, T., Decencière, E., Magneron, C.: Morphological segmentation applied to 3D seismic data. In Ronse, C., Najman, L., Decencière, E., eds.: Mathematical Morphology: 40 Years On. Computational Imaging and Vision (2005) 475–484
3. Taner, M.: Attributes revisited. Rock Solid Image Houston Texas (1992)
4. Bakker, P.: Image structure analysis for seismic interpretation. PhD thesis, Technische Universiteit Delft (2002)
5. Bouchereau, I.B.: Analyse d'images par transformée en ondelettes; application aux images sismiques. PhD thesis, Université Joseph Fourier Grenoble (1997)

6. Hale, D., Emmanuel, J.: Seismic interpretation using global image segmentation. 73th Annual International Meeting, Society of Exploration Geophysicists (2003)
7. Valet, L., Mauris, G., Bolon, P., Keskes, N.: Seismic image segmentation by fuzzy fusion of attributes. IEEE Transactions on Instrumentation and Measurement **50** (2001) 1014–1018
8. Moueddene, K.: Analyse d'images en sismique : pretraitement et extraction d'informations par la morphologie mathématique. PhD thesis, Université Paul Sabatier, Toulouse, France (1987)
9. N'Guyen, M.: Analyse multi-dimensionnelle et analyse par les ondelettes des signaux sismiques. PhD thesis, Institut National Polytechnique de Grenoble (2000)
10. Beucher, S., Lantuéjoul, C.: Use of watersheds in contour detection. In: International workshop on image processing, real-time edge and motion detection. (1979)
11. Meyer, F.: Minimal spanning forests for morphological segmentation. In Serra, J., Soille, P., eds.: Mathematical Morphology and its applications to signal processing (Proceedings ISMM'94), Fontainebleau, France, Kluwer Academic Publishers (1994)
12. S.Beucher, Decencière, E., Sandjivy, L., Magneron, C., Faucon, T.: (Demande de brevet français no 05 03793 pour un procédé de détermination hiérarchique d'événements cohérents dans une image sismique)
13. Kong, T., Rosenfeld, A.: Digital topology: Introduction and survey. Computer Vision, Graphics, And Image Processing **48**(1) (1989) 357–393
14. Svensson, S., Nyström, I., di Baja, G.S.: Curve skeletonization of surface-like objects in 3d images guided by voxel classification. Pattern Recognition Letter **23** (2002) 1419–1426

Watershed Segmentation with Chamfer Metric

Vasily Goncharenko[1] and Alexander Tuzikov[2]

[1] National Center of Information Resources and Technologies, National Academy of Sciences of Belarus, Akademicheskaja 25, 220072 Minsk, Belarus
vasily@mpen.bas-net.by
[2] United Institute of Informatics Problems, National Academy of Sciences of Belarus, Surganova 6, 220012 Minsk, Belarus
tuzikov@newman.bas-net.by

Abstract. Watershed transformation is introduced as a computation in image graph of a path forest with minimal modified topographic distance in $(\mathbb{R}^+)^2$. Two algorithms are presented for image segmentation that use a metric defined by a unit neighborhood as well as a chamfer (a, b)-metric. The algorithms use ordered queues to propagate over image pixels simulating the process of flooding. Presented algorithms can be applied to gray-scale images where objects have noticeable boundaries.

1 Introduction

Watershed transformation of gray-scale images often results in better segmentation and contour detection outcomes in comparison with other methods. Image segmentation by watershed transformation belongs to the region growing methods that combine pixels according to similarity of their properties relative to the properties of their local neighbors. This method works good for images with objects characterized by brightness or color characteristics rather than texture features.

Watershed transformation of gray-scale images was firstly described by S. Beucher and C. Lantuéjoul [1] in their paper devoted to contour detection of objects on metallographic pictures. The authors adopted geographic terminology for describing the contour detection process, based on the disclosure of the areas with the greatest absolute gradient values. The image was presented as a topographic surface the drops of water fall on, stream down and come into local minima. Each local minimum has its own set of points of the surface named **catchment basin**. If a drop falls on a point belonging to a catchment basin, it will come down into a corresponding local minimum. Several catchment basins may intersect - their common points form watersheds. Formal construction of watershed points was based on the detection of points equidistant from different catchment basins lying on a given level λ.

In a later work S. Beucher [2] considered two groups of watershed transformation algorithms. The first group contained algorithms which simulate the flooding process (or immersion into water). An algorithm based on morphological operations was taken as an example. The second group was made of procedures detecting watershed points directly.

A. Kuba, L.G. Nyúl, and K. Palágyi (Eds.): DGCI 2006, LNCS 4245, pp. 518–529, 2006.

L. Vincent and P. Soille in their work [3] presented an algorithm of watershed transformation based on immersion process described in terms of morphological operations and graph theory. The immersion process used FIFO queues to propagate over the pixels. Their algorithm was faster and more exact than any other algorithm presented at the moment of their publication.

In addition to morphological approach to calculation of watershed transformation of discrete images there is another way based on presentation of image in the form of a graph and calculation a shortest path forest for given local minima. S. Beucher and F. Meyer have developed an algorithm using a queue for finding the shortest path between arbitrary nodes of the graph in the process of building a tree of shortest paths [4]. However, if the image has "flat" regions (plateau), such an algorithm will result in errors at the regions. To use the algorithm in the presence of plateaus, the image should be corrected to remove them. A. Lotufo and R. Falcão proposed an algorithm using the ordered queue which allows to solve the problems with plateaus [5]. However, their algorithm has drawbacks: one pixel might be pushed into the queue more than once.

There is a number of other approaches to calculation of watershed transformation. J. Roerdink and A. Meijster in their review article [6] have discussed two groups of watershed transformation algorithms: based on immersion, and on topographical distance calculation. They also discussed problems of parallelization of watershed transform calculation and concluded that it is hard to parallelize because of its inherently sequential nature.

Another approach to watershed transform calculation is the topological one [7]. Topological watershed transform uses a graph representation of the image and is based on a notion of "simple" topology. Such a transformation of gray-scale image results in another gray-scale image preserving topological information. Topological transform calculation is based on detection of "simple" points on each cross section of gray-scale image.

In this paper an algorithm of watershed transform calculation is based on detection of minimal cost paths. The path cost consists of two components. The first characterizes maximal intensity of pixels along the path. The second component of the cost gives the length of the "flat" path part and is determined using a chosen metric. This allowed to extend the algorithm to the case of chamfer metric.

2 Basic Notions and Definitions

Given a set of pixels $X \subseteq Z^n$, where n is image dimension, let Y be a set of intensity values of the image. **Digital n-dimension gray-scale image** is a function $f : X \to Y$ where the value $f(p) \in Y$ defines intensity of pixel $p \in X$. We will consider a case when function f is discrete. Watershed transformation is interpreted as a result of immersion of surface f into water. It is assumed that at selected local minima $m_i \in M$ of function f the holes are pierced such that water will be gradually filling up "cavities" starting from the minima.

Graph of gray-scale image f is a weighted graph $G = (X, \Gamma, f, d)$, with function f defined on the set of graph nodes. Arc (p, q) with weight $d(p, q)$ between nodes $p, q \in X$ exists iff q belongs to the neighborhood $\Gamma(p)$ of p. Hence, image graph contains information about intensity of pixels and their adjacency. For each node p of graph G the value $\eta = |\Gamma(p)|$ is called **the connectivity** of node p. For our task the connectivities of all graph G nodes are the same: for 2D images given on rectangular lattice $\eta = 4$ or $\eta = 8$, for 3D images $\eta = 6, \eta = 14$, or $\eta = 26$.

Path $\pi = (p_0, p_1, \ldots, p_k)$ from node p_0 to node p_k of graph G is an ordered sequence of pixels $(p_0, p_1, \ldots, p_k)$, such that $p_{i+1} \in \Gamma(p_i)$ and $p_i \neq p_j$ for $j \neq i$, $0 \leq i \leq k-1$. Path $(p_0, p_1, \ldots, p_k)$ is denoted by π_k or simply π and its part $(p_i, p_{i+1}, \ldots, p_j)$, $0 \leq i \leq j \leq k$ from node p_i to node p_j in graph G is denoted by $\pi_{i,j}$. If $i = 0$, then the first index in the designation of path is omitted.

Let Π be a set of all paths in graph G. Function $\rho : \Pi \rightarrow (\mathbb{R}^+)^2$ is called **path function** of graph G if for any π_i, π_j such that π_i is a part of π_j (denoted as $\pi_i \subset \pi_j$), the following inequality is always true:

$$\rho(\pi_i) < \rho(\pi_j),$$

with $\rho(\pi_0) = (0, 0)$. In other words, appending at least one pixel to any path results in increasing its path function. Increasing path function is treated in the sense of **lexicographic ordering** in $(\mathbb{R}^+)^2$.

The value $C(p_i) = \rho(\pi_i), i \leqslant k, \pi_i \subseteq \pi_k$ is called **the cost of pixel** p_i on the path π_k. The value $C(p_i)$ is a vector in $(\mathbb{R}^+)^2$. The first and second coordinates of vector $C(p_i)$ are denoted by $C_x(p_i)$ and $C_y(p_i)$ respectively.

The distance $\delta(p, q)$ between any pixels $p, q \in X$ is defined as follows:

$$\delta(p, q) = \min_{\pi \in \Pi_{p,q}} \{\rho(\pi)\},$$

where $\Pi_{p,q}$ is a set of all paths between pixels p and q in graph G.

Let M be a set of selected pixels (markers) of graph G, $|M| > 1$, and I_M is a set of marker indices. **Catchment basin** $CB(m_i)$ of marker $m_i \in M, i \in I_M$ is defined as a set of points $x \in X$ which satisfy the following condition:

$$CB(m_i) = \{p \in X | \delta(m_i, p) < \delta(m_j, p), i \neq j\}.$$

Watershed $W(G)$ of graph G is a set of points not belonging to any catchment basin:

$$W(F) = X \backslash \left(\sum_{m_i \in M} CB(m_i) \right).$$

Catchment basins and watershed points define a segmentation of image. For simplicity it is possible to include watershed pixels into the relevant catchment basins in order to get segmentation just by catchment basins. Usually markers $m_i \in M$ are chosen as representative pixels of image objects as well as the background (i.e. not belonging to any object).

It is clear that catchment basins are Voronoi cells of the selected local minima. Therefore standard algorithms developed for Voronoi cells construction can be used for building catchment basins.

Let G be a graph of image f with selected markers $m_i \in M, i \in I_M$ and ω is an arbitrary label, $\omega \notin I_M$. **Watershed transformation** of image f is a mapping $\lambda : X \to I_M \cup \{\omega\}$ such that $\lambda(p) = i$ if $p \in CB(m_i)$, and $\lambda(p) = \omega$ if $p \in W(f)$.

3 Path Function

Choosing a path function having maxima at edges between different source image objects is an important task. It is based on the assumption that all objects of source gray-scale image f differ in intensity level, and there is a noticeable leap of intensity at edges between different objects.

As a simplest example of path function one can take a function $l(\pi)$ defined as follows:

$$l(\pi) = \left(0, \sum_{i=0}^{k-1} d(p_i, p_{i+1})\right),$$

where $p_{i+1} \in \Gamma(p_i)$. The function $l(\pi)$ is called **the length of flat path** π.

A drawback of the function $l(\pi)$ is that it does not depend on intensity values of pixels p_i lying at the path π and considers image f as being flat. Therefore, function $l(\pi)$ can not be used as a criterion to join pixels into one region based on their intensities.

Topographical distance takes care of relief of image [8]. Since path function should have maxima for pixels lying on edges between objects, **a gradient image** $|\nabla f|$ is often used instead of original one. Every pixel of the gradient image has intensity value equal to the absolute gradient value of the source image in this pixel. The value of absolute gradient of each pixel can be calculated based on discrete approximation. Later on the gradient image obtained from the source image f will be denoted by ψ.

The absolute gradient value can be approximated as follows:

$$\psi(p) = \max_{\gamma_p \in \Gamma(p)} \left\{ \left| f(p) - f(\gamma_p) \right| \right\}.$$

Watershed transformation of source image f can be computed in one of the following ways:

1. Transform source image f into gradient image ψ, and then make the watershed transform on gradient image ψ.
2. Make watershed transform directly on the source image f. The approximation of absolute gradient value $\psi(p)$ is calculated for each pixel p in the process of watershed transformation.

Regardless of the selected way the value $\psi(p)$ can be treated as an intensity of pixel p of the gradient image. Therefore one can build graph G of the image

ψ. All definitions that use image f are correct for gradient image ψ too. Taking this into account, we assume that graph G is built for image ψ unless otherwise specified.

Topographical length $l_\psi(\pi_k)$ of path π in G is defined as follows:

$$l_\psi(\pi_k) = \begin{cases} (0,0), & \text{if } k = 0, \\ \left(\max_{i=0,\dots,k}\{\psi(p_i)\}, \sum_{i=0}^{k-1} d(p_i, p_{i+1})\right), & \text{if } k > 0. \end{cases} \quad (1)$$

For $k > 0$ the expression (1) can be formulated as follows:

$$l_\psi(\pi_k) = \left(\max_{i=0,\dots,k}\{\psi(\pi_i)\}, 0\right) + l(\pi_k).$$

Gradient image usually has local maxima on object edges. At areas where maximal pixel intensity is constant, topographical length increases due to the second coordinate.

Topographical distance $T_\psi(p,q)$ between pixels $p, q \in X$ is equal to topographical length of the shortest path π from p to q in X. The lengths are compared in lexicographic sense.

The definition of topographical distance as a vector value allows to calculate correctly the distance between points in images with arbitrary intensities. The original definition of the topographical distance as a scalar value introduced in [8] was restricted to a class of images with non-marked pixels having at least one neighbor with lower intensity value. Such images do not contain non-minimal flat regions (plateaus).

As it was stated above, the watershed points are located on the same topographical distance from markers of at least two different objects. If topographical length l_ψ is used as path function, the watershed points are not always located on the place with maximal absolute gradient value indicating the edge of objects. Therefore we introduce a notion of a modified topographical distance which allows to create a set of watershed points located on the areas with maximal absolute gradient value.

Modified topographical length $l'_\psi(\pi_k)$ of path π_k is a path function defined as follows:

$$l'_\psi(\pi_0) = (0,0);$$
$$l'_\psi(\pi_1) = \begin{cases} (\psi(p_1), 0), & \text{if } \psi(p_1) > \psi(p_0); \\ (\psi(p_0), d(p_0, p_1)), & \text{if } \psi(p_1) \leqslant \psi(p_0). \end{cases}$$

And for $1 < i \leqslant k$:

$$l'_\psi(\pi_i) = \begin{cases} (\psi(p_i), 0), & \text{if } \psi(p_i) > \max_{j=0,\dots,i-1}\{\psi(p_j)\}; \\ l'_\psi(\pi_{i-1}) + (0, d(p_{i-1}, p_i)), & \text{if } \psi(p_i) \leqslant \max_{j=0,\dots,i-1}\{\psi(p_j)\}. \end{cases}$$

The expression for $l'_\psi(\pi_i), 0 < i \leqslant k$ can be rewritten in the following form:

$$l'_\psi(\pi_i) = \left(\max_{j=0,\dots,i}\{\psi(p_j)\}, 0\right) + l(p_i, p_{i-1}, \dots, p_{i-n}).$$

Here n is calculated as follows:

$$n = \max \left\{ t \mid 0 \leqslant t \leqslant i, \psi(p_{i-t}) = \max_{s=0,\dots,i} \{\psi(p_s)\} \right\}.$$

Hence, p_{i-t} is the first pixel of the path π_k which has maximal intensity value along the whole path.

For $0 < i \leqslant k$ the following relation is true:

$$C(p_i) = l'_\psi(\pi_i) = \begin{cases} (\psi(p_i), 0), & \text{if } \psi(p_i) > C_x(p_{i-1}), \\ C(p_{i-1}) + (0, d(p_{i-1}, p_i)), & \text{if } \psi(p_i) \leqslant C_x(p_{i-1}). \end{cases}$$

Modified topographical distance $T'_\psi(p,q)$ between arbitrary pixels p and q of graph G is equal to the modified topographical length of the shortest path π from p to q in G. Here like above the shortest path is treated in the lexicographic sense.

4 Algorithms Implementing Watershed Transformation

The algorithms implementing watershed transformation (see algorithms 1 and 2) define the shortest paths between arbitrary nodes of image graph using ordered queues [9,10]. Now let us define operations with ordered queues. The operation $Enqueue(p, C(p))$ pushes pixel p into the tail of queue with priority $C(p)$. The operation $DequeueMin()$ pops one pixel from the head of the first non-empty queue with minimal priority.

Let us discuss the algorithm of gray-scale image segmentation using a simple metric (the distances to adjacent pixels equal to 1). Priorities of the queues in this algorithm are scalars because ordered queues automatically sort pixels lying along every path by the second coordinate of their costs: the closer (in the sense of flat path) a pixel is from the object marker, the sooner it will be pushed into a queue with priority equal to the first coordinate of its cost, and the earlier it will be popped from the queue as compared to other pixels with the same priority.

The algorithm starts from the initialization step. At this step each marked pixel $p \in M$ gets a unique label $\lambda(p) > 0$ of the object it belongs to, and the cost $C_x(p) = 0$. After that the pixel is pushed into a queue with priority $C_x(p)$ (see lines 9-13 of algorithm 1). Each not marked pixel $p \notin M$ gets cost $C_x(p) = \infty$ and label $\lambda(p) = 0$. Not marked pixels are not pushed into queue (see lines 5-8). At the initialization step all pixels of graph G are considered to be non-examined, so they have flag $TEMP$ (see lines 2-4).

At the second step (propagation) one pixel p is popped from the head of the first non-empty queue with minimal priority by $DequeueMin()$ operation (see line 16). If it has flag $TEMP$ (not examined before), then it already has the minimal cost at the path from the marker, so it is now considered to be examined and gets flag $DONE$ (see lines 17-18). One distinction of the proposed algorithm from the Lotufo-Falcão algorithm [5] is that in the algorithm 1 each pixel being popped from the queue is checked to have flag $TEMP$. Therefore a not examined

Algorithm 1. Gray-scale image segmentation by watreshed transformation (using a metric defined by a unit neighborhood)

Require: Graph G of gradient image ψ; a set of markers M with indices in I_M. The priorities are scalars.

1: {Initialization}
2: **for all** $p \in X$ **do**
3: $\quad flag(p) \Leftarrow TEMP$;
4: **end for**
5: **for all** $p \notin M$ **do**
6: $\quad C_x(p) \Leftarrow \infty$;
7: $\quad \lambda(p) \Leftarrow 0$;
8: **end for**
9: **for all** $p \in M$ **do**
10: $\quad C_x(p) \Leftarrow 0$;
11: $\quad Enqueue(p, C_x(p))$;
12: $\quad \lambda(p_i) \Leftarrow i, i \in I_M, i > 0$;
13: **end for**
14: {Propagation}
15: **while** Queue not empty **do**
16: $\quad p \Leftarrow DequeueMin()$;
17: $\quad$ **if** $flag(p) == TEMP$ **then**
18: $\quad\quad flag(p) \Leftarrow DONE$;
19: $\quad\quad$ **if** $\lambda(p) \neq w$ **then**
20: $\quad\quad\quad$ **for all** neighbor q of p **and** $flag(q) == TEMP$ **do**
21: $\quad\quad\quad\quad C'_x(q) \Leftarrow \max\{C_x(p), \psi(q)\}$
22: $\quad\quad\quad\quad$ **if** $C'_x(q) < C_x(q)$ **then**
23: $\quad\quad\quad\quad\quad C_x(q) \Leftarrow C'_x(q)$;
24: $\quad\quad\quad\quad\quad \lambda(q) \Leftarrow \lambda(p)$;
25: $\quad\quad\quad\quad\quad Enqueue(q, C_x(q))$;
26: $\quad\quad\quad\quad$ **else**
27: $\quad\quad\quad\quad\quad$ **if** $C'_x(q) == C_x(q)$ **and** $\lambda(q) \neq \lambda(p)$ **then**
28: $\quad\quad\quad\quad\quad\quad \lambda(q) \Leftarrow w, w \notin I_M$;
29: $\quad\quad\quad\quad\quad$ **end if**
30: $\quad\quad\quad\quad$ **end if**
31: $\quad\quad\quad$ **end for**
32: $\quad\quad$ **end if**
33: $\quad$ **end if**
34: **end while**

pixel may have several "copies" in the queue. A pixel popped in that way will always have the minimal cost and, consequently, there is no necessity to check the queue for the presence of its "copy" each time it is pushed into the queue, as it is done in Lotufo-Falcão algorithm. If pixel p does not belong to a set of watershed points, then each its neighbor $q \in X$ having flag $TEMP$ is determined (see lines 19,20). Then the cost of each pixel q is calculated relative to the path passing through pixel p: $C'_x(q) = \max\{C_x(p), \psi(q)\}$ and if pixel q had cost $C'_x(q) < C_x(q)$, it gets a new cost $C_x(q) = C'_x(q)$ and label $\lambda(p) \in I_M$ showing that pixel q now belongs to the same object as pixel p. Besides, pixel q is pushed

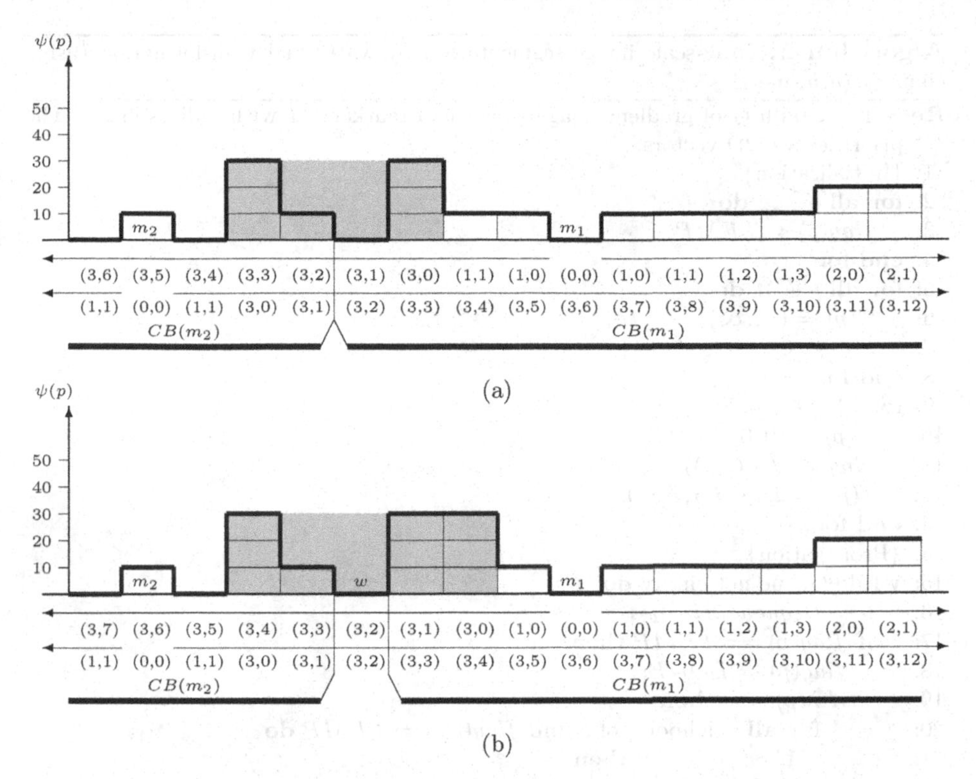

Fig. 1. Catchment basins for two selected markers using modified topographical distance T'_ψ. Gradient image ψ is shown by bold line. Modified topographical distance from marker m_1 to other pixels (top row) and from marker m_2 to other pixels (lower row) is shown in the brackets.

into the tail of queue with priority $C'_x(q)$ (see lines 21-25). If pixel q has been already pushed, the first "copy" of this pixel being popped from the queue has the least priority (and therefore the least cost). All other "copies" of this pixel having greater priority when popped from the queue will already have the flag $DONE$ and will not be examined (see line 17).

If pixel q is located on the same distance from markers of different catchment basins and has cost $C'_x(q) = C_x(q)$ and label $\lambda(p) \neq \lambda(q)$, then this means that the pixel belongs to a set of watershed points and gets new label $w \notin I_M$ (see lines 26-29). The algorithm finishes when the queue is empty. Each pixel p having label $\lambda(p) \in I_M$ belongs to the object corresponding to this label. If graph G is connected, all pixels not belonging to the watershed should have the label $\lambda(p) \in I_M$. If pixel p has label $\lambda(p) = w \notin I_M$, it belongs to the set of watershed points.

The set of watershed points, that is built by the proposed algorithm, not necessarily forms a closed contour lying at object edges (see fig. 1). If width of the segment of pixels with maximal cost (emphasized by gray color at the fig.1) is even, then there is no watershed point, since no pixels of the path lie on the

Algorithm 2. Gray-scale image segmentation by watreshed transformation (using chamfer (a, b)-metric)

Require: Graph G of gradient image ψ; a set of markers M with indices in I_M. The priorities are 2D vectors.

```
 1: {Initialization}
 2: for all p ∈ X do
 3:    flag(p) ⇐ TEMP;
 4: end for
 5: for all p ∉ M do
 6:    C(p) ⇐ (∞,∞);
 7:    λ(p) ⇐ 0;
 8: end for
 9: for all p ∈ M do
10:    C(p) ⇐ (0,0);
11:    Enqueue(p, C(p));
12:    λ(p_i) ⇐ i, i ∈ I_M, i > 0;
13: end for
14: {Propagation}
15: while Queue not empty do
16:    p ⇐ DequeueMin();
17:    if flag(p) == TEMP then
18:       flag(p) ⇐ DONE;
19:       if λ(p) ≠ w then
20:          for all neighbor q of p and flag(q) == TEMP do
21:             if ψ(q) > C(p) then
22:                C'(q) ⇐ (ψ(q),0);
23:             else
24:                C'(q) ⇐ C(p) + (0, d(p,q));
25:             end if
26:             if C'(q) < C(q) then
27:                C(q) ⇐ C'(q);
28:                λ(q) ⇐ λ(p);
29:                Enqueue(q, C(q));
30:             else
31:                if C'(q) == C(q) and λ(q) ≠ λ(p) then
32:                   λ(q) ⇐ w, w ∉ I_M;
33:                end if
34:             end if
35:          end for
36:       end if
37:    end if
38: end while
```

same distance from both markers m_1 and m_2 (see fig.1a). The path illustrated in the fig. 1b includes the watershed point lying on the same distance from the both markers. This feature of the algorithm makes information about watershed points almost useless for extraction of object contours. Watershed points are extracted merely to show that they may belong to any adjacent object. Therefore, user can put them to any object at his discretion using some extra information contained in the image. If watershed points are unnecessary, the algorithm can be modified not to extract them, but put them to any adjacent catchment basin. It can be done by just removing the lines 19, 26-29, 32 of the algorithm 1 and substituting ”$<$” by ”$\leqslant$” in the line 22.

Let us consider the algorithm of gray-scale image segmentation using chamfer (a, b)-metric (see fig. 2). In this case the priorities are vectors, because the distances to adjacent pixels are different and, therefore, different pixels get different second coordinate of the cost value. The segmentation algorithm (see algorithm 2) is similar to the algorithm 1. The only distinction is in using vector costs of pixels and increasing the second coordinate of the cost in accordance with chamfer (a, b)-metric.

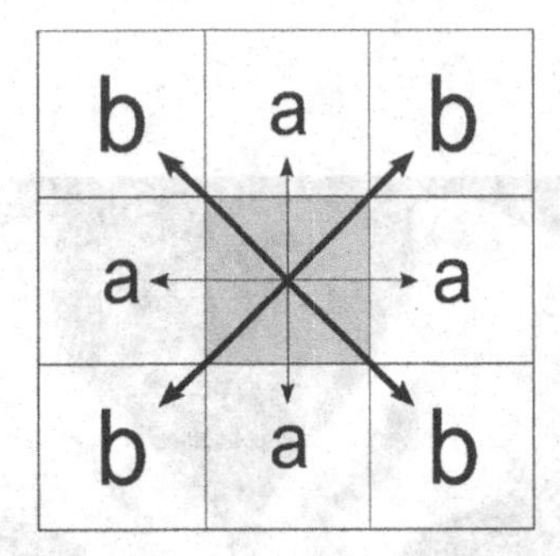

Fig. 2. The illustration of chamfer (a, b)-metric for two-dimensional rectangular lattice. Central pixel is emphasized with gray color. There are two groups of neighbor pixels: the distance to one group is a, and that to another group is b.

If watershed points are unnecessary, the algorithm 2 can be modified not to extract them, but put them to any adjacent catchment basin. It can be done by removing the lines 19, 30-33, 36 of the algorithm 2 and substituting ”$<$” by ”$\leqslant$” in the line 26.

5 Algorithm Discussion

Fig. 3 shows the results of a synthetic image segmentation containing an ellipsis against a homohenious background. This figure illustrates ”the reaction” of algorithms to ”the noise”. One can see that chamfer metric gives better results as compared with the metrics defined by a singular vicinity. Ideally the image should have been divided into halves.

Fig. 4 shows the result of segmentation of a 3D tomographic image of pelvis given by algorithm 1. Markers are selected automatically where intensities exceed some threshold value. The time of segmentation of the image with 512*512*136

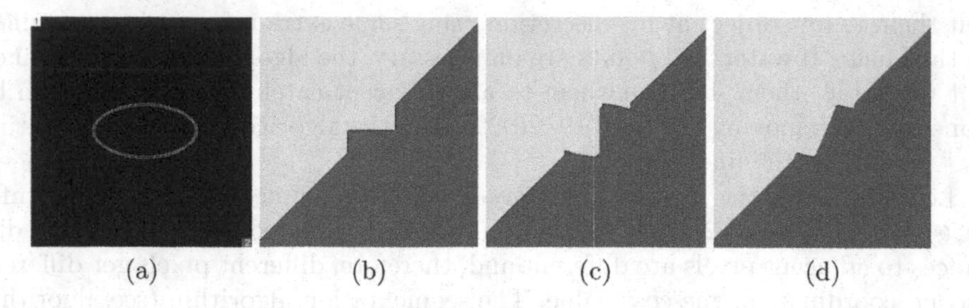

Fig. 3. The result of segmentation by the proposed algorithms
(a) Gradient image to be segmented. Markers are noted with numbers at image corners.
(b) Result of segmentation by algorithm 1 using "city block" metric.
(c) Result of segmentation by algorithm 1 using "chessboard" metric.
(d) Result of segmentation by algorithm 2 using chamfer (a, b)-metric with $a = 3$, $b = 4$.

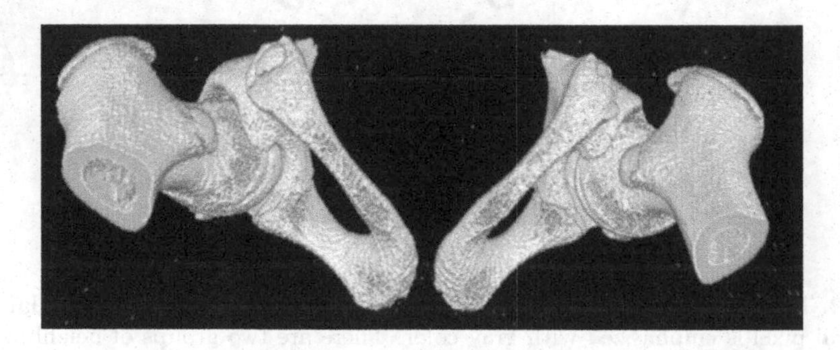

Fig. 4. CT image segmentation by a watershed algorithm

voxels, 16 bits per voxel by algorithm 1 without extracting watershed points is about 800ms. Not optimized algorithm 2 does it in about 4.5 seconds.

6 Conclusion

The paper formulates the notion of watershed transformation in terms of graph theory. Presented graph approach abstracts from specific metric and topology of the image, allowing to be applied to various data sets. The chamfer metric allows approximate more precisely Euclidean metric when calculating path lengths. Two algorithms of segmentation are presented: using a metric defined by singular neighborhood, and using chamfer (a, b)-metric. These algorithms use ordered queues permitting effective implementation. Unlike the similar algorithm proposed by Lotufo and Falcão, the presented in this paper algorithm gives more accurate result and allows extending for the Euclidean metric.

The work was partially carried out in the framework of INTAS N 04-77-7003 project.

References

1. Beucher, S., Lantuéjoul, C.: Use of Watersheds in Contour Detection. International Workshop on Image Processing, CCETT/IRISA (1979) 2.1–2.12
2. Beucher, S.: The Watershed Transformation Applied to Image Segmentation. Conference on Signal and Image Processing in Microscopy and Microanalysis (1991) 299–314
3. Vincent, L., Soille, P.: Watersheds in Digital Spaces: An Efficient Algorithm Based on Immersion Simulations. IEEE Trans. Pattern Analysis and Machine Intelligence, **13** (1991) 583–598
4. Beucher, S., Meyer, F., Dougherty, E. R.(ed.): The Morphological Approach to Segmentation: the Watershed Transformation. Mathematical Morphology in Image Processing. Marcel Dekker, New York (1993) 433–481
5. Lotufo, R., Falcao, A.: The Ordered Queue and the Optimality of the Watershed Approaches. In: Goutsias, J., Vincent, L., Bloomberg, D.S.(eds.): Mathematical Morphology and Its Applications to Image and Signal Processing, Vol. 18. Kluwer Academic Publishers, Boston Dordrecht London (2000)
6. Roerdink, J. B. T. M., Meijster, A.: The Watershed Transform: Definitions, Algorithms and Parallelization Strategies. Fundamenta Informaticae, **41** (2000) 187–228
7. Couprie, M., Bertrand, G.: Topological Grayscale Watershed Transformation. SPIE Vision Geometry V Proceedings, Vol. 3168 (1997) 136–146
8. Meyer, F.: Topographic Distance and Watershed Lines. Signal Processing, **38** (1994) 113–125
9. Cormen, T., Leiserson, C., Rivest, R., Stein, C.: Introduction to Algorithms. MIT Press, Cambridge (2001)
10. Moore, E.: The Shortest Path Through a Maze. Proceedings of an International Symposium on the Theory of Switching, Part II. Harvard University Press, Cambridge (1959) 285–292

Generalized Map Pyramid for Multi-level 3D Image Segmentation

Carine Grasset-Simon and Guillaume Damiand

SIC - Université de Poitiers
bât. SP2MI, Bvd M. et P. Curie
BP 30179, 86962 Futuroscope Chasseneuil Cedex - France
{simon, damiand}@sic.univ-poitiers.fr

Abstract. Graph pyramids are often used to represent an image with various levels of details. Generalized pyramids have been recently defined in order to deal with images in any dimension. In this work, we show how to use generalized pyramids to represent 3D multi-level segmented images. We show how to construct such a pyramid, by alternating segmentation and simplification steps. We present how cells to be removed are marked: by using an homogeneous criterion to mark faces and the cell degree to mark other cells. When the pyramid is constructed, the main problem consists in retrieving information on regions. In this work, we show how to retrieve two types of information. The first one is the low level cells that are merged into a unique high level cell. The second one is the inter-voxel cells that compose a given region.

Keywords: Irregular image pyramid, Inter-voxel elements, Generalized map, Hierarchical segmentation.

1 Introduction

To segment an image consists in partitioning this image into homogeneous connected regions given a criterion. A classical approach to region segmentation is the *split-and-merge* method and all its variants: a *bottom-up* approach [1,2] consists in taking small regions and merging them into bigger and bigger regions; a *top-down* approach [3, 4] is the opposite one, starting from big regions and cutting them into smaller and smaller regions; a *mixed* approach [5, 6] consists in combining the two previous ones.

For bottom-up approaches, it is important to be able to extract information on regions (for example mean, variance, ...), and to be able to retrieve adjacent regions of a given region. Graph based structures allow to retrieve such information, and this is why they are often used in image processing. However such structures have several drawbacks: they do not represent all the topological information and all the cells (the volumes representing the regions of the image and the faces, edges and vertices composing the region boundaries).

To solve these problems, structures based on combinatorial maps have been defined [7,8]. These structures have several advantages over graphs:

A. Kuba, L.G. Nyúl, and K. Palágyi (Eds.): DGCI 2006, LNCS 4245, pp. 530–541, 2006.

- they represent topological information, such as multi-adjacency or inclusion relations;
- they represent all the cells of the represented objects, and not only the regions as in the region adjacency graph;
- they allow to retrieve inter-voxel elements that composed the regions of the image and thus allow to compute geometrical features;
- they allow to compute topological characteristics of image regions.

Moreover, it is often necessary to represent a same image with different segmentation levels. For that, classical structures are extended in hierarchical structures [9, 10] in order to be able to represent a same object with different resolutions. In this work, we use a 3-G-map pyramid in order to deal with a 3D multi-level segmented image. Considered images are in grey level, and the segmentation method is a bottom-up approach based on a very simple criterion that uses the squared error.

The 3-G-map pyramid used in this work is similar to the one presented in [11]. With this structure, we represent various partitions of a same image as well as links between the levels in order to be able to run through the pyramid. Moreover, each cell and each adjacency and incidence relation are represented for each different level. This information allows multi-level operations, such as for example a local modification of a region at a given level, with propagations on neighbor levels in order to keep the coherence of the structure.

In order to compute topological or geometrical features on regions of the image, it is often necessary to retrieve:

- which regions of a fine segmentation (a segmentation composed of many small regions) were merged in a unique region at a higher level in the pyramid;
- inter-voxel elements in the initial image that composed a given cell of a given region in the pyramid (e.g. the voxels of a region or the surfels of a face).

In this work, we show that this information can be retrieved in the 3-G-map pyramid by using the notion of generalized cells (particular cases of generalized orbits defined in [12]). The main result of this work is the definition of the algorithms which allow to retrieve all the information concerning a given region.

This paper is organized as follows. Section 2 provides some recalls about pyramids of n-dimensional generalized maps and about the notion of generalized cells. In section 3 we examine the construction of the n-G-map pyramid representing different segmentation levels of a same image in grey level. In section 4 we show how to retrieve voxels and inter-voxel elements which compose a region at a higher level. Conclusion and further issues are discussed in section 5.

2 Recall: Pyramid of n-Dimensional Generalized Maps and Generalized Cells

An *n-dimensional generalized map* (n-G-map) allows to represent the topology of n-dimensional objects. For example a 3-G-map can represent the topology of a 3D image. An n-G-map is a set of abstract elements, called darts, together with $n+1$ involutions defined on this set of darts (an involution f on a finite set S is a one to one mapping from S onto S such as $f = f^{-1}$). Each involution α_i represents an adjacency relation between i-dimensional cells (c.f. Fig. 1 and definition in [13]).

The different *cells* of an image such as pointel, linel, surfel and voxel (or in general vertex, edge, face and volume corresponding to 0, 1, 2, and 3-dimensional cells) are implicitly represented as subset of darts by using the orbit notion. Intuitively, an orbit $< f_1, \dots f_k > (d)$ is the set of darts that we can reach by a breath first search algorithm starting from d and using any application f_i or f_i^{-1}. Each i-cell is defined by a particular orbit in n-G-maps, using all the involutions except α_i (see [13] for definition of i-cells).

The *degree* of an i-cell c is the number of distinct $(i+1)$-cells incident to c. For example, the degree of a vertex v is the number of edges incident to v. The *local degree* of an i-cell c is similar to the degree but computed locally without to run through the $(i+1)$-cells incident to c. So if an incident $(i+1)$-cell is incident to c twice, it is considered as two $(i+1)$-cells when we compute the local degree of c.

The operation of *cell removal* (defined in [14]) removes simultaneously various cells in an n-G-map. These cells can be removed if they respect two preconditions: they have to be disjoint, and the local degree of each cell is two. For removing an i-cell c, we delete the darts which form this cell and for each surviving neighbor dart of c, we redefine the value of involution α_i applied to c (see Fig. 1) in order to jump over the removed cells.

A *pyramid of n-G-maps* is a hierarchical data structure composed of several n-G-maps, where each map is a reduction deduced from the previous map (cf. de-

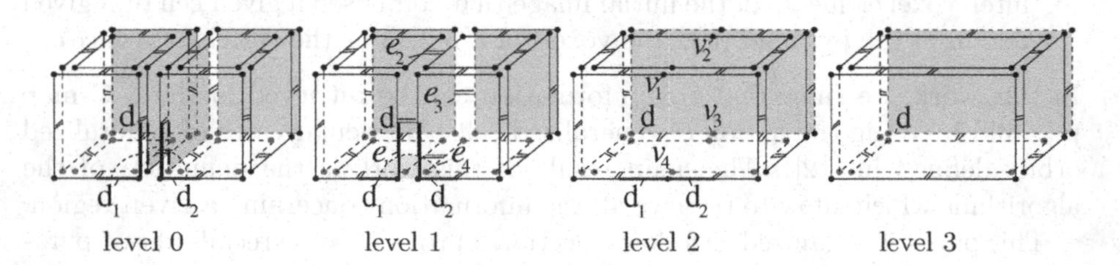

Fig. 1. Example of a 3-G-map pyramid composed of four levels. At each level darts are represented by black segments. Two darts linked by α_i are in the same j-cell (with $j \neq i$) but in two adjacent i-cells. In this pyramid, level 0 is composed of 2 volumes, level 1 is obtained by removing the face between the two volumes, level 2 is obtained by removing edges e_1, e_2, e_3 and e_4, and level 3 is obtained by removing degree two vertices v_1, v_2, v_3 and v_4. In level 2, the edges incident to d_1 and d_2 are adjacent and so linked by α_1. This link is deduced from those of level 1 which allow to go from d_1 to d_2 by jumping removed edge e_3. This principle is recursive. Generalized face $CG_{(2,0,3)}(d)$ (resp. $CG_{(2,1,3)}(d)$ and $CG_{(2,2,3)}(d)$) corresponding to the grey face incident to dart d at level 3 corresponds to the set of grey faces at levels 0 (resp.1 and 2).

finition in [11]). In the particular case of a region growing segmentation method, each n-G-map is deduced from the previous level by applying the general operation of cell removal. The choice of cells to remove depends on the application and is the result of an external process. Fig. 1 illustrates an example of a 3-G-map pyramid composed of four levels. An important property of this pyramid is that a one to one mapping exists between the surviving darts of a level (the darts which are not marked to remove) and the darts of the following level.

With an n-G-map pyramid it is possible to represent geometrical information by adding attributes on cells such as vertex coordinates, or region colors... Then, it is necessary to be able to retrieve this information which is generally kept in the initial level.

The notion of *generalized cell* (particular case of generalized orbit defined in [12]) allows to retrieve the set of the i-cells of a level which are concerned in the formation of an i-cell c given at a higher level. This set of i-cells is composed of:

- i-cells which have been merged into c by removing incident $(i-1)$-cells;
- i-cells incident to these $(i-1)$-cells, which have been removed.

We note $CG_{(i,l',l)}(d)$ the generalized i-cell at level l' that corresponds to the i-cell incident to dart d at level l. This generalized cell is a union of i-cells at level l' (see Fig. 1). The generalized cells of dimension 0, 1, 2 or 3 are respectively named generalized vertices, edges, faces or volumes.

3 Presentation of the Pyramid

3.1 Choice of the Structure

In this work, we use a pyramid of 3-G-maps in order to realize a multi-level segmentation of a 3D image. Indeed this structure have several advantages.

First, n-G-maps are defined for any dimension of the space, their definition is homogeneous for all dimensions and so the operations defined above them are generic (in particular cell removals). They allow to represent all the cells of an image (not only the regions and their boundaries) as well as the incidence, adjacency and inclusion relations.

Second, a pyramidal structure allows to keep in memory various segmentation levels of the same image and so to work at the best level according to each operation. Since the levels of such a structure are linked between them, it allows to work simultaneously at various levels or to retrieve the set of regions of a fine segmentation which have been merged into a region of a coarse segmentation.

To facilitate the retrieval of information in the pyramid, it is important to simplify each segmentation level. This type of simplification is often used in 2D with dual graph pyramids [15] and combinatorial pyramids [16]. To add a new segmentation level, we propose to use three different steps:

- first, the merge of similar adjacent regions of the previous level;
- second, a first simplification of the boundary of each region by merging the adjacent faces incident to two same regions;

– third, the second simplification of the boundary of each region by merging the adjacent edges incident to two same faces.

In the pyramid, these three steps are achieved by removing respectively the faces, edges or vertices between the elements to merge. They are successively applied and as many times as necessary. To represent a new segmentation level, three pyramid levels are constructed and noted level 0, 1 and 2 (mod 3).

3.2 Construction of the Pyramid

The construction of a new 3-G-map is achieved by applying a removal kernel (see [17] for all algorithms constructing a pyramid) based on the following principle:

– first, the cells to remove are marked. The choice of these cells depends on the level we want to add. For example, to construct a level 0 (mod 3), the faces separating two homogeneous regions according to a segmentation criterion are marked;
– second, the new G-map is constructed by copying each surviving dart of the previous level and by linking them by taking into account disappeared darts.

For all the levels, the method used to add a new level is the same. The unique difference existing between the levels is the criterion used to mark the cells.

Marking Faces to Remove. Two adjacent regions have to be merged if their union is homogeneous according to a criterion. In our application, the homogeneity of a region R is measured with the squared error which corresponds to the sum of the square distance of each grey level to the mean grey level ν of R. This criterion can be formulated with the moments of order zero, one and two of a region:

$$EQ(R) = M_2(R) - M_0(R)\nu(R)^2$$

where $M_0(R)$ is the number of voxels contained in R, $M_1(R)$ is the sum of voxel grey levels of R, $M_2(R)$ is the sum of squared voxel grey levels of R, and $\nu(R) = \frac{M_1(R)}{M_0(R)}$. Since the moments of the union of two regions can be incrementally computed from the moments of both regions, the squared error can be efficiently updated.

Two regions have to be merged if their union is homogeneous i.e. the squared error of their union is inferior to a threshold T. Otherwise, their union is non-homogeneous and both regions should not be merged.

To obtain coarser segmentations when we go up in the pyramid, the threshold have to be increased with the levels ($T^l > T^{l-1}$ with l the level we want to add). It is possible to compute a new threshold by multiplying the previous threshold by a constant k ($T^l = T^{l-1} \times k$), or to use a different formula like $T^l = (T^{l-1})^2$ (if $T^{l-1} > 1$) depending on how we want the segmentation to evolve.

To merge all homogeneous regions, the algorithm scans twice the 3-G-map:

– it considers each face of the G-map and marks it to remove if it separates two similar regions according to the homogeneity criterion;

- it considers each not marked face and marks it if it separates two regions which will be merged in the next level. This step is necessary to avoid inner faces (i.e. faces inside a region).

Since face removal can lead to volume disconnection, a region inclusion tree is added to each level (see [18, 19] for more details on disconnection problems and possible solutions).

Marking Edges to Remove. In the first step of the boundary simplification, the edges separating two faces between the same two regions (i.e. local degree two edges) are removed. Edge removal can lead to face disconnection and object disappearance (for example in Fig. 2-b the removal of edge e_3 leads to a face disconnection, and in Fig. 2-d the removal of edge e_5 removes the representation of the cube). A way to solve this problem is not to mark an edge if its removal leads to a disconnection or a disappearance. Algo. 1 considers successively each edge and marks it if:

- its local degree is two;
- its removing does not lead to disappearance;
- its removing does not lead to disconnection.

The two first points are realized by a direct test (achieved in $\mathcal{O}(1)$), and the third point is realized by running through an orbit (cost $\mathcal{O}(f)$ with f the number of darts of the face). Note that this last test can be optimized by using a union-find tree.

When an edge is marked to remove, the algorithm reconsiders possibly incident edges. For example, in Fig. 2 we want to simplify the boundary of the region represented by the cube. If we consider edge e_3 before e_4, we cannot mark it since its removal leads to a face disconnection. Then, when we consider e_4, it is marked to remove, and we can see that edge e_3 can now be marked to remove without face disconnection. This is the reason why e_3 needs now to be reconsidered. Note that this case occurs only when the marked edge is adjacent to a unique not marked edge. To solve this case, we just test both extremities of the current edge after it was marked to remove. If a single not marked edge is incident to an extremity, this edge is added to a list of edges to reconsider. Note that an edge can be treated at most twice since an already treated edge can be reconsidered only when it becomes the unique edge incident to a vertex.

Marking Vertices to Remove. In the second step of the boundary simplification, the vertices separating two edges between the same two faces (i.e. degree two vertices) are removed. In this step, it is important to test the degree and not the local degree as for edges since the removal of a degree two vertex does not lead to topological modification while the removal of a local degree two vertex can lead to the disappearance of the object (in the case where the vertex is only incident to a loop). Note that the vertex removal cannot lead to a disconnection, and so this step is achieved directly without problem. Each vertex is successively considered and marked to remove if its degree is two.

Algorithm 1. Edge marking.

```
Data: G: a 3-G-map,
Result: G in which edges to remove are marked.
e ← an edge of G ;
while e ≠ null do
    // processing of edge e
    if the local degree of e is 2 then
        if the removing of e leads neither to disconnection nor disappearance
        then
            mark e to remove ;
            foreach vertex v incident to e do
                if it exists a single not marked edge e' incident to v then
                    add e' in list_edge_to_treat ;
    // choice of the next edge to treat
    if list_edge_to_treat is not empty then
        e ← the first of list_edge_to_treat ;
    else e ← an edge of G not yet treated;
```

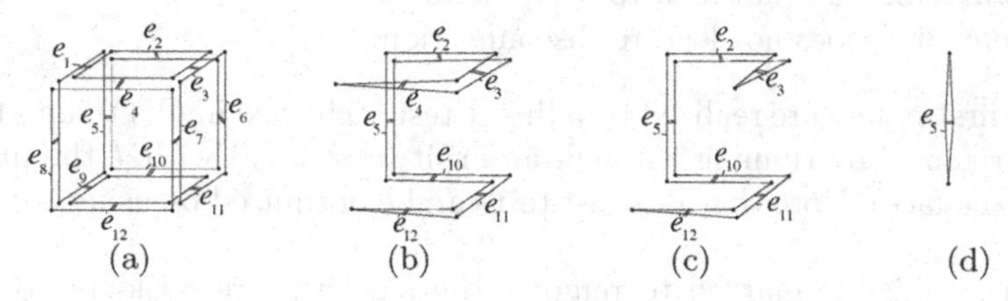

Fig. 2. Example of boundary simplification for a cube. (a) The initial 3-G-map representing a cube with a boundary composed of six faces. (b) The 3-G-map obtained by removing local degree two edges e_1, e_6, e_7, e_8 and e_9 from (a). (c) The 3-G-map obtained by removing edge e_4 from (b). (d) The 3-G-map composed of two vertices, one edge and one face, which is obtained after simplification of the boundary.

3.3 The First Level

The whole pyramid is built starting from a first level, and thus the question concerning the definition of this first level is important. Indeed, this level can either represent each voxel of the image or represent a fine segmentation of this image. Note that, in the first case, the best adapted structure in order to represent a regular subdivision seems to be a matrix. For our application, we have chosen the second possibility since the initial image is not the reality but a discretization of it, and so it can contain noise. Moreover, in image analysis it is usual to make a pre-segmentation before other steps.

To compute the first segmentation starting from the image, we have used here a semi-supervised classification based on a histogram analysis, but any method can be used. This first level is built in two scans of the image:

- a first scan, in order to construct the histogram, and then the classes;
- a second scan, in order to construct incrementally the 3-G-map of the first pyramid level. In this step, the voxels are added one by one to the G-map and merged with these neighbor voxels if they are similar and already constructed.

After this first pyramid level, the construction of the next levels follows the principle explained before: add both simplification levels and possibly other segmentation levels until the desired result (or an image composed of a unique region).

4 Retrieving Regions and Inter-region Elements

When we keep in memory various segmentation levels of a same image, we often want to work simultaneously at various levels, or we want to run through the same object at various levels. For a given cell c at a level, it is necessary to be able to retrieve information: the cells of a lower level which have been merged into c, and the inter-voxel elements which represent c in the image. With this information, we are for example able to:

- modify the segmentation of a part of the image (at a given level) without reconstruct all the levels. For that we modify a given region and propagate the modifications in all the pyramid but only for concerned regions;
- compute geometric criteria, for example characteristics of face curvature by using the surfels which compose it.

The notions of cells and generalized cells defined in an n-G-map pyramid allow to retrieve this information.

4.1 Retrieving the Cells

Given a cell c at a particular level l, we want to retrieve all the cells at a lower level that are merged into c at level l.

Volumes. For a given volume V, we want to retrieve the volumes at a lower level which have been merged into V. The idea is to use generalized volumes. When there is no disconnection, the result is directly given by the generalized volume computed between the two considered levels. Otherwise, we need to make the union of generalized volumes for each boundary of each volume obtained by the initial generalized volume (each volume is represented by an external boundary and possibly several internal boundaries, one for each cavity).

Algo. 2 gives the set of volumes at a given level which have been merged into a volume V at a higher level. First, it computes the generalized volume representing V at the lower level. The obtained set corresponds to the external boundary of V. Second, the generalized volumes are computed for all the internal boundaries.

Algorithm 2. Retrieve volumes(d, l, l').

Input: d: a dart of a volume V ;
l: the level containing V ;
l': the level where we want to retrieve the volumes merged into V ;
Output: the set of volumes of level l' which have been merged into V ;
$Res \leftarrow$ the volumes incident to $CG_{(3,l',l)}(d)$;
foreach *volume* $V' \in Res$ **do**
  foreach *internal boundary* B *of* V' **do**
    $d' \leftarrow$ a dart of B ;
    $Res \leftarrow Res \cup CG_{(3,l',l)}(d')$;
return Res ;

Faces. For a given face F, we want to retrieve the faces at a lower level which have been merged into F. By using generalized faces, we obtain, by definition, the faces which were concerned in the formation of F. As we can see in Fig. 3, this does not give immediately the expected result since we obtain too many faces. To solve this problem, we need to progressively go down in the pyramid

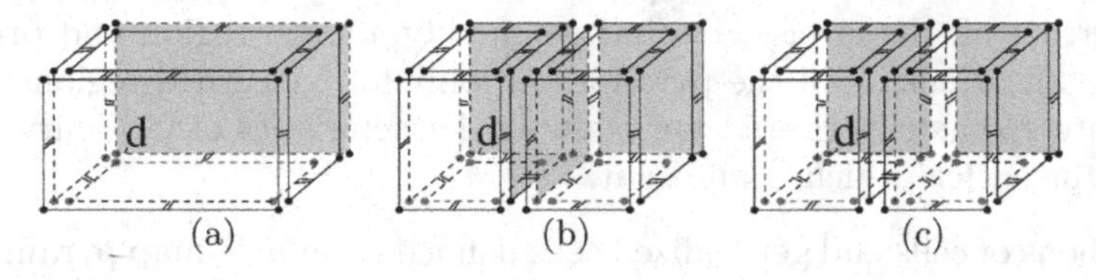

Fig. 3. Comparison between a generalized face and faces merged into a face. (a) Level 3 of the pyramid of Fig. 1. Face F is in grey. (b) and (c) Level 0 of the pyramid where grey faces respectively correspond to $CG_{(2,0,3)}(d)$ and the ones merged into face F.

and add different cells depending on the current level. Algo. 3 computes the set of faces of a given level l' which have been merged into face F incident to dart d at level l. Its principle is the following:

- if $l \equiv 1$ or $2 \pmod 3$, the set of faces which have been merged at the previous level in order to form F is the set of faces given by $CG_{(2,b-1,b)}(d)$. Indeed, the unique operations used to construct levels 1 or 2 (mod 3) are the edge and vertex removal but not face removal. So by definition, the generalized face gives us directly the set of faces which have been merged into face F;
- otherwise, $l \equiv 0 \pmod 3$. In this case, we cannot use the generalized face since this level is obtained from the previous one by removing faces. Since between both levels, only faces have disappeared, the surviving faces have not been modified. Thus, to obtain the face corresponding to F at the previous level, we only use the existing links between the darts of F and the darts of the previous level.

The algorithm stops when the current level is l' and in this case there is only one face incident to dart d.

Algorithm 3. Retrieve faces(d, l, l').

Input: d: a dart of a face F ;
l: the level containing F ;
l': the level where we retrieve the faces merged into F ;
Output: the set of faces of level l' which have been merged into F ;
if $l = l'$ **then** return the face incident to d ;
if $l \equiv 1$ *or* 2 (mod 3) **then**
  $F' \leftarrow$ the faces incident to $CG_{(2,l-1,l)}(d)$;
else $F' \leftarrow$ the face at level $(l-1)$ which corresponds to face F ;
foreach *face* $f \in F'$ **do**
  $d' \leftarrow$ a dart of f ;
  $Res \leftarrow Res \cup$ Retrieve faces($d', l-1, l'$) ;
return Res ;

Edges. For a given edge E, we want to retrieve the edges at a lower level which have been merged into E. We use exactly the same principle than for faces. We use the generalized edge when the current level is 0 or 2 (mod 3) since the unique operation used is the face removal or vertex removal, and we use the links existing in the pyramid for level 1 (mod 3) in order to directly retrieve the corresponding edge in the previous level.

Vertices. Since vertices cannot be merged in G-map pyramids, to retrieve the set of vertices of a given level l' which have been merged into a vertex V of a higher level l, amounts to take the unique vertex V' at level l' which corresponds to V. In order to do that, we only use the bijective links allowing to go down directly to level l', and then use the classical orbit notion in this level.

4.2 Retrieving the Inter-voxel Elements

To retrieve the voxels of the image which have been merged into a given region, or to retrieve the inter-voxel elements of an image which represent the boundary of a given region, we use the previous algorithms.

To retrieve the **surfels**, **linels** and **pointels** which respectively represent the faces, edges and vertices of the boundary of a region, it is enough to directly apply corresponding algorithms of the previous section between level l of the region and level 0. Indeed surfels, linels and pointels are directly represented in our level 0 G-map.

To retrieve the **voxels** which have been merged into a given region R, we need to:

- retrieve the regions of level 0 which have been merged into R by using Algo. 2;
- retrieve the surfels of these regions by running through their boundaries;
- use a classical flood-fill algorithm in order to reconstruct the voxels.

Retrieving voxels is more complex than for inter-voxel elements since voxels are not represented explicitly in the first pyramid level.

5 Conclusion and Perspectives

In this paper, we have presented the construction of a 3-G-map pyramid in the framework of multi-level segmentation of a 3D grey level image. Each new segmentation level is deduced from the previous level in the pyramid by applying a particular removal kernel which can use any homogeneous criterion. In order to facilitate the information retrieval, this level is simplified, and thus each new segmentation is represented by three successive levels in the pyramid. The first level is the new segmentation. The second level is obtained by removing all the degree two edges, and the third level is obtained by removing all the degree two vertices. Additional constraints are added in order to guaranty that no adjacency or incidence relation is lost during the simplifications.

We have shown how information can be retrieved in such a pyramid. We have given algorithms that allow to retrieve, given a region of a particular level, any cells that composed this region in a lower level. The methods used in these algorithms are based on the generalized orbit notion, and on the links between successive levels of the pyramid. When we are able to retrieve any cells between any levels, it is then easy to retrieve inter-voxels elements since it is just a particular case where the considered level is the first level of the pyramid.

Now, we want to study if it is possible to optimize the construction of the pyramid in order to keep only one level for each new segmentation. This construction is theoretically possible since the operation which removes simultaneously cells of various dimensions is defined in [14]. But we need to study how the generalized orbits can be used in such a case. Moreover, we are working to conceive operations for handling this pyramid. A first interesting operation consists in locally modifying a region at a given level and propagating the modifications without re-computing all the levels. After that, many other operations can be considered in order to propose a whole framework image processing.

Acknowledgements

The authors wish to thank Pascal Lienhardt for its encouragements and help.

References

1. Brice, C., Fennema, C.: Scene analysis using regions. Artificial Intelligence **1**(3-4) (1970) 205–226.
2. Fiorio, C., Gustedt, J.: Two linear time union-find strategies for image processing. Theoretical Computer Science **A: 154**(2) (1996) 165–181.
3. Lee, C.: Recursive region splitting at hierarchial scope views. Computer Vision, Graphics, and Image Processing **33**(2) (1986) 237–258.

4. Ohlander, R., Price, K., Reddy, R.: Picture segmentation by a recursive region splitting method. Computer Graphics and Image Processing **8** (1978) 313–333.
5. Pietikainen, M., Rosenfeld, A., Walter, I.: Split-and-link algorithms for image segmentation. Pattern Recognition **15**(4) (1982) 287–298.
6. Cheevasuvit, F., Maitre, H., Vidal-Madjar, D.: A robust method for picture segmentation based on a split-and-merge procedure. Computer Vision, Graphics, and Image Processing **34**(3) (1986) 268–281.
7. Braquelaire, J., Brun, L.: Image segmentation with topological maps and interpixel representation. Journal of Visual Communication and Image Representation **9**(1) (1998) 62–79.
8. Damiand, G., Bertrand, Y., Fiorio, C.: Topological model for two-dimensional image representation: definition and optimal extraction algorithm. Computer Vision and Image Understanding **93**(2) (2004) 111–154.
9. Montanvert, A., Meer, P., Rosenfeld, A.: Hierarchical image analysis using irregular tessellations. PAMI **13**(4) (1991) 307–316.
10. Kropatsch, W., Macho, H.: Finding the structure of connected components using dual irregular pyramids. In: Discrete Geometry for Computer Imagery, Clermont-Ferrand, France (1995) 147–158.
11. Grasset-Simon, C., Damiand, G., Lienhardt, P.: Pyramids of n-dimensional generalized maps. In: proceedings of 5th Workshop on Graph-based Representations in Pattern Recognition. Number 3434 in LNCS, Poitiers, France (2005) 142–152.
12. Grasset-Simon, C., Damiand, G., Lienhardt, P.: Receptive fields for generalized map pyramids: the notion of generalized orbit. In: Discrete Geometry for Computer Imagery. Number 3429 in LNCS, Poitiers, France (2005) 56–67.
13. Lienhardt, P.: N-dimensional generalized combinatorial maps and cellular quasi-manifolds. In: International Journal of Computational Geometry and Applications. (1994) 275–324.
14. Damiand, G., Lienhardt, P.: Removal and contraction for n-dimensional generalized maps. In: Discrete Geometry for Computer Imagery. Number 2886 in Lecture Notes in Computer Science, Naples, Italy (2003) 408–419.
15. Kropatsch, W.: Building irregular pyramids by dual-graph contraction. IEE Proceedings Vision, Image and Signal Processing **142**(6) (1995) 366–374.
16. Brun, L., Kropatsch, W.: Receptive fields within the combinatorial pyramid framework. In: Discrete Geometry for Computer Imagery. Number 2301 in LNCS, Bordeaux, France (2002) 92–101.
17. Simon, C., Damiand, G., Lienhardt, P.: nd generalized map pyramids: definition, representations and basic operations. Pattern Recognition **39**(4) (2006) 527–538.
18. Damiand, G.: Définition et étude d'un modèle topologique minimal de représentation d'images 2d et 3d. Phd thesis, Université Montpellier II, France (2001)
19. Damiand, G., Resch, P.: Split and merge algorithms defined on topological maps for 3d image segmentation. Graphical Models **65**(1-3) (2003) 149–167.

Topologically Correct Image Segmentation Using Alpha Shapes

Peer Stelldinger, Ullrich Köthe, and Hans Meine

University of Hamburg, 22527 Hamburg, Germany

Abstract. Existing theories on shape digitization impose strong constraints on feasible shapes and require error-free measurements. We use Delaunay triangulation and α-shapes to prove that topologically correct segmentations can be obtained under much more realistic conditions. Our key assumption is that sampling points represent object *boundaries* with a certain maximum error. Experiments on real and generated images demonstrate the good performance and correctness of the new method.

1 Introduction

A fundamental question of image analysis is how closely a computed image segmentation corresponds to the underlying real-world partitioning. Existing geometric sampling theorems are limited to binary partitionings, where the plane is split into (not necessarily connected) fore- and background components. In this case, the topology of the partition is preserved under various discretization schemes when the original regions are r-regular and the sampling grid has a maximum pixel radius of $r' \leq r$ [1, 2]. By making slighlty stronger assumptions, this property is preserved when the shapes are blurred by a disc or square of radius p prior to discretization [3, 4] or when regions may have corners [5].

However, these theorems have two important limitations: they are not applicable when there are more regions than just fore- and background, and they do not make any predictions about the consequences of measurement errors. One reason for these limitations is the assumption of a fixed sampling grid. We are dropping this assumption in favour of *adaptive sampling* where sampling points are placed roughly along the contour of the regions to be segmented.

Our treatment of adaptively placed sampling points is inspired by research on laser range scanning. Here, a number of isolated sampling points is scattered over the surface of the object of interest, and the task is to reconstruct the surface from the set of points. A successful solution of this problem is the concept of *α-shapes* [6, 7]. Under certain conditions, an α-shape is homeomorphic or at least homotopy equivalent to the desired object surface. By applying this idea to the problem of image segmentation, we are able to derive a new condition on object shape that ensures homotopy equivalence of the digital segmentation with the original analog plane partitioning. This means in particular that there is a 1-to-1 mapping between the computed and the ground-truth regions. By imposing slightly stronger requirements on region shape, these properties can even be guaranteed when the segmentation is subject to measurement errors.

A. Kuba, L.G. Nyúl, and K. Palágyi (Eds.): DGCI 2006, LNCS 4245, pp. 542–554, 2006.

2 Preliminaries

We consider the task of reconstructing a partition of the Euclidean plane from a sampled representation. The plane partition to be recovered is defined as follows:

Definition 1. *A* partition of the plane $\mathbb{R}^2$ *is defined by a finite set of* points $P = \{p_i \in \mathbb{R}^2\}$ *and a set of pairwise disjoint* arcs $A = \{a_i \subset \mathbb{R}^2\}$ *such that every arc is a mapping of the open interval* $(0,1)$ *into the plane, the start and end points* $a_i(0)$ *and* $a_i(1)$ *are in* P *(but not in* a_i*). The union of the points and arcs is the* boundary *of the partition* $B = P \cup A$*, and the* regions $R = \{r_i\}$ *are the connected components (maximal connected sets) of the complement of* B.

The partition is called *binary* when we can assign two labels (foreground and background) to the regions such that every arc is in the closure of exactly one foreground and one background region. A binary partition is called *r-regular*, when at every boundary point there exist two osculating discs of radius r which are entirely in the foreground and background respectively [2, 4]. This implies that regions cannot have corners, and junctions of three or more regions are impossible. These restrictions are somewhat relaxed by the notion of *r-halfregular partitions*, where an osculating r-disc must exist at least in the foreground *or* the background, and the number of regions must not change under either morphological opening or closing with a disc of radius $\leq r$ [5]. Corners are now possible, but the partition is still binary and has no junctions. The two notions of r-regularity and r-halfregularity have been central to all existing geometric sampling theorems. In this paper, the class of feasible plane partitions is extended as follows:

Definition 2. *A plane partition is* r-stable *when its boundary can be dilated with a closed disc of radius* s *without changing its homotopy type for any* $s \leq r$.

In other words, we can replace an infinitely thin boundary with a strip of width $2r$ such that the number and enclosure hierarchy of the resulting regions is preserved. In particular, "waists" are forbidden, whereas junctions are allowed, see Fig. 1. This includes r-regular and r-halfregular partitions, but also allows non-binary partitions and junctions and models real images much better. In particular, *polygonal partitions* (all arcs are straight lines) are always r-stable for some sufficiently small r. Unfortunately, the traditional way of proving a geometric sampling theorem (using a fixed grid at arbitrary position and angle, in connection with subset or supercover digitization) does not work for these partitions because topological equivalence cannot be guaranteed in general. Therefore, we consider another approach to digitization: we approximate the *boundary* of the partition with a finite set of *adaptively placed* sampling points. The sampling points are selected somehow "near" the boundary. We formalize this as follows:

Definition 3. *A finite set of sampling points* $S = \{s_i \in \mathbb{R}^2\}$ *is called a* (p,q)-sampling of the boundary B *when the distance of every point* $b \in B$ *to the nearest point in* S *is at most* p*, and the distance of every point* $s \in S$ *to the nearest point in* B *is at most* q*. The elements of* S *are called* edgels*. The sampling is said to be* strict *when all edgels are exactly on the boundary, i.e.* $q = 0$.

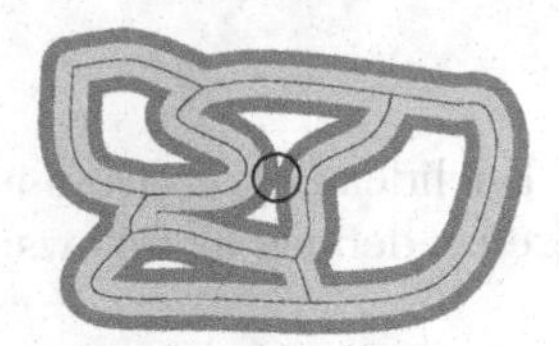

Fig. 1. The homotopy type of an r-stable plane partition does not change when dilated with a disc of radius of at most r (light gray), while dilations with bigger radius (dark gray) may connect different arcs as marked by the circle

The Hausdorff distance between the boundary and its sampling is $d_H(S, B) = \max(p, q)$. Obviously, $q < p$ is required because S is finite. Non-zero edgel shifts $q > 0$ can be caused by systematic or statistical measurement errors. Edgels may be determined in various ways (section 4), but this only matters in so far as it determines the accuracy of the sampling, i.e. the values of p and q. Once computed, we consider edgels as isolated points that somehow define the digital boundary and connect them by means of the Delaunay triangulation:

Definition 4. *The* Delaunay triangulation D *of a set of points* S *is the set of all triangles formed by triples* $t \subset S$ *such that the open circumcircle of every triangle does not contain any point of* S*. If the points are in general position, the Delaunay triangles, their edges and corners (also denoted as 2-, 1- and 0-cells in this context) form a uniquely defined, connected simplicial complex. The union of all cells* $|D| = \bigcup_{c \in D} c$ *is called the* polytope of D.

In order to approximate the boundary of the partition, we want to remove those edges and triangles from the Delaunay triangulation that are not related to the boundary. A useful subset is defined by the α-complex introduced in [6]:

Definition 5. *The* α-complex D_α *of a set of points* S *is the subcomplex of the Delaunay triangulation* D *of* S *which contains all cells* c *such that (a) the radius of the smallest open circumcircle of* c *is smaller than* α*, and this circle contains no point of* S*, or (b) an incident cell* c' *with higher dimension is in* D_α*.*

The polytope $|D_\alpha|$ is called *α-shape*. Since cells are removed from the Delaunay triangulation, the α-complex has holes which hopefully correspond to the regions we are trying to segment. In order to determine when this is the case, the following theorem is of fundamental importance (the proof can be found in [7]):

Theorem 1 (Edelsbrunner). *The union of closed* α*-discs with centers at the points* $s_i \in S$ *covers* $|D_\alpha|$*, and the two sets are homotopy equivalent.*

Consequently, the α-shape $|D_\alpha|$ is homotopy equivalent to the original plane partition if and only if the dilation of the edgels with α-discs is homotopy equivalent to the boundary of the partition. This requirement is indeed fulfilled in certain situations: In [8] it is proved that $|D_\alpha|$ is even homeomorphic to B if B is the boundary of an r-regular set with $p < \alpha < r$ and $q = 0$. Unfortunately, this no longer applies when the original partition is not r-regular and/or the edgels are not exactly on the original boundary. Fig. 2 shows an example where the r-dilation of the boundary is homotopy equivalent to the boundary (i.e. the partition is r-stable), but the dilation of the edgels is not. The rest of the paper is devoted to the question what can be said under these more general conditions.

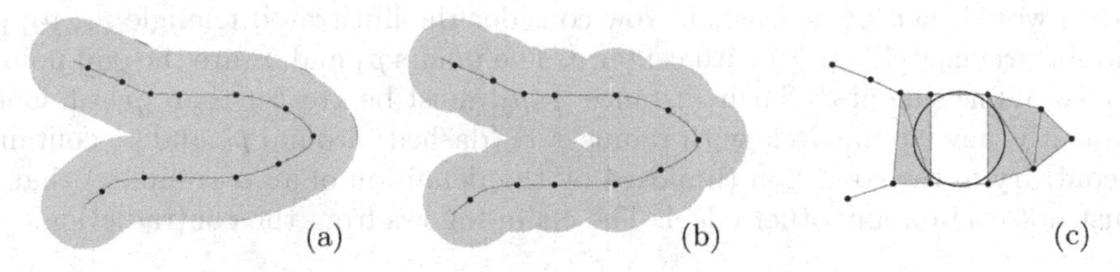

Fig. 2. The α-dilation (a) of the boundary of an α-stable plane partition may not be homotopy equivalent to the union (b) of the α-discs centered at the edgels. Thus the α-shape (c), which is always homotopy equivalent to the union of discs (b), may contain unwanted holes consisting of Delaunay triangles of radius greater than α. Thus there exists an α-disc centered in the hole which does not cover any edgel, as shown in (c).

3 Segmentation with Alpha-Shapes

Since holes of the α-complex do not necessarily correspond to regions of the original plane partition, we must characterize these holes in more detail:

Definition 6. *Consider the Delaunay triangulation D of a point set S and the complement $D_\alpha^C = \mathbb{R}^2 \setminus |D_\alpha|$ of the corresponding α-polytope with $\alpha > 0$. A connected component of D_α^C is called an* α-hole of $|D_\alpha|$. *When the radius of the circumcircle of the largest Delaunay triangle in an α-hole's closure is at least $\beta \geq \alpha$, we speak of an (α, β)-hole.*

For simplicity, we also use the term "hole" for the component which contains the infinite region. It is an (α, β)-hole for arbitrary large β. It follows from theorem 1 that there is a 1-to-1 relation between α-holes and the holes in the union of α-discs around the edgels. A similar relationship exists for (α, β)-holes:

Lemma 1. *An α-hole h is an (α, β)-hole if and only if it contains a point v whose distance from the nearest edgel is at least β.*

Proof. **I** ($d_H(v \in h, S) \geq \beta \Rightarrow h$ is an (α, β)-hole): when v is in the infinite region, the claim follows immediately. Otherwise, v is contained in some Delaunay triangle. By assumption, the corners of this triangle must have distance $\geq \beta$ from v. Hence, the triangle's circumradius must be at least β and the claim follows.

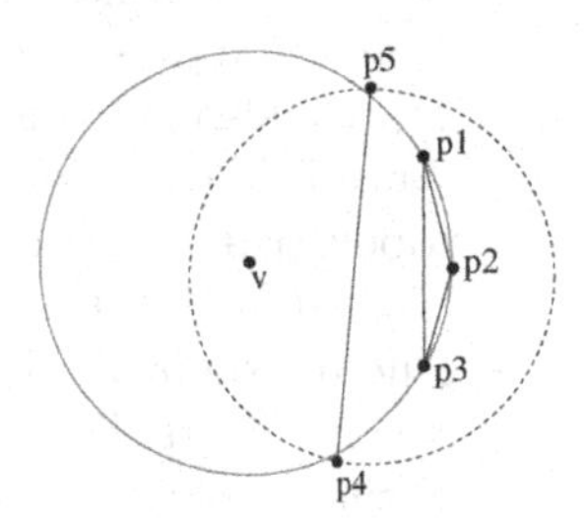

II (h is an (α, β)-hole $\Rightarrow \exists v \in h$ with $d_H(v, S) \geq \beta$): by assumption, the closure of h contains a Delaunay triangle t with circumradius of at least β. Consider the center v of its circumcircle. If it is within the triangle t, it is also in h and the claim follows. Otherwise, it is at least in some (α, β)-hole, and we must prove that t is in the same hole. Suppose to the contrary that v and t are in different α-holes. Then there exists a Delaunay triangle t' or a single edge e between t and v whose smallest circumcircle is smaller than α. The corners of t' or e cannot be inside t's circumcircle since it is a Delaunay triangle. Neither t' nor e can contain v because their circumcircle radius would then be at least β. Now consider the illustrated triangle p_1, p_2, p_3 and its circumcircle (gray) with center v. The points p_4 and p_5 are the end points of e or of one side of t'. Their distance $|p_4p_5|$ must be greater than $|p_1p_3|$. Consequently, any circumcircle with radius $\leq \alpha$ (dashed) around p_4 and p_5 contains t, contrary to the condition (imposed by the definition of an α-complex) that it must not contain any other edgel. The claim follows from the contradiction. □

Even for optimally chosen α, the α-complex does not necessarily reconstruct the homotopy type of the original boundary, since it may contain too many holes (see Fig. 2). This can be "repaired" by identifying (α, β)-holes:

Definition 7. *An (α, β)-boundary reconstruction from an edgel set S is defined as the union of the polytope $|D_\alpha|$ with all α-holes of D_α that are not (α, β)-holes.*

In other words, surplus holes are simply "painted over", and (α, β)-boundary reconstruction essentially amounts to hystheresis thresholding on the triangle size of a Delaunay triangulation. The following theorem shows that exactly the desired holes survive when α and β are properly chosen.

Theorem 2 (Boundary Sampling Theorem). *Let $\mathcal{P}$ be an r-stable plane partition, and S a (p, q)-sampling of $\mathcal{P}$'s boundary B. Then the (α, β)-boundary reconstruction $\mathcal{R}$ defined by S is homotopy equivalent to B, and the (α, β)-holes of $\mathcal{R}$ are topologically equivalent to the regions r_i of $\mathcal{P}$, if*

1. $p < \alpha \leq r - q$
2. $\beta = \alpha + p + q$
3. *every region r_i contains an open γ-disc with $\gamma \geq \beta + q > 2(p + q)$.*

Proof. Let U be the union of open α-discs centered at the points of S. Furthermore, let $B^\oplus = B \oplus \mathcal{B}^o_{\alpha+q}$ be the dilation of B with an open $\alpha + q$-disc, and $r_i^\ominus = r_i \ominus \mathcal{B}_{\alpha+q}$ the erosion of region $r_i \in \mathcal{P}$ with a closed $(\alpha + q)$-disc.

- According to the definition of a (p, q)-sampling, the dilation of B with a closed q-disc covers S. Consequently, $B^\oplus$ covers U. Therefore, U cannot have fewer connected components than $B^\oplus$. $B^\oplus$ has as many components as B due to r-stability of $\mathcal{P}$. Conversely, since $\alpha > p$, every open α-disc around a point of S intersects B, and the union U of these discs covers B. It follows that U cannot have more components than B. The number of components of B and U is thus equal. Due to homotopy equivalence of U and $|D_\alpha|$ (theorem 1), this also holds for the components of $|D_\alpha|$.
- Since $\mathcal{P}$ is r-stable with $r \geq \alpha + q$, each $r_i^\ominus$ is a connected set with the same topology as r_i. The intersection $r_i^\ominus \cap B^\oplus$ is empty, and $r_i^\ominus$ cannot intersect $|D_\alpha| \subset U \subset B^\oplus$. Hence, $r_i^\ominus$ is completely contained in a single α-hole of $|D_\alpha|$.
- Due to condition 3, r_i contains a point whose distance from B is at least $\gamma = \beta + q$. Its distance from S is therefore at least $\gamma - q = \beta$. Due to lemma 1, the α-hole which contains $r_i^\ominus$ is therefore also an (α, β)-hole.
- Since $B^\oplus$ covers U and U covers B, no (α, β)-hole can intersect both $r_i^\ominus$ and $r_j^\ominus$ $(i \neq j)$. It follows from this and the previous observation, that every region r_i can be mapped to exactly one (α, β)-hole which will be denoted h_i.

- An α-hole that does not intersect any region $r_i^\ominus$ must be completely contained within $B^\oplus$. Every point $v \in B^\oplus$ has a distance $d < \alpha + q$ to the nearest point of B. In turn, every point in B has a distance of at most p to the nearest point in S. Hence, the distance from v to the nearest point of S is $d' < \alpha + p + q = \beta$. According to lemma 1, this means that an α-hole contained in $B^\oplus$ cannot contain a triangle with circumradius β and cannot be an (α, β)-hole.
- The previous observation has two consequences: (i) All holes remaining in $\mathcal{R}$ intersect a region $r_i^\ominus$. Therefore, the correspondence between r_i and h_i is 1-to-1, and B and $|\mathcal{R}|$ enclose the same number of regions. (ii) All differences between $\mathcal{R}$ and D_α (i.e. all Delaunay cells re-inserted into $\mathcal{R}$) are confined within $B^\oplus$. This implies that $|\mathcal{R}|$ cannot have fewer components than $B^\oplus$ and B. Since all re-inserted cells are incident to D_α, $|\mathcal{R}|$ cannot have more components than $|D_\alpha|$, which has as many components as B (see first observation). Hence, B and $|\mathcal{R}|$ have the same number of components.
- Consider the components of the complement $(r_i^\ominus)^C$ and recall that $r_i^\ominus$ is a subset of both r_i and h_i for any i. Since B and $|\mathcal{R}|$ have the same number of components, it is impossible for h_i^C to contain a cell that connects two components of $(r_i^\ominus)^C$. This means that the sets r_i^C and h_i^C have the same number of components. This finally proves the topological equivalence of r_i and h_i, and implies homotopy equivalence of B and $|\mathcal{R}|$. □

If there exists no r fulfilling all conditions of theorem 2 for a given plane partition (or if the chosen α is too big), topology preservation is no longer guaranteed. Very small regions may get lost in the reconstruction. A region that is split into two or more parts by an s-erosion (i.e. has an s-waist) with $s < \alpha$ may also be split in the reconstruction. In case of very small waists, i.e. when $s+2p+2q \leq \alpha$, this is even guaranteed to happen. Thus, we can still apply our sampling theorem: we modify the original plane partition by connecting the different sides of small waists by a new arc. When the new partition fulfills our requirements, the modified topology is preserved, and the difference between the modified reconstruction and the original plane partition is well defined, see the second column of Fig. 3. When a thick boundary representation is undesirable, we apply *topology-preserving thinning*. An edge in the (α, β)-boundary reconstruction is called *simple* if its removal does not change the topology of the reconstructed regions. Simple edges always bound an (α, β)-hole on one side and a triangle in the boundary reconstruction on the other. Thinning removes simple edges until none are left:

1. Find all simple edges of the given (α, β)-boundary reconstruction and put them in a priority queue (the sorting is discussed below).
2. As long as the queue is not empty: Remove the topmost edge in the queue from the boundary reconstruction when it is still simple (it may have lost this property after removal of other edges). Put the edges in the triangle of the removed edge in the queue if they have now become simple.

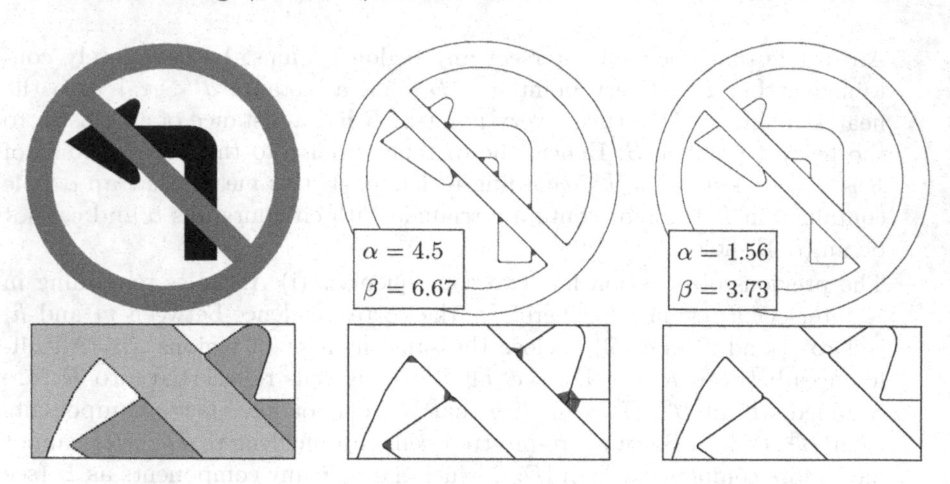

Fig. 3. Reconstructions before (red and black) and after (black only) thinning (note edgels in lower left image). Connectivity errors can occur when α is too big (center).

As far as region topology is concerned, the ordering of the edgels in the priority queue is arbitrary. For example, we can measure the contrast (image gradient) and remove weak edges first. Ordering by edge length is particularly interesting:

Definition 8. *A (not necessarily unique)* minimal boundary reconstruction *is obtained from an (α, β)-boundary reconstruction by means of topology-preserving thinning where the longest edges are removed first.*

The resulting boundaries are illustrated in Fig. 3. Since region topology is preserved, a minimal boundary reconstruction is homotopy equivalent to B. The two boundaries do not in general have the same topology, because the reconstruction may contain short edges, which end in the interior of a region.

Since a minimal boundary reconstruction can be shown to be a shortest possible one with correct topology, surviving edges connect edgels closest to each other. Neighboring edgels optimally align on the thinned boundary. The length $d_{\max}$ of the longest surviving edge is a measure of the density of the boundary sampling. The maximum distance p from the true boundary to the nearest edgel may be much larger than $d_{\max}/2$ if the displacements of neighboring edgels are highly correlated. This often occurs in practice: for example, Canny edgels along a circular arc are consistently biased toward the concave side of the curve. An (α', β) reconstruction of the edgel set with $\alpha' = d_{\max}/2 + \epsilon < p$ and arbitrarily small ϵ is still correct in the sense of theorem 2: since a minimal reconstruction is a subset of the (α', β) reconstruction, no true regions can get merged. Since $\alpha' < \alpha$, no region can get lost, and since β remains unchanged, no additional holes can be created. In fact, $\beta' = \alpha' + p + q < 2p + q$ would have been sufficient.

We found experimentally that undesirable holes (α-holes that are not (α, β)-holes) are actually quite rare, and their largest triangles are hardly ever as large as the maximal possible circumradius β allows. Therefore, an (α', β')-boundary reconstruction with β' even smaller than $\alpha' + p + q$ often produces the

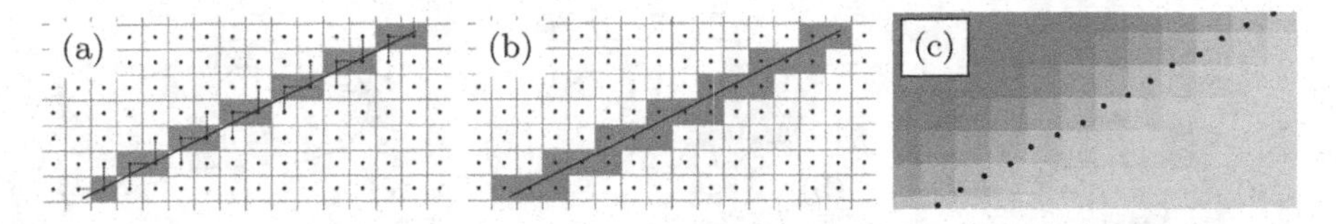

Fig. 4. (a) Where the boundary intersects the dual grid, the nearest sampling points form the *grid intersection digitization.* (b) The *supercover digitization* contains all sampling points whose pixel facets intersect the arc. (c) *Canny's algorithm* produces subpixel-accurate edgels from gray scale images.

correct region topology. We are currently investigating the conditions which permit weaker bounds. This is important, because a smaller β leads to a correspondingly reduced γ, i.e. the required size of the original regions is reduced, and more difficult segmentation problems can be solved correctly.

4 Application to Sampling and Segmentation Schemes

In theorem 2, p and q are assumed given. We now make their meaning and consequences more intuitive, by computing or estimating them for common sampling and segmentation schemes. Let's first look at grid intersection digitization:

Definition 9. *Consider a plane partition $\mathcal{P}$ with boundary B and a square grid. Compute all intersection points of B with the grid lines (i.e. with the lines connecting 4-adjacent sampling points) and round their coordinates to the nearest sampling point. The set of edgels thus defined is called* grid intersection digitization *of B, see Fig. 4a.*

For simplicity, let the grid size (i.e. the smallest distance from one sampling point to another) be unity. When each component of B crosses at least one grid line, the distance p of any point of B to the nearest selected grid point is less than $\sqrt{2}$, and the distance q of any grid intersection to its rounded coordinate cannot exceed $1/2$. Inserting this into the conditions of theorem 2, we get $\alpha \geq \sqrt{2}$, $r \geq \sqrt{2}+\frac{1}{2}$, $\beta \geq 2\sqrt{2}+\frac{1}{2} \approx 3.3$, and $\gamma \geq 2\sqrt{2}+1 \approx 3.8$. However, the worst case configurations giving rise to the values of β and γ in the theorem cannot actually occur in a square grid because Delaunay edges between grid points cannot have arbitrary length. It can be shown that the largest circumradius in an undesirable α-hole is below $\sqrt{34} \approx 2.9$, so that $\gamma \approx 3.4$ (circle area 37 pixels) is sufficient.

Generally the grid intersection digitization of a connected curve is an 8-connected digital curve. It is identical to Bresenham's digital straight line in case of a straight arc. Moreover the grid intersection digitization is a subset of the supercover digitization on a square grid, which produces a 4-connected digital curve for any connected curve:

Definition 10. *Let $\mathcal{P}$ be a plane partition with boundary B and $\mathcal{G}$ a finite set of sampling points such that the Voronoi cells of $\mathcal{G}$ have a radius of at most g. The* supercover digitization *of B is the set of all sampling points whose Voronoi cell intersects B, see Fig. 4b.*

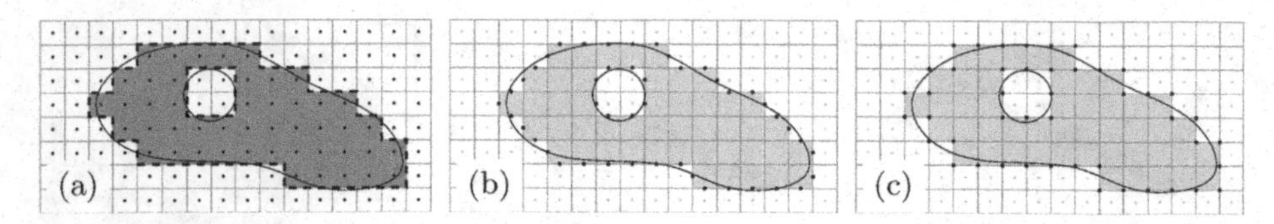

Fig. 5. The *interpixel boundary* (dashed) can be extracted from the subset digitization (a). It includes both the *midcrack digitization* (b) and the *endcrack digitization* (c).

The constraint on the size of the Voronoi cells implies that $p = g$ and $q < g$. Hence, $\alpha > g$, $r > 2g$, $\beta > 3g$ and $\gamma > 4g$ are required. For example, in a unit square grid we have $q < p = \sqrt{2}/2$ and $\gamma > 2\sqrt{2} \approx 2.8$. Thus, the supercover digitization imposes weaker constraints on the original plane partition $\mathcal{P}$ than the grid-intersection digitization. This is mainly due to the denser sampling of the boundary (smaller spacing of the edgels) in the former. As stated in [9], the supercover digitization is a Hausdorff discretization, i.e. a set of sampling points which minimizes the Hausdorff distance $\max(p, q)$ to the boundary B. Thus the given bounds for α, β and γ are sufficient for all Hausdorff discretizations. Another interesting question is what can be said about region based digitization:

Definition 11. *Let $\mathcal{P}$ be a plane partition with regions $R = \{r_i\}$ and $\mathcal{G}$ a finite set of sampling points such that the Voronoi cells (i.e. the pixels) of $\mathcal{G}$ have a radius of at most g. The* subset digitization *$\hat{r}_i$ of region r_i is the union of all Voronoi cells whose sampling point is in r_i, see Fig. 5a. The union of the boundaries of all $\hat{r}_i$ is called the* interpixel boundary. *A boundary digitization scheme where all edgels are on the interpixel boundary $\mathbb{B}$ is an* interpixel digitization. *Two examples are the the* midcrack digitization *(Fig. 5b) where the center points of all pixel edges inside $\mathbb{B}$ are chosen as edgels, and the* endcrack digitization *(Fig. 5c) where all pixel corner points inside $\mathbb{B}$ are used.*

Thus, *boundary-based* digitizations like endcrack and midcrack digitization can be derived from the *region-based* subset digitization. While the maximal distance q of any edgel to the nearest boundary point cannot exceed g, the distance p from any boundary point to the nearest edgel can be arbitrary large: An r-stable, but non-binary plane partition is never r-regular. Consequently, $\hat{r}_i$ is generally not topologically equivalent to the closure of r_i and may even be disconnected. The distance of the components of $\hat{r}_i$ may approach the diameter of r_i when r_i has a long narrow spike. Obviously, this is not a useful practical bound for the value of p. We need a restriction that is stronger than r-stability, but weaker than r-regularity and which prevents these undesirable spikes:

Definition 12. *Let $\mathcal{P}$ be a plane partition with boundary B. We say two points* $x1, x2 \in \mathcal{B}$ delimit a (θ, d)-spike, *if the distance from x_1 to x_2 is at most d and if every path on B from x_1 to x_2 contains at least one point with $\angle x_1 y x_2 < \theta$. We say that $\mathcal{P}$* has no (θ, d)-spikes *if for any pair of boundary points $x_1, x_2 \in B$ with distance of at most d, there exists a path $Y \subset B$ between x_1 and x_2 such that $\angle x_1 y x_2 \geq \theta$ for all points $y \in Y$.*

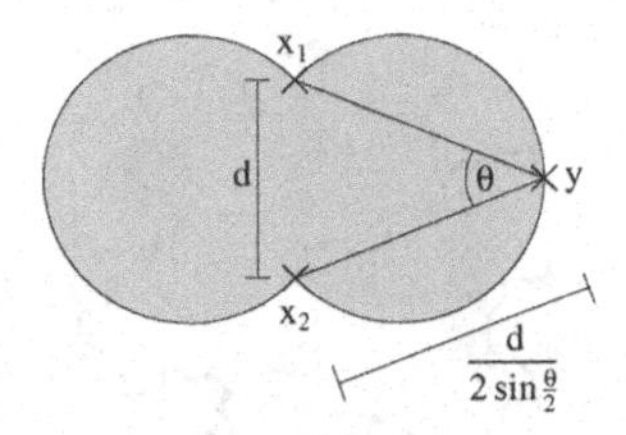

Fig. 6. Any point which encloses an angle of at most θ with x_1 and x_2 must lie inside the shaded region. The shown y is the one with the maximal distance to the nearer one of x_1 and x_2. Thus there is a path from x_1 to x_2 inside the shaded region and each of its points has a distance of at most $d/(2\sin\frac{\theta}{2})$.

Intuitively, two points delimit a (θ, d)-spike, if the shortest boundary path between them does not differ too much from a straight line, i.e. it lies inside the shaded region in Fig. 6.

Note that r-regular partitions have no (θ, d)-spikes for $d \leq r$ and $\theta = 2\arctan\left(d/(2r - \sqrt{4r^2 - d^2})\right)$ (e.g. for $\theta = 90°, 60°$ we get $d = r$ and $d = \sqrt{3}r$ respectively). By sampling dense enough one can enforce the angles to be arbitrarily flat. But in general, absence of (θ, d)-spikes does not imply r-stability, so we will require both. The fact that the boundary cannot be too far away from the edgels can be used for estimating p and q, e.g. in the midcrack and endcrack digitization case:

Theorem 3. *Let G be a square grid with sample distance h (pixel radius $g = h/\sqrt{2}$), and let $\mathcal{P}$ be a plane partition such that every region $r_i \in \mathcal{P}$ contains a closed g-disc and the boundary B has no (θ, d)-spikes. Then the endcrack digitization is a (p, q)-boundary sampling with $q = h/\sqrt{2}$ and $p = q + \left(\frac{h}{2} + q\right) / \sin\frac{\theta}{2}$, provided that $h \leq d/(1 + \sqrt{2})$, and the midcrack digitization is a (p, q)-boundary sampling with $q = \frac{h}{2}$ and $p = q + \left(\frac{h}{2} + q\right) / \sin\frac{\theta}{2}$, provided that $h \leq \frac{d}{2}$.*

Proof. First, we prove the bounds on q. Let x, y be two 4-adjacent square grid points. Their common pixel edge is in the interpixel boundary if and only if x and y lie in different regions r_i and r_j, i.e. the grid line between x and y intersects the boundary B in at least one point v. The endcrack edgels are exactly the end points of these pixel edges, and their distance to v is at most $h/\sqrt{2}$. It follows that $q = h/\sqrt{2}$ for the endcrack digitization. The midcrack edgels are the center points between x and y, so their maximum distance to v is $\frac{h}{2}$. Hence, $q = \frac{h}{2}$ for the midcrack digitization. The maximum distance between neighboring edgels is h in both cases.

Now, we prove the bound on p given q. By definition $B = \bigcup \partial r_i$, where ∂r_i is the boundary of region r_i. Since every region contains a closed disc of radius $g = h/\sqrt{2}$, and every such disc contains at least one grid point, every region r_i contains a grid point, i.e. $\hat{r}_i$ is not empty, and there exist at least four edgels near ∂r_i. Due to the nonexistence of (θ, d)-spikes, any two components $(\partial r_i)_j$ and $(\partial r_i)_l$ of the boundary ∂r_i must have a distance of more than $d \geq 4q$. So, for every component there exists a set of edgels which are closer to $(\partial r_i)_j$ than to any other component. Obviously every component $(\partial r_i)_j$ is a closed curve. Thus by mapping every edgel to the nearest point of B, one gets a cyclic list of points $[b_k]^{(ij)}$ for every component $(\partial r_i)_j$, and each point b_k has a distance of at most $h + 2q$ to its successor b_{k+1} in the list. For endcrack edgels, we have

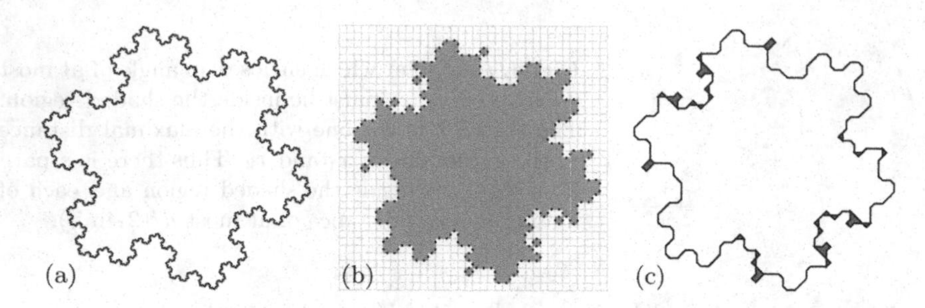

Fig. 7. (a) Koch Snowflake; (b) subset digitization (note the topology violations); (c) (α, β)-boundary reconstruction from marked midcrack edgels. Areas where the edgels do not unambigously determine the boundary shape pop out by remaining thick.

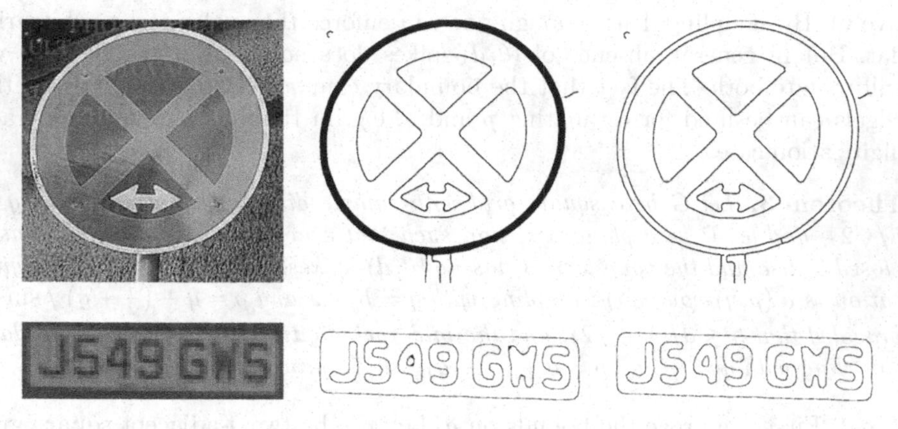

Fig. 8. Real images, (α, β)-boundary reconstruction (center) and minimal reconstruction after thinning (right). Edgels have been computed by Canny's algorithm on a color (top) and intensity (bottom) gradient.

$h + 2q = (1 + \sqrt{2})h \leq d$, and for midcrack edgels $h + 2q = 2h \leq d$. Thus, the boundary part between b_k and b_{k+1} includes no point with an angle smaller than θ. As shown in Fig. 6, this implies that the distance from any boundary point between b_k and b_{k+1} to the nearer one of these two points is at most $\left(\frac{h}{2} + q\right) / \sin\frac{\theta}{2}$. Thus, the maximum distance to the nearest of the two edgels which are mapped onto b_k and b_{k+1} is $p = q + \left(\frac{h}{2} + q\right) / \sin\frac{\theta}{2}$. □

E.g. if a plane partition has no $(60°, d)$-spikes for sufficiently big d, we get $q = h/\sqrt{2}$ and $p = (1 + 3\sqrt{2})\frac{h}{4} \approx 1.31h$ for endcrack digitization and $q = \frac{h}{2}$ and $p = h$ for midcrack digitization. It follows that midcrack digitization should be favoured over endcrack digitization.

The nonexistence of shape spikes allows us to topologically correctly digitize even objects having a fractal boundary like the Koch Snowflake (see Fig. 7): let K be the object bounded by the Koch Snowflake based on a triangle of sidelength 1. Then it can be shown that K is r-stable for all $r < 1/\sqrt{3}$ and it has no $(60°, d)$-spikes for $d < 1/\sqrt{3}$ and it contains a γ-disc for any $\gamma \leq 1/\sqrt{3}$. Thus the (α, β)-boundary reconstruction based on the midcrack digitization with a square grid of grid size h is correct for all $h < 1/\sqrt{27} \approx 0.192$.

Our sampling theorem can also be applied to commonly used edge detectors on real images (see Fig. 8), like Canny's algorithm [10]. In [11] we derive the following bounds: suppose the original partition is r-stable and free of $(60°, 2r)$-spikes, and the combined PSF and edge detector scale is at most $\sigma = 0.8r$, with pixel distance $h \leq r$. Then q does not exceed $0.9\sigma + 0.3 \approx 1.1$ pixels when the boundary contains corners or junctions and SNR $= 10$ (noise at this level is already quite visible), and $q \approx 0.2$ pixels when the partition is (4-pixel)-regular and SNR $= 30$. Note that the average error is much lower and approaches zero along straight edges. When the edgels are not represented with subpixel accuracy, a maximal round-off error of $\frac{h}{\sqrt{2}}$ must be added, and the average error cannot fall below $\frac{h}{\sqrt{6}} \approx 0.4h$ pixels even in case of straight edges.

5 Conclusions

To our knowledge, this paper proposes the first geometric sampling theorem that explicitly considers measurement errors. Moreover, the theorem applies to a much wider class of shapes (r-stable partitions) than existing ones (r-regular partitions). The situation in real images is thus modeled much more faithfully because shapes may now have corners and junctions, and standard segmentation algorithms can be used. The resulting segmentations are similar to what one gets from traditional heuristic edgel linking, but their properties can now be formally proven due to their theoretical basis in Delaunay triangulation. We showed that many known digitization and segmentation methods can be analysed and applied in the new framework by simply determining their error bounds. Our approach (including boundary thinning) provides a novel way for computing a combinatorial map representation [12] of the boundaries in real images.

References

1. Pavlidis, T.: Algorithms for Graphics and Image Processing. Computer Science Press (1982)
2. Serra, J.: Image Analysis and Mathematical Morphology. Academic Press (1982)
3. Latecki, L.J., Conrad, C., Gross, A.: Preserving Topology by a Digitization Process. J. Mathematical Imaging and Vision **8** (1998) 131-159
4. Stelldinger, P., Köthe, U.: Towards a General Sampling Theory for Shape Preservation. Image and Vision Computing **23**(2) (2005) 237-248
5. Stelldinger, P.: Digitization of Non-regular Shapes. In: Ronse, C., Najman, L., Decenciere, E. (eds.): Mathematical Morphology, ISMM (2005)
6. Edelsbrunner, H., Mücke, E.P.: Three-dimensional alpha shapes. ACM Trans. Graphics **13** (1994) 43-72
7. Edelsbrunner, H.: The union of balls and its dual shape. Discrete Comput. Geom. **13** (1995) 415-440

8. Bernardini, F., Bajaj, C.L.: Sampling and Reconstructing Manifolds Using Alpha-Shapes. Proc. 9th Canadian Conf. Computational Geometry (1997)
9. Ronse, C., Tajine, M.: Discretization in Hausdorff Space. J. Mathematical Imaging and Vision **12** (2000) 219-242
10. Canny, J.: A Computational Approach to Edge Detection. IEEE Trans. Pattern Analysis and Machine Intelligence **8**(6) (1986) 679-698
11. Stelldinger, P., Köthe, U., Meine, H.: Topologically Correct Image Segmentation using Alpha Shapes. University Hamburg, Computer Science Department, Technical Report FBI-HH-M-336/06 (2006)
12. Braquelaire, J.-P., Brun, L.: Image segmentation with topological maps and interpixel representation. J. Visual Communication and Image Repr. **9**(1) (1998) 62-79

New Removal Operators for Surface Skeletonization

Carlo Arcelli, Gabriella Sanniti di Baja, and Luca Serino

Institute of Cybernetics "E.Caianiello", CNR, Pozzuoli (Naples), Italy
(c.arcelli, g.sannitidibaja, l.serino)@cib.na.cnr.it

Abstract. New 3×3×3 operators are introduced to compute the surface skeleton of a 3D object by either sequential or parallel voxel removal. We show that the operators can be employed without creating disconnections, cavities, tunnels and vanishing of object components. A final thinning process, aimed at obtaining a unit-thick surface skeleton, is also described.

1 Introduction

Topology preserving removal operations have received much attention in the literature dealing with discrete objects, since their use is crucial for processes such as shrinking, thinning and skeletonization. In this respect, the interest is towards operations based on a small neighborhood of an object element p (the 3×3 neighborhood in 2D, and the 3×3×3 neighborhood in 3D). In fact, it has been proved that if removal of p does not alter the topology in the neighborhood of p, then p can be safely removed since topology is also globally maintained [1, 2].

To preserve topology in 2D, removal of p from an 8-connected object should not create holes or disconnections. In other words, p should be a *simple point*. Removal of p does not create holes, if p has at least a 4-adjacent neighbor in the background. Disconnections do not occur if p has exactly one 8-connected component of object neighbors. In 3D, topology preserving removal operations can still be based on the notion of simple point [3,4], but besides cavities and disconnections, also tunnels have to be taken into account in the definition of point simplicity. Similarly to the 2D case, cavities are not created by removing p from a 26-connected object if p has at least a 6-adjacent neighbor (face-neighbor) in the background. To this purpose, the number of 26-connected components of object voxels and the number of 6-connected components of background voxels have to be computed by processing the 3×3×3 neighborhood of p. These numbers can be used to determine whether removal of p causes disconnections. Moreover, removal of p should prevent creation of tunnels. In other words, removal of p from a tunnel-free solid object should not change the object into one that cannot be deformed to a single voxel [5]. To this aim, the number of 6-connected components of background voxels, having p as face-neighbor and computed in the 3×3×3 neighborhood of p deprived of the eight vertex-neighbors, has to be computed.

In this paper, we introduce a set of 3×3×3 topology preserving removal operations, based on the configurations of the voxels in the neighborhood of p. We show that

A. Kuba, L.G. Nyúl, and K. Palágyi (Eds.): DGCI 2006, LNCS 4245, pp. 555–566, 2006.

these operations can be used in the framework of 3D surface skeletonization. In particular, we introduce two surface skeletonization algorithms using respectively sequential and parallel voxel removal.

2 Notions and Definitions

We work with a binary finite voxel image in a cubic grid, where the object is the set of 1's and the background is the set of 0's. As implicitly stated in the previous section, we choose the 26-connectedness for the object and the 6-connectedness for the background. The object consists of border voxels, i.e., those having a face-neighbor in the background, and inside voxels, i.e., those having no face-neighbors in the background. Any voxel p has 6 face-, 12 edge- and 8 vertex-neighbors. We call N(p) the 3×3×3 set including p and its 26 neighbors, and N(p)* the set N(p) deprived of the vertex-neighbors.

A 6-connected path of voxels is termed *face-aligned* if no change of direction is allowed along the path. Three directions, each with two orientations, are possible.

We say that a *transition* exists for a border voxel p, if a face-aligned path consisting of three voxels can be found, where p is the intermediate voxel and the two extremes of the path are respectively an inside voxel and a background voxel.

A *protrusion* is a maximal connected set of border voxels, none of which has an inside voxel as face-neighbor. By this definition it follows that a protrusion is the at most 2-voxel thick union of surface-like and curve-like sets.

A *cavity* is a 6-connected component of the background fully surrounded by the object [6].

A *tunnel* exists if a path can be found in the object that cannot be deformed to a single voxel [5].

An object voxel p is *simple* if the object including p is homotopic to the object deprived of p. Simplicity of p means that the numbers of cavities, object components, and tunnels are the same, independently of whether p is in the object or in the background. Four main conditions are satisfied in N(p) by a simple voxel p: 1) p is not an inside voxel (cavity prevention condition); 2) the number of 26-connected object components in N(p) is equal to 1 (object connectedness condition); 3) the number of 6-connected components of background voxels in N(p) is equal to 1 (background connectedness condition); and 4) the number of 6-connected components of background voxels, having p as a face-neighbor and computed in N(p)* is equal to 1 (tunnel prevention condition).

We point out that voxel configurations, in correspondence of which tunnels risk to be created in the object by removal of p, may exist when in N(p)* two background face-neighbors of p belonging to distinct 6-connected components either form a face-aligned path with p (type 1), or are edge-neighbors of each other (type 2).

For completeness, we note that a voxel p whose 26 neighbors are all background voxels, i.e., an *isolated* voxel, is not simple.

The *distance transform* of an image is a labeled version of the image where the label of any voxel is the distance of that voxel from the background. In this paper we compute the distance transform DT by using the distance D^6, which is the 3D version of the well-known city-block distance in 2D. For the computation of DT, as well as of distance transforms based on other metrics, see [7]. Voxels in DT can be interpreted as centers of balls with radius equal to the distance label.

A *center of a maximal ball*, CMB, is an object voxel whose associated ball is not included by any other single ball in the object. In DT, a CMB is any voxel whose distance label is not smaller than the labels of its 6 face-neighbors. The union of all the maximal balls coincides with the object.

The *surface skeleton* of a 3D object is a set of surfaces and curves, symmetrically placed within the object, which has the same topology and reflects the shape of the object. See [8] and the references listed therein for more details on 3D surface skeletonization. If the surface skeleton includes all the CMBs, then the object can be fully reconstructed starting from its surface skeleton. Unit thickness and inclusion of all CMBs are not possible at the same time in presence of object regions whose thickness is given by an even number of voxels. Since the set of the CMBs is at most 2-voxel thick, a nearly-thin skeleton, at most 2-voxel thick, is obtained if full recoverability is desired. A process, generally referred to as *final thinning*, has to be performed if a unit-thick surface skeleton is requested.

3 New 3×3×3 Topology Preserving Removal Operations

As already pointed out, a voxel p should have a face-neighbor in the background, i.e., p should be a border voxel, to avoid creation of cavities in N(p) when p is removed. Moreover, since we are interested in removal operations to be used in the framework of skeletonization, protrusion voxels cannot be removed. In fact, their presence in the skeleton is necessary to reflect the geometrical properties of the object, e.g., to account for its elongated parts. Thus, to be removable, p should have an inside face-neighbor.

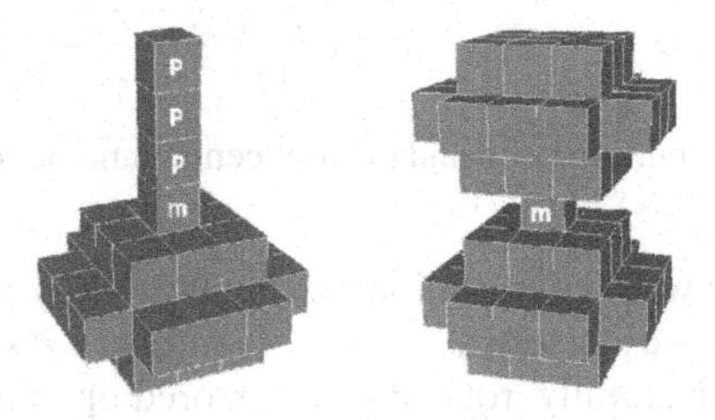

Fig. 1. Voxels denoted by p form a protrusion, voxels denoted by m are necessary to maintain connectedness between the protrusion and the inside voxels, left, and between two components of inside voxels, right

To consider p as a candidate to removal, we request that the above mentioned background and inside face-neighbors of p form with p a face-aligned path, i.e. we request that a transition exists for p. It is immediate to see that the existence of a transition for p guarantees that when p is removed: 1) it is not an isolated voxel, thus no vanishing of object components occurs, 2) cavities are not created, 3) protrusion voxels are preserved, 4) type 1 tunnels and background fusions are not created in the direction of the transition since the two face-neighbors of p forming a face-aligned path are not both background voxels, and 5) voxels connecting a protrusion and the set of inside voxels, or two components of inside voxels, are preserved, if they are the only voxels ensuring connection, since no transition exists for them. Thus, the remaining conditions to pose for safe removal of p should guarantee only object connectedness preservation in the remaining directions, and avoid the creation of type 2 tunnels as well as fusion of background components.

For the 3D objects shown in Fig.1, no transition exists for voxels (denoted by p) forming a protrusion and for voxels (denoted by m) necessary to maintain connectedness along the face direction between the protrusion and the inside voxels, Fig.1 left, and between two components of inside voxels, Fig.1 right.

Once the two face-neighbors of p necessary for the existence of a transition have been identified, we describe N(p) as consisting of three 3×3 windows. The 3×3 sets, respectively centered on p, on its background face-neighbor p_T, and on its inside face-neighbor p_B, are called *central, top* and *bottom* windows. The remaining elements in the windows are denoted by their cardinal directions with respect to p, p_T and p_B, as shown in Fig.2. In this interpretation, independently of the orientation of the actual neighborhood of p, the transition is always seen as occurring from top to bottom.

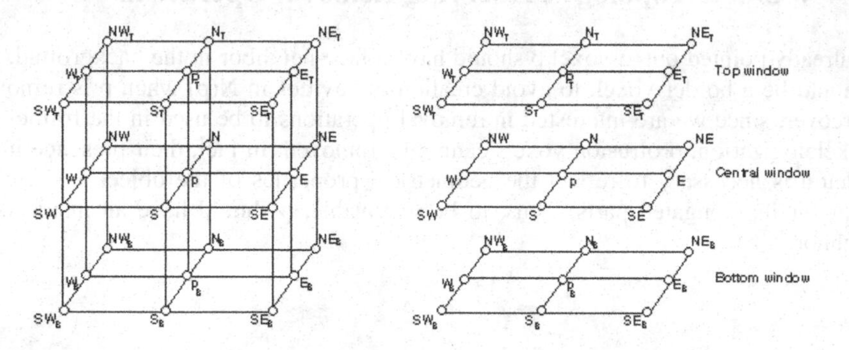

Fig. 2. The 3×3×3 neighborhood of p and the top, central and bottom windows constituting it

Since a voxel p for which a transition exists necessarily has a number k=1,2,..,5 of object face-neighbors, we suggest five basic operations for safe removal of p, depending on the value of k. Naturally, rotated and mirrored operations have also to be taken into account. The basic removal operations are discussed below with reference to the interpretation of N(p) into the top, central and bottom windows.

As far as object connectedness preservation is concerned, we note that the object neighbors of p in the central and bottom windows are all directly 26-connected to the inside voxel p_B. Thus, we should only consider the object neighbors of p in the top window (of course, we don't consider p_T, which is a background voxel). While vertex-neighbors of p in the top window play a role only in the object connectedness condition, edge-neighbors of p in the top window are also important to design a condition preventing background fusion and type 2 tunnel creation, whenever appropriate face-neighbors of p in the central window are in the background.

Case k=1

The only object face-neighbor of p is p_B and p can be removed if
$N_T=0 \wedge E_T=0 \wedge S_T=0 \wedge W_T=0 \wedge (NE_T=0 \vee NE=1) \wedge (SE_T=0 \vee SE=1) \wedge (SW_T=0 \vee SW=1) \wedge (NW_T=0 \vee NW=1)$

Note that $(NE_T=0 \vee NE=1)$, $(SE_T=0 \vee SE=1)$, $(SW_T=0 \vee SW=1)$ and $(NW_T=0 \vee NW=1)$ respectively prevent that object disconnections occur in any of the four possible vertex-directions. Moreover, no background fusion or tunnels are created and no object disconnections occur in the four possible edge-directions, if the edge-neighbors of p in the top window are in the background, i.e., if $N_T=0 \wedge E_T=0 \wedge S_T=0 \wedge W_T=0$. In Fig.3, an example of removable voxel is shown for case k=1.

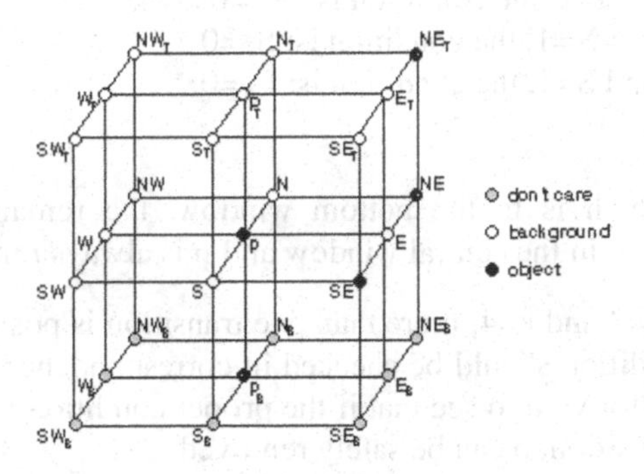

Fig. 3. Case k=1. The voxel p with one object face-neighbor, the inside voxel p_B, is removable. It would not be removable if, still being NE_T an object voxel, NE is a background voxel, or if any among N_T, E_T, S_T and W_T is an object voxel.

Case k=2

The two object face-neighbors of p, one of which is p_B, cannot form a face-aligned path with p, due to existence of a transition for p. Thus, they are edge-neighbors of each other. Since there exist four possible pairs of edge-connected face-neighbors of p, one of which is p_B, the removal condition is one of the following four conditions, depending on which is the second object face-neighbor of p.

If N=1, the condition is: $W_T=0 \wedge S_T=0 \wedge E_T=0 \wedge (SW=1 \vee SW_T=0) \wedge (SE=1 \vee SE_T=0)$
If W=1, the condition is: $N_T=0 \wedge S_T=0 \wedge E_T=0 \wedge (SE=1 \vee SE_T=0) \wedge (NE=1 \vee NE_T=0)$
If S=1, the condition is: $W_T=0 \wedge N_T=0 \wedge E_T=0 \wedge (NW=1 \vee NW_T=0) \wedge (NE=1 \vee NE_T=0)$
If E=1, the condition is: $W_T=0 \wedge S_T=0 \wedge N_T=0 \wedge (NW=1 \vee NW_T=0) \wedge (SW=1 \vee SW_T=0)$

Case k=3

Due to the transition from p_T to p_B, the remaining two object face-neighbors of p are necessarily placed in the central window and either are edge-neighbors of each other (four possible configurations), or form a face-aligned path with p (two possible configurations). Thus, the removal condition is one of the following six conditions, depending on which are the two object face-neighbors of p in the central window.

If W=1 and S=1, the condition is: $N_T=0 \wedge E_T=0 \wedge (NE_T=0 \vee NE=1)$
If W=1 and N=1, the condition is: $S_T=0 \wedge E_T=0 \wedge (SE_T=0 \vee SE=1)$
If N=1 and E=1, the condition is: $S_T=0 \wedge W_T=0 \wedge (SW_T=0 \vee SW=1)$

If E=1 and S=1, the condition is: $N_T=0 \wedge W_T=0 \wedge (NW_T=0 \vee NW=1)$
If W=1 and E=1, the condition is: $N_T=0 \wedge S_T=0$
If N=1 and S=1, the condition is: $E_T=0 \wedge W_T=0$

Case k=4
Except for p_B, which is in the bottom window, the remaining three object face-neighbors of p are all in the central window (four possible configurations). Thus, the removal condition is one of the following four conditions, depending on the positions of the three object face-neighbors of p in the central window.

If W=1, E =1, and S=1, the condition is: $N_T=0$
If N=1, E=1, and S=1, the condition is: $W_T=0$
If W=1, E =1, and N=1, the condition is: $S_T=0$
If W=1, N =1, and S=1, the condition is: $E_T=0$

Case k=5
Except for p_B, which is in the bottom window, the remaining four object face-neighbors of p are all in the central window and p is always removable.

In Cases k=2, k=3 and k=4, more than one transition is possible. In these cases, the relative above condition should be checked in correspondence with any existing transition. It is straightforward to see that if the proper condition is satisfied in correspondence with one transition, p can be safely removed.

In the following we will denote by *f* the removal operator that includes detection of transitions and involves the five cases k=1-5. A voxel p is removed if the operator finds out that at least a transition exists for p and the proper removal condition, based on the number of object face-neighbors of p, is satisfied. We remind that voxels constituting object protrusions are not removed by *f* since no transition exist for them. This is an important feature to use *f* in skeletonization, since it guarantees that no ad hoc criterion is necessary to prevent unwanted shortening of peripheral surfaces/branches of the skeleton.

For simplicity, we will refer to $(NE_T=0 \vee NE=1)$, $(SE_T=0 \vee SE=1)$, $(SW_T=0 \vee SW=1)$ and $(NW_T=0 \vee NW=1)$ as to *vertex conditions* and to the conditions $N_T=0$, $E_T=0$, $S_T=0$, and $W_T=0$ as to *edge conditions*. Note that the argument involved in an edge condition is the only one relative to the edge-neighbor in the top window of p such that the corresponding voxel vertically aligned with it, i.e., the face-neighbor of p in the central window, is a background voxel.

4 Surface Skeletonization by Sequential Voxel Removal

In surface skeletonization by sequential voxel removal, each currently inspected border voxel is removed, i.e., is set to 0, if the proper condition is satisfied. Of course, to guarantee that the surface skeleton is centrally placed within the object, an iterative process is necessary to candidate to removal at each iteration only voxels belonging to the current border of the object. To save computation time and have all borders directly available, the distance transform DT is used. The border of the object at the *l*-th iteration includes all voxels with distance label *l* in DT, as well as voxels with distance label smaller than l, if these latter voxels were not removed during previous iterations. Voxels with label greater than l are inside voxels at the l-th iteration. The background is, at all iterations, the current set of 0's. A transition exists for a voxel p at the l-th iteration, if p has two face-neighbors forming with p a face-aligned path such that one of these neighbors is 0 and the other has label l+1.

The algorithm can be sketched as follows.

Compute DT. Let m be the maximal distance label in DT
  for l = 1 to m do
  apply f to any voxel p with distance label l, and set p to 0, if removable.

It is easy to see that CMBs are voxels for which no transition occurs. In fact, none of the face-neighbors of a CMB can be an inside voxel. Thus, the obtained surface skeleton SK includes all the CMBs and complete recovery of the object is possible. SK is likely to be 2-voxel thick in some parts and final thinning is necessary if unit thickness is a desired feature.

An example of the performance of the algorithm is given in Fig.4.

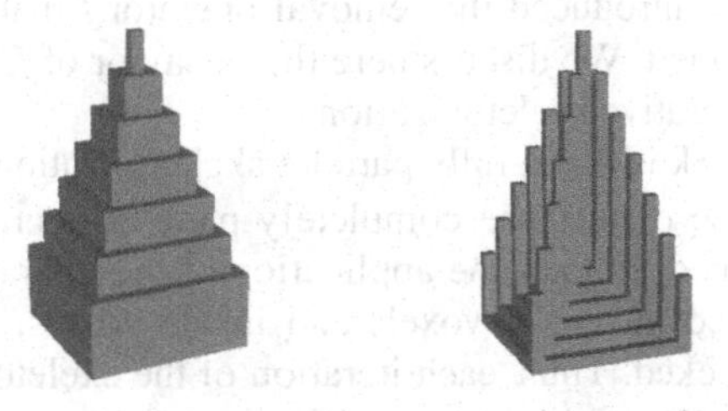

Fig. 4. Object, left, and its surface skeleton, right

We point out that small asymmetries could exist in the resulting SK. These asymmetries occur only in presence of pairs of face-connected border voxels such that only one of them is removable, depending on the order in which voxels are inspected. See Fig.5. There, the central section of a small 3D object is shown; gray voxels are inside voxels, all other voxels belong to the border. The two sections before and after the central section include only two voxels each; these voxels are the face-neighbors of the voxels shown in gray in the central section. Depending on the order in which voxels are examined, removal of the voxel k (h) will allow a transition for the voxel p (q). The voxel p (q) is thus removed since the proper removal condition is satisfied.

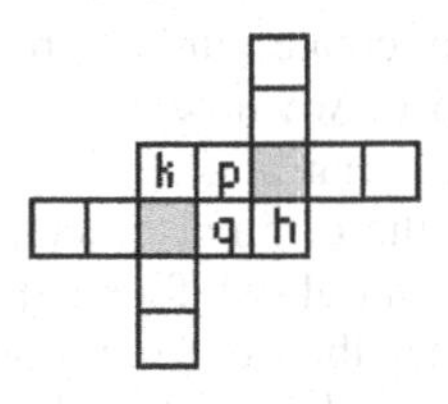

Fig. 5. Section of a 3D object, where inside voxels are gray. Voxels k, p and h (q, h and k) are removed if the section is examined in forward (backward) fashion.

To avoid creation of asymmetries, one could detect parallelwise, at each iteration, the voxels for which a transition exists. For the example in Fig.5, both p and q would remain in the skeleton. In fact, p and q belong to the same current border including k and h, and no transition would exist for them. We point out that a parallel detection of voxels for which a transition exists implies doubling the number of iterations necessary to obtain SK. Moreover, other asymmetries are likely to be created by final thinning when a unit-thick skeleton is desired. Thus, especially in this latter case, resorting to parallel detection of transitions is not convenient.

For the sake of completeness we point out that, due to the implicit selection of all CMBs as skeletal voxels, the surface skeleton may include unwanted peripheral branches/surfaces. These are originated in the presence of noisy border configurations in correspondence of which some CMBs, generally characterized by small distance values, are found. Thus, as it is always the case for skeletonization, some pruning should be applied to remove noisy peripheral branches/surfaces (see e.g., [9]).

5 Surface Skeletonization by Parallel Voxel Removal

In Section 3, we have introduced the removal operator *f* that does not alter topology when sequentially applied. We discuss here the behavior of *f*, when applied in parallel in the framework of iterative skeletonization.

An immediate remark is that a fully parallel skeletonization algorithm, i.e., an algorithm based on the use of just one completely parallel operator with 3×3×3 support [10], cannot be designed. In fact, the application of the operator *f* requires that object voxels be distinguished in border voxels and inside voxels, since the existence of a transition must be checked. Thus, each iteration of the skeletonization algorithm must include a sub-iteration devoted to the identification of the border voxels.

It is straightforward to observe that object vanishing and creation of cavities cannot occur, since no transition exists for voxels of object parts entirely consisting of border voxels and for inside voxels. A formal proof that *f* does not alter object connectedness and does not create fusion of background components, when two neighboring border voxels p and q are simultaneously removed, can be found in [11]. In fact, proving that *f* does not alter object and background connectedness in the set N(p)∪N(q) is enough to guarantee that object and background connectedness are also globally preserved. Here we just point out that, although our removal operations apparently have only a 3×3×3 support, the existence of the transition for a voxel p candidate to removal implies that at least a face-neighbor of p is an inside voxel and, hence, the operations implicitly use a support larger than 3×3×3. In [11], the worst case, occurring when a candidate voxel has only one inside face-neighbor, is discussed.

As regards the possibility to create tunnels by simultaneous removal of two neighboring object voxels p and q, we note that p and q must obviously be face-neighbors of each other. Indeed, it can be seen by examining Fig.6 that tunnels, not corresponding to real tunnels in the original object, can be created under some circumstances by simultaneous removal of two face-adjacent voxels p and q.

A suitable operator *g*, to detect the configurations that could create tunnels has to be designed. Such an operator can be employed only after the operator *f* has partitioned the set of border voxels into removable and non-removable voxels, since spurious tunnels can be eliminated only after the configurations that would cause them have been created. Thus, the identification on the current border of which voxels to remove should be done in two sub-iterations. During the first sub-iteration *f* is applied to mark non-removable border voxels as belonging to the skeleton. Obviously, border voxels that have already been assigned to the skeleton at some previous iteration do not need to be newly checked by *f*. During the second sub-iteration, the operator *g* is applied to the non-marked border voxels to mark as non-removable (and, hence, as belonging to the skeleton) all the voxels whose removal would create tunnels.

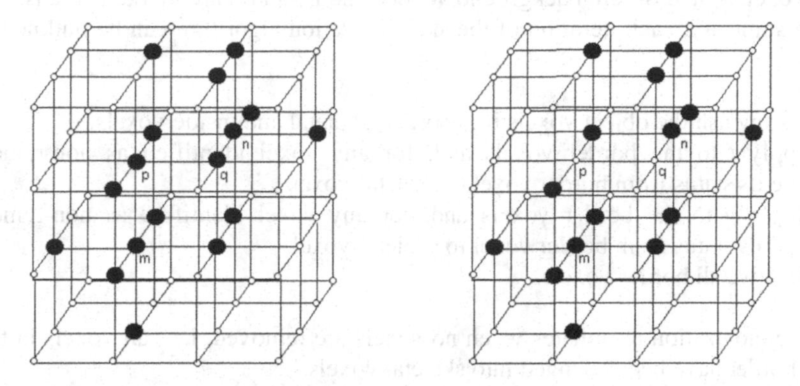

Fig. 6. The 3D configuration, where black dots are object voxels and m and n are inside voxels, before (left) and after (right) simultaneous removal of p and q

We have examined the various configurations embedding a pair of face-adjacent voxels p and q to identify the cases in which the operator *f* fails to preserve topology when applied in parallel, so as to find the adequate operator *g*. Actually, in [11] we show that only one basic configuration exists, where the operator *f* fails.

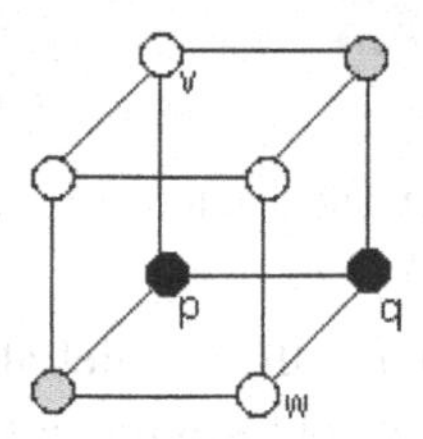

Fig. 7. Basic configuration where parallel removal of p and q causes spurious tunnel creation. Gray voxels are voxels already assigned to the skeleton

The basic configuration is shown in Fig.7, where the two voxels v and w are background voxels (by the way, they are necessary for the transitions relative to p and q), the two voxels which are face-neighbors of both p and w, and of both q and v, respectively (shown in gray) have already been assigned to the skeleton (during any previous iteration, or during the sub-iteration of the current iteration in which *f* has been applied) and the remaining two voxels (denoted for simplicity by white dots like v and w) are don't care. Since it is enough that one of the two voxels p and q is marked as non-removable to avoid creation of a spurious tunnel, the number of mirrored and rotated configurations figures up at 24. The 3×3×3 operator *g* is designed by taking into account the basic configuration shown in Fig.7, as well as the configurations derived from it.

For example, with reference to Fig.2, if p has q as its face-neighbor E in the central window, the neighbors of p that are actually checked by the operator *g* in N(p) are q, p_T, E_T, S, and SE. The voxel p is marked as non-removable if q is a non-marked border voxel, p_T and SE are background voxels, and E_T and S are marked voxels.

In summary, each iteration of the skeletonization algorithm can be outlined as follows.

Distinguish the object voxels into border, skeletal and inside voxels.

Apply *f* to the border voxels and, for any voxel identified as non-removable, change its status from border voxel to skeletal voxel.

Apply *g* to the border voxels and, for any voxel identified as non-removable, change its status from border voxel to skeletal voxel.

Remove all border voxels.

Skeletonization terminates when no voxels are removed, i.e., all voxels in the current border have been changed into skeletal voxels.

An example of the performance of the parallel algorithm is given in Fig.8.

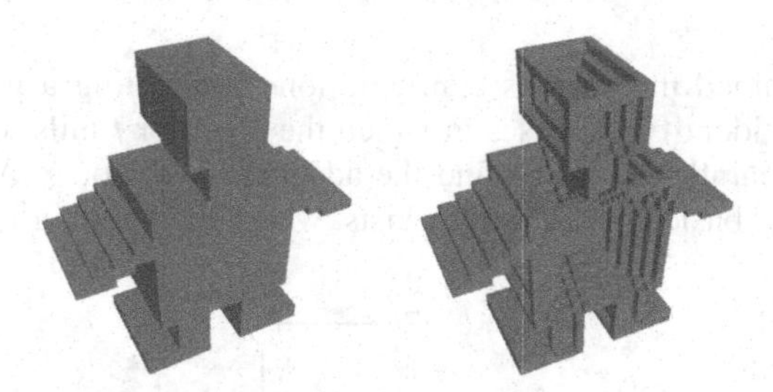

Fig. 8. Object, left, and its surface skeleton, right

In particular, when implementing the parallel algorithm on a standard sequential computer by scanning the 3D array in the top-bottom, left-right and front-back directions, as soon as the operator *g* identifies a voxel p, whose neighborhood satisfies any of the 24 configurations, p can be marked together with the relative face-neighbor q. In this way, both voxels in the pair matching the configurations derived from Fig.7 are marked, and a perfectly symmetrical surface skeleton is obtained. Moreover, DT can be conveniently used to reduce the computational effort. In fact, instead of including at each iteration the sub-iteration necessary to distinguish border voxels and inside voxels (which would require one scan of the image), distinction between border and inside voxels is available at all iterations once DT is computed (in two scans of the image). At the *l*-th iteration, the border includes all voxels with distance label *l* in DT as well as all voxels with distance label smaller than l and assigned to the skeleton during some previous iteration. The surface skeleton SK results to be centered within the object, includes all CMBs, is at most 2-voxel thick and is completely symmetrical.

6 Final Thinning

The surface skeleton obtained by using sequential or parallel voxel removal is expected to consist only of border voxels. However, inside voxels can still be present in SK. For example, this is the case for the object (coinciding with its surface skeleton) shown in Fig.9, which consists of three bricks, each with section 2×2 voxels and mutually crossing each other, where the kernel of the crossing region includes eight inside voxels. Another example is given by complex objects including irreducible sets, i.e., objects such that the only way to obtain a unit-thick SK would be that of forcing topology changes. We do not consider the case of irreducible sets here (for a discussion of this case in 2D see [12]) and refer to the standard case, where a few inside voxels exist only in the presence of crossings.

Both the algorithm based on sequential removal and the one based on parallel removal require final thinning if unit thickness is desired for SK. We describe below a final thinning algorithm based on sequential removal.

Fig. 9. A nearly-thin surface skeleton including eight inside voxels

Since the only voxels that should be removed by this process are those located where the object is 2-voxel thick in face-direction, we should detect the existence of such subsets of SK. To this purpose, we use six 4×1 masks, each consisting of four voxels forming a face-aligned path along one of the three principal directions (top-bottom or bottom-top, left-right or right-left, and front-back or back-front). In each mask, the two external voxels are background voxels and the two internal ones, say v_1 and v_2, are object voxels.

A simplified version f' of the operator f is used to perform final thinning along the direction where 2-voxel thickness is identified by one of the above masks. Of course, no transition exists in the direction of the mask, since SK is nearly-thin in that direction, but the voxel v_2 (v_1) plays the role of inside voxel in $N(v_1)$ ($N(v_2)$). The operator f' removes v_1 (v_2), depending on the number of object face-neighbors in $N(v_1)$ ($N(v_2)$).

Final thinning is done in two scans. During the first scan, the three masks detecting 2-voxel thickness along the top-bottom, left-right and front-back directions are used. During the second scan, the remaining three masks are used.

If, during the first scan, for the current voxel v_1 the mask is matched, but v_1 is not removed by the operator f', the coordinates of v_2 are recorded to directly access that voxel during the second scan. In this way, the mask does not need to be newly checked and the second scan is avoided.

Final thinning should be applied twice since, as said above, the nearly-thin skeleton might include some inside voxels, which could become removable border voxel, after removal of some of their neighbors.

7 Conclusion

We have introduced two 3×3×3 operators *f* and *g*. Only *f* is necessary if skeletonization is accomplished by sequential voxel removal. Both operators are used to design an iterative parallel algorithm for surface skeletonization. The obtained surface skeleton has all the expected properties except for unit thickness, as it is 2-voxel thick wherever the thickness of the object is expressed by an even number of voxels. In this respect, we have also introduced a sequential final thinning algorithm, which originates the unit-thick surface skeleton.

References

1. Rosenfeld, A.: A characterization of parallel thinning algorithms. Information and Control **29** (1975) 286-291
2. Tsao, Y.F., Fu, K.S.: A parallel thinning algorithm for 3-D pictures. Computer Graphics and Image Processing **17** (1981) 315-331
3. Saha, P.K., Chaudhuri, B.B.: Detection of 3D simple points for topology preserving transformations with application to thinning. IEEE Trans. PAMI **16** (1994) 1028-1032
4. Bertrand, G., Malandain, G.: A new characterization of three-dimensional simple points. Pattern Recognition Letters **15** (1994) 169-175
5. Kong, T.Y.: A digital fundamental group. Computers and Graphics. **13** (1989) 159-166
6. Kong, T.Y., Rosenfeld, A.: Appendix: Digital Topology – A brief introduction and bibliography. In: Kong, T.Y., Rosenfeld, A (eds.): Topological Algorithms for digital image processing. Elsevier, Amsterdam (1996) 263-292
7. Borgefors, G.: On digital distance transform in three dimensions. Computer Vision and Image Understanding. **64**(3) (1996) 368-376
8. Sanniti di Baja, G., Nystrom, I.: Skeletonization in 3D discrete binary images. In Chen, C.H., Wang, P.S.P. (eds.): Handbook of Pattern Recognition and Computer Vision. World Scientific, Singapore (2005) 137-156
9. Borgefors, G., Nyström, I., Svensson, S., Sanniti di Baja, G.: Simplification of 3D skeletons using distance information. In Latecki, L.J. et al. (eds.): Vision Geometry IX, Proc. SPIE 4117 (2000) 300-309
10. Hall, R.W.: Parallel connectivity-preserving thinning algorithms. In: Kong, T.Y., Rosenfeld, A (eds.): Topological Algorithms for digital image processing. Elsevier, Amsterdam (1996) 145-179
11. Arcelli, C., Sanniti di Baja, G., Serino, L.: Topology preservation in 3D" Internal Report R.I.161/05, Istituto di Cibernetica, December 2005
12. Eckhardt, U., Latecki, L., Maderlechner, G.: Irreducible and thin binary sets. In Arcelli, C. et al. (eds.): Aspects of Visual Form Processing. World Scientific, Singapore (1994) 199-208

Skeleton Pruning by Contour Partitioning

Xiang Bai[1,2], Longin Jan Latecki[1], and Wen-Yu Liu[2]

[1] CIS Dept., Temple University, Philadelphia, PA 19094, USA
{baixiang, latecki}@temple.edu
[2] Dept of Electronics & Information Engineering, Huazhong University of Sci. & Tech. Wuhan, Hubei. 430074 P.R. China
liuwy@hust.edu.cn

Abstract. In this paper, we establish a unique correspondence between skeleton branches and subarcs of object contours. Based on this correspondence, a skeleton is pruned by removing skeleton branches whose generating points are on the same contour subarc. This has an effect of removing redundant skeleton branches and retaining all the necessary visual branches. We show that this approach preserves skeleton topology, does not shift the skeleton, and it does not shrink the remaining branches.

Keywords: Skeleton, skeleton pruning, discrete curve evolution.

1 Introduction

The skeleton is important for object representation and recognition in different areas, such as image retrieval and computer graphics, character recognition, image processing, and analysis of biomedical images [1]. Skeleton-based representations are the abstraction of objects, which contain both shape features and topological structures of original objects. Due to the importance of the skeleton, many skeletonization algorithms have been developed to represent and measure different shapes. The important factor that constraint the matching of skeletons is the skeleton's sensitivity to object's boundary deformation: little noise or variation of boundary often generates redundant skeleton branches that may disturb the topology of skeleton's graph seriously. For example, the skeleton in Fig. 1(a) has many redundant skeleton branches generated by boundary noise.

The most common approaches to overcome skeleton's instability are based on skeleton pruning, i.e., eliminating redundant skeleton branches. Pruning can be either performed implicitly as a post processing step or implicitly integrated in the skeleton computation. However, none of the existing skeleton pruning methods yields satisfactory results without user interaction. Before we describe the existing skeleton pruning approaches, we characterize desirable properties of skeletons. The skeleton of a single connected shape that is useful for skeleton-based recognition should have the following properties: (1) it should preserve the topological information of the original object; (2) the position of the skeleton should be accurate; (3) it should be stable under small deformations; (4) it should contain the centers of maximal disks, which can be used for reconstruction of original object; (5) it should be invariant under

A. Kuba, L.G. Nyúl, and K. Palágyi (Eds.): DGCI 2006, LNCS 4245, pp. 567 – 579, 2006.

Euclidean transformations, such as rotations and translations, and (6) it should represent significant visual parts of objects.

The main goal of this paper is to present a novel skeleton-pruning method that achieves all the above properties. We stress that no existing method can provide a skeleton with all these properties. In addition the proposed method is easy to implement, and can be computed efficiently.

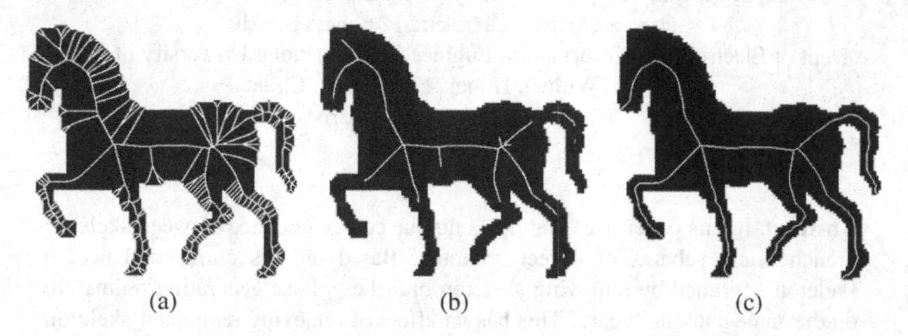

Fig. 1. The skeleton in (a) has many redundant branches. To remove them, usually skeleton pruning is applied. (b) illustrates the problems of actual pruning approaches (it is generated by a method in [7]). In particular, observe that pruning may change the topology of the original skeleton. (c) illustrates the pruning result of the proposed method that is guaranteed to preserve topology.

Now we present a brief overview of skeletonization and skeleton-pruning approaches. The skeletonization algorithms can broadly be classified into four types: thinning algorithms, e.g., [5, 6]; algorithms based on the Voronoi diagrams, e.g., [2, 8, 22]; algorithms based on distance maps, e.g., [4, 6, 7, 9, 15, 24, 25, 29]; algorithms based on mathematical morphology, e.g., [19-21]. All these algorithms require skeleton pruning, either as a postprocessing step or directly incorporated in the course of computation. There are mainly two ways of pruning methods: (1) based on significance measures assigned to skeleton points [2-4, 16, 22], and (2) based on boundary smoothing before extracting the skeletons [16, 28]. In particular, curvature flow smoothing still have some significant problems that make the position of skeletons shift and have difficulty in distinguishing noise from low frequency shape information on the boundary [16]. A different kind of smoothing is proposed in [10]. A great progress have been made in the type (1) of pruning approaches that define a significance measure for skeleton points and remove points whose significance is low. Shaked and Bruckstein [16] give a complete analysis and compare such pruning methods. To the common significance measures of skeleton point belong propagation velocity, maximal thickness, radius function, axis arc length, the length of the boundary unfolded. Ogniewicz and Kübler [2] present a few significance measures for pruning hairy Voronoi skeletons without disconnecting the skeletons. Siddiqi et al combine a flux measurement with the thinning process to extract a robust and accurate connected skeleton [20].

All presented methods have several drawbacks. First, many of them are not guaranteed to preserve topology. This is illustrated in Fig. 2, where the skeleton in (d) violates the topology of the input skeleton in (c). This skeleton was obtained by the method in [4]. However, any other method described above would lead to topology violation. In particular, all methods presented in [16] (including the method in [2]). Methods described in [16] are guaranteed to preserve topology for simply connected objects (objects with a single contour), but not for objects with more than one contour like the can in Fig. 2. There exist several methods that preserve topology [24, 30] and thinning algorithms (that are usually guaranteed to preserve topology). The topology preserving skeleton obtained by the proposed pruning method is illustrated in Fig. 2(e). Even if the input shape is simply connected many of methods described above are not guaranteed to preserve the original topology (e.g., the skeleton in Fig. 1(b), generated by pruning method in [7]).

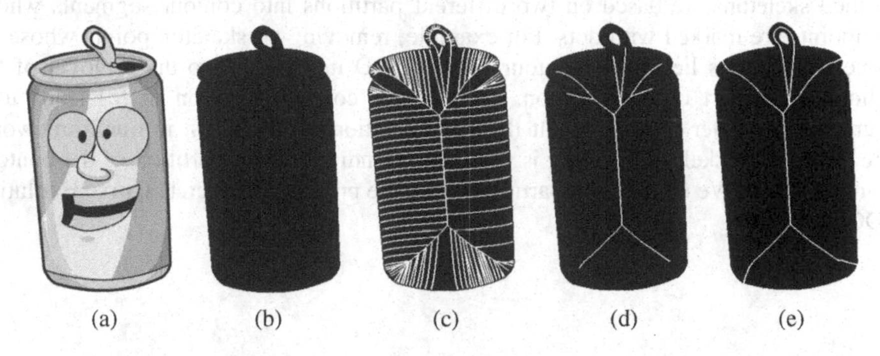

Fig. 2. (a) The input object. (b) Binary object mask. (c) The initial skeleton. (d) A pruned skeleton obtained by the method in [4]. (e) A pruned skeleton obtained by the proposed method. While skeleton in (d) violates the topology, the proposed method guaranteed to preserve the topology.

The second drawback of all methods described above is that main skeleton branches are shorten to some extent and short skeleton branches are not removed completely, which seriously complicates the structure of the skeletons. These effects are illustrated in Figs. 1(b) and 2(d).

To summarize, although the existing skeleton pruning methods have many drawbacks, they are definitely needed to remove spurious or redundant skeleton branches. The skeleton generating approaches suffer from the fact that a small protrusion on the boundary may result in a large skeleton branch, which is an intrinsic problem of the definition of the skeleton, since the mapping of boundary points to the skeleton points is not continuous. An obvious solution to this problem is to first remove the protrusions on the boundary and then compute the skeleton. As stated above, various smoothing approaches are either applied to the contour or to the distance map before the skeleton is computed. The problem is that isotropic (e.g., Gaussian) as well as anisotropic smoothing only reduces but does not remove the protrusions. A common characteristic of the above approaches is that they displace the boundary points, and consequently displace the location of skeleton points.

2 Main Ideas of the Proposed Approach

We propose an approach that completely removes protrusions without displacing the boundary points, and consequently, without displacing the remaining skeleton points.

Spurious or redundant branches are completely removed while the main branches are not shorten. The main observation of our approach is that it is possible to perform a topology preserving skeleton pruning based on a contour partition into curve segments. Returning to Blum's definition of the skeleton, every skeleton point is linked to boundary points that are tangential to its maximal circle, so called **generating points**. The main idea is to remove all skeleton points whose generating points all lie on the same contour segment. This works for any contour partition into segments, by some partitions yield better results than other. Fig. 3 illustrated two different pruned skeletons (b) and (c) obtained for the same input skeleton in (a). The pruned skeletons are based on two different partitions into contour segments whose endpoints are marked with dots. For example, removing all skeleton points whose all generating points lie on the contour segment CD in (b) leads to the removal of the whole lower part of the skeleton. Clearly, the contour partition in (c) leads to a significantly better pruning result than the partition in (b). Thus, in our framework, the question of skeleton pruning is reduced to finding a good partition of the contour into segments. We obtain such partitions with the process of Discrete Curve Evolution (DCE) [11-13].

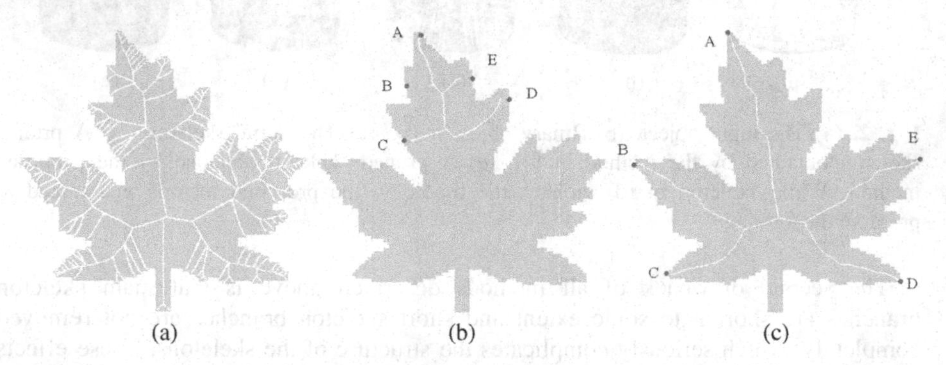

Fig. 3. Pruning the input skeleton (a) with respect to contour partition induced by five random points on the boundary in (b). The five points in (c) are selected with DCE.

First, observe that every object boundary in a digital image can be represented without loss of information as a finite polygon, due to finite image resolution. The process of DCE is proven experimentally and theoretically to eliminate the noisy points [11-13]. This process eliminates such points by recursively removing polygon vertices with the smallest shape contribution (which are the most likely to result from noise). As the result of DCE, we obtain a subset of vertices that best represent the shape of a given contour. This subset can also be viewed as a partition of the original contour polygon into contour segments defined by consecutive vertices of the simplified polygon. A hierarchical skeleton structure obtained by the proposed approach is illustrated in Fig. 4, where the (red) bounding polygon represents the contour simplified by DCE.

Because DCE can reduce the boundary noise without displacing the remaining boundary points, the accuracy of the skeleton position is guaranteed. The continuity, which implies stability in the presence of noise, of the proposed pruning methods follows from the continuity of the DCE. This means if a given contour and its noisy versions are close (measured by Hausdorff distance), the obtained pruned skeletons will be close too. A formal proof of DCE continuity with respect to Hausdorff distance of polygonal curves is given in [18]. Thus, our approach provides a solution to the instability of the classical skeleton pruning algorithms.

The proposed pruning method can be applied to any input skeleton. We only require that each skeleton point is the center of a maximal disk and that the boundary points tangent to the disk (generating points) are given. It is also possible to perform a skeleton growing that includes the proposed pruning method. The main idea is that the pruning is not done in postprocessing (after the skeleton is computed) but is integrated into the skeleton growing process. To implement this idea, we extended the skeleton growing algorithm in [4] based on the Euclidean distance map. First, we select a skeleton seed point as a global maximum of the Euclidean distance map. Then, the remainder of the skeleton points is decided by a growing scheme. In this scheme, the new skeleton points are added using a simple test that examines their eight connected points. During this process, the redundant skeleton branches are eliminated with respect to a given contour partition computed by DCE.

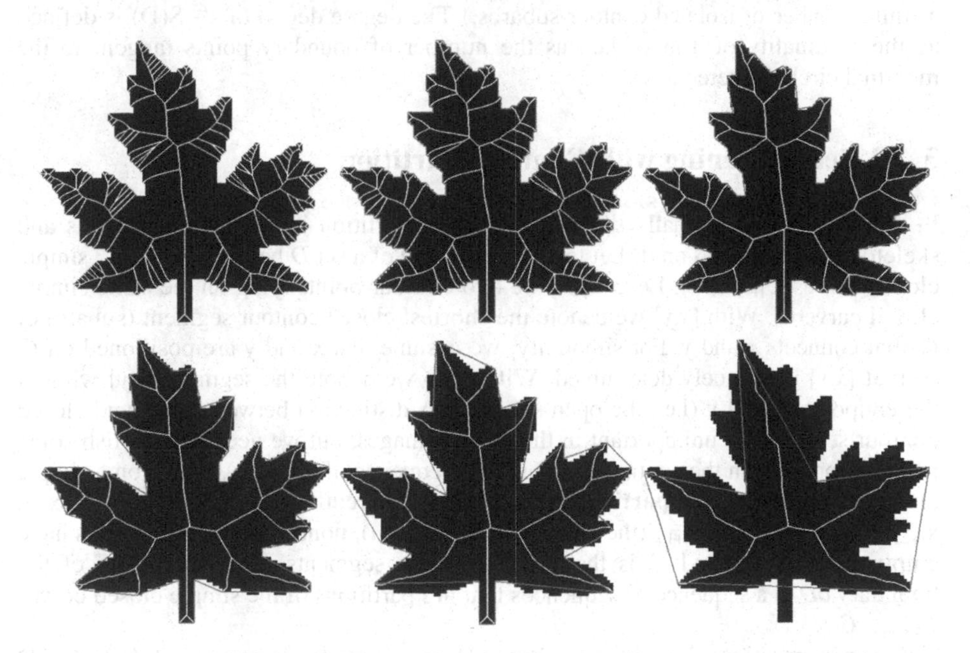

Fig. 4. Hierarchical skeleton of leaf obtained by pruning the input skeleton (top left) with respect to contour segments obtained by the Discrete Curve Evolution (DCE). The outer (red) polylines show the corresponding DCE simplified contours.

Before we formally define the proposed skeleton pruning, we need to characterize planar sets for which we can determine the skeleton. Following [23] we assume that a planer set D is the closure of a connected bounded open subset of R^2 whose boundary ∂D is composed of a finite number of mutually disjoint simple closed curves. Each simple closed curve in ∂D consists of finite number of pieces of real analytic curves. We further assume in this paper that each simple closed curve is a polygonal curve, i.e., the pieces they consist of are line segments. We make this assumption only to simplify some definitions, and we stress that all our results also hold for simple closed curves that consist of finite number of real analytic curves. Observe also that this assumption does not introduce any restriction on object contours in digital images, since each boundary curve in a digital image can be regarded as polygonal curve with vertices being the boundary pixels.

According to Blum's definition of the medial axis [1], the skeleton $S(D)$ of a set D is the locus of the centers of maximal disks. A maximal disk of D is a closed disk contained in D that is interiorly tangent to the boundary ∂D and that is not contained in any other disk in D. Observe that each maximal disc must be tangent to the boundary in at least two different points. We denote as **Tan(s)** the set of the boundary points tangent to the maximal disk B(s) centered at $s \in S(D)$. The points in Tan(s) are called **generating points** of the skeleton point s. Due to our assumption that each boundary curve is a simple closed polygonal curve, Tan(s) is composed of a final number of isolated boundary points, since B(s) can intersect each boundary line segment in at most one point. (Without this assumption, Tan(s) would be composed of a finite number of isolated contour subarcs.) The degree deg(s) of $s \in S(D)$ is defined as the cardinality of Tan(s), i.e., as the number of boundary points tangent to the maximal circle centered at s.

3 Skeleton Pruning with Contour Partition

In this section, we formally define the contour partition into contour segments and skeleton pruning based on it. Let the boundary ∂D of a set D be composed of k simple closed curves $C_1, \ldots, C_k$. Let x and y be two contour points lying on the same simple closed curve C_i. With $[x,y]$ we denote the shortest closed contour segment (subarc) of C_i that connects x and y. For simplicity, we assume that x and y are positioned on C_i so that $[x,y]$ is uniquely determined. With (x,y) we denote the segment $[x,y]$ without the endpoints x and y (i.e., the open subarc). (A distinction between open and closed contour segments is unimportant in the digital images, but we need to establish some formal properties in the continuous plane.) A sequence of points $x_0, \ldots, x_{n-1}$ on a simple closed curve C_i forms a **partition** of C_i if two consecutive segments $[x_i, x_{i+1}]$, $[x_{i+1}, x_{i+2}]$ that intersect in $\{x_{i+1}\}$ (the indices are modulo n), nonconsecutive segments have empty intersection, and C_i is the union of these segments. The partition Γ of the boundary ∂D is a sequence of sequences that are partitions of the simple closed curves $C_1, \ldots, C_k$.

Given a partition Γ of the boundary ∂D of a simply connected set D (i.e., ∂D consist of one simple closed curve), the **skeleton pruning** is defined as removal of all skeleton points $s \in S(D)$ whose all generating points lie in the same open segment (x,y) of partition Γ. More precisely, the **pruned skeleton** is composed of all points $s \in S(D)$ such that $Tan(s)$ is not contained in the same open segment of the partition Γ. This is a very simple definition of skeleton pruning, and it works with any contour partition. The key issue is to get reasonable partitions. As we will show, DCE provides a very good partition for the pruning. When D is not simply connected (i.e., ∂D consist of more than one simple closed curve), we additionally need to check whether all skeleton points removed with respect to the same partition segment have their generating points on the same simple closed curve C_i for some $i=1, \ldots, k$. It can be proven that the topology of pruned skeleton is preserved for pruned skeleton generated by any partition of the contour. The proof is omitted due to the limited space. It will appear in a forthcoming paper. For simplicity of the presentation, we assume in the following definitions that D is simply connected (i.e., ∂D consist of one simple closed curve).

Now we introduce a contour partition based on Discrete Curve Evolution (DCE). A hierarchical decomposition of the boundary of the set D obtained by DCE is the key component of the proposed skeleton pruning method. Given a skeleton $S(D)$ of a planar shape D and given a DCE simplified contour D given as polygon P^k with convex vertices $x_0, \ldots, x_{n-1}$, we perform skeleton pruning with respect to the open segments of the partition $[x_i, x_{i+1}]$ (modulo n). Thus, we remove all points $s \in S(D)$ such that the generating points of s, $Tan(s)$, are contained in a single open side of the DCE polygon. Observe that we ignore concave vertices of the DCE simplified polygon.

The simplification of the boundary contour with DCE corresponds to pruning complete branches of the skeleton. In particular, a removal of a single convex vertex v from P^{n-k} to obtain $P^{n-(k+1)}$ by DCE implies a complete removal of the skeleton branch that ends at v. We give an example illustrating this fact in Fig. 5. It shows a polygon with seven vertices obtained from the leaf contour by DCE and the skeleton obtained by pruning based on this polygon. Observe that there are only five skeleton branches ending in the five convex vertices of the simplified polygon. The pruned skeleton is computed with respect to the DCE segments (A,C), (C,D), (D,E), (E,F), (F, A). The pruning was applied to the leaf skeleton shown in the first image in Fig. 4. (The skeleton in Fig. 5 is the same as in the last image in Fig. 4.) We can illustrate the main idea of our approach by explaining why the green skeleton branch in Fig. 5 that ends at point C remained. It remained, since each of its points has maximal disks tangent to points on two different DCE segments, which are contour arcs (A,C) and (C,D).

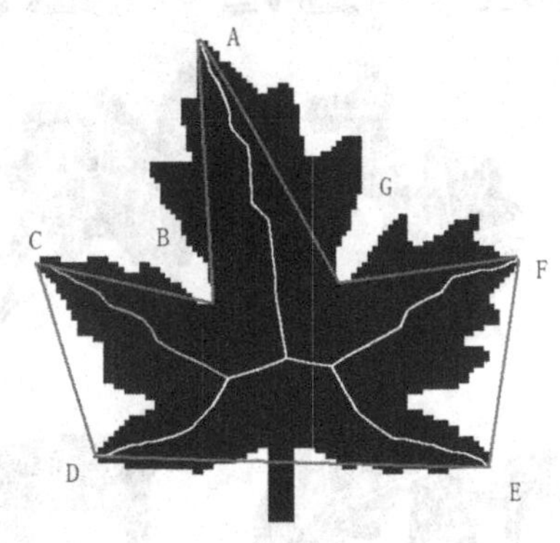

Fig. 5. A simplified polygon with 7 vertexes (in red) and the skeleton obtained based on this polygon. The green skeleton branch remained, since each of its points has generating points on two different arcs BC and CD of the original contour.

A very important property of DCE induced contour partition, and every partition that is restricted to vertices of the boundary polygon, is that fact that there is a skeleton branch ending at every partition point. If a partition point that is also a polygon vertex u_i is deleted in a DCE evolution step, i.e., $u_i \in P^{n-k} - P^{n-(k+1)}$, then the arc $[u_{i-1},u_{i+1}]$ replaces arcs $[u_{i-1},u_i]$, $[u_i,u_{i+1}]$ in the contour partition. Consequently, the whole skeleton branch that ends at vertex u_i is eliminated by the skeleton pruning.

4 Experimental Results and Comparison

4.1 Stability of Pruning with DCE

Some results on shapes from MPEG-7 Core Experiment CE-Shape-1 database [27] are showed in Fig. 6. For each shape class, we show pruned skeletons for several objects from the same class. Although the objects differ significantly from each other, the obtained pruned skeletons have the same structures. The final DCE simplified polygons are also showed overlaid on the shapes with red segments. The skeleton pruning is performed with respect to contour partition induced by the vertices of these polygons. In the first row in Fig. 6, the skeletons of the thin and long tails of rats remained complete, which cannot be achieved by other pruning methods, since they may shorten or disconnect the skeleton. Although the camels differ significantly in their shape, all obtained skeletons have similar global structure. The last row of Fig. 6 shows the DCE's stability to the same shapes in different scales.

Fig. 6. Our results on Mpeg 7 shape database illustrate extraordinary stability of pruned skeletons in the presence of significant shape variations and deformations

4.2 Analysis and Comparison

In this part, we describe our test results with the proposed approach on several binary shape images with the size 500×500. All the images tested have significant boundary distortions.

A hierarchy of pruned skeletons is shown for the walking human in Fig. 7. The pruning is preformed with respect to DCE simplified contours with N = 200, 100, 50, 30, and 12 vertices. We have also shown a hierarchy of pruned skeletons in Fig. 4 above. We can see that the results of our algorithm are in accord with human visual perception. Besides hierarchical and visual property, our skeleton has a unique property: the pruned branches are eliminated completely, i.e., the obtained skeletons are without the presence of remaining half-shortened small, short branches. For example in Fig. 7, each skeleton branch is removed, and no remaining fractions are left.

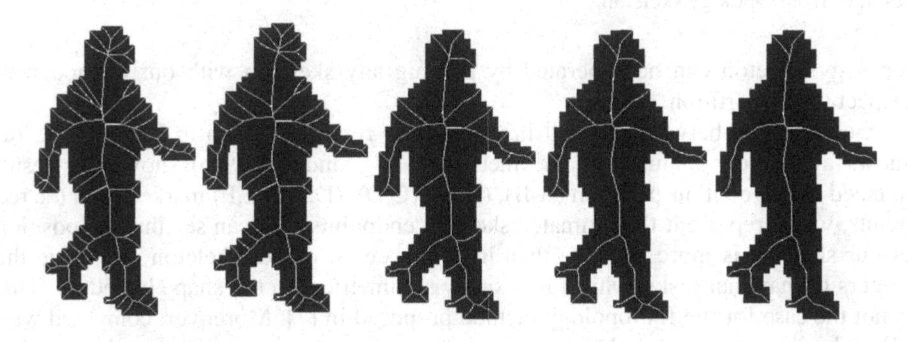

Fig. 7. Hierarchical skeleton of a walking human. The input image is similar to a walking human in [7].

The problem of spurious, half-shortened braches is clearly seen in Fig. 8(a), below, Figs. 1(b) and. 2(d), above, where we see several spurious branches that are not related to any obvious boundary features. Figs. 1 and 2 show a comparison of our method and the method in [4]. As can be clearly seen our method does not suffer from shortening main skeleton branches, and preserves topology of the skeleton. Moreover, the obtained skeletons seem to be in accord with human perception. The result obtained by the method in [4] exhibits clear problems with the skeleton topology and it shortens main skeleton branches in Figs. 1(b) and 2(d).

Our pruning method can also be used in pruning Voronoi skeleton branches. As Voronoi skeleton points are symmetrical to the boundary sample points, the generating boundary points of each skeleton point are known. Fig. 8 shows a comparison of our method to the method by Ogniewicz and Kübler [2]. Fig. 8(b) shows that the proposed approach is able to eliminate all the unimportant branches and still preserve all main structure. Fig. 8(c) shows an application of our method to generate a fixed topology skeleton introduced in Golland and Grimson [7]. It shows that the proposed pruning is not limited to the DCE induced contour partitioning. Once the positions of skeleton's endpoints is estimated along the boundary as in the method in [7], the endpoints induce a partition of the boundary curve, and the fixed

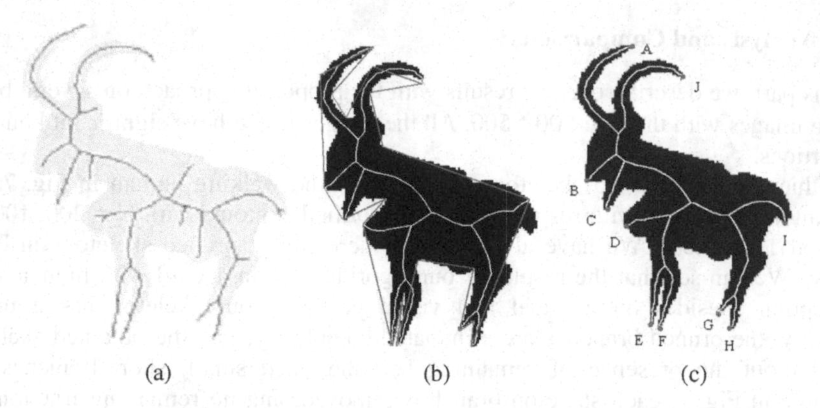

Fig. 8. Comparison between pruning result in [2] in (a) and our results in (b), and (c) is the result of fixed topology skeleton

topology skeleton can be generated by pruning any skeleton with our method with respect to this partition.

A comparison between a result in [7] and our result is shown in Fig. 9. Fig. 9(a) shows a skeleton obtained by the method in [7], and Fig. 9(b) shows our result induced by the contour partition (A,B), (B,C), (C,D), (D,E), (E,F) marked with the red points, which represent the estimated skeleton endpoints. We can see that the position of our skeleton is more accurate than in (a), since all of our skeleton points are the centers of maximal disks, which are exactly symmetrical to the shape boundary. This is not the case for the fix topology method proposed in [7]. Moreover, compared with [7], only the endpoints need to be estimated; we do not need to estimate the junction points of the skeleton. Another example of a fixed topology skeleton produced with our method is shown in Fig. 8(c).

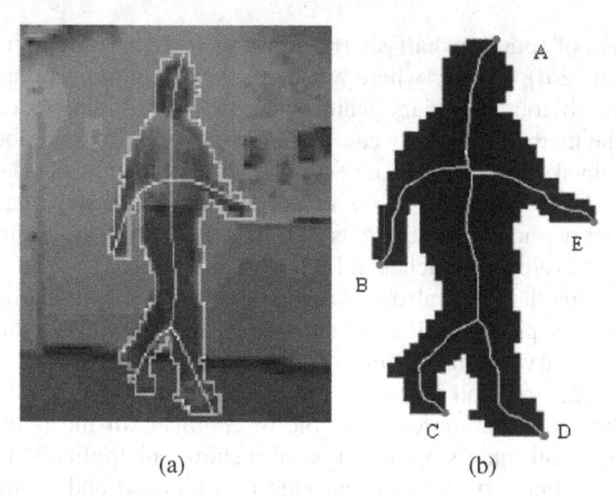

Fig. 9. Comparison between the fixed topology skeleton in [7] in (a) and our skeleton in (b)

We omitted here a formal prove that our method is guaranteed to preserve topology. We illustrated this fact in Fig. 2(e) above. Fig. 10 shows another example for a shape with three holes that has total of four contour curves. For comparison, the result of the method in [4] is shown in Fig. 10(b). Fig. 10(c) shows that the proposed approach preserves the original topology. In Fig. 10 (d), the contour partition is only composed of the four boundary curves, i.e., there are no segments on any of the four curves, so that the skeleton points must have their tangent points on the different boundary curves in order to remain.

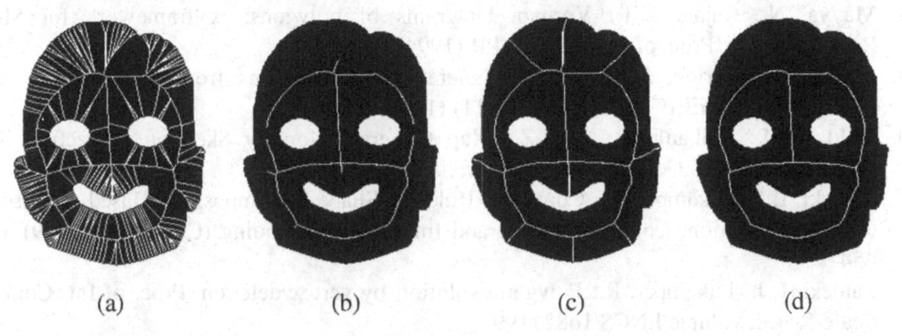

Fig. 10. (a) The input skeleton. (b) A pruned skeleton obtained by the method in [4] violates the topology. (c,d) Pruned skeletons obtained by the proposed method, which is guaranteed to preserve the topology.

5 Conclusions and Future Work

In this paper, we introduce a new skeleton pruning method based on contour partitioning. Any contour partition can be used, but the partitions obtained by Discrete Curve Evolution (DCE) yield very good results. The theoretical properties and the presented experiments demonstrate that the obtained skeletons are in accord with human visual perception and stable, even in the presence of significant noise, and have the same topology as the original skeletons. Thus, we provide a solution to the instability of the classical skeleton algorithms. The stability of skeletons is the key property required to measure shape similarity of objects using their skeletons. Moreover, we have shown that the proposed approach never produces spurious branches, so common to the known skeleton pruning methods, and that the proposed pruning method does not displace the skeleton points. Consequently, all skeleton points are centers of maximal disks. We have proven that our pruning method is guaranteed to preserve topology. The proof will appear in a forthcoming paper.

References

1. Blum, H.: Biological Shape and Visual Science (Part I). J. Theoretical Biology **38** (1973) 205–287
2. Ogniewicz, R.L., Kübler, O.: Hierarchic Voronoi skeletons, Pattern Recognition **28**(3) (1995) 343–359

3. Malandain, G., Fernandez-Vidal, S.: Euclidean skeletons. Image and Vision Computing **16** (1998) 317–327
4. Choi, W.-P., Lam, K.-M., Siu, W.-C.: Extraction of the Euclidean skeleton based on a connectivity criterion. Pattern Recognition **36** (2003) 721–729
5. Pudney, C.: Distance-Ordered Homotopic Thinning: A Skeletonization Algorithm for 3D Digital Images. Computer Vision and Image Understanding **72**(3) (1998) 404–413
6. Leymarie, F., Levine, M.: Simulating the grassfire transaction form using an active Contour model. IEEE Trans. PAMI **14**(1) (1992) 56–75
7. Golland, P., Grimson, E.: Fixed topology skeletons. In CVPR, Vol. 1 (2000) 10–17
8. Mayya, N., Rajan, V.T.: Voronoi Diagrams of polygons: A framework for Shape Representation. Proc. of the IEEE CVPR (1994) 638–643
9. Ge, Y., Fitzpatrick, J.M.: On the Generation of Skeletons from Discrete Euclidean Distance Maps. IEEE Trans. PAMI **18**(11) (1996) 1055–1066
10. Gold, C.M., Thibault, D., Liu, Z.: Map Generalization by Skeleton Retraction. ICA Workshop on Map Generalization, Ottawa, August 1999
11. Latecki, L.J., Lakämper, R.: Convexity Rule for Shape Decomposition Based on Discrete Contour Evolution. Computer Vision and Image Understanding (CVIU) **73** (1999) 441–454
12. Latecki, L.J., Lakamper, R.: Polygon evolution by vertex deletion. Proc. of Int. Conf. on Scale-Space, volume LNCS 1682 (1999)
13. Latecki, L.J., Lakamper, R.: Shape similarity measure based on correspondence of visual parts, IEEE Trans. Pattern Analysis and Machine Intelligence (PAMI) **22**(10) (2000) 1185–1190
14. Latecki, L.J., Lakamper, R.: Application of planar shape comparison to object retrieval in image databases. Pattern Recognition **35**(1) (2002) 15–29
15. Borgefors, G.: Distance transformations in digital images. Computer Vision, Graphics and Image Processing **34**(3) (1986) 344–371
16. Shaken, D., Bruckstein, A.M.: Pruning Medial Axes. Computer Vision and Image Understanding **69**(2) (1998) 156–169
17. Dimitrov, P., Damon, J.N., Siddiqi, K.: Flux Invariants for Shape. CVPR (2003)
18. Latecki, L.J., Ghadially, R.-R., Lakämper, R., Eckhardt, U.: Continuity of the discrete curve evolution. Journal of Electronic Imaging **9**(3) (2000) 317–326
19. Dimitrov, P., Phillips, C., Siddiqi, K.: Robust and Efficient Skeletal Graphs. In CVPR (2000) 1417–1423
20. Siddiqi, K., Bouix, S., Tannenbaum, A.R., Zucker, S.W.: Hamilton-Jacobi Skeletons. International Journal of Computer Vision **48**(3) (2002) 215–231
21. Vasilevskiy, A., Siddiqi, K.: Flux Maximizing Geometric Flows. IEEE Trans. PAMI **24**(12) (2002) 1565–1578
22. Brandt, J.W. Algazi, V.R.: Continuous skeleton computation by Voronoi diagram. Comput. Vision, Graphics, Image Processing **55** (1992) 329–338
23. Choi, H.I., Choi, S.W., Moon, H.P.: Mathematical Theory of Medial Axis Transform. Pacific Journal of Mathematics **181**(1) (1997) 57–88
24. Arcelli, C. Sanniti di Baja, G.: Euclidean skeleton via center of maximal disk extraction. Image and Vision Computing **11** (1993) 163–173
25. Kimmel, R., et al.: Skeletonization via Distance Maps and Level Sets. CVIU: Comp. Vision and Image Understanding **62**(3) (1995) 382–391
26. Sebastian, T.B., Klein, P.N., Kimia, B.B.: Recognition of shapes by editing their shock graphs. IEEE Trans. PAMI **26**(5) (2004) 550–571

27. Latecki, L.J., Lakamper, R., Eckhardt, U.: Shape Descriptors for Non-rigid Shapes with a Single Closed Contour. Proc. CVPR (2000)
28. Pizer, S.M., Oliver, W.R., Bloomberg, S.H.: Hierarchial shape description via the multiresolution symmetric axis transform. IEEE Trans. PAMI **9** (1987) 505–511
29. Borgefors, G., Ramella, G., Sanniti di Baja, G.: Hierarchical decomposition of multiscale skeletons. IEEE Trans. PAMI **13**(11) (2001) 1296–1312
30. Sanniti di Baja, G.: Well-shaped, stable and reversible skeletons from the (3,4)-distance transform. Visual Communication and Image Representation **5** (1994) 107–115

A New 3D Parallel Thinning Scheme Based on Critical Kernels

Gilles Bertrand and Michel Couprie

Institut Gaspard-Monge
Laboratoire A2SI, Groupe ESIEE
Cité Descartes, BP 99
93162 Noisy-le-Grand Cedex France
g.bertrand@esiee.fr, m.couprie@esiee.fr

Abstract. Critical kernels constitute a general framework settled in the category of abstract complexes for the study of parallel thinning in any dimension. We take advantage of the properties of this framework, and we derive a general methodology for designing parallel algorithms for skeletons of objects in 3D grids. In fact, this methodology does not need to handle the structure of abstract complexes, we show that only 3 masks defined in the classical cubic grid are sufficient to implement it. We illustrate our methodology by giving two new types of skeletons.

Introduction

Precisely forty years ago, in 1966, D. Rutovitz proposed an algorithm which is certainly the first parallel thinning algorithm [1]. Since then, many 2D parallel thinning algorithms have been proposed, and several 3D ones may be found in the literature, see for example [2, 3, 4, 5, 6, 7, 8, 9, 10, 11, 12]. A fundamental property required for such algorithms is that, they do preserve the topology of the original objects. In fact, such a guarantee is not obvious to obtain [13, 14]. In [15], see also [16], one of the authors introduces a general framework for the study of parallel thinning in any dimension in the context of abstract complexes. A new definition of a simple point has been proposed, this definition is based on the collapse operation which is a classical tool in algebraic topology and which guarantees topology preservation. The most fundamental result is that, if a subset Y of X contains the so-called *critical kernel* of X, then Y has the same topology as X.

In this paper, we focus on 3D objects. We introduce the notion of *crucial voxels*, which permits to make a link with the framework of digital topology [17]. This leads to a general methodology for designing 3D parallel thinning algorithms. We illustrate our methodology by giving two new types of skeletons. The first one corresponds to a "minimal" skeleton of an object, the second one to a skeleton which contains a part of the medial axis of an object. All these skeletons are obtained by a sequence of symmetric operators. To our best knowledge, they have no equivalent. All previously proposed symmetric thinning conditions

A. Kuba, L.G. Nyúl, and K. Palágyi (Eds.): DGCI 2006, LNCS 4245, pp. 580–591, 2006.

are not sufficiently "powerful" for removing enough points in order to obtain a skeleton such as the "minimal" skeleton of Fig. 7.

For the sake of space, proofs are not given in this paper, most of them will be available in an extended version of the paper.

1 Cubical Complexes

In this section, we give some basic definitions for cubical complexes, see also [18]. We consider here the three-dimensional case. Note that many of the notions introduced in the first sections make sense in arbitrary n-dimensional cubical spaces.

If T is a subset of S, we write $T \subseteq S$, we also write $T \subset S$ if $T \subseteq S$ and $T \neq S$.

Let $\mathbb{Z}$ be the set of integers. We consider the families of sets $\mathbb{F}_0^1$, $\mathbb{F}_1^1$, such that $\mathbb{F}_0^1 = \{\{a\} \mid a \in \mathbb{Z}\}$, $\mathbb{F}_1^1 = \{\{a, a+1\} \mid a \in \mathbb{Z}\}$. A subset f of $\mathbb{Z}^n$, $n \geq 2$, which is the Cartesian product of exactly m elements of $\mathbb{F}_1^1$ and $(n-m)$ elements of $\mathbb{F}_0^1$ is called a *face* or an *m-face* of $\mathbb{Z}^n$, m is the *dimension of* f, we write $dim(f) = m$.

We denote by $\mathbb{F}^3$ the set composed of all m-faces of $\mathbb{Z}^3$, $m = 0, 1, 2, 3$. An *m-face* of $\mathbb{Z}^3$ is called a *point* if $m = 0$, a *(unit) interval* if $m = 1$, a *(unit) square* if $m = 2$, a *(unit) cube* if $m = 3$.

Let f be a face in $\mathbb{F}^3$. We set $\hat{f} = \{g \in \mathbb{F}^3 \mid g \subseteq f\}$ and $\hat{f}^* = \hat{f} \setminus \{f\}$.

Any $g \in \hat{f}$ is a *face of* f, and any $g \in \hat{f}^*$ is a *proper face of* f.

If X is a finite set of faces in $\mathbb{F}^3$, we write $X^- = \cup\{\hat{f} \mid f \in X\}$, X^- is the *closure of* X.

A set X of faces in $\mathbb{F}^3$ is a *cell* or an *m-cell* if there exists an m-face $f \in X$, such that $X = \hat{f}$. The *boundary of a cell* $\hat{f}$ is the set $\hat{f}^*$.

A finite set X of faces in $\mathbb{F}^3$ is a *complex (in $\mathbb{F}^3$)* if $X = X^-$. Any subset Y of a complex X which is also a complex is a *subcomplex of* X. If Y is a subcomplex of X, we write $Y \preceq X$. If X is a complex in $\mathbb{F}^3$, we also write $X \preceq \mathbb{F}^3$.

Let $X \preceq \mathbb{F}^3$. A face $f \in X$ is *principal for* X if there is no $g \in X$ such that $f \in \hat{g}^*$. We denote by X^+ the set composed of all principal faces of X.

Observe that, in general, X^+ is not a complex, and that $[X^+]^- = X$. See illustrations Fig. 1.

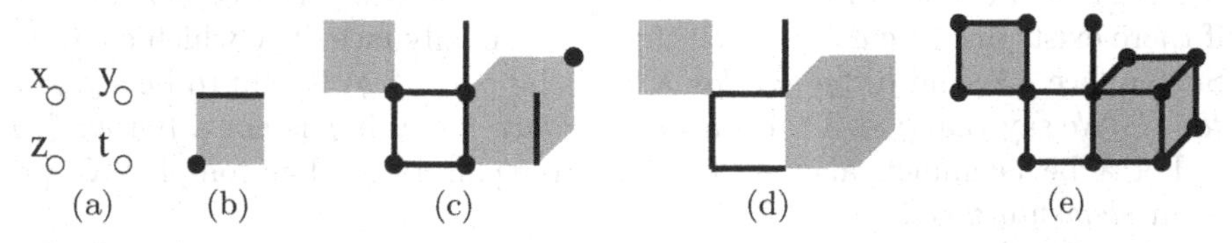

Fig. 1. (a): Four points x, y, z, t. (b): A graphical representation of the set of faces $\{\{x, y, z, t\}, \{x, y\}, \{z\}\}$. (c): A set of faces X, which is not a complex. (d): The set X^+, composed by the principal faces of X. (e): The set X^-, *i.e.* the closure of X, which is a complex.

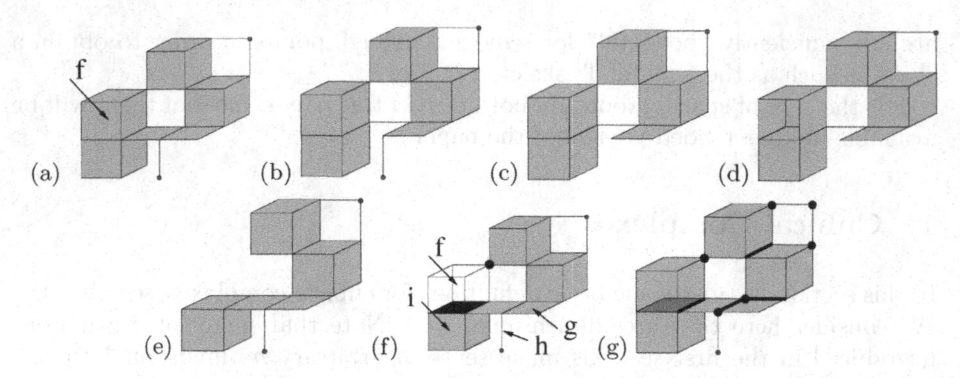

Fig. 2. (a) A complex X, (b), (c), and (d) three steps of elementary collapse of X, (e) the detachment of $\hat{f}$ from X, (f) the attachment of the 3-face f is highlighted in dark, the face f is not simple, whereas g, h, and i are simple, (g) the essential faces of X which are not principal are highlighted in dark

Let $X \preceq \mathbb{F}^3$, $dim(X) = \max\{dim(f) \mid f \in X^+\}$ is the *dimension of* X. We say that X is an m-complex if $dim(X) = m$.

We say that X is *pure* if, for each $f \in X^+$, we have $dim(f) = dim(X)$.

Let $X \preceq \mathbb{F}^3$ and $Y \preceq X$. If $Y^+ \subseteq X^+$, we say that Y is a *principal subcomplex of* X and we write $Y \sqsubseteq X$. Observe that, for any $X \preceq \mathbb{F}^3$, $\emptyset \sqsubseteq X$.

If $X \preceq \mathbb{F}^3$ and if X is a pure 3-complex, we also write $X \sqsubseteq \mathbb{F}^3$.

Let $X \preceq \mathbb{F}^3$ and let $Y \preceq X$. We set $X \oslash Y = [X^+ \setminus Y^+]^-$. The set $X \oslash Y$ is a complex which is the *detachment of* Y *from* X.

2 Simple Cells

Intuitively a cell $\hat{f}$ of a complex X is simple if its removal from X "does not change the topology of X". In this section we propose a definition of a simple cell based on the operation of collapse [19], which is a discrete analogue of a continuous deformation (a homotopy). Note that this definition is a rather general one, in particular, it may be directly extended to n-dimensional cubical complexes [15].

Let X be a complex in $\mathbb{F}^3$ and let $f \in X^+$. The face f is a *border face for* X if there exists one face $g \in \hat{f}^*$ such that f is the only face of X which contains g. Such a face g is said to be *free for* X and the pair (f, g) is said to be a *free pair for* X. We say that $f \in X^+$ is an *interior face for* X if f is not a border face.

Let X be a complex, and let (f, g) be a free pair for X. The complex $X \setminus \{f, g\}$ is an *elementary collapse of* X.

Let X, Y be two complexes. We say that X *collapses onto* Y if there exists a *collapse sequence from* X *to* Y, *i.e.*, a sequence of complexes $\langle X_0, ..., X_l \rangle$ such that $X_0 = X$, $X_l = Y$, and X_i is an elementary collapse of X_{i-1}, $i = 1, ..., l$. If X collapses onto Y, we also say that Y *is a retraction of* X. See illustration Fig. 2 (a), (b), (c), (d).

We give now a definition of a simple point, it may be seen as a discrete analogue of the one given by T.Y. Kong in [20] which lies on continuous deformations in the n-dimensional Euclidean space.

Definition 1. Let $X \preceq \mathbb{F}^3$. Let $f \in X^+$.
We say that $\hat{f}$ and f are *simple for* X if X collapses onto $X \oslash \hat{f}$.

The notion of attachment, as introduced by T.Y. Kong [13, 20], leads to a local characterization of simple cells.

Definition 2. Let $X \preceq \mathbb{F}^3$ and let $f \in X^+$. The *attachment of* $\hat{f}$ *for* X is the complex $Attach(\hat{f}, X) = \hat{f}^* \cap [X \oslash \hat{f}]$.

In other words, a face g is in $Attach(\hat{f}, X)$ if g is in $\hat{f}^*$ and if g is a (proper) face of a principal face h distinct from f.

The following proposition is an easy consequence of the above definitions.

Proposition 3. *Let* $X \preceq \mathbb{F}^3$, *and let* $f \in X^+$.
The cell $\hat{f}$ *is simple for* X *if and only if* $\hat{f}$ *collapses onto* $Attach(\hat{f}, X)$.

The attachment of a 3-face f of a complex X is highlighted Fig. 2 (f) and $X \oslash \hat{f}$ is depicted in (e). It may be seen that f is not simple: there is no collapse sequence from X (a) to $X \oslash \hat{f}$ (e). On the other hand the faces g, h, and i are simple.

3 Critical Kernels

Let X be a complex in $\mathbb{F}^3$. We observe that, if we remove simultaneously simple cells from X, we may obtain a set Y such that X does not collapse onto Y. In other words, if we remove simple cells in parallel, we may "change the topology" of the original object X. Thus, it is not possible to use directly the notion of simple cell for thinning discrete objects in a symmetrical manner.

In this section, we introduce a framework for thinning in parallel discrete objects with the warranty that we do not alter the topology of these objects. This method may be extended for complexes of arbitrary dimension [15]. As far as we know, this is the first method which allows to thin arbitrary complexes in a symmetric way.

This method is based solely on three notions, the notion of an *essential face* which allows to define the *core of a face*, and the notion of a *critical face*.

Definition 4. Let $X \preceq \mathbb{F}^3$ and let $f \in X$. We say that f is an *essential face for* X if f is precisely the intersection of all principal faces of X which contain f, *i.e.*, if $f = \cap\{g \in X^+ \mid f \subseteq g\}$. We denote by $Ess(X)$ the set composed of all essential faces of X. If f is an essential face for X, we say that $\hat{f}$ is an *essential cell for* X.

Observe that a principal face for X is necessarily an essential face for X, *i.e.*, $X^+ \subseteq Ess(X)$. The essential and non-principal faces of the complex X of Fig. 2 (a) are highlighted Fig. 2 (g).

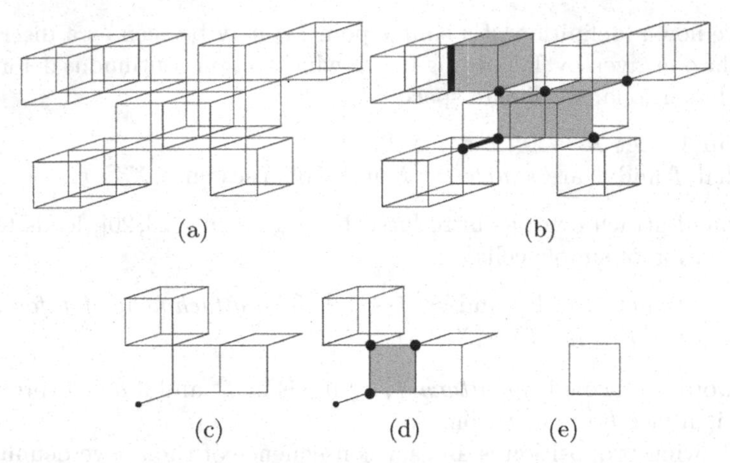

Fig. 3. (a): A complex X_0 in $\mathbb{F}^3$. (b): The critical faces of X_0 are highlighted. (c) The complex $X_1 = \mathit{Critic}(X_0)$. (d): The critical faces of X_1 are highlighted. (e) The complex $X_2 = \mathit{Critic}(X_1)$: X_2 is such that $\mathit{Critic}(X_2) = X_2$.

Definition 5. Let $X \preceq \mathbb{F}^3$ and let $f \in \mathit{Ess}(X)$. The *core of* $\hat{f}$ *for* X is the complex, denoted by $\mathit{Core}(\hat{f}, X)$, which is the union of all essential cells for X which are in $\hat{f}^*$, *i.e.*, $\mathit{Core}(\hat{f}, X) = \cup\{\hat{g} \mid g \in \mathit{Ess}(X) \cap \hat{f}^*\}$.

The preceding definition may be seen as a generalization of the notion of attachment for arbitrary essential cells (not necessarily principal).

Proposition 6. *Let* $X \preceq \mathbb{F}^3$ *and let* $f \in X^+$. *The attachment of* $\hat{f}$ *for* X *is precisely the core of* $\hat{f}$ *for* X, *i.e, we have* $\mathit{Attach}(\hat{f}, X) = \mathit{Core}(\hat{f}, X)$.

Definition 7. Let $X \preceq \mathbb{F}^3$ and let $f \in X$. We say that f and $\hat{f}$ are *regular for* X if $f \in \mathit{Ess}(X)$ and if $\hat{f}$ collapses onto $\mathit{Core}(\hat{f}, X)$. We say that f and $\hat{f}$ are *critical* for X if $f \in \mathit{Ess}(X)$ and if f is not regular for X.

We set $\mathit{Critic}(X) = \cup\{\hat{f} \mid f \text{ is critical for } X\}$, $\mathit{Critic}(X)$ is the *critical kernel of* X. A face f in X is *a maximal critical face, or an M-critical face for* X, if f is a principal face of $\mathit{Critic}(X)$.

Again, the preceding definition of a regular cell is a generalization of the notion of a simple cell. As a corollary of Prop. 6, a principal face of a complex $X \preceq \mathbb{F}^3$ is regular for X if and only if is simple for X.

The following theorem holds for complexes of arbitrary dimensions (see [15]). This is our basic result in this framework. See Fig. 3 where the successive critical kernels of a complex are depicted.

Theorem 8. *Let* $X \preceq \mathbb{F}^3$. *The critical kernel of* X *is a retraction of* X. *Furthermore, if* $Y \sqsubseteq X$ *is such that* Y *contains the critical kernel of* X, *then* Y *is a retraction of* X.

4 Crucial Kernels

If X is a complex in $\mathbb{F}^3$, the subcomplex $Critic(X)$ is not necessarily a principal subcomplex of X as illustrated Fig. 3. In this paper, we investigate thinning algorithms which take as input a pure 3-complex and which return a principal subcomplex of the input (thus also a pure 3-complex). For that purpose, we propose some notions which allow to recover a principal subcomplex Y of a pure complex X, with the constraint that Y is a retraction of X.

Definition 9. Let $X \sqsubseteq \mathbb{F}^3$, and let $f \in X^+$ be a 3-face for X.

We say that f and $\hat{f}$ are *3-crucial for* X if f is critical for X. We say that f and $\hat{f}$ are *k-crucial for* X, $k = 2, 1, 0$, if $\hat{f}^*$ contains a k-face which is M-critical for X and which is not a proper face of an l-crucial face, $l > k$. We say that f and $\hat{f}$ are *crucial for* X if f is k-crucial for some $k \in [0, 1, 2, 3]$.

Thus, a 3-face f is 2-crucial iff it contains a 2-face which is M-critical. A 3-face f is 1-crucial iff it contains a 1-face which is M-critical and which is not contained in a 2-crucial face. A 3-face f is 0-crucial iff it contains a 0-face which is M-critical and which is not contained in a 2- or 1-crucial face. Observe that a face f which is k-crucial cannot be l-crucial, with $l \neq k$.

In Fig. 4 (a), the M-critical faces of a complex are highlighted (see also Fig. 3 (b) where the critical faces of the same complex are given). The 3-face h is not crucial: it contains a 0-face which is critical but not M-critical. The 3-face f is 3-crucial, the 3-face g is 2-crucial. The 3-face i is not crucial: it contains a 1-face which is M-critical but which is a face of the 2-crucial face g.

Let $X \sqsubseteq \mathbb{F}^3$. We define the *crucial kernel of* X as the set $Cruc(X)$ which is the union of all cells of X which are crucial for X. By the very definition of a crucial face, $Cruc(X)$ contains the critical kernel of X. Thus, by Prop. 8, the crucial kernel of X is a retraction of X.

In Fig. 4 (a), a complex X_0 and its M-critical faces are depicted (5 faces). The complex $X_1 = Cruc(X_0)$ is given in (b) also with its M-critical faces (only one 2-face). Finally, in (c), the complex $X_2 = Cruc(X_1)$ contains also one M-critical face, and it may be seen that $X_2 = Cruc(X_2)$.

For thinning objects, we often want to keep other faces than the ones which are crucial. That is why we introduce the following definition in order to generalize the previous notions. Intuitively, the set K corresponds to a set which is preserved by a thinning algorithm (like extremities of curves, if we want to obtain a curvilinear skeleton).

Definition 10. Let $X \sqsubseteq \mathbb{F}^3$, let $K \subseteq X^+$ be a set composed of 3-faces of X, and let $f \in X^+$ be a 3-face of X.

We say that f and $\hat{f}$ are *3-crucial for* $\langle X, K \rangle$ if f is critical for X or if f is in K. We say that f and $\hat{f}$ are *k-crucial for* $\langle X, K \rangle$, $k = 2, 1, 0$, if $\hat{f}^*$ contains a k-face g which is M-critical for X and which is not a proper face of an l-crucial face for $\langle X, K \rangle$, with $l > k$. The set of 3-faces of X which contain such a face g is *a k-crucial clique* for $\langle X, K \rangle$.

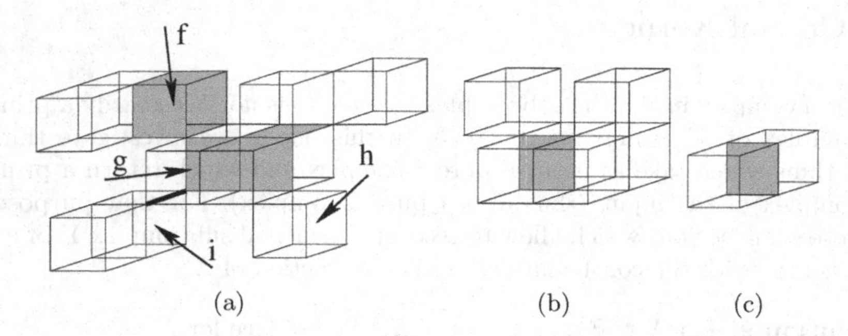

Fig. 4. (a): A complex X_0 and its M-critical faces (highlighted). (b): $X_1 = Cruc(X_0)$ and its M-critical faces. (c): The complex $X_2 = Cruc(X_1)$ contains only one M-critical face (highlighted), and $X_2 = Cruc(X_2)$.

We say that f and $\hat{f}$ are *crucial for* $\langle X, K\rangle$ if, for some $k \in [0,1,2,3]$, f is k-crucial for $\langle X, K\rangle$. We say that a set of 3-faces is *a crucial clique for* $\langle X, K\rangle$ if, for some $k \in [0,1,2]$, this set is a a k-crucial clique for $\langle X, K\rangle$.

Definition 11. Let $X \sqsubseteq \mathbb{F}^3$, and let $K \subseteq X^+$ be a set composed of 3-faces of X. The *crucial kernel of* X *constrained by* K is the set which is the union of all crucial cells for $\langle X, K\rangle$.

From the previous definitions and from Th. 8, we immediately deduce the following proposition which ensures that any constrained crucial kernel of an object preserves the topology of this object.

Proposition 12. *Let* $X \sqsubseteq \mathbb{F}_2^n$*, and let* $K \subseteq X^+$ *be a set composed of 3-faces of* X*. The crucial kernel of* X *constrained by* K *is a retraction of* X*.*

5 Crucial Voxels in the Cubic Grid

We introduce the following definitions in order to establish a link between pure complexes in $\mathbb{F}^3$ and objects in the cubic grid as often considered in image processing.

We define the *cubic grid* as the set $\mathbb{G}^3$ composed of all 3-faces of $\mathbb{F}^3$. A 3-face of $\mathbb{G}^3$ is also called a *voxel.* In the sequel, we consider only finite subsets of $\mathbb{G}^3$.

For any pure complex in $\mathbb{F}^3$, *i.e.*, for any $X \sqsubseteq \mathbb{F}^3$, we associate the subset X^+ of $\mathbb{G}^3$. In return, to each finite subset S of $\mathbb{G}^3$, we associate the complex S^- of $\mathbb{F}^3$. This will be our basic methodology to "interpret" a set of voxels. In particular, all definitions given for a principal face in X^+ have their counterparts for a voxel in $\mathbb{G}^3$. For example if $S \subseteq \mathbb{G}^3$ and $p \in S$, we will say that the voxel p is *simple for* S if p is simple for S^-. Border, interior, (k-) critical, and (k-) crucial voxels are defined in the same manner.

Some local characterizations of simple voxels in the cubic grid have been proposed [21,22,23], these characterizations may be used for detecting 3-crucial voxels (*i.e.*, non-simple voxels). We give now some simple local conditions, also

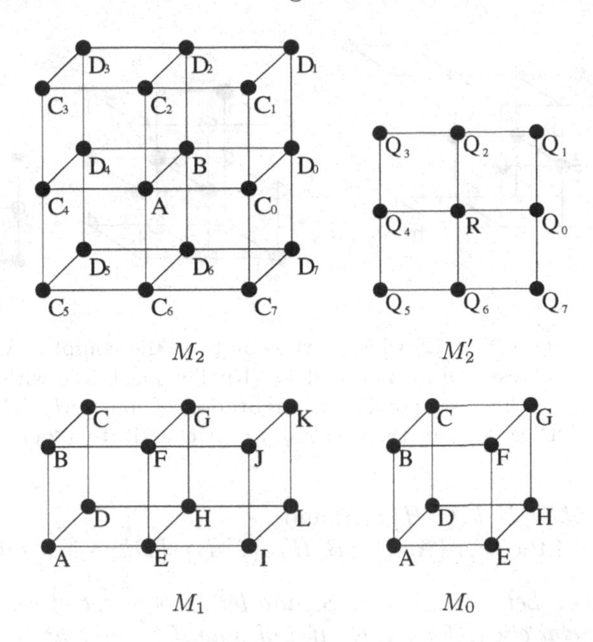

Fig. 5. Masks for 2-crucial (M_2), 1-crucial (M_1), and 0-crucial (M_0) voxels, M_2' is a configuration derived from M_2. Here, a voxel is represented by a point.

in the cubic grid, for k-crucial voxels, with $k = 2, 1, 0$. We express these local conditions by a set of masks, as in most papers related to parallel thinning in the digital topology framework. These masks M_2, M_1, M_0 are given Fig. 5. For each of these masks, we also consider all the masks obtained from them by applying $\pi/2$ rotations. We get 7 masks (3 for M_2, 3 for M_1, and 1 for M_0). The 2D configuration M_2', which appears also in Fig. 5, is derived from M_2 as explained in the following. Different characterizations for 2D simple configurations may be found in [17], they may be used for checking condition iii) for M_2'. See also Fig. 6 for an illustration of the use of mask M_2.

Definition 13. Let $S \subseteq \mathbb{G}^3$, and let M be a set of voxels of S.

1) The set M *matches the mask* M_2 if:
i) the voxels in M are simple for S; and
ii) $M = \{A, B\}$; and
iii) the 2D configuration M_2' obtained by setting $R \in M_2'$ and setting $Q_i \in M_2'$ if $\{C_i, D_i\} \cap S \neq \emptyset$, with $i \in [0, ..., 7]$, is such that R is non-simple in the 2D sense.

2) The set M *matches the mask* M_1 if:
i) the voxels in M are simple and not 2-crucial for S; and
ii) $M = \{E, F, G, H\} \cap S$; and
iii) the set $\{E, G\}$ or the set $\{F, H\}$ (or both) is included in M; and
iv) we have either $[\ U \cap S \neq \emptyset$ and $V \cap S \neq \emptyset\ ]$ or $[\ U \cap S = \emptyset$ and $V \cap S = \emptyset\ ]$, with $U = \{A, B, C, D\}$ and $V = \{I, J, K, L\}$.

3) The set M *matches the mask* M_0 if:
i) the voxels in M are simple and neither 2-crucial nor 1-crucial for S; and

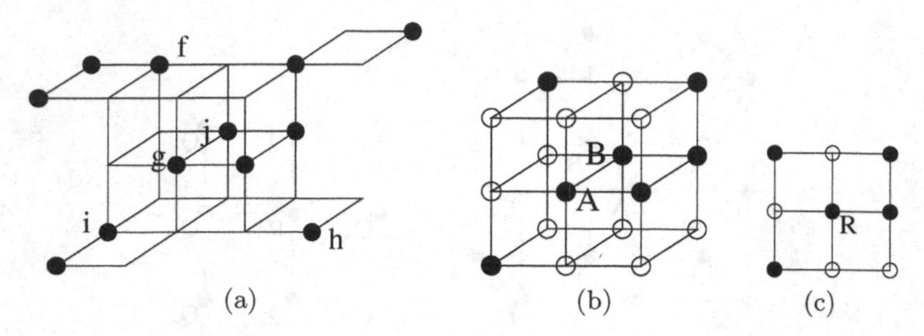

Fig. 6. (a): The subset S of $\mathbb{G}^3$ which corresponds to the complex X_0 of Fig. 4 (a). Each voxel of S is represented by a black disk. (b): The mask M_2, with A, B matching voxels g, j of S. (c): The corresponding configuration of mask M_2'. The element R is not simple in the 2D sense, thus the voxels g, j of S constitute a 2-crucial clique.

ii) $M = \{A, B, C, D, E, F, G, H\} \cap S$; and
iii) at least one of the sets $\{A, G\}, \{B, H\}, \{C, E\}, \{D, F\}$ is a subset of M.

Proposition 14. *Let $S \subseteq \mathbb{G}^3$, $K \subseteq S$, and let M be a set of voxels in $S \setminus K$.*
i) M is a 2-crucial clique for $\langle X, K \rangle$ if and only if M matches the mask M_2;
ii) M is a 1-crucial clique for $\langle X, K \rangle$ if and only if M matches the mask M_1;
iii) M is a 0-crucial clique for $\langle X, K \rangle$ if and only if M matches the mask M_0.

Observe that a voxel is k-crucial iff it belongs to a k-crucial clique, with $k \in [0, 1, 2]$. Thus, Prop. 14 also provides a method for detecting 0-, 1-, and 2-crucial voxels.

6 A Generic Thinning Scheme

We define the following notion of skeleton which is constrained to include a given set K. We then give an algorithm for computing this skeleton, this algorithm may be viewed as a generic thinning scheme where different kinds of skeletons may be obtained by considering different sets K. At last, we give two examples of skeletons derived from this scheme: one is a minimal skeleton, the other is constrained to contain some of the centers of the maximal balls included in the object. All these skeletons are obtained by a sequence of symmetric operators, thus they are invariant by $\pi/2$ rotations.

Definition 15. Let $S \subseteq \mathbb{G}^3$ and let $K \subseteq S$. We denote by $Cruc(S, K)$ the set composed of all voxels which are crucial for $\langle S, K \rangle$.
Let $\langle S_0, S_1, ..., S_k \rangle$ be the unique sequence such that $S_0 = S$, $S_k = Cruc(S_k, K)$ and $S_i = Cruc(S_{i-1}, K)$, $i = 1, ..., k$. The set S_k is the *$\mathcal{K}$-skeleton of S constrained by K.*

By Prop. 12, the $\mathcal{K}$-skeleton of a set S constrained by a set K is a retraction of S. By construction, the following algorithm computes the $\mathcal{K}$-skeleton of S constrained by K. It consists in the repetition of 5 steps, each step may be done

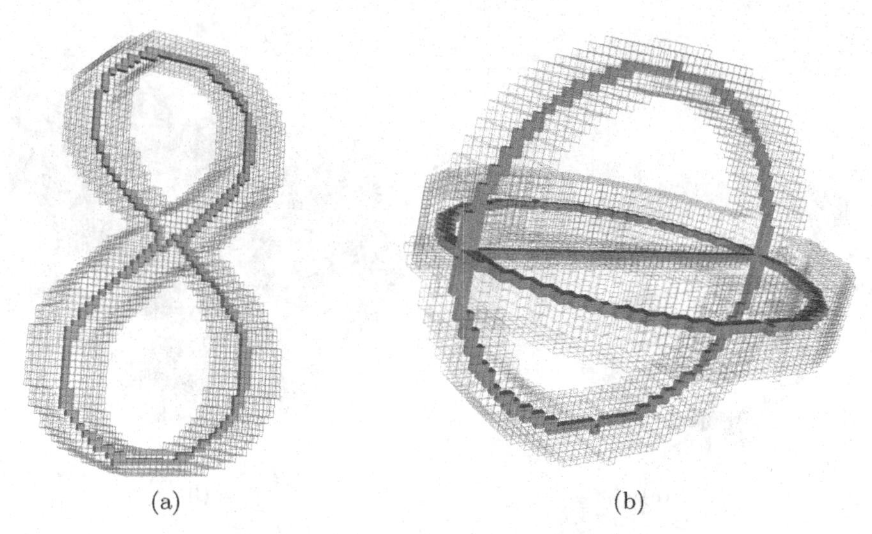

(a) (b)

Fig. 7. Two objects in $\mathbb{G}^3$ and their minimal $\mathcal{K}$-skeleton (in red)

in parallel, with the local characterization of critical (non-simple) voxels (step 02), and with the characterizations given Prop. 14 (steps 03, 04, 05).

SK^3 (Input: $S \subseteq \mathbb{G}^3$, $K \subseteq S$; Output: S)
01. Repeat Until Stability
02. R_3 := set of voxels which are critical for S or which are in K
03. R_2 := set of voxels which belong to a 2-crucial clique included in $S \setminus R_3$
04. R_1 := set of voxels which belong to a 1-crucial clique included in $S \setminus (R_3 \cup R_2)$
05. R_0 := set of voxels which belong to a 0-crucial clique included in $S \setminus (R_3 \cup R_2 \cup R_1)$
06. $S := R_3 \cup R_2 \cup R_1 \cup R_0$

A minimal skeleton of an object may be obtained by imposing no constraining set. Let $S \subseteq \mathbb{G}^3$. The *minimal $\mathcal{K}$-skeleton of S* is defined as the $\mathcal{K}$-skeleton of S constrained by K, with $K = \emptyset$.

Two examples of minimal $\mathcal{K}$-skeletons are given Fig. 7. As far as we know, SK^3 is the first thinning scheme which allows to compute such skeletons which are invariant by $\pi/2$ rotations. Furthermore, the result of SK^3 is an object which is well-defined. To our best knowledge, this is also the first attempt to give a precise definition of such a notion.

The quality of a skeleton is often assessed by the fact that it contains, approximately or completely, the medial axis of the shape.
Let $S \subseteq \mathbb{G}^3$. We consider the balls induced by the city-block distance. A ball is *maximal for S* if it is included in S and if it is not a proper subset of another ball included in S. We denote by K_r the set composed of the centers of all maximal balls which have a radius greater than or equal to r. The *medial axis of S* is precisely the set K_0. In [24], A. Rosenfeld and J.L. Pfaltz have proved that, for the city-block and the chessboard distance, the medial axis of a shape can be obtained by detecting the local maxima of its distance transform. This provides an algorithm for computing any set K_r. In Fig. 8, different $\mathcal{K}$-skeletons constrained

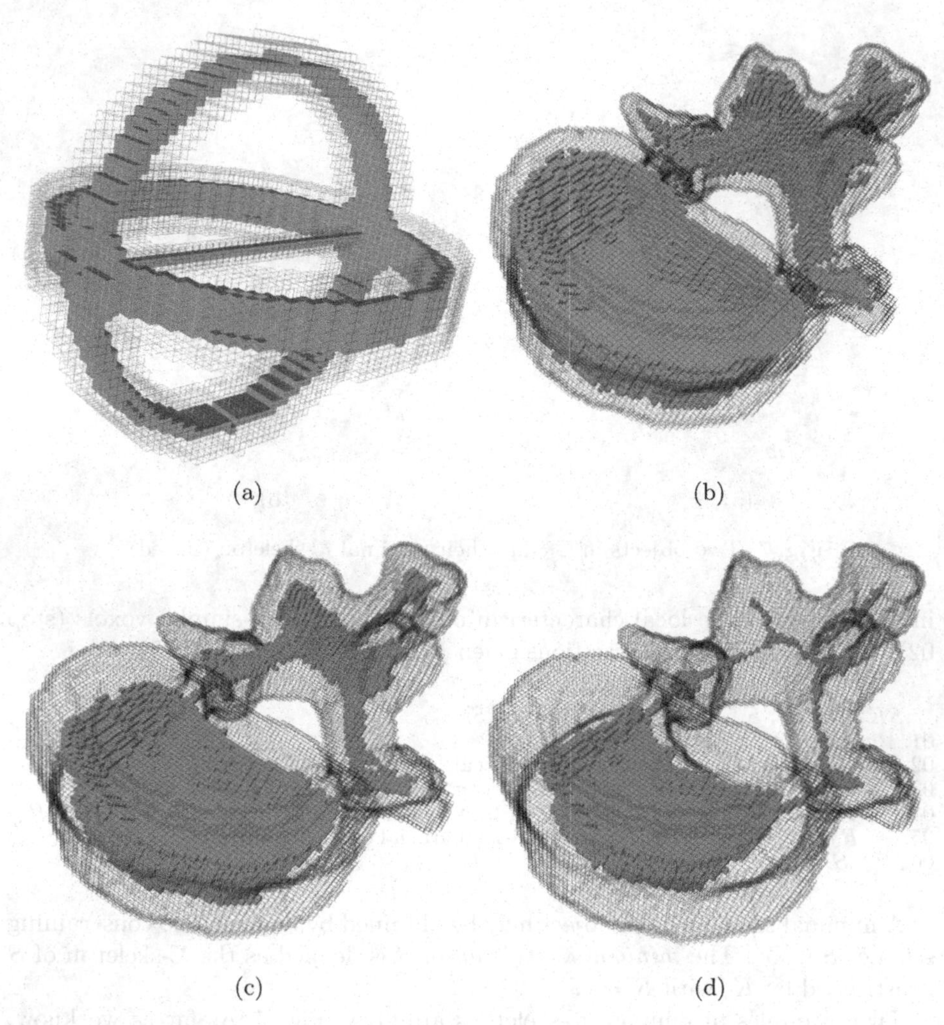

Fig. 8. (a): A subset S of $\mathbb{G}^3$ and its $\mathcal{K}$-skeleton constrained by K_3. (b): A subset T of $\mathbb{G}^3$ and its $\mathcal{K}$-skeleton constrained by K_5. (c): The $\mathcal{K}$-skeleton of T constrained by K_7. (d): The $\mathcal{K}$-skeleton of T constrained by K_9.

by K_r are given. As far as we know, this is the first thinning method which allows to obtain such skeletons constrained by a set and which are invariant by $\pi/2$ rotations.

References

1. Rutovitz, D.: Pattern recognition. Journal of the Royal Statistical Society **129** (1966) 504–530
2. Tsao, Y., Fu, K.: A parallel thinning algorithm for 3d pictures. CGIP **17**(4) (1981) 315–331

3. Saha, P., Chaudhuri, B., Dutta, D., Majumder, D.: A new shape-preserving parallel thinning algorithm for 3d digital images. PR **30**(12) (1997) 1939–1955
4. Bertrand, G.: A parallel thinning algorithm for medial surfaces. PRL **16** (1995) 979–986
5. Gong, W., Bertrand, G.: A simple parallel 3d thinning algorithm. In: ICPR90. (1990) 188–190
6. Ma, C.M.: A 3d fully parallel thinning algorithm for generating medial faces. Pattern Recogn. Lett. **16**(1) (1995) 83–87 FILE=ma1995.pdf.
7. Manzanera, A., Bernard, T., Prêteux, F., Longuet, B.: N-dimensional skeletonization: a unified mathematical framework. Journal of Electronic Imaging **11**(1) (2002) 25–37
8. Ma, C.M., Sonka, M.: A 3d fully parallel thinning algorithm and its applications. Computer Vision and Image Understanding **64**(3) (1996) 420–433
9. Palágyi, K., Kuba, A.: A 3d 6-subiteration thinning algorithm for extracting medial lines. Pattern Recognition Letters (19) (1998) 613–627
10. Palágyi, K., Kuba, A.: A parallel 3d 12-subiteration thinning algorithm. Graphical Models and Image Processing (61) (1999) 199–221
11. Lohou, C., Bertrand, G.: A 3d 12-subiteration thinning algorithm based on p-simple points. Discrete Applied Mathematics **139** (2004) 171–195
12. Lohou, C., Bertrand, G.: A 3d 6-subiteration curve thinning algorithm based on p-simple points. Discrete Applied Mathematics **151** (2005) 198–228
13. Kong, T.Y.: On topology preservation in 2-d and 3-d thinning. International Journal on Pattern Recognition and Artificial Intelligence **9** (1995) 813–844
14. Bertrand, G.: On P-simple points. Comptes Rendus de l'Académie des Sciences, Série Math. **I**(321) (1995) 1077–1084
15. Bertrand, G.: On critical kernels. Internal Report, Université de Marne-la-Vallée **IGM2005-05** (2005) Also submitted for publication.
16. Bertrand, G., Couprie, M.: Two-dimensional parallel thinning algorithms based on critical kernels. Internal Report, Université de Marne-la-Vallée **IGM2006-02** (2006) Also submitted for publication.
17. Kong, T.Y., Rosenfeld, A.: Digital topology: introduction and survey. Comp. Vision, Graphics and Image Proc. **48** (1989) 357–393
18. Kovalevsky, V.: Finite topology as applied to image analysis. Computer Vision, Graphics and Image Processing **46** (1989) 141–161
19. Giblin, P.: Graphs, surfaces and homology. Chapman and Hall (1981)
20. Kong, T.Y.: Topology-preserving deletion of 1's from 2-, 3- and 4-dimensional binary images. In: Lecture Notes in Computer Science. Volume 1347. (1997) 3–18
21. Bertrand, G.: Simple points, topological numbers and geodesic neighborhoods in cubic grids. Pattern Recognition Letters **15** (1994) 1003–1011
22. Bertrand, G., Malandain, G.: A new characterization of threedimensional simple points. Pattern Recognition Letters **15**(2) (1994) 169–175
23. Saha, P., Chaudhuri, B., Chanda, B., Dutta Majumder, D.: Topology preservation in 3d digital space. PR **27** (1994) 295–300
24. Rosenfeld, A., Pfaltz, J.: Sequential operations in digital picture processing. Journal of the Association for Computer Machinery **13** (1966) 471–494

Order Independence in Binary 2D Homotopic Thinning

Marcin Iwanowski and Pierre Soille

Institute of Control and Industrial Electronics, Warsaw University of Technology
ul.Koszykowa 75, 00-662 Warszawa, Poland
iwanowski@isep.pw.edu.pl
Joint Research Centre of the European Commission,
T.P. 262, I-21020 Ispra (VA), Italy
Pierre.Soille@jrc.it

Abstract. This paper investigates binary homotopic 2D thinning in view of its independence of the order of processing image pixels. Pixel removal conditions are provided leading to an order independent thinning. They are introduced for various types of connectivity. Two kinds of pixels to be removed are considered: simple and b-simple. Use of each of those pixels yields to different types of order independent thinnings: homotopic marking and local-SKIZ.

1 Introduction

A widely used approach for computing discrete skeletons consists in iteratively thinning the input pattern with a series of homotopic structuring elements until no more pixels can be removed. *Homotopic marking* is obtained when removable pixels are those which can be removed without modifying the homotopy of the input image. These removable pixels are called *simple pixels*. Another type of thinning result is obtained when the notion of homotopy that is the basis of the definition of simple pixels, is replaced by the notion of *background homotopy*, i.e., homotopy of connected components of the background. This notion leads to a different kind of simple pixels which we call the *b-simple* pixels. Thinning with b-simple pixels leads to kind of a skeleton which resembles the skeleton of influence zones of the complement of the input image, but is based on local pixel characterisation. It will be called in this paper *local-SKIZ*. The main developments of the paper refer to the *order independence* of the thinning. When this property is satisfied, the computed skeleton is the same no matter the order in which the image pixels are processed. Unfortunately, most algorithms produce different results for forward and backward scans of the image pixels. This also occurs when computing the skeleton of an object and its mirrored version or rotation by a multiple of 90 degrees. Recently [1, 2], the concept of order independent homotopic thinning was introduced. It was also shown that it leads to order independent skeletonisation when iterated until the algorithm reaches stability. In this paper, we introduce a general framework for order independent thinning leading to various types of result. Order independence is achieved

A. Kuba, L.G. Nyúl, and K. Palágyi (Eds.): DGCI 2006, LNCS 4245, pp. 592–604, 2006.

by introducing a supplementary condition that a removable pixel must meet, and thus guarantees its removal whatever the scanning order. These conditions refer to a neighborhood of removable pixels. All the developments described in this paper are presented for both the 8-foreground with 4-background and 4-foreground with 8-background connectivities. A case when removable pixels must be simultaneously simple (resp. b-simple) for both connectivities is also described.

The paper is organised as follows. In section 2, background notions are presented. Section 3 looks at the order independence. Section 4 illustrates the results. Finally, the conclusions are given in section 5.

2 Preliminaries

2.1 Adjacency, Connectivity, and Homotopy

A *binary image* f is represented as the *foreground* (set) $F = \{p : f(p) = 1\}$ consisting of *foreground pixels*. The complement of F, referred to as *background* (set), is defined as $\overline{F} = \mathcal{D} \setminus F = \{p : f(p) = 0\}$ and consists of *background pixels*[1]. $\mathcal{D}$ is referred to as the definition domain of the image f.

Two pixels $p, q \in \mathcal{D}$ can be either 4- or 8-adjacent. If, moreover, they have the same value, we say that they are 4- or 8-connected respectively. The k-*neighborhood* of a pixel p, denoted by $\mathcal{N}_k(p)$, is the set of all pixels q such that q and p are k-adjacent. Let $\mathcal{N}_k^0(p) = \mathcal{N}_k(p) \cap \overline{F}$ and $\mathcal{N}_k^1(p) = \mathcal{N}_k(p) \cap F = \mathcal{N}_k(p) \setminus \mathcal{N}_k^0(p)$. Let also $\mathcal{CC}_k(X)$ be the set of all k-connected components of a set X and $|\mathcal{CC}_k(X)|$ be the number of its elements (also called the cardinal number). In order to avoid the well-known connectivity paradoxes (see e.g. [3]) two combinations of connectivity are possible for the square grid: $\mathcal{G} = 8$ (and $\mathcal{G}' = 4$) or $\mathcal{G} = 4$ (and $\mathcal{G}' = 8$). *Homotopy* describes the adjacency relations between the elements of $\mathcal{CC}_\mathcal{G}(F)$ and $\mathcal{CC}_{\mathcal{G}'}(\overline{F})$. Two sets are said to be *homotopic* when they have the same homotopy tree. In this paper, we are interested in characterising homotopy between a set and an arbitrary thinning of this set so that we can restrict our attention to the notion of homotopy between ordered sets or functions due to the anti-extensivity of the thinning transformation. In this context, it was shown in [4] that two ordered sets F and G (in the sense of set inclusion) are homotopic if and only if there exists a one to one correspondence between the connected components of these sets as well as those of their background. We propose therefore to split the notion of homotopy between ordered sets into background and foreground homotopy:

Definition 1. Background and foreground homotopy. *Let F and G be two ordered sets for the inclusion relationship, then F is background homotopic (resp. foreground homotopic) to G if and only if there exists a one to one correspondence (i.e., bijective mapping) between the elements of $\mathcal{CC}_\mathcal{G}(F)$ and those of $\mathcal{CC}_\mathcal{G}(G)$ (resp. between $\mathcal{CC}_{\mathcal{G}'}(\overline{F})$ and $\mathcal{CC}_{\mathcal{G}'}(\overline{G})$).*

[1] Both kinds of notation are used in this paper to refer to binary images. An 'image F' refers to the set of foreground pixels, while an 'image f' refers to the mapping that generates F and $\overline{F}$.

Definition 2. Homotopy. *Two ordered sets are* homotopic *if and only if they are both background and foreground homotopic.*

2.2 Pixel Configurations

Topological numbers were introduced in [5] to characterise simple pixels. They are defined as: $\mathcal{T}_k(p)$ is the number of k-connected components of the 8-adjacent foreground neighbours of p that are k-adjacent to p. $\overline{\mathcal{T}_k}(p)$ is obtained by replacing foreground with background in the previous definition. We can define them also as: $\mathcal{T}_4(p) = |\{S \in \mathcal{CC}_4(\mathcal{N}_8^1(p)) : S \cap \mathcal{N}_4^1(p) \neq \emptyset\}|, \mathcal{T}_8(p) = |\mathcal{CC}_8(\mathcal{N}_8^1(p))|$, $\overline{\mathcal{T}_4}(p) = |\{S \in \mathcal{CC}_4(\mathcal{N}_8^0(p)) : S \cap \mathcal{N}_4^0(p) \neq \emptyset\}|, \overline{\mathcal{T}_8}(p) = |\mathcal{CC}_8(\mathcal{N}_8^0(p))|$. Using the topological numbers we can define the basic configurations of foreground pixels as follows: $\overline{\mathcal{T}_{\mathcal{G}'}}(p) = 0$ for interior (inner) pixels, $(\overline{\mathcal{T}_{\mathcal{G}'}}(p) > 0 \wedge \mathcal{T}_{\mathcal{G}}(p) > 0)$ for boundary pixels, and $\mathcal{T}_{\mathcal{G}}(p) = 0$ for isolated pixels. A pixel $p \in F$ is *simple* [6,3] if and only if its removal from the image does not change the homotopy of F (according to Definition 2). If p is simple, F and $F \setminus \{p\}$ are homotopic, as well as $\overline{F}$ and $\overline{F} \cup \{p\}$. It was proved in [6,3] that the simpleness of a pixel can be determined by analysing its 3×3 neighborhood. A definition of simple pixels based on topological numbers is as follows:

Definition 3. Simple pixel [5]. *A pixel p is simple* $\Leftrightarrow \mathcal{T}_{\mathcal{G}}(p) = \overline{\mathcal{T}_{\mathcal{G}'}}(p) = 1$.

If $p \in F$ the above definition is equivalent to:

Definition 4. Simple pixel [6]. *A pixel $p \in F$ is simple if and only if it satisfies the following three conditions:*

1. $\mathcal{N}_{\mathcal{G}}^1(p) \neq \emptyset$, *2.* $\mathcal{N}_{\mathcal{G}'}^0(p) \neq \emptyset$, *3.* $\exists S \in \mathcal{CC}_{\mathcal{G}}(\mathcal{N}_8^1(p))$ *such that* $\mathcal{N}_{\mathcal{G}}^1(p) \subseteq S$.

The simpleness property depends on the connectivity used for foreground and background. In [7], another type of simple pixels was considered: *{4,8}-simple*, which are simple according to both $\mathcal{G} = 4, \mathcal{G}' = 8$ and $\mathcal{G} = 8, \mathcal{G}' = 4$. This case will be referred to later in the text as $\mathcal{G} = \{4, 8\}$ and can be checked according to the following proposition[2].

Proposition 1. *A pixel p is {4,8}-simple if and only if* $\mathcal{T}_4(p) = \overline{\mathcal{T}_4}(p) = 1$.

When substituting homotopy with background homotopy in the definition of a simple pixel we get the *b-simple* ('*b*' refers to the background homotopy) pixels:

Definition 5. b-simple pixel. *A pixel p is b-simple if and only if* $\overline{\mathcal{T}_{\mathcal{G}'}}(p) = 1$.

The relationship between b-simple[3] and simple pixels is expressed by the following proposition:

Proposition 2. b-simple pixel. *A pixel p is b-simple if and only if one of the following is true: 1. p is simple or 2. p is isolated.*

Finally, we define a {4,8}-b-simple pixel as a pixel which is b-simple for both $\mathcal{G} = 4$ and $\mathcal{G} = 8$.

---

[2] Proofs of all propositions are presented in the appendix.

[3] By analogy, a *f-simple* pixels can be defined as those p for which $\mathcal{T}_{\mathcal{G}}(p) = 1$. Simple pixel can be thus defined as a pixel which is both b-simple and f-simple. The f-simpleness is however out of scope of this paper.

2.3 Thinning

In the simplest case, during the thinning process, all simple pixels are iteratively removed. This is usually done in two-stage iterative process. During the first stage the simple pixels are detected while during the second one they are removed. The removal of a simple pixel, by definition, preserves the homotopy of the input image F. This type of thinning is called a *homotopic marking* [8,2]. It reduces binary objects to single pixels or to loops surrounding the holes inside the figures (preserving the homotopy). Homotopic marking can be thus considered as a skeleton satisfying the topological constraints (preserving the homotopy) but not the geometrical ones [7][4].

The replacement of the simple pixels by the b-simple pixels in the thinning procedure leads to a (non-homotopic) skeleton preserving only those connected components of the homotopic marking which are not simply connected. Contrary to the previous method, in this one connected components without holes disappear. That is this type of a skeleton resembles the skeleton of influence zones (SKIZ) of the complement of the input image. Comparing to the distance-based SKIZ the result of thinning with b-simple pixels contains some lines which does not separate different influence zones, but are located inside the same one. This is due to the fact that the actual SKIZ cannot be obtained using local characterisation of pixel neighborhood [9]. To stress this difference the thinning by removal of b-simple pixels will be called *local-SKIZ* (SKIZ based on local pixel characterisation)[5].

3 Order Independence

When the property of *order independence* is satisfied, the result of thinning is the same no matter the order in which the image pixels were processed. Let us assume a generic two stage iterative thinning scheme. In the first (detection) phase, the image is scanned to find all simple pixels. In the second (removal) phase the values of the simple pixels should be removed. However, all cannot be removed at once because modification of one pixel can result in the neighbour becoming non-simple. A solution to this problem consists in finding a supplementary characterisation of the simple pixels. It is able to detect in the detection phase, only those pixels which can be removed no matter what scanning order is considered. These pixels are called *order independent simple pixels*. The order

[4] In order to get a skeleton that better characterises the input shape, an *anchored skeletonisation* can be performed [1,2]. It allows to pre-define a set of pixels which, by definition, cannot be removed during the thinning process.

[5] In order to get the actual SKIZ (which - contrary to local-SKIZ - is not b-homotopic to the input image) one needs to label the connected components of the background in advance and add one additional test when testing the b-simpleness. According to this test, if a non-b-simple pixel has among its background neighbours all pixels with the same label it can be removed during the thinning process. This approach requires also that when a pixel is removed, the closest background label must be propagated.

independence should be then incorporated into this generic scheme of thinning. In view of this scheme, order independence refers to the removal phase.

A separate subject of research available in the literature is *parallel thinning* [10]. Most of these studies aim at defining for binary images a minimal set of pixels which can be removed at once (in parallel) without modifying the homotopy of the original image [4,7,11]. The parallel approach itself is not sufficient to get a fully order independent algorithm, which requires also a two-stage scheme which guarantees that pixels are removed iteratively 'layer' by 'layer'. In an order independent algorithm, when processing in a given iteration, all pixels that are flagged as removable can be removed in parallel. In [1,2] the concept of order independent homotopic thinning was introduced, which leads to order independent skeletonisation when iterated until the algorithm reaches stability.

3.1 Parallel Removal of Pixels

Useful analysis of the set of simple binary pixels that can be removed in parallel has been proposed by Ronse [7]. This approach aims at finding the condition which a set of pixels has to meet to be *deletable*, i.e., to be able to be removed without changing the homotopy of the image. According to the results presented in [7] such a set cannot contain a *minimal non deletable set.* This set is defined as a minimal set of simple pixels which cannot be modified in parallel:

Proposition 3. Minimal non deletable set [7] (for proof see [7]) *A set $U \in F$ is a* minimal non deletable set *if and only if one of the following conditions holds:*

1. U consists of an isolated pixel, 2. U is a pair of 8-adjacent simple pixels which is not deletable, 3. for $\mathcal{G} = 8$, U is a triple or quadruple of pairwise 8-adjacent simple pixels, and U is an 8-connected component of F.

Proposition 3 allows us to define a set of conditions that a pixel must meet to be modified without influencing the simpleness of its neighbours. It is based on the assumption that such a pixel cannot belong to a minimal non deletable set. We can thus define an *order independent simple pixel:*

Definition 6. Order independent simple pixel. *A pixel p is* order independent simple *if and only if the three following conditions are met:*

1. p is simple, 2. p does not belong to any non deletable pair of 8-adjacent simple pixels, 3. for $\mathcal{G} = 8$, p belongs neither to an isolated (i.e., surrounded by background pixels) triple nor to an isolated quadruple of pairwise 8-adjacent simple pixels.

Comparing Definition 6 with the simpleness test (according to Definitions 4 or 3), two more conditions must be checked when analysing the given pixel: the deletability of a pair of simple pixels as well as whether they are part of the triple and quadruple configurations. Rules for finding these configurations are described further in this paper.

```
................. ................. ................. ................. ................. .................
....2223......22. ....2333......21  ....2222......21. ....2223......22. ....2333......21. ....2222......21.
..33111333121..1. ..33313111112..2. ..33111111111..1. ..33111333121..1. ..33313111112..2. ..33111111111..1.
.333111.33.1.11.. .113332.12.1.12.. .113111.12.1.11.. .333111.33.1.11.. .113332.12.1.12.. .113111.12.1.11..
.1..33.13..1.1.2. .2..33.31..1.2.1. .1..33.11..1.1.1. .1..33.13..1.1.2. .2..33.31..1.2.2. .1..33.11..1.1.2.
..1.213113311.... ..3.333333313.... ..1.313113311.... ..1.213113311.... ..2.333333313.... ..1.313113311....
..2.223223322.33. ..3.233333332.33. ..2.233333332.33. ..2.223223322.22. ..2.233333332.22. ..2.233333332.22.
................. ................. ................. ................. ................. .................
```

$\mathcal{G} = 8$ $\mathcal{G} = 4$ $\mathcal{G} = \{4, 8\}$ (a) $\mathcal{G} = 8$ $\mathcal{G} = 4$ $\mathcal{G} = \{4, 8\}$ (b)

Fig. 1. An example of binary order independent and order dependent simple (a) and b-simple pixels (b) in different types of connectivity. Dots represent background, digits stand for: '1' - a non-simple (resp. non-b-simple) foreground pixel, '2' - an independent simple (resp. b-simple) pixel, '3' - a dependent (resp. b-simple) simple pixel. Underlined pixels belong to the final thinning result.

In case of b-simpleness, a similar reasoning to the one described above leads to the formulation of the conditions a pixel must meet to be order independent b-simple. The difference between simpleness and b-simpleness relates to the fact that in the latter case we should also consider the configurations of isolated pixels. In the local-SKIZ transformation case, isolated pixels are removable and therefore removal of all configurations leading to an isolated pixel are allowed. Such configurations are possible based on the third item of Proposition 3: isolated triple and quadruple of simple pixels. Since they are isolated and have no 'holes' they do not belong to the local-SKIZ. For the same reason, an isolated pair of simple pixels can be removed. Consequently, the following definition can be formulated:

Definition 7. Order independent b-simple pixel. *A pixel p is* order independent b-simple *if and only if the following conditions are met:*
1. p is b-simple, 2. p does not belong to any deletable pair of b-simple pixels, which is not an isolated connected component.

The foreground image pixels can be classified into the following three classes: non-simple (resp. non-b-simple) pixels, dependent simple (resp. b-simple) pixels, and order independent simple (resp. b-simple) pixels. Only the pixels belonging to the third group can be removed by an order independent thinning algorithm. In order to determine whether a pixel is independent some additional tests must be performed in addition to the simpleness (resp. b-simpleness) tests. All the necessary tests are described below. Figure 3.1 shows a binary test image and the classification of its pixels.

3.2 A Pair of Simple and b-Simple Neighbours

First, let us consider the case when a simple pixel (resp. b-simple pixel) is adjacent to only one other simple (resp. b-simple) pixel. These pixels are pairwise independent if removal of one of them will not influence the removability of the second. The following proposition holds for two adjacent simple pixels.

Proposition 4. Order independent pair of simple pixels. *Let p and q be two 8-adjacent simple pixels. Let $\mathcal{G}$ be the foreground connectivity. p is order* *independent of q if and only if:* $|\mathcal{CC}_8(S \cap \mathcal{N}_8(q))| = 1$, *where* $S = \mathcal{N}_8^0(p)$ *if* $\mathcal{G} = 4$ *and* $S = \mathcal{N}_8^1(p)$ *if* $\mathcal{G} = 8$.

Contrary to the simple pixel case, in the case of b-simple pixels, an isolated pixel can be removed. Thus we have the following proposition:

Proposition 5. Order independent pair of b-simple pixels. *Let p and q be two 8-adjacent b-simple pixels. Let* $\mathcal{G}$ *be the foreground connectivity. Pixel p is order independent of q if and only if* $|\mathcal{CC}_8(S \cap \mathcal{N}_8(q))| = 1$ *or* $|\mathcal{CC}_8(\mathcal{N}_8^1(p) \cap \mathcal{N}_8(q))| = 0$, *where* $S = \mathcal{N}_8^0(p)$ *if* $\mathcal{G} = 4$ *and* $S = \mathcal{N}_8^1(p)$ *if* $\mathcal{G} = 8$.

When simple pixels for $\mathcal{G} = \{4, 8\}$ are investigated, p should be independent from q for $\mathcal{G} = 4$ and for $\mathcal{G} = 8$ simultaneously.

```
*0q  *0q  *1q  *00* *01* *10* *11* *01* *01* *10* *01* *10* *11*
*p0  *p1  *p1  *pq* *pq* *pq* *pq* *pq* *pq* *pq* *pq* *pq* *pq*
***  ***  ***  *00* *00* *00* *00* *10* *01* *10* *11* *11* *11*
 a    b    c    d    e    f    g    h    i    j    k    l    m
```

Fig. 2. Configurations of common 8-neighbours of two 8-adjacent pixels p and q, '$*$' stands for any pixel value

Figure 2 shows generic configurations of intersecting neighbourhood of two pixels. To get all possible configurations, one should add configurations that are symmetrical to *b,e,f,g,h,k,* and *l* with respect to the *pq* axis, as well as rotations through 90°,180°, and 270°. Configurations of independent pixels from Fig. 2 according to propositions 4 and 5 are given in table 1. Note that some configurations do not exist for $\mathcal{G} = 4$ and consequently for $\mathcal{G} = \{4, 8\}$.

Table 1. Independent configurations from Fig. 2

	$\mathcal{G} = 8$	$\mathcal{G} = 4$	$\mathcal{G} = \{4, 8\}$
independent simple	b,c,e,f,g	a,b,g,k,l	b,g
independent b-simple	a,b,c,d,e,f,g	a,b,d,g,k,l	a,b,d,g

In order to check whether a pixel p is independent simple, we must test its pairwise independence from all its simple neighbours. If it is independent from all of its simple neighbours, the second condition of Definition 6 is met and only the third condition has to be verified. However, according to Definition 7, this condition does not have to be tested for b-simple pixels.

One of the configurations (marked as c in Fig. 2) requires more attention when $\mathcal{G} = 4$ and $\mathcal{G}' = 8$. This is due to the fact that for some configurations of '$*$'-neighbours of p this pixel can be removed, even if it is not independent. This case is described in detail in the next section.

3.3 Multiple Simple Neighbours

There are two cases where multiple configurations of simple or b-simple pixels have to be analysed. The first one refers to the third condition of Definition 6 and concerns the isolated set of pixels for $\mathcal{G} = 8$. The second one refers to the configuration c from Fig. 2 for $\mathcal{G} = 4$.

```
0000 000 0000 0000 1000 0000 ****
0010 010 0110 0110 0110 0p11 *rq*
0100 010 0100 0110 0110 01q1 0ps*
0000 000 0000 0000 0001 0000 *0**
(a)  (b) (c)  (d)  (e)  (f)  (g)
```

Fig. 3. Multiple simple neighbours. (a,b) isolated pair. (c) isolated triple. (d) isolated quadruple (for $\mathcal{G} = 4$ and $\mathcal{G} = 8$). (e) isolated quadruple (for $\mathcal{G} = 4$ only). (f) example of configuration where p is dependent of q, but can be removed ($\mathcal{G} = 4$). (g) generic 'edge' configuration of neighbourhood ($\mathcal{G} = 4$, '*' stands for any pixel values, but such that r,q,s remain simple).

As far as the first case is concerned, for skeletonisation with $\mathcal{G} = 8$, isolated triple and quadruple configurations of simple pixels cannot be removed even if they are independent. This situation is similar to the case of an isolated pair of pixels as shown in Figs. 3a and 3b, where depending on the order of processing we can get different thinning results. Fortunately, a pair of simple pixels does not have to be considered, because such a pair is not pairwise independent and thus it cannot be removed. However, using pairwise tests, there is no obstacle to remove triple and quadruple configurations shown in Figs. 3c and 3d because every pixel belonging to these isolated connected components is independent. But such a removal would be order dependent because different pixel processing orders of the triple/quadruple result in different isolated pairs of pixels. Consequently, an additional test is required to protect pixels belonging to isolated triple or quadruple of simple pixels from being removed. The test can either follow the external boundary of the investigated triples or quadruples of simple pixels and check whether there are only background pixels or it can check the number of neighbours of every pixel belonging to it. If every pixel has exactly three (resp. four) foreground 8-neighbours then it means that the triple (resp. quadruple) is isolated.

For the configuration c of Fig. 2, for $\mathcal{G} = 4$, a supplementary analysis must be performed. This is due to the fact that for some combinations of the neighbourhood of the quadruple of pixels (consisting of p,q and two other pixels of value

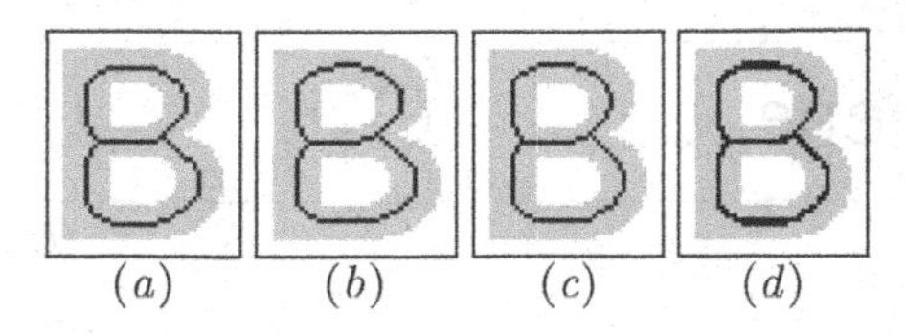

Fig. 4. (a–c) Order dependent thinning with scanning orders: (a) from top-left (TL) to bottom-right (BR), (b) from BL to TR, (c) from BR to TL. (d) shows the order independent thinning result.

1 in their common neighbourhood), pixel p is independent from q. An example of such a case is shown in Fig. 3*f*. Due to the dependence of p on q, p cannot be set to 0, based on the previously described conditions, whereas it should be, because the whole quadruple of simple (resp. b-simple) pixels can be removed without changing the homotopy of the image. The same result can be obtained for any other similar configuration when p is an 'edge' pixel. Therefore such an 'edge' configuration must be treated separately. The 'edge' pixel is characterised by a pixel having three adjacent simple (resp. b-simple) neighbours with a dependent diagonal neighbour and two independent non-diagonal ones. This produces the configuration shown in Fig. 3*g*. Since p is independent of r and s, the rest of its 4-adjacent neighbours must be background neighbours. Otherwise p would not be simple. In addition, in case of simple pixels (not b-simple), the quadruple being considered must be checked whether it is isolated. If so, p cannot be removed. Owing to that, a similar (to the first case described in this section) check of the third condition of Definition 6 must be executed. It this case however, to test whether a quadruple p, q, r, s is isolated must to check whether only 4-adjacent external neighbours belong to the background. The difference between non-deletable quadruples for $\mathcal{G} = 8$ and $\mathcal{G} = 4$ is illustrated with the two examples shown in Fig. 3*d* and 3*e*.

4 Results

The fast implementation of the algorithm according to the proposed method is queue-based. It consists of three phases. In the first one, all order independent simple (resp. b-simple) pixels are put into the queue. In the second phase, pixels are removed from the queue and set to 0. In the third phase, all the order-independent neighbours of the pixels processed in the second phase are put into the queue. Second and third phases are performed iteratively until there are new order independent simple pixels to process.

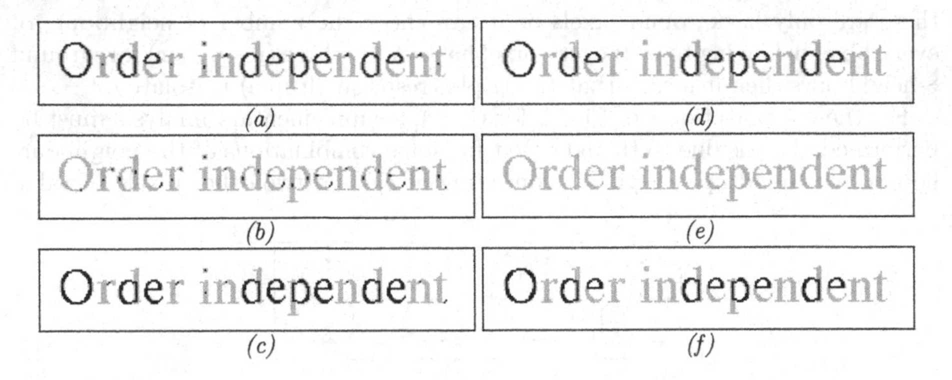

Fig. 5. Results of order independent thinning without anchor pixels. Left column: homotopic marking (skeleton): (*a*) $\mathcal{G} = 8$, (*b*) $\mathcal{G} = 4$, (*c*) $\mathcal{G} = \{4, 8\}$. Right column - local-SKIZ: (*d*) $\mathcal{G} = 8$, (*e*) $\mathcal{G} = 4$, (*f*) $\mathcal{G} = \{4, 8\}$.

The impact of the order of processing image pixels on the result of order-dependent thinning is shown in Fig. 4 and compared to the order independent result. Three scanning orders were applied: from top-left to bottom-right (Fig. 4*a*), from bottom-left to top-right (Fig. 4*b*) and finally from bottom-right to top-left (Fig. 4*c*). One can observe that every skeleton is different, although everyone of them was obtained using the successive ('layer' by 'layer') removal of simple pixels. The result of the order independent approach is shown in Fig. 4*d*. The same result was obtained for all tested scanning orders. All results have been obtained for $\mathcal{G} = 8$. The influence of the chosen connectivity on the thinning result is shown in Fig. 5. The test image was thinned in all three combinations of connectivity and produced two types of result: homotopic marking and local-SKIZ. Comparing homotopic marking for $\mathcal{G} = 8$ (Fig. 5*a*) and $\mathcal{G} = 4$ (Fig. 5*b*) we can observe that thin closed objects were not preserved in the latter case. This is due to the fact that closed paths on the original image are often one-pixel thick and sometimes two adjacent pixels are diagonal neighbours. In this case these pixels are not 4-connected; for $\mathcal{G} = 4$ they belong to different connected components. This is also a reason why local-SKIZ in case of $\mathcal{G} = 4$ (Fig. 5*d*) is an empty image - contrary to skeletonisation, the local-SKIZ preserves only the closed paths, which are not present in this image for $\mathcal{G} = 4$.

5 Conclusions

In this paper, the issue of order independence in thinning was discussed. The skeletonisation by thinning is based on the successive removal of simple pixels, which are defined as pixels which can be removed from the image without modifying its homotopy. We have introduced the notion of the b-simple pixel, which is defined as a pixel whose removal does not change the background homotopy of the image. Based on these types of simpleness, different skeletons can be obtained: homotopic marking (for simple pixels) and the skeleton by influence zones of the complement of an input image (local-SKIZ, for b-simple pixels). The order independence was also investigated in the paper. When this property is satisfied by an iterative thinning algorithm, the result does not depend on the order of processing image pixels. The conditions which are necessary to remove a pixel whatever is the scanning order was formulated. All kinds of thinning methods were analysed in the paper for three types of connectivity. Two of them were the classical 8-foreground with 4-background and 4-foreground with 8-background. The third type is the one where the removed pixel must be simultaneously simple (resp. b-simple) for both above mentioned connectivity combinations.

References

1. Ranwez, V., Soille, P.: Order independent homotopic thinning. Lecture Notes in Computer Science **1568** (1999) 337–346
2. Ranwez, V., Soille, P.: Order independent homotopic thinning for binary and grey tone anchored skeletons. Pattern Recognition Letters **23**(6) (2002) 687–702

3. Kong, T., Rosenfeld, A.: Digital topology: Introduction and survey. Computer Vision, Graphics, and Image Processing **48** (1989) 357–393
4. Ronse, C.: A topological characterization of thinning. Theoretical Computer Science **43** (1986) 31–41
5. Bertrand, J., Everat, J., Couprie, M.: Image segmentation through operators based upon topology. Journal of Electronic Imaging **6**(4) (1997) 395–405
6. Rosenfeld, A.: Connectivity in digital pictures. Journal of the ACM **17**(1) (1970) 146–160
7. Ronse, C.: Minimal test patterns for connectivity preservation in parallel thinning algorithms for binary digital images. Discrete Applied Mathematics **21** (1988) 67–79
8. Soille, P.: Morphological Image Analysis: Principles and Applications. corrected 2nd printing of the 2nd edn. Springer-Verlag, Berlin and New York (2004)
9. Couprie, M., Bertrand, G.: Tesselations by connection. Pattern Recognition Letters **23** (2002) 637–647
10. Rosenfeld, A.: A characterization of parallel thinning algorithms. Information and Control **29** (1975) 286–291
11. Bertrand, G.: On P-simple points. Comptes Rendus de l'Academie des Sciences, Serie Math. **I**(321) (1995) 1077–1084

Appendix - Proofs

The following lemmas will be used to prove the first two propositions.

Lemmas
1. $\mathcal{T}_{\mathcal{G}}(p) = 1 \Rightarrow \overline{\mathcal{T}_{\mathcal{G}'}}(p) \leq 1$, **2.** $\overline{\mathcal{T}_{\mathcal{G}'}}(p) = 1 \Rightarrow \mathcal{T}_{\mathcal{G}}(p) \leq 1$,
3. $\mathcal{T}_{\mathcal{G}}(p) = 0 \Rightarrow \overline{\mathcal{T}_{\mathcal{G}'}}(p) = 1$, **4.** $\mathcal{T}_4(p) = 1 \Rightarrow \mathcal{T}_8(p) \geq 1$; $\overline{\mathcal{T}_4}(p) = 1 \Rightarrow \overline{\mathcal{T}_8}(p) \geq 1$.

Proof of Lemma1. Let consider first the particular case where $\mathcal{G} = 4$ and therefore $\mathcal{G}' = 8$: $\mathcal{T}_4(p) = 1 \Rightarrow \overline{\mathcal{T}_8}(p) \leq 1$. Left side indicates that there can be one or more 4-conn. components of the foreground, but only one is 4-adjacent to p, so that all the other ones must be single diagonal neighbours of p. If $\mathcal{N}_8^0(p) \neq \emptyset$ then all its pixels must be 8-conn., so $\overline{\mathcal{T}_8}(p) = 1$. Otherwise (if $\mathcal{N}_8^0(p) = \emptyset$) $\overline{\mathcal{T}_8}(p) = 0$. Combining both we get $\overline{\mathcal{T}_8}(p) \leq 1$. In case of $\mathcal{T}_8(p) = 1 \Rightarrow \overline{\mathcal{T}_4}(p) \leq 1$ and $\mathcal{N}_4^0(p) = \emptyset$, we get an isolated pixel and $\mathcal{T}_4(p) = 0$. In case when $\mathcal{N}_4^0(p) \neq \emptyset$, it must contain only one 4-conn. component 4-adjacent to p (i.e., $\overline{\mathcal{T}_4} = 1$) otherwise there would be more than one 8-conn. component of p which are 8-adjacent to p.

Proof of Lemma 2. The proof is similar to that developed for Lemma 1.

Proof of Lemma 3. Left-hand side means that there are no foreground pixels in $N_{\mathcal{G}}(p)$, i.e., p is $\mathcal{G}$-isolated pixel. Consequently there exists exactly one $\mathcal{G}'$-conn. component of the background.

Proof of Lemma 4. Any 4-conn. component consists of one or more 8-conn. ones.

Proof of Proposition 1. A {4,8}-simple pixel must be both 4-simple and 8-simple. According to Definition 3 for both combinations of connectivity we

have four conditions which should be met, and therefore Proposition 1 can be re-written as follows:
$\overline{T_4}(p) = T_4(p) = 1 \Leftrightarrow \overline{T_4}(p) = T_8(p) = T_4(p) = \overline{T_8}(p) = 1$.
The implication'$\Leftarrow$' is obvious. For '$\Rightarrow$' we have: $\overline{T_4}(p) = 1 \wedge T_4(p) = 1 \Leftrightarrow$
$\Leftrightarrow$ [*Lemmas* 1, 2] $\overline{T_4}(p) = 1 \wedge T_8(p) \leq 1 \wedge T_4(p) = 1 \wedge \overline{T_8}(p) \leq 1 \Rightarrow$
$\Rightarrow$ [*Lemma* 4] $\overline{T_4}(p) = 1 \wedge T_8(p) = 1 \wedge T_4(p) = 1 \wedge \overline{T_8}(p) = 1$.

Proof of Proposition 2. Proposition 2 can rewritten using topological numbers as follows: $\overline{T}_{\mathcal{G}'}(p) = 1 \Leftrightarrow (\overline{T}_{\mathcal{G}'}(p) = 1 \wedge T_{\mathcal{G}}(p) = 1) \vee T_{\mathcal{G}}(p) = 0$. Let then perform the following transformations: $\overline{T}_{\mathcal{G}'}(p) = 1 \Leftrightarrow$ [*Lemma* 2] $\overline{T}_{\mathcal{G}'}(p) = 1 \wedge T_{\mathcal{G}}(p) \leq 1 \Leftrightarrow \overline{T}_{\mathcal{G}'}(p) = 1 \wedge (T_{\mathcal{G}}(p) = 1 \vee T_{\mathcal{G}}(p) = 0) \Leftrightarrow (\overline{T}_{\mathcal{G}'}(p) = 1 \wedge T_{\mathcal{G}}(p) = 1) \vee (\overline{T}_{\mathcal{G}'}(p) = 1 \wedge T_{\mathcal{G}}(p) = 0) \Leftrightarrow$ [*Lemma* 3] $(\overline{T}_{\mathcal{G}'}(p) = 1 \wedge T_{\mathcal{G}}(p) = 1) \vee T_{\mathcal{G}}(p) = 0$.

Proof of Proposition 4. To analyse precisely dependencies in the neighbourhood of such a pair of pixels, we must consider all possible cases. In order to guarantee that p is order independent of q, we must show, that when q is set to 0, p remains simple. Let us first consider the case where $\mathcal{G} = 8$ and $\mathcal{G}' = 4$ (this case was described in [2]). According to Proposition 4, the following configurations of pixels from Fig. 2 are order-independent: *b*,*c* and *e*,*f*,*g*. All the other are dependent, which means that once q is modified, p will not be simple any more. In the proof we will check if the conditions from Definition 4 (referred to as the 1st, the 2nd and the 3rd conditions) are met for pixel p when q is modified. Let us consider now all particular cases from Fig. 2. **Conf.*a*,*d*:** Since p is simple, then according to the 3rd cond. $\mathcal{N}_8^1(p) = \{q\}$ (otherwise it would not be conn.). If so, when q is modified, then due to the 1st cond., p will become non simple and p is dependent on q. **Conf.*b*,*e*,*f* and *g*:** When q is removed, there will still be pixels belonging to $\mathcal{N}_4^0(p)$ and $\mathcal{N}_8^1(p)$. The 1st and the 2nd cond. will then be met. The 3rd cond. as well since $\mathcal{N}_8^1(p)$ will still be connected when q is removed. Consequently p is independent of q. **Conf.*c*:** According to the 3rd cond., $\mathcal{N}_4^0(p)$ must consists of at least one pixel. In this configuration, there must exist a 4-adjacent neighbour of p marked by '*'. When q is modified, there will still be some pixels in $\mathcal{N}_4^0(p)$ and $\mathcal{N}_8^1(p)$ and consequently the 1st and 2nd conditions will be met. Also, $\mathcal{N}_8^1(p)$ will not become disconnected when q is modified, so that the 3rd cond. is met as well. Pixel p is therefore independent of q. **Conf.*h*,*i*,*k*:** When q is modified, $\mathcal{N}_8^1(p)$ will become non-connected and the 3rd cond. will not be met so that p is dependent on q. **Conf.*j*,*l*,*m*:** According to the 3rd cond., $\mathcal{N}_4^0(p)$ must consists of at least one pixel. In these configurations it must be a 4-adjacent neighbour of p marked as '*'. The only path connecting between 'upper-*' and 'lower-*' pixels in $\mathcal{N}_8^1(p)$ goes through q. If q is modified then $\mathcal{N}_8^1(p)$ will become non-connected and the 3rd cond. will not be met any more so that p is dependent on q.

A similar analysis is performed for the second case when $\mathcal{G} = 4$ and $\mathcal{G}' = 8$. According to Proposition 4, configurations of pixels *a*,*b*,*g*,*k*,*l* from Fig. 2 are order-independent. In the remaining configurations, when q is modified, p become non-simple. Considering particular cases from Fig. 2 and referring to the conditions of Definition 4 we get: **Conf.*a*:** Since p is simple, $N_4^1(p)$ consists of at least one pixel, which must be one (or both) of 4–adjacent '$*$'-pixels. $N_4^1(p)$ is then located within $*$-pixels, and q anyway does not belong to it. When q is modified, the 3rd cond. will still be met, as well as the 1st and 2nd conditions. Thus, p is independent of q. **Conf.*b*:** Pixel q is a 8–, but not 4–adjacent neighbour of p, so when q is removed the $N_4^1(p)$ does not change. Also, q does not belong to 4–path connecting pixels from $N_4^1(p)$. Consequently, when q is removed the 3rd cond. will still be met. Since the 1st and 2nd cond. are met as well, p is independent from q. **Conf.*c*,*m*:** Since p is simple there must exist at least one pixel in $\mathcal{N}_8^0(p)$ and it has to be one of those from $\mathcal{N}_8(p)$ which are marked as '$*$'. The only 4–path connecting pixels from $\mathcal{N}_4^1(p)$ goes then through q. If q is modified, the 3rd cond. will not be met and therefore p is dependent on q. **Conf.*d*:** Since p is simple, then according to the 3rd cond. $\mathcal{N}_8^1(p) = \{q\}$ (otherwise 3rd cond. would not be met). If so, when q is removed, then due to the 1st cond., p is becoming non simple, so p is dependent on q. **Conf.*e*,*f*,*h*–*j*:** these configuration cannot appear on an actual image, since either p (in *f*,*j*) or q (in *e*,*i*) or both (in *h*) cannot be simple according to the 3rd cond. **Conf.*g*,*k*,*l*:** The 1st and 2nd cond. are met no matter the value of q. When q is removed, also the 3rd cond. will be met, since it does not split the 4-conn. component of $N_8^1(p)$ due to the fact that q is the last pixel 4-adjacent to p in the 4-path in $N_8^1(p)$.

Proof of Proposition 5. Comparing the current case with the case of simple independent pixels, the difference in pixels' configuration which are independent is in configurations *a* and *d* (from Fig. 2). Contrary to the case of simple pixels, in the current one both configurations are independent. This is due to the fact that when q is modified, p becomes isolated. In case of b-simpleness, this is allowed (thanks to Proposition 2). In all other configurations, the isolated pixel cannot appear when q is modified, so for those configurations the proof is the same as the former one.

Exact Euclidean Medial Axis in Higher Resolution

André Vital Saúde[1,2,⋆], Michel Couprie[2], and Roberto Lotufo[1]

[1] State University of Campinas, School of Electrical and Computer Engineering, DCA-FEEC-UNICAMP, Caixa Postal 6101, 13081-970 Campinas/SP, Brazil
{andrevit, lotufo}@dca.fee.unicamp.br
[2] Institut Gaspard-Monge, Laboratoire A2SI, Groupe ESIEE
Cité Descartes, BP 99, 93162 Noisy-le-Grand Cedex, France
{vitalsaa, coupriem}@esiee.fr

Abstract. The notion of skeleton plays a major role in shape analysis. Some usually desirable characteristics of a skeleton are: sufficient for the reconstruction of the original object, centered, thin and homotopic. The Euclidean Medial Axis presents all these characteristics in a continuous framework. In the discrete case, the Exact Euclidean Medial Axis (MA) is also sufficient for reconstruction and centered. It no longer preserves homotopy but it can be combined with a homotopic thinning to generate homotopic skeletons. The thinness of the MA, however, may be discussed. In this paper we present the definition of the Exact Euclidean Medial Axis on Higher Resolution which has the same properties as the MA but with a better thinness characteristic, against the price of rising resolution. We provide an efficient algorithm to compute it.

1 Introduction

In 1961, Blum [1] introduced the notion of medial axis or skeleton, which has since been the subject of numerous theoretical studies and has also proved its usefulness in practical applications. Consider a subset X (called object) of a metric space. The medial axis of X is the set composed by the centers of the maximal balls for X, that is, the balls which are included in X but which are not included in any other ball included in X.

Although originally defined in the continuous plane, the medial axis can be defined using the same terms in the n-dimensional discrete grid $\mathbb{Z}^n$. The discrete medial axis is a set of points which is, by nature, centered in the object with respect to the distance which is used to define the notion of ball. To achieve a certain degree of rotation invariance, the Euclidean distance between points of $\mathbb{Z}^n$ may be used. Nevertheless even in this case, the medial axis has not, in general, the same nice properties as its continuous counterpart. In particular, the properties of thinness and homotopy are lost (see for example Figure 1(a), where the medial axis is "two-pixels thick" in some places, and has not the same number of connected components as the original object).

⋆ A.V. Saúde is thankful to the financial support from Fapesp and Capes, Brazil.

A. Kuba, L.G. Nyúl, and K. Palágyi (Eds.): DGCI 2006, LNCS 4245, pp. 605–616, 2006.

Topological aspects are out of the scope of this paper. Nevertheless, let us mention that in order to obtain an homotopic skeleton which contains the medial axis, the use of guided and constrained discrete homotopic transformations has been proposed by several authors (see *e.g.* [2,3,4]).

On the other hand, the problem of thickness of the skeleton has been tackled in different ways. Some authors use an asymmetric thinning step in order to reduce two-pixel thick configurations. But in this case, the result cannot be considered anymore as centered with respect to the original object. Furthermore the property of reconstruction is lost. In order to get a thin medial axis while preserving centeredness (in the sense of the so-called 8-distance), G. Bertrand introduced the notion of derived grid [5]. Also, in the study of discrete topology-preserving transformations, several recent works promote the use of the Khalimsky grid or its variations [6, 7, 8, 9, 10]. From a geometrical point of view, the Khalimsky grid $\mathbb{H}^n$ associated to $\mathbb{Z}^n$ can be embedded in $[\frac{1}{2}\mathbb{Z}]^n$, that is, by doubling the resolution of the grid for each dimension. Starting from an initial object in $\mathbb{Z}^n$, a "model" of this object in $\mathbb{H}^n$ can be computed and then thinned by a symmetrical algorithm [9], producing a result which is both centered and thin.

The drawback of the above approaches lies in their sensitivity to rotations. Since they implicitly or explicitly rely on the 4- or 8-distance, the skeletons can be quite different for an object and its rotation by, say, $\pi/4$. Our aim in this paper is to provide a notion of medial axis in a doubled resolution grid, based on the Euclidean distance, and an efficient algorithm to compute it. In addition to the Euclidean balls centered on points of $\mathbb{Z}^n$, we will also consider the balls centered on points of $[\frac{1}{2}\mathbb{Z}]^n$ (see Figure 1(b)).

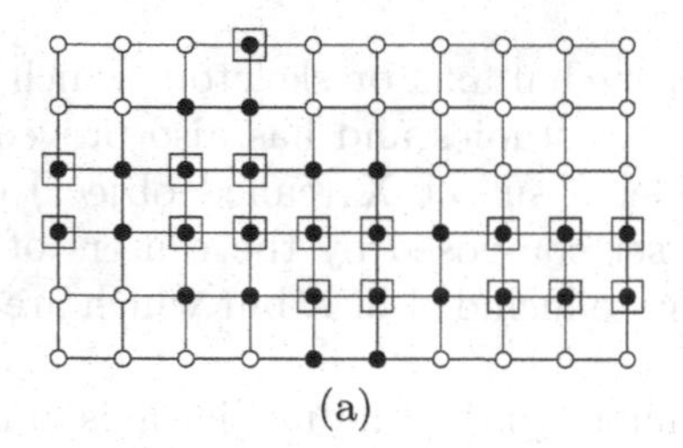
(a)

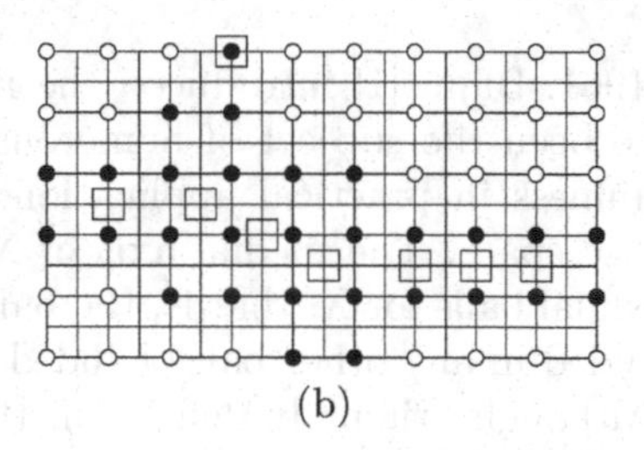
(b)

Fig. 1. (a)The object $X \subset \mathbb{Z}^2$ is depicted in black. The squares mark the points of the medial axis, based on the Euclidean distance; (b)The medial axis of X in the doubled resolution grid (squares).

Dealing with the Euclidean distance in $\mathbb{Z}^n$ is far from easy. Many algorithms found in the literature only compute approximations of the Euclidean distance transform (*e.g.* [11]) or medial axis (*e.g.* [12]). A simple to implement, optimal algorithm for building exact Euclidean distance maps was proposed only in 1994 [13,14,15], and efficient algorithms to compute exact Euclidean medial axes were not known before 2003 [16,17,18]. The proofs for our algorithm will be provided in an extended version of this paper; meanwhile they are available online in [19].

Let us briefly and informally sketch the proposed method. First, rather than using the grid $[\frac{1}{2}\mathbb{Z}]^n$, we transform the original object X by doubling the coordinates of all its points and consider the grid $\mathbb{Z}^n$ as the doubled resolution grid. This choice leads to simpler notations and proofs. Then, we consider the set X_h composed of all the points of $\mathbb{Z}^n$ which are in the neighborhood of a point of X (scaled). We compute the exact squared Euclidean distance map of X_h using a linear-time algorithm [13,14]. Finally, we propose an efficient algorithm to extract the higher resolution medial axis of X from this distance map in 2D and 3D. This algorithm is based on the same idea as the one proposed by Rémy and Thiel [16,18] and is also based on pre-computed lookup tables, following the approach originally proposed in [20]. The biggest of these tables is indeed shared by Rémy and Thiel's method and ours. Obtaining a medial axis for an object X from a distance map relative to a different object X_h is not obvious, and we present several intermediate properties to establish this fact. It should be noted that the naive solution, which consists of doubling the resolution and using the classical Euclidean medial axis, does not provide a satisfactory result.

2 Basic Notions

We denote by $\mathbb{Z}$ the set of integers, and by $\mathbb{N}$ the set of nonnegative integers. Let $\mathrm{X} \subset \mathbb{Z}^n$, we denote by $\overline{\mathrm{X}}$ the complementary of X. We denote by $(y-x)^2$ the squared Euclidean distance between two points $x \in \mathbb{Z}^n$ and $y \in \mathbb{Z}^n$.

Let $\mathrm{X} \subset \mathbb{Z}^n$, the *squared Euclidean distance transform* of X, denoted by D^2_{X}, associates to each point $x \in \mathrm{X}$ its squared Euclidean distance to the nearest point in $\overline{\mathrm{X}}$: $D^2_{\mathrm{X}}(x) = \min\{(y-x)^2, y \in \overline{\mathrm{X}}\}$.

Let x be a point in $\mathbb{Z}^n$ and let R be a positive integer. The set of points y of $\mathbb{Z}^n$ such that $(y-x)^2 = R$ will be of particular interest in what follows. Each such point y corresponds to a decomposition of R into a sum of n square integers. We introduce some notations to deal with square decompositions of integers.

Definition 1 (Square Decomposition). *Let $n \in \mathbb{N}$, $R \in \mathbb{N}$, the n-uple $(r_1, r_2, ..., r_n) \in \mathbb{N}^n$ is a* square n-decomposition of R *if $r_1 \geq r_2 \geq \ldots \geq r_n \geq 0$ and $(r_1^2 + r_2^2 + ... + r_n^2) = R$. We denote by $SQD_n(R)$ the set of all square n-decompositions of R.*

There are other decompositions of R into the sum of n squares, satisfying only the second condition. They may be obtained from the above decompositions by permutations and sign changes.

We set $\mathbb{D}_n = \{R \in \mathbb{N} \mid \mathrm{SQD}_n(R) \neq \emptyset\}$, the set of n-decomposable integers. It is known that for any $n \geq 4$, $\mathbb{D}_n = \mathbb{N}$ (Lagrange's theorem, see [21], Section 20.5).

Let $R \in \mathbb{N}$, we define:

$$\overline{R} = \min\{\delta \in \mathbb{D}_n \,|\delta \geq R\}; \qquad R^+ = \overline{R+1}$$
$$\underline{R} = \max\{\delta \in \mathbb{D}_n \,|\delta \leq R\}; \qquad R^- = \underline{R-1}$$

Let $R \in \mathbb{N}$. Observe that, if $R \in \mathbb{D}_n$ then $\overline{R} = \underline{R} = R$ and $(R^-)^+ = (R^+)^- = R$. In addition, if $R \notin \mathbb{D}_n$ then $\overline{R} = R^+$ and $\underline{R} = R^-$.

Definition 2 (Euclidean Ball). *Let $x \in \mathbb{Z}^n$, $R \in \mathbb{N}$, we denote by $B^{\leq}(x, R)$ the* Euclidean ball *centered in x with (squared) radius R and we denote by $B^{<}(x, R)$ the* Euclidean ball *centered in x with (squared) strict radius R, where $B^{\leq}(x, R) = \{y \in \mathbb{Z}^n, (x - y)^2 \leq R\}$ and $B^{<}(x, R) = \{y \in \mathbb{Z}^n, (x - y)^2 < R\}$.*

Note that $B^{\leq}(x, R) = B^{<}(x, R^+)$ and $B^{<}(x, R) = B^{\leq}(x, R^-)$.

Definition 3 (Maximal Ball). *Let $\mathrm{X} \subset \mathbb{Z}^n$, $x \in \mathrm{X}$, $R \in \mathbb{N}$,*
a) A ball $B^{<}(x, R) \subseteq \mathrm{X}$ is a maximal ball for x in X *if it is the largest ball centered in x and included in* X, i.e., *$\forall R' \in \mathbb{N}, B^{<}(x, R) \subseteq B^{<}(x, R') \subseteq \mathrm{X} \Rightarrow B^{<}(x, R) = B^{<}(x, R')$;*
b) A ball $B^{<}(x, R) \subseteq \mathrm{X}$ is a maximal ball for X *if it is not strictly included in any other ball included in* X, i.e., *$\forall R' \in \mathbb{N}, \forall y \in \mathrm{X}, B^{<}(x, R) \subseteq B^{<}(y, R') \subseteq \mathrm{X} \Rightarrow B^{<}(x, R) = B^{<}(y, R')$.*

Proposition 1. *Let $\mathrm{X} \subset \mathbb{Z}^n$, $x \in \mathrm{X}$ and $R \in \mathbb{D}_n$. The ball $B^{<}(x, R)$ is maximal for x in X if and only if $R = D^2_{\mathrm{X}}(x)$.*

Observe that, if $B^{<}(x, R)$, with $R \in \mathbb{N}$, is maximal for X, then it is maximal for x in X. Now let us recall the definition of the medial axis [1].

Definition 4 (Medial Axis). *Let $\mathrm{X} \subset \mathbb{Z}^n$, the* medial axis of X, *denoted by $MA(\mathrm{X})$, is the set of the centers of all the maximal balls for* X.

3 Euclidean Medial Axis in Higher Resolution

The goal of changing resolution is to extract a medial axis of the object X by considering a new family of Euclidean balls which are not necessarily centered on points of X. More precisely, we also take into account Euclidean balls centered on the vertices of a doubled resolution grid. For simplicity, instead of considering half integers for the coordinates in the higher resolution grid, we begin by doubling coordinates of the original object. Thus $\mathbb{Z}^n$ is used as the higher resolution grid, and the points with only even coordinates constitute the support of the scaled original image.

Definition 5. *Let $i \in \{0\,..\,n\}$. We define the set $E_i \subset \mathbb{Z}^n$ as the set of elements in $\mathbb{Z}^n$ with exactly i even coordinates, more precisely $E_i = \{(z_1, z_2, \ldots, z_n) \in \mathbb{Z}^n, \sum_{j=1}^{n}((z_j + 1) \mod 2) = i\}$.*

Let $\mathrm{X} \subset \mathbb{Z}^n$, we write $E_i(\mathrm{X}) = E_i \cap \mathrm{X}$. The family $\{E_i\}_{i=0..n}$ forms a *partition* of $\mathbb{Z}^n$, *i.e.*, $\bigcup_{i=0..n} E_i = \mathbb{Z}^n$ and $\forall i, j \in \{0\,..\,n\}, i \neq j \Rightarrow E_i \cap E_j = \emptyset$.

Definition 6 (Neighborhood). *Let* $x \in \mathbb{Z}^n$, *we define the* neighborhood of x *as the set* $N_n(x) = \{y \in \mathbb{Z}^n \mid \max_{i=1..n} |y_i - x_i| \leq 1\}$. *Let* $X \subset \mathbb{Z}^n$, *we define the set* $N_n(X) = \bigcup_{z \in X} N_n(z)$.

Definition 7. *Let* $X \subset \mathbb{Z}^n$, *we define* $\phi(X)$ *and* $\phi^{-1}(X)$ *by* $\phi(X) = \{2z, z \in X\}$ *and* $\phi^{-1}(X) = \{z, 2z \in X\}$.

Definition 8 (H-transform). *Let* $X \subset \mathbb{Z}^n$, *the* H-transform *of* X, *denoted by* $\mathcal{H}(X)$, *is defined by* $\mathcal{H}(X) = N_n(\phi(X))$.

Figure 2 illustrates the H-transform of a set in $\mathbb{Z}^2$. Let $X \subset \mathbb{Z}^n$, observe that $\phi(X) \subseteq E_n$, that $\phi^{-1}(\phi(X)) = X$ and that $E_n(\mathcal{H}(X)) = \phi(X)$, hence $\phi^{-1}(E_n(\mathcal{H}(X))) = X$. The following proposition is elementary.

Proposition 2. *The* H*-transform is increasing,* i.e., *for each* $A, B \subset \mathbb{Z}^n, A \subset B \Rightarrow \mathcal{H}(A) \subset \mathcal{H}(B)$.

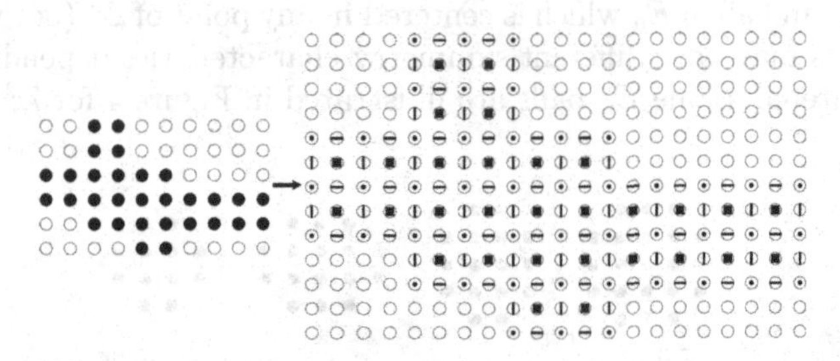

Fig. 2. H-transform of a set in $\mathbb{Z}^2$. On the left, the set X (in black), and on the right, the set $\mathcal{H}(X)$. Elements of $\mathcal{H}(X)$ are marked based on the sets of the partition: E_0 (with a dot), E_1 (with a vertical or horizontal line) and E_2 (with a square).

Definition 9 (E_n-tight). *Let* $X \subset \mathbb{Z}^n$, $x \in X$, $R \in \mathbb{D}_n$. *The ball* $B^{\leq}(x,R)$ *is said to be* E_n-tight *if there is no* n*-decomposable integer* $R' < R$ *such that* $E_n(B^{\leq}(x,R')) = E_n(B^{\leq}(x,R))$.

It can easily be seen that $B^{\leq}(x,R)$ is E_n-tight if and only if there exists $y \in E_n(B^{\leq}(x,R))$ such that $(x-y)^2 = R$. The following proposition will play an important role to prove the algorithms that we are going to propose. Put briefly, it says that any E_n-tight ball B is included in $N_n(E_n(B))$.

Proposition 3. *Let* $n \in \mathbb{N}, n \leq 3$. *Let* $B = B^{\leq}(x,R)$, *where* $x \in \mathbb{Z}^n, R \in \mathbb{D}_n$, *be an* E_n*-tight ball. For any* $z \in B$, *there exists a point* $w \in N_n(z)$ *such that* $w \in E_n(B)$, i.e., $B \subseteq N_n(E_n(B))$.

Although this proposition may seem simple, it is in fact false in general but true at least in dimensions 2 and 3. Furthermore, the counter-example of Figure 3 shows that the condition "E_n-tight" is indeed necessary for the proposition. With this proposition, we will be able to justify our algorithms, and the practical applications are mostly in dimensions 2 and 3.

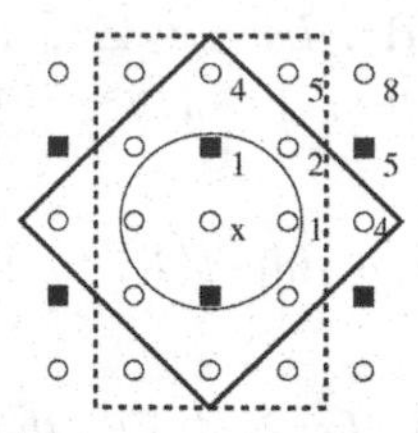

Fig. 3. Consider the ball $B = B^{\leq}(x, 4)$, surrounded by a square. The points of E_n are marked by black squares. We see that B is not E_n-tight, and that B is not included in $N_n(E_n(B))$ (surrounded by a dashed rectangle). On the other hand, the ball $B' = B^{\leq}(x, 1)$ (surrounded by a circle) is E_n-tight and included in $N_n(E_n(B'))$.

E_n-Balls and Higher Resolution Medial Axis

For any $x \in \mathbb{Z}^n, R \in \mathbb{D}_n$, the set $E_n(B^{<}(x, R))$ is called an *E_n-ball*, it can be seen as an Euclidean ball in E_n which is centered in any point of $\mathbb{Z}^n$ (not necessarily in E_n). E_n-balls may have different symmetry characteristics depending on where they are centered. Some E_n-balls are illustrated in Figure 4 for $\mathbb{Z}^2$.

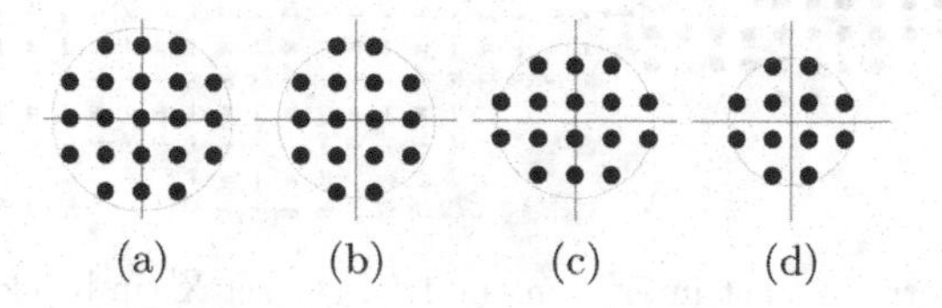

Fig. 4. E_n-balls in $\mathbb{Z}^2$. Only points of E_2 are represented. (a) E_n-ball centered in E_2, (b) and (c) E_n-balls centered in E_1, (d) E_n-ball centered in E_0.

Definition 10 (E_n-maximal balls). *Let* $X \subset \mathbb{Z}^n$, $x \in X$, $R \in \mathbb{N}$,
a) An E_n-ball $E_n(B^{<}(x, R)) \subseteq X$ is an E_n-maximal ball for x in X *if it is the largest E_n-ball centered in x and included in* X, i.e., $\forall R' \in \mathbb{N}, E_n(B^{<}(x, R)) \subseteq E_n(B^{<}(x, R')) \subseteq X \Rightarrow E_n(B^{<}(x, R)) = E_n(B^{<}(x, R'))$;
b) An E_n-ball $E_n(B^{<}(x, R)) \subseteq X$ is an E_n-maximal ball for the set X *if it is not strictly included in any other E_n-ball included in* X, i.e., $\forall R' \in \mathbb{N}, \forall y \in \mathbb{Z}^n, E_n(B^{<}(x, R)) \subseteq E_n(B^{<}(y, R')) \subseteq X \Rightarrow E_n(B^{<}(x, R)) = E_n(B^{<}(y, R'))$.

Definition 11 (HMA). *Let* $X \subset \mathbb{Z}^n$. *The* higher resolution medial axis *$HMA(X)$ is the set of centers of all E_n-maximal balls of $\mathcal{H}(X)$.*

The problem of extracting the HMA is to find the set of E_n-maximal balls for the object. Knowing the radii of the E_n-maximal balls for each point is necessary to test maximality for the object. Let $X \subset \mathbb{Z}^n$, $X_h = \mathcal{H}(X)$, $x \in X_h$, in nD the radius of the E_n-maximal ball for x can be obtained from $D^2_{X_h}(x)$ by looking for the first integer $R \geq D^2_{X_h}(x)$ such that $x + v \in E_n(\overline{X_h})$, with v being any vector satisfying $v^2 = R$. However, in 2D and 3D, we can avoid such iterations, based on the following proposition, which ensures that $D^2_{X_h}(x)$ is precisely the radius of the E_n-maximal ball for x in X_h. It can be proved with the help of Proposition 3.

Proposition 4. *Let $n \in \mathbb{N}, n \leq 3$. Let $X \subset \mathbb{Z}^n$, $X_h = \mathcal{H}(X)$, $x \in X_h$, $R \in \mathbb{D}_n$. If $B^<(x, R)$ is maximal for x in X_h, then $E_n(B^<(x, R))$ is E_n-maximal for x in X_h.*

4 Algorithm to Compute the HMA

In this section we present an algorithm to compute the higher resolution medial axis (HMA). Testing if an E_n-ball is maximal for the object is not trivial. One way of doing this is to test if it is not included in another E_n-ball. This can be done by an adaptation of the algorithm presented by Rémy and Thiel [18] for the extraction of the exact Euclidean medial axis (MA).

Euclidean balls have a number of symmetries that simplify the problem. An Euclidean ball can be reconstructed from only one of its cones (octant in 2D) by retrieving symmetries of each point (or vector) of that cone. We chose the *generator* cone to be the set of vectors $v^g \in \mathbb{Z}^n$ such that $v^g = (v_1, v_2, \ldots, v_n), v_1 \geq v_2 \geq \cdots \geq v_n \geq 0$. We distinguish two types of symmetries of v^g:

Type 1. A symmetry obtained by setting signs to the coordinates of v^g. It can be obtained by $S_1 v^g$, where S_1 is a matrix in which every element in the diagonal equals 1 or -1, with 0s everywhere else. A vector in $\mathbb{Z}^n$ has therefore 2^n Type 1 symmetries, which gives 4, 8 and 16 for $n = 2$,3 and 4 respectively. We denote by S^n_1 the set of all Type 1 matrices in nD.

Type 2. A symmetry obtained by a permutation of the coordinates of v^g. It can be obtained by $S_2 v^g$, where S_2 is a matrix obtained by permuting the rows of an identity matrix according to some permutation of the numbers 1 to n. Every row and column therefore contains precisely a single 1 with 0s everywhere else. A vector in $\mathbb{Z}^n$ has therefore $n!$ Type 2 symmetries, which gives 2, 6 and 24 for $n = 2$,3 and 4 respectively. We denote by S^n_2 the set of all Type 2 matrices in nD.

For the algorithm we are going to present in this section, we need all the combinations of symmetries of Types 1 and 2. Any symmetry of v^g in $\mathbb{Z}^n$ can be obtained by $S_1 S_2 v^g, S_1 \in S^n_1, S_2 \in S^n_2$. There are therefore $2^n n!$ symmetries in $\mathbb{Z}^n$, which gives 8, 48 and 384 for $n = 2$,3 and 4 respectively. We denote by S^n the set of all products $S_1 S_2$, with $S_1 \in S^n_1$, and $S_2 \in S^n_2$.

4.1 Algorithm for the Medial Axis (Rémy and Thiel)

Given a set $X \in \mathbb{Z}^n$ and its square Euclidean distance transform D^2_X, the algorithm proposed by Rémy and Thiel [18] tests, for each point $x \in X$, if the maximal ball for x in X, $B^<(x, D^2_X(x))$, is a maximal ball for X. This is done by testing if $B^<(x, D^2_X(x))$ is not included in another ball in X. If the maximal radius of all maximal balls for X is not greater than a previously known radius $R_{max} \in \mathbb{N}$, the inclusion test can be performed efficiently with the help of previously computed lookup tables. These lookup tables are described below and the algorithms to compute them shall be found in references [18,19].

- Let $X \in \mathbb{Z}^n$, $x \in X$, if the maximal radius of all maximal balls for the set X is not greater than a previously known radius $R_{max} \in \mathbb{N}$, it is possible to precompute a limited set of generator vectors $\mathcal{M}^g_{R_{max}}$ which is sufficient to ensure that, if $\forall v^g \in \mathcal{M}^g_{R_{max}}, \forall S \in S^n, B^<(x, D^2_X(x)) \not\subseteq B^<(x+Sv^g, D^2_X(x+Sv^g))$, then $x \in \text{MA}(X)$. $\mathcal{M}^g_{R_{max}}$ is the set of sufficient vectors for the radius R_{max}.
- Let $x \in \mathbb{Z}^n, v^g \in \mathcal{M}^g_{R_{max}}$, for any value of R_{max}, let $R \in \mathbb{D}_n$. The table $Lut[v^g, R]$ gives the minimal radius $R' \in \mathbb{D}_n$ necessary for having $\forall S \in S^n, B^<(x, R) \subseteq B^<(x + Sv^g, R')$. Note that $Lut[0, R] = R$.

To compute the MA of a set $X \in \mathbb{Z}^n$, it is sufficient to apply the IsMA function for every point $x \in X$. The correctness of Function IsMA lies on Proposition 5 and Proposition 6, proved in [18].

Function IsMA(x, R_{max}, D^2_X)

```
   // tests if x ∈ MA(X)
1  foreach v^g ∈ M^g_Rmax do
2      foreach S ∈ S^n do
3          v ← Sv^g;
4          R_v ← Lut[v^g, D²_X(x)];
5          if D²_X(x + v) ≥ R_v then return false
6  return true
```

Proposition 5. *Let $x \in \mathbb{Z}^n, v^g \in \mathcal{M}^g, S \in S^n, v = Sv^g, R \in \mathbb{D}_n, R' \in \mathbb{D}_n$, we have $B^<(x, R) \subseteq B^<(x + v, R') \Leftrightarrow R' \geq Lut[v^g, R]$.*

Proposition 6. *Let $X \subset \mathbb{Z}^n, x \in X, R_{max} = \max\{D^2_X(z) \mid z \in X\}$. The ball $B^<(x, D^2_X(x))$ is maximal for X if and only if $\forall v^g \in \mathcal{M}^g_{R_{max}}, \forall S \in S^n, B^<(x, D^2_X(x)) \not\subseteq B^<(x + v, D^2_X(x + v))$ where $v = Sv^g$.*

4.2 Algorithm to Extract the HMA

The MA algorithm presented above takes profit of the minimal radii given by the lookup table *Lut*. We need a notion of minimal radius for the E_n-balls.

Definition 12 (E_n-minimal radius). *Let $x \in \mathbb{Z}^n, R \in \mathbb{D}_n, v \in \mathbb{Z}^n$, the* E_n-minimal radius *relative to x, R and v, denoted by $R^\vee(x, R, v)$, is the strict radius of the smallest ball centered in $x + v$ which includes $E_n(B^<(x, R))$*, i.e., $R^\vee(x, R, v) = min\{R' \in \mathbb{D}_n, E_n(B^<(x, R)) \subseteq E_n(B^<(x + v, R'))\}$.

The following proposition is elementary.

Proposition 7. *Let* $x \in \mathbb{Z}^n, R \in \mathbb{D}_n, R' \in \mathbb{D}_n, v \in \mathbb{Z}^n$, $R' < R^{\vee}(x,R,v) \Leftrightarrow E_n(B^{<}(x,R)) \not\subseteq E_n(B^{<}(x+v,R'))$.

Unlike Euclidean balls in $\mathbb{Z}^n$, E_n-balls may not be invariant by symmetries of Type 2. Thus the value of $R^{\vee}(x,R,v)$ depends on to which subset E_i of $\mathbb{Z}^n$ belong the points x and $x+v$. The construction of a lookup table with E_n-minimal radii may be prohibitive. We propose to calculate $R^{\vee}(x,R,v)$ on runtime with the Function `EnRmin`.

Function `EnRmin`(x, R, v)

```
   // we consider S1 ∈ S1^n, S2 ∈ S2^n,
      v = S1 S2 v^g, where v^g is the
      generator of v
1  R_v ← Lut[v^g, R];
2  while R_v > 0 do
3      foreach r^g ∈ SQD_n(R_v^-) do
4          foreach S2' ∈ S2^n do
5              r ← S1 S2' r^g;
6              if (((x + v − r) ∈ E_n) and
               ((v − r)^2 < R) then return R_v
7      R_v ← R_v^-;
8  return 0
```

Proposition 8. *The value* R_v *returned by the* `EnRmin` *function is equal to* $R^{\vee}(x,R,v)$.

An example with Function `EnRmin` is given in [19] and may be useful for its comprehension.

Now we need to construct a table $\mathcal{M}^h$, which gives the set of vectors sufficient to compute the HMA. The construction of this table is done by the `BuildMhLut` procedure presented in [19]. This procedure, similar to the one of [18] to compute $\mathcal{M}^g$, is based on the observation that if $\mathcal{M}^h_{R_{max}}$ is sufficient to extract, from any ball with a radius less or equal to R_{max}, a medial axis which is reduced to a single point, then $\mathcal{M}^h_{R_{max}}$ enables to extract correctly the HMA from any squared distance map which values do not exceed R_{max}.

By construction of $\mathcal{M}^h$ and as a consequence of Proposition 4, Proposition 7 and Proposition 8, we have the two following propositions.

Proposition 9. *Let* $n \in \mathbb{N}, x \in \mathbb{Z}^n, v^g \in \mathcal{M}^h, S \in S^n, v = Sv^g, R \in \mathbb{N}, R' \in \mathbb{N}$. *We have* $E_n(B^{<}(x,R)) \subseteq E_n(B^{<}(x+v,R')) \Leftrightarrow R' \geq$ `EnRmin`(x,R,v).

Proposition 10. *Let* $n \in \mathbb{N}, n \leq 3$. *Let* $\mathrm{X} \subset \mathbb{Z}^n, \mathrm{X}_h = \mathcal{H}(\mathrm{X}), x \in \mathrm{X}_h, R_{max} = max\{D^2_{\mathrm{X}_h}(x'), x' \in \mathrm{X}_h\}$. *The* E_n*-ball* $E_n(B^{<}(x, D^2_{\mathrm{X}_h}(x)))$ *is* E_n*-maximal for* X_h *if and only if* $\forall v^g \in \mathcal{M}^h_{R_{max}}, \forall S \in S^n, E_n(B^{<}(x, D^2_{\mathrm{X}_h}(x))) \not\subseteq E_n(B^{<}(x+v, D^2_{\mathrm{X}_h}(x+v)))$, *where* $v = Sv^g$.

We conclude from Proposition 9 and Proposition 10 that, given a set $\mathrm{X} \subset \mathbb{Z}^n$, the computation of HMA(X) may be done by the computation of the `IsHMA` function for every point $x \in \mathcal{H}(\mathrm{X})$. Note the direct use of $D^2_{\mathrm{X}_h}$, valid only in 2D and 3D, thanks to Proposition 4.

Function `IsHMA`$(x, R_{max}, D^2_{\mathrm{X}_h})$

// tests if $x \in$ HMA(X) in $\mathbb{Z}^n$, $\mathrm{X}_h = \mathcal{H}(\mathrm{X})$

1 **foreach** $v^g \in \mathcal{M}^h_{R_{max}}$ **do**
2   **foreach** $S \in S^n$ **do**
3     $v \leftarrow Sv^g$;
4     $R_v \leftarrow \mathrm{EnRmin}(x, D^2_{\mathrm{X}_h}(x), v)$;
5   **if** $D^2_{\mathrm{X}_h}(x+v) \geq R_v$ **then** **return** *false*
6 **return** *true*

To efficiently calculate R^+, R^- for any integer R, another precomputed lookup table is used: the square decompositions table SQD_n defined in Section 2. The computation of such table is very simple in any dimension, but the algorithm is available in [19, 4].

Let $R \in \mathbb{N}, x \in \mathbb{Z}^n$, calculations are expressed as follows:

$\overline{R}$: (**while** $(\mathrm{SQD}_n(R) = \emptyset)$ **do** $R \leftarrow R+1$); R^+: $(R \leftarrow R+1; R \leftarrow \overline{R})$
$\underline{R}$: (**while** $(\mathrm{SQD}_n(R) = \emptyset)$ **do** $R \leftarrow R-1$); R^-: $(R \leftarrow R-1; R \leftarrow \underline{R})$

5 Results

Besides the didactic example given in Figure 1(a), in Figure 5 we present a practical 2D example of the HMA, comparing it to the MA.

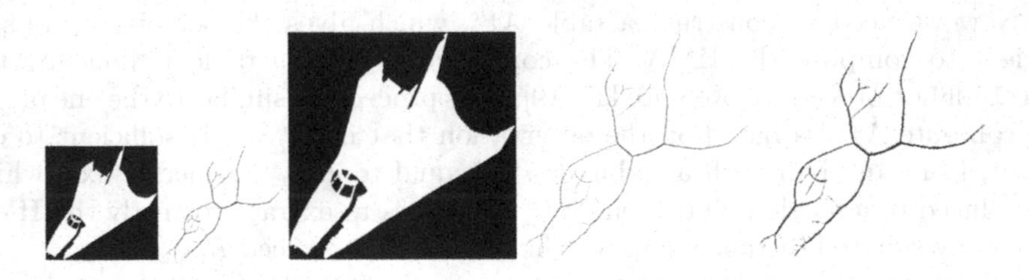

Fig. 5. 2D HMA example. From left to right: the original object, its MA, the object in doubled resolution, its HMA and, for comparison, the MA scaled to a double resolution.

As a practical evaluation of the HMA computation time, we have executed MA and HMA on several images similar to those presented in Figure 5 and also on 3D segmentations of a cerebral structure called hippocampus. The estimations were performed on an AMD Athlon XP 2400+, 2.0 GHz, running Linux, without compiler optimizations.

Let X be the object. We measured the times $t(\mathrm{X})$ and $t_h(\mathrm{X})$ for computing MA(X) and HMA(X), respectively.

For the 2D images, we have zoomed the images by factors of 0.5, 1 and 2 in order to have different ball sizes in similar shapes. The object sizes are 12445, 50157 and 112842 pixels. We obtained for $t(X)$: 0.187, 0.426, 0.662 and for $t_h(X)$: 0.487, 1.281, 2.906 (in seconds). The ratios between $t_h(X)$ and $t(X)$ are respectively 2.6, 3.0 and 4.4. The fact that we obtained ratios below four, whereas doubling the resolution multiplies the data volume by four in 2D, is mainly due to the relative cost of loading the lookup tables: this cost is significant for small images but is better amortized for large images. Notice, however, that only the parts of the lookup tables which are really needed are loaded.

For the 3D images, we used zooming factors of 1, 1.5 and 2, and the object sizes are 3530, 11799 and 24771 voxels. We obtained for $t(X)$: 0.077, 0.148, 0.330, and for $t_h(X)$: 0.422, 2.01, 5.36; hence the ratios: 5.45, 13.5, 16.2. Notice that in 3D, doubling the resolution multiplies the data volume by eight.

We also evaluated and compared the "thinness" of the HMA and the one of the MA. Let us denote by $T(X)$ the ratio: size(MA(X))/size(X), and by $T_h(X)$ the ratio: size(HMA(X))/size($\mathcal{H}$(X)). In 2D, we obtained for the three zoomed images: $T(X) = 0.091, 0.054, 0.042$, and $T_h(X) = 0.035, 0.022, 0.017$ respectively. In 3D, we obtained for the three zoomed images: $T(X) = 0.33, 0.26, 0.24$, and $T_h(X) = 0.067, 0.072, 0.077$ respectively.

It was interesting to note in the 2D case that the bigger is X, the thinner is MA(X), and that the HMA offers better improvements for small objects. In 3D, the same observation may be done, and the improvements brought by the HMA are more sensible than in 2D.

6 Conclusion and Perspectives

We have defined the HMA - the exact Euclidean medial axis in higher resolution in a completely discrete framework. We showed that the HMA presents better thinness characteristics than the classical discrete Euclidean MA. We explained how to compute such axis in n-dimensions and we provided an efficient algorithm to compute it in 2D and 3D. Although the proofs could not be presented here, the algorithm has been systematically proved. We presented a practical evaluation of the algorithm's behaviour in terms of speed. A more precise complexity analysis will be given in the extended version of this paper. The HMA is based on a transformation to higher resolution which permits the application of further homotopic thinning for the computation of homotopic skeletons. In a forthcoming paper [22], we define the notion of a symmetric homotopic skeleton which contains the Euclidean HMA, and provide a parallel algorithm to compute it, based on the framework of critical kernels (see [10]). We also intend to exploit a different approach (see [17]) in a future work, in order to obtain an even more efficient algorithm, generalized to n dimensions.

References

1. Blum, H.: An associative machine for dealing with the visual field and some of its biological implications. Biological prototypes and synthetic systems **1** (1961) 244–260

2. Davies, E., Plummer, A.: Thinning algorithms: a critique and a new methodology. Pattern Recognition **14** (1981) 53–63
3. Talbot, H., Vincent, L.: Euclidean skeletons and conditional bisectors. In: Procs. VCIP'92, SPIE. Volume 1818. (1992) 862–876
4. Couprie, M., Coeurjolly, D., Zrour, R.: Discrete bisector function and euclidean skeleton in 2d and 3d. Image and Vision Computing (2006) accepted.
5. Bertrand, G.: Skeletons in derived grids. In: procs. Int. Conf. Patt. Recogn. (1984) 326–329
6. Kovalevsky, V.: Finite topology as applied to image analysis. Computer Vision, Graphics and Image Processing **46** (1989) 141–161
7. Khalimsky, E., Kopperman, R., Meyer, P.: Computer graphics and connected topologies on finite ordered sets. Topology and its Applications **36** (1990) 1–17
8. Kong, T.Y., Kopperman, R., Meyer, P.: A topological approach to digital topology. American Mathematical Monthly **38** (1991) 901–917
9. Bertrand, G.: New notions for discrete topology. In: procs. DGCI, LNCS, Springer Verlag. Volume 1568. (1999) 216–226
10. Bertrand, G., Couprie, M.: New 3d parallel thinning algorithms based on critical kernels. In Kuba, A., Palágyi, K., Nyúl, L., eds.: DGCI. LNCS, Springer (2006)
11. Danielsson, P.: Euclidean distance mapping. Computer Graphics and Image Processing **14** (1980) 227–248
12. Meyer, F.: Cytologie quantitative et morphologie mathématique. PhD thesis, École des Mines de Paris, France (1979)
13. Saito, T., Toriwaki, J.: New algorithms for euclidean distance transformation of an n-dimensional digitized picture with applications. Pattern Recognition **27** (1994) 1551–1565
14. Hirata, T.: A unified linear-time algorithm for computing distance maps. Information Processing Letters **58**(3) (1996) 129–133
15. Meijster, A., Roerdink, J., Hesselink, W.: A general algorithm for computing distance transforms in linear time. In J. Goutsias, L.V., Bloomberg, D., eds.: Mathematical morphology and its applications to image and signal processing 5th. Volume 18 of Computational Imaging and Vision., Kluwer Academic Publishers (2000) 331–340
16. Rémy, E., Thiel, E.: Look-up tables for medial axis on squared Euclidean distance transform. In: procs. DGCI, LNCS, Springer Verlag. Volume 2886. (2003) 224–235
17. Cœurjolly, D.: d-dimensional reverse Euclidean distance transformation and Euclidean medial axis extraction in optimal time. In: procs. DGCI, LNCS, Springer Verlag. Volume 2886. (2003) 327–337
18. Rémy, E., Thiel, E.: Exact medial axis with euclidean distance. Image and Vision Computing **23**(2) (2005) 167–175
19. Saúde, A.V., Couprie, M., Lotufo, R.: Exact Euclidean medial axis in higher resolution. Technical Report IGM2006-5, IGM, Université de Marne-la-Vallée (2006)
20. Borgefors, G., Ragnemalm, I., di Baja, G.S.: The Euclidean Distance Transform: finding the local maxima and reconstructing the shape. In: Seventh Scandinavian Conference on Image Analysis. Volume 2., Aalborg, Denmark (1991) 974–981
21. Hardy, G., Wright, E.: An Introduction to the Theory of Numbers. 5th edn. Oxford University Press (1978)
22. Couprie, M., Saúde, A.V., Bertrand, G.: Euclidean homotopic skeleton based on critical kernels. In: Procs. SIBGRAPI. (2006) to appear.

Skeletonization and Distance Transforms of 3D Volumes Using Graphics Hardware

M.A.M.M. van Dortmont, H.M.M. van de Wetering, and A.C. Telea

Department of Mathematics and Computer Science
Technische Universiteit Eindhoven, the Netherlands
m.a.m.m.van.dortmont@student.tue.nl, wstahw@win.tue.nl, alext@win.tue.nl

Abstract. We propose a fast method for computing distance transforms and skeletons of 3D objects using programmable Graphics Processing Units (GPUs). We use an efficient method, called distance splatting, to compute the distance transform, a one-point feature transform, and 3D skeletons. We efficiently implement 3D splatting on GPUs using 2D textures and a hierarchical bi-level acceleration scheme. We show how to choose near-optimal parameter values to achieve high performance. We show 3D skeletonization and object reconstruction examples and compare our performance with similar state-of-the-art methods.

1 Introduction

The skeleton of a three-dimensional object is the set of interior points that have at least two closest points on the object surface. Alternative definitions use the set of centers of maximal contained balls [1] or first order singularities of the object surface's distance transform (DT). The skeleton points, together with their distance to the 3D surface, define the Medial Surface Transform (MST), which can be used for volumetric animation [2], surface smoothing [3], or topological analysis used in shape recognition, registration, or feature tracking.

While 2D skeletonization of raster images is a well-studied problem, skeletonization of 3D volumes still has some open issues. First, 3D skeletons tend to be far more complex than their 2D counterparts. Second, there exist several 2D criteria used to detect and/or simplify the skeleton in a noise-resistant way, e.g. the collapsed boundary length criterion [4, 5, 6]. However, there are hardly any similar 3D criteria that comply with the same requirements, e.g. prune and/or detect the skeleton starting from its less important points inwards, prevent skeleton disconnection during pruning, and are robust to noise. Last but not least, computing skeletons for large 3D volumes like nowadays medical scans can be a time-consuming process.

In this paper, we show how to compute 3D skeletons and distance transforms by extending to the 3D case a recent 2D skeletonization method that uses a new idea of computing skeletons by splatting distance textures [7]. We show how to efficiently implement the non-trivial 3D distance splatting on GPUs. Next, we show how to integrate a well-known 3D skeletonization criterion [8] in our splatting approach in order to compute 3D skeletons fully on the graphics card. We

A. Kuba, L.G. Nyúl, and K. Palágyi (Eds.): DGCI 2006, LNCS 4245, pp. 617–629, 2006.

keep the attractive features of the original 2D method (speed, implementation simplicity, arbitrary distance metrics). We demonstrate our approach with examples of skeletonizing and surface smoothing of real-world complex 3D objects.

The structure of this paper is as follows. Section 2 briefly overviews related work. Section 3 outlines the 2D splatting proposed by [7]. Section 4 details how we extended splatting to compute 3D skeletons. Section 5 presents our results, discusses the method, and compares it with its main competitor [8]. Finally, Section 6 concludes this paper.

2 Background

The methods for computing medial axes and skeletons can be algorithmically classified into three groups: thinning [9], Voronoi-based methods [4], and distance field methods [3, 7, 6]. In 3D, many such methods still have limitations. First, there is no generally accepted skeleton detection and/or pruning criterion that yields noise-resistant and connected 3D skeletons. For example, the θ-SMA method [10] detects skeleton points by thresholding the angle between the so-called feature points, or anchor points. This can yield skeletons with holes or even disconnections and is sensitive to noise. Euclidean Skeletons [11] improves upon θ-SMA by using a combined angle and feature point distance criterion. Other local criteria, e.g. divergence-based (Siddiqi *et al.* [12]) and moment-based (Rumpf and Telea [3]) have the same problem, i.e. can yield disconnected skeletons, unless homotopy is explicitly enforced, e.g. as in [13]. In this paper, we do not consider homotopy preservation as this is not efficiently implementable on GPUs. A second problem of 3D skeletonization is its relatively low speed. Recent GPU-based methods are one up to two magnitude orders faster than CPU-based skeletonization methods. Sud *et al.* [8] extract 3D skeletons on the GPU using the θ-SMA detector and Voronoi-based clamping techniques to limit overdraw. A related method [14] computes 3D signed distance transforms on the GPU, but not 3D skeletons. Strzodka and Telea [7] use the GPU to compute 2D skeletons using the collapsed boundary length, or anchor point distance, detector [4, 5, 6]. The skeleton and the boundary's distance transform (DT) are computed by a simple idea, called distance splatting, which is efficiently implemented on GPUs. Besides being simple, this method allows using any L_p metric, like Manhattan or (an)isotropically weighted Euclidean. Finally, we mention the important class of 3D thinning methods that compute skeletons by iteratively removing voxels from the object boundary in a given order [9]. Although simple to implement, and yielding connected skeletons, such methods can generate ill-centered and/or noisy skeletons, unless voxel removal is done in a true distance-to-boundary order, e.g. as proposed by [15].

3 Distance Splatting in 2D

Our aim is to generalize the 2D method described in [7] to perform 3D DT computation and skeletonization on the GPU, preserving its attractive points:

simplicity, accomodation of several distance metrics, and efficiency. The extension is not trivial, as the 3D case introduces specific difficulties, not present in 2D. We detail these (and our solution) in the following, starting with some definitions. Given an object $\Omega \in \mathbb{R}^3$ with surface $\partial\Omega$, the distance transform $DT : \Omega \rightarrow \mathbb{R}$ of $\partial\Omega$ can be defined as

$$DT(p) = \min_{q \in \partial\Omega} (dist(p, q)) \tag{1}$$

where $dist(p, q)$ is a distance metric (e.g. Euclidean or Manhattan). For a $p \in \Omega$, the feature transform $FP(p)$ yields the boundary points at distance $DT(p)$ from p

$$FP(p) = \{q \in \partial\Omega | dist(p, q) = DT(p)\} \tag{2}$$

The skeleton of Ω can be defined as

$$S(\Omega) = \{p \in \Omega | \exists q, r \in \partial\Omega, q \neq r : dist(p, q) = dist(p, r) = DT(p)\} \tag{3}$$

The tuples $(p, DT(p))$ with $p \in S(\Omega)$ form the medial surface transform (MST). Using the MST, one can reconstruct the surface $\partial\Omega$. To allow us to easily measure distances at any point $q \in \Omega$ from a given point $p \in \partial\Omega$, we introduce the Point Distance Function (PDF)

$$PDF_p(q) = dist(p, q) \tag{4}$$

For typical distances, we also have that

$$PDF_p(q) = PDF_0(q - p), \tag{5}$$

i.e. we can compute PDF_p by translating the PDF centered at the origin, PDF_0.

The 2D splatting method [7] we shall extend to 3D works on a discrete (image) sampling (V, V_S) of $(\Omega, \partial\Omega)$. Splatting computes 2D skeletons on the GPU in two steps. First, $DT(V_S)$ is computed by drawing PDF_0, sampled in a 2D texture, centered on all pixels $p \in V_S$. The actual distance minimization (Eqn. 1) is done during the drawing, by assigning the luminance-encoded distance values to the depth channels of the drawn pixels, and using the depth (Z buffer) test to mask pixels with greater distance values. The implementation takes a single texture draw with the pixel shaders functions of modern GPUs. Besides distance, splatting also propagates a second signal U, which encodes an arc-length boundary parameterization, so the method effectively computes a one-point feature transform of V_S. Next, the (pruned) skeleton $S(V, \tau)$ is computed as

$$S(V, \tau) = \{(i, j) \in \Omega | max(U_{i+1,j} - U_{i,j}, U_{i,j+1} - U_{i,j}) > \tau\} \tag{6}$$

The above gives the so-called collapsed boundary length at every pixel [4, 5, 6], i.e. all skeleton points where more than τ boundary units have collapsed. Increasing τ values prune the skeleton inward from its outer branches, yielding a connected, noise-free skeleton.

4 Distance Splatting in 3D

4.1 New Algorithm

A first problem of extending the above 2D algorithm to 3D is finding a suitable 3D replacement for the collapsed boundary length. A 'collapsed surface area' criterion would be a good candidate. However, we do not know how to (easily) compute such a measure. Hence, we use some simpler, though arguably less robust, local skeletonization criteria. Unlike global criteria, like the collapsed boundary length, local criteria, e.g. the θ-SMA angle [10], the divergence-based [12] or the moment-based criterion [3] use only information in a small neighbourhood of the considered point. These are more vulnerable to noise and can yield gaps or even disconnections in the skeleton. However, local criteria are simple and very efficient to implement on GPUs. After several experiments, we found the combined measure of angle between feature points and distance between feature points [11] the most robust in 3D and chose it as basis for our GPU skeletonization. A second problem is how to efficiently extend the 2D distance splatting [7] to 3D. In 2D, splatting could directly implement Eqn. 1, as explained in Sec. 3. However, though modern GPUs have 3D (volumetric) textures, they cannot render 3D primitives. To perform 3D splatting, we must find efficient ways to render volumetric primitives as a set of 2D (polygonal) primitives.

In our algorithm, we first generate the DT similarly to the 2D algorithm [7]. For all points p in the discretely sampled (voxelized) volume V counterpart of Ω, we compute the distance $DT(p)$ to the voxelized surface V_S counterpart of $\partial\Omega$, as well as *one* of its feature points $FP(p)$

```
1 Initialize DT to ∞
2 forall p in V_S
3   forall q in V
4   if (PDF_p(q) < DT(q))
5     DT(q) = PDF_p(q)
6     FP(q) = p
```

Listing 1.1. *Splatting-based DT computation*

This yields a one-point feature transform of V_S [16]. Next, we compute a skeleton detector $f(p)$ similar to [11]. In detail, we use

$$f(p) = angle(q)^a * DT(q)^b \quad (7)$$

where a=1, b=3/2, $angle(q)$ is the maximum angle between feature vectors $r-p$ at p, where $r \in FP(p)$ and $dist(q)$ is the maximum distance between feature points $FP(p)$ at p. Since we compute a single feature point $FP(p)$ instead of all potentially many feature points, we actually compute $angle(q)$ and $dist(q)$ using the neighbours $n(p)$ of p. Indeed, if p is near or on the skeleton, it will have a neighbour $n(p)$ that has a feature point $FP(n(p))$ in a significantly different location than $FP(p)$, yet with a similar DT as p (see Eqn. 3). Another property to check for skeleton points is whether they are centers of maximal balls. If q is such a point, no ball centered at a neighbor p of q, of radius $DT(p)$, can completely contain a ball centered at q with radius $DT(q)$, i.e. $\forall p, q \in \Omega : p \in n(q) : DT(q) + \|q - p\| > DT(p)$. This property holds, among others, for the city block, chessboard, D^6 and D^{26} distance metrics. If a neighbour p of q fails this test, q is not the center of a maximal ball, so is not part of the skeleton. The complete detector computation is shown in Listing 1.2.

```
1  forall q in V
2    detector(q) = dist(q) = angle(q) = 0
3  forall q in V
4    forall p in n(q)∩V
5      if (DT(p) ≤ DT(q) + ||q−p||)
6        angle(q) = max(∠ (FP(p)−p,FP(q)−q),angle(q))
7        dist(q) = max(||FP(q)−FP(p)||,dist(q))
8      else
9        angle(q) = dist(q) = 0
10       break out of loop
11   detector(q) = f(angle(q),dist(q))
```

Listing 1.2. *Pseudocode for angle and distance-based skeleton detector*

For $n(p)$, we use the 6-neighbour set. [11] states that this suffices for accurately computing the detector in Eqn. 7. The skeleton $S(p, \tau, \alpha, \beta) = \{p \in \Omega | f(p) > \tau \wedge angle(q) > \alpha \wedge DT(q) > \beta\}$ is obtained by thresholding the detector f as well as the maximal feature angle *angle* and maximal inter-feature distance *dist*. Similar to [11], typical thresholds values are $\tau \approx 180$, $\alpha \in [45, 100]$ degrees, and $\beta \in [0.05\mathcal{D}, 0.15\mathcal{D}]$ where $\mathcal{D} = 2\max(DT)$ is the object diameter.

4.2 Implementation

We implemented our method in C++ using OpenGL and Cg (C for graphics) [17] as our shader language. We splat the 3D PDF texture (Listing 1.1, Sec. 4.1) using only 2D rendering primitives. We splat several 2D textures on an xy-axis-aligned, slice-by-slice basis, as described next (see also Fig. 1; line numbers refer to Listing 1.1). For every xy slice, the initialization (line 1) is done by clearing the depth and color buffers. We implement the loops in lines 2 and 3 by drawing quadrilaterals on the current slice (the thick vertical line in Fig. 1), textured with a 2D slice from the 3D PDF function (Eqn. 5). The distance minimization (line 4) is done by assigning the PDF value from the texture to the depth value of the drawn pixels, using a pixel shader. We use the depth test, so this implicitly does the minimization and yields the minimal DT value in the depth (Z) buffer. We save the DT (line 5) by copying it to the alpha channel of the drawn pixel.

Finally, we store the feature point (line 6) by writing the splatted point p's coordinates to the RGB color channels of the drawn pixel. The drawn image thus holds the DT in the alpha channel and the one-point feature transform FP in the RGB channels. The efficiency of our implementation depends critically on the PDF texture size. We store the 3D PDF as 2δ 2D texture slices of size $(2\delta)^2$, where δ is the PDF radius. Such a slice is shown in Fig. 1 with gray values. We do not use 3D textures as these lack the high numerical precision needed and also do not allow non-power-of-two sizes, which would increase δ unnecessarily. When splatting, we do not iterate over the entire set V_S, but over the smaller 'band' V'_S (thick line in Fig. 1), which includes the points on slices at most δ pixels from the current slice, since these are the only ones that can influence the DT result on the current slice.

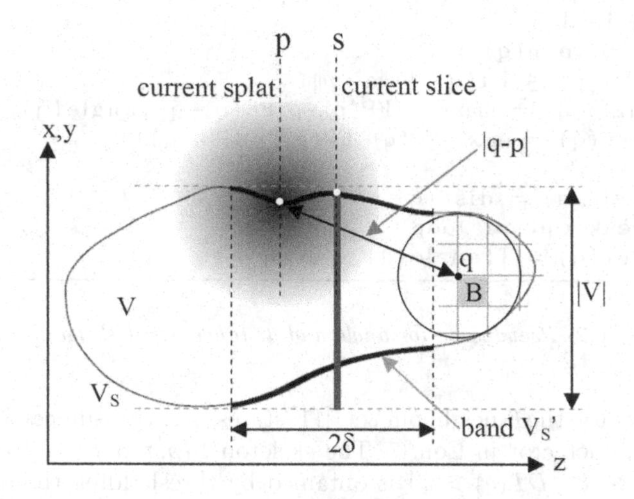

Fig. 1. 3D distance splatting principle

A (naive) upper bound for δ is $|V|/2$, i.e. half of the shortest axis of V's bounding box. However, this leads to many draw operations that do not affect the final DT. We reduce the overdraw by two techniques: a hierarchical optimization and a tighter upper bound estimation for δ, as follows. We implement a 3D version of the adaptive hierarchical optimization proposed by [7], as follows. We divide V in equally sized blocks B of c^3 voxels. We construct a coarse-scale version V_c of V, where V_c contains one sample of every block in V. We compute (also by GPU splatting) the coarse-scale distance transform DT_c of V_c, where the distance between two samples in V_c is is the maximal distance between any two voxels from their blocks $B \subset V$. For every block B, we splat only those boundary voxels that can affect its DT. These are all $p \in V_S$ that are closer to B than $DT_c(B)$. This bi-level hierarchical scheme has three advantages. First, we can quickly skip splatting the blocks B which are outside V. Second, we check if the minimal distance from B to the surface point p undergoing splatting ($|p-q|$ in Fig. 1) exceeds $DT_c(B)$ (shown by the radius of the circle centered at B).

If so, p cannot affect the DT of any voxel in B, so we skip splatting p over B. Finally, DT_c upper-bounds the radius at which a surface point p can influence the DT, so we use it as a tighter upper bound for the PDF size d than $|V|/2$. Our improved PDF size δ' is

$$\delta' = \min\left(\frac{|V|}{2}, \max_{B \subset V_c}(DT_c(B)) + 1\right). \quad (8)$$

This is a globally optimized PDF size (GPDF). We also tried a locally optimized PDF size (LPDF) that changes for every block B. However, this was slower than the GPDF, as detailed further in Sec. 5.

The second stage of the algorithm (Listing 1.2) is also implemented by rendering xy-aligned slices. The initializations (lines 1,2) are done by clearing the color buffer before drawing a slice. Next, we draw a rectangle for every volume slice (loop at line 3). The inner loop (line 4) is done using a vertex shader to generate the texture coordinates of the neighbours $n(q)$ so that the pixel shader can use these to access the relevant textures. If the fragment fails the ball containment test (line 5), it is discarded, since not part of the skeleton (lines 9,10). If the fragment passes the test, the maximum distance and angle are calculated (lines 6,7). We then use these to evaluate the detector f (Eqn. 7) and store it in a texture (line 11). Finally, we threshold this texture on-the-fly with the user-chosen values τ, α, β (Sec. 4.2), yielding the desired pruned 3D skeleton.

5 Discussion

We tested our method on both synthetic volumes and volumes segmented from real 3D scans (see Fig. 3). We used an Athlon 3.4GHz PC with 1 GB RAM and tested on two different GPUs, i.e. a GeForce 6800 with 128 MB and a GeForce 6600 with 256 MB graphics memory. We first compared our results with a software-only implementation based on the Euclidean Feature Transform method [16], which efficiently computes a feature transform (Table 1,column SW) and uses the same skeleton detector (Eqn. 7). Both methods yielded identical skeletons. We also used the pruned skeletons to reconstruct smoothed objects, by splatting the skeleton voxels with PDF functions equal to their corresponding MST values. It is well known that this replaces small-scale boundary details, corresponding to pruned skeleton points, with spherical surface segments. Figure 4 shows reconstructions for several objects. Our skeletons are indeed exact, as shown by the cube reconstructed from a non-pruned skeleton (Fig. 4 b), which is identical to the original cube (Fig. 3 f). We can easily handle noisy objects with highly complex 3D skeletons, e.g. the CT-scanned frog intestin (Fig. 3 a) or the MRI-scanned colon (Fig. 3 h). Reconstructing the colon from a highly pruned skeleton yields the smooth shape shown in Fig. 4 d.

We stress that our 3D distance splatting is exact by construction. Splatting propagates the distance from a boundary point directly, thus exactly, to the interior points. The depth test guarantees that the minimal distance is always correctly kept. This is not the case for incremental methods, e.g. level-set based [3,6],

Table 1. Benchmarks of splatting-based skeletonization

model	volume size	PDF size	object voxels	surface voxels	SW time	6600 time	6800 time	skel. voxels	recon. time	voxels/ sec.
cube	128x128x128	45	91125	11618	18	3.8	2.5	4961	0.8	4647
box	151x101x101	37	67392	9592	11.1	2.3	1.7	4032	0.5	5642
sphere 1	128x128x128	85	324157	18642	145.2	23.2	10.0	1	0.5	1864
sphere 2	256x256x256	171	2627271	75942	N/A	452.7	199.3	1	2.0	381
cylinder 1	51x51x51	31	61590	8138	5.9	1.7	1.2	4303	0.7	6781
cylinder 2	129x129x213	61	1674880	72043	781	188.1	19.2	37461	42.7	3752
cow	165x107x64	53	190041	21152	30	13.3	6.7	6402	2.6	3157
ellipse	100x100x100	25	23094	4164	3.2	0.9	0.7	288	0.1	6940
spring	100x100x100	15	38978	14013	2.2	1.7	1.5	2289	1.2	9342
ice 2	80x80x80	23	29880	5948	3.6	1.2	1.0	1255	0.6	5948
ice 3	80x80x80	29	41964	8104	4.7	2.3	1.7	1551	1.3	4767
rings	100x100x100	33	264784	28272	28.4	9.1	6.1	3222	1.6	4634
duo	72x69x90	23	36931	8636	3.1	1.7	1.4	2261	1.2	6168
intestin	60x71x94	17	13599	5724	3	0.9	0.8	1611	0.6	7155
colon	256x256x311	43	653170	81308	350.7	42.4	26.1	65120	35.7	3115
bent	150x150x150	49	429307	34211	92.9	21.8	11.6	10706	7.9	2949

that propagate information (e.g. distance, feature points) from point to point. Unless special measures are taken, such methods accumulate errors yielding visibly incorrect DTs and skeletons [16].

We would like to compare the performance of our GPU-based skeletonization with other methods, e.g. [3], [10], [11], [13], [8], and [12]. Unfortunately, this is far from trivial. These methods use different input and/or skeleton data models and skeleton detectors; have non-trivial, non-available implementations and/or test datasets; and performance is reported for different platforms. For example, we use a voxel-based model for both the input object and the computed skeleton, just as [11] and [12]. In contrast, [13], [10] and [3] use polygonal surface models for either or both.

The most interesting method to compare against is probably DiFi [8]. DiFi also uses GPUs to compute a DT and skeletons, and has a very similar skeleton detector (θ-SMA). DiFi handles both polygonal and volumetric objects. Since we do not have a DiFi implementation, nor its test objects, we shall compare our method with DiFi using the number of input object surface points processed per second. Comparing Table 1 (rightmost column) with Table 2), we see a large performance overlap between our method and DiFi. Our method skeletonizes objects at a rate of [3157..9342] surface voxels/second, with an average of 5356 (we left out the two spheres from this benchmark, since they are special absolute worst-case situations for any skeletonization method, also not present in DiFi's benchmarks). For DiFi, these figures are [1500..10500] voxels/second, with an average of 5516. As our method, DiFi can also handle many distance metrics, e.g. all L_p norms, if the Voronoi regions of the surface elements are connected. However, it is much easier to change the distance metric with our method than with DiFi. We can use a specific distance metric by providing its sampled version as a 3D PDF texture. We can do this globally, but also locally. Every surface point can use another PDF function just by using another texture. For example, we can easily compute the so-called Johnson-Mehl or Apollonius diagrams [18], also called generalized skeletons, using additively, respectively multiplicatively, weighted Euclidean PDF functions, by scaling or multiplying the PDF texture at every point [7]. Doing this with DiFi appears to be significantly more complex [8].

Table 2. Skeletonization performance, DiFi method (from [8])

model	surface (voxels)	time (sec.)	voxels/sec.
octahedron	4862	0.85	5720
brain 1	18944	1.82	10408
brain 2	4988	0.64	7793
sinus 1	34507	22.1	1561
sinus 2	104154	49.7	2095

As Table 3 c shows, using our globally optimized PDF size (GPDF) calculation (Sec. 4.2) has a major performance impact for relatively elongated objects (e.g. 'bent', 'colon', 'intestin') where it massively reduces the amount of GPU overdraw during splatting. For objects tightly fitting their bounding-box, e.g. 'sphere' or 'cube', the optimization has no impact. Since the optimization itself does not cost extra time, it is always an efficient, valuable mechanism. Finally, we see that reconstruction is clearly faster than skeletonization (Table 1, column 'recon'). This is as expected, since a (pruned) skeleton has less points than the surface it comes from, and its MST values are exactly equal to the distance-to-boudary at every point, i.e. they match the absolute optimal PDF size value (Sec. 4.2).

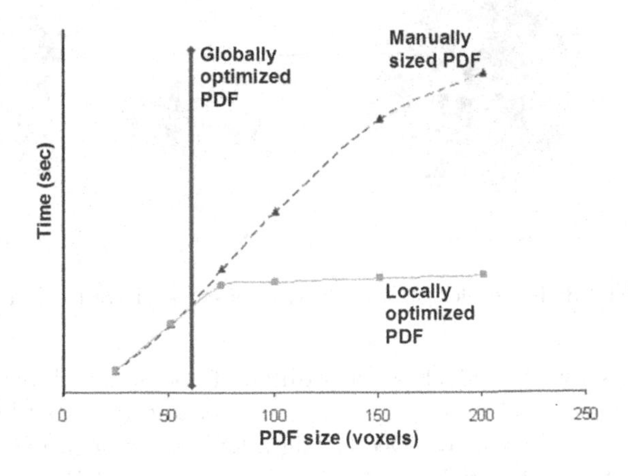

Fig. 2. Performance of local versus global PDF size choices

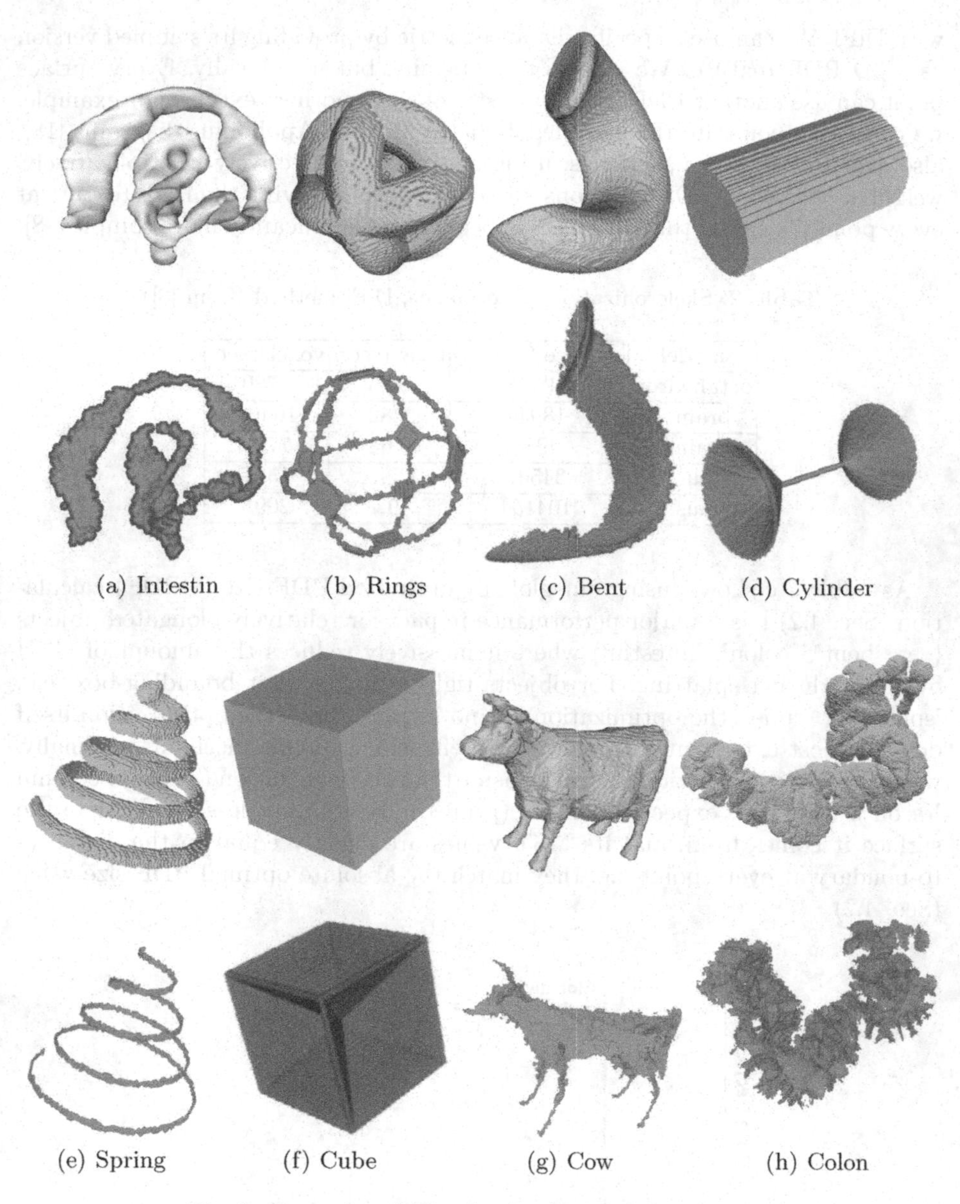

Fig. 3. Examples of 3D splatting-based skeletonization

Table 3 (a,b) shows the effect of using different coarse grid block sizes c in our bi-level hierarchical acceleration (Sec. 4.2). Increasing c means less CPU overhead, but more GPU overdraw. Decreasing c has the opposite effect. Varying c also implicitly affects the PDF size (Table 3). An optimal PDF size estimation would be obtained for the minimal block size $c = 1$. However, decreasing c increases the time needed to compute DT_c as well as the GPDF (Eqn. 8). For $c \leq 9$

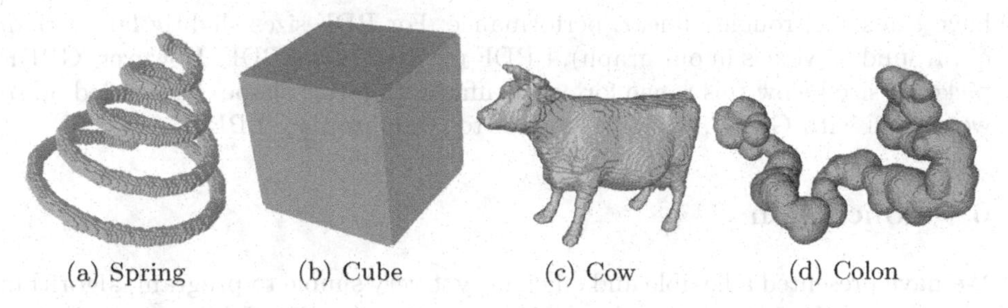

(a) Spring (b) Cube (c) Cow (d) Colon

Fig. 4. Reconstruction of smoothed objects by splatting pruned skeletons

Table 3. Benchmarks for variable coarse grid size (a,b); Naive versus globally-optimized PDF size performance (c)

(a)

model	grid size	time (sec)	PDF size
rings	6	10.0	49
	7	9.9	53
	8	10.9	55
	9	12.8	63
	10	12.7	61
	11	14.6	67
	100	23.3	97
cow	6	7.4	53
	7	6.3	53
	8	6.5	53
	9	6.7	53
	10	6.7	53
	11	6.8	53
	100	6.8	53
bent	6	19.7	65
	7	16.1	65
	8	17.1	73
	9	17.4	74
	10	18.2	75
	11	21.0	83
	100	49.4	131

(b)

model	grid size	time (sec)	PDF size
spring	6	3.4	37
	7	4.0	43
	8	4.6	49
	9	5.2	55
	10	5.9	61
	11	6.6	67
	100	8.7	83
duo	6	2.0	37
	7	2.4	43
	8	2.6	49
	9	2.9	55
	10	3.3	61
	11	3.3	61
	100	3.5	61
colon	6	76.0	55
	7	57.7	57
	8	47.9	61
	9	45.9	67
	10	44.4	69
	11	51.9	77
	100	112	119

(c)

	naive		globally optimized	
model	PDF size	time	PDF size	time
cube	45	2.5	45	2.5
box	37	1.7	37	1.7
sphere 1	85	10.0	85	10.0
sphere 2	171	199.3	171	199.3
cylinder 1	31	1.2	31	1.2
cylinder 2	129	72.3	61	19.2
cow	53	6.7	53	6.7
ellipse	25	0.7	25	0.7
spring	83	7.4	15	1.5
ice 2	43	1.6	23	1.0
ice 3	79	4.5	29	1.7
rings	97	19.7	33	6.1
duo	61	3.3	23	1.4
intestin	43	1.5	17	0.8
colon	119	179.9	43	26.1
bent	131	50.1	49	11.6

voxels, this cost is no longer negligible. After extensive testing on several models, we found the optimal coarse block size c to lie between 7 and 10, so we chose 10 as a default value. Finally, we compared the efficiency of local (per-block) and global PDF size optimizations (see Sec. 4.2). We timed our method using the locally optimized PDF size (LPDF), globally optimized PDF size (GPDF), and also, for comparison purposes, a fixed-size PDF (FPDF) manually set to values ranging from 25 to 201. As the graph in Fig. 2 shows, GPDF picks a PDF size δ' for which the FPDF (ascending graph) and LPDF (leveled graph) have the same, roughly linear, performance. For PDF sizes slightly larger than δ' (around 60 voxels in our graph), LPDF clearly beats FPDF. However, GPDF picked a size below this range for any configuration (3D shape) we availed of, so we settled with GPDF, which is simpler to compute than LPDF.

6 Conclusion

We have presented a flexible and efficient, yet very simple to program, algorithm to compute 3D skeletons on the GPU. We generalize the 2D distance splatting idea presented in [7] to the 3D case, and combine it with a different skeleton detector. Similar to [7], we use a bi-level hierarchical scheme to speed up our method by reducing the overdraw amount. Additionally, we use the coarse-scale distance transform (DT) to estimate an optimal size for our splat radius (PDF size), and thus reduce the overdraw even further. Since the optimal PDF size is highly object-dependent, and the GPU drawing performance is at least linearly dependent on the PDF size, this optimization can drastically improve the overall performance, as shown by our experiments. We performed extensive testing to evaluate our method on a range of volumetric objects, deduce optimal parameter values, and validated our results by performing (smoothed) object reconstructions from the skeleton. Overall, our simple splatting-based DT computation and skeletonization is as efficient as more complex methods, such as DiFi [8], and also lets one quite easily customize the distance metric used just by defining a 3D texture.

A more challenging subject, however, is finding efficient global criteria for noise-resistant detection and hole-free pruning of 3D skeletons. What such criteria might be, and whether they can efficiently be implemented on GPUs, is a subject for further research.

References

1. Blum, H.: A Transformation for Extracting New Descriptors of Shape. In: Models for the Perception of Speech and Visual Form. MIT Press (1967) 362–380
2. Gagvani, N., Kenchammana-Hosekote, D., Silver, D.: Volume animation using the skeleton tree. Proc. IEEE Volume Visualization (1998) 47–53
3. Rumpf, M., Telea, A.: A Continuous Skeletonization Method Based on Level Sets. Proc. VisSym (2002) 151–158
4. Ogniewicz, R.L., Kübler, O.: Hierarchic Voronoi skeletons. Pattern Recognition **28**(3) (1995) 343–359
5. Costa, L., Cesar, R.: Shape Analysis and Classification: Theory and Practice. CRC Press, Inc. (2000)
6. Telea, A., van Wijk, J.J.: An Augmented Fast Marching Method for Computing Skeletons and Centerlines. Proc. IEEE VisSym (2002) 251–258
7. Strzodka, R., Telea, A.: Generalized Distance Transforms and Skeletons in Graphics Hardware. Proc. VisSym (2004) 221–230
8. Sud, A., Otaduy, M.A., Manocha, D.: DiFi: Fast 3D Distance Field Computation Using Graphics Hardware. Computer Graphics Forum **23**(3) (2004) 557–566

9. Palágyi, K., Kuba, A.: Directional 3D Thinning Using 8 Subiterations. Proc. DGCI (1999) 325–336
10. Foskey, M., Lin, M.C., Manocha, D.: Efficient Computation of a Simplified Medial Axis. Proc. ACM Symp. Solid Modeling (2003) 96–107
11. Malandain, G., Fernández-Vidal, S.: Euclidean Skeletons. Image and Vision Computing **16**(5) (1998) 317–327
12. Siddiqi, K., Bouix, S., Tannenbaum, A., Zucker, S.: Hamilton-Jacobi Skeletons. IJCV **48**(3) (2002) 215–231
13. Sud, A., Foskey, M., Manocha, D.: Homotopy-Preserving Medial Axis Simplification. Proc. ACM Symp. Solid Modeling (2005) 39–50
14. Sigg, C., Peikert, R., Gross, M.: Signed Distance Transform Using Graphics Hardware. Proc. IEEE Visualization (2003) 83–90
15. Pudney, C.: Distance-ordered homotopic thinning: A skeletonization algorithm for 3d digital images. Computer Vision and Image Understanding **72**(3) (1998) 404–413
16. Reniers, D., Telea, A.: Quantitative comparison of tolerance-based distance transforms. Proc. VISAPP'06 (2006) 57–65
17. Pharr, M., Fernando, R.: GPU Gems 2: Programming Techniques for High-Performance Graphics. Addison-Wesley (2005)
18. Aurenhammer, F.: Voronoi diagrams: A survey of a fundamental geometric data structure. SIAM J. Comp. (27) (1998) 654–667

How to Tile by Dominoes the Boundary of a Polycube

Olivier Bodini and Sandrine Lefranc

LIRMM, 161, rue ADA, 34392 Montpellier Cedex 5, France
bodini@lirmm.fr, sandrine.lefranc2@libertysurf.fr

Abstract. We prove that the boundary of a *polycube* (finite union of integer unit cubes) has always a tiling by *foldable dominoes* (two edge-adjacent unit squares on the boundary). Moreover, the adjacency graph of the unit squares in the boundary of a spherical polycube has a Hamiltonian cycle.

1 Introduction

Polyominoes (finite and simply-connected union of unit squares in the plane) have been introduced in the sixties by S. Golomb [1]. Some natural questions arise for polyominoes:

- How many different polyominoes can be build with n squares? [17, 18, 20, 2]
- Is it possible to tile the plane (or a rectangle) with a given set of polyominoes? [14, 15, 16, 19, 21, 7, 23, 24, 25, 26, 3]
- Is it possible to tile a given polyomino with a given set of polyominoes? [4]
- How many different ways to do that? [22, 2]

All these questions have been studied by several authors as mentioned. To solve the third question below, J.H. Conway and J.C Lagarias [5] introduced a new point of view. They encoded the boundary of a polyomino by a word and obtained some important properties by using algebraic considerations of this word. In his paper, W. Thurston using this idea has shown that we can tile in linear time a polyomino by *dominoes* (two edge-adjacent unit squares). Successively, many authors have extended this tools: T. Chaboud [6], R. Kenyon [7], J.C. Fournier [8], K. Ito [9], N. Thiant [10], E. Remila [11] and myself with M. Latapy [12]. We introduce here 3-dimensional extension of the notion of polyominoes and another way to deal with, considering a set of "squares" drawn on a surface of $\mathbb{R}^3$. In particular, we study the boundary of a *spherical polycube* (finite union of unit cubes which is a topological ball). In a first section, we prove that the boundary of a polycube can always be tiled by foldable dominoes (i.e. a couple of edge-adjacent and not necessarily coplanar unit squares). In the second section, we show a new criterion of graphtheoretical h-connectivity and prove that the adjacency graph of external unit squares of a spherical polycube admits a Hamiltonian cycle. We deduce this result from a famous theorem of W. Tutte [13]. Let us now introduce our definitions.

A. Kuba, L.G. Nyúl, and K. Palágyi (Eds.): DGCI 2006, LNCS 4245, pp. 630–638, 2006.

2 Definitions

All objects that we consider in the sequel are built from the nodes of the lattice $\mathbb{Z}^3$. A *3-cell* (unit cube) in position (i, j, k) is the set:

$$\{(x, y, z) \in \mathbb{R}^3; i \leq x \leq i + 1, j \leq y \leq j + 1, k \leq z \leq k + 1\}.$$

With this definition, a polycube is a finite union of 3-cells. Two other classical objects are 1-cells (unit edges) and 2-cells (unit squares). A *1-cell* (resp. *2-cell*) is a face of dimension 1 (resp. of dimension 2) of a 3-cell. For instance the set $\{(1, 0, z) \in \mathbb{R}^3; 2 \leq z \leq 3\}$ is a 1-cell and the set

$$\{(3, y, z) \in \mathbb{R}^3; -1 \leq y \leq 0, 2 \leq z \leq 3\}$$

is a 2-cell. Naturally, we generalize the notion of domino in the following way: A *domino* is the union of two 2-cells sharing a 1-cell which is called *inner edge.* We say that the boundary of a polycube can be *tiled* by dominoes if and only if it is possible to cover without overflow nor overlap (i.e. the intersection of two dominoes is of dimension 1) this boundary with dominoes. In other words, a polycube can be tiled by dominoes if there exists a complete tiling of its boundary with non overlapping dominoes.

We denote by $G_{\mathbb{Z}^3}$ the non directed graph such that the vertex set is $V(G_{\mathbb{Z}^3}) = \mathbb{Z}^3$ and the edge set $E(G_{\mathbb{Z}^3})$ is:

$$\{\{v, v'\}; v \in V, v' \in V \text{and } d(v, v') = 1)\}$$

where d denotes the Euclidian distance in $\mathbb{R}^3$.

We notice that the 1-cells are in one-to-one correspondence with the edges of the graph $G_{\mathbb{Z}^3}$.

Let P be a polycube and f be a 2-cell, f is an *external 2-cell* of P if and only if f belongs to exactly one 3-cell of P. The union of the external 2-cells of P is the *topological boundary* of P. We denote it by ∂P.

Now, the *graphtheoretical boundary* of P, denoted by $\partial P_{\mathcal{G}}$, is the graph whose vertices are the external 2-cells and such that $\{e_1, e_2\}$ is an edge of $\partial P_{\mathcal{G}}$ if and only if:

- the 2-cells e_1 and e_2 share a 1-cell a.
- the 3-cells c_1, c_2 of P which contain respectively e_1 and e_2 do not verify $c_1 \cap c_2 = a$.

This second condition asserts that a 1-cell is the inner edge of at most one domino.

3 Tiling on the Boundary

Let H be the convex hull of all the permutations of the vector $(-1, -1/2, 0, 1/2, 1)$ and P be the image of H under the canonical projection $\pi : (x_1, x_2, x_3,$

$x_4) \longmapsto (x_1, x_2, x_3)$. P is a *permutohedron* (see fig.1). It is known that P tiles the space. Let $\mathcal{T}$ be the tiling of the space by translated copies of permutohedra which has a permutohedron centered on the origin (see fig. 2). Let $\mathcal{E}$ be the 1-skeleton (the union of its 1-cells) of this tiling. The set $\mathcal{E}$ allows us to bijectively associate each 3-cell c to a hexagon $(c \cap \mathcal{E}\ )$ in such a way that two hexagons share a 1-cell if and only if their associated 3-cells share a 2-cell. In a sense, we build a derived cellular complex.

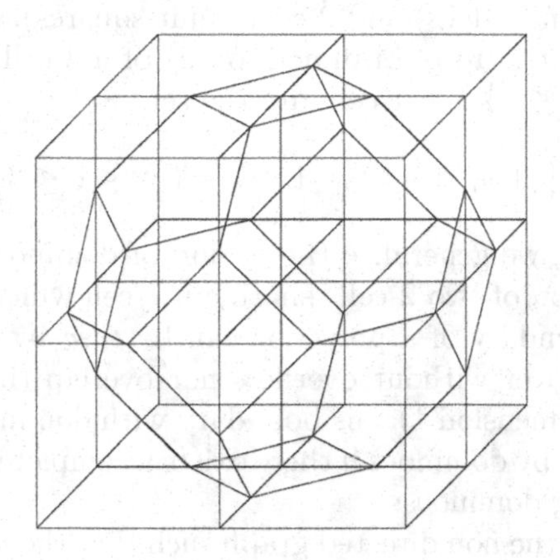

Fig. 1. A centered permutohedron

More explicitly, we can define $\mathcal{E}$ from the following 16 points (on the top of the figure):

$p_1 = (-1/2, 0, 1)$, $p_2 = (1, 1/2, 0)$, $p_3 = (1/2, 1, 0)$,
$p_4 = (-1, 0, 1/2)$, $p_5 = (1, 0, 1/2)$, $p_6 = (-1, 1/2, 0)$,
$p_7 = (0, 1, 1/2)$, $p_8 = (-1/2, 1, 0)$, $p_9 = (1/2, 0, 1)$,
$p_{10} = (-1, -1/2, 0)$, $p_{11} = (-1/2, -1, 0)$, $p_{12} = (0, 1/2, 1)$,
$p_{13} = (1/2, -1, 0)$, $p_{14} = (1, -1/2, 0)$, $p_{15} = (0, -1/2, 1)$,
$p_{16} = (0, -1, 1/2)$.

First of all, we build the following four hexagons:

$$h^1 = [p_5, p_9] \cup [p_9, p_{12}] \cup [p_{12}, p_7] \cup [p_7, p_3] \cup [p_3, p_2] \cup [p_2, p_5]$$

$$h^2 = [p_{12}, p_7] \cup [p_7, p_8] \cup [p_8, p_6] \cup [p_6, p_4] \cup [p_4, p_1] \cup [p_1, p_{12}]$$

$$h^3 = [p_5, p_9] \cup [p_9, p_{15}] \cup [p_{15}, p_{16}] \cup [p_{16}, p_{13}] \cup [p_{13}, p_{14}] \cup [p_{14}, p_5]$$

$$h^4 = [p_1, p_4] \cup [p_4, p_{10}] \cup [p_{10}, p_{11}] \cup [p_{11}, p_{16}] \cup [p_{16}, p_{15}] \cup [p_{15}, p_1] .$$

Then, using the following map t from $\mathbb{Z}^3$ to $\{1, 2, 3, 4\}$:

$$t((a_1, a_2, a_3)) = \begin{cases} 1 \text{ if } a_1 + a_3 \text{ and } a_2 + a_3 \text{ even} \\ 2 \text{ if } a_1 + a_3 \text{ odd and } a_2 + a_3 \text{ even} \\ 3 \text{ if } a_1 + a_3 \text{ even and } a_2 + a_3 \text{ odd} \\ 4 \text{ if } a_1 + a_3 \text{ and } a_2 + a_3 \text{ odd} \end{cases}$$

We can write: $\mathcal{E} = \bigcup_{(a_1,a_2,a_3)\in Z^3} \tau_{(a_1,a_2,a_3)} h^{t((a_1,a_2,a_3))}$ where $\tau_{(a_1,a_2,a_3)}$ denotes the translation of vector (a_1, a_2, a_3) (see fig. 2). As we have done for the boundary, we can put a structure of graph on $\mathcal{E}$. The vertices of $\mathcal{E}_\mathcal{G}$ are the vertices of all the hexagons. Let v and v' be two vertices of $\mathcal{E}_\mathcal{G}$, the set $\{v, v'\}$ is an edge of $\mathcal{E}_\mathcal{G}$ if and only if the segment $[v, v']$ belongs to a hexagon.

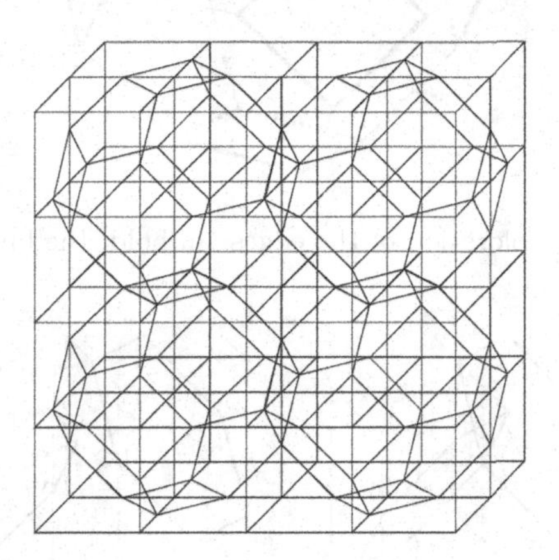

Fig. 2. Space tiling by permutohedra

In the following theorem, we deal with the natural bicoloration of the 3-cells. In order to do that, we recall that the adjacency graph of the 3-cells is bipartite and so, we can partition its vertices (3-cells) into two classes: the *white* 3-cells which are those in position (i, j, k) where $i + j + k = 0 \bmod 2$ and the *black* 3-cells which are the others. Moreover, we color each edge e of $\mathcal{E}_\mathcal{G}$ in blue if e belongs to a cycle of length 4, otherwise we color the edge in yellow (see fig 3.). This is the *cycle edge-coloration.*

We notice that all the permutohedra of the tiling are colored identically.

The set $\mathcal{E} \cap \partial P$ is composed by the edges of the hexagons of $\mathcal{E}$ "visible" on the boundary of P. Let G be the graph whose edges are the edges of $\mathcal{E}_\mathcal{G}$. We denote by $(\mathcal{E} \cap \partial P)_\mathcal{G}$ the graph where we have split each vertex of degree 4, in two new vertices of degree 2 in the following way: if a vertex v has degree 4, it is the middle of a 1-cell which is the intersection of two 3-cells c_1 and c_2. In the new graph, the 1-cells of c_1 linked to v are now linked to v_1 and the 1-cell of c_2 are now linked to v_2 (see fig. 4).

We have the following theorem:

Theorem 1. *For every polycube P, the graph $(\mathcal{E} \cap \partial P)_\mathcal{G}$ is a union of disjoint elementary cycles of even length (fig. 5).*

Proof. By definition, each edge a of $(\mathcal{E} \cap \partial P)_\mathcal{G}$ belongs to exactly one 3-cell c of P. We are going to define from the bicoloration of the 3-cells and the coloration of $\mathcal{E}_\mathcal{G}$, a new coloration in green and red for the edges of $\mathcal{E}_\mathcal{G} \cap \partial P_\mathcal{G}$ as follows:

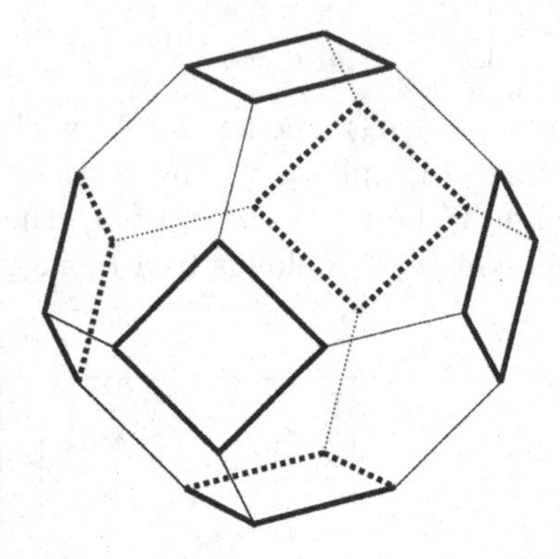

Fig. 3. Coloration of the edges (in bold, the blue edges)

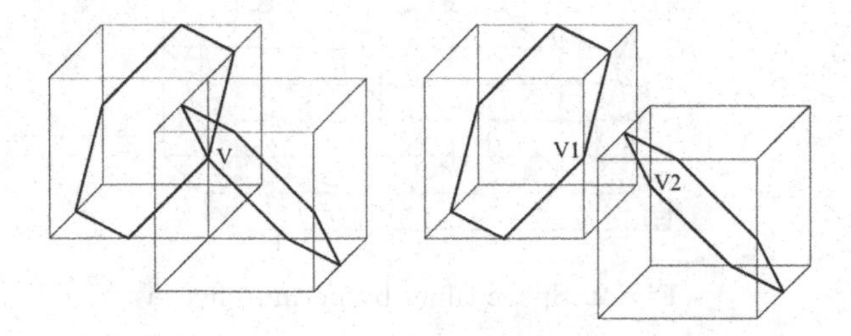

Fig. 4. Split of the vertex v

- if the 3-cell c is black and a is yellow (resp. blue) then a is red (resp. green) for the new coloration.
- if the 3-cell c is white and a is yellow (resp. blue) then a is green (resp. red).

Then, Two adjacent edges e_1 and e_2 have always different colors. Indeed, if they belongs to a same 3-cell, it is clear. Otherwise, they belongs to different 3-cells sharing a 1-face f. We denote by e the edge belonging to f. Clearly, e_1 (resp. e_2) has not the same color than e for the cycle edge-coloration. So, e_1 and e_2 have the same color for the cycle edge-coloration. As they belongs to adjacent 3-cells, it follows that they have different colors for the new coloration. Finally, this graph is 2-regular, the theorem is proved.

Corollary 1. *The boundary of a polycube P is tilable by dominoes.*

Proof. As $(\mathcal{E} \cap \partial P)_{\mathcal{G}}$ admits a perfect matching, by duality edge-vertex, ∂P admits a tiling by dominoes. Actually, there are at least 2^c different tilings where c is the number of even cycles on the boundary. Indeed, there is at least two different ways to tile the squares covered by each even cycle.

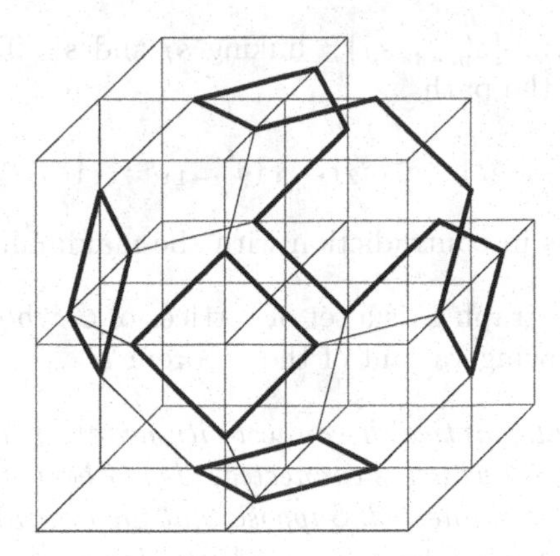

Fig. 5. Even cycles on the boundary of a polycube

4 Graphtheoretical h-Connectivity Conditions and Main Theorem

Let us recall that a graph G is h-connected if the deletion of every set of $h-1$ vertices does not disconnect G. Let X be a subset of vertices of G. We denote by $\Gamma(X)$ the set of adjacent vertices of X. We give in this section a new criterion of graphtheoretical h-connectivity which uses only connected subgraphs of G.

Theorem 2. *(Sufficient condition of graphtheoretical h-connectivity). Let G be a connected graph such that for every $1 \leq i \leq h-1$ and every connected subgraph induced by a set X of i vertices, the graph induced by $\Gamma(X)\backslash X$ is connected, then G is h-connected.*

Proof. Let G be a graph verifying the previous hypotheses and let us suppose that there exists a set $D = \{v_1, ..., v_{h-1}\}$ which disconnects this graph. Let p, q be two vertices which do not belong to the same connected component of $G\backslash D$. There exists in G a path:

$$C = (\{p = s_1, s_2\}, ..., \{s_{k-1}, q = s_k\})$$

linking p and q such that:

$$I = \min_{1 \leq i \leq k-1} \{i \text{ such that } s_i \notin D \text{ and } s_{i+1} \in D\}$$

is maximal. Let

$$J = \min_{I+1 \leq i \leq k} \{i \text{ such that } s_{i-1} \in D \text{ and } s_i \notin D\},$$

then s_{I+1} and s_{J-1} belong to the same connected component D' of the subgraph induced by D. From the hypotheses, the graph induced by $\Gamma(D')\backslash D'$ contains a path $C' = (\{s_I, s'_2\}, ..., \{s'_{m-1}, s_J\})$ linking s_I and s_J. This path has no vertex belonging to D. So, the path

$$(\{p, s_2\}, ..., \{s_{I-1}, s_I\}, \{s_I, s'_2\}, ..., \{s'_{m-1}, s_J\}, \{s_J, s_{J+1}\} ..., \{s_{k-1}, q\})$$

links p and q. This is in contradiction with the maximality of I.

A *separating set* of a graph G is a set of vertices of G whose deletion disconnects G. We have the following variant of the theorem 2:

Theorem 3. *(graphtheoretical h-connectivity criterion for the graphs having a minimum separating set which is connected). Let G be a graph having a minimum separating set which is connected. Suppose that for every $1 \leq i \leq h-1$ and every connected subgraph induced by a set X of i vertices, the graph induced by $G \backslash X$ is connected, then G is h-connected.*

Proof. Let us suppose that there exists a minimum connected separating set $D = \{v_1, ..., v_{h-1}\}$ for G. Let p, q be two vertices which do not belong to the same connected component of $G \backslash D$. There exists in G a path $C = (\{p = s_1, s_2\}, ..., \{s_{k-1}, q = s_k\})$ linking p with q and this path uses at least one vertex of D. Let s_I be the first vertex of C such that $s_{I+1} \in D$ and s_T the last vertex of C such that $s_{T-1} \in D$. As s_I and s_T belong to $G \backslash D$, there exists a path C' linking s_I and s_T. This path has no vertex belonging to D. So, we obtain a path linking p and q. This is a contradiction.

Now, let us come back to the boundary of a polycube. In fact, to solve our problem, we deal with a variant of this criterion.

Let F be a set of external 2-cells of a polycube P, we denote by $N_T(F)$ the set of all external 2-cells which touch F.

Lemma 1. *Let G be the adjacency graph of the external 2-cells of a polycube, then for every i, $1 \leq i \leq 3$, and every topologically connected subgraph F with i 2-cells, the graph induced by $N_T(F) \backslash F$ is connected.*

Proof. To disconnect the boundary, a connected subset S has to contain a closed curve which is non contractible on S (It separates the components). Clearly, this is not possible with less than four 2-cells.

Corollary 2. *The graph $\partial P_{\mathcal{G}}$ is 4-connected.*

Proof. Let P be a polycube verifying the previous hypotheses and let us suppose that there exists a set $D = \{v_1, v_2, v_3\}$ which disconnects it. Let p, q be two external 2-cells which do not belong to the same topological connected component of $\partial P/D$. There exists in $\partial P_{\mathcal{G}}$ a path $C = (\{p = s_1, s_2\}, ..., \{s_{k-1}, q = s_k\})$ linking p and q such that $I = \min_{1 \leq i \leq k-1} \{i \text{ such that } s_i \notin D \text{ and } s_{i+1} \in D\}$ is maximal. Let

$$J = \min_{I+1 \leq i \leq k} \{i \text{ such that } s_{i-1} \in D \text{ and } s_i \notin D\},$$

then s_I and s_J belong to the same topological connected component D' of the set D. From the Lemma 1, $N_T(D') \setminus D'$ contains a path $C' = (\{s_I, s'_2\}, ..., \{s'_m, s_J\})$ linking s_I and s_J. This path does not have any vertex belonging to D. Then, the path

$$(\{p, s_2\}, ..., \{s_{I-1}, s_I\}, \{s_I, s'_2\}, ..., \{s'_m, s_J\}, \{s_J, s_{J+1}\} ..., \{s_{k-1}, q\})$$

links p and q. This is in contradiction with the maximality of I.

Lemma 2. *Let P be a spherical polycube, then the graph $\partial P_{\mathcal{G}}$ is planar.*

Proof. By definition, we can represent this graph on a sphere. It suffices to do a stereographic projection from the center of a 2-cell of the polycube.

Theorem 4. *The adjacency graph of the external 2-cells of a spherical polycube has a Hamiltonian cycle.*

Proof. This graph is 4-connected and planar. So, the Theorem of Tutte [13] asserts that there exists a Hamiltonian cycle.

Let us notice that the theorem 3.2 is an easy consequence of this result. Indeed, we have a even number of external 2-cells. So, we can split the Hamiltonian cycle into couples of adjacent external 2-cells. This clearly involves a tiling of the boundary by dominoes.

Now, let us generalize this statement. We consider a sphere S subdivided into simply connected subsets called *cells* such that:

- each cell has four neighbors.
- two non-disjointed cells share a simple arc of curve.

We call this cellular complex a *4-regular subdivision of the sphere.* So, the adjacency graph of a 4-regular subdivision of the sphere is without loop nor multiple edge. Moreover, it is 4-regular.

Lemma 3. *Less than 4 cells can not disconnect the sphere.*

Proof. It is clear that one or two cells can not disjointed the sphere. Let C be the union of 3 connected cells which disconnects the sphere. Let T be the adjacency graph of one of the connected components of $S \setminus C$. The sum of the degree of T is $4n - 3$ (n the number of vertices of T) which obviously can not be equal to $2m$ (m the number of edges). By contradiction, we have the lemma.

Theorem 5. *The adjacency graph of the cells of a 4-regular subdivision of the sphere has a Hamiltonian cycle.*

Proof. This graph is 4-connected and planar. So, the Theorem of Tutte [13] asserts that there exists a Hamiltonian cycle.

Remark 1. The proofs of Theorem 4.6 and 4.8 do not provide any hint to algorithmically build the Hamiltonian cycle. This problem seems to be NP-complete as in the general case.

Conjecture 1. The adjacency graph of the external 2-cells of a connected polycube has a Hamiltonian cycle.

References

1. Golomb, S.W.: Tiling with polyominoes, J.C.T. Series A **1** (1966) 280–296.
2. Krattenthaler, C., Okada, S.: The number of rhombus tilings of a "punctured" hexagon and the minor summation formula. Adv. Appl. Math. **21** (1998). 381–404.
3. Walkup, D.W.: Covering a rectangle with T-tetraminoes. Amer. Math. Monthly **72** (1965). 986–988.
4. Bodini, O., Nouvel, B.: Z-tilings of Polyominoes and Standard Basis. ICWIA'04. (2004). 137–150.
5. Conway, J.H., Lagarias, J.C: Tiling with polyominoes and combinatorial group theory. JCT Series A **53**. (1990). 183–208.
6. Chaboud, T.: Pavage par des dominos dans des graphes planaires biréguliers. C. R. Acad. Sci. Paris **318**. Série I. (1994) 591–594.
7. Kenyon, R.: A note on tiling with integer-sided rectangles. J.C.T. Series A **74**. (1996) 321–332.
8. Fournier, J.C.: Tiling pictures of the plane with dominoes. Discrete Mathematics **165/166** (1997) 313–320.
9. Ito, K.: Domino tilings on planar regions. J.C.T. Series A **75** (1996) 173–186.
10. Thiant, N.: An O(n log n)-algorithm for finding a domino tiling of a plane picture whose number of holes is bounded. Theoretical Computer Science. **303**. Issues 2-3. (2003).353–374.
11. Rémila, E.: The lattice structure of the set of domino tilings of a polygon. Theoretical Computer Science. **322**. Issue 2. (2004). 409–422
12. Bodini, O., Latapy. M.: Generalized Tilings with Height Functions. Morfismos **7** volume 3. (2003). 47–68.
13. Tutte, W.T.: A theorem on planar graphs. Trans. Amer. Math. Soc. **82** (1956). 99–116.
14. Barnes, F.W.: Best packing of rods into boxes. Discrete Mathematics **142** (1995) 271–275.
15. Berger, R.: The undecidability of the domino problem. Mem. Amer. Math. Soc. **66** (1966).
16. Bodini, O.: Tiling a Rectangle with Polyominoes. DMCS'03. (2003). 89–102.
17. Bousquet-Mélou, M.: Codage des polyominos convexes et équations pour l'énumération suivant l'aire. Discrete Applied Mathematics. **48**. Issue 1. 4 January 1994. 21–43.
18. Bousquet-Mélou, M.: q-énumération de Polyominos Convexes. J.C.T. Series A. **64**. Issue 2. November 1993. 265–288.
19. Dahlke, K.A.: The Y-hexomino has order 92. J.C.T. Series A **51** (1989) 125–126.
20. Feretic, S.: A q-enumeration of convex polyominoes by the festoon approach. Theoretical Computer Science. **319**. Issues 1-3. (2004). 333–356.
21. Golomb, S.W.: Polyominoes which tile rectangles. J.C.T. Series A **51** (1989) 117–124.
22. Kenyon, R.W., Sheffield, S.: Dimers, tilings and trees. J.C.T. Series B. **92**. Issue 2. (2004). 295–317
23. Klarner, D.A.: Packing a rectangle with congruent n-ominoes. J.C.T. Series A **7** (1969) 107–115.
24. Marshall, W.R.: Packing rectangles with congruent polyominoes. J.C.T. Series A **77** (1997). 181–192.
25. Reid, M.: Tiling rectangles and half strips with congruent polyominoes. J.C.T. Series A **80** (1997). 106–123.
26. Rhoads, G.C.: Planar tilings by polyominoes, polyhexes, and polyiamonds. Journal of Computational and Applied Mathematics. **174** Issue 2 (2005). 329–353.
27. Thurston, W.P.: Conway's tiling groups. Amer. Math. Monthly **95** (1990). 757–773.

A Generalized Preimage for the Standard and Supercover Digital Hyperplane Recognition

Martine Dexet and Eric Andres

Laboratoire SIC, Université de Poitiers
Bât. SP2MI, bvd Marie et Pierre Curie, BP 30179
86962 Futuroscope Chasseneuil Cedex, France
{dexet, andres}@sic.univ-poitiers.fr

Abstract. A new digital hyperplane recognition method is presented. This algorithm allows the recognition of Standard and Supercover hyperplanes by incrementally computing in a dual space the generalized preimage of a given hypervoxel set. Each point in this preimage corresponds to a Euclidean hyperplane which intersects all given hypervoxels. An advantage of the generalized preimage is that it does not depend on the hypervoxel locations. Moreover, the proposed recognition algorithm does not require the hypervoxels to be connected or ordered in any way.

1 Introduction

In digital geometry, objects are usually considered as digital point or hypervoxel (*pixels* in 2D and *voxels* in 3D) sets. Indeed, this is the structural decomposition mostly used to store digital information. A drawback of this kind of representation is that it does not provide any information on the shape or topology of digital objects. Another way of obtaining the description of digital objects is the hyperplane decomposition. This process, called *digital hyperplane recognition*, consists of determining if a digital point set forms a hyperplane segment, that is a hyperplane bounded region.

The recognition problem has so far mainly been studied in dimensions 2 and 3 (see [1] for an overview on 2D recognition algorithms), with various approaches such as linear programming techniques [2, 3], computational geometry methods [4,5,6] or *preimage* computation based algorithms [7]. Very few papers handle the problem in arbitrary dimensions [8,9]. Computational and efficiency aspects of digital hyperplane recognition problems are investigated in [10].

In this paper, we are interested in preimage based approaches for the recognition of Standard and Supercover analytical hyperplanes. Informally, the preimage [11] of a hypervoxel set consists of all Euclidean hyperplanes the digitization of which contains the given hypervoxels. More precisely, the preimage of a hypervoxel set is computed in a dual space where each point is mapped onto a Euclidean hyperplane. Preimage computation algorithms depending on the hypervoxel locations have been proposed in dimensions 2 and 3 [7,12].

In this work, we perform the recognition of Standard and Supercover digital hyperplanes by computing the set of Euclidean hyperplanes which intersect a

A. Kuba, L.G. Nyúl, and K. Palágyi (Eds.): DGCI 2006, LNCS 4245, pp. 639–650, 2006.

given hypervoxel set. In order to do that, we incrementally compute the *generalized preimage* of the hypervoxels, which is a preimage defined in any dimension and independent of the hypervoxel connectivity and location. This preimage is computed from the dual of each hypervoxel. Indeed, each point in this dual object corresponds to a Euclidean hyperplane which cuts the hypervoxel. Hence, a major part of this paper is devoted to determining the formulas describing the dual of a hypervoxel. First, a positive and a negative extrusion are defined. Then, we show that the dual of a hypervoxel can be computed from the extrusions of the dual of its vertices. Finally, the intersection of all hypervoxel duals forms the generalized preimage. The recognition process consists therefore simply in computing the generalized preimage of a hypervoxel set (i.e. computing the dual of a hypervoxel set). More precisely, we start with a hypervoxel dual and add hypervoxel duals as long as the generalized preimage is not empty.

In Section 2, we introduce our notations and definitions as well as the Standard and Supercover analytical hyperplane descriptions. In Section 3, we determine the dual of a hypervoxel. The generalized preimage of a hypervoxel set is introduced in Section 4. Then, we explain how the Standard and Supercover hyperplane recognition algorithm works. We also provide some hints on how the computational efficiency can be increased thanks to a particular hypervoxel dual description. Conclusion and future works are proposed in Section 5.

2 Preliminaries

In this section, we first propose some notations and give the definition of a hypercube. Then, we present the two digitization models considered in this work that are the Standard model and the Supercover model.

2.1 Notations and Definitions

Let $n \in \mathbb{Z}$, $n > 0$. In the following, we will denote by $\mathcal{E}_n$ the classical n-dimensional Euclidean space, and by $[\![1, k]\!]$ the subset of integer values $\{1, \ldots, k\} \subset \mathbb{Z}$. We define an *$\alpha$-hypercube*, $\alpha \in \mathbb{R}$, as follows:

Definition 1 (α-Hypercube). *The hypercube (or n-dimensional cube) centered on $(c_1, \ldots, c_n) \in \mathbb{R}^n$, with size $\alpha \in \mathbb{R}$, is the set of points $(x_1, \ldots, x_n) \in \mathbb{R}^n$ verifying*

$$\forall i \in [\![1, n]\!], c_i - \frac{\alpha}{2} \leq x_i \leq c_i + \frac{\alpha}{2}$$

Let $p \in \mathbb{Z}^n$ be a point with integer-valued coordinates, also called *digital point*. We call *hypervoxel* a unit-size hypercube centered on a digital point. Hypervoxels in dimensions 2 and 3 are respectively called *pixels* and *voxels*.

2.2 The Standard and Supercover Analytical Models

The *Standard model* [13] and *Supercover model* [14,15] are both digital analytical models, defined in any dimension, which provide a digitization of Euclidean objects. Moreover, the Standard model is the only one that allows, in dimension n, the $(n-1)$-connected digitization of any linear object of $\mathbb{R}^n$.

Standard and Supercover hyperplanes (or n-dimensional planes) are defined analytically as follows (see Figure 1):

Definition 2 (Standard Hyperplane [13]). *The Standard hyperplane with parameters* $(c_0,\ldots,c_n)\in\mathbb{R}^{n+1}$ *is the set of points* $(x_1,\ldots,x_n)\in\mathbb{Z}^n$ *verifying*

$$-\frac{\sum_{i=1}^{n}|c_i|}{2}\leq c_0+\sum_{i=1}^{n}c_ix_i<\frac{\sum_{i=1}^{n}|c_i|}{2}$$

where $c_1\geq 0$, *or* $c_1=0$ *and* $c_2\geq 0$, *or* $\ldots$, *or* $c_1=c_2=\ldots=c_{n-1}=0$ *and* $c_n\geq 0$.

Definition 3 (Supercover Hyperplane [15]). *The Supercover hyperplane with parameters* $(c_0,\ldots,c_n)\in\mathbb{R}^{n+1}$ *is the set of points* $(x_1,\ldots,x_n)\in\mathbb{Z}^n$ *verifying*

$$-\frac{\sum_{i=1}^{n}|c_i|}{2}\leq c_0+\sum_{i=1}^{n}c_ix_i\leq\frac{\sum_{i=1}^{n}|c_i|}{2}$$

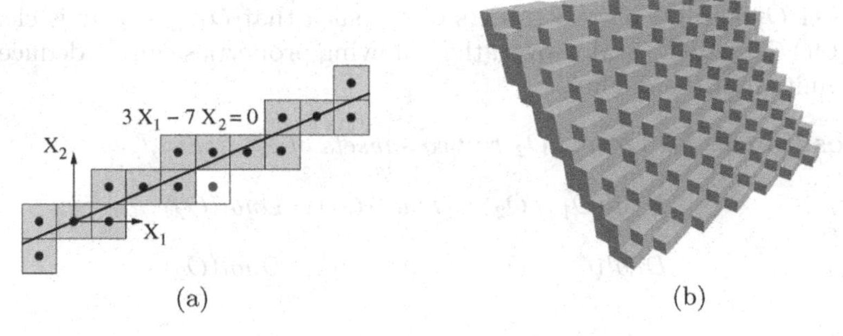

Fig. 1. Standard and Supercover hyperplane examples in dimensions 2 and 3: (a) The Supercover and Standard lines with parameters $(0,3,-7)$. The pixel colored in white does not belong to the Standard line (pixels colored in grey) but belongs to the Supercover one, (b) The Standard plane with parameters $(0,3,-1,2)$

Remark 1. The Supercover digitization of a Euclidean hyperplane also consists of all hypervoxels which are intersected by the hyperplane, whereas the Standard one consists of all hypervoxels cut by the hyperplane except when a hypervoxel vertex is intersected (see Figure 1a). In this case, several hypervoxels adjacent to this vertex do not belong to the Standard digitization. This is due to the fact that one inequality in Definition 2 is strict.

3 Dual of a Hypervoxel

We use a dual transformation similar to the well known Hough transform which is an efficient tool usually used in image processing to recognize parametric shapes in an image. A review on existing variations of this method is presented in [16].

In the two following sections, we first define the parameter space in which our dual transformation is performed. Then, we describe the dual of a hypervoxel, which is the basis of the recognition algorithm presented in Section 4.

3.1 Parameter Space: Definition and Properties

In this work, we use the n-dimensional parameter space $\mathcal{P}_n \subset \mathbb{R}^n$, and define the two functions $\mathcal{D}_\mathcal{E} : \mathcal{E}_n \to \mathcal{P}_n$ and $\mathcal{D}_\mathcal{P} : \mathcal{P}_n \to \mathcal{E}_n$ by:

$$\mathcal{D}_\mathcal{E}(x_1, \ldots, x_n) = \left\{ (y_1, \ldots, y_n) \in \mathcal{P}_n | y_n = -\sum_{i=1}^{n-1} x_i y_i + x_n \right\}$$

$$\mathcal{D}_\mathcal{P}(y_1, \ldots, y_n) = \left\{ (x_1, \ldots, x_n) \in \mathcal{E}_n | x_n = \sum_{i=1}^{n-1} y_i x_i + y_n \right\}$$

Informally, each point in $\mathcal{E}_n$ (resp. $\mathcal{P}_n$) is transformed by $\mathcal{D}_\mathcal{E}$ (resp. $\mathcal{D}_\mathcal{P}$) into a hyperplane in $\mathcal{P}_n$ (resp. $\mathcal{E}_n$). In the rest of this paper, we will generically write *Dual* for $\mathcal{D}_\mathcal{E}$ or $\mathcal{D}_\mathcal{P}$.

Let O be a subset of $\mathbb{R}^n$. Then, $Dual(O) = \bigcup_{p \in O} Dual(p)$ is called the *dual* of O. Let O_1 and O_2 be two subsets of $\mathbb{R}^n$ such that $O_1 \subseteq O_2$. It is clear that $Dual(O_1) \subseteq Dual(O_2)$. Moreover, the following properties can be deduced from our definition of the duality.

Proposition 1. *Let O_1 and O_2 be two subsets of $\mathbb{R}^n$. Then,*

$$Dual(O_1 \cup O_2) = Dual(O_1) \cup Dual(O_2) \tag{1}$$

$$Dual(O_1 \cap O_2) \subseteq Dual(O_1) \cap Dual(O_2) \tag{2}$$

Proof. (1): $Dual(O_1 \cup O_2) = \bigcup_{p \in O_1 \cup O_2} Dual(p) = \left[\bigcup_{p \in O_1} Dual(p)\right] \cup \left[\bigcup_{p \in O_2} Dual(p)\right] = Dual(O_1) \cup Dual(O_2)$.

(2): Since $O_1 \cap O_2 \subseteq O_1$ and $O_1 \cap O_2 \subseteq O_2$, we deduce that $Dual(O_1 \cap O_2) \subseteq Dual(O_1)$ and $Dual(O_1 \cap O_2) \subseteq Dual(O_2)$. Thus, $Dual(O_1 \cap O_2) \subseteq Dual(O_1) \cap Dual(O_2)$. □

Property 1. Let $p \in \mathbb{R}^n$ be a point. The dual of each point which lies in $Dual(p)$ is a hyperplane which passes through p.

3.2 Hypervoxel Dual Representation

In this work, we need to define the dual of a hypervoxel. We first define the positive and negative extrusions of a point.

Definition 4 (Positive and Negative Extrusions). *Let $p = (x_1, \ldots, x_n) \in \mathbb{R}^n$ be a point. The positive extrusion of p is defined by:*

$$p^+ = \{p' = (x'_1, \ldots, x'_n) \in \mathbb{R}^n | \forall i \in [\![1, n-1]\!], x_i = x'_i \text{ and } x_n \leq x'_n\}$$

In the same way, the negative extrusion of p is defined by:

$$p^- = \{p' = (x'_1, \ldots, x'_n) \in \mathbb{R}^n | \forall i \in [\![1, n-1]\!], x_i = x'_i \text{ and } x_n \geq x'_n\}$$

Let O_1 and O_2 be two subsets of $\mathbb{R}^n$ such that $O_1 \subseteq O_2$. Then, $O_1^+ \subseteq O_2^+$ and $O_1^- \subseteq O_2^-$. Moreover, the following properties can be deduced from Definition 4.

Proposition 2. *Let O_1 and O_2 be two subsets of $\mathbb{R}^n$. Then,*

$$(O_1 \cup O_2)^+ = O_1^+ \cup O_2^+$$

In the same way, $(O_1 \cup O_2)^- = O_1^- \cup O_2^-$.

Proof. $(O_1 \cup O_2)^+ = \bigcup_{p \in O_1 \cup O_2} p^+ = \left[\bigcup_{p \in O_1} p^+\right] \cup \left[\bigcup_{p \in O_2} p^+\right] = O_1^+ \cup O_2^+$. The proof of $(O_1 \cup O_2)^- = O_1^- \cup O_2^-$ is obtained in the same way. □

Proposition 3. *Let $p \in \mathbb{R}^n$ be a point. Then,*

$$Dual(p)^+ = Dual(p^+)$$

In the same way, $Dual(p)^- = Dual(p^-)$.

Proof. Let us consider $p = (x_1, \ldots, x_n) \in \mathcal{E}_n$. Then, $Dual(p^+) = \mathcal{D}_{\mathcal{E}}(p^+) =$

$$\bigcup_{p' \in p^+} Dual(p') = \bigcup_{p' = (x'_1, \ldots, x'_n) \in p^+} \{(y_1, \ldots, y_n) \in \mathcal{P}_n | y_n = -\sum_{i=1}^{n-1} x'_i y_i + x'_n\} =$$

$$\{(y_1, \ldots, y_n) \in \mathcal{P}_n | y_n \geq -\sum_{i=1}^{n-1} x_i y_i + x_n\} = \bigcup_{p' \in \mathcal{D}_{\mathcal{E}}(p)} p'^+ = \mathcal{D}_{\mathcal{E}}(p)^+ = Dual(p)^+.$$

The proof of $Dual(p)^- = Dual(p^-)$ can be obtained in the same way. □

Proposition 3 is illustrated in Figure 2.

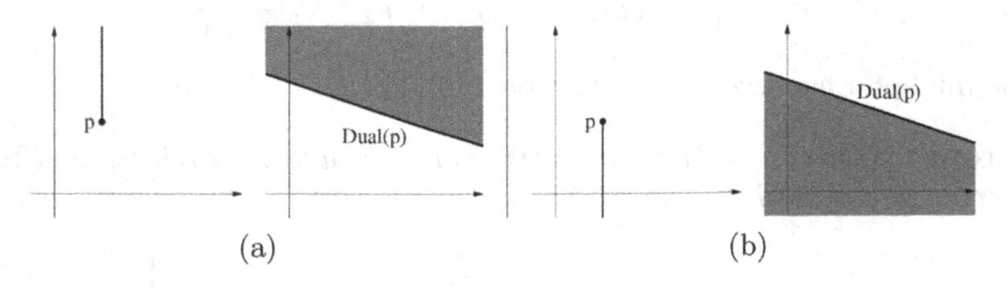

Fig. 2. Positive and negative extrusions of a point p (half-lines) and their dual object: a half-space, (a) Positive extrusion of p, (b) Negative extrusion

Proposition 4. *Let H be a hypervoxel. Then,*

$$Dual(H) = Dual(H)^+ \cap Dual(H)^-$$

Proof. In the following, we assume that $H \in \mathcal{E}_n$. Since $Dual(H) \subseteq Dual(H)^+$ and $Dual(H) \subseteq Dual(H)^-$, we deduce that $Dual(H) \subseteq Dual(H)^+ \cap Dual(H)^-$.

We now prove that $Dual(H)^+ \cap Dual(H)^- \subseteq Dual(H)$. Consider a point $p = (x_1, \ldots, x_n) \in Dual(H)^+ \cap Dual(H)^-$. Then, there exists $p' = (x'_1, \ldots, x'_n) \in Dual(H)$ such that $p \in p'^+$ and there exists $p'' = (x''_1, \ldots, x''_n) \in Dual(H)$ such that $p \in p''^-$. We deduce that $\forall i \in [\![1, n-1]\!]$, $x'_i = x_i = x''_i$ and $x'_n \leq x_n \leq x''_n$.

Next we prove that $Dual(p) \cap H \neq \emptyset$, which would imply $p \in Dual(H)$. Since $p' \in Dual(H)$ and $p'' \in Dual(H)$, we have $Dual(p') \cap H \neq \emptyset$ and $Dual(p'') \cap H \neq \emptyset$. Let $q' = (q'_1, \ldots, q'_n) \in Dual(p') \cap H$ and $q'' = (q''_1, \ldots, q''_n) \in Dual(p'') \cap H$. Then, we have

$$q'_n = \sum_{i=1}^{n-1} x_i q'_i + x'_n \text{ and } q''_n = \sum_{i=1}^{n-1} x_i q''_i + x''_n$$

Since $x'_n \leq x_n \leq x''_n$, we deduce that

$$q'_n \leq \sum_{i=1}^{n-1} x_i q'_i + x_n \text{ and } q''_n \geq \sum_{i=1}^{n-1} x_i q''_i + x_n$$

Thus, $Dual(p) \cap [q', q''] \neq \emptyset$. Finally, since H is convex we know that $[q', q''] \subset H$. We then deduce that $Dual(p) \cap H \neq \emptyset$. □

Let us now describe the dual of a hypervoxel from its vertices. Let H be a hypervoxel centered on $(c_1, \ldots, c_n) \in \mathbb{Z}^n$ and let $\mathcal{V}$ be the set of its 2^n vertices. We define two subsets of $\mathcal{V}$ with cardinality 2^{n-1} as follows:

$$\mathcal{V}_+ = \left\{ v = (x_1, \ldots, x_n) \in \mathcal{V} | x_n = c_n + \frac{1}{2} \right\}$$

$$\mathcal{V}_- = \left\{ v = (x_1, \ldots, x_n) \in \mathcal{V} | x_n = c_n - \frac{1}{2} \right\}$$

The dual of a hypervoxel can then be defined by:

Theorem 1 (Dual of a Hypervoxel). *Let H be a hypervoxel, $\mathcal{V}_+$ and $\mathcal{V}_-$ the two vertex sets defined previously. Then:*

$$Dual(H) = \left[\bigcup_{v \in \mathcal{V}_-} Dual(v)^+ \right] \cap \left[\bigcup_{v \in \mathcal{V}_+} Dual(v)^- \right]$$

Proof. We assume that H is centered on $(c_1, \ldots, c_n) \in \mathbb{Z}^n$. Let us first prove that $Dual(H^+) = Dual(\mathcal{V}_-^+)$, and then that $Dual(H)^+ = Dual(\mathcal{V}_-)^+$. The proof of $Dual(H^-) = Dual(\mathcal{V}_+^-)$ can be obtained in the same way.

The convex hull of the vertices in $\mathcal{V}_-$ is defined by

$$\mathcal{C} = \{(x_1, \ldots, x_n) | \forall i \in [\![1, n-1]\!], c_i - \tfrac{1}{2} \leq x_i \leq c_i + \tfrac{1}{2} \text{ and } x_n = c_i - \tfrac{1}{2}\}.$$

We can deduce from Definition 4 that $H^+ = \mathcal{C}^+$. Then, let us prove that $Dual(\mathcal{C}^+) = Dual(\mathcal{V}_-^+)$. Specifically, we know that $\mathcal{V}_- \subset \mathcal{C}$. Then, $\mathcal{V}_-^+ \subset \mathcal{C}^+$ and thus $Dual(\mathcal{V}_-^+) \subseteq Dual(\mathcal{C}^+)$. Next we show that $Dual(\mathcal{C}^+) \subseteq Dual(\mathcal{V}_-^+)$.

Let $p \in Dual(\mathcal{C}^+)$. Then, $Dual(p) \cap \mathcal{C}^+ \neq \emptyset$. We need to prove that there exists $v \in \mathcal{V}_-$ such that $Dual(p) \cap v^+ \neq \emptyset$. We proceed by contradiction and assume that $\forall v \in \mathcal{V}_-, Dual(p) \cap v^+ = \emptyset$. Since $\mathcal{C}$ is the convex hull of the vertices in $\mathcal{V}_-$, we can deduce that $\mathcal{C}^+$ is the convex hull of the points in $\mathcal{V}_-^+$. Hence, since $Dual(p)$ is a hyperplane, we deduce that $Dual(p) \cap \mathcal{C}^+ = \emptyset$. □

An illustration of Proposition 1 and Theorem 1 is given in Figure 3.

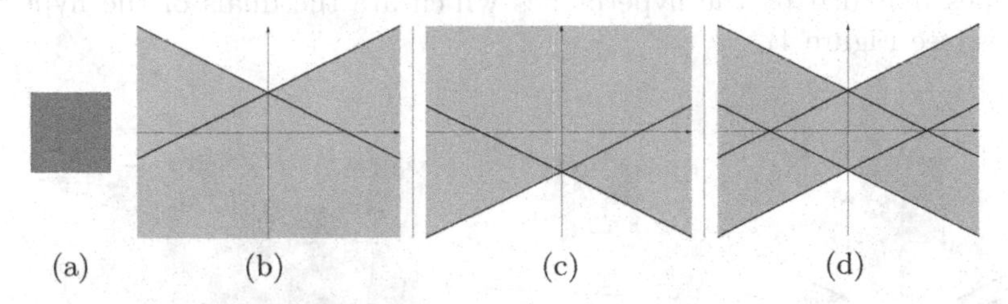

Fig. 3. (a) A pixel H, (b) Illustration of $Dual(H)^-$ ($= Dual(H^-)$), (c) Illustration of $Dual(H)^+$ ($= Dual(H^+)$), (d) Dual of H

Theorem 1 allows us to compute the dual of a hypervoxel from its vertices. The dual of a hypervoxel can also be described as the union of several open polytopes.

Let $\mathcal{S} = \{-1, 1\}$ and let $\mathcal{R} : \mathcal{S} \to \{\mathbb{R}^-, \mathbb{R}^+\}$ be the function defined by: $\mathcal{R}(-1) = \mathbb{R}^-$ and $\mathcal{R}(1) = \mathbb{R}^+$. Moreover, let $s = (s_1, \ldots, s_{n-1}) \in \mathcal{S}^{n-1}$. We denote by $\mathcal{R}^s$ the Cartesian product $\mathcal{R}(s_1) \times \ldots \times \mathcal{R}(s_{n-1})$. A partition of $\mathbb{R}^n$ can thus be defined by:

$$\mathbb{R}^n = \bigcup_{s \in \mathcal{S}^{n-1}} (\mathcal{R}^s \times \mathbb{R})$$

Let H be a hypervoxel centered on $(c_1, \ldots, c_n) \in \mathbb{Z}^n$. We denote by v_-^s the vertex of H with coordinates $(c_1 + \frac{1}{2}s_1, \ldots, c_1 + \frac{1}{2}s_{n-1}, c_n - \frac{1}{2})$ and by v_+^{-s} the vertex with coordinates $(c_1 - \frac{1}{2}s_1, \ldots, c_1 - \frac{1}{2}s_{n-1}, c_n + \frac{1}{2})$. Then, we have the following property:

Corollary 1 (Hypervoxel Dual Decomposition). *Let H be a hypervoxel. Then,*

$$Dual(H) = \bigcup_{s \in \mathcal{S}^{n-1}} \left[(\mathcal{R}^s \times \mathbb{R}) \cap Dual(v_-^s)^+ \cap Dual(v_+^{-s})^-\right]$$

Proof. Let us assume that $H \subset \mathcal{E}_n$. We know that
$\forall s \in \mathcal{S}^{n-1}$, $\left[(\mathcal{R}^s \times \mathbb{R}) \cap Dual(v_-^s)^+ \cap Dual(v_+^s)^-\right] \subset Dual(v_-^s)^+ \cap Dual(v_+^s)^-$.
Hence, since $Dual(v_-^s)^+ \cap Dual(v_+^s)^- \subset Dual(\mathcal{V}_-)^+ \cap Dual(\mathcal{V}_+)^-$, we deduce from Theorem 1 that
$\forall s \in \mathcal{S}^{n-1}$, $\left[(\mathcal{R}^s \times \mathbb{R}) \cap Dual(v_-^s)^+ \cap Dual(v_+^s)^-\right] \subset Dual(H)$. Let us prove the reverse inclusion.

Let $p = (x_1, \ldots, x_n) \in Dual(H)$. Then, there exists $s = (s_1, \ldots, s_{n-1}) \in \mathcal{S}^{n-1}$ such that $p \in \mathcal{R}^s$. Moreover, by Theorem 1 it follows that $\forall v = (v_1, \ldots, v_n) \in \mathcal{V}_-, x_n \geq \sum_{i=1}^{n-1} -v_i x_i + v_n$, and $\forall v = (v_1, \ldots, v_n) \in \mathcal{V}_-, x_n \leq \sum_{i=1}^{n-1} -v_i x_i + v_n$. However, since $p \in \mathcal{R}^s$, and thus $x_i \in \mathcal{R}(s_i)$, we deduce that $-(c_i + \frac{1}{2}s_i)x_i \leq -(c_i - \frac{1}{2}s_i)x_i$. Hence, p verifies the following inequalities: $x_n \leq \sum_{i=1}^{n-1} -(c_i - \frac{1}{2}s_i)x_i + c_n + \frac{1}{2}$ and $x_n \geq \sum_{i=1}^{n-1} -(c_i + \frac{1}{2}s_i)x_i + c_n - \frac{1}{2}$. We get that $p \in Dual(v_-^s)^+ \cap Dual(v_+^{-s})^-$. □

Corollary 1 allows the decomposition of the dual of a hypervoxel into 2^{n-1} open polytopes bounded by the hyperplanes which are the duals of the hypervoxel vertices (see Figure 4).

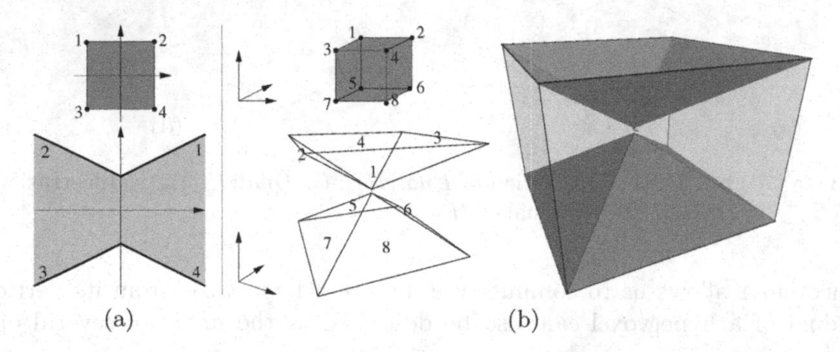

Fig. 4. Hypervoxel dual examples. Numbering shows the correspondence between the vertices of a pixel (resp. voxel) and the lines (resp. planes) forming the border of its dual: (a) Dual of the pixel centered on $(0, 0)$, (b) Dual of the voxel centered on $(0, 0, 0)$.

4 Standard and Supercover Hyperplane Recognition

In this section, we present our Standard and Supercover hyperplane recognition algorithm. The aim is to determine if a hypervoxel set belongs to a Standard or Supercover digital hyperplane. The idea of our algorithm is to compute the set of Euclidean hyperplanes (if it exists) which cross the given hypervoxels. Each hyperplane in this set is the dual of a point in the parameter space which belongs to a particular polytope. We call this polytope the *generalized preimage* of the hypervoxels. Then, based on the shape (empty or not) of this preimage, we can deduce if the hypervoxel set belongs or not to a Standard or Supercover hyperplane. In the following, we first give the definition of the generalized preimage.

Then, we detail our recognition algorithm, and give some simplifications we can apply on it in order to improve its complexity.

4.1 The Notion of Generalized Preimage

As said previously, each point in the generalized preimage of a hypervoxel set $\mathcal{H}$ is the dual of a Euclidean hyperplane which cuts all hypervoxels of $\mathcal{H}$. We define the generalized preimage of a hypervoxel set as follows:

Definition 5 (Generalized Preimage). *Let $\mathcal{H} = (H_1, \ldots, H_k)$ be a set of k hypervoxels, and let $Dual(H_i)$, $i \in [\![1, k]\!]$, be the dual of H_i in the parameter space. The generalized preimage $\mathbb{G}_P$ of $\mathcal{H}$ is defined by:*

$$\mathbb{G}_P(\mathcal{H}) = \bigcap_{i=1}^{k} Dual(H_i)$$

Note that the Standard digitization of many hyperplanes in the dual of the generalized preimage does not contain the given hypervoxels (see Remark 1). However, we know that incorrect hyperplanes are located on the border of the generalized preimage (because these hyperplanes cross hypervoxel vertices). Thus, in order to obtain a correct hyperplane, it is sufficient to choose a point which is not on the generalized preimage border.

In the following, we will assume that a generalized preimage which is not empty is not either reduced to m-dimensional subspace segments, $m < n$.

4.2 Recognition Algorithm

Let $\mathcal{H} = \{H_1, \ldots, H_k\}$ be a set of k hypervoxels. The Standard and Supercover hyperplane recognition (see Algorithm 1) is simply performed by computing the generalized preimage $\mathbb{G}_P$ of $\mathcal{H}$. First, $\mathbb{G}_P(H_1)$, i.e. the dual of H_1, is computed according to the hypervoxel dual definition given by Theorem 1. Then, $\mathbb{G}_P(\{H_1, H_2\})$ is computed from the intersection of $\mathbb{G}_P(H_1)$ and $Dual(H_2)$. And so on until $\mathbb{G}_P(\{H_1, \ldots, H_k\})$ is computed or $\mathbb{G}_P$ becomes empty. Note that the hypervoxels can be considered in any order, and do not need to be connected. Figure 5 illustrates the recognition process in dimension 2. A 3D generalized preimage example is shown in Figure 6.

To perform the intersection operations, a first naive approach is to intersect directly the generalized preimage and each hypervoxel dual. It is not an efficient method since the dual of a hypervoxel is an open concave polytope. However, Corollary 1 allows us to compute the generalized preimage of a hypervoxel set by simply computing intersections of convex polytopes and hyperplanes.

Let k be the number of given hypervoxels. These improvements lead to a complexity for our algorithm of $\mathcal{O}(k)$ in dimension 2 when applied on 4-connected curves [17]. Indeed, the generalized preimage of a pixel set is a polygone with at most four edges [11,18]. In dimension 3 [19,20] and higher, the complexity is $\mathcal{O}(k^2)$ in the worst case.

Algorithm 1. Standard and Supercover hyperplane recognition algorithm

---

Data: A set $\mathcal{H}$ of k hypervoxels $H_1, \ldots, H_k$.
begin
  $GP \longleftarrow Dual(H_1)$;
  $i \longleftarrow 2$;
  while $GP \neq \emptyset$ **and** $i \leq n$ **do**
    $GP \longleftarrow GP \cap Dual(H_i)$;
    $i \longleftarrow i + 1$;
  if $GP \neq \emptyset$ **then**
    $\mathcal{H}$ belongs to a Standard and Supercover hyperplane.
  else
    $\mathcal{H}$ does not belong to a Standard or Supercover hyperplane.
end

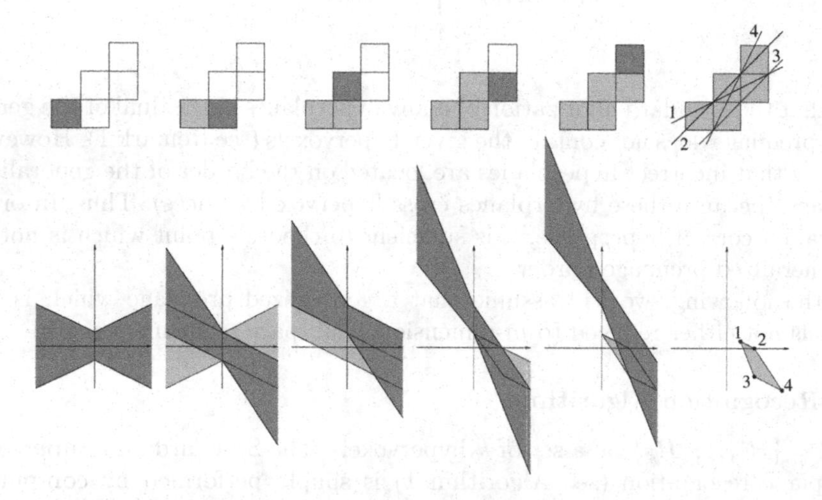

Fig. 5. Example of 2D generalized preimage computation

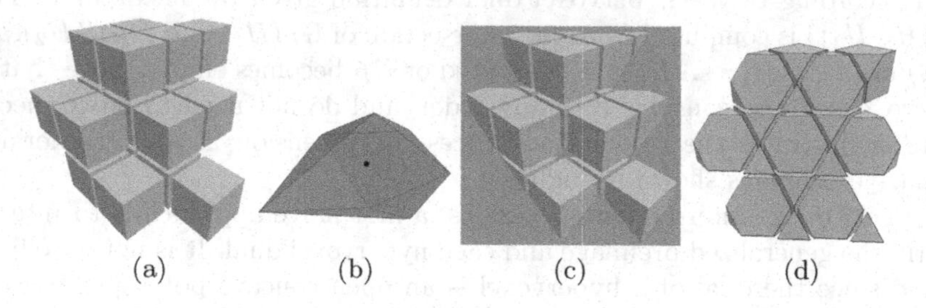

Fig. 6. Example of 3D generalized preimage: (a) A voxel set V, (b) Corresponding generalized preimage $\mathbb{G}_P(V)$ and its barycenter B (back point), (c) and (d) Result of the intersection operation between V and $Dual(B)$. Remark that $Dual(B)$ does not cross any voxel vertex of V since B does not lie on the border of $\mathbb{G}_P(V)$

5 Conclusion and Future Works

In this article, a new Standard and Supercover hyperplane recognition algorithm in arbitrary dimension has been presented. This algorithm determines if a given hypervoxel set belongs to a Standard or Supercover hyperplane by providing the set of Euclidean hyperplanes which cut all hypervoxels. This set is deduced from the computation in a dual space of the generalized preimage of the hypervoxels. This preimage is defined as the intersection of the duals of the hypervoxels. A description of the dual of a hypervoxel is given in order to increase the preimage computation efficiency. The recognition algorithm does not require given hypervoxels to be connected. Moreover, during the recognition process, hypervoxels can be considered in any order.

The results proposed in this paper are very general. They allow many extensions. For instance, the hypercubes in the recognition process are not required to have the same size. This can easily lead to recognition algorithms in multi-scale grids or heterogeneous grids. We focused in this paper on the dual of a hypervoxel but as long as the cell is convex we can compute its dual object. We can therefore also propose recognition algorithms in grids which are not based on hypercubes. The recognition algorithm can also be adapted to recognize other types of hyperplanes than Standard or Supercover analytical hyperplanes. Indeed, for each digital analytical model, there is a distance and/or a unit ball associated to the definition. For instance, for the Naive model [21] in 3D, it is the distance d_1 and the ball is an octahedron. This makes the general definition of the dual in dimension n a bit more difficult but nonetheless feasible.

References

1. Klette, R., Rosenfeld, A.: Digital straightness – a review. Discrete Applied Mathematics **139**(1–3) (2004) 197–230
2. Françon, J., Schramm, J.M., Tajine, M.: Recognizing arithmetic straight lines and planes. In: Discrete Geometry for Computer Imagery. Volume 1176 of LNCS. (1996) 141–150
3. Buzer, L.: A linear incremental algorithm for Naive and Standard digital lines and planes recognition. Graphical models **65**(1–3) (2003) 61–76
4. Kim, C.E., Stojmenović, I.: On the recognition of digital planes in three-dimensional space. Pattern Recognition Letters **12**(11) (1991) 665–669
5. Debled-Rennesson, I., Reveillès, J.: A linear algorithm for segmentation of digital curves. International Journal of Pattern Recognition and Artificial Intelligence **9**(6) (1995) 635–662
6. Gerard, Y., Debled-Rennesson, I., Zimmermann, P.: An elementary digital plane recognition algorithm. DAMATH: Discrete Applied Mathematics and Combinatorial Operations Research and Computer Science **151** (2005)
7. Vittone, J., Chassery, J.M.: Recognition of digital Naive planes and polyhedrization. In: Discrete Geometry for Computer Imagery. Volume 1953 of LNCS. (2000) 296–307
8. Brimkov, V.E., Dantchev, S.S.: Complexity analysis for digital hyperplane recognition in arbitrary fixed dimension. In: Discrete Geometry for Computer Imagery. Number 3429 in LNCS, Poitiers, France (2005) 287–298

9. Stojmenović, I., Tošić, R.: Digitization schemes and the recognition of digital straight lines, hyperplanes, and flats in arbitrary dimensions. In: Vision Geometry. Volume 119 of Contemporary Mathematics Series., American Mathematical Society (1991) 197–212
10. Cœurjolly, D., Brimkov, V.: Computational aspects of digital plane and hyperplane recognition. In: 11th International Workshop on Combinatorial Image Analysis. Volume 4040 of LNCS., Berlin, Germany (2006) 291–304
11. Dorst, L., Smeulders, A.W.M.: Discrete representation of straight lines. IEEE Transactions on Pattern Analysis and Machine Intelligence **6**(4) (1984) 450–463
12. Cœurjolly, D.: Algorithmique et géométrie discrète pour la caractérisation des courbes et des surfaces. PhD thesis, Université Lumière Lyon 2, Lyon, France (2002)
13. Andres, E.: Discrete linear objects in dimension n: the Standard model. Graphical Models **65** (2003) 92–111
14. Cohen-Or, D., Kaufman, A.: Fundamentals of surface voxelization. Graphical Models and Image Processing **57**(6) (1995) 453–461
15. Andres, E., Nehlig, P., Françon, J.: Tunnel-free Supercover 3D polygons and polyhedra. Computer Graphics Forum **16**(3) (1997) 3–14 Proceedings of Eurographics '97. ISSN 1067-7055.
16. Matre, H.: Un panorama de la transformation de Hough – a review on Hough transform. Traitement du Signal **2**(4) (1985) 305–317
17. Dexet, M., Andres, E.: Linear discrete line recognition and reconstruction based on a generalized preimage. In: 11th International Workshop on Combinatorial Image Analysis. Volume 4040 of LNCS., Berlin, Germany (2006) 174–188
18. McIlroy, M.D.: A note on discrete representation of lines. AT&T Technical Journal **64**(2) (1985) 481–490
19. Brimkov, V.E., Coeurjolly, D., Klette, R.: Digital planarity - a review. Technical Report CITR-TR-142, Center for Image Technology and Robotics, University of Auckland, New Zealand (2004) http://citr.auckland.ac.nz/techreports/.
20. Cœurjolly, D., Sivignon, I., Dupont, F., Feschet, F., Chassery, J.: On digital plane preimage structure. DAMATH: Discrete Applied Mathematics and Combinatorial Operations Research and Computer Science **151** (2005)
21. Andres, E., Acharya, R., Sibata, C.: Discrete analytical hyperplanes. Graphical Models and Image Processing **59**(5) (1997) 302–309

Distance Transforms on Anisotropic Surfaces for Surface Roughness Measurement

Leena Ikonen[1], Toni Kuparinen[1], Eduardo Villanueva[1], and Pekka Toivanen[2]

[1] Laboratory of Information Processing, Department of Information Technology
Lappeenranta University of Technology, P.O.Box 20, 53851 Lappeenranta, Finland
{leena.ikonen, toni.kuparinen}@lut.fi, villanue@gmail.com
[2] Tampere University of Technology / Digital Media Institute / EPANET
FRAMI, Kampusranta 9C, 60320 Seinäjoki, Finland
pekka.toivanen@tut.fi

Abstract. The Distance Transform on Curved Space (DTOCS) calculates distances along a gray-level height map surface. In this article, the DTOCS is generalized for surfaces represented as real altitude data in an anisotropic grid. The distance transform combined with a nearest neighbor transform produces a roughness map showing the average roughness of image regions in addition to one roughness value for the whole surface. The method has been tested on profilometer data measured on samples of different paper grades. The correlation between the new method and the arithmetic mean deviation of the roughness surface, S_a, for small wavelengths was strong for all tested paper sample sets, indicating that the DTOCS measures small scale surface roughness.

1 Introduction

The roughness of a surface is a property that needs to be measured in many applications, and the application motivating our research is in quality assessment of paper. Paper roughness has a significant effect on printability, which eventually defines the quality of the printed product. The roughness of paper is measured using profilometers, which acquire the real topography of a surface. The basic idea of profilometers is that they have a stylus, which travels on a surface measuring its height [1]. The Distance Transform on Curved Space (DTOCS) measures distances along surfaces represented as gray-level height maps, or range images, and can be used directly on the profilometer data. Here, the DTOCS and its locally Euclidean modification, the Weighted DTOCS (WDTOCS) [2] are generalized to anisotropic profilometer data, and used for estimating surface roughness based on the fact that distance values calculated along a highly varying surface are larger than distances calculated along a smoother surface. The presented method combines the DTOCS with the nearest neighbor transform (NNT), and produces a roughness map, which can be used to compare the roughness of different regions in the same image. In addition, an average roughness value can be calculated to characterize the whole surface. Other approaches to roughness inspection utilize statistical features, like kurtosis [3], roughness parameters, Fourier analysis [4], wavelets [5], and fractal dimension [6, 7].

A. Kuba, L.G. Nyúl, and K. Palágyi (Eds.): DGCI 2006, LNCS 4245, pp. 651–662, 2006.

2 DTOCS for Anisotropic Grids

The DTOCS calculates distances along gray-level surfaces, when gray-levels are understood as height values. Local distances, which are summed along digital paths to calculate the distance transform, are defined as $d(p_i, p_{i-1}) = |\mathcal{G}(p_i) - \mathcal{G}(p_{i-1})| + 1$, where $\mathcal{G}(p)$ denotes the gray-value of pixel p, and p_{i-1} and p_i are subsequent pixels on a path. The WDTOCS produces more accurate distance values by using the locally Euclidean distance between pixels, and the Optimal DTOCS [8] improves the distance approximation even further. The DTOCS and the WDTOCS can be used for profilometer data consisting of real height values without any changes in the distance definitions. The integer gray-values defining the height differences are replaced with the floating point altitude data. Alternatively, the height data could be represented using units, which can be scaled to integers, for example, data represented in micrometers could be scaled by 10 and then rounded. However, the accuracy of the measuring device can be fully exploited by using the floating point data directly. Converting the data to a volume image and using well known distance transforms in 3D [9] would also require rounding of the height measurements, and result in an increased problem size.

The scaling of the surface must be carefully considered, when using the DTOCS. The values of the pixels represent the height, or the z-coordinate, of the surface represented as a range image. If the resolution in the xy-plane differs from the resolution in the z-direction, the height differences must be scaled in order to obtain approximations of true distances along the surface. Scaling in the horizontal image plane is needed, if the grid of the range image is anisotropic or rectangular, that is, if the resolution in the x-direction differs from the resolution in the y-direction. Interpolating additional values in the direction with the lower resolution would inevitably introduce some error compared to measured data, and lead to a multifold increase in the image size. Instead, the DTOCS local distances are generalized as follows:

$$d(p_i, p_{i-1}) = \begin{cases} r_z|\mathcal{G}(p_i) - \mathcal{G}(p_{i-1})| + r_x \ , \ p_{i-1} \text{ neighbor of } p_i \text{ in } x\text{-dir.} \\ r_z|\mathcal{G}(p_i) - \mathcal{G}(p_{i-1})| + r_y \ , \ p_{i-1} \text{ neighbor of } p_i \text{ in } y\text{-dir.} \\ r_z|\mathcal{G}(p_i) - \mathcal{G}(p_{i-1})| + \max(r_x, r_y) \ , \ p_{i-1} \text{ diag. neighbor of } p_i \end{cases} \tag{1}$$

where r_z is the scaling factor for the height differences, and r_x and r_y are the distances between neighbor pixels in the x- and y-direction, as visualized in Fig. 1. The factors r_x, r_y and r_z may have any non-negative values, not necessarily integers. Similarly, the WDTOCS, in which the local distance is calculated using the Pythagoras' theorem from the height difference and the horizontal displacement between the neighbor pixels, can be generalized to rectangular grids as follows:

$$d(p_i, p_{i-1}) = \begin{cases} \sqrt{r_z^2|\mathcal{G}(p_i) - \mathcal{G}(p_{i-1})|^2 + r_x^2} \ , \ p_{i-1} \text{ neighbor of } p_i \text{ in } x\text{-dir.} \\ \sqrt{r_z^2|\mathcal{G}(p_i) - \mathcal{G}(p_{i-1})|^2 + r_y^2} \ , \ p_{i-1} \text{ neighbor of } p_i \text{ in } y\text{-dir.} \\ \sqrt{r_z^2|\mathcal{G}(p_i) - \mathcal{G}(p_{i-1})|^2 + r_x^2 + r_y^2} \ , \ p_{i-1} \text{ diag. neighbor of } p_i \end{cases} \tag{2}$$

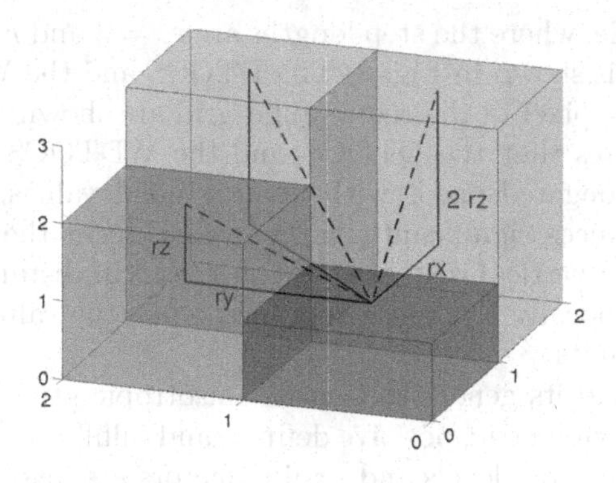

Fig. 1. Local distance definitions for the DTOCS (solid lines) and the WDTOCS (dashed lines) in an anisotropic grid visualized on a surface of 2×2 pixels

0	0	0	0	0
0	0	1	0	0
0	2	2	2	0
0	0	0	0	0
0	0	0	0	0

(a) Original image

3	1	3	6	9
3	0	4	6	9
3	3	5	8	9
4	6	8	10	10
5	7	9	11	11

(b) DTOCS

3.16	1.00	3.16	6.16	9.16
3.00	0.00	3.16	6.32	9.32
3.16	2.24	3.74	6.48	9.49
4.16	4.47	5.98	7.48	10.22
5.16	5.47	6.98	8.48	10.65

(c) WDTOCS

3	1	3	6	9
3	0	3	6	9
3	1	3	6	9
4	2	4	6	10
5	7	5	7	11

(d) Proj. dist. (DTOCS)

3.16	1.00	3.16	6.16	9.16
3.00	0.00	3.00	6.32	9.32
3.16	1.00	3.16	6.16	9.49
4.16	2.00	4.16	6.32	9.32
5.16	3.00	5.16	7.32	9.49

(e) Proj. dist. (WDTOCS)

Fig. 2. Example of the DTOCS, the WDTOCS and the corresponding projection distances in a rectangular grid, where $r_x = 3$ and $r_y = 1$

12.30	1.32	3.37	1.32
7.91	−0.44	4.25	1.61
1.32	1.61	4.83	6.01
−1.61	7.76	6.15	8.79
2.78	9.08	5.86	10.99

(a) Original image

17.74	2.76	8.81	15.86
13.35	0.00	9.69	15.57
6.76	3.05	10.27	16.45
10.69	10.20	12.59	19.23
16.08	12.52	13.88	22.43

(b) DTOCS

13.73	2.02	6.36	11.77
9.73	0.00	6.85	11.76
5.39	2.28	7.34	12.25
8.31	8.51	8.99	13.79
12.82	10.17	10.03	16.02

(c) WDTOCS

5	1	5	10
5	0	5	10
5	1	5	10
6	2	6	10
7	3	7	11

(d) Proj. dist (DTOCS)

5.10	1.00	5.10	10.10
5.00	0.00	5.00	10.20
5.10	1.00	5.10	10.10
6.10	2.00	6.10	10.20
7.10	3.00	7.10	11.20

(e) Proj. dist (WDTOCS)

Fig. 3. Example of the DTOCS, the WDTOCS, and the corresponding projection distances on anisotropic profilometer data, where $r_x = 5$ and $r_y = 1$

A small example, where the step lengths are $r_x = 3$ and $r_y = 1$, and the height scaling is $r_z = 1$, is shown in Fig. 2. The DTOCS and the WDTOCS calculated from one reference pixel in the rectangular grid are shown in Fig. 2 (b) and (c). Fig. 3 demonstrates that the DTOCS and the WDTOCS can be applied also to images with elongated pixels with floating point values. It can be seen that the DTOCS produces significantly larger values than the WDTOCS. Adding the horizontal and vertical displacement in the local distance definition clearly overestimates the locally Euclidean distance, when the values of neighbor pixels differ by several units.

The DTOCS and its generalization to anisotropic grids are metrics, that is, the distances are symmetric, positive definite and fulfill the triangle inequality, as long as only integer gray-levels and scaling factors are involved [10]. Calculating floating point distance values, either by using the WDTOCS definition or by having floating point input data, may result in violations of the metrics criteria, due to the limited precision available. For example, the least significant bits of the distance values calculated from pixel p to pixel q and from q to p may differ, as they are the result of several floating point operations.

The roughness measurement method utilizes normalized distance values. Distances are divided by the so called projection distance [11]. The projection distance value of pixel p is the length of the shortest path from p to the nearest reference pixel projected into the image plane. The local distances are obtained by removing the effect of the height differences from the DTOCS or WDTOCS local distances, that is, by setting $r_z = 0$ in Eq. (1) and (2). Fig. 2 (d) and (e), and Fig. 3 (d) and (e) show the projection distances of the corresponding DTOCS and WDTOCS images. Dividing the DTOCS or WDTOCS distance values with the corresponding projection distances result in values indicating the average height variation along the shortest path to the nearest reference pixel.

3 Distance and Nearest Neighbor Transformation

An efficient priority pixel queue transformation algorithm for calculating the DTOCS is presented in [14]. It is very similar to the Fast Marching algorithm for calculating forward propagating level sets [15], but the calculations are simpler, as the DTOCS algorithm is developed directly for the discrete geometry inherent for digital image processing. The reference pixels, from which distances are calculated, are enqueued into a minimum heap, from which they are dequeued in priority order. New distance values are calculated for neighbors of the dequeued pixel, and subsequently enqueued. The best first approach ensures that distance values are final when they are dequeued, and propagated further. The projection distance values are calculated simultaneously with the DTOCS or the WDTOCS values. The propagation order also enables easy implementation of the nearest neighbor transformation, which assigns the identity of the nearest feature pixel to each pixel in the image. The distances and the nearest site are determined according to the DTOCS, as described in [16]. Unique seed values assigned to each reference pixel are propagated simultaneously with the distance values, so that each pixel gets the seed value of the pixel from which the distance propagated to it. A similar region growing algorithm for tessellation of 3D volumes is presented in [17]. The complexity of the pixel queue algorithm is in $\mathcal{O}(n \log n_q)$, where n_q is the length of the queue, which varies throughout the transformation. As $n_q \ll n$, the algorithm is near-linear, with running times increasing only slightly with increasing surface complexity [14].

4 Roughness Measurement Using the DTOCS

In the new roughness evaluation method, a distance map is calculated using an evenly spaced set of reference pixels, or seeds, on the original image. A nearest neighbor transform is calculated simultaneously to attach each pixel to the nearest reference pixel. The curved distances within each region are divided by the corresponding projection distances. The more variation there is around the seed pixel, the larger are the distances. The averages of the normalized distance values within each region produce a roughness map of local roughness values.

Fig. 4 illustrates how the roughness measurement method proceeds on a topography image. The original surface image, Fig. 4 (a), is 250 pixels wide and 50 pixels high, but represents a square surface, a piece of paper of size 2.5 $mm \times 2.5$ mm.

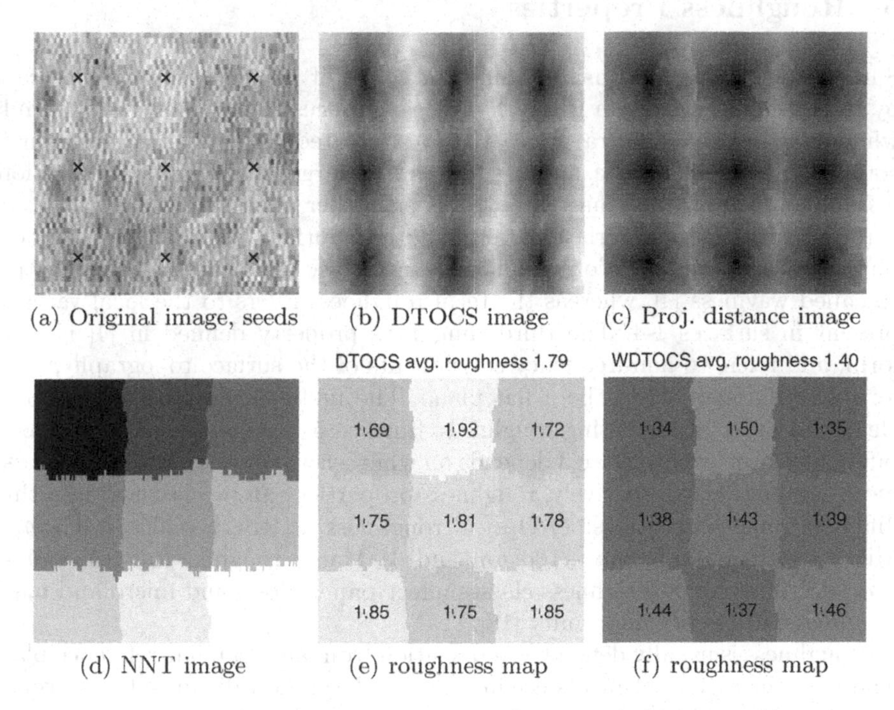

Fig. 4. The phases in the roughness measurement method (a)–(e). Image (f) shows that WDTOCS roughness values are consistently smaller than DTOCS roughness values.

This means that one pixel represents a surface area of size 10 $\mu m \times 50$ μm. The height values are measured in micrometers, but as the height variation is very small compared to the horizontal displacements, a factor 10 is added to the height component to emphasize the variation. The resulting scaling factors for the local distances are $r_x = 10$, $r_y = 50$, and $r_z = 10$, or in practise, $r_x = 1$, $r_y = 5$, and $r_z = 1$.

As local distances based on gray-values can vary significantly, the nearest neighbor transformation can result in any shapes of regions around each site. The region sizes also vary, as the distance propagation covers more pixels in a smoother area. Seed pixels in areas with higher variation are typically surrounded by smaller nearest neighbor regions. On a highly varying surface, some seed values may not propagate at all, if each neighbor of the seed pixel is closer, or equally close, to one of the surrounding reference pixels. In such cases, the roughness value is approximated using the average of the distance values in the 8-neighborhood of the seed pixel, plus one. The idea is that distances from the reference pixel to its neighbors must be greater or equal to the distance between each neighbor and some other reference pixel. This approximation can make the method more robust against noise, as a reference pixel differing significantly from its neighborhood can "borrow" its roughness value from its neighborhood rather than cause a peak in the roughness map.

5 Roughness Properties

Surface roughness can consist of different scales of variations, as demonstrated by the example surfaces in Fig. 5. Synthetic surfaces 2, 3 and 4 are very similar when examined at close range, as they are created by adding the same noise component to a flat surface, and to surfaces with larger scale Gaussian variations, or bumps. It is obvious that surface 2 is smoother than surfaces 3 and 4, and surface 1 with less local variation even smoother. Surface 5 is locally very smooth, but the larger scale variation is similar as in surface 4. Surfaces 3-5 demonstrate so called waviness [4], whereas the term roughness refers to the local variation present in surfaces 1-4. The third roughness property defined in [4] is called form, and refers to non-frequency components of the surface topography, which for paper surfaces should be a flat plane. The first four surfaces in Fig. 5 are clearly in order of increasing roughness. Surface 5 is smoother than surface 4, but comparison with surface 3 depends on whether waviness or local roughness is more significant. Alternatively, roughness properties can be classified into three different roughness classes: 1) Optical roughness at length scales < 1 μm, 2) Micro roughness at 1 μm - 100 μm and 3) Macro roughness at 0.1 mm - 1 mm. All these three roughness classes affect paper gloss, and micro and macro roughness also paper uniformity [18].

Roughness is usually defined as a deviation from an ideal, flat reference plane, where all the surface elements are in the same level. Our distance transform approach is designed for measuring local or smaller scale roughness, and additional measures will be needed to extract waviness properties. In the experiments, the DTOCS roughness measure is compared to the arithmetic mean deviation, S_a,

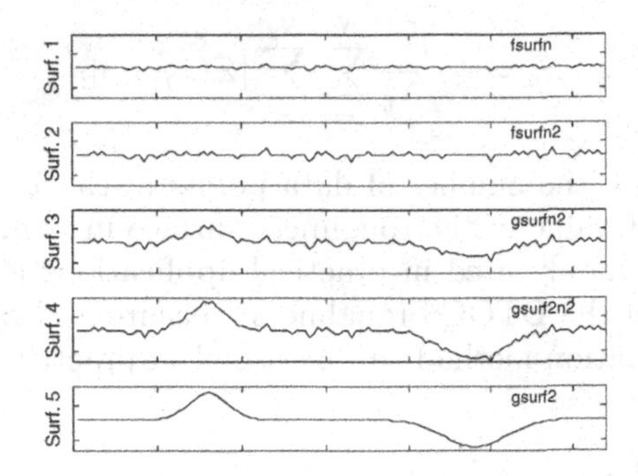

Fig. 5. Profiles of example surfaces with different roughness properties

calculated from the so called roughness surface, from which the effect of the waviness has been removed by filtering. The measured topographies are filtered using a Gaussian filter in order to extract roughness and waviness surfaces from the inspected surface. The result of low-pass Gaussian filtering is the waviness surface, and the roughness surface is extracted from the original surface by subtracting the waviness surface from the original surface, see Fig. 6. The filter is calculated by a direct convolution of the surface topography with a Gaussian weighting function $S(x, y)$, which is given by

$$S(x,y) = \frac{1}{\beta\lambda_{xc}\lambda_{yc}} exp\left\{-\frac{\pi}{\beta}\left[\left(\frac{x}{\lambda_{xc}}\right)^2 + \left(\frac{y}{\lambda_{yc}}\right)^2\right]\right\}, \tag{3}$$

where x and y are the positions from the center of the weighting function, (λ_{xc},λ_{yc}) are the cutoff wavelengths at 50% attenuation ratio and $\beta = ln2/\pi$ [4]. ISO standards recommend cutoff wavelengths 0.08, 0.25, 0.8, 2.5, and 8 mm [19].

The roughness surface can be characterized by statistical analysis, 2-D spectral analysis and time series analysis. A statistical roughness parameter, the arithmetic mean deviation of the surface, S_a, is defined as

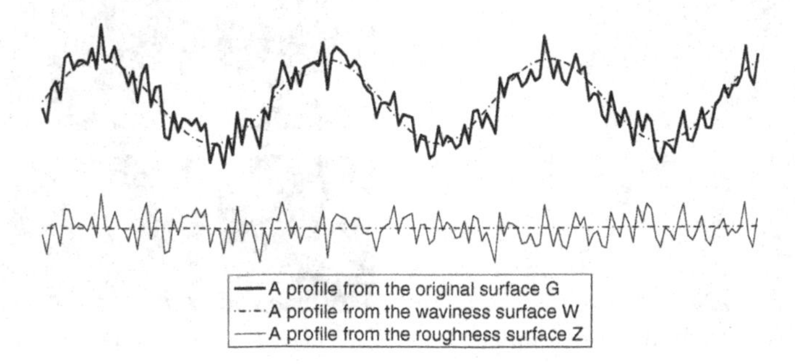

Fig. 6. Profiles from the original, the waviness and the roughness surface

$$S_a = \frac{1}{N_x N_y} \sum_{i=1}^{N_y} \sum_{j=1}^{N_x} |Z(x_j, y_i)|, \tag{4}$$

where N_x and N_y are the number of data points in the x- and y-direction and $Z(x, y)$ is the height value of the roughness surface in point (x, y). The S_a parameter is very commonly used in practical applications [4], so obtaining high correlations between the DTOCS roughness measure and S_a indicates that the new roughness evaluation method can be useful in practice.

6 Experiments

The roughness of synthetic images, some of which were used as examples in Fig. 5, were evaluated using the DTOCS and the WDTOCS. The nine original surfaces in Fig. 7 a) are arranged so that the waviness increases from top to bottom. The first surface is flat, the second one contains Gaussian bumps, and in the third surface the Gaussian bumps are twice as high or deep. The local roughness increases from left to right, so surfaces to the left are locally smooth, surfaces in the middle contain a noise component, and to the right, the noise component is doubled. The standard deviation of surface height values is indicated above each image. Fig. 7 (b) shows the DTOCS roughness maps of the corresponding test surfaces. The WDTOCS roughness maps, not shown due to lack of space, are visually similar, but with consistently lower roughness values. The intensity of each region indicates the local roughness value, that is, darker regions lie in smoother areas of the image. Areas with only local roughness without waviness can have equally high DTOCS roughness values as areas containing Gaussian

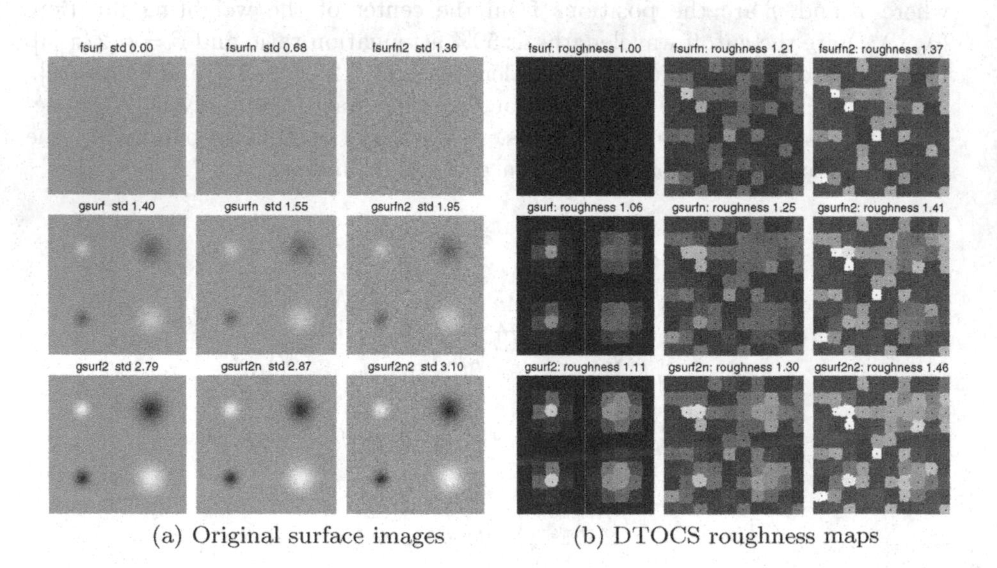

(a) Original surface images (b) DTOCS roughness maps

Fig. 7. Synthetic images and their roughness maps based on a 10×10 grid of seeds

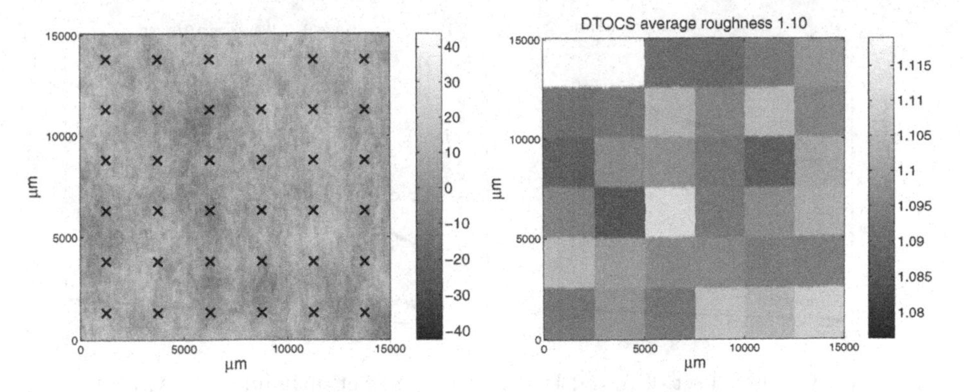

Fig. 8. A sample from test set C, and its DTOCS roughness map

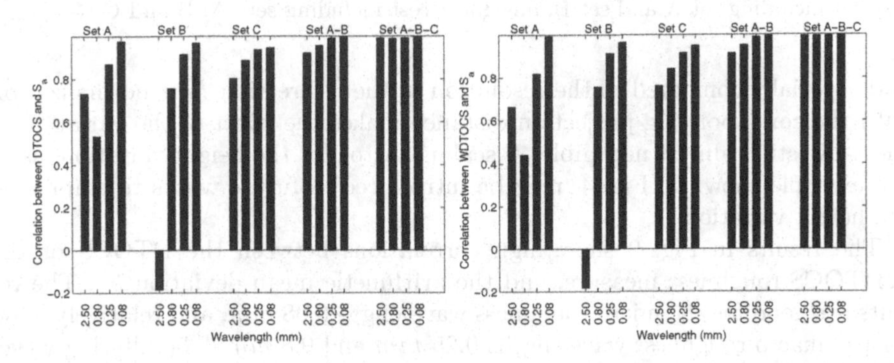

Fig. 9. Correlation values for comparison between S_a and DTOCS or WDTOCS

bumps, as a smooth slope produces only slightly larger distance values than a flat surface. The average roughness values shown above each surface image increase only slightly with increasing waviness.

Experiments on real profilometer data were performed using sample sets, which include a variety of paper samples and cardboard samples. Sample set A consists of 8 light weight coated paper samples, sample set B of 11 supercalendered paper samples and sample set C of 8 base cardboard samples. The first two sample sets, A and B are similar in roughness compared to set C, which is significantly rougher. Each sample was marked with a $15mm \times 15mm$ measurement area, on which the profilometer measurements were performed. The resolution in the x-direction is $10\mu m$ and the resolution in the y-direction $50\mu m$. The height value is given in micrometers, that is, $r_x = 10$, $r_y = 50$ and $r_z = 1$. One sample from the roughest test set C is shown in Fig. 8, visualizing also the 6×6 grid of seed pixels used in the experiments. It can be seen that the roughness variation is very small within the image, resulting in almost square regions in the roughness map. The variation in the corresponding WDTOCS roughness map is even smaller. This is due to the fact that the height variation in the profilometer data is small compared to the resolution in the x-direction,

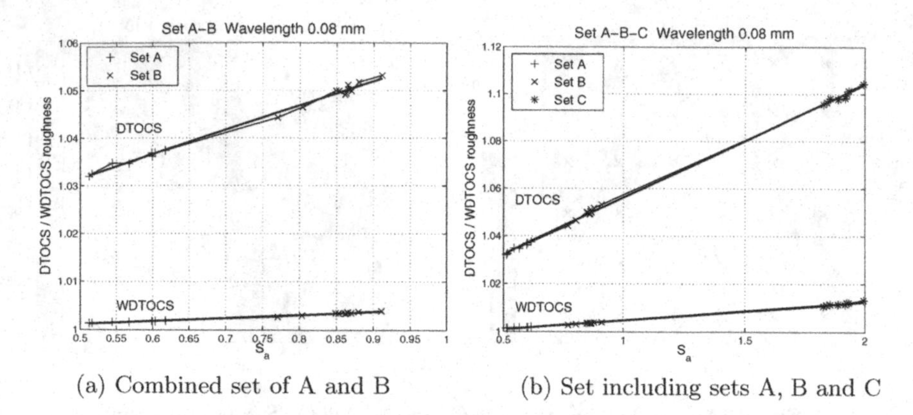

(a) Combined set of A and B

(b) Set including sets A, B and C

Fig. 10. Scatter plots of S_a and DTOCS or WDTOCS roughness values for a combined test set including set A and set B, and for a test including sets A, B and C

and especially compared to the resolution in the y-direction. The normalization with the corresponding projection distance makes the effect of the variation in the y-direction almost negligible. A scaling factor for the height variation, as in the example shown in Fig. 4, may be introduced in future works to emphasize the height variation.

The results in Fig. 9 show high correlations between the DTOCS or the WDTOCS roughness measure, and the arithmetic mean deviation S_a. The results are good for the micro roughness wavelength 0.08 mm, and relatively good for the macro roughness wavelengths 0.25 mm and 0.8 mm. The filtering wavelength 2.5 mm approaches the size of the measurement area, so the resulting S_a value does not characterize the surface accurately. The correlations calculated using the micro roughness wavelength are strong for all tests sets A, B and C separately, and also for a combined set containing the samples from A and B, and a set containing all samples from A, B and C. Since the samples in set C are clearly rougher than the samples in A and B, the correlation is very strong, but it can been seen from Fig. 10 (b) that there is a clear linear dependency between the DTOCS or WDTOCS roughness values, and the S_a roughness parameter calculated using the micro roughness wavelength 0.08 mm. Fig. 10 also illustrates, that the DTOCS roughness values are consistently higher than the WDTOCS roughness values, as the DTOCS local distance overestimates the locally Euclidean distance between neighbor pixels.

7 Discussion

A new distance transform method has been developed for measuring surface roughness. The distance transforms, the DTOCS and the WDTOCS, which measure distances along a surface, have been generalized to anisotropic grids. Previously, the distance transforms have been applied to gray-level images, but the new modifications are applicable to floating point altitude data. In the new roughness measurement method, the distance transforms are combined with a nearest neighbor transform to produce a roughness map characterizing local roughness within regions of the image. Experiments were performed on synthetic surfaces, and on real topographies of paper surfaces obtained using a profilometer. The correlations between the arithmetic mean deviation of the surface, S_a, for the micro roughness wavelength, and the new methods were strong for all test sets. The results suggest that the roughness can be calculated directly from the topography using the DTOCS or the WDTOCS, without first extracting the roughness surface using Gaussian filtering. Further experiments and comparisons to other methods will be performed in future work.

In future work, the distance transform method will be tested using different numbers of seed pixels for the nearest neighbor transform. Alternative configurations, like a hexagonal grid of seeds, may also be beneficial compared to the square grid used here. Also, more tests are needed to determine whether the WDTOCS method provides more accurate results compared to the DTOCS, which is slightly simpler and faster to calculate. Furthermore, as it is well known that the piecewise Euclidean distance overestimates true distances (see e.g. [20]), the more accurate Optimal DTOCS [8] will be generalized to images of unequal resolution, for example, by using weights derived by Sintorn and Borgefors [12] for distance transforms of binary images in rectangular grids. Alternatively, ideas behind weighted 3D distance transforms for elongated voxel grids [13] could be utilized in order to obtain more accurate approximations of true distances along anisotropic gray-level surfaces. However, in measuring surface roughness, using the most accurate approximations of true distances may not be necessary, as long as the distance values increase in proportion with the surface variation. As shown by the experiments, the effect of the height variation may need to be emphasized by using a scaling factor for the height component of the local distance. In this work, the distance values approximate the true distance measured along the representation of the surface. However, the representation is not entirely accurate, as the sparse 50 μm resolution data misses some small scale variation in the surface. In future work, data measured at the more dense resolution, 10 μm, in both directions will be available for comparison.

This work is part of a research project aiming to develop a machine vision system for measuring the roughness of paper, which could be used in paper industry during processing. The new method was shown to be a simple and efficient approach to characterize the micro roughness of paper surfaces. Particularly, the roughness maps, which provide interesting information about the roughness properties of different areas of the surface, will be investigated further. If the DTOCS roughness values and the S_a values calculated from the same nearest neighbor regions are found to correlate, the DTOCS provides an easy way to divide the surface to areas with different roughness properties. The distance transform measurements could be part of a larger pattern recognition system, providing features to be used together with, for example, fractal dimension or statistical features in classifying any surfaces, not just paper, by their roughness.

References

1. Wagberg, P., Johansson, P. A.: Surface profilometry - a comp. between optical and mechanical sensing on printing papers. Tappi Journal **76**(12) (1993) 15–121
2. Toivanen, P.: New geodesic distance transforms for gray-scale images. Pattern Recognition Letters **17** (1996) 437–450
3. Johansson, J-O.: Measuring homogenity of planar point-patterns by using kurtosis. Pattern Recognition Letters **21**(13-14) (2000) 1149–1156
4. Stout, K. J. (ed.): Development of Methods for the Characterization of Roughness in Three Dimensions. Penton Press, revised reprint (original 1993) edition (2000)
5. Yuan, C., Peng, Z., Yan, X.: Surface characterization using wavelet theory and confocal laser scanning microscopy. Journal of Tribology **127**(2) (2005) 394–404
6. Kent, H. J.: The fractal dimension of paper surface topography. In: TAPPI/CPPA International Printing and Graphics Arts Conference, Vancouver, Canada (1990) 73–78
7. Kuparinen, T., Rodionov, O., Toivanen, P., Mielikainen, J., Bochko, V., Korkalainen, A., Parviainen, J., Vartiainen, E.: Fractal dimension analysis and statistical processing of paper surface images towards surface roughness measurement. In: Scandinavian Conference on Image Analysis (SCIA) Joensuu, Finland (2005) 1218–1227
8. Ikonen, L., Toivanen, P.: Shortest routes on varying height surfaces using gray-level distance transforms. Image and Vision Computing **23**(2) (2005) 133–141
9. Borgefors, G.: On digital distance transforms in three dimensions. Computer Vision and Image Understanding **64**(3) (1996) 368–376
10. Ikonen, L.: Distance Transforms on Gray-Level Surfaces. PhD thesis, Lappeenranta University of Technology (2006)
11. Ikonen, L., Toivanen, P.: Distance and nearest neighbor transforms of gray-level surfaces. Submitted to Pattern Recognition Letters, September (2005)
12. Sintorn, I-M., Borgefors, G.: Weighted distance transforms in rectangular grids. In: International Conference on Image Analysis and Processing (ICIAP), Palermo, Italy (2001) 322–326
13. Sintorn, I-M., Borgefors, G.: Weighted distance transforms for images using elongated voxel grids. In: Discrete Geometry for Computer Imagery (DGCI), Berlin, Germany (2002) 244–254
14. Ikonen, L.: Pixel queue algorithm for geodesic distance transforms. In: Discrete Geometry for Computer Imagery (DGCI), Poitiers, France (2005) 228–239
15. Sethian, J. A.: Level Set Methods and Fast Marching Methods. Cambridge University Press, 2nd edition (1999)
16. Ikonen, L., Toivanen, P.: Distance and nearest neighbor transforms of gray-level surfaces using priority pixel queue algorithm. In: Advanced Concepts for Intelligent Vision Systems (ACIVS), Antwerp, Belgium (2005) 308–315
17. Kapoutsis, C. A., Vavoulidis, C. P., Pitas, I.: Morphological iterative closest point algorithm. IEEE Transactions on Image Processing **8**(11) (1999) 1644–1646
18. Niskanen, K. (ed.): Paper Physics. Book 16 - Papermaking Science and Technology. Fabet Oy and Tappi Press, Jyväskylä (1998)
19. National Institute of Standards and Technology. Internet-based surface metrology algorithm testing system. WWW-page. Available from http://ats.nist.gov/VSC/jsp/index.jsp, accessed 2006-03-31
20. Borgefors, G.: Distance transformations in digital images. Computer Vision, Graphics, and Image Processing **34** (1986) 344–371

A 3D Live-Wire Segmentation Method for Volume Images Using Haptic Interaction

Filip Malmberg[1], Erik Vidholm[2], and Ingela Nyström[2]

[1] Centre for Image Analysis, Swedish University of Agricultural Sciences, Uppsala, Sweden
[2] Centre for Image Analysis, Uppsala University, Uppsala, Sweden
{filip, erik, ingela}@cb.uu.se

Abstract. Designing interactive segmentation methods for digital volume images is difficult, mainly because efficient 3D interaction is much harder to achieve than interaction with 2D images. To overcome this issue, we use a system that combines stereo graphics and haptics to facilitate efficient 3D interaction. We propose a new method, based on the 2D live-wire method, for segmenting volume images. Our method consists of two parts: an interface for drawing 3D live-wire curves onto the boundary of an object in a volume image, and an algorithm for connecting two such curves to create a discrete surface.

1 Introduction

Fully automatic segmentation of non-trivial images is a difficult problem, and despite decades of research there are still no robust methods for automatically segmenting arbitrary images. This is mainly due to the fact that it is hard to identify objects from the image data only. Often, we also need some high-level knowledge about the type of objects we are interested in. This is recognized in [1], where the segmentation process is divided into two steps: recognition and delineation. Recognition is the task of roughly determining where in the image the objects are located, while delineation consists of determining the exact extent of the object. Human users outperform computers in most recognition tasks, while computers are often better at delineation. Semi-automatic segmentation methods take advantage of this fact by letting a human user guide the recognition, while the computer performs the delineation. A successful semi-automatic method minimizes the user interaction time, while maintaining tight user control to guarantee the quality of the result.

Many semi-automatic segmentation methods exist for two-dimensional (2D) images, but in most cases it is not obvious how to extend these methods to work efficiently for three-dimensional (3D) images. One problem is that efficient interaction with a 3D image is much harder to achieve than interaction with a 2D image. In this project, we have used a system that simplifies 3D interaction. The system supports 3D input, stereo graphics, and haptic feedback through a sensing probe. The haptic feedback allows the user to feel where surfaces and structures in the volume are located, making it easier to navigate within the

A. Kuba, L.G. Nyúl, and K. Palágyi (Eds.): DGCI 2006, LNCS 4245, pp. 663–673, 2006.

volume. Haptic feedback has previously been used in the context of interactive image segmentation in, e.g. [2] and [3]. Our system gives us new possibilities to create efficient user interfaces for interacting with volume images. The aim is to investigate how these new possibilities can be used to extend an existing semi-automatic segmentation method to also handle volume images efficiently.

Live-wire [1, 4] is a popular semi-automatic segmentation method that has been used successfully for many 2D problems. Various ways of extending this method to segment volume images have been proposed in the literature. Most of these methods are based on using the 2D live-wire method on a subset of 2D slices in the volume, and then reconstructing the entire object using this information. Examples of such approaches can be found in [5, 6, 7, 8]. While the reconstruction algorithms might take 3D information into account, all user interaction is performed in 2D. This restriction gives rise to problems, e.g., how to handle cases where a single object has different numbers of connected components in consecutive slices. The method presented in this paper is a step towards a more direct 3D approach that may overcome these problems.

In our method, the user still draws live-wire curves, but these curves are not restricted to planes. Pairs of such curves are then connected by a surface in the following way: using the image foresting transform (IFT) [9], a large number of live-wire curves are computed between the original curves. The areas between these curves are covered by polygons which are then rasterized to produce a tunnel-free discrete surface between the two curves that well approximates the boundary of the underlying object.

2 Environment

Our setup consists of a Reachin desktop display [10] which combines a PHANToM desktop haptic device [11] with a stereo-capable monitor mounted over a

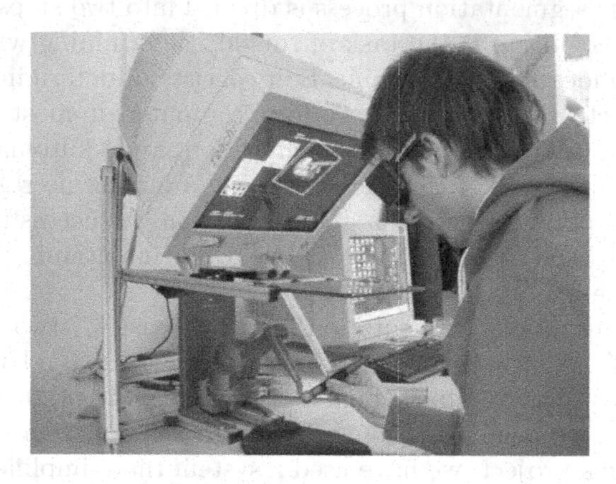

Fig. 1. The Reachin desktop display [10]. The PHANToM device [11] is positioned beneath a semi-transparent mirror. The graphics are projected through the mirror to obtain co-location of graphics and haptics.

semi-transparent mirror. See Figure 1. The PHANToM is a 3D input device that also provides the user with haptic feedback. It is designed as a stylus, and the haptic feedback is given at one point, the tip. It provides six degrees of freedom (DOF) for input and three DOF for output, i.e., the input is the position and orientation of the stylus, and the output is a force vector.

We have implemented our software using the Reachin API 3.2 [10], which is a C++ API that combines graphics and haptics in a scene-graph environment. The workstation we use is equipped with dual 2.4 GHz Pentium IV processors and 1GB of RAM.

3 The 2D Live-Wire Method

The basic idea of the live-wire method is that the user places a *seed-point* on the boundary of an object. As the user moves the cursor in the image, a path (live-wire) is displayed in real-time from the current position of the cursor back to the seed-point. The wire is attracted to edges in the image and it will snap onto the boundary of the object. If the user moves too far from the original seed-point, the wire might snap onto an edge that does not correspond to the desired object. When this happens the user can move the cursor back a little and place a new seed-point. The wire from the old seed-point to the new then freezes, and the tracking continues from the new seed-point. In this way, the entire object boundary can be traced with a rather small number of live-wire segments. See Figure 2.

The live-wire algorithm is based on graph-searching techniques. Therefore, we consider the image to be a graph where the nodes correspond to pixels and the

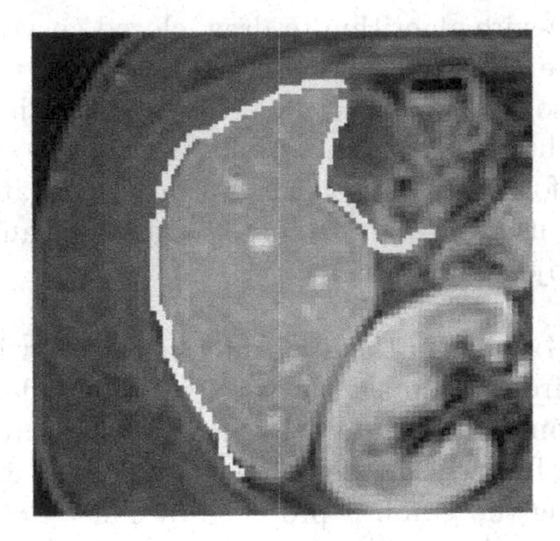

Fig. 2. In the live-wire method, the user segments an image by interactively placing seed-points along the boundary of an object. As the user moves the cursor through the image, a path from the last seedpoint to the current cursor position is displayed in real-time. The path attempts to follow edges in the image, and will thus snap onto the boundary of the object. Here, the liver in a slice of a magnetic resonance (MR) image is being segmented.

graph edges correspond to paths between pixels. For every edge, a cost is assigned to represent the "likelihood" that the edge belongs to a desired boundary in the image. How this cost function should be chosen is discussed in, e.g., [1]. When the user places a seed-point, Dijkstra's algorithm [12] is used to compute the optimal path to the seed-point from all other points in the image. Once these paths have been computed, it is possible to display the live-wire in real-time with virtually no computational cost as the user moves the cursor through the image.

4 Our 3D Live-Wire Method

In the 2D live-wire method, the user interacts with the program by placing seed-points which are then connected by curves in a 2D plane. Our 3D analogy of this is to let the user create curves, which are then automatically connected by discrete surfaces in 3D space, a process we call *bridging*. An entire object in a volume image can be segmented by drawing a relatively small number of live-wire curves on the boundary of the object.

After a tunnel-free surface representing the boundary of the desired object has been created, the entire region occupied by the object can be found by taking the complement of all connected background components that touch the border of the volume. To create this segmentation method, we need to solve two problems: (1) how to draw live-wire curves in 3D, and (2) how to connect two closed curves to form a tunnel-free discrete surface.

4.1 Drawing Live-Wire Curves in 3D

To extend the live-wire algorithm to draw closed curves in 3D, there are again two issues that we need to consider. First, we need to transform the volume image to a graph so that we can apply the shortest path calculation. This can be achieved in a number of ways, and we will see that the choice of graph structure will affect the performance of the shortest-path calculation. The second issue is how to design the user interface so that seed-points can easily be placed on the boundary of an object.

Defining a 3D Graph. In the literature on 2D live-wire, various ways of constructing a graph from the image have been used by different authors. Mortensen and Barrett [13] construct the graph by placing a node at the center of each pixel, with edges passing from each node to the nodes of all neighboring pixels. Another way of constructing the graph is proposed by Falcão *et al.* [1] where the nodes are instead placed at the pixel vertices, and the pixel edges are used as directed graph edges. The latter graph formulation has the advantage that thin (even one pixel wide) structures can be traced as a closed boundary, and the directed nature of the graph allows us to use cost functions that depend on the direction in which the edge is traversed. Both these 2D graph models are illustrated in Figure 3.

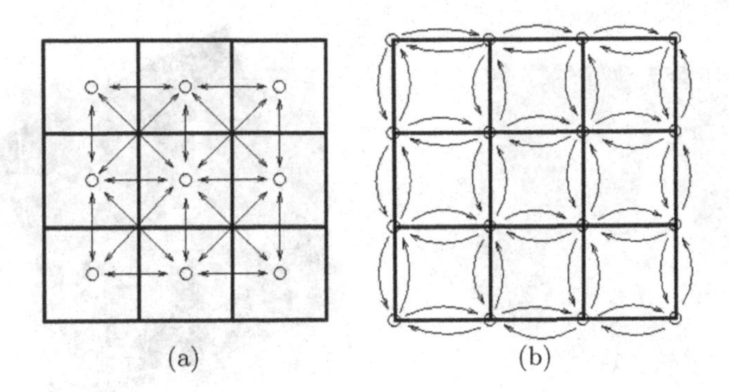

Fig. 3. Two different ways of transforming a 2D image to a graph. (a) The graph definition used in [13]. (b) The graph definition used in [1].

For our 3D graph, we have chosen to place a node at the center of each voxel and create edges to all its 26 neighbors. Although this definition does not use oriented edges, it has the advantage that the resulting graph has relatively few nodes. The bottleneck of Dijkstra's algorithm is the maintenance of a sorted heap containing all nodes that are currently examined, which means that fewer nodes lead to faster computations. Using a smaller neighborhood to define the edges of the graph would also result in faster computations, but the penalty in computational cost of increasing the number of edges is not as large as the penalty for increasing the number of nodes. Our experiments have shown that good realtime performance is achieved even if we use the full 26-neighborhood, which is advantageous since we then obtain thin and less "jagged" live-wire curves.

User Interface. We have implemented a user interface with two options for drawing live-wire curves onto the boundary of an object.

The first option is to place seed-points freely in the volume. In this case, the volume is displayed by maximum intensity projection (MIP). See Figure 4a. To help the user locate the boundary of the object, we have used a volume haptics technique described in [2]. The gradient of the volume is used as a surface normal that defines a virtual surface that can be felt with the haptic device.

The second option is to "slice" the volume with a plane that can be moved and rotated freely relative to the volume, and draw the curve onto this arbitrary plane. See Figure 4b. The plane is a haptic surface so that the user can feel where it is located and easily place seed-points onto it. Drawing a live-wire curve onto the plane feels similar to drawing on a real surface. The shortest-path computations are still performed in 3D, but since we want the curve to stay in the vicinity of the plane we also add a term to our cost function that grows quickly as we move away from the chosen plane.

The two methods have different advantages. To draw a curve around an object using the first method, the user might have to rotate the volume while drawing the curve in order to avoid twisting the wrist too much. Using the second method, however, it is easy to do this by placing the guiding plane at the desired

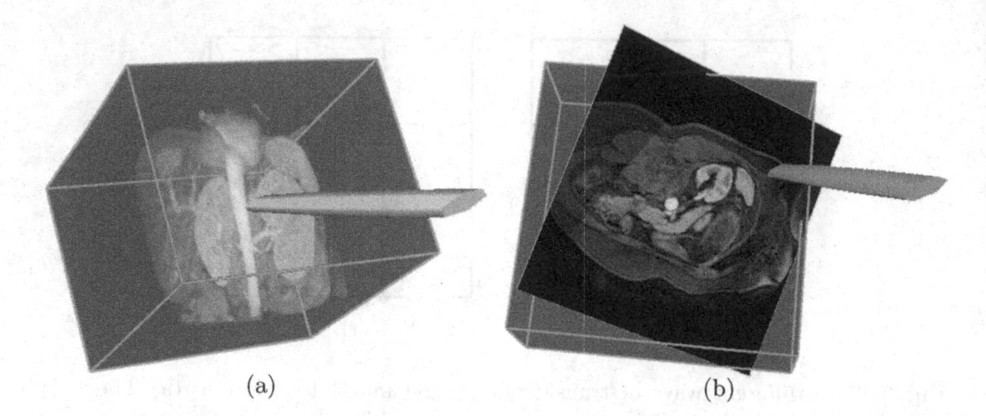

Fig. 4. (a) Placing seed-points freely in the volume using volume rendering and volume haptics to locate the boundary of the object. (b) Drawing a live-wire curve onto a guiding plane.

cross-section of the object. When testing the application we have mainly used the second method for drawing curves, while the first method has been used for placing single points (which are also considered to be closed curves).

4.2 Bridging Algorithm

We have based our bridging algorithm on the image foresting transform (IFT) [9]. The IFT is essentially Dijkstra's algorithm, modified to allow an arbitrary number of seed-points as input for the computations. The result of using more than one seed-point is that we find the shortest path from all pixels in the image to *any* of the seed-points.

Using the IFT, it is quite straightforward to create a rough "wire-frame" of the surface we are looking for. To do this, we compute the optimal curves from each voxel in one of the curves, using all voxels of the other curve as seed-points. An illustration of the result is shown in Figure 5c.

Since the curves we compute are independent of each other, the result is generally not a tunnel-free surface. To fill the gaps between the curves we define a polygonal surface between adjacent curves, as shown in Figure 6. These polygons are then rasterized using the method described in [14] to obtain a discrete surface. The result is exemplified in Figure 5d. Algorithm 1. shows the pseudo-code for our bridging algorithm.

Since all points on the second curve are not necessarily the "closest" point to a point on the first curve, a surface generated using our method is unfortunately not guaranteed to fit the "goal" curve exactly. In the worst case, all points on the first curve have the same closest point on the second curve, in which case the second curve will be approximated by this point only. Our current solution to this is to apply the algorithm from both directions and return the union of the results, i.e., run the algorithm starting from one of the curves and then run

Algorithm 1. Bridging Algorithm

Input: Two sets of voxels, *curve*1 and *curve*2, that represent the initial curves.
Output: A set of voxels, *output*, that represents the generated surface.
Additional variables: Two sets of voxels, *current_curve* and *next_curve* that will be used to store intermediate curves. A voxel *next_voxel*.

- *current_curve* ← *curve*1

while *next_curve* contains voxels that are not in *curve*2 **do**
- *next_curve* ← *empty*

  for every voxel v in *current_curve* **do**
  - *next_voxel* ← v

    if v is not in *curve*2 **then**
    - calculate the optimal path from v to *curve*2
    - set *next_voxel* to be the next voxel along this path

    end if

    if *next_curve* is not empty **then**
    - rasterize the polygon between the last voxel of *next_curve* and the two last voxels of *current_curve*
    - add the resulting voxels to *output*
    - rasterize the polygon between *next_voxel* and the last voxels of *next_curve* and *current_curve*
    - add the resulting voxels to *output*

    end if
  - add *next_voxel* to *next_curve*

  end for
- rasterize the polygon between *next_voxel*, the first voxel of *next_curve* and the last voxel of *current_curve*
- add the resulting voxels to *output*
- rasterize the polygon between the first voxels of *current_curve* and *next_curve* and the last voxel of *current_curve*
- add the resulting voxels to *output*
- *current_curve* ← *next_curve*

end while

it again starting from the other curve. The result then always contains the two original curves. However, if the shapes of the two curves differ to much "pockets" may be produced on the surface, as illustrated in Figure 7. In most cases when this happens, it is possible to reduce the error by drawing additional curves inbetween the two original curves.

4.3 Performance

For a moderately sized volume (128x128x100) our algorithm runs in approximately 1–20 seconds. The exact times depend on the length of the curves we are bridging and the distance between them. A number of things can be done to improve the performance. For example, our current implementation uses a heap data structure to manage the sorted queue when computing the IFT, but

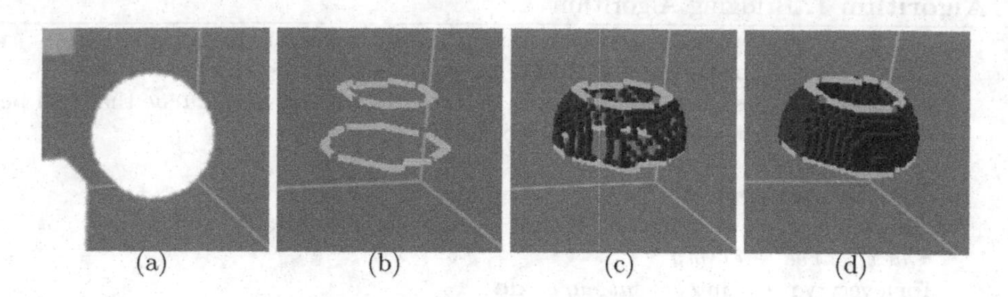

Fig. 5. (a) A synthetic object. (b) Two closed curves on the boundary of the object. (c) Result of connecting the two curves by live-wires. (d) Result of using our proposed algorithm.

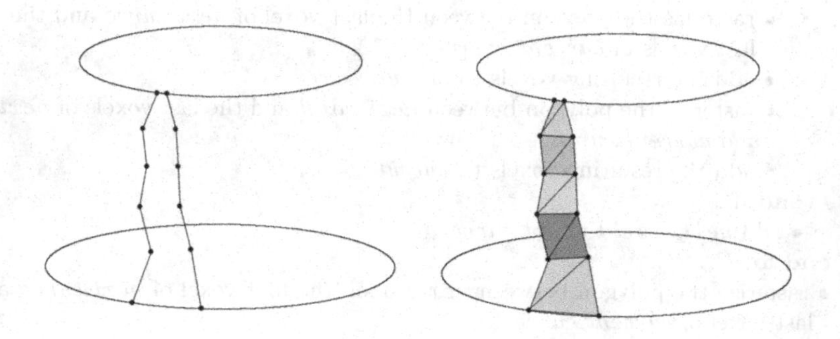

Fig. 6. To obtain a tunnel-free surface, all live-wire curves generated between the two user-defined curves are connected to their adjacent curves by triangular polygons

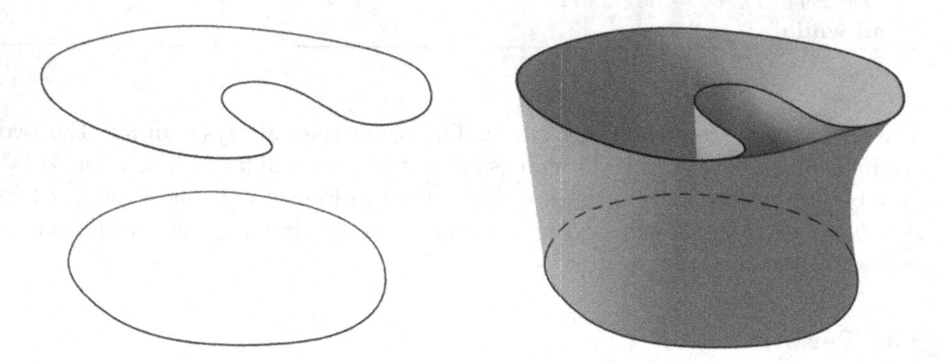

Fig. 7. If the shapes of the two curves differ too much, the surface produced by our bridging algorithm may contain pockets

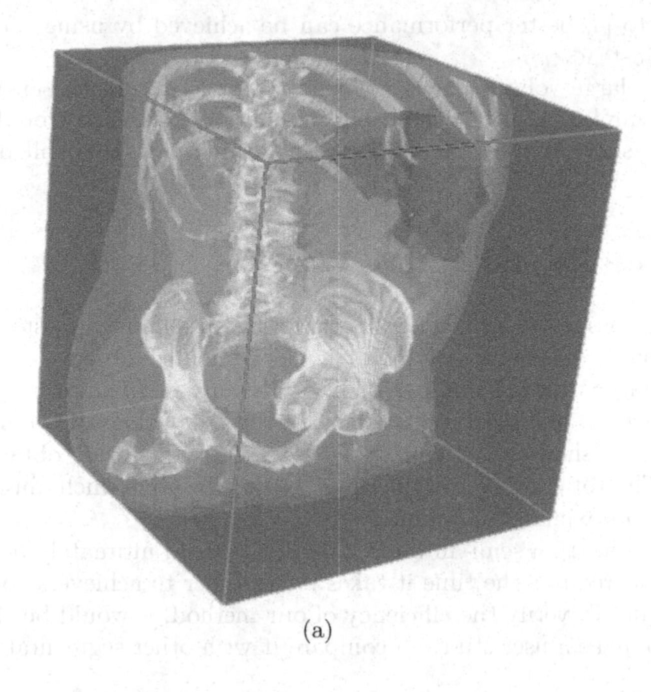

(a)

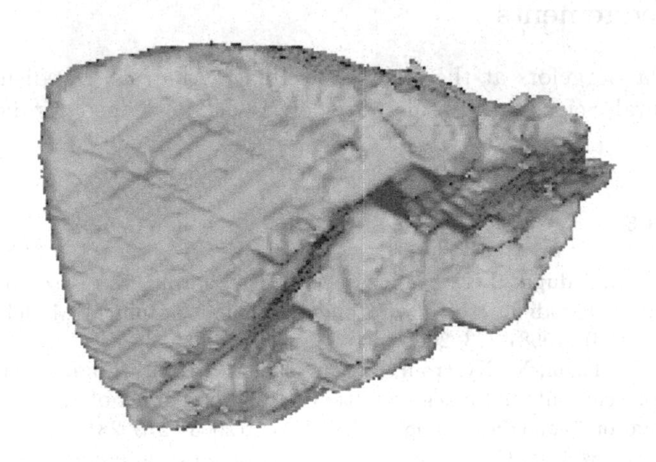

(b)

Fig. 8. (a) Segmentation of the liver in a 153x153x161 CT image of a human, obtained with our method by drawing 7 live-wire curves onto the boundary of the liver. (b) Surface rendering of the segmentation result.

according to [4] better performance can be achieved by using a circular queue structure instead.

Tests on large volumes have not been possible due to the current memory-inefficient implementation where big amounts of data are stored redundantly during the shortest-path calculations. In future work, this will be made more efficient.

5 Discussion and Future Work

Despite the issues mentioned in Section 4.2, preliminary tests of our application show promising results. We have segmented objects of varying complexity using a varying number of live-wire curves. In most cases, the closed discrete surfaces obtained are satisfactory representations of the underlying object. Figure 8 shows an example of a segmentation result obtained with our method. The total segmentation time for this example, including user interaction, was approximately 3 minutes.

As stated in [1], a semi-automatic method should ultimately be measured by how much it reduces the time it takes for the user to achieve a correct segmentation result. To verify the efficiency of our method, it would be therefore be of interest to make a user study to compare it with other segmentation methods.

Acknowledgements

Prof. Gunilla Borgefors at the Centre for Image Analysis, Swedish University of Agricultural Sciences, Uppsala, Sweden is acknowledged for her scientific support.

References

1. Falcão, A.X., Udupa, J.K., Samarasekera, S., Sharma, S.: User-steered image segmentation paradigms: Live wire and live lane. Graphical Models and Image Processing (60) (1998) 223–260
2. Vidholm, E., Tizon, X., Nyström, I., Bengtsson, E.: Haptic guided seeding of MRA images for semi-automatic segmentation. In: Proceedings of IEEE International Symposium on Biomedical Imaging (ISBI'04). (2004) 278–281
3. Harders, M., Székely, G.: Enhancing human-computer interaction in medical segmentation. Proceedings of the IEEE, Special Issue on Multimodal Human Computer Interfaces **9**(91) (2003) 1430–1442
4. Falcão, A.X., Udupa, J.K., Miyazawa, F.K.: An ultra-fast user-steered image segmentation paradigm: Live wire on the fly. IEEE Transactions on Medical Imaging **19**(1) (2000) 55–62
5. Falcão, A.X., Udupa, J.K.: A 3D generalization of user-steered live-wire segmentation. Medical Image Analysis **4**(4) (2000) 389–402
6. Knapp, M., Kanitsar, A., Groller, M.E.: Semi-automatic topology independent contour based 2 1/2 D segmentation using live-wire. Journal of WSCG **20**(1–3) (2004)

7. Schenk, A., Prause, G., Peitgen, H.O.: Efficient semiautomatic segmentation of 3D objects in medical images. In: Proceedings of MICCAI 2000, Springer-Verlag (2000) 186–195
8. Hamarneh, G., Yang, J., McIntosh, C., Langille, M.: 3d live-wire-based semi-automatic segmentation of medical images. In: Proceedings of SPIE Medical Imaging: Image Processing. Volume 5747. (2005) 1597–1603
9. Falcão, A.X., Bergo, F.P.G.: Interactive volume segmentation with differential image foresting transforms. IEEE Transactions on Medical Imaging **23**(9) (2004)
10. Reachin Technologies: (http://www.reachin.se) Accessed 23 January 2006.
11. Technologies, S.: (http://www.sensable.com) Accessed 23 January 2006.
12. Dijkstra, E.W.: A note on two problems in connexion with graphs. Numerische Mathematik **1** (1959) 269–271
13. Mortensen, E.N., Barrett, W.: Interactive segmentation with intelligent scissors. Graphical Models and Image Processing **60**(5) (1998) 349–384
14. Kaufman, A.: Efficient algorithms for scan-converting 3D polygons. Computers and Graphics **12**(2) (1988) 213–219

Minimal Decomposition of a Digital Surface into Digital Plane Segments Is NP-Hard

Isabelle Sivignon and David Coeurjolly

Laboratoire LIRIS - Université Claude Bernard Lyon 1
Bâtiment Nautibus - 8, boulevard Niels Bohr
69622 Villeurbanne cedex, France
{isabelle.sivignon, david.coeurjolly}@liris.cnrs.fr

Abstract. This paper deals with the complexity of the decomposition of a digital surface into digital plane segments (DPS for short). We prove that the decision problem (does there exist a decomposition with less than k DPS?) is NP-complete, and thus that the optimisation problem (finding the minimal number of DPS) is NP-hard. The proof is based on a polynomial reduction of any instance of the well-known 3-SAT problem to an instance of the digital surface decomposition problem. A geometric model for the 3-SAT problem is proposed.

1 Introduction

Digital objects are defined as sets of grid points in $\mathbb{Z}^n$. Those objects carry redundant geometrical information due to their discrete structure: an object is represented as a set of elementary cells (called pixels in 2D, voxels in 3D). The definition of digital linear structures like digital lines [1] and digital planes [2, 3] originated a lot of works dealing with the decomposition of the contour of a digital object into digital linear primitives. Such a decomposition actually apprehends global geometrical properties of those objects. Many decomposition strategies may be designed and the number of parts computed by the algorithms may be a first criterion to compare the results. In this work, we focus on the complexity of the optimal (minimal number of parts) decomposition problem. In the 2D case, it has been shown that the minimal decomposition of a digital curve into digital line segments can be computed in linear time [4]. In the 3D case of surfaces, many decomposition algorithms have been proposed [5, 6, 7, 8], offering comparisons on the number of faces recognised by different algorithms. Nevertheless, no optimality results exist, and no complexity study has been carried out. Related results have been recently proposed in [9] concerning the NP-completeness of the construction of an integer lattice polyhedron P with minimal number of convex facets such that $P \cap \mathbb{Z}^3$ corresponds to the input 3D digital object.

In computational geometry, the decomposition of a shape (*e.g.* a polygon) into a minimal number of elements (*e.g.* convex polygons) usually leads to NP-complete problems [10]. A problem is in the NP class of algorithms if it can

A. Kuba, L.G. Nyúl, and K. Palágyi (Eds.): DGCI 2006, LNCS 4245, pp. 674–685, 2006.

only be solved in polynomial time by a non-deterministic machine [11]. As a corollary, the problem is in NP if a solution to the decision problem can be verified in polynomial time on a deterministic machine. A problem is said to be NP-complete if it is at least as difficult as any NP problem. In other words, if a problem is NP-complete, an old conjecture is that no time efficient solution exists to solve it. The remaining option is to consider approximation algorithms with or without heuristics.

Prior to a complexity study, the problem has to be formalised. In the sequel, we consider 6-connected sets of voxels which surface $\mathbb{S}$ is defined as the set of object voxels sharing a face with the background. The surface is a set of 18-connected voxels, and maximal digital naive planes [12, 13, 2, 3] are used for the decomposition. A digital plane segment (DPS for short) is maximal if no surface voxel may be added to it. In the following, we consider a sequential decomposition algorithm: given a voxel on the surface (called a *seed*), we construct the maximal digital naive plane segment adding iteratively voxels that are 18-connected to the DPS initialised with the seed. Then, a new seed is considered from the set of remaining voxels in $\mathbb{S}$. In this algorithm, both the propagation process during the DPS growing and the initialisation of seeds must be taken into account.

The optimisation problem we consider is defined as follows:

Min-DSD (Digital Surface Decomposition): Given a digital object surface $\mathbb{S}$, find the minimal decomposition of $\mathbb{S}$ into maximal digital naive plane segments using a sequential algorithm.

In order to study the complexity of an optimisation problem, the related decision problem has to be considered:

k-DSD: Given a digital object surface $\mathbb{S}$ and a number $k \in \mathbb{N}^*$, does there exist a decomposition of $\mathbb{S}$ into k maximal digital naive plane segments using a sequential algorithm?

In this article, we prove that $k-$DSD is NP-complete whatever the propagation heuristic. Furthermore, the only requirement on the digital plane segments topology is connectivity.

To prove that a problem $\mathcal{P}$ is NP-complete, a classical scheme is to exhibit a polynomial reduction of any instance of a classical NP-complete problem, denoted $\mathcal{P}_{NP}$ into an instance of $\mathcal{P}$. Then, we have to prove that a solution of $\mathcal{P}$ also leads to a solution of $\mathcal{P}_{NP}$. Since $\mathcal{P}_{NP}$ is known to be NP-complete, we could conclude that $\mathcal{P}$ is also NP-complete [11]. In the literature, the Boolean Satisfiability Problem (SAT) is a decision problem classically used in complexity theory since it was the first known NP-complete problem. An instance of SAT is a boolean expression written using only AND, OR, NOT, variables and parentheses. The decision problem is: given an expression, is there an assignment of the variables such that the expression is TRUE? The problem remains NP-complete even if the expression is written in conjunctive normal form with three variables per clause, yielding the 3-SAT problem. An expression ϕ has the form:

$$(x_1 \vee x_2 \vee \neg x_3) \wedge (\neg x_1 \vee x_4 \vee x_5) \wedge (\neg x_6 \neg \vee x_3 \vee \neg x_5) \wedge \dots , \quad (1)$$

where each x_i is a binary variable (and $\neg x_i$ its negation) that can appear several times in the expression.

In the following we define a polynomial reduction of any instance of the 3-SAT problem to an instance of the k-DSD problem. The construction process, defining geometrical objects for variables, variable instances and clauses, is presented in Section 2, while the NP-completeness proof derived from this construction is given in Section 3.

2 A Geometric Model for 3-SAT

Given a 3-SAT expression ϕ, we show how to construct a geometric discrete object. This construction is a two steps process: after defining geometric elements for variables, instances of variables and clauses, we see how these basic components are organised and linked together in the 3D space.

2.1 General Considerations

All the basic elements we define further are composed of two parts:

- **idle part**: the surface of this part will be made of planes parallel to axis planes and only aims at defining a 6-connected object. The minimal number of DPS needed to cover the idle part will be fixed for each basic element, and any decomposition of this part into maximal DPS will exactly cover the same voxels, with no possible extension. This part is not used in the encoding of a 3-SAT expression;
- **active part**: this part consists of the remaining voxels after the decomposition of the idle part. It takes advantage of digital planes properties to geometrically encode the 3-SAT elements.

For each basic element, we provide an illustration[1] of an example set of seeds (black voxels) which may be used in a decomposition algorithm. Moreover, those sets have the remarkable property that any two seeds cannot be covered by a single DPS.

The underlying basic idea for this construction is the following: the optimal decomposition chosen for a variable object generates a "signal" sent to clause objects through "wires" representing instances of variables. This kind of geometric construction of 3-SAT is a classic way to prove NP-completeness of geometric problems (see [14,15] for instance). The construction of basic elements relies on several properties of digital planes structure that we set forth here (Figure 1):

Property 1. For the four configurations (a), (b), (b') and (c) represented in Figure 1, one DPS cannot simultaneously cover the two light-coloured voxels and all the other ones. In the configuration represented in Figure 1(d), the three voxels cannot be covered by one DPS, but any two voxels can.

---

[1] Most illustrations of this paper are originally colour artworks. To make the understanding of the paper easier from B&W printings, colour images are available on `http://liris.cnrs.fr/isabelle.sivignon/SatDSD.html`

Proof. The proofs of these properties are straight forward using either digital naive plane structural basic properties or their arithmetical definition [12, 6, 2].

In the following, we refer to these configurations as Property 1(a), (b), (b'), (c) and (d).

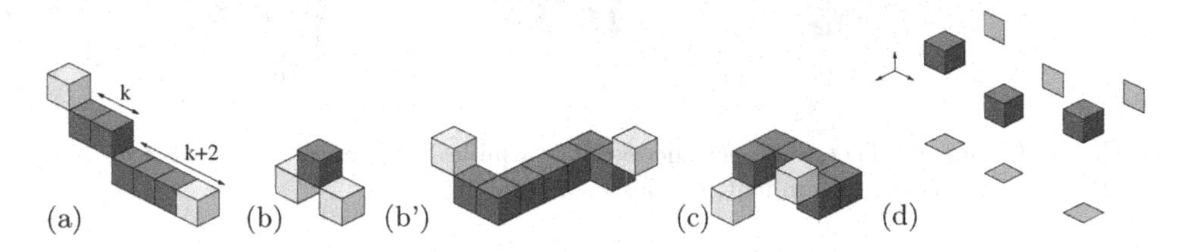

Fig. 1. Five configurations used for the reduction process

2.2 Variable and Clause Objects

An illustration of a variable element is given in Figure 2. Optimal decompositions are represented in Figure 2 (b) and (c): five DPS are used to cover the idle part of the object, and two more DPS are required to cover the remaining active part (upper part on the figure). Indeed, regardless of the idle part decomposition, the only two optimal decompositions for the active part (composed of the "bump" and the five voxels on each side of this bump) consists of two DPS, with two possible configurations (Figure 2 (b) and (c)): the "bump" voxels are either covered by the left or the right DPS (see Property 1(b)). Actually, any other decomposition is either not optimal or not composed of maximal DPS. We set that these two possibilities respectively encode true and false assignments of the variable. From these decomposition schemes and Property 1(b), seven seeds can be defined on each variable object.

Variable elements are linked to clauses thanks to wires that are connected on the area circled in red on Figure 2(a). The first part of these wires, described in details in Section 2.3, aims at generating a "signal" encoding the assignment of the variable. This signal is then "sent" to clause objects (see Section 2.3 for the transmission process). Figure 3 illustrates the signal generation: the number of voxels covered by the two active DPS differ by one according to the truth value of the variable - in the case of a true value, one more voxel is covered.

For a positive instance of variable, the wire is connected to the variable on the right-hand side of the variable, as depicted on Figure 3: the signal corresponding to the value of the variable is generated. For a negative instance of variable, the wire is connected on the left-hand side of the variable: in this case, the signal corresponding to the negated value of the variable is generated.

Finally, and to handle multiple instances of the same variable in a boolean expression, the length (along y axis) of the variable v depends on the maximum number of positive or negative instances of v in an expression, so that all the connections can be made. Note that the length of the variable does not change the optimal number of DPS required for the decomposition.

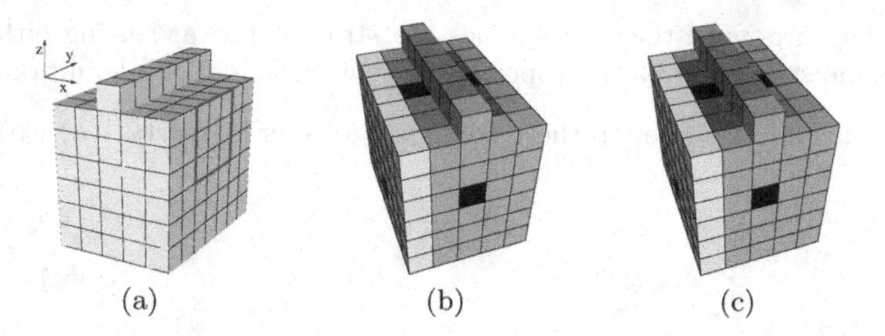

Fig. 2. Geometric discrete object encoding a variable: (a) general view with wire plugging area, (b) truth assignment, (c) false assignment

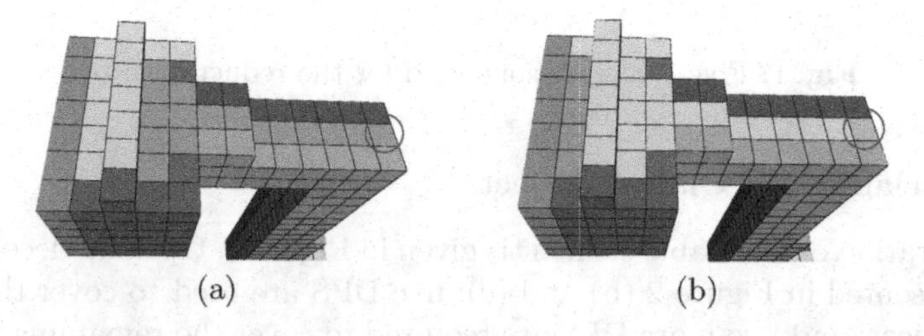

Fig. 3. Generation of a "signal" according to the variable assignment: when the variable is set to TRUE (a), the voxel circled in red is covered by a DPS of the variable, otherwise (b), this voxel cannot be covered by one of the two DPS of the variable (Property 1 (a))

A clause element is depicted in Figure 4. It is composed of a transversal rectangular parallelepiped on which three terminals are plugged. Since each clause has three literals (recall that 3-SAT is considered), each clause element has three incoming wires. The active part consists of the upper part of the object as depicted in Figure 4. Five idle DPS are required in order to decompose five out of the six faces of the rectangular parallelepiped. Idle planes related to the three terminals will be taken into account in the wire definition.

The active part of a clause can be entirely covered by a single DPS, except one out of the three terminal extremities (see Figure 4(b)): indeed, following Property 1(d), the three terminal extremities cannot be covered by a single DPS whereas any couple of terminals can be entirely covered by a single DPS. To sum up, if one terminal extremity can be covered by a DPS of the wires active part, then only one DPS is required to cover the clause active part. Otherwise, two DPS are necessary. The link between a boolean clause and the geometric object we propose can be drawn up as follows: a boolean clause is true if and only if at least one literal is true; one DPS is enough to cover the whole active part of a clause if and only if one terminal extremity is covered by another DPS. From the optimal decomposition, six seeds can be defined on the surface of each clause object. The wires linking variable elements to a clause are plugged on the areas circled in red on Figure 4(a), one on each terminal extremity.

To end with the variable and clause geometric objects, Figure 5 illustrates how these objects are put together in the 3D space when many objects are involved

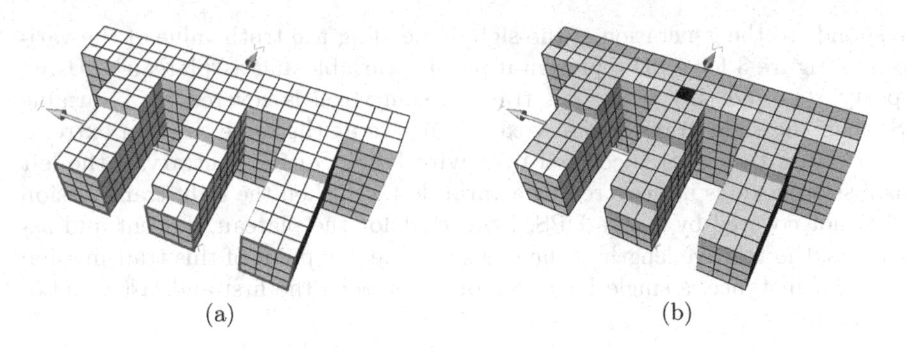

Fig. 4. Geometric discrete object encoding a clause: (a) general view, (b) optimal decomposition

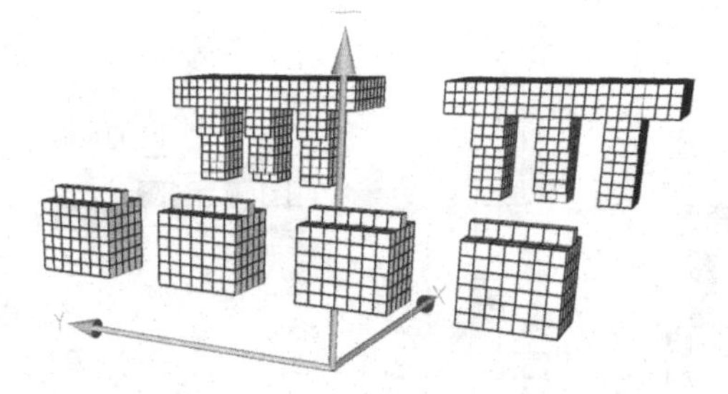

Fig. 5. Positions of (four) variables and (two) clauses objects in the 3D space

in a boolean expression: variables and clauses are lined up on two axis parallel to the y axis. The definition of wires connecting variables and clauses relies on this spatial construction.

2.3 Linking Variables and Clauses

Variable elements are connected to clause elements through wires, that represent variable instances: if a variable v appears in a clause c, a wire links the geometric elements of v and c. Those wires aim at "transmitting" the truth value of a variable to the clause it belongs to. Before defining the geometric construction of wires, we describe the transmission process.

Figure 6 illustrates how the truth value of a variable is transmitted to a clause through a wire. This Figure represents a vertical cut of the active part of a variable, a wire and a clause terminal. Figure 6(a) illustrates the propagation of a true value while Figure 6(b) shows how a false value is transmitted to a clause. From the construction we propose, the vertical cut of a variable-wire-clause connection can be thought of as a 2D digital curve that we decompose in digital straight segments, using their properties.

We call "transmission voxels" the two voxels circled in red in Figure 6 (intermediate transmission voxels are circled in black). We consider an optimal decomposition of the surface into DPS. The left transmission voxel actually corresponds to the generation of the signal encoding the truth value of the variable (see Figure 3 for a 3D representation of a variable and a "plateau"). Using Property 1(a) and (b'), if the left transmission voxel is covered by a variable DPS, then the right transmission voxel (which is at the same time an extremity of a clause terminal) is covered by a wire DPS. On the contrary, if the left transmission voxel is not covered by a variable DPS, then the right transmission voxel is not covered by a wire DPS. Note that for the plateau, descent and ascent parts, the relative length of the steps are the key point of this transmission process: for instance, a single DPS cannot cover both the first and last voxel of the descent.

Since the left transmission voxel is covered by a variable DPS if and only if the variable instance is set to the value "true" (Property 1(a)), the clause terminal extremity is covered by a wire DPS if and only if the variable instance is set to the value "true".

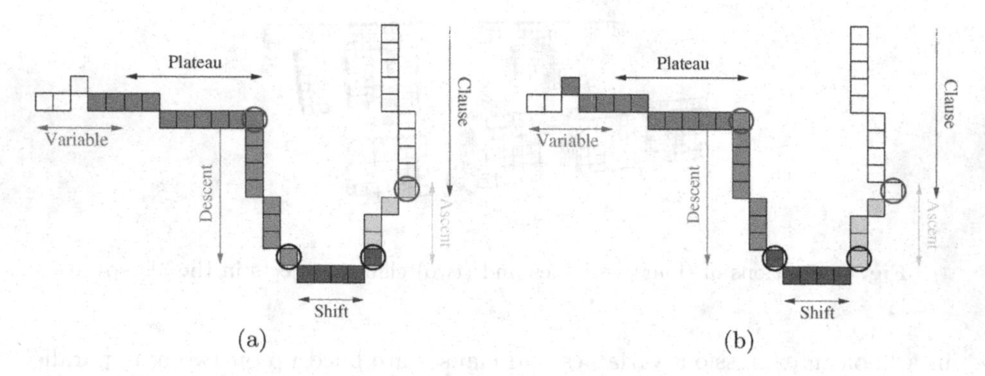

Fig. 6. Vertical cut illustration of the transmission of a truth assignment through a wire: (a) the value "true" is transmitted; (b) the value "false" is transmitted

Following the spatial arrangement of variable and clause elements (see Figure 5), and the rules defined for the connections of negative and positive instances of variables (Section 2.2), wires standing for positive instances are plugged on the variable side closest to clauses, while wires corresponding to negative instances are plugged on the opposite side. We see that in the case of a negative wire, a U-turn towards clause objects is required. As depicted in 2D in Figure 6, wires are basically composed of four parts, that are depicted in 3D in Figure 7(a) for a positive wire:

- a plateau (blue) generates the "signal" corresponding to the truth value of the variable;
- a descent (green) to a given level L: two distinct variable instances descend on two different levels to ensure an intersection free construction;
- a shift movement (red) on the level L to reach the clause position;
- an ascent (yellow) from the level L to the clause terminal extremity.

The optimal decomposition of the plateau is made of one DPS only, for the idle part (bottom of the plateau). Indeed, the sides are covered with descent idle

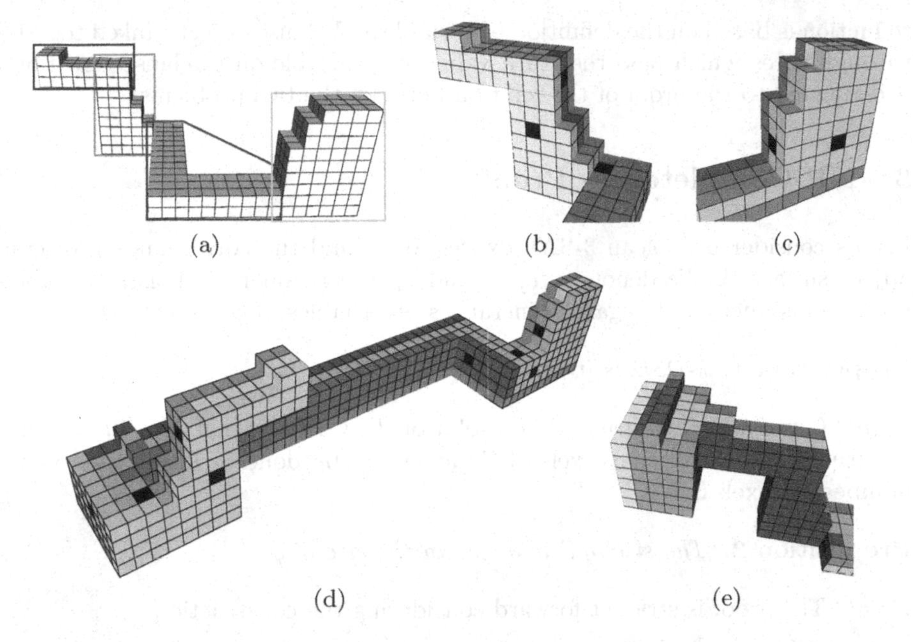

Fig. 7. Wire between a variable and a clause: (a) case of a positive variable instance, with zooms on (b) the descent and (c) ascent parts; (d) case of a negative variable instance

DPS, and the upper part (active) is covered by a DPS coming from the variable decomposition. Figure 7(e) illustrates how a wire is connected to a variable.

Concerning the descent and ascent parts, optimal decompositions are depicted in Figure 7(b) and (c). Seven idle and one active DPS are required for both positive and negative wires. Moreover, the active part is a three steps surface such that the first step is k voxels long ($k \geq 3$), the second one is $k - 2$ voxels long and the third one is made of one voxel. In addition to the function of this construction in the transmission process (see previous paragraph), the parameter k is used to ensure that every wire descend on a different level so that wires do not intersect.

The shift parts of positive and negative wires are different. In the case of a positive wire, only one DPS covers the active part (upper part). Five more idle planes are necessary for a negative wire, and two DPS are needed to cover the active part (see Figure 7(d)). The transmission of the truth value through the U-turn part of a negative wire is ensured thanks to the small "bump" on the shift part and Property 1(c).

All in all, optimal decomposition of wires leads to the definition of 11 seeds (eight for the idle part, three for the active part) for each positive wire and 17 (13 for the idle part, four for the active part) for each negative wire (see Figure 7(b), (c) and (d)).

To summarise the construction, we have proposed a polynomial reduction of any instance of the 3-SAT problem into an instance of the k-DSD problem. This reduction is based on the definition of variable and clause objects linked together through wires which pass the truth value of a variable on to clauses. Section 3 is dedicated to the proof of the relation between the two problems.

3 NP-Completeness Proof

Let us consider a boolean 3-SAT expression ϕ and the corresponding discrete object surface $\mathbb{S}$. We denote c, v, v_p and v_n the number of clauses, variables, positive instances and negatives instances of variables in ϕ respectively.

Proposition 1. *k-DSD is in NP.*

Proof. Given a digital surface $\mathbb{S}$ and solution D, verifying that $|D| \leq k$ and that it actually covers all the voxels of $\mathbb{S}$ can easily be done in linear time in the number of voxels $\mathbb{S}$. □

Proposition 2. *The size of* $\mathbb{S}$ *is linear in the size of* ϕ.

Proof. The proof is straight forward considering the construction.

We shall now prove that the construction is a reduction of 3-SAT to k-DSD, *i.e.* that the expression ϕ is satisfiable if and only if $\mathbb{S}$ admits a decomposition with k maximal DPS. We prove the two implications one after the other.

Lemma 1. *If the expression* ϕ *is satisfiable, then* $\mathbb{S}$ *admits a decomposition with* k *maximal DPS.*

Proof. Assume that ϕ is satisfiable under some truth assignment T. The following algorithm builds a decomposition of the surface of $\mathbb{S}$ into k maximal DPS:

1. label all the voxels belonging to a construction DPS regardless of T: $5v + 5c + 8v_p + 13v_n$ DPS are used to cover the entire idle part of $\mathbb{S}$;
2. decompose each variable according to its truth assignment in T: this decompositions requires $2v$ DPS;
3. use $3v_p$ and $4v_n$ DPS to decompose the wires active parts, which may leave the tips of some wires (which are also the clause terminal extremities) uncovered;
4. since T satisfies ϕ, every clause has at least one incoming wire with a covered tip. Thus, every clause has at least one covered terminal extremity. Consequently, each clause active part can be covered with one single DPS.

All in all, $(5v+5c+8v_p+13v_n)+2v+3v_p+4v_n+c = 7v+6c+11v_p+17v_n = k$ DPS are used in this decomposition. □

In order to prove the reverse implication, we need to show that there is only one way of decomposing $\mathbb{S}$ into k DPS. Next, we show that this unique solution leads to a satisfactory assignment of ϕ's variables.

Lemma 2. *Consider a decomposition of $\mathbb{S}$ with k DPS. Then the decompositions of variable, positive wire, negative wire and clause objects are respectively covered by 7, 11, 17, 6 DPS.*

Proof. Consider a decomposition D of $\mathbb{S}$ with $|D| = k$. Suppose that there exist a variable object with a decomposition D_v such that $|D_v| > 7$. Extra DPS are either idle or active. Using more than five idle DPS has no effect on the number of DPS required to cover other variables, wires and clauses. Thus, $|D| = 7(v-1) + |D_v| + 6c + 11v_p + 17v_n > k$, which is a contradiction.

Now suppose that extra DPS are used for the active part. With these DPS, one can at best ensure that the "true" value is transmitted to every clause objects linked to this variable. Nevertheless, the number of DPS required to cover wires and clauses does not change, and we still have $|D| = 7(v-1)+|D_v|+6c+11v_p+17v_n > k$, which is a contradiction.

On the contrary, if less than seven DPS are used to cover a variable object, it is easy to check that some voxels will remain uncovered even if more DPS are used for wires or clause objects. Similar arguments can be used to show that positive and negative wires, and clause decompositions have to be composed of 11, 17 and 6 DPS. □

Lemma 3. *If $\mathbb{S}$ admits a decomposition into k maximal DPS, then ϕ is satisfiable.*

Proof. Suppose that $\mathbb{S}$ admits a decomposition D with k DPS. Since $|D| = k$, from Lemma 2 the decomposition of every variable object is made of seven DPS. Variable objects can only be decomposed two ways into seven DPS, each of which encodes a truth assignment. This decomposition is made of 5 DPS for the idle part and 2 DPS for the active part (regardless of the algorithm used). Thus, covering all variables requires $7v$ DPS. In the same way, using Lemma 2 covering wires uses $11v_p + 17v_n$ DPS. All in all, $k - 7v - 11v_p - 17v_n = 6c$ DPS remain for covering clause objects. The idle part of clause objects requires 5 DPS regardless of the rest of the decomposition. Thus c DPS remain to cover the clauses active parts. Since there are c clause objects, and c DPS remain, we know that the clause active parts are covered by one DPS only in D. This is possible if and only if every clause is satisfied, and thus ϕ is satisfied too. □

Theorem 1. *k-DSD is a NP-complete problem.*

Proof. The result is derived from Lemma 1 and 3.

This theorem proves that the decision problem associated to Min-DSD is NP-complete. Thus, according to the theory of complexity, Min-DSD is said to be NP-hard.

4 Example

A software that generates a 3D object from a 3-SAT boolean expression is available on `http://liris.cnrs.fr/isabelle.sivignon/code.html`. This program also generates the seeds of the object, and a simple surface decomposition algorithm into maximal DPS is also provided to compute the decomposition derived from those seeds.

Figure 8 is an illustration of the digital surface encoding the expression $\phi = (a \vee \neg b \vee c)$. The optimal decomposition into maximal DPS is composed of 49 idle DPS and 17 active DPS. In Figure 8(a), the variable objects encode the assignment ($a = true$, $b = true$, $c = false$), and the optimal decomposition is represented. In Figure 8(b), the variable objects encode the assignment ($a = false$, $b = true$, $c = false$): in this case, since ϕ is not satisfied, the optimal decomposition cannot be achieved, and an extra DPS (in red) is added.

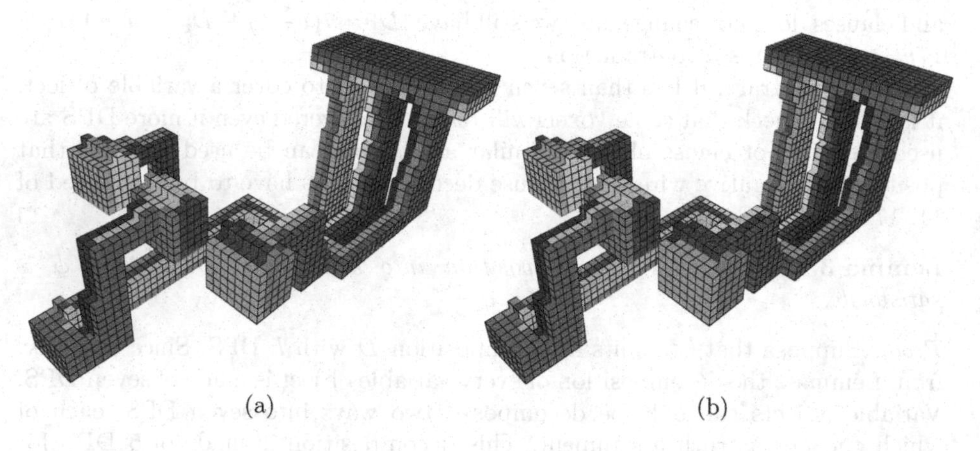

Fig. 8. Discrete object encoding the expression $\phi = (a \vee \neg b \vee c)$: (a) optimal decomposition corresponding to the satisfaction of ϕ; (b) ϕ is not satisfied and one more DPS is required to achieve a complete decomposition

5 Conclusion and Future Works

In this article, we have proved that the decomposition of a digital object surface into a minimal number of maximal DPS using a sequential algorithm is NP-complete. This theoretical result concludes an important open problem in the discrete geometry community: no efficient algorithms exist to solve the Min-DSD problem. A logical consequence of this answer is that only heuristics can be used.

Among possible heuristics, important theoretical future works exist: does there exist a polynomial-time approximation scheme for the Min-DSD problem? More precisely, is there a polynomial in time approximation of Min-DSD that produces a solution that is within ϵ factor of the optimal solution?

By construction of variables, clauses and links, the genus of the obtained binary object depends on the number of cycle in the 3-SAT instance. Is $k-$DSD still NP-complete, and thus Min-DSD still NP-hard for hole-free objects?

References

1. Rosenfeld, A., Klette, R.: Digital straightness. In: Int. Workshop on Combinatorial Image Analysis. Volume 46 of Electronic Notes in Theoretical Computer Science., Elsevier Science Publishers (2001)
2. Klette, R., Rosenfeld, A.: Digital Geometry: Geometric Methods for Digital Picture Analysis. Series in Comp. Graph. and Geom. Modeling. Morgan Kaufmann (2004)
3. Brimkov, V., Coeurjolly, D., Klette, R.: Digital planarity - a review. Technical Report CITR-TR-142, CITR - Univerity of Auckland (2004)
4. Feschet, F., Tougne, L.: On the min dss problem of closed discrete curves. Discrete Applied Mathematics **151**(1-3) (2005) 138–153
5. Françon, J., Papier, L.: Polyhedrization of the boundary of a voxel object. In: 8th DGCI. Volume 1568 of LNCS., Springer-Verlag (1999) 425–434
6. Debled-Rennesson, I.: Etude et reconnaissance des droites et plans discrets. PhD thesis, Université Louis Pasteur (1995)
7. Sivignon, I., Dupont, F., Chassery, J.M.: Decomposition of a three-dimensional discrete object surface into discrete plane pieces. Algorithmica **38**(1) (2003) 25–43
8. Sivignon, I., Dupont, F., Chassery, J.M.: Reversible polygonalization of a 3D planar discrete curve: Application on discrete surfaces. In: 12th DGCI. (2005) 347–358
9. Brimkov, V.: Discrete volume polyhedrization is srongly NP-hard. Technical report, CITR-RR 179 (2006)
10. Goodman, J.E., O'Rourke, J., eds.: Handbook of Discrete and Computational Geometry. CRC Press (1997)
11. Cormen, T.H., Leiserson, C.E., Rivest, R.L., Stein, C.: Introduction to Algorithms. 2nd edn. MIT Press, Cambridge, MA (2001)
12. Reveillès, J.P.: Géométrie discrète, calcul en nombres entiers et algorithmique. PhD thesis, Université Louis Pasteur - Strasbourg (1991)
13. Andres, E., Acharya, R., Sibata, C.: Discrete analytical hyperplanes. Graphical Models and Image Processing **59**(5) (1997) 302–309
14. Worman, C.: Decomposing polygons into diameter bounded components. In: Canadian Conference on Computational Geometry (CCCG'03). (2003) 103–106
15. Chazelle, B., Dobkin, D.P., Shouraboura, N., Tal, A.: Strategies for polyhedral surface decomposition: An experimental study. In: Symp. on Compututational Geometry. (1995) 297–305

Author Index

Lecture Notes in Computer Science

For information about Vols. 1–4175

please contact your bookseller or Springer

Vol. 4270: H. Zha, Z. Pan, H. Thwaites, A.C. Addison, M. Forte (Eds.), Interactive Technologies and Sociotechnical Systems. XVI, 547 pages. 2006.

Vol. 4269: R. State, S. van der Meer, D. O'Sullivan, T. Pfeifer (Eds.), Large Scale Management of Distributed Systems. XIII, 282 pages. 2006.

Vol. 4267: A. Helmy, B. Jennings, L. Murphy, T. Pfeifer (Eds.), Autonomic Management of Mobile Multimedia Services. XIII, 257 pages. 2006.

Vol. 4265: N. Lavrač, L. Todorovski, K.P. Jantke (Eds.), Discovery Science. XIV, 384 pages. 2006. (Sublibrary LNAI).

Vol. 4264: J.L. Balcázar, P.M. Long, F. Stephan (Eds.), Algorithmic Learning Theory. XIII, 393 pages. 2006. (Sublibrary LNAI).

Vol. 4254: T. Grust, H. Höpfner, A. Illarramendi, S. Jablonski, M. Mesiti, S. Müller, P.-L. Patranjan, K.-U. Sattler, M. Spiliopoulou (Eds.), Current Trends in Database Technology – EDBT 2006. XXXI, 932 pages. 2006.

Vol. 4253: B. Gabrys, R.J. Howlett, L.C. Jain (Eds.), Knowledge-Based Intelligent Information and Engineering Systems, Part III. XXXII, 1301 pages. 2006. (Sublibrary LNAI).

Vol. 4252: B. Gabrys, R.J. Howlett, L.C. Jain (Eds.), Knowledge-Based Intelligent Information and Engineering Systems, Part II. XXXIII, 1335 pages. 2006. (Sublibrary LNAI).

Vol. 4251: B. Gabrys, R.J. Howlett, L.C. Jain (Eds.), Knowledge-Based Intelligent Information and Engineering Systems, Part I. LXVI, 1297 pages. 2006. (Sublibrary LNAI).

Vol. 4249: L. Goubin, M. Matsui (Eds.), Cryptographic Hardware and Embedded Systems - CHES 2006. XII, 462 pages. 2006.

Vol. 4248: S. Staab, V. Svátek (Eds.), Engineering Knowledge in the Age of the Semantic Web. XIV, 400 pages. 2006. (Sublibrary LNAI).

Vol. 4247: T.-D. Wang, X. Li, S.-H. Chen, X. Wang, H. Abbass, H. Iba, G. Chen, X. Yao (Eds.), Simulated Evolution and Learning. XXI, 940 pages. 2006.

Vol. 4245: A. Kuba, L.G. Nyúl, K. Palágyi (Eds.), Discrete Geometry for Computer Imagery. XIII, 688 pages. 2006.

Vol. 4243: T. Yakhno, E.J. Neuhold (Eds.), Advances in Information Systems. XIII, 420 pages. 2006.

Vol. 4241: R.R. Beichel, M. Sonka (Eds.), Computer Vision Approaches to Medical Image Analysis. XI, 262 pages. 2006.

Vol. 4239: H.Y. Youn, M. Kim, H. Morikawa (Eds.), Ubiquitous Computing Systems. XVI, 548 pages. 2006.

Vol. 4238: Y.-T. Kim, M. Takano (Eds.), Management of Convergence Networks and Services. XVIII, 605 pages. 2006.

Vol. 4236: L. Breveglieri, I. Koren, D. Naccache, J.-P. Seifert (Eds.), Fault Diagnosis and Tolerance in Cryptography. XIII, 253 pages. 2006.

Vol. 4234: I. King, J. Wang, L. Chan, D. Wang (Eds.), Neural Information Processing, Part III. XXII, 1227 pages. 2006.

Vol. 4233: I. King, J. Wang, L. Chan, D. Wang (Eds.), Neural Information Processing, Part II. XXII, 1203 pages. 2006.

Vol. 4232: I. King, J. Wang, L. Chan, D. Wang (Eds.), Neural Information Processing, Part I. XLVI, 1153 pages. 2006.

Vol. 4229: E. Najm, J.F. Pradat-Peyre, V.V. Donzeau-Gouge (Eds.), Formal Techniques for Networked and Distributed Systems - FORTE 2006. X, 486 pages. 2006.

Vol. 4228: D.E. Lightfoot, C.A. Szyperski (Eds.), Modular Programming Languages. X, 415 pages. 2006.

Vol. 4227: W. Nejdl, K. Tochtermann (Eds.), Innovative Approaches for Learning and Knowledge Sharing. XVII, 721 pages. 2006.

Vol. 4225: J.F. Martínez-Trinidad, J.A. Carrasco Ochoa, J. Kittler (Eds.), Progress in Pattern Recognition, Image Analysis and Applications. XIX, 995 pages. 2006.

Vol. 4224: E. Corchado, H. Yin, V. Botti, C. Fyfe (Eds.), Intelligent Data Engineering and Automated Learning – IDEAL 2006. XXVII, 1447 pages. 2006.

Vol. 4223: L. Wang, L. Jiao, G. Shi, X. Li, J. Liu (Eds.), Fuzzy Systems and Knowledge Discovery. XXVIII, 1335 pages. 2006. (Sublibrary LNAI).

Vol. 4222: L. Jiao, L. Wang, X. Gao, J. Liu, F. Wu (Eds.), Advances in Natural Computation, Part II. XLII, 998 pages. 2006.

Vol. 4221: L. Jiao, L. Wang, X. Gao, J. Liu, F. Wu (Eds.), Advances in Natural Computation, Part I. XLI, 992 pages. 2006.

Vol. 4219: D. Zamboni, C. Kruegel (Eds.), Recent Advances in Intrusion Detection. XII, 331 pages. 2006.

Vol. 4218: S. Graf, W. Zhang (Eds.), Automated Technology for Verification and Analysis. XIV, 540 pages. 2006.

Vol. 4217: P. Cuenca, L. Orozco-Barbosa (Eds.), Personal Wireless Communications. XV, 532 pages. 2006.

Vol. 4216: M.R. Berthold, R. Glen, I. Fischer (Eds.), Computational Life Sciences II. XIII, 269 pages. 2006. (Sublibrary LNBI).

Vol. 4215: D.W. Embley, A. Olivé, S. Ram (Eds.), Conceptual Modeling - ER 2006. XVI, 590 pages. 2006.

Vol. 4213: J. Fürnkranz, T. Scheffer, M. Spiliopoulou (Eds.), Knowledge Discovery in Databases: PKDD 2006. XXII, 660 pages. 2006. (Sublibrary LNAI).

Vol. 4212: J. Fürnkranz, T. Scheffer, M. Spiliopoulou (Eds.), Machine Learning: ECML 2006. XXIII, 851 pages. 2006. (Sublibrary LNAI).

Vol. 4211: P. Vogt, Y. Sugita, E. Tuci, C. Nehaniv (Eds.), Symbol Grounding and Beyond. VIII, 237 pages. 2006. (Sublibrary LNAI).

Vol. 4210: C. Priami (Ed.), Computational Methods in Systems Biology. X, 323 pages. 2006. (Sublibrary LNBI).

Vol. 4209: F. Crestani, P. Ferragina, M. Sanderson (Eds.), String Processing and Information Retrieval. XIV, 367 pages. 2006.

Vol. 4208: M. Gerndt, D. Kranzlmüller (Eds.), High Performance Computing and Communications. XXII, 938 pages. 2006.

Vol. 4207: Z. Ésik (Ed.), Computer Science Logic. XII, 627 pages. 2006.

Vol. 4206: P. Dourish, A. Friday (Eds.), UbiComp 2006: Ubiquitous Computing. XIX, 526 pages. 2006.

Vol. 4205: G. Bourque, N. El-Mabrouk (Eds.), Comparative Genomics. X, 231 pages. 2006. (Sublibrary LNBI).

Vol. 4204: F. Benhamou (Ed.), Principles and Practice of Constraint Programming - CP 2006. XVIII, 774 pages. 2006.

Vol. 4203: F. Esposito, Z.W. Raś, D. Malerba, G. Semeraro (Eds.), Foundations of Intelligent Systems. XVIII, 767 pages. 2006. (Sublibrary LNAI).

Vol. 4202: E. Asarin, P. Bouyer (Eds.), Formal Modeling and Analysis of Timed Systems. XI, 369 pages. 2006.

Vol. 4201: Y. Sakakibara, S. Kobayashi, K. Sato, T. Nishino, E. Tomita (Eds.), Grammatical Inference: Algorithms and Applications. XII, 359 pages. 2006. (Sublibrary LNAI).

Vol. 4200: I.F.C. Smith (Ed.), Intelligent Computing in Engineering and Architecture. XIII, 692 pages. 2006. (Sublibrary LNAI).

Vol. 4199: O. Nierstrasz, J. Whittle, D. Harel, G. Reggio (Eds.), Model Driven Engineering Languages and Systems. XVI, 798 pages. 2006.

Vol. 4198: O. Nasraoui, O. Zaiane, M. Spiliopoulou, B. Mobasher, B. Masand, P. Yu (Eds.), Advances in Web Minding and Web Usage Analysis. IX, 177 pages. 2006. (Sublibrary LNAI).

Vol. 4197: M. Raubal, H.J. Miller, A.U. Frank, M.F. Goodchild (Eds.), Geographic, Information Science. XIII, 419 pages. 2006.

Vol. 4196: K. Fischer, I.J. Timm, E. André, N. Zhong (Eds.), Multiagent System Technologies. X, 185 pages. 2006. (Sublibrary LNAI).

Vol. 4195: D. Gaiti, G. Pujolle, E. Al-Shaer, K. Calvert, S. Dobson, G. Leduc, O. Martikainen (Eds.), Autonomic Networking. IX, 316 pages. 2006.

Vol. 4194: V.G. Ganzha, E.W. Mayr, E.V. Vorozhtsov (Eds.), Computer Algebra in Scientific Computing. XI, 313 pages. 2006.

Vol. 4193: T.P. Runarsson, H.-G. Beyer, E. Burke, J.J. Merelo-Guervós, L.D. Whitley, X. Yao (Eds.), Parallel Problem Solving from Nature - PPSN IX. XIX, 1061 pages. 2006.

Vol. 4192: B. Mohr, J.L. Träff, J. Worringen, J. Dongarra (Eds.), Recent Advances in Parallel Virtual Machine and Message Passing Interface. XVI, 414 pages. 2006.

Vol. 4191: R. Larsen, M. Nielsen, J. Sporring (Eds.), Medical Image Computing and Computer-Assisted Intervention – MICCAI 2006, Part II. XXXVIII, 981 pages. 2006.

Vol. 4190: R. Larsen, M. Nielsen, J. Sporring (Eds.), Medical Image Computing and Computer-Assisted Intervention – MICCAI 2006, Part I. XXXVVIII, 949 pages. 2006.

Vol. 4189: D. Gollmann, J. Meier, A. Sabelfeld (Eds.), Computer Security – ESORICS 2006. XI, 548 pages. 2006.

Vol. 4188: P. Sojka, I. Kopeček, K. Pala (Eds.), Text, Speech and Dialogue. XV, 721 pages. 2006. (Sublibrary LNAI).

Vol. 4187: J.J. Alferes, J. Bailey, W. May, U. Schwertel (Eds.), Principles and Practice of Semantic Web Reasoning. XI, 277 pages. 2006.

Vol. 4186: C. Jesshope, C. Egan (Eds.), Advances in Computer Systems Architecture. XIV, 605 pages. 2006.

Vol. 4185: R. Mizoguchi, Z. Shi, F. Giunchiglia (Eds.), The Semantic Web – ASWC 2006. XX, 778 pages. 2006.

Vol. 4184: M. Bravetti, M. Núñez, G. Zavattaro (Eds.), Web Services and Formal Methods. X, 289 pages. 2006.

Vol. 4183: J. Euzenat, J. Domingue (Eds.), Artificial Intelligence: Methodology, Systems, and Applications. XIII, 291 pages. 2006. (Sublibrary LNAI).

Vol. 4182: H.T. Ng, M.-K. Leong, M.-Y. Kan, D. Ji (Eds.), Information Retrieval Technology. XVI, 684 pages. 2006.

Vol. 4180: M. Kohlhase, OMDoc – An Open Markup Format for Mathematical Documents [version 1.2]. XIX, 428 pages. 2006. (Sublibrary LNAI).

Vol. 4179: J. Blanc-Talon, W. Philips, D. Popescu, P. Scheunders (Eds.), Advanced Concepts for Intelligent Vision Systems. XXIV, 1224 pages. 2006.

Vol. 4178: A. Corradini, H. Ehrig, U. Montanari, L. Ribeiro, G. Rozenberg (Eds.), Graph Transformations. XII, 473 pages. 2006.

Vol. 4177: R. Marín, E. Onaindía, A. Bugarín, J. Santos (Eds.), Current Topics in Artificial Intelligence. XV, 482 pages. 2006. (Sublibrary LNAI).

Vol. 4176: S.K. Katsikas, J. Lopez, M. Backes, S. Gritzalis, B. Preneel (Eds.), Information Security. XIV, 548 pages. 2006.